John Buridan, *Quaestiones super octo libros Physicorum Aristotelis* (*secundum ultimam lecturam*)

History of Science and Medicine Library

VOLUME 55

Medieval and Early Modern Philosophy and Science

VOLUME 27

The titles published in this series are listed at *brill.com/memps*

John Buridan
Quaestiones super octo libros Physicorum Aristotelis (secundum ultimam lecturam)

Libri III–IV

Edited by

Michiel Streijger
Paul J.J.M. Bakker

Guide to the Text by

Edith D. Sylla

BRILL

LEIDEN | BOSTON

Front cover illustration: manuscript Kraków, Biblioteka Jagiellońska, cod. 1771, f. 106vb (detail)
Back cover illustration: manuscript Kraków, Biblioteka Jagiellońska, cod. 1771, f. 142vb (detail)

Library of Congress Cataloging-in-Publication Data

Names: Buridan, Jean, 1300-1358, author. | Streijger, Michiel, 1974- editor. | Bakker, Paul J. J. M., editor. | Sylla, Edith Dudley, writer of supplementary textual content.
Title: Quaestiones super octo libros Physicorum Aristotelis (secundum ultimam lecturam) : Libri III-IV / John Buridan ; edited by Michiel Streijger, Paul J.J.M. Bakker ; a guide to the text by Edith D. Sylla.
Other titles: John Buridan, Quaestiones super octo libros Physicorum Aristotelis (secundum ultimam lecturam) libri III-IV | History of science and medicine library ; v. 55. | History of science and medicine library. Medieval and early modern philosophy and science ; v. 27.
Description: Leiden ; Boston : Brill, 2016. | Series: History of science and medicine library ; volume 55 | Series: Medieval and early modern philosophy and science ; volume 27 | Text in Latin; introduction and guide to the text in English. | Includes bibliographical references and index.
Identifiers: LCCN 2016030305 (print) | LCCN 2016032154 (ebook) | ISBN 9789004131873 (hardback) : alk. paper) | ISBN 9789004322356 (e-book)
Subjects: LCSH: Aristotle. Physics. | Physics–Early works to 1800. | Motion–Early works to 1800. | Infinite–Early works to 1800. | Place (Philosophy)–Early works to 1800. | Vacuum–Early works to 1800. | Time–Early works to 1800.
Classification: LCC Q151.A73 B976 2016 (print) | LCC Q151.A73 (ebook) | DDC 530–dc23
LC record available at https://lccn.loc.gov/2016030305

Typeface for the Latin, Greek, and Cyrillic scripts: “Brill”. See and download: brill.com/brill-typeface.

ISSN 2468-6808
ISBN 978-90-04-13187-3 (hardback)
ISBN 978-90-04-32235-6 (e-book)

This book is printed on acid-free paper and produced in a sustainable manner.

To the memory of Charles H. Lohr SJ (1925–2015)

Contents

Introduction

Paul J.J.M. Bakker and Michiel Streijger

The present volume contains the first complete critical edition of Books III and IV of John Buridan's *Quaestiones super octo libros Physicorum Aristotelis (secundum ultimam lecturam)* (henceforth: *Quaestiones Physicorum*).[1] Buridan's commentary has been preserved in the following thirty-two manuscripts:[2]

1. Berlin, Staatsbibliothek zu Berlin—Preußischer Kulturbesitz, cod. lat. fol. 852 [= Be]
2. Bratislava, Archív mesta Bratislavy, cod. E.L.5 [= Br]
3. Buenos Aires, Biblioteca Nacional, cod. 342R [= J]
4. Carpentras, Bibliothèque Inguimbertine, cod. 293 (L. 289) [= A]
5. Erfurt, Universitätsbibliothek, cod. CA F. 300 [= Er]
6. Frankfurt am Main, Stadt- und Universitätsbibliothek, cod. Praed. 52 [= B]
7. København, Kongelige Bibliotek, cod. Ny kgl. Samling 1801 fol. [= C]
8. Kraków, Biblioteka Jagiellońska, cod. 659 [= D]
9. Kraków, Biblioteka Jagiellońska, cod. 660 [= E]
10. Kraków, Biblioteka Jagiellońska, cod. 661 [= F]
11. Kraków, Biblioteka Jagiellońska, cod. 1771 [= G]
12. Kremsmünster, Bibliothek des Benediktinerstiftes, cod. CC 169 [= H]
13. Lambach, Bibliothek des Benediktinerstiftes, cod. Ccl. 175 [= La]

---

1 The edition of Books I and II appeared in 2015: John Buridan, *Quaestiones super octo libros Physicorum Aristotelis (secundum ultimam lecturam), Libri I–II*, ed. M. Streijger & P.J.J.M. Bakker, Leiden [etc.] 2015 (Medieval and early modern science, 25). The questions on the infinite (III.14–19) and the questions on time (IV.12–16) have been previously edited by J.M.M.H. Thijssen, *John Buridan's Tractatus de infinito. Quaestiones super libros Physicorum secundum ultimam lecturam, Liber III, quaestiones 14–19*, Nijmegen 1991 (Artistarium. Supplementa, 6), and D.-J. Dekker, *De tijdfilosofie van Johannes Buridanus († ca. 1360). Een historisch-wijsgerige studie met editie van Buridanus'* Quaestiones super octo libros *Physicorum* Aristotelis (secundum ultimam lecturam), *IV, 12–16*, Ph.D. dissertation, Radboud University, Nijmegen 2003, respectively.

2 For descriptions of these manuscripts, see J.M.M.H. Thijssen, 'Introduction,' in: John Buridan, *Quaestiones super octo libros Physicorum Aristotelis (secundum ultimam lecturam). Libri I–II*, ed. M. Streijger & P.J.J.M. Bakker, Leiden 2015 (Medieval and early modern science, 25), XIII–XLII, at XXI–XXXIII.

14. Liège, Bibliothèque de l'Université, cod. 114 C [= I]
15. Oxford, Balliol College Library, cod. 97 [= Ox]
16. Paderborn, Erzbischöfliche Akademische Bibliothek, cod. VVa 12 [= Pb]
17. Paris, Bibliothèque Nationale de France, cod. lat. 14723 [= L]
18. Salzburg, Stiftsbibliothek St. Peter (Erzabtei), cod. b.IX.24 [= M]
19. Salzburg, Universitätsbibliothek, cod. M.II.311 [= N]
20. Stralsund, Stadtarchiv der Hansestadt Stralsund, cod. 1050 [= K]
21. Torino, Biblioteca Nazionale Universitaria, cod. G.IV.10 [= O]
22. Vaticano (Città del), Biblioteca Apostolica Vaticana, cod. Vat. lat. 2163 [= P]
23. Vaticano (Città del), Biblioteca Apostolica Vaticana, cod. Vat. lat. 2164 [= Q]
24. Wien, Bibliothek des Dominikanerkonvents, cod. 107/73 [= R]
25. Wien, Österreichische Nationalbibliothek, cod. 5112 [= S]
26. Wien, Österreichische Nationalbibliothek, cod. 5332 [= T]
27. Wien, Österreichische Nationalbibliothek, cod. 5338 [= U]
28. Wien, Österreichische Nationalbibliothek, cod. 5367 [= V]
29. Wien, Österreichische Nationalbibliothek, cod. 5424 [= W]
30. Wien, Österreichische Nationalbibliothek, cod. 5458 [= X]
31. Wien, Österreichische Nationalbibliothek, cod. 5481 [= Y]
32. Zaragoza, Biblioteca Capitular de la Seo, cod. 15–61 [= Z]

In addition, the *ultima lectura* of the *Physics* was edited by John Dullaert and published in 1509:

> Iohannes Buridanus, *Subtilissimae Quaestiones super octo Physicorum libros Aristotelis*. Paris: Pierre le Dru for Denis Roce, 1509 [= p]

For the present edition of Books III and IV we used all the above mentioned manuscripts (with the exception of manuscript Be, which contains only questions I.2–8) as well as the 1509 edition by John Dullaert. Manuscript Pb offers an incomplete text of book III. The text in this manuscript breaks off abruptly in question II.13.[3] Questions III.1–11 are missing (just as is most of III.12).[4] Ques-

3 The last line on f. 33vb runs: 'diceret quod ab illa sanitate dependeret eius operatio' (cf. John Buridan, *Quaestiones Physicorum*, II.13, ed. Streijger & Bakker, 345^{15}).

4 Insofar as we are able to read the heavily mutilated text, the first line of f. 34ra seems to correspond with the answer to the second principal argument of question III.12 (cf. *infra*, 113^{3}). The preceding text of question III.12 is missing.

tion III.13 seems to be complete.[5] The text of question III.14 breaks off abruptly in the middle of the answer, and questions III.15–19 are missing.[6] Apart from the arguments 'quod non' and 'in oppositum' and the beginning of the answer of question IV.1, Book IV seems to be complete. In manuscript N the text of Book IV is incomplete: it breaks off in question IV.14, and question IV.15 is missing.[7] In John Dullaert's 1509 edition (p), there is an irregularity in the numbering of the folia: 60, 61, **61**, **62**, 64, 65. In our edition we restore the correct numbering (using 62 and 63 instead of 61 and 62).

For the present edition of Books III and IV of Buridan's *Quaestiones Physicorum*, we follow the same method as used for Books I and II.[8] The edition is based on manuscript C, which has been fully collated with manuscripts G and P, as well as with John Dullaert's edition (p). We take into account Dullaert's edition mainly for historical reasons. Given that substantial parts of manuscript G have become unreadable due to severe (water) damage on ff. 59–69, we use manuscript I instead of manuscript G for our edition of questions III.6–16.[9] We chose manuscript I because it contains a relatively small number of accidents and individual readings, and because of its close relationship with manuscript G: manuscripts G and I belong to the same branch of the stemma.[10]

In our edition, we follow manuscript C wherever possible. As a rule, we indicate all variant readings of G (or I), P, and p. The only variants we systematically ignore are the following: 'ergo'/'igitur', 'ille'/'iste' (etc.), 'item'/'iterum', variant ways of writing numbers (e.g., 'decimo tertio'/'tredecimo'/'13°'), and variation in spelling (e.g., 'intrinsecitas'/'intrinseitas').[11] In all these cases, we follow the

5 Question III.13 starts in the middle of f. 34ra and ends on the last line of f. 34vb ('est actio et passio; ad quod sequitur quod actio est passio;' cf. *infra*, 121^{19}). The first line of f. 34va reads: 'ad motus nocivos vel tristes ut patet quinto *Metaphysicae*' (cf. *infra*, 118^{2-3}).

6 Question III.14 starts on the first line of f. 35ra. The last line of f. 35vb runs: 'occuparet totum spatium etiam imaginabile, ideo non permitteret' (cf. *infra*, 128^{18-19}). The first line of f. 36ra ('quod non sit pars illius. Sic enim dicit Aristoteles locum esse') corresponds with the beginning of Buridan's answer to question IV.1 (cf. *infra*, 205^{20-21}).

7 The (hardly readable) text on f. 80vb seems to correspond with the first proof of the fifth conclusion of question IV.14 (*infra*, 327^{9}).

8 For the following, see Thijssen, 'Introduction,' XXXVI–XL.

9 Cf. *infra*, 60–152.

10 See Dekker, *De tijdfilosofie van Johannes Buridanus*, 134–158 (for the stemma, see 156–157).

11 Here we would like to make a correction to the edition of Books I and II: we have considered 'difformis'/'deformis' to be a variation of spelling of one and the same word, using 'deformis' in the edition. As it turns out, 'difformis' and 'deformis' are two different words with (slightly) different meanings. See H. Antony (ed.), *Mittellateinisches Wörterbuch bis*

reading of manuscript C (or the most natural reading of an abbreviation in manuscript C). In those instances in which the reading of manuscript C does not seem to make sense, we follow G (or I) and/or P (and p), and move the reading of C to the critical apparatus. It should be noted that C is not very precise in the mood of the verbs (at least according to the standards of classical Latin) and at times offers less elegant readings than G (or I) and/or P (and p). Nonetheless, also in those cases we follow C throughout. We indicate all (marginal) corrections in C in the apparatus, except in cases in which the corrected text is the same as the text of the three other witnesses (G[I]Pp). We also indicate in the apparatus the 'variant readings' proposed by C (in the margins and above the line).[12] Occasionally, when none of the four witnesses CG[I]Pp seems to provide a reading that makes sense, or when *only* the reading of the 1509 edition p seems meaningful, we consulted seven additional manuscripts: ABHLMTU.

For seven passages in Book III and four in Book IV it was necessary to consult all surviving manuscripts. In some of these cases, other manuscripts provide a possible reading. In these cases, the edition follows the reading of these manuscripts. In other cases, none of the other manuscripts provides either a possible reading or the reading that one would expect on the basis of the content of the text. In these cases, the edition retains the reading of the manuscripts, and our conjecture is indicated in the critical apparatus. The seven passages in Book III for which all manuscripts have been consulted are the following (in places that are missing in manuscript Pb, the apparatus indicates: '*deest* Pb'):

---

zum ausgehenden 13. Jahrhundert, III, München 2007, 204–205 (deformis etc.), 618 (difformis etc.); J.W. Fuchs, O. Weijers & M. Gumbert-Hepp, *Lexicon Latinitatis Nederlandicae medii aevi. Woordenboek van het middeleeuws Latijn van de noordelijke Nederlanden*, III, Leiden 1986, 1308–1309 (deformis etc.), 1462 (difformis etc.); R.E. Latham, *Dictionary of Medieval Latin from British Sources*, I, Oxford 1997, 594 (deformis etc.), 659 (difformis etc.). Accordingly, the following passages in our edition of Books I and II need to be corrected as follows: 126^{7-8} (read 'difformiter' instead of 'deformiter'); 126^{17} (read 'difformiter' instead of 'deformiter'); 127^{1} (and the corresponding lemma in the critical apparatus: read 'difformiter' instead of 'deformiter'); 127^{10} (read 'difformitates' instead of 'deformitates'); 129^{17} (read 'difformiter' instead of 'deformiter'); 129^{22} (critical apparatus, variant reading of 'uniformiter': read 'difformiter' instead of 'deformiter').

12 On some folia, the marginalia in C are no longer visible because part of the folium has been cut off. This will be indicated as follows: †...†.

Question III.6 (66^{16-20})

Et secundum istum usum verum est quod omne quod movetur movebatur prius, omne quod movetur movebitur posterius et omne quod est fuit prius et erit posterius, si nihil est per solum instans indivisibile (**'si' pro 'quia'**, cum nihil sit instans indivisibile).

Question III.9 (92^{4-5})

Item diversas et contrarias **habent** (one would expect: 'haberet') operationes, ut forte quia terminus motus, scilicet deorsum, est frigefaciens et motus naturaliter calefacit.

Question III.11 (107^{13-19})

De argumento autem Aristotelis secundo *Posteriorum* dicendum est quod, cuius est demonstratio tamquam conclusionis scitae per demonstrationem, eius non est definitio. Conclusio ista enim, ut quod triangulus habet tres etc., non definitur, licet hoc nomen '**conclusio**' (one would expect: 'triangulus') definiatur. Quando autem dicit Aristoteles quod accidentis est demonstratio, sensus debet esse quod conclusionis in qua accidens, id est passio, praedicatur de subiecto est demonstratio; et iam dictum est quod eius non est definitio.

Question III.17 (154^{23}–155^{10})

Item si 'numerum' capiamus pro rebus numeratis vel numerabilibus, hoc potest esse dupliciter. Uno modo quod solum loquimur de rebus ab invicem separatim existentibus, ita quod nec una sit pars alterius nec plures sint partes eiusdem totius, ita quod nihil loqueremur de partibus continui. Et tunc statim esset dicendum quod aliquis est numerus rerum quo nullus est maior secundum multitudinem. Non enim est infinita multitudo rerum separatarum a magnitudine. De rebus autem magnitudinem habentibus inter eas quae separatim existunt de facto est aliqua minima, scilicet qua nulla alia sit minor separatim existens; aliter continuum esset divisum in infinitum. Minima autem concessa non est possibile esse infinitatem secundum multitudinem **rerum quarum** quaelibet esset tanta vel maior, nisi ex eis poneremus resultare magnitudinem actu infinitam, quam non ponimus.

Question III.18 (176^{3-8})

Quarta conclusio de infinito secundum multitudinem, scilicet quod haec est falsa 'infinitae secundum multitudinem sunt partes in continuo', quia si in continuo sunt duae partes, tamen **non** ('non' should perhaps be omitted) sunt in eo plures partes quam illae duae, quoniam illae duae sunt et centum et mille, sicut ante dictum est. Et ita etiam concluditur secundum dicta quod haec est falsa 'infinitus est numerus secundum multitudinem'.

Question III.18 (185^{10-16})

Ad aliam dictum fuit prius quod non sunt plures partes in lineis *a* et *b* quam in linea *b*, immo nec quam in una centesima lineae *b*. Unde in continuo bene sequitur 'tantundem et amplius, igitur maius', quia illud amplius reddet maiorem extensionem. Sed non sequitur 'tot et adhuc alia, igitur plura', quia ista non reddunt plurem **multitudinem** quam essent illa tot, immo nec quam esset unum istorum, quia quodlibet est infinitum secundum multitudinem.

Question III.19 (197^{26}–198^{12})

Non enim sequitur, si verum est quod hoc album potest esse nigrum, quod haec est possibilis 'hoc album est nigrum' vel 'album est nigrum', sed sufficit quod ponatur in esse per pronomen demonstrativum demonstrato eo pro quo subiectum supponebat aliis quae in subiecto implicabantur circumscriptis. Et ita credo de praedicato esse, quoniam sicut propositioni de possibili verae debet correspondere propositio **de inesse possibilis, ita propositioni de futuro verae debet correspondere propositio** de praesenti quae erit vera, ut si haec est vera 'ego curram', sequitur quod haec erit vera 'ego curro', si tunc proponatur; et tamen non oportet quod illi de futuro correspondeat illa de praesenti salvata tota forma praedicati, quoniam haec est vera 'ego bibam cras', et tamen haec numquam erit vera 'ego bibo cras'; ita igitur de possibili. Nam demonstrato puero qui de ventre matris nascitur, haec est vera 'iste puer potest currere in tempore futuro'.

The four passages in Book IV are the following:

Question IV.2 (221^{6-11})

Ad quartam dicendum est eodem modo quod non sunt differentiae specificae locorum secundum suas essentias sursum et deorsum, quia idem locus cum locato suo posset ascendere sursum, cum prius esset deorsum. Sed sunt differentiae locorum, quantum ad sursum et deorsum, per distantiam vel propinquitatem ad caelum et sunt diversae naturales potentiae **communicantes**, quia aliter caelum influit propinque et remote.

Question IV.7 (259^{1-8})

Tunc igitur arguitur sic: cum ille ignis genitus occupet multo maiorem locum quam facerent stramina, igitur oportet corpora circumstantia cedere, nisi fiat penetratio corporum. Et illis cedentibus oportet alia iterum cedere. Et sic tandem oportet caelum cedere, nisi sint in corporibus aliquae vacuitates in quas corpora sic cedentia recipiantur, vel nisi dicatur quod necesse est **quod**, quantumcumque generatur hic de denso rarum et ex straminibus ignis, tantundem alibi oportet generari ex raro densum et simul horum utrumque fieri, ut semper remaneat totum aequale.

Question IV.10 (295^{9-12})

Et tunc est septima conclusio quod adhuc ille lapis non moveretur naturaliter per suam gravitatem, quia non esset **supra gravius nec supra levius** nec recedendo ab aere haberet aliquod corpus superius vel inferius sibi proximum grave vel leve, gravius vel levius.

Question IV.11 (302^{21-28})

Et forte quod in violentis incurvationibus lignorum habent locum huiusmodi condensatio et rarefactio. Nam cum arcus quasi **est** rectificatus et habens superficiem concavam quasi aequalem secundum longitudinem superficiei convexae, tamen quando multum incurvatur, oportet superficiem concavam fieri multo breviorem et superficiem convexam longiorem. Quod forte est per violentam condensationem partium interiorum et violentam rarefactionem exteriorum; ideo remoto incurvante revertitur velocissime et impetuosissime ad naturalem rectitudinem.

Just as in the edition of Books I and II, the apparatus is, in principle, negative. Only in cases in which we consulted other manuscripts besides CG[I]Pp is the apparatus positive. In those cases, the apparatus indicates immediately after the square bracket which of the manuscripts offer(s) the reading retained in the edition. Punctuation and paragraph divisions are according to modern conventions. We put a colon after the word 'quia' whenever it does not have the meaning 'because,' but rather serves the function of a colon (for example in 'probatur quia' or 'arguitur quia'). The spelling is homogenized according to classical standard orthography (as, for instance, codified in the *Oxford Latin Dictionary*). The medieval *e* for *ae* or *oe* is not retained, and *u* and *v* are always distinguished. All abbreviations are resolved, including 'Socrates' for 'Sor,' with the exception of 'etc.' (for example in 'igitur etc.' at the end of an argument). Literal quotations are put between single quotation marks. Double quotation marks are only used if they occur within single quotation marks.

In addition to the critical apparatus, which contains the variant readings, there is a second apparatus identifying the references to the sources quoted in the text. Only explicit references have been identified. As is to be expected, most references are to Aristotle's texts, which we identify by title, book, chapter, and pagination of the Bekker edition. If references also occur in the *Auctoritates Aristotelis*, we indicate this by *AA*, followed by the number of the Aristotelian work and the number of the relevant *auctoritas* according to the edition of Jacqueline Hamesse (e.g. '*AA* 2: 6' refers to *auctoritas* 6 on the *Physics*: 'Contra negantem principia non est disputandum'). We identify references to Averroes by work, book, commentum, and folium according to the Giunta edition of 1562–1574, unless a modern critical edition is available (as in the case of the long commentaries on *De anima* and *De caelo*). We indicate references to the Bible according to the system of abbreviation in the modern edition of the Vulgate.

We also indicate Buridan's internal references to his own *Quaestiones Physicorum*, unless they are to a passage in the same *quaestio* in which they occur. We traced references to books V to VIII of the *Physics* in John Dullaert's 1509 edition with book, *quaestio*, and folium. Some references are ambiguous, such as the announcement that one should look in such-and-such a book by Aristotle (*debet videri*). In these cases, it is not immediately clear whether Buridan is referring to his own text (an *expositio* or *quaestiones* on the relevant work by Aristotle) or to the appropriate passage in Aristotle in which the issue at stake is discussed. Only if the reference can, indeed, be found in one of Buridan's own commentaries, have we identified the passage. We have not tracked down references to anonymous 'some' (*aliqui*/*quidam*), and parallel passages in other texts.

The edition refers to the following sources:

Anonymus, *Liber sex principiorum*, ed. L. Minio-Paluello, Brugge, Paris 1966 (Aristoteles Latinus, 1/6–7).

Averroes, *Aristotelis opera cum Averrois commentariis*, ed. Iuntina, Venezia 1562–1574 (repr. Frankfurt am Main 1962).

Averroes, *Commentarium magnum in Aristotelis De anima libros*, ed. F.S. Crawford, Cambridge (MA) 1953 (Corpus commentariorum Averrois in Aristotelem. Versionum latinarum 6/1).

Averroes, *Commentum magnum super libro De celo et mundo Aristotelis*, ed. F.J. Carmody & R. Arnzen, 2 vols, Leuven 2003 (Recherches de Théologie et Philosophie médiévales. Bibliotheca, 4).

Chartularium Universitatis Parisiensis, ed. H. Denifle & A. Chatelain, 4 vols, Paris 1889–1897.

Hamesse, J., *Les Auctoritates Aristotelis. Un florilège médiéval*, Leuven, Paris 1974 (Philosophes médiévaux, 17).

Iohannes Buridanus, *In Metaphysicen Aristotelis quaestiones argutissimae*, Paris 1518 (repr. Frankfurt am Main 1964).

Iohannes Buridanus, *Quaestiones in Praedicamenta*, ed. J. Schneider, München 1983 (Veröffentlichungen der Kommission für die Herausgabe ungedruckter Texte aus der mittelalterlichen Geisteswelt, 11).

Iohannes Buridanus, *Quaestiones super De sensu et sensato*, in: G. Lokert (ed.), *Quaestiones et decisiones physicales insignium virorum*, Paris 1516, ff. 28va–40vb.

Iohannes Buridanus, *Quaestiones super libros De caelo et mundo*, ed. E.A. Moody, Cambridge (MA) 1942.

Iohannes Buridanus, *Quaestiones super libros De generatione et corruptione Aristotelis*, ed. M. Streijger, P.J.J.M. Bakker & J.M.M.H. Thijssen, Leiden [etc.] 2010 (Medieval and early modern science, 14).

Iohannes Buridanus, *Quaestiones super octo libros Physicorum Aristotelis (secundum ultimam lecturam), Libri I–II*, ed. M. Streijger & P.J.J.M. Bakker, Leiden [etc.] 2015 (Medieval and early modern science, 25).

Iohannes Buridanus, *Quaestiones super octo Physicorum libros Aristotelis*, Paris 1509 (repr. Frankfurt am Main 1964).

Iohannes Buridanus, *Sophismata*, ed. T.K. Scott, Stuttgart-Bad Canstatt 1977 (Grammatica speculativa, 1).

Iohannes Buridanus, *Summulae de practica sophismatum*, ed. F. Pironet, Turnhout 2004 (Artistarium, 10/9).

Iohannes Buridanus, *Summulae de suppositionibus*, ed. R. van der Lecq, Turnhout 1998 (Artistarium, 10/4).

Porphyrius, *Isagoge*, ed. L. Minio-Paluello, Brugge, Paris 1966 (Aristoteles Latinus, 1/6–7).

Thomas de Aquino, *In octo libros Physicorum Aristotelis expositio*, ed. M. Maggiòlo, Torino, Roma 1965.

Guide to the Text

Edith D. Sylla

1 Introduction

Books III and IV of Aristotle's *Physics* cover concepts or topics that, for the most part, remained relevant to modern physics, namely, motion, infinity, place, vacuum, and time. Volume VII of Pierre Duhem's *Le système du monde. Histoire des doctrines cosmologiques de Platon à Copernic* covered medieval scholastic treatments of most of these topics, saving 'motion' to be considered with time, adding three chapters on the latitudes of forms, and leaving 'vacuum' to be covered in Volume VIII. In 1985, Roger Ariew published an edition and English translation of Duhem's coverage of these topics, omitting what Duhem wrote on motion and adding relevant sections from Volume X, covering the fifteenth century, and a new Part IV, on the Plurality of Worlds, making use of sections from Duhem's Volumes I, IX, and X.

Anneliese Maier had covered the late scholastic treatments of these and other topics in her magisterial five-volume *Studien zur Naturphilosophie der Spätscholastik*. More recently Silvia Donati and Cecilia Trifogli have been providing fuller treatments of the background to Maier's work by studying the many anonymous and attributed thirteenth-century questions commentaries on Aristotle's *Physics* found mostly in Cambridge, Oxford, and Parisian libraries. In 2000, Trifogli published *Oxford Physics in the Thirteenth Century (ca. 1250–1270): Motion, Infinity, Place and Time*, after which she published two large repertories to the works, divided into a volume on the thirteenth century and a volume on the fourteenth century. She had previously published four lengthy papers on the earlier English commentators on Books III and IV in the *Documenti e studi sulla tradizione filosofica medievale* (see bibliography for details). These prior surveys of the topics covered in Books III and IV of Aristotle's *Physics* provide a foundation for this Guide to Books III and IV of John Buridan's *Quaestiones super octo libros Physicorum Aristotelis (secundum ultimam lecturam)* (henceforth: *Quaestiones Physicorum*).

If these earlier works provide a foundation, a model for thinking about Buridan's *Quaestiones* as a unit may be found in Jürgen Sarnowsky's *Die Aristotelisch-Scholastische Theorie der Bewegung. Studien zum Kommentar Alberts von Sachsen zur Physik des Aristoteles*, published in 1989. Sarnowsky discusses the whole of Albert of Saxony's questions on the *Physics*, book by book and chapter by chapter, often footnoting parallel discussions in the works of the other fourteenth-century Parisians including Buridan and Oresme. There is much too much valuable detail in Sarnowsky's book for it to make sense to refer in this Guide to all the places it casts light on Buridan's work. Hence I will just say here that for anyone wanting to work on Buridan's questions it always is worthwhile to look at Sarnowsky's book, particularly, for the present volume of Buridan's *Physics*, to chapter 4 ('Die *passiones rerum naturalium*'), which covers systematically what Albert of Saxony has to say on Aristotle's Books III and IV. It would be difficult to decide whether looking at Albert of Saxony's *Quaestiones* casts more light on Buridan's *Quaestiones* by the ways in which it is *similar* or by the ways in which it is *different*. Both perspectives are essential.

Pierre Duhem famously wrote that the condemnation of 1277 by Etienne Tempier, Bishop of Paris, constituted the 'birth certificate' of modern science.[1] In the opening paragraph of Ariew's translation from Duhem's *Le système du monde*, Duhem wrote:

> From the start of the fourteenth century, the grandiose edifice of Peripatetic physics was doomed to destruction. Christian faith had undermined all its essential principles; observational science, or at least the only observational science which was somewhat developed—astronomy—had rejected its consequences. The ancient monument was about to disappear; modern science was about to replace it. The collapse of Peripatetic physics did not occur suddenly; the construction of modern physics was not accomplished on an empty terrain where nothing was standing. The passage from one state to the other was made by a long series of partial transformations, each one pretending merely to retouch or to enlarge some part of the edifice without changing the whole. But when all these minor modifications were accomplished, man, encompassing at one glance the result of his lengthy labor, recognized with surprise that nothing remained of the old palace, and that a new palace stood on its place.[2]

---

1 P. Duhem, *Medieval Cosmology. Theories of Infinity, Place, Time, Void, and the Plurality of Worlds*, ed. and tr. R. Ariew, Chicago [etc.] 1985, xxii.

2 Duhem, *Medieval Cosmology*, 3. Cf. P. Duhem, *Le système du monde. Histoire des doctrines cosmologiques de Platon à Copernic*, VII, Paris 1956, 3: 'La destruction de la Physique péripatéticienne ne fut pas un subit écroulement; la construction de la Physique moderne ne se fit pas sur un terrain où rien n'était plus debout. De l'une à l'autre, le passage se fit par une longue suite de transformations partielles, dont chacune prétendait seulement retoucher ou agrandir quelque pièce de l'édifice sans rien changer à l'ensemble. Mais lorsque toutes ces modifications de détail eurent été faites, l'esprit humain, embrassant d'un regard le résultat

Although Duhem's historical framing of the work of fourteenth-century natural philosophers including John Buridan no longer fits the scope and detail of what is known, his comprehensive survey of relevant sources is still a reasonable place to begin. One point that emerges in looking at the primary sources that Duhem successively discusses is that he assumes more connection, continuity, and influence between the particular sources he describes than the documentary evidence solidly supports. So, for instance, when he sees a similarity between the ideas of two thinkers, he tends to infer that the later author took up his ideas from the earlier, whereas in historical fact there may have been a different chain of influence and numerous other thinkers expressing similar thoughts that we happen not to know about.

The catalog of thirteenth-century questions commentaries on Aristotle's *Physics* compiled by Cecilia Trifogli shows how many multifarious commentators existed with similarities and differences from each other.[3] When these commentaries are anonymous, there is little evidence to distinguish between what might be versions authored by the same individual at different times and versions made by different individuals at roughly the same time. In the current state of our knowledge, it has been noted that there are many passages in the questions on the *Physics* of Thomas Wylton that correspond to similar passages in the commentaries on the *Physics* attributed to Walter Burley. On the other hand, to Burley himself are attributed at least three versions of questions and/or expositions on the *Physics*. Likewise, for John Buridan and Albert of Saxony, there are many passages or questions that are quite similar, while for Buridan himself, there are thought to be collections of questions from two, three, or more stages of his career. Similarities between the work of Wylton and Burley, or between the work of Buridan and Albert of Saxony, could result not because one relied on the other, but because both were making use of the text of Aristotle, or of the commentaries of Averroes or Thomas Aquinas, or of John Duns Scotus' commentary on the *Sentences* of Peter Lombard.

The present Guide, compiling lists of authors whose work contains discussion of the same or similar questions, produces only a limited entry point to the earlier work which might have influenced Buridan's *Quaestiones Physicorum* and, likewise, a limited entry point to the later work that might have been

---

de ce long travail, reconnut avec surprise qu'il ne restait rien de l'ancien palais et qu'un palais neuf se dressait à sa place.'

3 C. Trifogli, *Liber Tertius* Physicorum *Aristotelis: Repertorio delle Questioni. Commenti inglesi ca. 1250–1270*, Firenze 2004 (Corpus philosophorum medii aevi. Subsidia, 13); Ead., *Liber Quartus* Physicorum *Aristotelis: Repertorio delle Questioni. Commenti inglesi ca. 1250–1270*, Firenze 2007 (Corpus philosophorum medii aevi. Subsidia, 16).

influenced by Buridan's *Quaestiones*. Two examples may suffice to illustrate the limitations of the methodology of this Guide. On the one hand, considering only questions commentaries on Aristotle's *Physics* fails to catch evidence of the influence of the work of Duns Scotus as exemplified by his commentary on Peter Lombard's *Sentences*. On the other hand, the methodology of Pierre Duhem and Anneliese Maier, making heavy use of commentaries on the *Sentences*, may exaggerate the influence of Thomas Aquinas or of Gregory of Rimini's theological work on Buridan. Buridan refers to Thomas Aquinas by name, but he does not refer to Gregory of Rimini by name. How are we to understand this? How are we to understand the relations between natural philosophy and theology at Paris in the fourteenth century insofar as they are manifested in Buridan's questions on the *Physics*?

In the introduction to *Die Vorläufer Galileis im 14. Jahrhundert*, Anneliese Maier writes:

> Die Schöpfer und Hauptvertreter der neuen Physik des 14. Jahrhunderts sind vor allem die Pariser Nominalisten, d.h. Johannes Buridan und seine Schule, die hauptsächlich durch drei hervorragende Denker repräsentiert wird: Nicolaus von Oresme, der auf mehr als einem Gebiet einer der genialsten Geister seiner Epoche gewesen ist, und die beiden Deutschen Albert von Sachsen, der später der Gründer der Universität Wien wurde, und Marsilius von Inghen, dem die Universität Heidelberg ihre Entstehung verdankt. Sie sind die eigentlich spekulativen Köpfe der neuen Bewegung, die sich um die physikalischen Theorien als solche, um exakte Begriffsbestimmungen und Herausarbeitung der Grundprinzipien bemühen, ohne sich zu sehr in Einzelprobleme zu verlieren.[4]

Recognizing in a footnote Duhem's groundbreaking work in the sources, Maier noted that Duhem's work required correction as well as expansion, since he had emphasized the Parisian terminists, while neglecting other forerunners of Galileo.[5] As already indicated in the Guide to Books I and II of Buridan's

4 A. Maier, *Die Vorläufer Galileis im 14. Jahrhundert* (Studien zur Naturphilosophie der Spätscholastik, 1), Roma 1949 (Storia e letteratura, 22), 3.

5 Maier, *Die Vorläufer Galileis*, 3, n. 2: 'Auf die Buridan-Schule und ihre Leistungen hat Duhem als erster hingewiesen und sie zum Gegenstand eingehender, auf gründliche Erforschung der Quellen gestützter Studien gemacht, freilich mit Resultaten, die, wie wir schon sagten, in manchen Punkten der Korrektur bedürfen. Und nicht nur der Korrektur, auch der Ergänzung. Für Duhem waren die Pariser Terministen, und nur sie, die "précurseurs de Galilée." Das ist entschieden zu einseitig gesehen. Den Naturphilosophen der Buridan-Schule kommt

Physics, scholars no longer easily accept the idea of 'John Buridan and his school', given that Buridan, Oresme, and Albert of Saxony, at least, might better be considered as a network of interacting contemporaries than as a school composed of a master and his students.[6] The history of natural philosophy in medieval universities does not fit the 'relay-runner' model of the history of philosophy or science: there was not one great philosopher or scientist or theologian passing on the baton to the next great scientist, but rather crowds of actors interacting with each other, about whose work we have only very partial evidence. One point of studying the questions on the *Physics* of Buridan and his predecessors, contemporaries, and successors is to investigate how they worked as a community, to practice 'historical epistemology': how did these individuals (and the communities they formed) work together and with what relative progress, stasis, or regression?[7]

Given this historiographical background, in the Guide to this second volume of the critical edition of Buridan's *Quaestiones Physicorum*, in my list of comparable collections of questions on Aristotle's *Physics*, I will refer readers to the work of Silvia Donati and Cecilia Trifogli for thirteenth-century works, giving brief summaries at the start of each major topic. For each topical section, I will then begin by making brief mention of the authors that Duhem surveys in his examination of the given topic. Duhem's method of work, and in particular the format of Ariew's edition and translation of it, leaves to footnotes in the back of the book any description of the particular contexts of given arguments, whether in commentaries on the *Physics* or other Aristotelian works or, very frequently, in commentaries on the *Sentences* of Peter Lombard. Anneliese Maier's work will, where appropriate, be used to supplement Duhem's.

---

zweifellos der Hauptanteil an der Schaffung der neuen Physik der Spätscholastik zu, aber sie sind nicht die einzigen gewesen, die zu ihr beigetragen haben. Wenn wir von Vorläufern Galileis im 14. Jahrhundert sprechen wollen, dann gehören auch andere dazu.'

6 See E.D. Sylla, 'Guide to the Text', in: John Buridan, *Quaestiones super octo libros Physicorum Aristotelis (secundum ultimam lecturam). Libri I–II*, ed. M. Streijger & P.J.J.M. Bakker, Leiden [etc.] 2015 (Medieval and early modern science, 25), XLIII–CLXXV, at XLVI–XLVII. See also J.M.M.H. Thijssen, 'The Buridan School Reassessed. John Buridan and Albert of Saxony', *Vivarium*, 42 (2004), 18–42.

7 Cf. J. Büttner, P. Damerow & J. Renn, 'Galileo's Unpublished Treatises. A Case Study on the Role of Shared Knowledge in the Emergence and Dissemination of an Early Modern "new science"', in: C.R. Palmerino & J.M.M.H. Thijssen (eds), *The Reception of the Galilean Science of Motion in Seventeenth-Century Europe*, Dordrecht 2004 (Boston studies in the philosophy of science, 239), 99–117.

It is, I believe, very well established, that a leading characteristic of the work of Buridan and of the other Parisian nominalists was their logical, analytical, or second intentional approach to physical problems as well as to the text of Aristotle's *Physics*. This point was made by Anneliese Maier, by Ernest Moody, and by John Murdoch, among others. In a chapter entitled 'Die erkenntnistheoretische Wendung und die neuen Strömungen des 14. Jahrhunderts', Maier writes:

> Eine neue Note kommt in die Betrachtung dadurch herein, dass die Probleme in anderer Weise angefasst, die Argumente von anderer Seite hergeholt werden: nämlich von der logischen. Das Problem der intensio und remissio hat ja auch eine logische Seite, denn die qualitas wird nicht nur vom Subjekt partizipiert, sie denominiert es auch ... Bei der Logisierung des Problems durch den Ockhamismus geht es auch nicht eigentlich darum, dass nun auf diese Seite der Frage der Akzent gelegt würde. Es handelt sich vielmehr um einen Ausfluss der allgemeinen Tendenz des Nominalismus, ontologische Probleme als logische anzusehen, und zwar als Probleme einer rein sermocinalen Logik. Die Frage, ob etwa der Satz 'forma habet latitudinem' wahr ist de virtute sermonis oder nicht, wird wichtiger als die, wie diese latitudo ontologisch zu deuten sei ...
>
> Neben das Interesse an dem Wie tritt nun das an dem Dass. Oder richtiger gesagt: der Umstand, dass das Wissen um die intensio und remissio der Formen nur aus der Erfahrung stammt, wird jetzt ausdrücklich zum Bewusstsein gebracht; der Akzent rückt von der ontologischen auf die erkenntnistheoretische Seite des Problems hinüber. So erklärt Ockham im Sentenzenkommentar: si quaeratur: unde est, quod una forma augmentatur et alia non?, dico quod nulla est ratio nisi quia ista natura est talis et alia est talis ...[8]

Of the questions that Buridan asks concerning Books III and IV of the *Physics*, some are exegetical and inherited from a long line of previous commentators. Buridan's question III.1, for instance, *Utrum necesse sit ignorato motu ignorare naturam*, had been asked by many previous authors of questions on the *Physics* going back to the thirteenth century. To make this clear, I report the appearance of this question in the questions commentaries cataloged by Trifogli or studied by Donati. On the other hand, Buridan's response to question III.1 makes abundant use of logic, as do the questions commentaries of his Parisian contempo-

8 A. Maier, *Zwei Grundprobleme der scholastischen Naturphilosophie* (Studien zur Naturphilosophie der Spätscholastik, 2), Roma 1968 (Storia e letteratura, 37), 75, 77.

raries. To investigate this side of Buridan's answer I make use of the editions (from more than one manuscript) or transcriptions from one manuscript of similar questions made by Benoît Patar.[9]

Other questions were new in the fourteenth century and sometimes stemmed from Ockhamism or nominalism. To show Ockham's likely role in their appearance I have included parallel passages not only from Ockham's *Quaestiones* on the *Physics*, but also from his other works. Buridan himself includes questions on alteration coming directly or indirectly from Walter Burley's work *De intensione et remissione formarum*, and this is echoed in later related works.[10] The inclusion of long sections of works involving 'analytical languages' in later questions commentaries on the *Physics* is demonstrated by referring to some works from later centuries.

2 Authors of Questions on Books III and IV of the *Physics* Related to Buridan's Questions

Group 1a: Commentaries Written before the 1320s

In this volume, I include only selected questions from Roger Bacon's *Questiones supra libros octo Physicorum*. Complete lists of questions in this work may be found in Roger Bacon, *Questiones supra libros octo Physicorum*, ed. F.M. Delorme & R. Steele, Oxford 1935 (Opera hactenus inedita, 13). I have not used here Roger Bacon, *Questiones supra libros quatuor Physicorum*, ed. F.M. Delorme & R. Steele, Oxford 1928 (Opera hactenus inedita, 8). In the *Questiones supra libros octo Physicorum*, all the questions on Book III (pp. 144–174) concern infinity and none concern motion.

Because of the extensive work of Cecilia Trifogli and Silvia Donati on thirteenth-century commentaries on Books III and IV of the *Physics* of English and Parisian origin, including lists of questions and a CD-ROM for selected texts, I mainly treat thirteenth-century named or anonymous authors as a group and

9 B. Patar, *La Physique de Bruges de Buridan et le Traité du Ciel d'Albert de Saxe*, 2 vols, Longueuil 2001. Patar transcribes or edits from more than one manuscript parallel questions from the following manuscripts: Brugge, Stedelijke Openbare Bibliotheek, cod. 477; Cesena, Biblioteca Malatestiana, cod. S.VIII.5, and Toulouse, Archives départementales de la Haute-Garonne, cod. 6; Erfurt, Universitätsbibliothek, cod. CA F. 298, and Città del Vaticano, Biblioteca Apostolica Vaticana, cod. Chigi lat. E.VI.199; and London, Wellcome Institute for the History of Medicine, cod. L.15.

10 See S. Caroti, 'Some Remarks on Buridan's Discussion on Intension and Remission,' *Vivarium*, 42 (2004), 58–85.

include only a selection of the works mentioned in Trifogli's repertories. Aside from and preceding the works of Trifogli and Donati are lists of questions found in Albert Zimmermann, *Verzeichnis ungedruckter Kommentare zur Metaphysik und Physik des Aristoteles aus der Zeit von etwa 1250–1350*, Leiden [etc.] 1971 (Studien und Texte zur Geistesgeschichte des Mittelalters, 9).

Included in my grouping of thirteenth-century named or anonymous authors (using their sigla in bold type) are:

A. Earlier (ca. 1250–1270) commentaries of likely English origin as studied in Cecilia Trifogli's *Liber Tertius* Physicorum *Aristotelis: Repertorio delle Questioni. Commenti inglesi ca. 1250–1270*, Firenze 2004 (Corpus philosophorum medii aevi. Subsidia, 13) and *Liber Quartus* Physicorum *Aristotelis: Repertorio delle Questioni. Commenti inglesi ca. 1250–1270*, Firenze 2007 (Corpus philosophorum medii aevi. Subsidia, 16). Anonymous questions are identified by Trifogli's boldface sigla. The following are a selection from the longer list studied by Trifogli:

1. **S** = manuscript Siena, Biblioteca Comunale degli Intronati, cod. L.III.21. Anonymous, *Quaestiones in Physicam I–VIII* (Book III, ff. 39vb–46vb; Book IV, ff. 46vb–64ra).
2. **P** = manuscript Cambridge, Peterhouse Library, cod. 157. William of Clifford, *Quaestiones in Physicam I–V, VII* (Book III, ff. 65rb–75va; Book IV, ff. 75vb–96va). I list these questions separately under Clifford's name rather than the siglum **P**. According to Trifogli, **S** is a source of **P**.[11]
3. **G¹** = manuscript Cambridge, Gonville and Caius College Library, cod. 367. Anonymous, *Quaestiones in Physicam I–IV* (Book III, ff. 139rb–144ra; Book IV, ff. 144ra–151vb).
4. **M²** = manuscript Oxford, Merton College Library, cod. 272. Anonymous, *Quaestiones in Physicam I, III.1–3, IV.10–14* (Book III, ff. 125ra–130rb; Book IV, ff. 130rb–$135C^{rb}$).
5. **M³** = manuscript Oxford, Merton College Library, cod. 272. Anonymous, *Quaestiones in Physicam I–V* (fragm.) (Book III, ff. 152rb–158rb; Book IV, ff. 158va–173rb). According to Trifogli, **G¹** is the main source of **M³**, and both show knowledge of Richard Rufus of Cornwall's commentary on the *Physics*.[12]

11 Cf. C. Trifogli, *Oxford Physics in the Thirteenth Century (ca. 1250–1270): Motion, Infinity, Place and Time*, Leiden [etc.] 2000 (Studien und Texte zur Geistesgeschichte des Mittelalters, 72), 28.

12 Cf. Trifogli, *Oxford Physics in the Thirteenth Century*, 32.

B. Anonymous Parisian questions from the 1270–1300s, listed in Silvia Donati, 'Commenti parigini alla Fisica degli anni 1270–1300 ca.', in: A. Speer (ed.), *Die Bibliotheca Amploniana. Ihre Bedeutung im Spannungsfeld von Aristotelismus, Nominalismus und Humanismus*, Berlin [etc.] 1995 (Miscellanea mediaevalia, 23), 136–256. These will be identified by the boldface sigla given to them by Donati:

1. **Er349(1)** = manuscript Erfurt, Universitätsbibliothek, cod. CA F. 349, ff. 1ra–68vb. Anonymous (Peter of Auvergne?), *Quaestiones in Physicam I–VIII.*[13]
2. **Er349(2)** = manuscript Erfurt, Universitätsbibliothek, cod. CA F. 349, ff. 75ra–117rb. Anonymous (Siger of Brabant?), *Quaestiones in Physicam I–VI.*[14] Cf. **Vat6758** = manuscript Città del Vaticano, Biblioteca Apostolica Vaticana, cod. Vat. lat. 6758, ff. 1ra–43vb. Anonymous (Siger of Brabant?), *Quaestiones in Physicam I–VIII.*[15] Of these two commentaries, only **Vat6758** has questions on time, and they are not similar to Buridan's questions.

C. Other, anonymous late thirteenth-century Parisian works that show familiarity with the two *Physics* commentaries in Erfurt, Universitätsbibliothek, cod. CA F. 349 and with the Paris condemnations of 1277, as follows:

1. **Ka11** = manuscript Kassel, Gesamthochschul-, Landes- und Murhardsche Bibliothek, cod. Phys. 2° 11, ff. 1ra–35rb. Anonymous, *Quaestiones in Physicam I–VIII.*[16] The same work is found in manuscript Paris, Bibliothèque Nationale de France, cod. lat. 14698, ff. 83ra–$129A^{r}$.[17]
2. **L1386(1)** = manuscript Leipzig, Universitätsbibliothek, cod. 1386, ff. 41ra–76ra. Magistri ⟨Pa⟩ (eras.), *Quaestiones in Physicam I–VIII.*[18]
3. **L1386(2)** = manuscript Leipzig, Universitätsbibliothek, cod. 1386, ff. 77vb–91ra. Magistri /// de ⟨C.idene (?)⟩ (eras.), *Quaestiones in Physicam I–VI.*[19]

13 For the list of questions, see Donati, 'Commenti parigini', 220–227.
14 For the list of questions, see Donati, 'Commenti parigini', 228–234.
15 For the list of questions, see Donati, 'Commenti parigini', 249–256.
16 For the list of questions, see Donati, 'Commenti parigini', 234–240.
17 For the list of questions, see Zimmermann, *Verzeichnis*, 284–291, and Donati, 'Commenti parigini', 179.
18 For the list of questions, see Donati, 'Commenti parigini', 240–245.
19 For the list of questions, see Donati, 'Commenti parigini', 245–249.

After listing many very similar parallel questions in these works for question III.1, for later questions I list parallels selectively, mainly to show the presence of the questions before Buridan. I refer to the following works:

1. William of Clifford (d. 1306), *Compilationes super librum Physicorum Aristotelis*, manuscript Cambridge, Peterhouse Library, cod. 157, I, ff. 43ra–104va (= William of Clifford). Arts master at Oxford by 1265. According to Trifogli, this work (which she gives the siglum **P**) probably dates from the decade 1250–1260. Trifogli lists 62 questions for Book III and 141 questions for Book IV.[20] Most are short. Zimmermann lists 46 questions from Book III and 126 questions from Book IV.[21] Because of including more or fewer question titles, the numbers assigned to the questions differ between Zimmermann and Trifogli. In this Guide to the text of Books III and IV, William of Clifford replaces William of Chelveston, who was included in the Guide for Books I and II (there is some uncertainty about the questions attributed to William of Chelveston).
2. Geoffrey of Aspall, *Quaestiones in Physicam*, manuscript Oxford, Merton College Library, cod. 272, ff. 88ra–118vb (given the siglum **M¹** by Trifogli) (= Geoffrey of Aspall). According to Trifogli, this work probably dates from 1255–1265. Another source suggests his MA ca. 1262. Trifogli lists 23 questions on Book III and omits Book IV.[22] Zimmermann lists 17 questions on Book III and 8 questions on Book IV.[23]
3. Boethius of Dacia (1260s and 1270s; MA ca. 1270), *Quaestiones super libros Physicorum*, ed. G. Sajó, København 1974 (Corpus philosophorum Danicorum medii aevi, 5/2) (= Boethius of Dacia). There are 37 questions on Book III, but only 5 on Book IV.[24]
4. Giles of Rome (d. 1316), *Commentaria in octo libros Phisicorum Aristotelis*, Venezia 1502 (repr. Frankfurt am Main 1968) (= Giles of Rome). As in the case of Walter Burley's final *Expositio*, Giles' commentary is an exposition rather than a questions commentary, but there are 'doubts' (*dubia*), which are listed in the '*Tabula huius operis*' (ff. 226^{r}–227^{r}). Since the statements

20 For the list of questions, see Trifogli, *Liber Tertius* Physicorum, 121–166, and Ead., *Liber Quartus* Physicorum, 197–299.

21 Zimmermann, *Verzeichnis*, 174–180.

22 For the list of questions, see Trifogli, *Liber Tertius* Physicorum, 369–370.

23 Zimmermann, *Verzeichnis*, 161.

24 For the list of questions, see pp. 323–326 in the critical edition. See also Zimmermann, *Verzeichnis*, 152–156.

of the questions are rarely like those of Buridan, I report Giles' doubts selectively. There are 51 doubts or other points listed for Book III, and 137 for Book IV.

5. Radulphus Brito, *Quaestiones in Physicam I–VIII*, manuscript Paris, Bibliothèque Nationale de France, cod. lat. 18160, ff. 3ra–79vb (= Radulphus Brito). According to W.J. Courtenay, Brito began the study of theology by 1299 and incepted as Master of Theology in 1313–1314. He would have continued to teach in the Faculty of Arts while studying theology.[25] In Book III, Brito has 16 questions related to the infinite and only 6 related to motion. In Book IV, he has 16 questions on place, 10 on vacuum, and 17 on time.[26]
6. Thomas Wylton, *Quaestiones in Physicam*, manuscript Cesena, Biblioteca Malatestiana, cod. S.VIII.2, ff. 2^r–141^v (= Thomas Wylton). Wylton has 16 questions on Book III and 32 questions on Book IV, so that more than one of his questions may relate to one of Buridan's questions.[27] Without a more detailed study, exact matches cannot be made with assurance. Cecilia Trifogli has published several articles based on Wylton's questions on the *Physics* and I include references to her publications where relevant.

It appears that Walter Burley wrote his early *Expositio et quaestiones librorum Physicorum* at Oxford before going to Paris to study theology. I list this work of Burley's and the separate *Quaestiones super libros Physicorum* immediately or shortly after the *Quaestiones in Physicam* of Thomas Wylton, to which they are closely related. I then list after the works of Ockham, to which it responds, Burley's final *Expositio in libros octo De phisico auditu*, which was the product of his teaching at Paris and which was completed after he returned to England from his study of theology in Paris:

1. Walter Burley, *Expositio et quaestiones librorum Physicorum*, manuscript Cambridge, Gonville and Caius College Library, cod. 448 (409), 172–543 (= Walter Burley [*Expositio et quaestiones*]).[28] This work was likely composed at Oxford before the works of John of Jandun and William of Ockham. The question titles listed here are from the single manuscript (Cambridge, Gonville and Caius College Library, cod. 448 [409]) and/or from Rega Wood's

25 Cf. W.J. Courtenay, 'Radulphus Brito, Master of Arts and Theology', *Cahiers de l'Institut du Moyen-Age Grec et Latin*, 76 (2005), 131–158.

26 For the list of questions, see Zimmermann, *Verzeichnis*, 182–190.

27 Unpublished list of questions distributed by John E. Murdoch.

28 See E.D. Sylla, 'Walter Burley's Practice as a Commentator on Aristotle's *Physics*', *Medioevo*, 27 (2002), 301–372.

list, where the exact wording is sometimes taken from the list of questions in manuscript Basel, Universitätsbibliothek, cod. F.V.12, rather than from the Gonville and Caius manuscript.[29] Questions from this work will be noted here only where they are relevant to showing Burley's views before he went to Paris to study theology.

2. Walter Burley, *Quaestiones super libros Physicorum*, manuscript Basel, Universitätsbibliothek, cod. F.V.12, ff. 108^{r}–171^{v} (= Walter Burley [*Quaestiones*]). Titles are taken from a list of questions that once were present in the manuscript, as published by Rega Wood.[30] Some of the questions listed exist as part of the *Expositio et quaestiones* in manuscript Cambridge, Gonville and Caius College Library, cod. 448 (409) and will be listed as Walter Burley (*Expositio et quaestiones*).[31] I am not convinced that any of these questions represent Burley's work after he went to Paris. Wood calls these the 'pre-1316 questions', but this, while true, is misleading, because I think they quite possibly were written ten or more years before 1316.

Another pre-1320 work is the following:

> Bartholomew of Bruges, *Quaestiones super libros Physicorum* (= Bartholomew of Bruges). According to Sten Ebbesen and Jan Pinborg, Bartholomew taught in the Faculty of Arts at Paris in the years 1307–1310.[32] J.M.M.H. Thijssen includes some of his questions on infinity in his *Johannes Buridanus over het oneindige*.[33]

Group 1b: Commentaries Written in the Period Just before Buridan

For what questions were asked on the *Physics* in the period just before Buridan began his teaching career, I have looked at:

> John of Jandun, *Quaestiones super 8 libros Physicorum Aristotelis*, Venezia 1551 (repr. Frankfurt am Main 1969) (= John of Jandun). This commentary

29 See R. Wood, 'Walter Burley's *Physics* Commentaries', *Franciscan Studies*, 44 (1984), 275–327.

30 See Wood, 'Walter Burley's *Physics* Commentaries', 312–313.

31 See Wood, 'Walter Burley's *Physics* Commentaries', 307.

32 See S. Ebbesen & J. Pinborg, 'Bartholomew of Bruges and his Sophisma on the Nature of Logic', *Cahiers de l'Institut du Moyen-Age Grec et Latin*, 39 (1981), iii–xxvi, 1–76.

33 J.M.M.H. Thijssen, *Johannes Buridanus over het oneindige. Een onderzoek naar zijn theorie over het oneindige in het kader van zijn wetenschaps- en natuurfilosofie*, Ph.D. dissertation, Radboud University, Nijmegen 1988, 2, 209–236.

gives an indication of questions asked at Paris before the logical turn associated with William of Ockham.

There is a profusion of works by Ockham that might be cited here. Up to now, historians have not mapped out which Continental authors evince knowledge of which works by Ockham, and whether transmission of his ideas and approaches came first or mainly through logic, theology, or natural philosophy. The works on physics listed here might be expected to have had the greatest likelihood to be used in Buridan's teaching on Aristotle's *Physics*, but that cannot be taken for granted. Even Ockham's theological works might have influenced Buridan indirectly.

1. William of Ockham, *Quaestiones in librum secundum Sententiarum* (*Reportatio*), ed. G. Gál & R. Wood, St. Bonaventure (NY) 1981 (Opera theologica, 5) (= William of Ockham [*Sent.* II]).
2. William of Ockham, *Summa logicae*, ed. Ph. Boehner, G. Gál & S. Brown, St. Bonaventure (NY) 1974 (Opera philosophica, 1) (= William of Ockham [*Summa logicae*]).
3. William of Ockham, *Expositio in libros Physicorum Aristotelis*. Prologus et libri I–III, ed. V. Richter & G. Leibold, St. Bonaventure (NY) 1985 (Opera philosophica, 4); *Expositio in libros Physicorum Aristotelis*, Libri IV–VIII, ed. R. Wood, R. Green, G. Gál, J. Giermek, F. Kelley, G. Leibold & G.J. Etzkorn, St. Bonaventure (NY) 1985 (Opera philosophica, 5) (= William of Ockham [*Expositio*]). Ockham's *Expositio* provides the best match with Buridan, but it is not organized in questions. Consequently, for Books III and IV, I have also checked for relevant question titles in Ockham's *Quaestiones* and in the *Tractatus de successivis* (see below). In reference to Ockham's *Expositio*, his chapter numbers will be used, rather than the standard chapter numbers in Aristotle's text.
4. William of Ockham, *Quaestiones in libros Physicorum Aristotelis*, ed. S. Brown, St. Bonaventure (NY) 1984 (Opera philosophica, 6) (= William of Ockham [*Quaestiones*]).

Cited in cases when the relevance seems especially obvious and there is not a better text in the *Expositio* or in the *Quaestiones*:

5. William of Ockham, *Tractatus de successivis*, ed. Ph. Boehner, *The Tractatus de successivis attributed to William Ockham*, St. Bonaventure (NY) 1944 (Franciscan Institute Publications. Philosophy Series, 1) (= William of Ockham [*De successivis*]). Mostly made up of excerpts from the *Expositio*.

6. William of Ockham, *Brevis summa libri Physicorum*, ed. S. Brown, St. Bonaventure (NY) 1984 (Opera philosophica, 6) (= William of Ockham [*Brevis summa*]).
7. William of Ockham, *Summula philosophiae naturalis*, ed. S. Brown, St. Bonaventure (NY) 1984 (Opera philosophica, 6). I have cited the early printed edition of this work because it contains passages at the end that are not in Brown's 1984 edition: William of Ockham, *Philosophia naturalis Guilielmi Occham*, Roma 1637 (repr. Vaduz-Liechtenstein 1963) (= William of Ockham [*Philosophia naturalis*]).

After Ockham's work are the following works by Walter Burley that reflect his time at the University of Paris:

1. Walter Burley, *Expositio in libros octo De phisico auditu*, Venezia 1501 (repr. Hildesheim 1972) (= Walter Burley [*Expositio*]).[34] At the end of the volume there is a list of *dubia* (ff. 266vb–268ra). Book III has 23 *dubia* and Book IV has 50. The numbering here is my own. I reserve the title *Expositio et quaestiones* for the early Oxford version found in manuscript Cambridge, Gonville and Caius College Library, cod. 448 (409). I list passages from this final *Expositio* after the works of Ockham because they contain evidence of Burley's reaction to Ockham's work.
2. Walter Burley, *Tractatus secundus de intensione et remissione formarum*, Venezia 1496 (= Walter Burley [*Tractatus secundus*]). This work is an extension of Burley's *Tractatus primus*, which derived from his activity as a bachelor of the *Sentences*. This work appears to be used in Buridan's questions III.3–5.

It is not certain what intermediaries there might have been between Burley's *Tractatus secundus* and Buridan's *Quaestiones Physicorum*. One place to look would be in the questions on the *Physics* of Richard Kilvington:

> Richard Kilvington, *Quaestiones quattuor super Physicam*, manuscript Venezia, Biblioteca Nazionale Marciana, cod. lat. VI.72, ff. 81ra–112rb, 168ra–

---

34 The 1972 reprint of this volume by Georg Olms Verlag has been given the title *In Physicam Aristotelis Expositio et Quaestiones*, although this title does not appear in the work itself, the closest approximation being the explicit: 'Finit expositio Gualterii de Burlei Anglici in libros octo de phisico auditu Aristotelis una cum questione eiusdem Gualterii de primo instanti et ultimo' (266vb).

169va (= Richard Kilvington).[35] The titles of the four questions are as follows:

1. *Utrum in omni motu potentia motoris excedat potentiam rei motae.* This question might fit into Aristotle's Book VII.
2. *Utrum qualitas suscipiat magis et minus.*[36] This question fits with Buridan's questions on alteration in Book III.
3. *Utrum aliquod corpus simplex possit moveri aeque velociter in vacuo et in pleno.*[37] This question addresses an issue of Book IV.
4. *Utrum omne transmutatum in transmutationis initio sit in eo ad quod primitus transmutatur.* This question addresses an issue of Book VIII.

The second and third questions will be discussed briefly in their proper places below in Books III and IV. There are probably other works that might have served as intermediaries between Burley's *Tractatus secundus* and Buridan's *Quaestiones Physicorum*, but they are not listed here because they are not question commentaries on the *Physics*.

Although Buridan's work has been taken by Jack Zupko as evidence for the origins of secular philosophical culture, it turns out that there is more influence of theology and ecclesiastical rulings than first meets the eye. As a resource for looking at intertextualities between the Faculty of Arts and the Faculty of Theology, I have looked at citations of theological authors in the following work:

---

35 See E. Jung, 'Works by Richard Kilvington,' *Archives d'histoire doctrinale et littéraire du Moyen Age*, 67 (2000), 181–223, esp. 203–217; E.D. Sylla, *The Oxford Calculators and the Mathematics of Motion. Physics and Measurement by Latitudes*, New York [etc.] 1991, 435–446.

36 This question is also found in other manuscripts, where it is either anonymous (manuscript Paris, Bibliothèque Nationale de France, cod. lat. 16401, 149^{v}–166^{v}) or ascribed to Walter Burley (manuscript Città del Vaticano, Biblioteca Apostolica Vaticana, cod. Vat. lat. 2148, 71^{r}–77^{v}) or Thomas Wylton (manuscript Oxford, Bodleian Library, Canon. Misc. 226, 38ra–42ra, incomplete). See also manuscript Città del Vaticano, Biblioteca Apostolica Vaticana, cod. Vat. lat. 4429, 64^{r}–70^{v}. The question is also (wrongly) included in some catalogues of works by Marsilius of Inghen. For this attribution, see Z. Kaluza, *Nicolas d'Autrécourt. Ami de la vérité*, Paris 1995 (Histoire littéraire de la France, 42/1), 197, n. 61, and P.J.J.M. Bakker, 'Syncatégorèmes, concepts, équivocité. Deux questions anonymes, conservées dans le ms. Paris, B.N., lat. 16.401, liées à la sémantique de Pierre d'Ailly (c. 1350–1420),' *Vivarium*, 34 (1996), 76–131, esp. 85, n. 26.

37 A related question appears also in manuscript Bologna, Biblioteca comunale dell'Archiginnasio, cod. lat. A.985 (*Utrum corpora gravia et levia in suis motibus requirant medium*).

Francesc Marbres (*alias* John the Canon), *Quaestiones super octo libros Physicorum*, Venezia 1520 (= Francesc Marbres).[38] This commentary was composed ca. 1325. The printed editions of this work differ from most of the other commentaries considered here in that they frequently list by name contemporary holders of opinions such as are usually referred to as '*aliqui*' or '*quidam*.' Among the authors cited by Francesc Marbres are Francis of Marchia, William of Ockham, Walter Burley, Landulph Caracciolo, Francis of Meyronnes, Thomas Anglicus, John Duns Scotus, Gerard of Odo, and others.

Group 2: Works of Buridan's Approximate Contemporaries in Paris

In what follows, I have listed questions from the printed versions of Albert of Saxony and from the *Abbreviationes Physicorum* of Marsilius of Inghen, but I have only occasionally attempted to investigate the content of these questions, nor of the earlier versions of Buridan's questions in relation to the *ultima lectura* edited in the present volume. I list the questions of Hugolinus of Orvieto because they appear to be closely related, although I have not seen the body of the questions. The recently published edition of Nicole Oresme's *Questiones super Physicam*, contained in manuscript Sevilla, Biblioteca Capitular y Colombina, cod. 7-6-30, is, however, so rich in evidence relevant to Buridan's *Quaestiones Physicorum* that I have made some initial forays into its content:

1. Nicole Oresme, *Questiones super Physicam* (*Books I–VII*), ed. S. Caroti, J. Celeyrette, S. Kirschner & E. Mazet, Leiden [etc.] 2013 (Studien und Texte zur Geistesgeschichte des Mittelalters, 112) (= Nicole Oresme).
2. Hugolinus of Orvieto, *Quaestiones super quattuor libros Physicorum*, ed. W. Eckermann, *Der Physikkommentar Hugolins von Orvieto OESA. Ein Beitrag zur Erkenntnislehre des spätmittelalterlichen Augustinismus*, Berlin [etc.] 1972 (Spätmittelalter und Reformation, 5) (= Hugolinus of Orvieto). This work is contained in manuscript Casale Monferrato, Biblioteca del Seminario Vescovile, cod. d. 17, ff. 1ra–76va, dated 1352. Since they are conveniently available in print, questions and conclusions from this work are noted as

38 Titles of questions from manuscript Città del Vaticano, Biblioteca Apostolica Vaticana, cod. Vat. lat. 3013, as in P.J.J.M. Bakker & D.-J. Dekker, 'Antoine Andrée ou Jean le Chanoine? A propos de l'authenticité du commentaire de la *Physique* conservé dans le Ms. Cambridge, Gonville et Caius College, 368 (590)', *Bulletin de philosophie médiévale*, 42 (2000), 101–131. On the identity of the author, see C. Schabel, 'Francesc Marbres, a.k.a. *Iohannes Canonicus*', *Bulletin de philosophie médiévale*, 56 (2014), 195–200. See also Sylla, 'Guide to the Text', LXVIII.

showing some influence of the interaction between realist (e.g., Walter Burley) and conceptualist (e.g., Ockham and Gregory of Rimini) approaches around 1350. According to Eckermann, Hugolinus' epistemology ('Erkenntnislehre') was influenced by Gregory of Rimini.[39] Although I have seen only Eckermann's outline of Books III and IV, since the whole work has 29 questions on Books I–IV in 77 folios, each question is fairly substantial. The first two questions of Book I are fully edited and fill pages 36–99 of Eckermann's book. Hugolinus does not name the authors of the opinions he discusses (he says, e.g., 'communis imaginatio modernorum'), but Eckermann is confident in naming William of Ockham and Walter Burley as the main proponents of opposed opinions concerning questions I.1 (*Utrum obiectum scientificum scientiae naturalis sit aliqua entitas extra animam*) and I.2 (*Utrum prima notitia rei naturalis sit notitia universalis vel notitia singularis*). Eckermann numbers the questions on all four books of the *Physics* sequentially; I provide separate numbering for each book.

3. [Albert of Saxony], *Quaestiones in Aristotelis octo libros Physicorum*, ed. B. Patar, *Expositio et quaestiones in Aristotelis Physicam ad Albertum de Saxonia attributae*, 3 vols, Louvain-la-Neuve, Leuven 1999 (Philosophes médiévaux, 39–41) (= Albert of Saxony).[40] The question titles in Patar's edition are by and large like those in the 1516 Lokert edition: *Quaestiones eximii Doctoris*

---

39 Compare also Hugolinus of Orvieto, *Commentarius in quattuor libros Sententiarum*, I, Prologus, q. 1, a. 2, ed. W. Eckermann, Würzburg 1980 (Cassiciacum, Supplement, 8), 68–69, 73: 'Consequenter quaeritur: Utrum omne verum complexum theologicum sit aliquid productum a creatura, ut propositio formata, vel quid sit ... Circa istud quaesitum sunt duo modi dicendi oppositi. Unus est Ockham rudis, qui ponit, quod obiectum scitum vel creditum est propositio, sed non absolute, ut est quaedam qualitas, sed potius ut est significativa ... Alia est via veritatis, quam scriptura innuit, Augustinus tenuit et Aegidius meminit ... et moderni declarant praecipue Gregorius [Ariminensis]. Pro primo igitur ostendetur quod significabile complexe sive obiectum cognitum non est propositio prout est significativa. Secundo quid sit. Sequendo igitur pro primo viam veritatis sit ista prima conclusio: quod esse verum, esse per se notum, esse per experientiam certum, esse scitum, esse necessarium, esse creditum, esse speratum per prius tempore et causalitate et consequentia et formaliter, non denominatione extrinseca convenit significato quam propositioni ... Omissa igitur hac inutili via restat nunc videre quid sit significabile complexe.'

40 For the list of questions on Book III, see vol. 2, XIII–XIV. For the list of questions on Book IV, see vol. 3, 1095–1096. For a deeper look at the content of Albert of Saxony's *Quaestiones Physicorum*, see J. Sarnowsky, *Die Aristotelisch-Scholastische Theorie der Bewegung. Studien zum Kommentar Alberts von Sachsen zur Physik des Aristoteles*, Münster 1989 (Beiträge zur Geschichte der Philosophie und Theologie des Mittelalters, n. F., 32).

Magistri Alberti de Saxonia in octo libros Physicorum Aristotelis, in: G. Lokert (ed.), *Quaestiones et decisiones physicales insignium virorum*, Paris 1516.[41] There are two irregularities in Books III and IV: question III.4, *Utrum ista sit concedenda 'motus est'*, and question IV.16, *Utrum tempus sit ab anima*, both in Patar's edition, are not in the 1516 edition, which leads to differences in the question numbers following these questions. In what follows, I shall give titles and numbers according to Patar's edition.

4. Marsilius of Inghen, *Abbreviationes super octo libros Physicorum Aristotelis*, Venezia 1521 (= Marsilius of Inghen). I have used the *tabula quaestionum* at the end of the volume. All the questions for Books III and IV appear on ff. 8^{v}–21^{v}, and very differing attention is paid to the questions, while more space is devoted to additional questions that arise, but are not included in the *tabula*. Marsilius writes at the start (f. 2ra) that to the extent that he can, he will abbreviate the books of natural philosophy as they are usually read at Paris ('pro meo posse philosophie naturalis libros Parisius legi solitos abbreviabo').

Finally, for question III.1, I will compare what is contained in the parallel questions in manuscripts that have been attributed by some to John Buridan (at various stages of his career) and by others to Albert of Saxony. For this I will use the transcriptions found in Benoît Patar's 2001 *La* Physique *de Bruges de Buridan et le* Traité du Ciel *d'Albert de Saxe*.

Group 3: Commentaries Written Some Decades or More after Buridan

Finally, I shall list parallel questions from later works, mainly to show that the influence of Buridan's *Quaestiones Physicorum* continued in later centuries.

1. Johannes Marsilii (?), *Quaestiones subtilissimae Johannis Marcilii Inguen super octo libros Physicorum secundum nominalium viam*, Lyon 1518 (repr. [ascribing the work to Marsilius of Inghen] Frankfurt am Main 1964) (= Johannes Marsilii [?]).[42] Most now consider the author of this set of questions—once published, with some doubt as to the author, in the *Opera*

41 Edition available online through the website of the Bayerische Staatsbibliothek in Munich (URL: http://www.mdz-nbn-resolving.de/urn/resolver.pl?urn=urn:nbn:de:bvb:12-bsb10195484-7).

42 Edition available online through the website of the Bayerische Staatsbibliothek: (URL: http://www.mdz-nbn-resolving.de/urn/resolver.pl?urn=urn:nbn:de:bvb:12-bsb10139261-9).

omnia of John Duns Scotus—as unknown. The Vivès edition of Scotus' works contains references to other works dealing with the topic of each question. According to Harald Berger, the author might be one Johannes Marsilii (or Marcilii), who determined in the Arts at Prague in 1396.[43] This seems to be the best (only) hypothesis yet offered, which would put a date for this set of questions at the very end of the fourteenth century, about the same time as Lawrence of Lindores, the author of the following work. One caveat is that in some cases the commentary by Johannes Marsilii (?) seems closer to the work of Buridan and Albert of Saxony than the abbreviation credited to Marsilius of Inghen.

2. Lawrence of Lindores, *Quaestiones Physicorum*, as in Th. Dewender, *Das Problem des Unendlichen im ausgehenden 14. Jahrhundert. Eine Studie mit Textedition zum Physikkommentar des Lorenz von Lindores*, Amsterdam 2002 (Bochumer Studien zur Philosophie, 36) (= Lawrence of Lindores). Thomas Dewender has the titles of the questions for Lindores' entire book (the questions for Books III and IV are on pp. 415–417) and has edited questions I.1–5, and 10; III.13–18; VI.9–10, and VIII.3. Lindores taught in the Faculty of Arts at Paris in the years 1395–1403. In 1405, he returned to Scotland, where he taught at St. Andrews. As late as 1438, Buridan's work was the basis of logic teaching at St. Andrews. In 1438, after protests, masters were allowed to teach using the work of Albert the Great or other philosophers accepted by the Church.
3. Benedictus Hesse, *Quaestiones super octo libros Physicorum Aristotelis*, ed. S. Wielgus, Wrocław [etc.] 1984 (= Benedictus Hesse). Essentially all of Buridan's question titles are contained in this work, along with numerous references to Buridan by name, together with many additional questions. Whereas Buridan has 19 questions on Book III, Hesse has 40; and whereas Buridan has 16 questions on Book IV, Hesse has 42. Of course, in many cases a single question in Buridan's commentary may be divided into several questions in Hesse. In many cases, Hesse may have learned about Buridan's questions through Lindores, whose work he cites even more often than Buridan's. According to Stanislaw Wielgus, Hesse's questions can be dated to 1421 in Cracow.

---

43 Personal communication in a letter of 8.10.1996. Berger refers to the *Liber decanorum Facultatis Philosophicae Universitatis Pragensis, ab anno Christi 1367 usque ad annum 1585*, Pars I, Praha 1830 (Monumenta historica Universitatis Carolo-Ferdinandeae Pragensis, 1/1), 314. Berger also sent other mentions of this Johannes Marsilii. He suggests that the 'Inguen' in the 1518 printing is an addition by the printer 'due to the association with the famous Marsilius.' I thank Prof. Berger for his kindness.

4. John Mair, *Propositum de infinito*, in: H. Elie, *Le traité* De l'infini *de Jean Mair*, Paris 1938 (= John Mair).[44] Although this is not a question commentary on the *Physics*, it will be included here in relation to Buridan's questions on infinity because it cites Buridan, Albert of Saxony, and Marsilius of Inghen by name.
5. John of Celaya, *Expositio in octo libros Physicorum Aristotelis cum quaestionibus eiusdem secundum triplicem viam beati Thomae, realium et nominalium*, Paris 1517 (= John of Celaya).[45] This work was composed some time before ca. 1520. In Celaya's questions on Book III, there is one reference to Buridan, but many to other authors, including long sections of the calculatory work of William Heytesbury and the Calculator (Richard Swineshead). For instance: 'Beatus Thomas' and 'Thomista' (60va); 'Opinio Scoti et aliorum realium' (60vb); 'Secundum Paulum Venetum in xxvii ca. sue *Metaphysice*' (60vb). Celaya refers to Gregory of Rimini: 'Alia est opinio nominalium non ponentium talia accidentia. Qui, veluti reales, divisi sunt. Nam Gregorius de Arimino in primo articulo quarte questionis, prime distinctionis secundi *Sententiarum* tenet quod ipsa res quam mobile continue acquirit secundum quod est actus eiusdem incompletus cum continua tendentia ad complementum est motus. Ex qua opinione sequuntur aliqua correlaria. Primum est quod in motu locali spacium supra quod mobile movetur est motus. Sequitur secundo quod in motu alterationis illa qualitas que continuo acquiritur aut deperditur dicitur motus. Sequitur tertio quod in motu augmentationis sive dimunitionis (!) illa pars que continuo acquiritur vel deperditur dicitur motus' (62vb). Celaya also refers to Albert of Saxony (63va). There is an entire *Tractatus proportionum* (63vb–68rb); then a work entitled *De motu penes causam* (68rb–76ra), in which the Calculator is frequently cited. This work is followed by a treatise *De potentia continuo variata in medio uniformiter dif-*

44 Mair's text opens as follows (2): 'Circa materiam de inifito sic procedam. Primo queram an sit aliquid infinitum extensive vel intensive. Secundo an implicet contradictionem Deum infinitum posse producere.' See also J. Biard, 'La logique de l'infini chez Jean Mair,' *Les études philosophiques*, 3 (1986), 329–348.

45 Edition available online through Google Books. Question III.1 is entitled *An ignorato motu necesse sit ignorare naturam* (59va). In this question, Celaya, *Expositio Physicorum*, III.1, 59vb, writes: 'Secundo est notandum quod ad hoc quod mobile moveatur requiruntur duo, scilicet mobile et dispositio secundum quam movetur. Verbi gratia, in motu locali requiruntur mobile et dispositio spacii per quod mobile debet moveri, et ita est de aliis motibus.' Like Buridan, in answering this question, Celaya distinguishes between the divided and the compounded senses depending on where the word 'necesse' appears in the proposition.

formi non variato (76ra–78ra) and a treatise *De motu in medio non resistente* (78ra–81rb), both based on the *Liber calculationum*. Then there is a question entitled *Penes quid debet attendi velocitas motus localis penes effectum* (81rb–88rb), in the course of which Albert of Saxony, William Heytesbury, Paul of Venice, and Jacob of Forlì are named. Next are the conclusions on motions difform in time ascribed to Nicole Oresme by Bernardus Tornius Florentinus, commentator on Heytesbury (88rb–91va). On the velocity of augmentation, one finds the opinions of the Calculator and of Paul of Venice (91va–94ra). Suddenly one has arrived at question III.8, which concerns the measures of alteration with respect to effect according to William Heytesbury and Gaetano of Thiene (94rb–97rb). After this long diversion into the works of the Oxford Calculators, Celaya returns to topics found in Buridan's *Quaestiones Physicorum*, such as the question whether the intension of forms occurs by the addition of degree to degree, with the prior remaining with the latter (Buridan's question III.5) (97rb–102rb). Here the authors whose opinions are mentioned are Thomas Aquinas, Johannes Capreolus, and Walter Burley (quoting conclusions from his *Tractatus secundus de intensione et remissione formarum*), where Celaya includes a series of arguments concerning Burley's view, some from James of Forlì's *Treatise on intension*. Then follow the opinions of Scotus and of all the nominalists, including the Calculator (100ra–102rb). The next topic is the intension of forms *in difformibus*, where figures showing configurations of form like those of Oresme appear, and the opinions of the Calculator and Jacob of Forlì are mentioned (102rb–108vb). Celaya then continues with further sections of the *Liber calculationum*, such as 'On the intension of mixed things having two contrary qualities' (108vb–110ra). This leads Celaya to another question appearing in Buridan (III.3) and in the other fourteenth-century Parisians: *An forme contrarie possunt esse adequate in eodem subiecto* (110ra–112va). Opinions Celaya discusses on this topic include those of Thomas Aquinas and Gregory of Rimini. This is followed by a question like Buridan's question III.4: *Utrum qualitas que intenditur successive acquiratur vel tota simul.* In his question III.15 (*An aliquod agens naturale valet producere effectum suum in instanti*), Celaya introduces two additional names: 'Pro huius dubii solutione est advertendum quod Robertus Holkot in primo *Sententiarum* et Hibernicus tenent partem negativam huius questionis' (115rb). I have not looked at Celaya's Book IV.

6. Domingo de Soto, *Super octo libros Physicorum Aristotelis subtilissimae quaestiones*, Venezia 1582 (= Domingo de Soto).[46] This work was written

46 Edition available online through the website of the Bayerische Staatsbibliothek (URL:

ca. 1546. Soto was educated at Paris and taught at Salamanca. He was a Dominican connected to the Thomist revival and a reformer of logic. His work has four questions on Book III (170–211) and four questions on Book IV (212–259).

7. Collegium Conimbricense, *Commentarii in octo libros Physicorum Aristotelis*, Lyon 1594 (repr. Hildesheim 1984) (= Conimbricenses). The Conimbricenses often report names of individuals who held various positions. Their work is very different from Buridan's *Quaestiones Physicorum*. Sometimes they make references like: 'Ochamus in q. 9. ac omnis fere schola Nominalium.'

3 The Questions on Book III

One of the obvious if implicit goals of Buridan's *Quaestiones Physicorum* is to be a model and provide resources for students and other readers who, immediately or in the future, might play roles in disputations concerning similar questions. This means that Buridan, as a master in the Faculty of Arts, not only attempts to answer or settle questions arising from Aristotle's *Physics*, but also tries to record and respond to arguments for and against alternative opinions on the questions. Beyond the interpretation of Aristotle's text, Buridan takes resources from Averroes' comments, other authors writing in Arabic (often via Averroes' comments), and earlier Latin commentators such as Thomas Aquinas. He also uses tools of the liberal arts, especially logic up through Aristotle's *Posterior analytics*, as well as mathematics and astronomy. Finally, it is not uncommon in Books III and IV for Buridan to make use of accepted truths of theology such as the general proposition that God could do anything that does not involve a logical contradiction and the specific doctrine that the body of Christ as present in the transubstantiated Eucharist is extended in itself, but not in relation to the visible extended parts of the bread or wine.

As can be seen below, some of Buridan's questions seem to cover overlapping topics. In his replies, Buridan might, in one case, answer only from a natural point of view, and in another case, bring in supernatural possibilities. This does not mean that Buridan has changed his opinions, but only that he is assuming the point of view of a different discipline. In question IV.8, edited here, Buridan makes an often-quoted remark to explain why he sometimes raises theological issues in his physical works.

---

http://www.mdz-nbn-resolving.de/urn/resolver.pl?urn=urn:nbn:de:bvb:12-bsb10150499-5) and through Google Books.

> Some of my lords and masters in theology have already blamed me for the fact that sometimes in my physical questions I mix some theological matters, since this does not pertain to those in the Arts Faculty. But I, with humility, respond that I very much wish not to be constrained only to this. All masters when they incept in arts vow that they will dispute no purely theological question, as for instance about the Trinity or the incarnation. And they vow further that if it happens to them that they dispute or determine some question which touches both faith and physics, they will determine it for the faith, and that they will resolve the arguments [for the other side] accordingly as it seems to them they should be resolved. It is evident, however, that if any question touches faith and theology, this is one of them, namely whether it is possible that there be a vacuum. Therefore, if I want to dispute it, I must say what it appears to me should be said according to theology or else be perjured. And I must turn aside the arguments for the other side accordingly as it seems possible to me. And I cannot resolve them unless I pose them. Therefore I am compelled to do this.
>
> I say therefore that we can imagine a vacuum in two ways, as was said in another question, and it is possible that a vacuum exist in either way by divine power. And this is believed by me and not proved by natural reason, and so I do not intend to prove it, but only to state the way in which it seems possible to me (268^5–269^2).[47]

In the *logica moderna* assumed by Buridan, the truth or falsity of propositions depends not only on the meaning of terms, but on their supposition, what they 'stand for' depending on their place in propositions.[48] Buridan assumes that terms in the categories of substance, quality, and (as he argues) quantity stand for *supposita* in the external world, whereas terms falling into other Aristotelian categories, such as action, passion, and relation, may not refer to individual separable things in the external world, but rather involve ways of looking at or conceiving things (connotations). In addition, propositions contain terms that do not refer directly to, or supposit for, things in the external world, but

---

47 See E.D. Sylla, '*Ideo quasi mendicare oportet intellectum humanum*: The Role of Theology in John Buridan's Natural Philosophy', in: J.M.M.H. Thijssen & J. Zupko (eds), *The Metaphysics and Natural Philosophy of John Buridan*, Leiden [etc.] 2001 (Medieval and early modern science, 2), 221–245, at 221–222.

48 For a good introduction to this topic, see Buridan's own compendium of logic: *Summulae de Dialectica*, ed. G. Klima, New Haven (CT) and London 2001 (Yale library of medieval philosophy).

rather affect the supposition of other terms. The term 'infinite' in particular may be used either categorematically, having supposition for things in the world, or syncategorematically, affecting the supposition of other terms. Technically (*de virtute sermonis*) the word order of propositions determines whether 'infinite' is to be understood categorematically or syncategorematically. Coming first in a proposition 'infinite' is to be understood syncategorematically. Coming later, and depending on the other words in the proposition, 'infinite' is to be taken categorematically. In Buridan's questions, the contrast between syncategorematic and categorematic replaces the distinction between potential and actual in analyzing propositions concerning infinity, as will be explained below.

Of the five main subjects of Books III and IV, namely motion, infinity, place, vacuum, and time, arguments might be made that none of them are things that exist separately in the external world. Earlier sections of the *Physics* discussed the status of artificial things, such as a house.[49] Is a house a single thing or a combination of things? In Books III and IV, questions arise about what were called at the time 'successive entities.' When water is heated to the boiling point, is that a single motion? And if so, is there a last instant of boiling water or a first instant when a given bit of water has been corrupted to be replaced by air (steam)? If time is divided into units by the rotation of the last sphere, might a day or a year be considered something that exists in the external world, or is it partly a human construct? Would there be days or years without human consciousness (the soul)?

Book III of Aristotle's *Physics* has two parts. The first part is on motion, and the second part on infinity. In chapters 1–2, Aristotle discusses the nature of motion, and in chapter 3 he asks where motion is. How is motion related to action and passion among Aristotle's categories? Averroes' fourth comment on Book III had a long lasting influence on medieval discussions of motion. In it he said that motion can be understood in two ways: either as the gradual taking on of the form eventually reached (*forma fluens*) or as the way to the final form (*via ad formam*). Another way of describing the difference was to ask whether what was real in a process of alteration was just the degrees of the new quality taken on (and so in the same genus or species as the final form) or whether it should be considered the process of taking on the form (*fluxus formae*), which might put it in the categories of action or passion rather than quality. According

49 See John Buridan, *Quaestiones super octo libros Physicorum Aristotelis* (*secundum ultimam lecturam*), II.1–2, ed. M. Streijger & P.J.J.M. Bakker, Leiden [etc.] 2015 (Medieval and early modern science, 25), 243–255.

to Averroes, the *fluxus formae* position was more common or popular (famous) in Aristotle's time or his own time, but the *forma fluens* position was truer.[50]

3.1 *Buridan's Questions on Motion: Questions III.1–13*

Buridan's questions on motion appear somewhat disorganized because, whereas in Book III Aristotle attempted to discuss motion in general, including motion in place, quality, and quantity (and motion in substance if sudden mutations are considered as motions in a wider sense), Buridan suddenly in questions III.2–5 turns to alteration or motion in quality in particular before returning to more standard questions following the order of Aristotle's writing.

In her *Oxford Physics in the Thirteenth Century (ca. 1250–1270): Motion, Infinity, Place and Time*, Cecilia Trifogli traces early English reactions to the first part of Book III of Aristotle's *Physics*, mainly in the period 1250–1270, but continuing on through Thomas Wylton and Walter Burley. In Trifogli's reading the early English commentators largely rejected Averroes' assertion that the truer way of understanding motion is to say that motion is the *forma fluens*. To the contrary, the English commentators take the realist position that motion is the *via ad formam*.[51] The arguments of the mostly anonymous commentators, none of whom seems to have known Thomas Aquinas' commentary on the *Physics*, are largely ontological and metaphysical.[52] In contrast, Aquinas criticizes Averroes' theory on physical grounds, saying, for instance, that in the case of heating water, the motion cannot be the form of heat gradually gained over the whole motion, because initially the body has degrees of cold, and then temperate, before gaining degrees of heat. In the *Quaestiones super Physicam* probably to be ascribed to William de Bonkes, an Oxford Master of Arts in the 1290s, Aquinas' argument about the heating of water is repeated.[53] The temperate degree that a body gains while being heated is not of the same essence as the heat finally attained. Further developments in the physical critique of Averroes' ideas of motion are then found in the commentaries on the *Physics* of Thomas Wylton and Walter Burley.[54]

---

50 For a comprehensive study of Averroes' view and its later medieval reception, see A. Maier, *Zwischen Philosophie und Mechanik* (Studien zur Naturphilosophie der Spätscholastik, 5), Roma 1958 (Storia e letteratura, 69), 59–143 ('Forma fluens oder fluxus formae?').

51 Cf. Trifogli, *Oxford Physics in the Thirteenth Century*, 51–59.

52 Cf. Trifogli, *Oxford Physics in the Thirteenth Century*, 59.

53 Cf. Trifogli, *Oxford Physics in the Thirteenth Century*, 63–64.

54 Cf. Trifogli, *Oxford Physics in the Thirteenth Century*, 65–66. On Wylton and Burley, see further C. Trifogli, 'Due questioni sul movimento nel commento alla *Physica* di Thomas Wylton', *Medioevo*, 21 (1995), 31–73; Ead., 'Thomas Wylton on Motion', *Archiv für Geschichte*

While Trifogli represents the views of Burley by his final *Expositio* of the *Physics*, more relevant to Buridan here would be Burley's *Tractatus secundus de intensione et remissione formarum*, which Buridan uses in his questions III.3–5. It is notable that Burley implicitly overcomes some of the arguments of Aquinas and Bonkes by arguing in this work and in its predecessor, the *Tractatus primus*, that contraries like hot and cold (and the temperate between them) are of the very same species, so that one particular argument against Averroes' *forma fluens* theory of motion would no longer hold, even though Burley continues, at the same time, to support the *fluxus formae* way of understanding motion.

Trifogli thus bridges the gap in the understanding of motion between her many Oxford questions commentaries on the *Physics*, on the one hand, and Buridan, on the other, by describing the contributions of Thomas Aquinas, William de Bonkes, Thomas Wylton, and Walter Burley. She writes:

> These physical objections against Averroes' *forma*-theory play a very marginal role in the discussion of our [early English] commentators. Nevertheless, they are historically important, since they indicate the direction in which the examination of this theory will be developed later in the thirteenth and fourteenth centuries.[55]

In her much earlier survey of the nature of motion ('Die Wesensbestimmung der Bewegung'), Anneliese Maier emphasized the sometimes neglected contribution of Albert the Great to the issue (much of the focus and language of the discussion about the nature of motion derives from Albert the Great rather than directly from Averroes).[56] Coming to the fourteenth century, Maier writes (in the translation of Steven Sargent):

> The question now arose, Should some further categorical element be postulated in addition, some kind of flux that constitutes the true essence of the reality we call motion? The new approach was less concerned with choosing between *forma fluens* and *fluxus formae* than with understanding how both *forma fluens* and *fluxus formae* may be involved. Doubts of this kind began to appear as early as the 1320s among a variety of Scotists,

---

der Philosophie, 77 (1995), 135–154; and Ead., 'The Reception of Averroes' View on Motion in the Latin West. The Case of Walter Burley', in: P.J.J.M. Bakker (ed.), *Averroes' Natural Philosophy and its Reception in the Latin West*, Leuven 2015 (Ancient and Medieval Philosophy, Series I, 50), 127–139.

55 Trifogli, *Oxford Physics in the Thirteenth Century*, 60.

56 See Sarnowsky, *Die Aristotelisch-Scholastische Theorie der Bewegung*, 136–137.

> who decided in favor of *fluxus formae* [Maier cites Francis of Meyronnes and William of Alnwick]. But these efforts remained isolated at first and did not lead to a unified theory of motion. Such an analysis only emerged in the natural philosophy of the Parisian school of 'terminists', that is, in the 'new physics' taught by Jean Buridan and his students and followers. Here as elsewhere, Ockham's theory of motion did not find ready acceptance, but instead stimulated its opponents to ask whether motion might indeed be something real and distinct from the object and its states or, in other words, whether another flux ought to be postulated in conjunction with the *forma fluens*.[57]

Thus Maier and Trifogli provide two possible historical backgrounds to Buridan's questions on motion in Book III. Some comments will be made, in their place, about the evidence presented here, without trying to determine how well it fits the pictures presented by previous historians. It is worth repeating here, however, the remark of Gyula Klima already quoted in volume 1:

> It was Buridan's careful attention to theoretical detail, coupled with his prudent practical judgment and pedagogical skill, that in his hands could turn Ockham's innovations into relatively uncontroversial, viable textbook material, capable of laying the foundations of a new, paradigmatically different conception of the relationships between language, thought and reality. And this is what renders the emergence of nominalist semantics the most significant development of late medieval philosophy. In the subsequent two centuries, the new theoretical conflicts that inevitably arose between practitioners of the nominalist 'modern way' (*via moderna*) and those of the realist 'old way' (*via antiqua*) were different in kind from the theoretical conflicts between members within each camp. Conflicts of this kind, to use Wittgenstein's happy analogy, are no longer about who wins the game, but rather about whose game everybody ought to play.[58]

Nominalist semantics differs from what had been the scholastic norm in the thirteenth century by emphasizing the supposition of terms in propositions,

---

57 A. Maier, *On the Threshold of Exact Science. Selected Writings of Anneliese Maier on Late Medieval Natural Philosophy*, ed. and tr. S.D. Sargent, Philadelphia 1982, 32–33 (translating Maier, *Die Vorläufer Galileis*, 18–19).

58 G. Klima, 'Nominalist Semantics', in: R. Pasnau & C. van Dyke (eds), *The Cambridge History of Medieval Philosophy*, 1, Cambridge 2010, 159–172, at 171–172.

and, within supposition, concentrating on personal supposition for things (*res*), that is substances, qualities, and perhaps quantities in the external world. None of the major subjects of Aristotle's Books III and IV turns out to be, in a simple sense, a substance or quality in the external world. Thus the game that Ockham invited nominalists to play was to explain the supposition of terms in propositions (or in the propositions into which exponible propositions were resolved) in terms of substances and qualities together with connotations reflecting the meaning of terms according to their definitions '*quid nominis*', as opposed to '*quid rei*.' Of course, not all of Buridan's questions are taken up with this nominalist approach or game. Some are more traditional natural philosophy. But it is relevant to look for the balance that Buridan represents between the traditional and the nominalist frameworks.

John of Jandun's question IV.27 (*Utrum tempus sit extra animam humanam*) begins with the following principal arguments:

> Arguitur primo quod sic quia: quod est in praedicamento est ens extra animam, ut videtur velle Aristoteles in sexto *Metaphysicae*; sed tempus est ens in praedicamento, ut vult Aristoteles in *Praedicamentis*; ergo etc.
>
> Item si tempus haberet esse ab anima, tunc esset de consyderatione logici—logicus enim consyderat entia rationis, ut communiter dicitur; sed hoc falsum est, quia ipsum tempus pertinet ad philosophum naturalem; ergo etc.[59]

Ultimately, Buridan was interested in the things in the external world, but he held that the truths of science were expressed in propositions, which were linked to the world through the suppositions of the terms in those propositions.[60]

In volume VII of *Le système du monde*, entitled 'La physique Parisienne au XIVe siècle', Duhem starts with the second part of Book III of the *Physics*, devoted to the infinitely small and the infinitely large. He reserves discussion

59 Jandun, *Quaestiones Physicorum*, IV.27, 72va. Cf. C. Trifogli, 'Il problema dello statuto ontologico del tempo nelle *Quaestiones super Physicam* di Thomas Wylton e di Giovanni di Jandun', *Documenti e studi sulla tradizione filosofica medievale*, 1 (1990), 491–548, esp. 530.

60 In his *Quaestiones Physicorum*, I.1, ed. Streijger & Bakker, 10^{8-12}, Buridan writes: 'Immo manifestum est quod non quaerimus habere scientiam tribus primis modis nisi propter habere scientiam illo quarto modo. Non enim curaret artifex de propositionibus et terminis, nisi per hoc crederet habere scientiam de rebus circa quas intendit agere et sibi utilia procurare.'

of what Aristotle says about motion in Book III to Chapter IV on motion and time. There he discusses the views on motion of John Duns Scotus, Nicholas Bonet, Gerard of Odo, John of Bassols, Francis of Meyronnes, Peter Aureol, Francis Bleth, John the Canon (i.e. Francesc Marbres), Graziadei of Ascoli, Gregory of Rimini, William of Ockham, and, finally, John Buridan 'and his students.'[61] Concerning Buridan, Duhem turns directly to Buridan's question III.7, 'Whether local motion is a thing distinct from place and from that which is locally moved.' In his discussion, Duhem emphasizes Buridan's rejection of Ockham's view that local motion is not something separate from the permanent things, i.e. the mobile and the distances gained. Near the end of his discussion of motion, Duhem says: 'After Buridan and Albert of Saxony, the Scholastics did not find anything else new to say on the nature of motion; as almost always happens, reading the works of Marsilius of Inghen [in which he includes the questions we here ascribe tentatively to Johannes Marsilii (?)] announces the decline of the School of Paris.'[62] In volume X, Duhem scorns the arguments of Paul of Venice on the question whether local motion differs from the mobile and the space traversed, Paul representing a conservative return towards the views of Aristotle and Averroes, with which Buridan had broken.[63] Nevertheless, Paul of Venice's final position is not unlike Buridan's, saying that local motion is a successive flowing (*fluxibile*) accident in its subject.[64]

Question III.1—Whether it is necessary that if motion is unknown, nature is unknown (*Utrum necesse sit ignorato motu ignorare naturam*). Aristotle, *Physics*, III, 1, 200b13–14; Averroes, *In Physicam*, III, comm. 1.

---

61 Duhem, *Le système du monde*, VII, 303–461. Ariew omits this whole section of Duhem's book from his English translation.

62 Duhem, *Le système du monde*, VII, 361: 'Après Buridan et Albert de Saxe, la Scolastique ne trouve plus rien de nouveau à dire sur la nature du mouvement; comme il advient presque toujours, la lecture des œuvres de Marsile d'Inghen nous annonce le déclin de l'Ecole de Paris.'

63 P. Duhem, *Le système du monde. Histoire des doctrines cosmologiques de Platon à Copernic*, X, Paris 1959, 412–413.

64 Duhem, *Le système du monde*, X, 414: 'Ce mouvement local, qu'est-il donc? "C'est un accident successif qui s'écoule au sein d'un sujet—*Motus localis est accidens successivum fluxibile in subjecto*." C'est la réponse qu'Avicenne avait donnée, que Jean Buridan avait formellement reprise. Cette réponse, Paul de Venise ne la veut pas attribuer à de tels auteurs; il faut, pour qu'elle lui paraisse recevable, qu'il la mette au compte d'Aristote.'

The table of questions for Book III has for the high points of the discussion: On the distinction of modalities between the composite sense and the divided sense. How [the propositions] 'I am ignorant of motion' and 'motion I am ignorant of' differ both on account of the negation implicit in a privative term and with regard to the appellation of a term following the verb 'I know'.[65]

Buridan's *tertia lectura* (which Patar calls the *secunda lectura*) also has 'necesse' before the rest of the proposition: *Utrum necesse sit ignorato motu ignorare naturam*, as do manuscripts Cesena, Biblioteca Malatestiana, cod. S.VIII.5 and Toulouse, Archives départementales de la Haute-Garonne, cod. 6. The manuscripts Brugge, Stedelijke Openbare Bibliotheek, cod. 477 and London, Wellcome Institute for the History of Medicine, cod. L.15, both ascribed to Albert of Saxony, have: *Utrum ignorato motu necesse sit ignorare naturam*.[66]

1. *Auctoritates Aristotelis*: 'Ignorato motu necesse est ignorare naturam' (2: 95)
2. **Er349(1)**: *Utrum, ignoto motu, necesse est ignorare naturam* (III.2)
3. **Er349(2)**: *Utrum, ignorato motu, necesse sit naturam ignorari* (III.1)
   Vat6758: *Utrum, ignoto motu, necesse sit naturam ignorari* (III.1)
4. **Ka11**: *Utrum, ignoto motu, necesse sit ignorare naturam* (III.1)
5. **L1386(1)**: *Utrum, motu ignoto, natura ignoratur* (III.1)
6. **L1386(2)**: *Utrum, ignorato motu, oportet ignorare naturam* (III.1)
7. **M²**: *Queritur de hoc quod dicit, quod ignoto motu necesse est ignoscere naturam* (III.1)
8. Manuscript Paris, Bibliothèque Mazarine, cod. 3493 (Anonymous): *Utrum habeat veritatem* ⟨*scil. quod ignorato motu necesse sit ignorare naturam*⟩[67]
9. Manuscript Paris, Bibliothèque Nationale de France, cod. lat. 14698 (Anonymous): *Utrum ignorato motu necesse sit ignorare naturam* (III.1)[68]
10. William of Clifford: *An ignorato motu, necesse est ignorare naturam* (III.4)
11. Boethius of Dacia: *Utrum ignoto motu necesse sit ignorare naturam* (III.1)
12. Giles of Rome: *Utrum ignorato motu ignoratur natura* (III.1)[69]

---

65 Cf. the following passage: 'De distinctione modalium inter sensum compositum et sensum divisum. Quomodo differt "ignoro motum" et "motum ignoro" tam ex parte negationis implicitae in nomine privativo quam de appellatione termini sequentis hoc verbum "cognosco"' (4^{4-7}).

66 Cf. Patar, *La Physique de Bruges*, 2: 37 (Annexe I); 2: 255 (Annexe III); 2: 293 (Annexe IV); and 427 (Appendice III).

67 Zimmermann, *Verzeichnis*, 277.

68 Zimmermann, *Verzeichnis*, 286.

69 Giles writes (*Commentaria in octo libros Phisicorum*, 50ra): 'Dicit ergo quod, quia natura

13. John of Jandun: *An necessarium sit, ignorantes motum ignorare naturam* (III.1)[70]
14. William of Ockham (*Expositio*): III, cap. 1 (t. 1, 200b12–15)[71]
15. Walter Burley (*Expositio*): *Utrum ignorato motu necesse sit ignorare naturam* (III.D.1)[72]
16. Nicole Oresme: *Utrum ignorato motu necesse sit ignorare naturam* (III.1)
17. Albert of Saxony: *Utrum ignorato motu necesse sit ignorare naturam* (III.1)[73]
18. Marsilius of Inghen: *Utrum ignorato motu necesse sit ignorare naturam* (III.1.1)[74]

---

est principium motus et mutationis et scientia nobis est de natura, igitur (.i. *ed.*) oportet nos naturales non ignorare quid sit motus. Necessarium est enim ignorato ipso motu ignorare naturam. Dubitaret forte aliquis quod, cum cognita causa non oporteat cognosci effectum, poterit sciri natura secundum id quod est absque eo quod cognoscatur motus. Non ergo oportet quod ignoto motu ignoretur natura. Dicendum quod absolute loquendo non oportet quod cognita causa secundum id quod est, quod cognoscatur effectus. Ad cognitionem enim effectus non sufficit scire causam nisi sciatur quoniam illius est causa vel sub ratione qua est causa.' Note that Giles has a translation of Aristotle, putting 'necessarium' at the start of the proposition, although the list of questions does not reflect this completely (leaving out the word 'necesssary').

70 Note that John of Jandun, like Buridan, places the word 'necessary' early in the proposition.

71 Ockham, *Expositio*, III, cap. 1, §1, 411: 'Circa primam partem primo ostendit quod ad naturalem pertinet determinare de motu, et hoc sic. Considerantem de natura oportet considerare de motu. Quod patet, quia natura est principium motus et status et mutationis, et per consequens ignorato motu ignoratur definitio naturae, cum motus ponatur in eius definitione; ergo cuius est considerare naturam et definitionem eius, ipsius est cognoscere motum. Sed scientia naturalis habet cognoscere naturam et definitionem naturae. Ergo scientia naturalis habet cognoscere motum. Sciendum est quod, licet illud quod est natura posset cognosci non cognito motu, tamen …'

72 Burley, *Expositio in Physicam* (1501), 60rb: 'Sed dubitatur an ignorato motu necesse sit ignorare naturam. Et videtur quod non, quia naturam esse est per se manifestum … Dicendum est quod naturam ignorari contingit dupliciter, scilicet vel si est, vel quid est. Dico igitur quod ignorato motu quid est non est necesse ignorare naturam si est.'

73 Cf. G. Geréby, '*Ignorato motu, ignoratur natura*. Logic and Physics in *Sophisma* II.110 of Albert of Saxony', in: J. Biard (ed.), *Itinéraires d'Albert de Saxe. Paris-Vienne au XIVe siècle*. Actes du Colloque organisé le 19–22 juin 1990 dans le cadre des activités de l'URA 1085 du CNRS à l'occasion du 600e anniversaire de la mort d'Albert de Saxe, Paris 1991 (Etudes de philosophie médiévale, 69), 175–189.

74 Marsilius of Inghen, *Abbreviationes Physicorum*, 8^{va-vb}: 'Hic queri solet utrum ignorato motu necesse est ignorare naturam, quod Philosophus videtur velle … Ad primam dicitur quod hec dictio "ignoro" includit negationem negantem terminum in quem tran-

19. Johannes Marsilii (?): *Utrum ignorato motu necesse est ignorare naturam* (III.1)
20. Lawrence of Lindores: *Utrum ignorato motu necesse sit ignorare naturam* (III.1)
21. Benedictus Hesse: *Utrum ignorato motu necesse est ignorare naturam* (III.2)

---

sit actus verbi, ut ignoro motum, id est non cognosco motum. Ex quo sequitur hanc esse impossibilem "ignoratur motus", quia infert illam "non cognoscitur motus", que est impossibilis propter cognitionem divinam de necessitate omnia cognoscentem. Secundo sequitur quod ad illam "ignoratur motus" sequitur quodlibet aliud. Patet, quia antecedens est simpliciter impossibile, per precedens. Tertio sequitur quod secundum hanc mentem hec est necessaria et necessaria "si ignoratur motus, ignoratur natura". Patet per precedens, quia cum antecedens sit impossibile, ad ipsum quodlibet sequitur. Sed istud est preter intentionem Philosophi, quia non loquitur de cognitione Dei. Et ideo aliter dicitur quod Philosophus voluit intelligere quod ly "ignorato motu" exponatur sic: "si aliquis ignorat motum". Et secundum hoc tria dicuntur. Primum quia hec est falsa "si aliquis ignorat motum, necesse est ipsum ignorare naturam". Patet, quia quamvis omnis homo ignoret motum, adhuc contingens est et non necessarium ignorare naturam. Secundo quod cognitione actuali stat aliquando hominem naturam cognoscere et omnem motum qui est accidens ignorare cognitione actuali. Patet, quia si aliquis solum attenderet ad conceptum substantie, talis naturam cognosceret et tamen omnem motum qui est accidens ignoraret, eo quod actualiter non attendit nisi ad conceptum substantie; et ideo non representat accidentia. Tertio dicitur quod illa conditionalis est necessaria: si aliquis ignorat motum, talis ignorat naturam. Et hanc voluit Philosophus. Probatur quia: ille terminus "natura" supponit pro materia et forma inquantum talia sunt principia activa motus vel passiva, ut dictum est in primo notabili secundi. Item dictum est in nono notabili eiusdem quod verba significantia actum anime interiorem faciunt terminum sequentem se appellare suam rationem. Et sic est sensus eius: ignorat motum, id est ignorat rem que est motus secundum conceptum seu rationem (rem *ed.*) secundum quam dicitur motus; et quicumque ignorat sic motum, ignorat rem que est natura secundum conceptum secundum quem dicitur natura. Patet, quia natura significat materiam et formam connotando quod sunt principia motus, ut dictum est. Ergo in conceptu nature intelligitur conceptus motus. Igitur si aliquis cognoscit naturam secundum conceptum nature, talis cognoscit motum. Ergo ex opposito qui non cognoscit motum, non cognoscit naturam. Quare necessaria est conditionalis: si aliquis ignorat motum, talis ignorat naturam. Tamen non intelligas quod, si aliquis ignorat diffinitionem motus, quod talis etiam ignorat diffinitionem nature. Non enim est hoc verum, quia appellatio rationis non dicitur respectu diffinitionis, sed dicitur respectu conceptus secundum quem terminus suum significat significatum. Et ideo necesse est non habentem conceptum motus non habere conceptum nature, eo quod in conceptu nature includitur conceptus motus.'

Buridan's first question on Book III repeats a question that had been asked by many previous commentators, as can be seen in the anonymous thirteenth-century works listed above, reflecting the first paragraph of Aristotle's Book III, which in English translation states:

> Nature is a principle of motion and change, and it is the subject of our inquiry. We must therefore see that we understand what motion is; for if it were unknown, nature too would be unknown.[75]

The apparent purpose of Aristotle's statement here is simply to explain why he will devote the first part of Book III of the *Physics* to motion in general—he is not making a controversial point. Averroes' comment on this passage in his long commentary suggests that nature in its material sense might be understood before understanding motion, while the prime mover cannot be understood before knowing motion.[76]

Many commentators before Buridan follow Averroes in saying that nature cannot be understood before motion, because motion is a part of the definition of nature, as when it is said that nature is a principle of motion and rest. Some say that nature as substance can in the present be understood without motion: although our knowledge of nature began with knowledge of motion, this may not be in our minds at present.[77]

---

75 Aristotle, *Physics*, III, 1, 200b12–14, tr. R.P. Hardie & R.K. Gaye, in: J. Barnes (ed.), *The Complete Works of Aristotle*, The Revised Oxford Translation, Princeton 1984 (Bollingen Series 71/2), 1: 315–446, 342.

76 Cf. Averroes, *In Physicam*, III, comm. 1, in: *Aristotelis opera cum Averrois commentariis*, 4, Venezia 1562–1574 (repr. Frankfurt am Main 1962), 85H–I: 'Incoepit notificare causam propter quam contingit ei loqui in hoc libro de motu et de rebus que sequuntur motum, quoniam, cum consyderatio eius sit de natura, cuius definitio est principium motus et quietis, fuit necesse ei declarare secundum quot modos dicitur principium et quod dicitur de materia et forma, scilicet primum principium materiale et formale. Sed primum principium materiale potuit declarare, antequam perscrutaretur de motu et accidentibus motus. Principium autem quod est primus motor non potuit perscrutari de eo nisi post perscrutationem de motu. Et incoepit hic declarare quod ex definitione naturae apparet quod perscrutatio de motu est necessaria.'

77 In his middle commentary, Averroes, in Latin translation, states: 'Dicamus quod, cum in definitione naturae accipiatur motus, cum dictum iam sit naturam esse principium motus et quietis cogaturque physicus tractare de omnibus partibus huius definitionis et de eius consequentibus, ergo cogitur tractare de definitione motus' (*Expositio media super libros Physicorum Aristotelis*, III, comm. 1, in: *Aristotelis opera cum Averrois commentariis*, 4, Venezia 1562–1574 [repr. Frankfurt am Main 1962], 449C).

Most commentators before Thomas Aquinas, and many after, state this question in a form such as that found in the *Auctoritates Aristotelis*, as quoted above: 'Whether if motion is unknown, it is necessary that nature is unknown.' On the other hand, Buridan in his question, and a few others, use the text as found in the *translatio vetus* made by James of Venice, and in the Moerbeke translation, putting the word 'necesse' earlier in the sentence. James of Venice has:

> Quoniam autem natura quidem est principium motus et status et mutationis, scientia autem nobis de natura est, oportet non ignorare quid sit motus. Necessarium enim est ignorato ipso ignorari et naturam.[78]

The translations printed with Averroes' commentary, as well as with the commentaries of Thomas Aquinas and Giles of Rome, may not be the translations that the authors themselves used. Be this as it may, some of the printed Latin texts of Aristotle in commentaries put the word 'necesse' earlier in the question.[79] Buridan not only has a word order unlike that in the *Auctoritates Aristotelis*, putting the word 'necesse' at the start of the relevant proposition, but he

78 Aristotle, *Physica*, III.1, 200b13 14, *translatio vetus*, ed. F. Bossier & J. Brams, Leiden [etc.] 1990 (Aristoteles Latinus, VII/1), 96–97.

79 Apparently the translation published with the commentary of Thomas Aquinas in 1965 is not an accurate representation of the translation Aquinas used. It also places the word 'necessary' before the key clause: 'Quoniam autem natura est principium motus et mutationis, scientia autem nobis de natura est, oportet non ignorare quid sit motus: necessarium enim est ignorato ipso, et ignorari naturam' (Thomas Aquinas, *In octo libros Physicorum Aristotelis expositio*, III, l. 1, ed. M. Maggiòlo, Torino, Roma 1965, 139). The English translation of Aquinas' comments by Blackwell, Spath and Thirkel reads: 'Thus it is clear that if one does not know motion, one does not know nature, since motion is placed in the definition of nature. Since, then, we intend to set forth the science of nature, it is necessary to know about motion' (Thomas Aquinas, *Commentary on Aristotle's Physics*, tr. R.J. Blackwell, R.J. Spath & W.E. Thirkel, revised edition, Notre Dame [IN] 1999 [first ed. New Haven (CT) 1963], 141). Thomas Aquinas' commentary runs as follows: '*Natura est principium motus et mutationis*, ut ex definitione in secundo posita patet (quomodo autem differant motus et mutatio, in quinto ostendetur): et sic patet quod ignorato motu, ignoratur natura, cum in eius definitione ponatur. Cum ergo nos intendamus tradere scientiam de natura, necesse est notificare motum' (Aquinas, *In octo libros* Physicorum *expositio*, III, l. 1, 139–140). The Latin in Giles of Rome's commentary on the *Physics* is the same as that in the 1965 edition of Thomas Aquinas, namely that of the *translatio vetus*. The fourteenth-century *moderni* are exceptional in attaching so much importance to word order.

also makes word order a focus of his attention, because according to his logical analysis, it changes the meaning of the proposition. The fact that, after Buridan, most authors of questions on the *Physics*, continue to use the time-honored word order of the question, putting 'necesse' in the middle of the proposition, rather than at the start, shows that Buridan did not succeed in changing the established habit, although in their discussions many, like Albert of Saxony, might note the difference in meaning of the proposition depending on the location of the word 'necesse.'

Obviously, question III.1 was originally an exegetical question, because it arises from Aristotle's text, the claims of which might be doubted. Reading Averroes' commentary as well as Aristotle's text, Walter Burley notes the falsity of Aristotle's claim, reported by Averroes, that the first material principle can be known without motion.[80] Despite the exegetical origin of Buridan's first question, he answers it mainly using the tools of logic. He applies many of the methods of the *logica moderna*, including attention to the word order of propositions and how that affects the supposition of terms. Most of the seven principal arguments, with which Buridan starts, had occurred in earlier authors. In his parallel question Albert of Saxony includes versions of five of Buridan's seven principal arguments. The arguments are:

1. What is known in itself (*per se notum*) does not require to be known by something else; but Aristotle says that it is ridiculous to demonstrate that nature exists in that it is known in itself (8^{7-9}).[81]

80 Burley, *Expositio in Physicam* (1501), 60rb: 'Notandum secundum Commentatorem hic quod Philosophus primum principium materiale potuit declarare antequam perscrutaretur de motu et accidentibus eius, sed primum principium, scilicet quod est primus motor, non potuit perscrutari de eo nisi post perscrutationem de motu. Sed hoc non videtur verum, videlicet quod primum principium materiale possit declarari sine perscrutatione de motu, quoniam primum principium materiale non cognoscitur a nobis nisi per transmutationem. Ex hoc enim quod est transmutatio inter formas oppositas et unum oppositorum non sic aliud scimus quod est unum subiectum quod manet subiective sub formis oppositis, et sic per transmutationem cognoscimus materiam primam. Dicendum igitur quod materia prima non cognoscitur a nobis sine cognitione motus si est.'

81 Cf. Jandun, *Quaestiones Physicorum*, III.1 (first principal argument), 41rb: 'Arguitur primo quod non quia: illud est per se notum quod non ignoratur ignorato quolibet alio; quod enim ignoratur ignorato alio cognoscitur per illud et sic non per se. Sed natura est quid per se notum, ut videtur velle Aristoteles in secundo *huius*, ubi dicit quod ridiculum est demonstrare naturam esse; quod non est pro alio nisi quia est per se notum. Quare etc.' Argument also in Burley, *Expositio in Physicam* (1501), 60rb: 'Sed dubitatur an ignorato motu necesse sit ignorare naturam. Et videtur quod non, quia naturam esse est per

2. Nature can be known without that without which it can exist; but nature can exist without motion, for example the nature of the earth (8^{10-11}).[82]
3. As nature is a principle of motion, so it is a principle of rest. Therefore, as it can be known through motion, so it can be known by rest, i.e., without motion (8^{12-14}).[83]
4. Substance is prior to accident in cognition (*notitia*), time, and definition, as is said in *Metaphysics* VII. Therefore substance can be known without cognition of accidents. But nature is substance and motion is accident. Therefore nature can be known without motion (8^{15-18}).[84]
5. As a thing is to being so it is to truth or knowledge. This is clear in *Metaphysics* II. Since therefore accident is not the cause of substance in being, so it also is not the cause of knowledge of it (8^{19-21}).
6. Effects are suited to be known by their causes. Therefore cause is more known than effect, and what is better known is not dependent in being known on what is less well known. Therefore, since nature is the cause of motion, it does not depend on motion in being known (8^{22-24}).[85]
7. What is simply contingent never becomes necessary on account of the positing of some possible thing in existence. But it is simply contingent that nature is unknown, and it is possible that motion is unknown. Therefore no matter how much motion is unknown, it is not necessary for nature to be unknown. Hence nature can be known without motion being known (9^{1-4}).[86]

---

se manifestum; sed illud quod per se est manifestum non ignoratur quocumque alio ignorato; ergo.' Also in Oresme, *Questiones super Physicam*, III.1 (principal argument 3), 293; and in [Albert of Saxony], *Quaestiones Physicorum*, III.1 (principal argument 2), 2: 458.

82 [Albert of Saxony], *Quaestiones Physicorum*, III.1 (principal argument 5), 2: 459: 'Quinto. Natura potest esse non existente motu; ergo potest cognosci non cognito motu.'

83 [Albert of Saxony], *Quaestiones Physicorum*, III.1 (principal argument 6), 2: 460: 'Sexto. Natura potest cognosci per quietem; ergo non oportet quod ignorato motu ignoretur natura.'

84 Jandun, *Quaestiones Physicorum*, III.1 (principal argument 3), 41rb: 'Item ignorato accidente non est necesse ignorare substantiam, cum substantia sit prior accidente notitia, definitione et tempore, septimo *Metaphysicae*; sed motus est accidens naturae et natura est substantia, scilicet materia et forma; quare etc.'

85 Jandun, *Quaestiones Physicorum*, III.1 (principal argument 2), 41rb: 'Item effectu ignorato non est necesse ignorare causam; causa enim non videtur dependere ex effectu nec in esse, nec in cognosci, sed e converso. Sed motus est effectus naturae. Quare etc.' Also in Oresme, *Questiones super Physicam*, III.1 (principal argument 2), 293.

86 Oresme, *Questiones super Physicam*, III.1 (first principal argument), 293: 'Et arguitur quod

After the seven principal arguments and noting that Aristotle at the beginning of Book III took the opposite position, Buridan continues:

> In this question there are many difficulties, both logical and natural. Immediately I exclude from this question the cognition by which God and the intelligences know other entities. I wish to speak only of our normal human cognition (9^{6-9}).

It makes a big difference, he argues, where the word 'necessary' appears in the proposition. If it appears first, as in 'necesse est ignorato motu ignorare naturam', then the proposition is to be taken in the compounded sense, and the sense is: 'this is necessary: motion being unknown, nature is unknown', and the proposition is true. On the other hand, the word order in 'ignorato motu necesse est ignorare naturam' means that the proposition is to be taken in the divided sense, making it equivalent to: 'if motion is unknown, then it is necessary that nature be unknown.' The latter, however, is false, because it is contingent and not necessary that nature is unknown. Buridan's first conclusion is that the proposition taken in the divided sense is simply false if it is taken literally (*de virtute seu proprietate sermonis*). If something possible is posited to exist, it does not follow that something in itself contingent is necessary (9^{10-18}).

Buridan here assumes that logical rules enable the reader to distinguish false from true propositions. For instance, the verb 'ignore' (*ignoro*) is a privative word opposed to 'I know' (*cognosco*). As such it implies a negation, which distributes a term following it, but not one preceding it. Therefore 'motion I ignore' (*motum ignoro*) must be distinguished from 'I ignore motion' (*ignoro motum*), just as to say 'motion I do not know' differs from 'I do not know motion'—in the first statement the word 'motion' is not distributed, but in the second it is distributed. For example, if someone saw Socrates running and did not know of the flux and reflux of the tides, then there would be [some] motion that that person was ignorant of, without being ignorant of [other] motion. On the basis of such distinctions, Buridan addresses the question at issue, distinguishing between different cases. He says that words like 'I desire' (*appeto*), 'I know', etc., cause the terms that follow them in a grammatical construction to appellate the meaning by which they signify what they signify (*appellare rationes secundum quas significant ea quae significant*) (12^{2-3}). He

non, quia semper est contingens quod natura ignoretur a Sorte vel ab alio homine vel sciatur, ergo nullo posito fiet necessarium.'

illustrates this with the common distinction between senses of knowing a person who is approaching—you might know a person, but not know that the person now approaching is that particular person whom you otherwise know.[87] In the case at hand, the word 'nature' does not simply mean the natural world, but that world as related to motion, as in the proposition that nature is a source of motion. It is in this sense that the proposition is necessary 'he who is ignorant of motion is ignorant of nature' if there is someone who is ignorant of motion (*illa est necessaria 'ignorans motum est ignorans naturam', si aliquis est ignorans motum*) (13^{5-8}).

Buridan's near contemporaries at Paris answer the question in similar fashion, although with their own twists. Albert of Saxony is even more logical than Buridan. He has the same point as Buridan about the different meanings of the proposition depending on the location of the word 'necesse.'[88] He goes on to explain other logical technicalities and then states a number of logical suppositions and conclusions.[89] The third supposition is that words about knowledge cause terms following them to 'appellate' the concepts connected to them.[90]

87 Cf. Buridan, *Summulae de Dialectica*, 294–295.

88 [Albert of Saxony], *Quaestiones Physicorum*, III.1, 2: 461–462: 'Sed est advertendum quod istae propositiones multum differunt: *ignorato motu necesse est ignorare naturam* et *necesse est ignorato motu ignorare naturam*. Prima enim est divisa et secunda est composita.'

89 [Albert of Saxony], *Quaestiones Physicorum*, III.1, 2: 462: 'Adhuc istae multum differunt: *necesse est ignorato motu ignorare naturam* et *necesse est motu ignorato ignorare naturam*, quia in prima hic terminus *motu* supponit distributive, postquam sequitur istum terminum *ignorato* includentem negationem, et in secunda hic terminus *motu* non supponit distributive, postquam non sequitur aliquem terminum ipsum distribuentem. Et adhuc istae multum differunt: *necesse est ignorato motu ignorare naturam* et *necesse est ignorato motu naturam ignorare*, nam in prima hic terminus *naturam* non supponit pro re quam significat absolute sed cum appellatione rationis vel conceptus istius termini *natura*, in secunda autem supponit pro illa re quam significat absolute sine tali appellatione.'

90 [Albert of Saxony], *Quaestiones Physicorum*, III.1, 2: 463–464: 'Tertia suppositio quod ista verba *cognosco, intelligo*, etc. faciunt terminum sequentem se supponere non absolute pro re quam significat, sed cum appellatione rationis vel conceptus illius termini. Et propter hoc solet concedi ista in casu: *venientem cognosco*, et negari ista: *cognosco venientem*, et similiter concedi ista: *hominem cognosco*, et negari ista: *cognosco hominem*, vel ⟨concedi⟩: *capram cognosco*, et negari: *cognosco capram*, quia aliquis habens in mente sua conceptum entis communem omnibus entibus per illum conceptum quodlibet ens mundi cognoscit, et per consequens per illum conceptum capram cognoscit; et tamen, si talis non cognosceret definitionem caprae, ille non cognosceret capram, quamvis capram cognosceret, nam ista: *cognosco capram* significat quod cognoscam aliquid quod est capra secundum rationem secundum quam dicitur capra.'

The rest of Albert's treatment of the question states conclusions and replies to the principal arguments using the tools of logic.

Nicole Oresme makes use of the same logical point about the divided vs. compounded senses of propositions depending on their form.[91] It is the whole inference from not knowing motion to not knowing nature that is necessary, not just the second part of the statement. Unlike Buridan and Albert of Saxony, however, Oresme devotes most of his attention to empirical questions—how, for instance, does one know motion through the senses. There are two kinds of knowledge, one intuitive of individual things, and another through complexes or propositions. In another sense, to know something it is not enough to know a single proposition about it, but rather one must know a whole set of propositions, as in a scientific discipline.[92] Oresme concludes that nature can be known intuitively and incomplexly without motion. But nature, under the concept of nature, cannot be known without knowing motion.[93] Then the whole second part of Oresme's determination concerns sight and perception of motion, making many references to experience. Experiencing motion, since it takes time, requires not only sight but memory and reason, comparing one image to another.[94]

Although Oresme's answer to this first question makes much greater use of theories of perspective and observation than Buridan's answer or that of Albert of Saxony, all three could be said to bring in auxiliary disciplines to answer a question that had been asked for many decades by working within the disciplines of physics and of exegesis of texts, with perhaps some help from metaphysics or epistemology. What is distinctive of Buridan's work is not so much the conclusion he reaches but the methods he uses to reach that conclusion.

It is misleading, however, to say that this is distinctive of Buridan's work in particular as opposed to saying that it is distinctive of the Parisian nominalists

91 Oresme, *Questiones super Physicam*, III.1, 295: 'Tertio notandum quod questio debet intelligi tamquam condicionalis, scilicet: si motus non cognosceretur vel fuisset incognitus, hoc est quantum ad esse, non cognosceretur natura; vel si non sciretur motum esse vel posse esse, nesciretur naturam esse, ita quod condicionalis est necessaria. Ideo "necessarium" non determinat aliquam partem sed totam. Ideo debet preponi dicendo: "necesse est ignorato motu ignorare naturam".'

92 Oresme, *Questiones super Physicam*, III.1, 294.

93 Oresme, *Questiones super Physicam*, III.1, 294.

94 Oresme, *Questiones super Physicam*, III.1, 301: 'Ultima conclusio est quod nihil videtur moveri aut iudicatur aliter se habere nisi a virtute conservativa specierum quasi in memoria, licet non sit in memoria proprie.'

or fourteenth-century *moderni*. Benoît Patar has transcribed or edited from two or more manuscripts this question as found in Buridan's *tertia lectura* (called *secunda lectura* by Patar), contained in manuscripts Erfurt, Universitätsbibliothek, cod. CA F. 298, and Città del Vaticano, Biblioteca Apostolica Vaticana, cod. Chigi lat. E.VI.199 (*Utrum necesse sit ignorato motu ignorare naturam*).[95] He has also transcribed it as found in manuscript Cesena, Biblioteca Malatestiana, cod. S.VIII.5, which he calls a 'cas particulier' (*Utrum necesse sit ignorato motu ignorare naturam*).[96] And then he has transcribed it from two manuscripts in which it is ascribed to Albert of Saxony: manuscripts Brugge, Stedelijke Openbare Bibliotheek, cod. 477 (*Utrum ignorato motu necesse sit ignorare naturam*[97]) and London, Wellcome Institute for the History of Medicine, cod. L.15 (*Utrum ignorato motu necesse sit ignorare naturam*).[98]

It might be noted that the two versions of the question commonly ascribed to Albert of Saxony (Brugge 477 and the Wellcome 15 manuscript) have the common word order (*Utrum ignorato motu necesse sit ignorare naturam*), whereas Buridan's *tertia lectura* and *ultima lectura* as well as the version ('cas particulier') in the Cesena manuscript have the new word order with 'necesse' at the start of the proposition. All the versions of the question listed here explain that the question can be solved either logically or physically. If these versions are compared with the similar questions in John of Jandun, Walter Burley, and other earlier authors, then it can be seen that the traditional way of answering the question is being called physical, although it includes also epistemology or issues related to the *Posterior analytics*.

---

95 Cf. Patar, *La Physique de Bruges*, 2: 37–45.

96 Cf. Patar, *La Physique de Bruges*, 2: 255–259. In a chart on p. 361, Patar lists the question as *Utrum ignorato motu necesse sit ignorare naturam*, but this seems to be in error.

97 [Albert of Saxony], *Quaestiones Physicorum*, III.1, 2: 457–472 (see also Patar, *La Physique de Bruges*, 1: 351*–354*).

98 Cf. Patar, *La Physique de Bruges*, 2: 293–297 (see also 1: 138*–139*, 147*, 431*–436*). At least Books I–V of the *Physics* in this manuscript, if not all eight books, are ascribed to Albert of Saxony. Contrary to Patar, I find the text in the Wellcome manuscript of a high quality, if more terse than some other versions. The determination begins as follows (2: 294): 'Pro evidentia quaestionis est notandum quod ista quaestio potest solvi logice et physice. Logice potest solvi si quaeratur utrum ista propositio sit concedenda: *ignorato motu necesse est ignorare naturam*. Pro quo notandum est quod, quando ille modus *necesse* ponitur in medio propositionis, talis propositio est in sensu diviso, id est termini supponunt significative; quando autem ponitur ante totam propositionem vel post, tunc propositio capitur in sensu composito, id est termini supponunt materialiter, ut dicendo: *necesse est ignorato motu ignorare naturam*.'

Patar argues that manuscript Brugge 477 transmits the oldest known version of Buridan's *Quaestiones Physicorum*, as Albert of Saxony might have used it as the basis for his early teaching.[99] If this were the case, would the later versions of Buridan's questions have represented a development beyond the Brugge 477 version, and what would count as development? In the case of question III.1 just discussed, I do not see obvious differences of doctrine. One difference is that, whereas most of the Parisian versions of question III.1 make use of logic without first setting out the relevant rules, the text in manuscript Brugge 477 has separate suppositions:

> Et ad videndum ista clarius primo ponam aliquas suppositiones logicales; secundo ponam conclusiones de quaesito.
>
> Prima suppositio sit ista: omnis negatio, vel omnis terminus includens negationem, confundit terminum sequentem se confuse, distributive, mobiliter, nisi aliud syncategorema impediat. Ista suppositio patet ex logica.
>
> Secunda suppositio est quod hoc verbum *ignoro* includit negationem. Patet hoc, quia *ignoro* est idem quod *non cognosco*. Ecce clare negationem.
>
> Tertia suppositio quod ista verba *cognosco*, *intelligo*, etc. faciunt terminum sequentem se supponere non absolute pro re quam significat, sed cum appellatione rationis vel conceptus illius termini ...[100]
>
> Quarta suppositio: quod li *necesse* aliquando reddit sensum divisum, aliquando compositum. Et quando sic vel sic, hoc scitur ex logica.
>
> Quinta suppositio: oratio in qua ponuntur duo ablativi in designatione consequentiae potest exponi per unam conditionalem, sicut ista: *ignorato motu, ignoratur natura*, potest exponi: *si ignoratur motus, ignoratur natura*.[101]

If at approximately the same time, before which logic had not been so widely deployed, one person producing a polished version of questions on the *Physics* sets out logical suppositions like these and another person simply uses these logical rules without first listing them as suppositions, does this provide any evidence as to which work is earlier or later? Actually, a list of logical suppositions appears not only in Bruges 477, but also in London, Wellcome 15, and

99 Patar, *La Physique de Bruges*, 1: 243*–446*.

100 For the continuation of this passage, see *supra*, n. 90.

101 [Albert of Saxony], *Quaestiones Physicorum*, III.1, 2: 462–465.

then there are logical suppositions too in manuscripts Cesena, Malatestiana S. VIII.5 and Toulouse 6 (Patar's 'cas particulier').[102]

If the tools of the *logica moderna* were introduced into teaching and disputing on Aristotle's natural works and then continued to be used over time, would it be more likely that the logical tools would be explicitly stated in the edited questions on the *Physics* when they were first introduced, or might they be included later to make the resulting text self-contained as it might be used in other universities? Further consideration of the relation between different versions of questions on the *Physics* by Buridan or Albert of Saxony is included in the discussion of question III.2. An alternate explanation for the listing of suppositions in manuscript Brugge 477, but not in Buridan's *ultima lectura*, might be to suggest that this approach is typical of Albert of Saxony, but not of Buridan. As in the case of question III.1, Patar transcribes or edits from several manuscripts a question on Book IV of the *Physics*, entitled *Utrum possibile sit esse vacuum* (or *vacuum esse*), and a similar inquiry could be made comparing other fourteenth-century Parisian treatments of that question.

After question III.1, which goes back to something Aristotle had written, Buridan's second question takes up a distinction initiated in Averroes' comment 4 on Book III, which distinguished between motion as a *forma fluens* and motion as a *via ad formam*. Averroes made a similar point in comment 9 on Book V. These comments had been elaborated by Albert the Great, leading to the frequent distinction between *forma fluens* and *fluxus formae*.[103]

Question III.2—Whether for alteration is required a flux distinct from the alterable and from the quality according to which the alteration occurs (*Utrum ad alterationem requiratur fluxus distinctus ab alterabili et a qualitate secundum quam est alteratio*).[104] Aristotle, *Physics*, III, 1, 200b32–201a3; Averroes, *In Physicam*, III, c. 4.[105]

---

102 Patar, *La Physique de Bruges*, 2: 256–258. For instance, there is the supposition common to several of these texts (257): 'Nota secundum grammaticos quod ablativus absolutus potest resolvi per *dum*, per *si* et per *quia* …'

103 Cf. Maier, *Zwischen Philosophie und Mechanik*, 74 ff.

104 The table of questions mentions no subtopics (4^{8-9}).

105 Averroes, *In Physicam*, III, comm. 1, 87C–E, 'Motus enim nihil aliud est quam generatio partis post aliam illius perfectionis ad quam intendit motus, donec perficiatur et fit in actu … Secundum autem quod est via ad perfectionem, quae est alia ab ipsa perfectione, necesse est ut sit genus per se. Via enim ad rem est aliud ab ipsa re. Et secundum hoc fuit positum praedicamentum per se. Et iste modus est famosior, ille autem est verior. Et ideo

Aristotle writes:

> There is no such thing as motion over and above the things. It is always with respect to substance or to quantity or to quality or to place that what changes changes. But it is impossible, as we assert, to find anything common to these which is neither 'this' nor quantity nor quality nor any of the other predicates. Hence neither will motion and change have reference to something over and above the things mentioned, for there *is* nothing over and above them.[106]

1. M²: *Cum mobile sit in duplici potentia passiva, scilicet respectu viae et respectu termini, quaeritur quae istarum est eius potentia secundum quod motus est eius actus* (III.8)
2. G¹: *Cum motus dicitur de istis duobus ⟨sc. de forma et de via ad formam⟩, de quo istorum dicitur verius* (III.7)
3. **Er349(2)**: *Utrum motus sit de genere perfectionis ad quam vadit* (III.2)
4. **L1386(1)**: *Utrum motus sit de genere perfectionis illius ad quod est motus* (III.4)
5. William of Clifford: *In quo genere naturae est motus, scilicet utrum sit de genere successivorum vel permanentium* (III.5)
6. Geoffrey of Aspall: *Propter hoc quod Commentator ponit hic quod motus sit ipsa forma inducenda non differens nisi secundum magis et minus, quaeritur utrum motus sit illa forma* (III.5)
7. Thomas Wylton: *Utrum motus sit aliquo modo in genere termini ad quem est motus* (III.7)[107]
8. John of Jandun: *An motus sit essentialiter idem cum termino ad quem* (III.6)
9. William of Ockham (*Sent.* II): *Utrum motus sit vera res extra animam differens realiter a mobili et a termino* (7)[108]

Aristoteles induxit illum modum famosum in *Praedicamentis* et istum modum verum in hoc libro.'

106 Aristotle, *Physics*, III, 1, 200b32–201a3, tr. Hardie & Gaye, 342.

107 Trifogli, 'Due questioni sul movimento', contains texts of Wylton's questions III.6 and III.7.

108 Ockham, *Sent.* II, q. 7, 100–103: 'Hic est una opinio communis omnibus quod motus addit aliquid distinctum ex natura rei super mobile et terminum, et illud secundum aliquos est absolutum, secundum alios est respectivum ... Contra istam opinionem arguo ponendo sex conclusiones. Quarum prima est quod motus alterationis non dicit aliquid positivum ultra res permanentes absolutas et respectivas, si ponantur respectus.' Ockham's discussion of this question continues on to page 151!

10. William of Ockham (*Expositio*): III, cap. 2 (t. 4, 200b32)[109]
11. William of Ockham (*Quaestiones*): *Utrum motus alterationis importet aliquam rem distinctam a rebus permanentibus* (15)
12. Walter Burley (*Expositio*): *Utrum motus sit res distincta a mobili* (III.D.6)[110]
13. Francesc Marbres: *Utrum motus sit aliqua realis entitas distincta essentialiter a termino ad quem est* (III.1)[111]

---

109 Ockham's text starts as follows (419): 'In ista parte ponit Philosophus quartam divisionem quae est quod cum motus non sit aliquid praeter res ad quas est motus, semper illud quod mutatur, aut mutatur secundum substantiam aut secundum qualitatem aut secundum quantitatem aut secundum locum. Et quia aliquis posset credere quod motus esset aliquid unum distinctum ab istis quattuor generibus et commune eis, sicut aliqui imaginantur quod motus est quaedam via ad rem distincta a re realiter, ideo hoc removet Philosophus ibi: *Commune autem in his*, dicens quod non est aliqua res communis istis ad quam terminatur motus, distincta ab eis …' The editors of Ockham's *Expositio*, insert the heading *Quod mutatio subita non est alia res a rebus permanentibus* for Section 5 (421–430) and the heading *Quod motus non est alia res a rebus permanentibus* for Section 6 (430–436). There Ockham writes (430): 'Istis visis de mutatione probandum est quod nullus motus est aliqua res secundum se totam distincta a rebus permanentibus.' He continues (433–434): 'Et si dicatur quod ista non-simultas partium est aliquid quando dicitur quod partes non sunt simul, dicendum est quod talis fictio nominum abstractorum de adverbiis, coniunctionibus, praepositionibus et verbis et syncategorematibus facit multas difficultates inexplicabiles et multos ducit in errores. Imaginantur enim multi per hoc quod sicut sunt nomina distincta ita sint res distinctae corre⟨s⟩pondentes ut tanta videlicet sit distinctio inter res significatas quanta est inter nomina significantia, quod tamen non est verum, sed aliquando eaedem res sunt significatae et tantum est diversitas in modo logicali vel grammaticali significandi. Et ideo non-simultas non est aliqua alia res a rebus quae possunt esse simul, sed significat quod res non sunt simul. Et ideo in modernis temporibus propter errores subortos ex usu talium abstractorum melius esset propter simplices in philosophia non uti talibus abstractis, sed tantum verbis, adverbiis, praepositionibus, coniunctionibus et syncategorematibus sicut primario fuerunt instituta quam fingere talia abstracta et uti eis. Immo nisi esset usus talium abstractorum "motus", "mutatio", "mutabilitas", "simultas", "successio", "quies" et huiusmodi, parva esset respective difficultas de motu et mutatione, tempore, instanti et huiusmodi.'

110 Cf. Burley, *Expositio in Physicam* (1501), 64rb: 'Sed circa naturam motus, id est cuius⟨modi⟩ ens sit motus, contingit dubitare, quoniam quidam de modernis philosophantibus dicunt quod nec motus nec mutatio est res secundum se totam distincta a rebus permanentibus. Et primo pono rationes eorum per quas probant quod motus non est res secundum se totam distincta a rebus permanentibus. Secundo ponam rationes probantes idem de mutatione subita' (manuscript Basel, Universitätsbibliothek, cod. F.II.30, 65ra *in marg.*: 'Hockam'). See also Trifogli, 'The Reception of Averroes' View on Motion.'

111 The authors mentioned are Aristotle, Averroes, Avicenna. Cf. Francesc Marbres, *Quaestiones Physicorum*, III.1, 34^{ra-vb}: 'Pono aliquas propositiones. Prima est ista: quod motus

14. Nicole Oresme: *Utrum motus sit res successiva sive fluxus distinctus a rebus permanentibus cuiusmodi sunt mobile et res acquisita ad quam est motus* (III.6) *Utrum moveri sit aliter se habere continue quam prius* (III.7)
15. Albert of Saxony: *Utrum de numero eorum quae sunt alia sint permanentia et alia successiva* (III.3) *Utrum motus alterationis sit res distincta a qualitate quae acquiritur et a qualitate quae deperditur et ab alterabili cui talis qualitas acquiritur et deperditur* (III.5)
16. Marsilius of Inghen: *Utrum ad alterationem requiratur aliquis fluxus distinctus ab alterabili* (III.1.2)
17. Lawrence of Lindores: *Utrum in alteratione sit ponendus fluxus distinctus ab alterabili et a qualitate secundum quam fit alteratio* (III.2)
18. Benedictus Hesse: *Utrum de motu alterationis praeter alterabile et qualitatem quae per alterationem acquiritur, oportet ponere aliquem fluxum esse distinctum ab alterabili et a qualitate, id est: utrum alteratio sit alia a subiecto et a qualitate acquisita* (III.9)
19. Conimbricenses: *An motus a termino in quem tendit et a mobili re ipsa distinguatur necne* (III.3)[112]

---

formaliter sumptus distinguitur essentialiter et realiter a termino ad quem est ... Secunda propositio est quod motus nihil aliud est nisi successiva generatio partis post partem ... Tertia propositio est ista quod motus non est ipse fluxus forme. Hec probatur quia: fluxus forme non est aliud quam successio secundum prius et posterius secundum quod pars posterior dicitur advenire priori; sed prius et posterius non sunt ipse motus, cum sint subiective in motu secundum Philosophum et Commentatorem quarto *huius*; ergo etc. ... His premissis pono duas propositiones. Prima est ista: quod motus est formaliter in genere passionis ... Secunda propositio est quod motus ut dicit accessum ad terminum ad quem est de genere termini ad quem vadit, saltem reductive ...' See also Bakker & Dekker, 'Antoine Andrée ou Jean le Chanoine?', 121–122.

112 Collegium Conimbricense, *Commentarii Physicorum*, III.2.3, 340–341: 'In hac disceptatione de duplici distinctione agendum nobis est. Primo de distinctione motus a termino in quem fertur. Secundo de distinctione eiusdem a subiecto cui inhaeret. Quod ad priorem attinet Aureolus apud Capreolum in 2, d. 1, q. 2, art. 2, Argentinas in 2, d. 19, q. 1, Venetus ad cap. 3 huius libri, multique alii opinati sunt motum distingui re a termino qui per ipsum acquiritur ... Quod spectat ad collationem motus cum subiecto in quo insidet, non distingui re ipsa ab eo motum probatur quia: motus localis angeli non videtur esse aliud quam ipsa angeli substantia prout aliis atque aliis partibus spatii successive respondet, et tamen hic motus recipitur in eadem substantia angeli ut in subiecto; idemque videtur dicendum de lationibus physicis respectu suorum mobilium. Non igitur inter motum et subiectum realis distinctio intercedit. *Articulus II. Totius controversiae enodatio.* Sit tamen conclusio motum a termino quem petit non re sed formaliter tantum distingui. Haec conclusio est D. Thomae 5 huius operis ad text. 9, Alexandri, Averrois et Themistii. Item Ochami in 2, q. 9, ac caeterorum Nominalium.'

Anneliese Maier made a thorough study of the issue lying behind this question, tracing opinions from Aristotle through Averroes, Avicenna, Albert the Great, John Duns Scotus, John of Jandun, Peter Aureol, William of Alnwick, Richard of Mediavilla, Antonius Andreae, William of Ockham, Walter Burley, John Buridan, Nicole Oresme, Albert of Saxony, Marsilius of Inghen, and Blasius of Parma.[113] In the preceding chapter of the same book ('Motus est actus entis in potentia …'), Maier had counted Buridan, along with Burley, as an opponent of Ockham's theory of motion, Buridan preferring to follow Thomas Aquinas and Giles of Rome with regard to local motion, if not with regard to alteration, augmentation, and diminution.[114]

At this point in Book III of the *Physics*, Aristotle was thinking about local motion and motion of augmentation and diminution, and not only about alteration. Buridan is changing the structure of Aristotle's reasoning by treating local motion, alteration, augmentation and diminution, generation and corruption, and mutation separately. Questions III.3–5 will cover alteration in particular. But something else is happening, namely that the *logica moderna* will be applied to the long-standing question of *fluxus formae* vs. *forma fluens*.

It appears that Ockham's works on the *Physics* provided a special impetus in this direction. Ockham's *Quaestiones* on the *Physics* are not arranged according to the order of the books of Aristotle's *Physics*. They begin with seven questions related to concepts, and then they turn to a long list of questions which might as a group be paired with Buridan's second question, except that, with the exception of question 15, they do not concern alteration in particular:

8. *Utrum mutatio subita sit aliqua res absoluta totaliter distincta ab omnibus rebus permanentibus*
9. *Utrum mutatio subita sit res respectiva*
10. *Utrum mutatio subita sit aliqua res*
11. *Utrum secundum intentionem Philosophi mutatio subita sit res distincta a rebus permanentibus*
12. *Utrum motus successivus importet respectus distinctos a rebus absolutis*
13. *Utrum motus sit aliquid absolutum distinctum a rebus permanentibus*
14. *Utrum motus localis importet rem aliam distinctam a rebus permanentibus*
15. *Utrum motus alterationis importet aliquam rem distinctam a rebus permanentibus*

113 Maier, *Zwischen Philosophie und Mechanik*, 59–143 ('Forma fluens oder fluxus formae?').

114 Maier, *Zwischen Philosophie und Mechanik*, 53.

16. *Utrum motus augmentationis importet aliquam rem distinctam ab omnibus rebus permanentibus*
17. *Utrum possit probari ratione sufficienti quod motus importet aliud a rebus permanentibus*
18. *Utrum intentio Aristotelis sit ponere motum aliam rem a rebus permanentibus*
19. *Utrum Commentator posuerit motum esse rem distinctam a rebus permanentibus*

What differentiates this list of questions from earlier discussions of the question whether motion was more truly a *forma fluens* or a *fluxus formae* is the constant use of the term 'thing' (*res*), indicating that real things are now a focus of attention.

Buridan concludes question III.2 as follows: to this question I respond that in true alteration, as in heating, there is not some other flux besides that heat which is continuously acquired part by part and that cold which is, conversely, continuously given off part by part (18^{4-7}). Among the arguments for this position are the following:

3. The Commentator concedes that the *forma fluens*, namely the quality which is acquired part by part, is the motion in its truer sense. This appears to be true. The *forma fluens* agrees with the definition of motion, for as long as the water is heated something is acquired of that heat and something remains to be acquired. So that heat is entirely the act of what is alterable insofar it is acquired, while it is still in potency to what remains to be acquired. Now if it is conceded that this quality is truly motion, then there is no reason why, to save motion, it is necessary to posit something else in addition (17^{6-14}).
4. Also if there were such an added flux, it would be either an action or a passion, as the Commentator says. But it would be worthless if we posited in alteration an action or passion added to the quality that is acquired, because without such an addition we could save everything, saying that the quality according to different points of view is said to be action and passion: that is, it is called action insofar as it is produced by the agent and it is called passion insofar as it is received in the body affected. But nothing should be posited uselessly in nature. Therefore there is no need to posit an added flux (17^{15-21}).

What is the meaning of 'flux' for Buridan? If Buridan's question is compared to Oresme's similar question, one sees that in Oresme's questions on the *Physics* the notion of 'flux' has more than one meaning. One can mean by a 'successive entity' a form existing in the mobile at any moment, or the whole process of change over time. In reply to his question III.7, 'Whether to be moved is continually to be otherwise than before' (*Utrum moveri sit aliter se habere continue quam prius*), Oresme writes:

> Lastly, from a comparison of the said opinions, it seems that any one of them touches on the truth, but, poorly understood, deviates from what is right. The first opinion says that motion is nothing (*nihil*). Taking something (*aliquid*) for something that really is some thing, this should be conceded, as Aristotle says that an accident is not a thing but of a thing (*non est ens, sed est entis*), also because it is not permanent but successive; about the latter [i.e. the successive] 'to be' is said equivocally. Secondly, the opinion that says that motion is many, although it is not worth much, nevertheless the truth is that for motion to exist, it is necessary that many exist, because an indivisible is not moved. The third opinion, which says that [motion] is the mobile—because it is imagined that to be so and so (*taliter*) is nothing but the thing being in this way (*res sic se habens*)—is plausible, because this condition or flux is not something superadded (*tale superadditum*), as many imagine. Nor is it a thing separable by any power, as white is separable, as is clear in the Sacrament. The fourth opinion, positing that it is what is acquired, also is true, because motion in one way is taken for what is acquired; but it is false in that it denies the other way of taking it (*acceptionem*). The other opinion, which posits that it is a flux in the way of (*ad modum*) a distinct form, as would be whiteness or the soul or something similar (*aliquod tale*) is the worst opinion of all (*omnium pessima*); however, if it is understood that it is not such a form or such a thing, but a mode or a condition of the mobile, then it is the most true (*verissima*) and more probable and easier than the other opinions, and it agrees with what Aristotle and philosophers say.[115]

115 Oresme, *Questiones super Physicam*, III.7, 340–341: 'Ultimo potest apparere de comparatione opinionum dictarum. Unde videtur quod quelibet tangat veritatem, tamen male intellecta deviat a rectitudine. Prima que dicit quod motus nichil est: capiendo "aliquid" pro eo quod vere est aliquid, concedendum est, sicut dicit Aristoteles quod accidens non est ens, sed est entis; etiam quia non est permanens sed successivum; et de istis dicitur "esse" equivoce. Secunda, que dicit quod motus est multa, licet parum valeat, tamen veritas est quod ad hoc quod motus sit, oportet multa esse, quia indivisibile non movetur. Tertia, que dicit quod est mobile, quia ymaginatur quod taliter se habere non sit nisi res sic se habens, secundum hoc habet apparentiam, quia etiam ista condicio vel fluxus non est tale superadditum, sicut multi ymaginantur, nec est res separabilis quacumque vir-

So far I have not noticed evidence that would clearly indicate whether Buridan knew this passage of Oresme's *Questiones super Physicam* or Oresme knew Buridan's answer to questions about the nature of motion, but there is plenty of evidence that Buridan, Oresme, Albert of Saxony, and Marsilius of Inghen were part of an interconnected community or network, so that their various works can shed light on each other. If the editors of Oresme's *Questiones super Physicam* are correct, after Mirecourt's condemnation in 1347, Oresme and others shied away from what had been Oresme's practice in his *Physics* questions of explaining reality using concepts of 'modi rerum' and 'condiciones', which assumed levels of ontological reality between real physical things and mere mental conceptions, then Buridan too in his *Quaestiones Physicorum* may well have been guarded in his ways of expressing himself about the ontological status of motion.

But it should also be noted that Ockham had argued that a 'mode' should not be understood as if it were something distinct from a thing. A qualitative form might have one mode during alteration, namely to be in flux, and another mode once alteration is complete and it is at rest. A similar distinction between the mode in which Christ's body is in place in heaven (circumscriptively) and the mode in which it is in place in the Eucharist (definitively) does not involve something distinctive existing in the body of Christ in the Eucharist.[116]

---

tute, sicut est albedo, ut patet in sacramento. Quarta, ponens quod est acquisitum, etiam est vera, quia motus uno modo capitur pro acquisito, sed falsa est in eo quod negat aliam acceptionem. Alia, que ponit quod est fluxus ad modum unius forme distincte, sicut esset albedo vel anima vel aliquod tale, est omnium pessima; tamen, si intelligatur quod non sit talis forma vel talis res, sed modus vel condicio ipsius mobilis, tunc est verissima, et probabilior, et facilior inter omnes, et concordat dictis Aristotelis et philosophorum.'

116 Ockham, *Summa logicae*, pars III-4, cap. 6, 781–783: 'Item tales propositiones "una potentia habet diversos modos operandi", "diversi modi essendi possunt eidem competere sine variatione rei" et huiusmodi; quia si accipiantur proprie, intelligendo per "modum" aliquid distinctum a re, sic tales propositiones sunt falsae. Quando enim dicimus quod anima respectu intellectionis et volitionis habet diversum modum operandi, non intelligimus quod sint aliqui modi distincti ab anima et ab actibus productis, quasi essent quaedam media; hoc enim simpliciter falsum est; et ideo sub tali intellectu tales propositiones sunt falsae. Alius sensus est iste "eadem res diversimode operatur", puta necessario elicit intellectionem et contingenter et libere elicit volitionem. Similiter si dicatur quod alius modus essendi competit corpori Christi in caelo et in sacramento Altaris, non est imaginandum quod modus essendi qui competit corpori Christi in caelo sit aliquid adveniens corpori Christi, distinctum ab uno alio quod advenit corpori Christi in sacramento. Sed per talem propositionem non intelligimus nisi quod corpus Christi est circumscriptive in loco in caelo et non in sacramento Altaris. Et si queras, quid est illa circumscriptivitas, dico

Is there any way to track the lines of influence in these arguments against the separate existence of motion beyond permanent things? Buridan's *Quaestiones Physicorum* obscure the question to some extent by starting off with alteration, concerning which there was less motive to conclude that motion is a flux beyond whatever form may be gained in motion (Buridan says there is not in the case of alteration). Moreover, he then goes on to three further questions on alteration derived directly or indirectly from Walter Burley's *Tractatus secundus de intensione et remissione formarum*. What might have been among the initial questions on Book III of the *Physics*, if Buridan had not taken this alternate route?

Interestingly, Albert of Saxony has in his *Quaestiones* on the *Physics*, after a second question on the definition of motion, the third question: 'Whether among the number of things that are, some are permanent and others successive' (*Utrum de numero eorum quae sunt alia sint permanentia et alia successiva*). This seems to be a question not found in any of the versions of Buridan's questions on the *Physics*, nor is it in the London Wellcome manuscript, although in that manuscript there is the question (not transcribed by Patar), entitled *Utrum motus sit aliqua res distincta a mobili et a rebus permanentibus* (III.3), which might have a similar content. Also, in the above list of related questions, for William of Clifford there is a similar question, so it might exist in other earlier sets of questions as well. Albert of Saxony starts with the interesting point that he is inquiring about the things that are (*de numero eorum quae sunt*) and not about entities.[117] If the word 'motion' is considered to have

quod est vox non-significativa, sicut bu-ba. Unde fingere talia abstracta de talibus adverbiis, verbis, praepositionibus et huiusmodi, est simplicibus multorum errorum occasio; tamen aliquando utilis potest esse intelligentibus, quia per tales fictiones frequenter brevius loqui possunt. Item omnis oratio in qua ponitur modus infinitivus pro supposito, sicut sunt tales "legere est bonum", "currere est moveri", "calefacere est agere" et huiusmodi, distingui possunt, quia unus sensus potest esse per quem denotetur quod praedicatum tale competat alicui quod nec est agens nec patiens nec effectus, quasi tales modi infinitivi importarent res distinctas ab agente et patiente et effectu et ceteris rebus quae possunt esse agentia et patientia et effectus producti. Et talis sensus simpliciter falsus est secundum principia Aristotelis. Alius sensus potest esse ut tales propositiones ponantur loco orationum in quibus participium praedicatur de participio vel verbum de participio vel duo verba correspondentia de eodem, ut talis oratio "calefacere est agere" habeat istum sensum "quod calefacit, agit" sive "calefaciens est agens;" et ista "legere est bonum" habeat istum sensum "legens facit bonum opus" sive "qui legit, facit bonum" et huiusmodi. Et per istum modum evacuantur multae difficultates.'

117 [Albert of Saxony], *Quaestiones Physicorum*, III.3, 2: 482–483, 'Notandum est tamen primo quod notanter formo quaestionem sub his verbis: "Utrum de numero eorum quae sunt,

supposition for an entity of some sort, this would limit the ways in which kinematics or dynamics could be developed. Anneliese Maier long ago pointed out that Buridan's famous concept of 'impetus', which had been said to be a step in the direction of the idea of inertia, actually was conceived as a qualitative form inhering in the mobile, and in some sense causing it to move, which is not what Newton's law of inertia calls for.[118]

Beyond the question of aiming for a minimalist ontology or applying Ockham's razor, there was the problem for Ockham's program of analyzing the truth of propositions by the things in the external world for which terms in propositions supposit (or have personal supposition). A presupposition of this program was that terms have supposition for presently existing things unless syncategorematic terms cause 'ampliation', so that a term might also stand for things existing in the past or future. This was a prominent issue in the analysis of propositions including the terms 'it begins' and 'it ceases' (*incipit* and *desinit*). To distinguish continuous motion from rest or sudden mutation, it seemed to be necessary to consider not only what is true at the present instant (if instants are accepted as existing), but what might be true immediately before or after a given time. Albert of Saxony's question on successive and permanent things immediately opens a vista in which many things thought to be permanent, might actually be successive, if for instance God not only creates things but also conserves them according to the theory of continuous recreation. The principal arguments in Albert's question III.3 begin as follows:

> Et arguitur primo quod nihil sit permanens, sed quodlibet sit successivum praeter Primam Causam, quia, sicut prius dicebatur, omnia praeter Primam Causam dependent a Prima Causa in fieri et in conservari, sicut

---

alia sint etc.", et non sub his verbis: "Utrum de numero entium, quaedam sint permanentia etc.", quia non de omni termino significative accepto de quo est verificabile esse, est verificabile ens. Statim patet hoc, quia de hoc nomine *exercitus* significative accepto bene verificatur esse, sed non ens, quia, quamvis haec sit vera: *exercitus est*, tamen haec est falsa: *exercitus est ens*. Et ita secundum ponentes motum non esse aliud a rebus permanentibus, sed esse ipsas res permanentes taliter vel taliter se habere ad invicem, non concederetur motus esse ens, quamvis bene concederetur esse, et per consequens bene concederetur esse de numero eorum quae sunt, sed non de numero entium.'

118 A. Maier, 'Die naturphilosophische Bedeutung der scholastischen Impetustheorie,' in: Ead., *Ausgehendes Mittelalter. Gesammelte Aufsätze zur Geistesgeschichte des 14. Jahrhunderts*, 1, Roma 1964, 353–379 (= Ead., *On the Threshold of Exact Science*, ed. and tr. Sargent, ch. 4: 'The Significance of the Theory of Impetus for Scholastic Natural Philosophy,' 76–102).

> lumen a corpore luminoso; sed modo sic est quod continue fit aliud et aliud lumen a corpore luminoso, et nullum est permanens …
>
> Quarto, omne quod est, est in instanti praesenti; sed entibus successivis repugnat esse in instanti praesenti propter hoc quod partibus ipsorum repugnat simultas (similitas *ed.*), ex eo quod una eorum est praeterita et alia futura; modo praeterito et futuro repugnant esse simul.[119]

Aside from light, which may seem to endure, but in fact is continuously renewed, many other entities are partly successive—for example animals and plants continuously change while remaining the same thing. Beyond light, other seemingly enduring things could be totally successive. An example would be a river such as the Seine.[120]

Thus, according to Albert, we can understand that some things are in part permanent and in part successive. In this way, the parts of quality that continuously add up in an alteration can be understood as—what? The term to be expected here is '*forma fluens*', though Albert does not use this expression:

> Quinta conclusio: similiter aliqua accidentia dicuntur successiva secundum quid. Patet hoc, sicut si aliqua albedo vel caliditas ex una parte fieret et ex alia parte desineret.
>
> Sexta conclusio: aliqua accidentia dicuntur simpliciter successiva. Patet de sono, quia sonus est una qualitas sensibilis quae consequitur motum et non generatur nisi mediante motu, et sic se habet quod pars primo generata non manet cum sequente, recte sicut nec pars motus praeterita manet cum futura. Similiter sic est de curvitate, cum aliquis baculus successive continue incurvatur, plus et plus continue est alia et alia curvitas, nec praeexistens manet cum sequente.
>
> Septima conclusio quod in aliquibus permanentia sunt eiusdem speciei cum successivis. Patet, quia si aliqua curvitas esset permanens, sicut sunt multae, illa esset eiusdem speciei cum curvitate successiva.
>
> Octava conclusio: aliqua sunt successiva quae non habent aliqua eiusdem speciei permanentia. Patet hoc de sono, motu et tempore: nullius

119 [Albert of Saxony], *Quaestiones Physicorum*, III.3, 2: 481–482.

120 [Albert of Saxony], *Quaestiones Physicorum*, III.3, 2: 483–484: 'Secunda distinctio: aliquid potest imaginari simpliciter successivum tam secundum substantiam quam secundum eius dispositionem. Verbi gratia, sicut si Socrates a Prima Causa continue fieret et fieret, recte ad modum ad quem continue Sequana fluit et fluit, ita quod nihil de praeexistenti maneret.'

> enim est motus permanens neque tempus: unde tempus sic fluit quod impossibile esset ipsum arrestare.[121]

By creating a range of types of things between permanent things and successive things, Albert of Saxony creates a space for thinking about motion, not as a form inhering in the mobile, but as something spread out over time. Immediately, this raises the issue for alteration whether the degrees of quality gained over time are continuous with each other and form a growing qualitative form. This could explain why Buridan thought it appropriate to import three questions at this point from Walter Burley's *Tractatus secundus*, questions addressing how qualities vary in alteration.

Does it make any sense to evaluate the various opinions on motion as *forma fluens* or *fluxus formae* in relation to the new physics of the seventeenth century? Sarnowsky raises the question at the close of his discussion of Albert of Saxony's questions on the nature of motion in Book III of his *Quaestiones* on the *Physics*. Sarnowsky reports that Anneliese Maier suggested that the concept of motion as *fluxus formae* might have been progressive in the sense of moving in the direction of the conception of inertia, but Sarnowsky is skeptical:

> Albert ist zwar besonders Buridan verpflichtet, hat aber auch dessen Position nicht immer übernommen. Das wichtigste Beispiel dafür aus dem Kontext der hier behandelten Quaestionen is die Entscheidung über die Realität der Bewegung neben *mobile* und *locus*. Dabei ist es zweifelhaft, ob diese Diskussion einen Beitrag zur Entwicklung der Klassischen Physik seit dem 17. Jahrhundert leisten konnte. Anneliese Maier hat die Hypothese aufgestellt, dass die scholastischen Philosophen mit der Lehre über die Bewegung als *fluxus distinctus* einen Schritt vor einem Analogon zum Trägheitsprinzip stehengeblieben seien, das *Bewegung als eine Art von Eigenschaft oder Zustand* gefasst hätte. Doch wenn auch die Erfassung der Bewegung als Zustand ein konstitutives Element des Trägheitsprinzips ist, sind damit alle anderen Elemente noch nicht gegeben. Mit der gleichen Berechtigung könnte man auch in der gegensätzlichen Auffassung Ockhams eine gewisse Vorbereitung des Bewegungsbegriffs der klassischen Physik sehen, zumal sich diese auf eine Analyse der tatsächlich gegebenen Faktoren (Raum, Zeit, Körper) beschränkt.
>
> Alle Überlegungen in dieser Richtung besitzen jedoch nur geringe Bedeutung, weil die Klassische Physik die Frage nach dem Wesen der

121 [Albert of Saxony], *Quaestiones Physicorum*, III.3, 2: 485–486.

> Bewegung nicht mehr explizit stellt, während sie in den hier behandelten Zusammenhängen im Mittelpunkt steht.[122]

Historians may find one view of motion or the other more attractive, but the goal here remains to understand Buridan in his own context and not from the backward-looking perspective of seventeenth-century physics.

3.1.1 Questions III.3–III.5: New Questions on Quality and Alteration

After the very traditional first question on Book III, albeit with an answer full of the tools of logic, and a second question picking up on comments of Averroes and possibly also of Albert the Great and William of Ockham, Buridan turns to questions on alteration that had not appeared in earlier questions on the *Physics*, apparently motivated by the recent appearance within the context of theology of the so-called 'addition-of-part-to-part theory' of qualitative change or of the intension and remission of forms, and its comparison to the so-called 'succession of forms' theory. Several of the arguments in these questions are found in Walter Burley's *Tractatus secundus de intensione et remissione formarum*. Similar questions appear in works by other, later authors not in Book III, but in Book V.

Question III.3—Whether contrary qualities such as whiteness and blackness, heat and cold, may be compatible, according to some degrees of them, at the same time, in the same subject (*Utrum qualitates contrariae, ut albedo et nigredo, caliditas et frigiditas, possint se compati simul in eodem subiecto secundum aliquos gradus ipsarum*).[123] Cf. Aristotle, *Physics*, VI, 4, 234b10–21; Averroes, *In Physicam*, VI, c. 32.

1. William of Ockham (*Expositio*): VI, cap. 6 (t. 32, 234b10–17)[124]
2. William of Ockham (*Quaestiones*): *Utrum contraria possint simul naturaliter exsistere in eodem subiecto* (84) *Utrum contraria maneant in eodem*

122 Sarnowsky, *Die Aristotelisch-Scholastische Theorie der Bewegung*, 148–149.

123 The list of questions for Book III continues: 'De triplici contrarietate, scilicet terminorum, propositionum et rerum aliarum. Quae caliditas cui frigiditati sit contraria. Quomodo contraria maxime distant' (4^{11-13}).

124 In his *Expositio*, VI, cap. 6, § 2, 499, Ockham writes: 'Sciendum est quod ex ista littera aliqui volunt habere quod contraria sub gradibus remissis sunt simul. Quia dicit hic Philosophus quod omne mobile partim est in termino a quo et partim in termino ad quem, ergo quando aliquid dealbatur, partim est ibi de albedine et partim de nigredine, ergo simul sunt albedo et nigredo. Sed ista non videtur mihi esse mens Aristotelis. Sed dico, sicut praedictum est,

subiecto sub gradibus remissis (85) *Utrum intentio Philosophi sit quod formae contrariae sint in eodem subiecto primo* (86)

3. William of Ockham (*Philosophia naturalis*): Pars III, c. 22[125]
4. Walter Burley (*Expositio*): VI, c. 32[126]

---

quod nulla contraria sunt simul sub quibuscumque gradibus. Tunc enim idem simul esset calidum et frigidum, album et nigrum, amarum et dulce, quae omnia sunt contra principia Aristotelis.'

125 Ockham, *Philosophia naturalis*, Pars III, c. 22, 75^a–76^a (ed. Brown, III, 23, 318–320): 'Subiectum alterari secundum qualitates sensibiles non potest aliter quam per experientiam probari. Unde per experientiam scimus quod aliquid de calore alteratur ad frigus et e converso, et similiter de albedine ad nigredinem et e converso, et sic de aliis consimilibus qualitatibus. Et secundum illas qualitates non solum est alteratio qualiscumque, sed est secundum eas successio et alteratio quae est de contrario in contrarium, sicut de dulcedine ad amaritudinem et de albedine ad nigredinem. Sed de ista alteratione est una difficultas specialis, quae in aliis motibus et mutationibus non habet locum. Est autem difficultas haec: quando aliquid alteratur de contraria qualitate in contrariam qualitatem, an simul expellantur aut una qualitas introducatur in alia, ita quod illae qualitates secundum gradus remissos simul maneant. Et videtur aliquibus quod sic, pro quibus possunt adduci rationes et auctoritates ... Sed istis non obstantibus oppositum tenendum est, nam si remaneant tales gradus remissi, puta gradus albedinis, quando dealbatum denigratur, sequeretur quod simul et semel esset album et nigrum. Nam ille gradus est vera albedo, quia si sit aliquod oportet quod sit qualitas, sive gradus, et non nisi albedo; igitur est vera albedo et per consequens subiectum habens ipsum erit vere album. Et eodem modo gradus nigredinis est vera nigredo et per consequens subiectum informatum illo gradu est vere nigrum, et ita simul et semel esset album et nigrum.' See also William of Ockham, *Scriptum in librum primum Sententiarum—Ordinatio*, d. 17, q. 6, ed. G.I. Etzkorn, St. Bonaventure (NY) 1977 (Opera theologica, 3), 512: 'Ideo dico quod sicut in augmentatione quantitativa, de qua loquitur Philosophus, primo *De generatione*, nihil augmentatur nisi per adventum alicuius realiter distincti a priori et remanentis cum eo, ita est in augmentatione qualitatis; quod cum non possit salvari per motum localem nec per coexsistentiam cum maiori loco, oportet quod aliquid realiter addatur priori de novo. Et si quaeratur: unde est quod una forma potest augmentari et alia non, dico quod nulla est ratio nisi quia ista natura est talis et alia est talis. Unde universaliter quaelibet natura, sive sit perfecta sive imperfecta, si potest facere per se unum cum alia natura eiusdem rationis sine maiori extensione quantitativa, in illa natura potest esse augmentatio.'

126 Burley, *Expositio in Physicam* (1501), 185rb: 'Ex illo loco accipiunt aliqui quod contraria sunt simul in eodem, quia si aliquid mutetur ab albedine in nigredinem, partim est sub albedine, partim est sub nigredine. Sed sine dubio intentio Philosophi non est ad illud propositum, sed magis ad oppositum, ut patet per textum. Dicit enim sic: quod non oportet, cum aliquid movetur ab albo in nigrum, quod illud sit partim sub albedine et partim sub nigredine. Et si dicatur quod ad minus oportet concedere quod alterum contrariorum simul est cum forma media, quoniam cum aliquid movetur de albo in

5. Walter Burley (*Tractatus secundus*): *Quod forma non intenditur neque remittitur per maiorem vel minorem admixtionem cum contrario* (Cap. II)[127]
6. Nicole Oresme: *Utrum contraria sint simul in eodem* (V.6) *Utrum media inter contraria componantur ex extremis* (V.8) *Utrum unum contrarium intendatur altero non remisso* (V.9)
7. Albert of Saxony: *Utrum qualitates contrariae secundum aliquos gradus earum possint se simul compati in eodem subiecto* (V.9)
8. Marsilius of Inghen: *Utrum contraria possint se simul compati in eodem subiecto* (III.1.5)
9. Johannes Marsilii (?): *Utrum forme contrarie possunt esse simul* (III.5)
10. Lawrence of Lindores: *Utrum qualitates contrariae possunt se compati in eodem subiecto secundum aliquos gradus ipsarum* (III.3)
11. Benedictus Hesse: *Utrum qualitates contrariae, ut albedo et nigredo, caliditas et frigiditas, possunt se compati in aliquo subiecto secundum aliquos gradus ipsarum* (III.10)
12. John of Celaya: *An forme contrarie possunt esse adequate in eodem subiecto* (III.12)

This is the first of three questions that stem directly from Walter Burley's *Tractatus secundus de intensione et remissione formarum*. This question, or similar ones, does not seem to have appeared in earlier questions commentaries on the *Physics*. Buridan replies to question III.3: things called contraries that are not significative terms, nor propositions, but real contraries, should be those that cannot exist together at the same time in the same thing. In other words, it should be impossible for them, or for things like them, to be in the same subject at the same time (27^{15-18}). With this principal condition, there are other conditions required for contrariety in the proper sense. The second property is that those things, or things like them, can be in the same subject successively, such that there can be a transmutation from one to the other in the same

pallidum, partim est sub albedine et partim sub pallore, et ita contraria in contrarietate extremi sunt simul in eodem—dicendum quod impossibile est alterum contrariorum esse simul in eodem subiecto adequato cum forma media, quia inter extremum contrarium et formam mediam potest esse motus et termini motus sunt incompossibiles.'

127 Burley, *Tractatus secundus*, 5rb: 'Ostenso quod intensio forme non fit per additionem partis ad partem utraque parte remanente, nec remissio per ablationem partis a parte, sequitur declarare quod causa intensionis et remissionis forme non est per admixtionem maiorem vel minorem cum contrario. Istud autem potest primo ostendi ostendendo quod contraria non possunt esse simul in eodem subiecto adequato.'

subject (27^{18-19}). Third it is required that they be maximally distant from each other—not meaning distance in the proper quantitative sense, but distance in degrees, in the sense that the most intense heat would be said to be maximally distant from the most intense cold, in that the transmutation from one to the other would take a longer time than a transmutation from a remiss degree to a remiss degree (27^{22-26}). If other conditions may be required for contrariety in the most proper sense, nevertheless these are the main ones. It is true that sometimes middle degrees are said to be contrary to each other in some sense, and middle degrees contrary to the extreme, as remiss heat to remiss cold, or a tepid degree to complete heat, but this is not precise speech nor perfect contrariety (28^{1-4}).

As noted in the list above, Nicole Oresme and Albert of Saxony have similar questions in Book V, rather than book III. This question is a complement to the following questions that advocate for the addition-of-part-to-part theory of intension of qualities in a version that includes the idea of admixture of qualities in the same subject as contributing to intension and remission. The admixture theory likely derives ultimately from Galenic medicine, i.e., from the theory of health as linked to a balance of the humors, and from the theory of compounding medicines from ingredients of varying degrees of hot, cold, wet, and dry.[128] The discussion here reflects Walter Burley's *Tractatus secundus*. In his *Quaestiones* 84–86, Ockham discusses whether contraries can be in the same subject at the same time and denies it. The topic is mentioned in several other places in Ockham's work.

Question III.4—Whether when there is proper and per se alteration, continuous and in time, the quality with respect to which there is alteration is acquired all at once or part after part (*Utrum qualitas secundum quam est alteratio per se et proprie dicta, continua et temporalis, acquiratur tota simul vel pars post partem*).[129]

1. Walter Burley (*Expositio*): *Utrum cuilibet parti sensibili temporis correspondeat pars sensibilis rei acquisite per motum* (III.D.4)

128 See J. Kaye, *A History of Balance, 1250–1375. The Emergence of a New Model of Equilibrium and its Impact on Thought*, Cambridge 2014, ch. 3–4, for the suggestion that patterns of reasoning were transferred in the Middle Ages from medicine to physics rather than in the other direction.

129 The table of questions for Book III continues: 'Quomodo qualitas est divisibilis quantitative et gradualiter. Quid debet intelligi per "gradum qualitatis intensibilis"' (4^{16-17}).

2. Nicole Oresme: *Utrum forma intendatur per additionem gradus ad gradum sive partis ad partem* (V.7)
3. Hugolinus of Orvieto: *Utrum motus ad formam sit in mobili subiective* (17 = III.2)[130]
4. Albert of Saxony: *Utrum in intensione qualitas quae acquiritur, acquiratur tota simul vel secundum partem post partem* (V.10)
5. Marsilius of Inghen: *Utrum qualitas quae acquiritur in alteratione per se acquiratur tota simul vel per partem post partem* (III.1.3)
6. Johannes Marsilii (?): *Utrum in alteratione qualitas acquiratur subito* (III.3)
7. Lawrence of Lindores: *Utrum qualitas secundum quam fit alteratio per se et proprie dicta, continua et temporalis, acquiratur tota simul vel pars post partem* (III.4)
8. Benedictus Hesse: *Utrum qualitas secundum quam est alteratio per se proprie dicta, temporalis et continua, acquiritur tota simul vel pars post partem* (III.11)
9. John of Celaya: *Utrum qualitas quae intenditur, successive acquiratur vel tota simul* (III.13)[131]

The arguments in this question directly or indirectly reflect Walter Burley's defense of the succession of forms theory of intension and remission (37[9]–38[3]). In the *oppositum*, Buridan quotes Averroes as saying that motion is the generation of part after part of that perfection towards which the mobile continually tends (38[4–6]). Quality is imagined to be divided in two ways, Buridan argues, either according to the quantity of the subject or into gradual parts existing together in the same part of the subject according to which it is said to be more or less such (38[15]–39[9]).

Before stating his own position, Buridan mentions another position which he then rejects:

130 Hugolinus, *Quaestiones Physicorum*, 17 = III.2, 25: 'In quarum prima ⟨decisione⟩ videbitur in generali quid sit motus ad formam. Prima conclusio est quod motus ad formam non est aliquis fluxus formae distinctus a forma et a partibus formae. Secunda conclusio est quod motus ad formam est acquisitio continua ipsius formae et partium eius in subiecto. Tertia conclusio est quod motus ad formam potest etiam dici forma fluens. In secunda ⟨decisione⟩ videbitur de quaesito videlicet an motus ad formam sit in mobili subiective. Prima conclusio est ista: quod omnis motus ad formam, sive sit acquisitus sive deperditus, est forma inhaerens. Secunda est quod omnis talis motus est in mobili subiective.'

131 Celaya, *Expositio Physicorum*, III.13, 111[va]: 'Ad hanc questionem respondetur per aliquas conclusiones. Quarum prima est ista: ponendo minimum naturale quoad extensionem necesse est dicere quod aliqua qualitas tota simul acquiritur in aliqua parte alicuius subiecti.'

> Many have doubted whether heat is divisible and whether one part of it is acquired after another. Some have posited two conclusions. The first was that quality is acquired all at once essentially; the second was that it is not acquired all at once gradually (39^{10-13}).

This position may have been the position advocated by Thomas Wylton.[132] After a long recitation of the arguments for these conclusions, Buridan rejects them and turns to his own view:

> This opinion does not seem to me to be true unless it is expounded in an improper sense. It seems to be imagined that there are degrees distinct from the essence of heat and that continually there is another and another degree and there is not another and another essence. This I do not believe to be true, and therefore I posit some conclusions.
>
> The first conclusion is that in calefaction something of heat is acquired earlier and something of heat is acquired later in the same part of the same subject, because it is necessary, if A is continually heated, that it continually becomes [earlier] less warm and [later] more warm. But being otherwise and otherwise in this way (*sic aliter et aliter se habere*) cannot be saved except by some disposition existing later which was not earlier, or vice versa, because it may not be saved by the relation of this A to something extrinsic, because everything extrinsic may be taken away in thought, retaining only what is continuously heated, and still it would be otherwise and otherwise. Also it may not be saved by a diverse relation and position of the parts in relation to each other. For in the question concerning figure and the figured (II.3), it was seen that it cannot happen except in one of the said ways that something is otherwise and otherwise later than before. But it is conceded that it is necessary that if A is warmer than before, something exists which did not exist before, or vice versa, not only extrinsically, but in A itself. All would concede that this something else is heat, or something of heat, namely parts or degrees or some such.

---

132 Stephen Dumont has been working on Wylton's theory of intension and remission of forms as found in particular in manuscript New Haven, Yale University: Beinecke Library, General cod. 470, 21rb–24va. For similar questions in manuscript Tortosa, Biblioteca de la Catedral y del Cabildo de la Sanctísima Iglesia Catedral (Archivo Capitular de Tortosa), cod. 88, 1ra–5ra, see Sylla, *The Oxford Calculators*, 429–435.

> The second conclusion is that a degree of heat is not something different from heat, that is, if it is posited that in calefaction that which is acquired first and that which is acquired later are called degrees of heat (41^{10}–42^{7}).

Buridan goes on to argue extensively for this point of view before replying to the principal arguments.

It is clear from the histories of theories of the intension and remission of forms that the addition-of-part-to-part theory, which Ockham and Buridan support, had its most significant source in the commentary on the *Sentences* of John Duns Scotus.[133] It is widely known that a key locus for discussions of the intension and remission of forms was in commentaries on the *Sentences*, Book I, d. 17, which dealt with the increase of *caritas* in the human soul. Walter Burley's *Tractatus secundus*, on which Buridan draws here, had its roots also in Burley's work as a student of theology, although Burley then traced the implications of his theories in many disciplines. To understand how Buridan's *Quaestiones Physicorum* fit into his intellectual context, it is important to consider how ideas from one genre of written work moved into another genre of written work. Was there an oral transmission of ideas from one faculty to another? The works that Anneliese Maier used to trace the history of intension and remission were predominantly theological. Gregory of Rimini's commentary on the *Sentences* (1343–1344) was an important locus of transmission of nominalist ideas on intension and remission.[134] In this context the version of the addition-of-part-to-part theory of increase of form that included admixture of contraries was often included in the discussion.

Question III.5—Whether in alteration the part of quality that is acquired earlier remains with the part that is acquired later (*Utrum in alteratione pars qualitatis quae prius acquiritur manet cum parte quae posterius acquiritur*).[135]

1. William of Ockham (*Quaestiones*): *Utrum haec sit concedenda de virtute sermonis 'alteratio est terminus ad quem partibiliter adquisitus vel deperditus'* (21)
2. Walter Burley (*Tractatus secundus*): *Quod forma non suscipit magis et minus per additionem partis ad partem utraque parte remanente* (Cap. I)

---

133 See Maier, *Zwei Grundprobleme*, 50 ff.

134 See Maier, *Zwei Grundprobleme*, 80.

135 The table of questions continues: 'Utrum caliditas dicatur intensior. Quid intenditur' (4^{19-20}).

3. Richard Kilvington: *Utrum qualitas suscipiat magis et minus* (2)
4. Nicole Oresme: *Utrum forma intendatur per additionem gradus ad gradum sive partis ad partem* (V.7)
5. Albert of Saxony: *Utrum in intensione alicuius qualitas quae primo acquiritur maneat cum qualitate quae posterius acquiritur* (V.11)
6. Marsilius of Inghen: *Utrum in alteratione proprie dicta pars qualitatis quae prius acquiritur maneat cum parte qualitatis quae post acquiritur* (III.1.4)
7. Johannes Marsilii (?): *Utrum in intensione pars primo acquisita maneat cum parte secundo acquisita* (III.4)
8. Lawrence of Lindores: *Utrum in alteratione per se et proprie dicta qualitas prius acquisita maneat cum qualitate posterius acquisita* (III.5)
9. Benedictus Hesse: *Utrum in alteratione proprie dicta pars prius acquisita maneat cum parte posterius acquisita* (III.12)

Here again, Buridan's argument is mainly against something like the succession of forms theory, and in particular against arguments found in Walter Burley's *Tractatus secundus*. The principal arguments in favor of the view that the part that is acquired earlier does not remain with the part that is acquired later are the following. In local motion the place that is acquired first does not remain with the place acquired later, and something similar should be the case in alteration (45[7–9]). Moreover, if the part of quality attained first remained with the part of quality attained later, one could not say either that the parts remain distinct, nor that they do not (45[10–11]). Finally, everyone who concedes that the part acquired earlier remains with the part acquired later says that by the accumulation of part to part there results a more intense quality, while the subtraction of part from part leads to a more remiss quality. If this could be disproved, one would need to concede that the parts do not remain (45[20–26]).

Buridan then turns to an elaboration of the third principal argument into nine points. All or most of the arguments come from Burley's *Tractatus secundus*. Briefly summarized, the arguments are the following:

1. It would follow that light would be infinitely intense in this house, which is absurd (46[1–9]).
2. It would follow that white would be of infinite perfection (46[10]–47[4]).
3. As A heated B, which is most cold at the start, the resistance of B to A's heating would repeatedly be cut in half, and so the heating would become twice as fast ... If A and B remained together, at the end of the day the heat would be infinite (47[5–18]).
4. As parts are added together, neither the prior part nor the new part would be more intense (47[19]–48[3]).

5. It would follow that a remiss warm body could produce in a cold body very intense heat and corrupt all its cold (48^{4-13}).
6. One would need to say that the whole hot was hotter than its gradual half (48^{14-17}).
7. A falling body accelerates, but one cannot say that a later degree is added to an earlier one (48^{18-21}).
8. Light increases in a house, but the earlier light does not remain with the later, because the air is continually moved by the wind (48^{22-27}).
9. Intension cannot occur by the corruption of degrees of the contrary, because all of the degrees are of the same nature (so why should one be corrupted rather than another?) (49^{1-6}).

Many of these arguments are in chapter 1 of the *Tractatus secundus*.[136] There are related arguments in Richard Kilvington's question *Utrum qualitas suscipiat magis et minus*. He replies, for instance, to arguments that if qualities undergo more and less, then the maximum heat will be infinite.[137] In her article on the *Tractatus secundus*, Anneliese Maier points out that in manuscript Città del

136 Cf. Sylla, *The Oxford Calculators*, 521–523.

137 Cf. Sylla, *The Oxford Calculators*, 442, and Jung, 'Works by Richard Kilvington,' 205, n. 112. See in particular the following passage (manuscript Venezia, Biblioteca Nazionale Marciana, cod. lat. VI.72, 89vb–90ra): 'Ad istud argumentum potest responderi septem modis quorum quilibet est specialis difficultatis. Prima responsio est quae concedit conclusionem, videlicet quod caliditas in summo et latitudo caliditatis sit infinita. Secunda responsio est quod argumentum fundatur super falsam imaginationem, videlicet quod caliditas habeat unam talem latitudinem, ita quod eadem caliditas continue intendatur et nulla nova generatur. Et causa istius responsionis est quia continue erit alia caliditas et alia, et sic quod in quolibet instanti generaretur una nova caliditas, et quaelibet praecedens corrumpitur. Tertia responsio est quod in quolibet instanti fiet nova caliditas et omnes caliditates praecedentes manebunt cum caliditate ultimo inducta. Quarta responsio est ista, quod nulla caliditas generatur in instanti, sed in quolibet tempore praecise generabitur una nova caliditas quae manebit cum caliditatibus praecedentibus. Quinta responsio est quod caliditas continue manebit eadem, nulla alia adducta, et ipsa in sua essentia nullo modo intendetur nec remittetur, sed solum intenditur quoad virtutem agendi vel patiendi, quia subiectum magis vel minus disponitur per ipsam. Sexta responsio est quod eadem caliditas continue intendetur et tamen nulla est nec esse potest talis latitudo caliditatis habens medietates distinctas. Septima et ultima responsio est, quod in intensione caliditatis praecise semper manet eadem caliditas, quae praecise substantialiter intenditur, sed illius caliditatis latitudo est idem realiter cum illa caliditate sicut tempus et coelum, sed alia erit ratio illius ut est latitudo et alia ut est caliditas.'

Vaticano, Biblioteca Apostolica Vaticana, cod. Vat. lat. 817, some of the arguments mentioned in this chapter of the *Tractatus secundus* are attributed to John Duns Scotus, William of Ockham, and Francis of Marchia.[138] Elżbieta Jung and Robert Podkoński argue that Kilvington was aware of Burley's arguments and that Buridan was likely aware of Kilvington's questions on the *Physics*, but until publication of Kilvington's work, this remains difficult to judge.[139]

Buridan answers the argument he lists about illumination by saying that the illuminations of the two candles add together in the room. If the previous quality did not remain, the quality in common when one element is changed to another would not remain (49^{8-22}). In augmentation the previous quantity remains and adds to the new quantity (49^{27}–50^{2}). If qualities did not remain, moral habits would be easily corrupted, which is false (50^{3-5}). Based on these three arguments that the prior part of quality remains, Buridan writes:

> I say briefly that in true and proper alteration, as in heating, the part of quality which is acquired earlier remains with that which is acquired later, because otherwise it would follow that a given resting quality would have been instantaneously acquired, such that nothing of it existed before the whole quality, and in this case there would be no temporal and continuous motion. But everyone concedes that there is continuous and temporal alteration to a quality such as heat. This would be just like the intellective soul which is acquired all at once (50^{6-15}).

Moreover, Buridan and those agreeing with his view assume (*supponimus*) that an indivisible instant is nothing, so that if something existed for only an indivisible instant, it would never exist (50^{20-23}). Moreover, if there were indivisible instants they would not be continuous with each other nor next to each other in time and then alteration would not be continuous, which is false, because all motion is continuous (50^{24}–51^{5}). Looking ahead, Buridan says that the opponents' view also will not hold for local motion, because even a purely successive thing does not exist in an instant (51^{24}–52^{3}). Again Buridan argues that the opposing view will not work for remission. It would follow from that view that even the most intense degree would be corrupted instantaneously

138 A. Maier, 'Zu Walter Burleys Traktat *De intensione et remissione formarum*', in: Ead., *Zwei Grundprobleme der scholastischen Naturphilosophie* (Studien zur Naturphilosophie der Spätscholastik, 2), Roma 1968 (Storia e letteratura, 37), Nachträge, 315–352, at 327.

139 Cf. E. Jung & R. Podkoński, 'The Transmission of English Ideas in the Fourteenth Century. The Case of Richard Kilvington,' *Mediaevalia Philosophica Polonorum*, 37/3 (2008), 59–69.

and all at once. This means that it would have no resistance to its corruption (52^{8-16}). Other arguments for Buridan's position follow.

In replying to the principal arguments at the start Buridan replies to the nine arguments that the prior part cannot remain with the later parts. Against the first argument concerning illuminations, Buridan's criticism is that the argument assumes there are indivisible instants, which do not exist (56^{8-9}). To the fourth argument, Buridan replies that neither the earlier part of quality nor the later part becomes more intense, but the subject of the quality has a greater degree (57^{24-26}). To some of the arguments, Buridan characterizes his replies as 'probable' (*probabiliter*) or 'persuasive' (*persuasio*).

In this question, as in the next on local motion, Buridan seems to report heated debates, and especially those also found in Burley's *Tractatus secundus*, without putting all the material into clear order. Albert of Saxony's related question covers nearly all the same material in a more organized way, beginning with the statement that there are two main theories, what we call the succession of forms theory and what we call the addition-of-part-to-part theory, and going on to give the arguments for and against each theory.[140]

Although Burley's *Tractatus secundus* can be said to be a source of the arguments in Buridan's questions III.3–5, where did the ideas in the *Tractatus secundus* come from? It seems to me that they came from Burley's *Tractatus primus*, which in turn reflects the *disputationes collativae* associated with the work of bachelors of the *Sentences* in this time period at Paris.[141] Again, the addition-of-part-to-part theory of intension of forms was developed by John Duns Scotus in his commentaries on Peter Lombard's *Sentences*, both at Oxford and at Paris. If one tries to understand the history of the development of the addition-of-part-to-part theory of qualitative change through an examination of commentaries on Aristotle's *Physics*, there are strange discontinuities as ideas formed and developed in the Faculty of Theology make their way into the Faculty of Arts. There is no commentary on the *Physics* by Scotus, who apparently decided to make known his ideas through his commentary on the *Sentences*.[142]

140 [Albert of Saxony], *Quaestiones Physicorum*, V.11, 3: 849–864.

141 Cf. E.D. Sylla, '*Disputationes Collativae*: Walter Burley's *Tractatus Primus* and Gregory of Rimini's *Lectura super primum et secundum Sententiarum*', *Documenti e studi sulla tradizione filosofica medievale*, 22 (2011), 383–464.

142 See R. Cross, *The Physics of Duns Scotus. The Scientific Context of a Theological Vision*, Oxford 1998, for an attempt to remedy this lack.

3.1.2 Questions III.6–8: Questions on Local Motion

Question III.6—Whether local motion is (or exists) or whether this is true 'local motion is' (*Utrum motus localis sit vel utrum haec sit vera 'motus localis est'*). Aristotle, *Physics*, III, 1–2, 201b6–202a3; Averroes, *In Physicam*, III, comm. 11–15.

The table of questions for Book III supplies a good deal of information for question III.6. As can be seen by this information, Buridan's discussion strays away from the original question:

> That by 'present' should not be understood an indivisible instant. That it is necessary to expound 'to be changed' by 'to be in differing conditions (*aliter se habere*) earlier and later', that is in both cases in the present. That this is true: 'Socrates is sitting and Socrates is not sitting.' From an affirmative infinite predicate does not follow the negative of a finite predicate. How the present is said simply with respect to the past and the future. How much time may be used for the present. Concerning the terms 'tomorrow', 'today', 'yesterday.' That it is a good inference (*consequentia*): 'In all present time Socrates is moved, therefore in all time Socrates is moved.'[143]

1. **Er349(1)**: *Utrum motus sit* (III.3)
2. **L1386(1)**: *Utrum motus sit* (III.2)
3. William of Clifford: *An motus sit* (III.1)
4. Boethius of Dacia: *Utrum motus sit* (III.5)[144]
5. Geoffrey of Aspall: *Utrum motus sit* (III.1)
6. Radulphus Brito: *Utrum motus sit* (III.1)
7. William of Ockham (*Quaestiones*): *Utrum haec sit concedenda de virtute sermonis 'motus est ens'* (20)

---

143 Cf. *infra*, 4^{22}–5^{3}: 'Quod per "praesens" non debet intelligi instans indivisibile, sed tempus divisibile. Quod oportet exponere "mutari" per aliter et aliter se habere prius et posterius, scilicet utrobique de praesenti. Quod haec est vera "Socrates est sedens et Socrates est non sedens." Quod ad affirmativam de praedicato infinito non sequitur negativa de finito. Quomodo praesens simpliciter dicitur respective praeteritum vel futurum. Quanto tempore liceat uti pro praesenti. De istis terminis "cras", "hodie", "heri." Quod est bona consequentia "in omni praesenti tempore Socrates movetur, igitur in omni tempore Socrates movetur".'

144 Cf. Boethius of Dacia, *Quaestiones Physicorum*, III.5, 264: 'Est autem motus medium inter purum actum et puram potentiam, est enim actus medius. Nec debet dici quod motus non sit, quia non est in instanti: nullum enim continuum successivum in instanti habet esse, sed est in tempore, et motum esse non significat motum esse in instanti, quia sic non esset motus. Et ideo non debet negari motum esse.'

8. William of Ockham (*Expositio*): III, cap. 4 (t. 15, 201b31–202a3)[145]
9. William of Ockham (*De successivis*): *Opinio propria*[146]

---

145 Ockham, *Expositio*, III, cap. 4, § 6, 472–474: 'Et si dicas quod non quaeris de illo quod movetur an sit in actu vel in potentia, sed de ipso motu qui non est illud quod movetur, dicendum est quod ista quaestio procedit ex falsa imaginatione antiquorum de motu in qua concordant aliqui moderni. Imaginabantur enim aliqui antiqui sicut aliqui moderni quod motus esset aliquod ens vel non ens distinctum ab omni re permanente et ab omnibus permanentibus, sicut albedo distinguitur ab omni substantia et ab omnibus substantiis coniunctim sumptis. Sed ista imaginatio est simpliciter falsa et contra intentionem Aristotelis, sicut dictum est. Quia ipse ponit motum nihil esse praeter res permanentes, sicut dixit supra, ita quod hoc nomen "motus" nihil penitus importat praeter res permanentes praesentes, praeteritas et futuras, sed importat res permanentes vel praecise praesentes, sicut patet in motu locali, vel praesentes et futuras, sicut in motu alterationis. Et ideo non potest quaeri de motu an sit in actu vel in potentia, nisi quaeras de aliqua re praesente vel praeterita vel futura vel de uno aggregato ex talibus ... Et ita patet quod non est hic magna difficultas nisi in verbis propter falsam imaginationem qua imaginantur aliqui quod, sicut hoc nomen "motus" est distinctum ab aliis nominibus rerum permanentium, quod ita importet unum secundum se totum distinctum ab importatis per alia nomina, quod tamen non est verum. Immo non plus motus importat aliquid secundum se totum distinctum ab aliis importatis per alia nomina quam tales coniunctiones "si", "et", "vel", "quia", et tales praepositiones "ab", "in", "de", "ex" et huiusmodi, et adverbia et syncategoremata important alia quam sint importata per alia nomina. Et quia sicut dictum est quod motus non importat aliquod ens vel non ens secundum se totum distinctum ab importatis per alia nomina, sed importat multa talia, unum in recto et aliud in obliquo, vel unum in uno obliquo et aliud in alio obliquo, vel talibus modis diversis grammaticalibus, vel saltem importat plura, hinc est quod aequivoce accipitur aliquando pro uno illorum, aliquando pro alio, aliquando pro omnibus simul et coniunctim, aliquando ponitur loco unius complexi, aliquando totum quod sequitur hoc nomen "motus" resolvendum est in alia vocabula et etiam ipsummet "motus", sicut tactum est prius. Et per ista possunt solvi omnes contrarietates quae videntur esse in dictis Philosophi et Commentatoris circa motum.'

146 Ockham, *Tractatus de successivis*, 45–47: 'Ideo dicendum est quod motus non est talis res distincta secundum se totam a re permanente, quia frustra fit per plura, quod potest fieri per pauciora; sed sine omni re tali possumus salvare motum et omnia quae dicuntur de motu; igitur talis res alia frustra ponitur. Quod autem sine tali re addita possumus salvare motum et omnia quae dicuntur de motu, patet discurrendo per singulas partes motus. De motu enim locali patet. Ponendo enim quod corpus sit in uno loco et postea in alio loco, sic procedendo sine omni quiete et omni re media alia ab ipso corpore et ipso agente quod movet, vere habemus motum localem; igitur frustra ponitur talis res alia ... Et si dicatur quod ista non-simultas partium est aliquid, quando dicitur "partes non sunt simul"—dicendum quod talis fictio nominum abstractorum ab adverbiis, coniunctionibus, praepositionibus, verbis, syncategorematibus facit multas difficultates, et multos

10. Walter Burley (*Expositio*): *Utrum motus sit res distincta a mobili* (III.D.6)[147]
11. Francesc Marbres: *Utrum motus sit aliqua realis entitas distincta essentialiter a termino ad quem est* (III.1)
12. Nicole Oresme: *Utrum motus sit aliquid* (III.2)
13. Hugolinus of Orvieto: *Utrum motus localis sit fluxus mobilis existens in mobili subiective* (16 = III.1)[148]
14. Albert of Saxony: *Utrum ista sit concedenda 'motus est'* (III.4)[149]
15. Marsilius of Inghen: *Utrum motus localis sit* (III.2.2)
16. Johannes Marsilii (?): *Utrum 'motus est' sit concedenda* (III.6)
17. Lawrence of Lindores: *Utrum haec sit concedenda 'motus localis est'* (III.6)
18. Benedictus Hesse: *Utrum motus localis sit, supposito quod motus localis nec sit locus, nec mobile localiter motum* (III.13)

This is a long inquiry, but it barely discusses the ostensible question. Buridan begins by saying that he will assume that local motion is neither the mobile nor the place reached, although this is not discussed until the seventh question (62^{9-10}). If motion is a successive entity, how are successives to be defined or understood? But this is not the whole difficulty. Buridan also assumes that indivisbles such as instants, points, and *mutata esse* do not exist, although this will not be dealt with directly until Book VI. There is still more trouble: if there

---

ducit in errores. Imaginantur enim multi per hoc quod, sicut sunt nomina distincta, ita sint res distinctae correspondentes, ut tanta videlicet sit distinctio inter res significatas, quanta est inter nomina significantia. Quod tamen non est verum. Sed aliquando eaedem res sunt significatae, et tamen est diversitas in modo logicali vel grammaticali significandi. Et ideo non-simultas non est aliqua alia res a rebus quae possunt simul esse, sed significat quod res non sunt simul. Et ideo in modernis temporibus propter errores subortos ex usu talium abstractorum melius esset in philosophia propter simplices non uti talibus abstractis, sed tantum verbis, adverbiis, coniunctionibus, praepositionibus, syncategorematibus sicut primario fuerunt instituta quam fingere talia abstracta et uti eis. Immo si non esset usus talium abstractorum "motus", "mutatio", "mutabilitas", "simultas", "successio", "quies" et huiusmodi, parva esset respective difficultas de motu, mutatione, tempore et instanti et huiusmodi'.

147 Burley, *Expositio in Physicam* (1501), 64rb. See *supra*, n. 110.

148 Hugolinus, *Quaestiones Physicorum*, 16 = III.1, 24: 'In prima ⟨decisione⟩ videbitur de veritate istius propositionis "motus est", an de virtute sermonis sit concedenda. Prima conclusio est ista: quod de virtute sermonis haec propositio est concedenda tamquam vera "motus est". Secunda est quod secundum viam Aristotelis haec est necessaria et non solum vera "motus est". Tertia conclusio est quod ponendo motum fore quiddam fluxum, ut ipsi ponunt, haec propositio est impossibilis "motus est".'

149 This question is missing in the 1516 Lokert edition.

are no instants, then the present, at or in which the truth of a proposition is to be assessed, cannot be an instant, but must be a shorter or longer length of time. Within different parts of that length of time contradictories can be true. That Buridan's answer to the question immediately diverges to such issues can be seen, above, by what the table of questions adds concerning the question.

After six arguments *quod non* (60^{9}–61^{30}) and three rather weak arguments *in oppositum* (62^{1-8}) Buridan writes that he assumes that local motion is not place (*locus*) nor what is moved locally, but that local motion exists:

> It is not necessary to prove this [that motion exists], because it is apparent to sense and should be supposed in natural science as a principle of doctrine. And because by 'present' we should not understand an indivisible thing in time, because it will be shown in Book VI that there is no indivisible in a continuum or magnitude, whether in motion or in time, therefore by 'present' one must understand divisible time. And nevertheless, time exists, because neither a grammarian nor a logician differentiates among present, past, and future except by being, having been, and being in the future (*esse, fuisse et fore*). Therefore time is, and as a consequence, motion is. Otherwise one could not say that some vocal proposition would be, and thus none would be true or false, which is absurd to say (62^{10-21}).
>
> There only remains to see how such successive things are said to be. And Aristotle in Books III and IV of the *Physics* immediately determines that the being of such [successives] does not consist in all of their parts being at the same time, but in one part after another existing, that is, in always becoming otherwise and otherwise. Such are days and years, as he says (62^{28}–63^{2}).

Buridan begins to try to expound propositions concerning motion without making use of indivisible instants, but he almost immediately turns to the objections and doubts that arise. One objection is that in any period of time contradictories can both be true, such as that Socrates runs and Socrates sits, although whoever sits does not run (64^{11-17}). A second objection is that Aristotle's argument that anything in motion was previously moving will not hold of a motion entirely contained within the present time taken as divisible (64^{18-20}). The problem of contradictories being true at the same time, if the present is not indivisible, is elaborated by talking about participants in a disputation: it will follow that the truth value of a proposition may change between the replies of two opponents or even as one disputant enunciates his proposition (67^{10-24}). Buridan suggests that to get around this problem, perhaps disputants should not direct their arguments to the present time of the disputation but to a different time (67^{25}–68^{7}). Others suggest that, if there is no indivisible instant, one should assume the minimum time that can be sensed (68^{21-23}). Buridan's view is that disputants should be allowed to use as the present whatever length of time they choose, even, say, a year (69^{4-13}). Many incongruities would result from this approach, however. Buridan tries to argue that one could reasonably accept some apparently self-contradictory propositions that would follow from this proposal. To the fourth principal argument at the start of the question, he replies:

> A purely successive thing does not come into being and is not corrupted except when it exists, and its being is the same as its becoming and its being corrupted. Aristotle says that its being consists in always becoming otherwise and that such beings are in generation and corruption. Such generations and corruptions are not contrary to each other, but occur in the same time (72^{9-13}).

All in all, this appears at best to be a work in progress, not a convincing response to the question. It is evidence, however, that Buridan is trying to advance the logical program often associated with William of Ockham, that is, to assess the truth or falsity of propositions by the suppositions of their terms and by their connotations while restricting real things to substances and qualities. The real question at issue is to be answered in question III.7.

Nicole Oresme's question related to Buridan's is on a similar topic, but is not very like Buridan's. Albert of Saxony's related question makes arguments much like Buridan's question, but it is more tightly edited. Rather than going off on a tangent to discuss determinations of what is meant by the 'present' in relation to disputations and the enunciation of propositions, Albert sticks to the subject at hand—how a successive entity like motion can be said to exist. Marsilius of Inghen's treatment is something like an abstract of Buridan's question, with some additions. Johannes Marsilii's (?) treatment is different, and has sophismata-like elements.

Question III.7—Whether local motion is a thing distinct from place and from that which is locally moved (*Utrum motus localis est res distincta a loco et ab eo quod localiter movetur*).[150]

150 The table of questions continues: 'Quod ultima sphaera potest moveri motu quo movetur sine loco. Quod motus ultimae sphaerae est res pure successiva' (5^{5-6}).

Whereas Aristotle and Averroes had asked a question like this for motion in general, and Buridan had already asked in question III.2 whether a flux is required in alteration, here he raises a similar issue for the case of local motion.

1. William of Ockham (*Quaestiones*): *Utrum motus successivus importet respectus distinctos a rebus absolutis* (12) *Utrum motus sit aliquid absolutum distinctum a rebus permanentibus* (13) *Utrum motus localis importet rem aliam distinctam a rebus permanentibus* (14) *Utrum possit probari ratione sufficienti quod motus importet aliud a rebus permanentibus* (17) *Utrum intentio Aristotelis sit ponere motum aliam rem a rebus permanentibus* (18) *Utrum Commentator posuerit motum esse rem distinctam a rebus permanentibus* (19) *Utrum haec sit concedenda de virtute sermonis 'motus localis est mobile successive adquirens spatium vel spatium partibiliter adquisitum'* (22) *Utrum octava sphaera sit in loco per se aut per accidens* (79) *Utrum octava sphaera moveatur per se* (80)
2. Nicole Oresme: *Utrum motus sit res mota vel ipsum mobile* (III.3) *Utrum motus localis sit illud quod acquiritur mobili tali motu, scilicet situs vel locus in quo et circa quod mobile movetur, vel utrum motus localis sit acquisitum* (III.5) *Utrum motus sit res successiva sive fluxus distinctus a rebus permanentibus cuiusmodi sunt mobile et res acquisita ad quam est motus* (III.6) *Utrum moveri sit aliter se habere continue quam prius* (III.7)
3. Hugolinus of Orvieto: *Utrum motus localis sit fluxus mobilis existens in mobili subiective* (16 = III.1)[151]
4. Albert of Saxony: *Utrum secundum Aristotelem et eius Commentatorem ad hoc quod aliquid moveatur localiter requiratur aliqua res quae sit quidam fluxus distinctus a mobili et a loco* (III.6) *Utrum admittentes casus divinos oporteat concedere quod motus localis sit alia res a mobili et a loco* (III.7)
5. Marsilius of Inghen: *Utrum motus localis sit res distincta a mobili et a loco* (III.2.1)
6. Johannes Marsilii (?): *Utrum motus localis sit res distincta a mobili* (III.7)
7. Lawrence of Lindores: *Utrum motus sit res distincta a mobili et a loco* (III.7)

---

151 Hugolinus, *Quaestiones Physicorum*, 16 = III.1, 24–25: 'In secunda ⟨decisione⟩ videbitur an motus localis sit aliqua res media distincta a mobili et partibus spatii ... Quarta ⟨conclusio⟩ est quod nulla est entitas successiva quae ab omni permanente realiter sit distincta ... In tertia vero ⟨decisione⟩ perscrutabitur de quaesito an motus localis sit in mobili subiective. Prima conclusio est quod motus localis non est agens existens in mobili subiective. Secunda est quod motus localis non est fluxus mobilis existens in mobili subiective. Tertia est quod moveri est accidens praedicabile de mobili tamquam de subiecto.'

8. Benedictus Hesse: *Utrum motus localis sit, supposito quod motus localis nec sit locus, nec mobile localiter motum* (III.13) *Utrum motus localis sit res distincta a mobili et a loco* (III.15)
9. Domingo de Soto: *Utrum motus sit res distincta et a mobili et a forma seu termino* (III.2)[152]
10. Conimbricenses: *Utrum latio in mobili an in corpore circumiacente inhaereat* (III.3.2)[153]

---

152 Soto cites many earlier authors by name or position. See for instance (*Quaestiones physicorum*, III.2, 182^b–183^b): 'Et ideo S. Thom. & antiqui Philosophi in hoc sensu appellabant has distinctiones reales, non quod seclusa operatione intellectus essent duae res, sed quod seclusa operatione intellectus est in rebus unde sumitur illa distinctio. Unde non inepte posset distingui duplex distinctio realis: una quae est inter rem et rem, ut distinguuntur substantia & qualitas, alia quae est inter illa quae distinguuntur tanquam res & modus realis eius, ut distinguitur sedens et sessio, fundamentum & relatio, motus & qualitas etc. … De motu autem locali est difficultas nova, propterea quod id quod acquiritur per motum localem non est forma intrinseca quae recipitur in ipso mobili, sicut per motum alterationis recipitur qualitas & per motum augmentationis quantitas, sed solum acquiruntur successive diversa loca, quae sunt res extrinsicae continentes locatum. Quapropter licet Ocham & Gregorius [Ariminensis] conveniant de motu alterationis & augmentationis, differunt tamen de motu locali. Gregorius enim existimat motum localem esse spacium per quod mobile movetur. Quam opinionem sustinens nonnulli Thomistae credentes illam opinionem esse S. Thomae. Ocham vero putat motum localem non distingui realiter a mobili, cui ego potius fidem adhibeo. Ratio & analogia Gregorii est haec: quemadmodum se habet qualitas in motu alterationis, se habet spacium in motu locali; & qualitas est idem realiter cum motu alterationis, ut modo dicebamus; ergo & spacium est idem cum motu locali … Ita latitudo qualitatis a non gradu usque ad summum est quodammodo distantia quae pertransitur per motum alterationis.'

153 Collegium Conimbricense, *Commentarii Physicorum*, III.3.2, 350–351: 'Articulus I. Opinio existimantium non inhaerere in mobili. Proposita quaestio non parum controversia est. Sunt enim qui existiment lationem non in mobili, sed in corpore circumfuso ut in subiecto inesse ac rem ipsam quae movetur extrinsecus tantum denominare … Huius ergo opinionis fuit D. Thomas in 4, d. 44, q. 2, art. 2 et 3, *Contra gentes* c. 82, Gregorius in 2, dist. 1, q. 4, art. 1, Sonc. 5 *Metaph.* q. 40, Hervaeus *Quodl.* 1, q. 9 … Articulus II. Lationem inesse in mobili nec adversariorum argumenta concludere. Contraria sententia existimantium lationem inhaerere in ipso mobili verisimilior ac philosophorum communior est. Eam defendit Sotus hoc in l. q. 2, Toletus q. 3, Iandunus 4 *Metaph.*, q. 4, Durandus in 2, d. 15, q. 3, Ioannes Maior d. 2, q. 2, Ochamus in q. 9, ac omnis fere schola Nominalium; in eademque videtur fuisse auctor *Sex principiorum* in explicatione categoriae ubi, cum doceat ubi esse in re ipsa quae movetur; quod item asserit Scotus in 4, d. 49, q. 16. Probatur vero haec opinio hunc in modum …'

After listing five arguments *quod non* (73^{6-23}) and eleven arguments *in oppositum* (73^{24}–75^{8}), Buridan remarks that the ancients did not doubt about this question but agreed that local motion is something different from the mobile and the place. Now, however, he claims, the 'later moderns',[154] because of some of the arguments just stated, have posited that motion is not another thing than the mobile (75^{9-12}). But everyone is agreed that motion is some kind of mutation, and to move is to be changed (75^{16-17}). On this basis he states the following conclusions:

1. It is possible for the last sphere to move with the motion by which it moves without a place (76^{2-3}).
2. The last sphere moves not only because it continually has another relation to the earth or to some other body (77^{1-3}).
3. For the last sphere to move is for it to be otherwise intrinsically before and after. This is proved because of the definition of the term 'to move' as 'to be otherwise before and after' (*aliter et aliter se habere prius et posterius*). Nevertheless the last sphere moves without having a different relation before and after to something extrinsic (77^{18-22}).
4. The motion of the last sphere is not that sphere itself, nor its place (78^{9-10}). It cannot be its place because it does not have a place. The sphere itself does not change substantially, so it cannot be the sphere itself either.
5. The motion of the last sphere is distinct from the last sphere and its place (79^{1-3}).
6. The motion of the last sphere is a purely successive thing (*res pure successiva*), of which, that is, there is an earlier part and a later part that do not remain together (79^{4-5}).

Do these conclusions amount to a rejection of Ockham's view on motion? It is usually said that they do, but it is not certain that by conclusion 6, Buridan is committed to making 'motion' correspond to an accidental form of the last sphere. He writes:

> The motion of the last sphere is not its substantial change, nor is it a change in relation to something extrinsic, unless this occurs in relation to being otherwise intrinsically, as was said before. Therefore it is a change according to a disposition (*dispositio*) other than the substance of the sphere and inhering in it (78^{23-26}).

---

154 *Posteriores moderni*: this would be Ockham and those influenced by him.

The word 'disposition' has many nuances, and it could mean something not unlike Oresme's notion of '*modi rerum*' or '*condiciones*'.[155] To decide exactly how Buridan intended '*dispositio*' here would require surveying his use of the word in many other contexts, as well as the use of the word in earlier, contemporary, or later authors.[156] It seems that a disposition must be real, but something less than a full-fledged form. The word is used, for instance, to describe what happens, for those who believe that contrary forms can be in the same subject at the same time, as one moves away from one contrary and towards another—the disposition to the later form is present before the form itself can be sensed. The word is also used when one talks of the preparation of matter for the reception of a new form, followed by God (or the *dator formarum*) causing the new form to exist in the matter (in embryology, the embryo might develop until the point when God causes the immortal soul to exist in it).

Albert of Saxony, as indicated above, has two conclusions in this place, one with regard to what Aristotle and the Commentator say, and the other 'admitting divine cases', such as that God might move the entire cosmos as one body. With regard to natural cases, some think it is necessary to posit a flux distinct from the mobile to save local motion, he says, but others think it is not.[157] Of the latter group, some think it is necessary that motion be judged with reference to a body at rest, while others think it is sufficient for the mobile to enter a place other than the one it was in at first. Albert thinks that both of these positions are deficient. In the case of the last sphere, which is not in place, there cannot be motion in the usual sense, but there can be motion of the parts of the sphere without requiring an added flux.[158]

If, however, you want to admit divine cases, Albert goes on, such as saying that God moves the whole world as a single body, then you do have to admit that the body will move with respect to something intrinsic (*secundum aliquod intrinsecum*), which is not the body, but inheres in the body. This something inherent is the motion or flux. God can move a body after annihilating everything outside it, without it acquiring some other disposition than motion (*absque hoc quod acquiratur sibi aliqua alia dispositio a motu*).[159] Albert's point

155 Cf. S. Kirschner, 'Oresme on Intension and Remission of Qualities in His Commentary on Aristotle's *Physics*', *Vivarium*, 38 (2000), 255–274.

156 The word 'dispositio' is contrasted with 'habitus' in chapter 8 of Aristotle's *Categories*, saying that dispositions are easily changed, whereas habits are not, the whole discussion having to do with qualities of humans rather than of other sorts of substance.

157 [Albert of Saxony], *Quaestiones Physicorum*, III.6, 2: 508.

158 [Albert of Saxony], *Quaestiones Physicorum*, III.6, 2: 510–511.

159 [Albert of Saxony], *Quaestiones Physicorum*, III.7, 2: 516–517.

is that in such a divine case, God can cause something of the same species as local motion, without anything extrinsic to the moving body and without having to create some kind of new thing in the body that is not present in a natural case.[160] Buridan seems to fit Albert's description of those who want to admit divine cases. If there were nothing to distinguish a single solid spherical body rotating in empty space from the same body at rest, what could be meant by saying that God can cause the body to rotate?

In his question III.6, Oresme describes the position that motion is a flux saying:

> From what has been said, the fifth conclusion emerges, namely that motion is a successive thing absolutely (*simpliciter*) distinct from permanent things. This can be understood in two ways. First that motion is one inherent thing that can be signified incomplexly, like a form. In this sense it is not true. Second, it can be understood that it is a condition or mode of the mobile, and in this sense it is true. This is proved as follows: when two things do not suffice as a basis for the truth of a proposition, one must posit another thing or at least another mode of a thing. This is immediately clear, because if it were sufficient before, it would now be true. But positing the mobile and the space is not enough to make the proposition 'this moves' true. Therefore when it was true, another thing is posited.[161]

Oresme then raises six arguments against the conclusion that motion is a successive thing, a condition, or a mode of the mobile distinct from permanent

---

160 [Albert of Saxony], *Quaestiones Physicorum*, III.7, 2: 517: 'Septima conclusio: in omni mobili quod movetur localiter, volentes admittere casus divinos oportet ponere fluxum seu motum inhaerentem mobili qui successive illi mobili acquiritur. Probatur quia: tales habent admittere quod Deus posset movere aliquod mobile, quocumque sibi extrinseco annihilato, consimili motu in specie absque productione alicuius rei quae nec ipsa vel ei consimilis praeexistebat in illo mobili; sed hoc non posset, si, omnibus annihilatis, illud mobile moveretur aliquo fluxu sibi superaddito, et non prius, sicut patet intuenti; igitur etc.'

161 Oresme, *Questiones super Physicam*, III.6, 334, 'Ex predictis potest elici quinta conclusio (opinio *ed.*), scilicet quod motus est res successiva distincta simpliciter a permanentibus. Et potest dupliciter intelligi: primo, quod sit una res inherens significabilis incomplexe sicut una forma, et sic non est verum; secundo, quod sit condicio vel modus ipsius mobilis, et sic est verum. Et probatur quia: quando due res non sufficiunt ad hoc quod aliqua propositio sit vera, oportet ponere aliam rem vel saltem alium modum rei; patet statim, quia, si sufficiebat ante, iam fuisset vera; sed posito mobili et spatio non sufficit ad hoc quod hec sit vera "hoc movetur"; ergo quando fuit vera, aliud ponitur'.

things. The fifth argument is that, if motion were a flux spread out over time, then it would have no parts, because what is only in the past or future does not exist:

> Fifth, the parts of such a flux do not exist, because one is past and the other is future. It is replied that in the same way one could argue that sound does not exist. Therefore one responds by distinguishing what Aristotle says, as Ockham does: some things are said to exist not in a perfect act, but in an incomplete act, of which there is always something in the future. Successive things are of this sort.[162]

One may understand that Oresme and Buridan conclude that local motion must be some inherent condition, because otherwise God could not cause a rotation of the world taken as a whole; but the case posits that God does rotate the whole sphere, and this is not logically impossible. If in a 'divine case' something inherent is required, so in natural cases it must also be required. Is this something inherent a form? Albert of Saxony makes the argument this way:

> Breviter in ista quaestione, admittentes casus divinos coguntur aliter respondere quam cogerentur non volentes admittere casus divinos, sicut essent solum volentes loqui et admittere in persona Aristotelis et Commentatoris. Nos ergo, volentes admittere casus divinos, supponamus primo quod moveri est aliter et aliter se habere secundum prius et posterius continue. Secundo supponamus quod Deus posset facere totalem mundum, scilicet orbes caelestes et ista inferiora, unum continuum, et ipsum volvere ab oriente versus occidens, vel aliter, sicut sibi placeret.[163]

In this supernatural case, the whole world as a continuous body could not be rotating with respect to something external, because there is nothing external, and supposing that the reference body could be a hypothetical body that does not actually exist is not sufficient. So there must be something intrinsic to the mobile:

---

162 Oresme, *Questiones super Physicam*, III.6, 335: 'Quinto partes talis fluxus non sunt, quia una est preterita et alia est futura. Respondetur quod eodem modo arguitur de sono quod non est. Ideo respondetur distinguendo ad Aristotelem, sicut Ocham: aliqua dicuntur esse non in actu perfecto, sed incompleto, cuius semper aliquid est futurum, cuiusmodi sunt successiva.'

163 [Albert of Saxony], *Quaestiones Physicorum*, III.7, 2: 515–516.

Quarta conclusio: illud intrinsecum non est illud mobile, licet sit aliquod sibi inhaerens ...

Quinta conclusio: illud intrinsecum secundum quod mobile aliter et aliter se habet est ipse motus seu fluxus. Probatur quia: cum mundus sic movetur, nihil aliud sibi acquiritur nisi sit talis motus. Unde Deus potest movere totalem mundum absque hoc quod acquiratur sibi aliqua alia dispositio a motu.

Sexta conclusio: motus talis est res distincta a mobili. Patet ex quarta et quinta conclusione.

Septima conclusio: in omni mobili quod movetur localiter, volentes admittere casus divinos oportet ponere fluxum seu motum inhaerentem mobili qui successive illi mobili acquiritur. Probatur quia: tales habent admittere quod Deus posset movere aliquod mobile, quocumque sibi extrinseco annihilato, consimili motu in specie absque productione alicuius rei quae nec ipsa vel ei consimilis praeexistebat in illo mobili; sed hoc non posset, si, omnibus annihilatis, illud mobile moveretur aliquo fluxu sibi superaddito, et non prius, sicut patet intuenti; igitur etc.[164]

If in a 'divine case' there has to be some flux or motion in the mobile, so must the same be true in a natural case. What is this flux or motion? In reply to a principal argument, Albert writes:

Ad quartam, cum dicebatur *motus est de essentia termini ad quem*: hic respondeatur per distinctionem Commentatoris intelligendo ipsam, sicut verba iacent, quod 'motus' accipitur dupliciter: uno modo pro fluxu, alio modo pro *forma fluente*. Primo modo non est de essentia termini ad quem, sed bene secundo modo. Unde in motu locali forma fluens dicitur locus vel spatium cuius una pars acquiritur mobili post aliam; in motu autem alterationis forma fluens dicitur qualitas quae successive acquiritur alterabili; et in motu alterationis idem est fluxus et forma fluens, sed non in motu locali, sicut patet ex praecedentibus.[165]

Before concluding this way, Albert had imagined various cases, such as one in which a stone is falling in a medium, and then everything outside the stone is annihilated. Could it be that in the natural case there was no intrinsic flux required, but suddenly when the rest of nature is annihilated, a flux has to be

164 [Albert of Saxony], *Quaestiones Physicorum*, III.7, 2: 517.

165 [Albert of Saxony], *Quaestiones Physicorum*, III.7, 2: 520.

created in the stone? What if God chose not to create the new flux? Comparing this case to that of the rotation of the outermost sphere, Albert argues that there cannot be local motion without place, but there could be motion of the same species as local motion. In such discussions, the Parisians are struggling with the problems that nominalist semantics poses. If they multiply entities, they do not do so without what they perceive as compelling arguments.

3.1.3 Questions III.8–9: The Termini of Motion

It was commonly said that motion and time are successive as opposed to permanent entities. A single motion has no time gaps. Within a single motion, continuity depends on the continuity of what is gained or lost, whether in place, quality, or quantity. If a motion leaves an old place or quality and gains a new one, the instant dividing the time when the mobile is in the old place or quality, and the time when it is in the new place or quality, should be assigned to the later place or quality, as Aristotle argues in Book VIII of the *Physics*.

Aristotle assumed the existence of indivisibles such as points, lines, surfaces, instants, and indivisible ends of motions (*mutata esse* or *momenta*). William of Ockham and Buridan reject the existence of such indivisibles in nature, however much they are a part of geometry. In the section of Book IV on time, where other commentators deal with questions on instants, Buridan includes no such questions. Up to the 1320s, logicians discussing the proper usage of the words '*incipit*' (it begins) and '*desinit*' (it ceases) came to the conclusion that, when permanent things begin, they have a first instant of being and no last instant of non-being, whereas successive things have a last instant of non-being and no first instant of being. This would mean that of any motion there is no first instant of being. Likewise, motions end with the acquisition of a permanent place, quality, quantity, or other form, and therefore have no last instant of being.

In questions III.8–9, Buridan deals with the termini of local motion. It seems that, like alterations, local motions should have termini (otherwise would they not be infinite?). But, if so, what could they be? Buridan assumes that motion extends throughout the mobile, so the limits of motion in space would be the same as the limits of the body. But what might be said about the limits of the motion in time? In particular, what might be said about the termini of a day, if the physical basis of time and of measures of time is the rotation of heavens? Days are in this sense continuous with each other, as the rotations are continuous, even if humans mark off days, whether they choose the turn of the day to be midnight, or sunset, or sunrise, etc. The problem is even more complicated if it is assumed that instants or indivisibles of time or motion do not exist.

Question III.8—Whether it is necessary for local motion to have positive termini other than the flux, namely the terminus from which and the terminus to which (*Utrum de necessitate motus localis sit habere terminos positivos praeter fluxum, scilicet terminum a quo et terminum ad quem*).[166]

1. Boethius of Dacia: *Utrum omnis motus habeat duas mutationes terminos suos* (III.20)[167]
2. William of Ockham (*Quaestiones*): *Utrum sit dare primam partem mobilis mutatam vel motam* (81) *Utrum sit dare primam partem formae adquisitae per motum* (82) *Utrum sit dare primam partem motus vel temporis* (83) *Utrum in successivis sit dare primum intrinsecum et extrinsecum* (98) *Utrum successivum habeat ultimum intrinsecum et ultimum extrinsecum sui esse* (99) *Utrum permanentia habeant primum intrinsecum sui esse* (100) *Utrum permanentia habeant ultimum intrinsecum et ultimum extrinsecum sui esse* (101)
3. Hugolinus of Orvieto: *Utrum omnis motus sit idem cum termino suo ad quem* (18 = III.3)[168]

---

166 The table of questions continues: 'De duplici divisione et magnitudine motus. Quod omnis continui finiti termini sunt prima et ultima partes eius. Quod eiusdem lineae infiniti sunt termini et in infinitum termini est terminus. Quod saepe in motu recto intenduntur termini extrinseci' (5^{8-11}). The word order used in the table of questions indicates that syncategorematic infinites are intended; cf. *infra*, n. 170.

167 Cf. Boethius of Dacia, *Quaestiones Physicorum*, III.20, 289–290: 'Ad primam quaestionem dico quod omnis motus habet duas mutationes terminos suos, quia sicut motus non terminatur nisi per mutationem—inter tempus motus et tempus quietis est instans, in quo instanti nec movetur mobile nec quiescit, ergo mutatur—ita mutatio similiter debet esse a parte inceptionis, quia motus non incipit per motum, quia tunc, cum mobile incipit moveri (movere *ed.*), movetur. Tunc esset bonum argumentum: incipit moveri, ergo movetur, quod est falsum. Si quaeras utrum omnis motus finitus habeat duas mutationes terminos suos, dico quod sic ... Ad aliud dico quod motus habet duplicem terminum finalem: unum qui fit per motum, ut albatio, et talis terminus divisibilis est, quia est motus successivus; alius est terminus qui non fit per motum, et talis non est divisibilis.'

168 Hugolinus, *Quaestiones Physicorum*, 18 = III.3, 26: 'Primus articulus. In quorum primo videbitur in generali quid sit esse terminus motus. Accipiendo "terminum" primo modo, sic pono istam conclusionem: quod terminus alicuius motus aliqua entitas est, et terminus alicuius motus nulla entitas est. Sed accipiendo "terminum" secundo modo, sic dico quod terminus motus penitus nihil est. Secundus articulus. In secundo videbitur de identitate motus cum suo termino ad quem. Prima conclusio est quod non omnis motus est idem realiter cum termino suo ad quem. Secunda est quod nullus motus est idem essentialiter cum termino suo ad quem. Tertia est quod aliquis (quis *ed.*) motus est idem realiter cum

4. Marsilius of Inghen: *Utrum in motu locali termini positi distincti sint necessarii* (III.2.3)
5. Lawrence of Lindores: *Utrum in motu locali sint ponendi termini distincti a fluxu et a mobili et a loco* (III.8)
6. Benedictus Hesse: *Utrum de necessitate motus localis sit quod habeat terminos positivos praeter fluxum, scilicet terminum a quo et terminum ad quem* (III.16)[169]

Buridan's question III.8 has few precedents or successors, but there is one in the questions on the *Physics* of Boethius of Dacia, as indicated in the list above. It seems that the place from which a local motion departs and the place where it ends are termini, but they are not intrinsic termini of motion as distinct from what is gained by local motion. Also if a motion is finite, it must be terminated at the beginning and the end (81^{15-16}). The opposite is argued, however, because it is not apparent that in local motion, beyond the flux that is motion, there ought to be posited anything beyond the indivisible places or moments, as would be said of points terminating lines. But places are not required, as has been said, nor indivisibles of motion (*momenta* or *mota esse*, i.e. having been moved), because, Buridan writes, we do not concede that they exist any more than we concede indivisible points in a line (82^{15-20}).

A motion is extended throughout the body moved and has termini at the end of the body, and it is also extended in time. Buridan writes:

> Thus I suppose that it is the same magnitude by which the mobile is extended and by which the motion is extended, as I would say that it is the same magnitude by which the stone and whiteness and hardness and other of its accidents are extended (83^{3-6}).
>
> In another way motion has magnitude, extension and divisibility according to the parts of time before and after each other in the same mobile and succeeding each other in the same part of it (83^{13-15}).

In this way, Buridan goes on, we say that the motion of the heaven is infinite. Thus the motion would be said to be infinite which is always or which is perpetual. And it would be said to be finite in the earlier part (*a parte ante*), if it

termino suo ad quem. Quarta conclusio est quod accipiendo "terminum" secundo modo, nullus motus est idem quocumque modo cum suo termino ad quem.'

169 Benedictus Hesse, *Quaestiones Physicorum*, III.16, 333: 'Nostra responsio est contra primam conclusionem Buridani, secundi articuli. Argumentum probatur: Buridanus dicit quod non est necesse in motu locali habere terminos, et responsio dicit: necesse est.'

was not always, but at some time earlier nothing of it existed (83^{15-19}). Against this background, Buridan formulates the following conclusion:

> Perpetual motion, if there is some perpetual motion, does not have termini of itself, that is, termini by which it would be terminated according to duration, because it is infinite in duration, and the infinite, if it is something, is unterminated and does not have termini (83^{21-24}).

Buridan goes on, however:

> However it is not incongruous that of this kind of motion, infinite are the termini of its parts (83^{24-25}).

So, suppose that the cosmos is perpetual. Then there will have been and will continue to be infinitely many days. But this has to be in the syncategorematic sense: no matter how many termini of days there may have been in the past, there will be more days in the past and more termini of those days.[170] This is also expressed in the following conclusions:

> The second conclusion is that there would be termini of the parts of it [of a perpetual motion], because finite parts of it can be designated, as a diurnal revolution or two, and of every finite magnitude, indeed of every finite continuum, it is necessary that there be termini on every side on which it is finite (83^{26}–84^{2}).
>
> Take a part of the motion of the last sphere finite in both directions. Call it A, and let it be as large as a diurnal rotation of the sun. Then the third conclusion will be that the motion of A has a terminus on the earlier part and on the later part, because it is finite on both sides (84^{3-6}).
>
> The fourth conclusion is that those termini are not extrinsic to that sphere, and likewise for other local motions, because the motion would be no less finite and terminated if there were nothing extrinsic (84^{7-9}).
>
> The fifth conclusion is that also of the finite motion of the last sphere, say of one revolution, the terminus is not the heaven, nor the motion of the heaven, nor some part of the heaven, or the magnitude, whether on the earlier or on the later side, because any of these is the same before and after, and in the middle (84^{24-27}).

---

170 This was indicated by the word order in the table of questions. See *supra*, n. 166.

And here comes the upshot. The termini of the parts of the motion in time have to be periods of time, not indivisible instants, because there are no such instants. And the parts are not any exact given part, but any one of an unlimited number of smaller and smaller intervals, as small as one chooses:

> But then it may reasonably be asked what are the termini by which the revolution is terminated?—This question should be answered as the question concerning the termini of a line or of a magnitude. Suppose that there were a rod ten feet long and that there were no other body in the world. It would be finite and terminated in length, but not by anything external. Nor would its termini be indivisible points or surfaces indivisible in depth, beause we assume that such do not exist. Nor would the whole line or rod be the terminus ... All this is conceded. From which it follows that its termini will be quantitative parts, that is its first and last parts (85$^{1-13}$).

Various objections can be raised against this position, for instance that it is inconsistent with Aristotle's statements, but these objections may be solved easily, according to Buridan (86$^{13-15}$).

What about rectilinear motion? According to Buridan, rectilinear motion need not have a terminus just because it is rectilinear, because, just as God could rotate the heaven infinitely, so God could move it in an infinite straight line (88$^{2-5}$). Naturally, it is necessary for every rectilinear motion to have termini because all such motion must be finite and not perpetual. Motion cannot be in a straight line except below the orb of the moon: the celestial orbs are rigid spherical shells, leaving no permeable dimensions for motion in a straight line. And inside the orb of the moon, the only straight line distance is finite (*ibi non est spatium rectum nisi finitum*) (88$^{6-11}$).

Buridan further concludes that every heavy or light body moved naturally by its gravity or levity has as a goal (*intendit*) an extrinsic terminus, namely its natural place (88$^{12-14}$). Another conclusion is that often animals or humans intend by their local motions to acquire something extrinsic, such as temperate air or better health or the joy of playing games (88$^{17-23}$).

Among the authors included in the list above, Benedictus Hesse wrote that he disagreed with Buridan on this last subject: Buridan argues that it is not necessary to posit intrinsic termini in local motion, but Hesse maintains that it is necessary.[171]

---

171 See *supra*, n. 169.

Question III.9—Whether motion is of the essence of the terminus to which it is (*Utrum motus sit de essentia termini ad quem est*).[172] Aristotle, *Physics*, III, 1, 200b32–33; Averroes, *In Physicam*, III, comm. 4.

1. **Er349(1)**: *Utrum motus sit in genere perfectionis ad quam est motus* (III.5)
2. **Er349(2)**: *Utrum motus sit de genere perfectionis ad quam vadit* (III.9)
3. **Ka11**: *Utrum motus sit in genere eius ad quod vadit* (III.2)[173]
4. **L1386(1)**: *Utrum motus sit de genere perfectionis illius ad quod est motus* (III.4)
5. **L1386(2)**: *Utrum motus sit in genere eius ad quod vadit* (III.2)
6. **Vat6758**: *De motu utrum sit de genere eius ad quod vadit* (III.2)
7. Geoffrey of Aspall: *Utrum motus sit illa forma ⟨inducenda⟩* (III.4)
8. Boethius of Dacia: *Utrum motus sit idem in essentia cum suo termino* (III.4)
9. Thomas Wylton: *Utrum motus sit aliquo modo in genere termini ad quem est motus* (III.7)
10. John of Jandun: *Utrum motus sit essentialiter idem cum termino ad quem* (III.6)[174]
11. William of Ockham (*Quaestiones*): *Utrum haec sit concedenda de virtute sermonis 'alteratio est terminus ad quem partibiliter adquisitus vel deperditus'* (21)
12. Nicole Oresme: *Utrum motus sit res acquisita mobili dum movetur* (III.4)
13. Hugolinus of Orvieto: *Utrum omnis motus sit idem cum termino suo ad quem* (18 = III.3)[175]

---

172 The table of questions continues: 'Hoc declaratur de motu locali, de alteratione, de augmentatione, diminutione, generatione, corruptione et instantanea mutatione' (5^{13-14}).

173 This question is also found in manuscript Paris, Bibliothèque Nationale de France, cod. lat. 14698.

174 Jandun, *Quaestiones Physicorum*, III.6, 44vb: 'Ad quaestionem intelligendum est quod, sicut communiter dicitur et accipitur a Commentatore, "motus" potest accipi dupliciter. Uno modo prout non differt a termino ad quem est motus nisi sicut magis perfectum et minus perfectum. Et hoc modo describit Commentator motum, quod est generatio unius partis post aliam partem illius perfectionis ad quam tendit motus donec perficiatur et sit in actu. Alio modo accipitur "motus" pro fluxu continuo huius formae, quo fluit et vadit ad aliquid ulterius. Et dicit Commentator quod prima opinio est verior, scilicet quae ponit motum esse formam imperfectam et fluentem, alia autem opinio est famosior, scilicet quod motus est fluxus et via quaedam huius formae ad aliquid ulterius acquirendum. Et tunc dicendum quod motus primo modo sumptus est eiusdem essentiae specificae cum termino ad quem vadit, secundo modo non.'

175 This question was already reported under Buridan's question III.8 (cf. *supra*, XCVII and n. 168). In what Eckermann includes, it is unclear whether Hugolinus deals in this question

14. Marsilius of Inghen: *Utrum omnis motus sit de essentia termini ad quem* (III.2.4)
15. Lawrence of Lindores: *Utrum omnis motus sit de essentia sui termini ad quem* (III.9)
16. Benedictus Hesse: *Utrum motus sit de essentia sui termini ad quem* (III.17)

At the start, Buridan notes in passing that, contrary to the impression one might derive from the question, Aristotle and Averroes do not use the terminology of being of the essence of the *terminus ad quem*, although the commentators (*expositores*) interpret them in this way (92^{17-19}). Before Buridan, I see only John of Jandun asking this question in these terms, but it really corresponds to asking about Averroes' 'truer' (*verior*) description of motion, that it is the *forma fluens* or acquisition part by part of the ultimate terminus of the motion. By adding questions III.3–5, picking up on the work of Walter Burley, to the questions asked by previous commentators, Buridan essentially took the question in a different direction. So what remains here may be what Buridan might otherwise have asked about alteration.

Throughout Aristotle's *Physics* and what commentators made of it in the Middle Ages, there is the argument that motion is not just local motion, as might be thought in early modern physics, but also alteration and augmentation and diminution at least, if not also generation and corruption. What Averroes had written about *forma fluens* and *fluxus formae* really applies best to alteration and not as well to local motion. Perhaps in recognition of the possible disparity of different types of motion, Buridan includes in this question separate discussions of local motion, alteration, generation and corruption (both in time and sudden), and augmentation and diminution (both simple rarefaction and condensation, and nutrition leading to growth). For local motion, there are special issues resulting from Aristotle's text. For instance, Aristotle had argued that during local motion the mobile is always partly in the *terminus a quo* and partly in the *terminus ad quem*. To this, Buridan rightly points out that it would be true if an extended body started in one extended place and then moved to the very next non-overlapping place. It would not, however, be true if there were a long gap between the original place and the final place (93^{18-33}). Buridan is also thinking, for instance, of a heavy body moving from the place of fire to the natural place of earth. During the motion, when the body is in air or water, that place would not be of the essence of the natural place of earth (92^{1-3}).

---

with motions in the categories of place, quality, and quantity, or primarily with one of these categories.

Turning to alteration, Buridan takes as his example alteration from cold to hot, assuming the admixture theory of intension and remission. If the extremes of the motion are the most intense cold and the most intense hot, then, while the subject of alteration is partly through the process, what qualifies it will not be precisely of the essence of the *terminus ad quem*, but it will include parts that later become parts of the *terminus ad quem* (94^{1-11}). In reply to the third principal argument, Buridan explains that, while local motion and its intrinsic termini (understood to be parts of the motion) are purely successive things, in alteration the final form existing at rest was successively generated one part after another during the alteration (98^{12-15}).

Buridan has nothing particularly notable to say about generation and corruption, or augmentation and diminution, only explaining how his conception of motion can be applied in these cases.

3.1.4 Questions III.10–11: The Definition of Motion

Question III.10—Whether all motion is the act of a being in potency (*Utrum omnis motus sit actus entis in potentia*).[176] Aristotle, *Physics*, III, 1–2, 201a8–202a12; Averroes, *In Physicam*, III, comm. 6–11.

1. **Ka11**: *Utrum* ⟨*motus*⟩ *sit actus entis in potentia secundum quod huiusmodi* (III.4)
2. Boethius of Dacia: *Utrum motus sit actus* (III.8)
3. Geoffrey of Aspall: *An motus sit actus* (III.6)
4. Radulphus Brito: *Utrum diffinitio motus sit bene assignata* (III.3)
5. Giles of Rome: *Utrum motus sit actus* (III.D.2) *Utrum motus sit entis in potentia* (III.D.3) *Utrum motus sit actus entis in potentia secundum quod huiusmodi* (III.D.4)
6. Thomas Wylton: *Utrum motus sit actus simpliciter* (III.2)
7. John of Jandun: *An motus sit actus* (III.2)
8. William of Ockham (*Expositio*): III, cap. 3 (*De definitione motus*) (tt. 6–11, 201a9–b15)[177]
9. Nicole Oresme: *Utrum motus bene diffiniatur quando dicitur quod est actus entis in potentia secundum quod in potentia* (III.8)
10. Marsilius of Inghen: *Utrum motus sit actus imperfectus* (III.2.5)
11. Lawrence of Lindores: *Utrum motus sit actus entis in potentia* (III. 10)

176 The table of questions continues: 'Quod est actus moventis et eius quod movetur et entis in actu et entis in potentia et imperfectus actus' (5^{16-17}).

177 Ockham, *Expositio*, III, cap. 3, §1–7, 452–467.

12. Benedictus Hesse: *Utrum omnis motus sit actus entis in potentia* (III.19)
13. Conimbricenses: *Rectene Aristoteles motum definierit* (III.1)

In question III.10, Buridan begins to discuss the first parts of Aristotle's definition of motion and in question III.11, he extends his consideration to the whole definition. It might be argued, as was done in early modern times, that Aristotle's definition of motion makes matters more rather than less obscure. What is the act of a being in potency? Isn't this an oxymoron? Explaining the significance of words, Buridan notes:

> It should be noted that of all the categories (*praedicamenta*), the predicates are sometimes said, explicitly or implicitly, with the verb 'is' (*est*) and sometimes with the verb 'can' (*potest*) ... By 'to be' we understand being in act (unless there is a predicate that amplifies to the past or future) and by 'can be' (*posse esse*) we understand being potentially (100^{11-17}).
>
> It should also be noted that, taking 'potentially' broadly, it follows from something being in act that it can be (*ad esse in actu sequitur esse in potentia*), but strictly speaking being in act and being in potentiality are opposed and cannot be true of the same thing in the same way and at the same time (100^{18-23}).

Against the background of these comments, Buridan states the following five conclusions:

1. Motion is the act of that which is moved and of the mover, because what previously could be moved and was not moved is moved by the present motion. And also what could move, moves with this motion. Likewise the agent acts with the action which it could act with before, and what suffers the action suffers the action (*patiens patitur*) when before it could suffer the action (*pati*). Therefore we say that the action is the act of the agent and the passion of the patient. Note that even though the heaven always was moved and thus never was said properly to be in potency to be moved, nevertheless the heaven was not always moved with today's revolution, but it could be moved by it and now it is moved by it. Thus it is also the act of God as mover, because God previously could move it with the revolution with which God now moves it (101^{3-14}).
2. Motion is the act of a being in act, because it is the act of that which moves and of that which is moved. But these verbs 'to move' and 'to be moved' signify to be in act according to such a disposition (*secundum talem dispositionem*) just as 'to be able to move' and 'to be able to be moved' signify

to be potentially. And universally, by the definition of the meaning of the term it is clear that all act by which something is said to be in act, when before it would be said to be in potency, is the act of a being in act (101^{15-20}).

3. Every motion is the act of a being in potency. This is clear first of perpetual motion ... But it is also clear of all non-perpetual motion, because everything that is moved with such a motion is not completely (*perfecte*) moved, but can be completely moved; therefore it is still in potency to be completely moved (101^{21-26}).
4. Every motion is an incomplete act as long as it is motion, because in those things which are able to become or to have been made (*fieri vel facta esse*) nothing is said to be complete unless it is totally made, because in the word 'perfect' the prefix 'per' signifies completeness. But whenever motion is totally accomplished (*factum*), it is no longer motion, because every motion consists in becoming one after another. Therefore it may well happen that motion is perfected and completed, but then it is not motion, indeed it was motion (102^{1-7}).
5. Motion is a second act and not a first act ... Aristotle calls science a first act and 'to consider' a second act, although the soul is in act before this science. But since names are a matter of choice (*ad placitum*), philosophers have called a 'first act' every form or habit from which procedes, or from which is able to procede, an operation, and they call a 'second act' the operation in relation to the form or habit from which it procedes ... Thus we call gravity and levity the first acts with respect to the local motions of the elements, and we call the latter second acts. For every motion is an operation proceding from some form, or from some agent in act, and therefore we call every motion a second act with respect to the act of the agent from which it procedes (102^{8-24}).

Buridan further explains that perfection can be understood in different ways. Motion can be called a perfection in the sense of a disposition inhering in a subject or coming from an agent, suitable to (*conveniens*) that subject or that agent (103^{1-8}). Thus in question III.10, Buridan has explained the first parts of Aristotle's definition of motion, and in doing so has said that motion is a perfection of a sort which is also called a disposition (*dispositio*).

Question III.11—Whether the definition of motion is good in which it is said that motion is the act of a being in potency or insofar as it is in potency (*Utrum definitio motus sit bona in qua dicitur quod motus est actus entis in potentia secundum quod in potentia*).[178] Aristotle, *Physics*, III, 1, 201a10–11.

---

178 The table of questions adds: 'Non convenit mutationi instantaneae' (5^{19}).

1. William of Clifford: *An motus possit habere diffinitionem* (III.3) *An motus sit actus* (III.4) *An motus sit actus entis in potentia secundum quod huiusmodi* (III.5)
2. Boethius of Dacia: *Utrum motus sit actus entis secundum quod ipsum est in potentia* (III.9)
3. Geoffrey of Aspall: *Utrum motus sit actus in potentia* (III.7)
4. Radulphus Brito: *Utrum diffinitio motus sit bene assignata* (III.3)
5. Thomas Wylton: *Utrum diffinitio motus sit bene data* (III.1)
6. Walter Burley (*Expositio et quaestiones*): *An diffinitio motus sit bene data* (III.1)[179]
7. John of Jandun: *An motus definitio sit conveniens* (III.3)
8. William of Ockham (*Expositio*): III, cap. 3 (*De definitione motus*) (tt. 6–11, 201a9–b15)[180]
9. Nicole Oresme: *Utrum motus bene definiatur quando dicitur quod est actus entis in potentia secundum quod in potentia* (III.8)
10. Albert of Saxony: *Utrum definitio motus, scilicet motus est actus entis in potentia secundum quod in potentia, sit bona* (III.2)
11. Benedictus Hesse: *Utrum definitiones motus et specierum motus sint bonae* (III.18) *Utrum definitio motus sit bene posita* (III.20)
12. Domingo de Soto: *Utrum diffinitio motus cum reliquis adiacentibus sit ab Aristotele bene assignata* (III.1)

In this question, Buridan uses different principal arguments than the ones he used in the previous question, in particular considering what is proper for a definition. His replies offer another chance to try to understand his exact position on what motion is, in the guise of his explication of and support for Aristotle's definition of motion. He precedes his main conclusion with a few clarifications:

> It should be noted that for motion is required a mobile and a form or disposition according to which the mobile differs earlier and later (*forma vel dispositio secundum quam mobile se habet aliter et aliter prius et posterius*). The mobile may be related to the form or disposition in three ways. In one way, the mobile is in pure potency with respect to the form or disposition, because it has nothing of it. In another way, the mobile has the form or disposition perfectly, and then it is no longer moved according

179 In the *Quaestiones*, this is question 27.

180 Cf. *supra*, n. 177.

> to it, but it has been moved. In a third way, the mobile has or has acquired something of the form or disposition, but it has not acquired it wholly, and in this way it is necessary for the mobile to be in relation to the disposition when it is moved in respect to it (105$^{14-22}$).
>
> Furthermore, it should be noted that the word 'motion' (*motus*) does not have supposition for the mobile ... but for the disposition according to which the mobile differs earlier and later (*se habet aliter et aliter prius et posterius*) (105$^{23-25}$).
>
> On this basis, it should be concluded that motion, namely that disposition according to which motion occurs, is the act of the mobile according to that which the mobile has of that disposition. And the mobile is still in potency to that which remains to be acquired further. Therefore motion is the act of a being in potency (105$^{26-29}$).
>
> But, again, it should be noted that, when the mobile was partly in act and partly in potency according to the disposition according to which it is or could be moved, this can happen in two ways. In one way that it is in a permanent state (*in permanentia*), i.e., without a current tendency to that of the disposition to which it is still in potency. In another way that it has a current tendency to that. If it is in the first way, then it is not moved according to that disposition ... But if there is still a process ... farther, then it is still moved. It is to exclude the possibility of remaining in the middle disposition that it is necessary to add to the definition of motion another short clause, namely 'according as it is in potency' (106$^{1-14}$).

After raising and responding to a further doubt, Buridan concludes that Aristotle's definition of motion is sound, because it states explicitly and convertibly the significance and every connotation of the term '*motus*'. It excludes the possibility that the mobile will not be moved further in respect to the disposition at issue and it includes the connotation of a tendency to continue (106^{29}–107^{5}). This definition of motion does not include sudden mutations (108$^{6-8}$). On the other hand, the definition fits time, because time is motion (109$^{1-2}$). In his overall answer to this question Buridan seems to be fairly traditional. If his answer were to be compared to the views of John of Jandun and Walter Burley, on the one hand, and to those of Nicole Oresme and Albert of Saxony, on the other, his position within a spectrum of positions could be more clearly defined.

3.1.5 Questions III.12–13: On Motion as Action and Passion

These questions arise from Aristotle's text. In Greek, the verb for 'to move' is transitive, so one expects to consider that A moves B, and not, intransitively, that 'B moves' in the sense of 'B is moving' (present participle). Interestingly, much of the discussion in Latin uses, not an active form of the verb, '*movere*', but a middle or passive form, '*moveri*'. If A moves B, is A's moving of B (A's action) in A or in B? If A's moving is an action and B's being moved is a passion, is A's action in B and is it the same as B's passion? Interestingly, teaching and learning are used to illustrate this problem. Has professor A done any actual teaching if B has not learned anything? Or are teaching and learning, so to speak, two sides of the same coin? Related questions about action and passion may have become pertinent because of the habit of expounding the truth of propositions on the basis of the personal supposition of the terms, i.e., what they stand for in the external world.

Question III.12—Whether all motion is in the mobile or in the mover or in both as in a subject (*Utrum omnis motus sit subiective in mobili vel movente vel in utroque*).[181] Cf. Aristotle, *Physics*, III, 3, 202a13–14; Averroes, *In Physicam*, III, comm. 18.

1. William of Clifford: *Utrum actio sit in eodem subiecto cum passione* (III.6)
2. Boethius of Dacia: *Utrum motus sit in movente ut in subiecto* (III.12) *Utrum idem sit actus moventis et mobilis* (III.13)
3. Geoffrey of Aspall: *Utrum motus sit in movente ut in subiecto* (III.8)
4. Radulphus Brito: *Utrum actio sit in agente* (III.6)
5. Giles of Rome: *Si actio et passio sint idem realiter, quomodo possunt facere diversa praedicamenta* (III.14)
6. Thomas Wylton: *Utrum motus sit in movente ut in subiecto* (III. 8) *Utrum idem sit actus motivi et mobilis* (III.9) *Utrum actio sit in passo vel in patiente sicut in subiecto, et loquor de actione inquantum tale ut est distinctus actus a passione* (III.10)
7. Walter Burley (*Expositio et quaestiones*): *Utrum actio sit in agente vel in passo* (III.4)
8. Walter Burley (*Quaestiones*): *Utrum actio sit in agente subiective* (III.30)
9. John of Jandun: *An motus sit in movente* (III.4)
10. William of Ockham (*Expositio*): III, cap. 6 (tt. 18–22, 202a13–b29)[182]
11. William of Ockham (*Quaestiones*): *Utrum haec sit concedenda de virtute sermonis 'actio et passio sunt idem motus'* (25)
12. Walter Burley (*Expositio*): *Utrum ambo, scilicet agere et pati, sint in agente vel patiente* (III.D.11) *Utrum actio sit in agente vel patiente* (III.D.13)

---

181 No subtopics are mentioned in the table of questions.

182 Ockham, *Expositio*, III, cap. 6, §1–10, 479–497.

13. Hugolinus of Orvieto: *Utrum actio aliqua elicita ab agente sic sit in ipso formaliter et subiective* (20 = III.5)[183]
14. Albert of Saxony: *Utrum motus sit subiective in mobili vel in movente* (III.8)
15. Marsilius of Inghen: *Utrum omnis motus sit in mobili* (III.3.1)
16. Johannes Marsilii (?): *Utrum omnis actio sit in agente* (III.8)
17. Lawrence of Lindores: *Utrum motus est subiective in movente vel in mobili vel in utroque simul* (III.11)
18. Benedictus Hesse: *Utrum motus sit subiective in mobili vel in movente vel in utroque* (III.22)
19. Conimbricenses: *Utrum actio transiens in agente an in patiente insit* (III.3.1)[184]

In his replies to this question, Buridan also refers to the next question, which probes more deeply in some respects. Some of the principal arguments state that motion is in the mover and some assert that it is in what is moved. For instance, the second principal argument is that the form or disposition naming (*denominans*) some subject is in that subject; but from that motion the mover is said to move; therefore the motion is in the mover (110^{8-9}). The

183 Hugolinus, *Quaestiones Physicorum*, 20 = III.5, 27: 'In secundo articulo respondebitur ad quaesitum. Prima conclusio est quod omnis actio elicita naturaliter est subiective et formaliter in patiente. Secunda est quod aliqua actio elicita est subiective et formaliter in agente.'

184 Collegium Conimbricense, *Commentarii Physicorum*, III.3.1, 347–348: 'Quod ad transeuntes attinet, utrum in agente insint an non, controversia est. Scotus in 4, d. 13, q. 1, Caietanus in commentariis 1 part. ad quaest. 25, art. 1, Raymundus Lullius 2 libro suorum proverbiorum, aliique nonnulli arbitrantur actionem transeuntem formaliter spectatam, hoc est ipsam emanationem formae ab agente, ut caloris ab igni, non in patiente, sed in agente recipi ... Contraria tamen sententia existimantium non solum formam, quae motu gignitur, sed ipsam quoque formae productionem sive emanationem inhaerere in patienti, Peripatetica et vera est, eamque amplectuntur praeter alios Themistius, Alexander, Philoponus, Simplicius, Averroes, Magnus Albertus, ut constat ex iis, quae annotant ad text. 20 et 23 huius libri, et ad text. 139 libri tertii *De anima*. Item Ferrariensis hoc loc., q. 4, et ca. 1 lib. 2 *Contra gentiles*, Sonc. 5 *Metaph.* q. 37, Iavellus lib. 9 *Metaph.* q. 15, Hervaeus *Quodlib.* 4, q. 4, Capr. in 2, d. 1, q. 2, art. 3. Secuti communem Magistrum D. Thomam, huiusce opinionis, quicquid alii velint, assertorem, tum aliis in locis, tum lib. 2 *Contra gent.*, c. 1. Ac quod ita senserit Aristoteles liquet ex c. 3 huius libri, ubi docet actionem et passionem esse unum eundemque motum, hunc vero in re, quae movetur, esse, et actionem esse actum huius in hoc, id est, in patiente; passionem vero huius ab hoc. Quod non minus luculente confirmat libro 9 *Metaph.*, cap. 9, text. 16, cum ait actionem esse in eo quod efficitur, ut aedificationem in eo quod aedificatur, et omnino motum in re quae movetur.'

third principal argument, however, is that the disposition according to which something differs earlier and later is in that which differs in this way, namely in that which is moved (110^{10-16}). The fifth principal argument has five sub-proofs to the effect that motion is in the agent, because to move is to act, and action is in the agent (110^{21}–111^{13}).

Buridan follows Aristotle and Averroes in claiming that motion is in the mobile (111^{14-20}), not in the mover: the action of the mover is of the mover, but not in the mover (*actio motoris est eius et non in eo*). However, he provides two answers, one for those who hold that local motion is the mobile (this would include Ockham and some of his followers) and another for those who claim that motion differs from the mobile and place (111^{21-23}). Those who hold that local motion is the same as the mobile cannot say that local motion is *subiective* in the mobile according to real inherence. They argue that, when Aristotle and other philosophers say that motion is subjectively in the mobile (*motum esse subiective in mobili*), what they mean is only that the mobile can be labelled as moving (not that there is a form of motion inhering in the mobile). This is called 'naming predication' (*praedicatio denominativa*). If the heaven is moved, the motion of the heaven is in the heaven, i.e., the term 'motion' is truly affirmed of the name 'heaven', saying 'the heaven is moved' (*caelum est motus*). This is contrasted with what this first group holds about alteration, where they would concede that there is a form in the alterable body as a subject by real inherence, and taking the terms significantly (111^{24}–112^{7}).

As opposed to the first group, the second group, in which Buridan includes himself (*nos autem dicimus*), claims that all motion is subjectively in the mobile, i.e., in that which is moved, by real inherence, as white is in the wall, because there does not seem to be any exception to this unless possibly in local motion; but it has been shown previously that local motion is not an exception. However, Buridan continues, speaking indefinitely (*indefinite loquendo*) one can say that motion is in the mover, because it may happen that the mover is moved, as will be shown in questions VII.1 and VIII.4. This is not always true, however, because the prime mover is not moved: the prime mover is immutable (112^{8-16}). Motion is not in the mover because it moves (something else), but because it is itself moved (112^{17-20}). In reply to the fourth sub-argument for the fifth principal argument, Buridan explains:

> Things in the external world do not name each other, but terms signifying things. Now the term 'to act' signifies action connotatively, although it does not supposit for action. Consequently, it also signifies passion, because action and passion are the same. And thus the term 'to suffer' (*pati*) also signifies action and passion. Therefore it is manifest that a term

> signifying action and passion names a term suppositing for the agent, saying that A acts, and a term signifying action and passion names a term suppositing for the patient, saying that B suffers (113^{27}–114^{6}).

Thus Buridan's reply ends, having, perhaps, said more about the relation of terms to things in the *logica moderna* than about the relation of things in the external world. He has, however, distinguished himself from those unnamed philosophers (Ockham and his followers) who say that motion is the mobile, instead identifying motion with a *dispositio*. It should be noted, however, that, according to Buridan, parts of the relevant disposition are acquired one after another in local motion. Does question III.13 clarify this issue further?

Question III.13—Whether all action is passion and vice versa (*Utrum omnis actio sit passio et econtra*).[185] Cf. Aristotle, *Physics*, III, 3, 202a15–20 (and III, 1, 201a23–25).

1. **Er349(1)**: *Utrum actio et passio sint motus unus* (III. 9)
2. **Er349(2)**: *Utrum actio et passio sint motus unus* (III.3)
3. **Ka11**: *Utrum actio et passio sint motus unus* (III.5)
4. **L1386(1)**: *Utrum actio et passio sint unus motus in numero* (III.8)
5. William of Clifford: *An actio et passio sunt idem secundum rem* (III.7) *An actio et passio sint diversa praedicamenta* (III.10)
6. Radulphus Brito: *Utrum actio et passio sint unus motus* (III.4) *Utrum actio et passio differant realiter* (III.5)
7. Thomas Wylton: *Utrum motus sit in genere passionis* (III.6) *Utrum actio sit in passo vel in patiente sicut in subiecto, et loquor de actione inquantum tale ut est distinctus actus a passione* (III.10) *Utrum actio et passio sint duo praedicamenta* (III.11)
8. Walter Burley (*Expositio et quaestiones*): *Utrum actio et passio sint idem motus numero* (III.5)[186]
9. John of Jandun: *An actio et passio sint duo diversa ac distincta praedicamenta* (IV.5)
10. William of Ockham (*Expositio*): III, cap. 6 (tt. 18–22, 202a13–b29)[187]
11. William of Ockham (*Quaestiones*): *Utrum haec sit concedenda de virtute sermonis 'actio et passio sunt idem motus'* (25)

---

185 The table of questions adds: 'De quo praedicamento est ille terminus "motus". An simplex corruptio est actio' (5^{23-24}).

186 In the *Quaestiones*, this is question 31.

187 Ockham, *Expositio*, III, cap. 6, §1–10, 479–497.

12. Walter Burley (*Expositio*): *Utrum agere quod est actio motoris et agi quod est actio motus sint eadem vel diversa* (III.D.10) *Utrum actio et passio sint idem motus* (III.D.12)
13. Francesc Marbres: *Utrum actio sit existens formaliter in agente* (III.2)[188]
14. Hugolinus of Orvieto: *Utrum actio et passio sint duae entitates distinctae realiter inter se* (19 = III.4)
15. Albert of Saxony: *Utrum omnis actio sit passio* (III.9)
16. Marsilius of Inghen: *Utrum omnis actio sit in passo* (III.3.2)
17. Lawrence of Lindores: *Utrum omnis actio sit passio vel e contra* (III.12)
18. Benedictus Hesse: *Utrum actio sit passio et e converso* (III.23) *Utrum omnis actio sit subiective in agente* (III.24) *Utrum motus sit separabilis ab actione et passione* (III.25)

---

188 Francesc Marbres, *Quaestiones Physicorum*, III.2, 35rb–36rb: 'Quantum ad primum sciendum est quod est una opinio Scotiçantium ponens ipsam actionem esse formaliter formam respectivam et dicere quendam respectum, extrinsecus tamen advenientem. Unde premittitur quod quidam est respectus intrinsecus adveniens, alius extrinsecus adveniens ... Franciscus de Marchia de isto respectu intrinseco et extrinseco modum sibi alium adinvenit dicens quod dicuntur ista sex predicamenta respectus extrinsecus advenientes, eo quod alterum extremum est perfectio extrinseca alterius ... Secundo est alia opinio cuiusdam doctoris qui ponit quod ipsa actio est forma simpliciter absolute (*in marg*: Opinio Tho. Cap. Ad aliud) ... Tenendo autem opinionem Scoti potest responderi ad rationes alterius doctoris ... Sic igitur potest dici ad illas rationes, tenendo istam opinionem. Verumtamen quia alia opinio est satis probabilis, potest dici ad rationes Scoti. Pro cuius evidentia et rationibus solvendis est sciendum quod tripliciter est genus forme ... Secundo modo est motus qui quidem est entitas partim absoluta et partim respectiva ... Quantum ad secundum articulum: ubi erant videnda duo puncta. Primum: utrum actio fundetur in motu tanquam in proprio fundamento vel subiecto. Respondeo. Ubi sciendum quod est opinio cuiusdam doctoris qui ponit quod ipsa actio fundatur in motu tanquam in proprio fundamento vel subiecto. Ad hoc adducit auctorem *Sex principiorum* qui hoc plane videtur dicere ... De secundo puncto: utrum actio sit distincta a motu. Respondeo. Ubi est una opinio cuiusdam doctoris qui ponit quod actio non est distincta realiter a motu ... Quantum ad tertium articulum videndum est in quo sit actio subiective, utrum scilicet vel in agente vel in patiente. Respondeo. Ubi primo distinguo de actione quod actio est dupliciter: quedam que nihil aliud est nisi quidam respectus quo agens refertur ad patiens, tenendo quod sit formaliter respectiva, vel saltem fundamentum quo mediante et ratione cuius agens refertur ad patiens; et ista est actio que est predicamentum distinctum a passione. Est alia actio que est res acta, que quidem est quidam effectus prime actionis et idem cum passione. Istam autem distinctionem videtur ponere Simplicius super *Predicamenta* ... Hoc premisso pono duas conclusiones. Prima conclusio est quod actio acta est in passo ... Secunda conclusio est ista: quod actio que est de genere actionis, ut dictum est, est formaliter existens in agente.'

Buridan begins his reply to question III.13 by expounding the terms of the question and their multiple meanings. Averroes, he claims, includes in 'action' everything in formal act whether existing in permanence and quiet, or in flux, as the soul is said to be the act of the body. In another way, action is called pure act, as it signifies a second act (*actus secundus*), as we would say that in God to think and to want, or even intellection and will, are actions immanent in God. Another way action is said to be what is produced by an agent, although it remains unchanging after the agent stops acting. These senses of action are not at issue here (117^{3-10}). Another way action is said to be an operation that comes from and is produced by an agent, which, when it is present, the agent is said to act it (*qua praesente agens ea dicitur agere*) (117^{11-12}).

Likewise, the word 'passion' is taken in many different ways. In one way, any accident is said to be a passion of what it inheres in, as whiteness in a wall, science in the soul, light in the sun, magnitude in a corporeal body, etc. Sometimes passion is limited to sensible qualities. Sometimes the word 'passion' is said of connotative terms in relation to substantial terms, or of more connotative terms in relation to less connotative terms. One term is said to be the subject of another if they both supposit for the same thing, and we say that passion is predicated of a subject denominatively (*denominative*). Speaking in a material sense, being capable of laughing (*risible*) is a passion of man, even and odd are passions of number, and time is a passion of motion. However, none of these senses of the notion of 'passion' are relevant here, where passion is taken to mean the operation by which the thing affected is said to be affected (*operatione qua passum dicitur pati*), as by action an agent is said to act (117^{19}–118^{5}). Against the background of these terminological clarifications, Buridan's conclusions are the following:

1. It is possible for there to be action without passion, as when God, without a receptive subject, creates some creature (118^{7-8}).
2. Conversely, is impossible for passion to exist without action (118^{19}).
3. It is impossible naturally for there to be action without passion, because nothing can be made naturally unless from a pre-existing subject or in a presupposed subject (118^{23-25}).
4. All natural action is passion and all passion is action, because all action is an operation which is produced by the agent, and all such is received in some subject, and thus it is said to be passion (118^{28-30}).
5. Every motion is action and passion, because all motion is the act of the mobile as that in which it is received, and thus it is passion. And all motion is also action of the mover as that by which it is produced, and thus it is action ... Then the same thing is motion, action, and passion, but the

names change according to reason. It is called 'motion' or 'mutation' because of succession, and it is called 'action' insofar as it is from an agent, and 'passion' insofar as it is received in a patient. The words 'action' and 'passion' connote different things, i.e., the agent and the patient, and because of this the names are predicated denominatively (*denominative*) of each other and not quidditatively (*quidditative*) (119^{3-12}).

A doubt is then raised about what category (*praedicamentum*) the terms 'motion' and 'mutation' belong to. Buridan replies:

> It may be said that, since the categories are proposed by Aristotle to distinguish among the diverse modes of predicating of the first substances, i.e., of discrete names signifying substances standing by themselves absolutely and without foreign connotation unless a grammatical one, and since many abstract terms suppositing for accidents are not predicated truly of first substances, as 'white', 'heat', 'magnitude', as I believe, and 'motion' and 'heating' etc., therefore such abstract terms ought to be posited in the categories according to the requirements of their concrete terms. But the concrete terms are 'motion' and 'mutation', 'to alter', 'to be altered', etc., of which now truly and quidditatively are said 'to act' and 'to suffer'. Therefore 'motion' would be said to be in the category of action to the extent that the mover is said to move, and in the category of passion to the extent that by it the mobile is said to be moved. And we may say that to the extent that we understand that the mobile is moved by motion, this word 'motion' connotes the mobile and it is quidditative predication 'motion is passion', and to the extent that we understand that by motion the mover moves, this word 'motion' connotes the mover and this predication is quidditative 'motion is action' (119^{20}–120^{5}).

In reply to a principal argument, Buridan argues that one category cannot be predicated of another quidditatively, but it can be predicated denominatively (119^{15-19}). To another principal argument, Buridan answers that, setting aside substance, connotative terms from different categories can supposit for the same thing (120^{29}–121^{2}).

3.2 *Buridan's Questions on Infinity: Questions III.14–19*

In the second part of Book III of the *Physics*, chapters 4–8, Aristotle discusses infinity. In chapter 4, he discusses the opinions of the early philosophers. In chapter 5, he criticizes the views of the Pythagoreans and Platonists. He says (204a34) that there is no infinite sensible body. In chapter 6, Aristotle concludes that the infinite exists, and inquires how it exists or what the infinite is (206b33). Chapter 7 describes the various kinds of infinite, and asks which of the four conditions of change the infinite is to be referred to, saying it is matter (207b34). In chapter 8, Aristotle refutes the arguments for the existence of an actual infinite.

The infinite has a place in Book III, because Aristotle connects the infinite division of magnitude to the continuity of motion—so infinity is of interest because of its connection to continuity and motion. As is well known, Aristotle believed that the cosmos is finite but eternal. For medieval authors, this raised questions. It says in the Bible that 'in the beginning God created the heavens and the earth', which seems to imply that time is not infinite in the past, although it might be infinite in the future. On the other hand, Aristotle rejected the existence of minimum corporeal quantities, or atoms, instead arguing that no matter how many times a body has been divided, it could be divided further. While Buridan's question III.14 has a clear basis in Aristotle's text, all the other questions on infinity go beyond Aristotle.

Cecilia Trifogli divides the issues addressed by thirteenth-century English commentators on the *Physics* into (1) the metaphysical grounds for rejecting an actual infinite in magnitude; (2) the actual infinite and the potential infinite by addition in magnitude; and (3) the infinite in number.[189] With regard to number, some of the early English commentators are finitists and some infinitists. Their differences to some extent derive from their views on the nature of number as much as from ideas about infinity. All of the commentators agree that for Aristotle number is only potentially infinite, but some maintain that there is a more general kind of number which is actually infinite.[190] For Aristotle, the infinity of number is based on the divisibility *ad infinitum* of the continuum.[191] A question concerns whether to be counted the parts of a continuum need to be separated or whether it is enough to mark them off mentally. Since most of the accounts in the early questions commentaries are quite short, there is little opportunity to develop any position in detail.

For the fourteenth century, Pierre Duhem discusses the views of the infinitist Gregory of Rimini, followed by John Buridan and Albert of Saxony ('Les adversaires de Grégoire de Rimini'), on the one hand, and by Nicole Oresme and Marsilius of Inghen ('Les partisans de Grégoire de Rimini'), on the other.[192]

189 See Trifogli, *Oxford Physics in the Thirteenth Century*, 87–132.

190 Trifogli, *Oxford Physics in the Thirteenth Century*, 114–115.

191 Trifogli, *Oxford Physics in the Thirteenth Century*, 116.

192 Duhem, *Le système du monde*, VII, 3–157.

In Duhem's view, the fourteenth-century Parisians took the discussion of infinity to a point that would not be exceeded even up to modern times. Duhem wrote (as translated by Ariew):

> After the departure of Marsilius of Inghen, the classrooms of the Sorbonne, the schools on the Rue de Fouarre, did not hear any new opinion worth noting on the infinitely small and the infinitely large; the teachings of the old masters—of William of Ockham, Gregory of Rimini, John Buridan, and Albert of Saxony—were forgotten or served as fodder for unintelligent, rote repetitions. The fate befalling the problem of infinity also befell all the cosmological problems that were the subjects of impassioned debate in Paris during the fourteenth century. The hour marking the start of the Western Schism also marks the end of the mission to initiate modern science that the University of Paris had received.[193]

In *Die Vorläufer Galileis im 14. Jahrhundert*, Anneliese Maier discusses infinity in connection with the continuum ('Kontinuum, Minima und aktuel Unendliches'). Among those in the fourteenth century who held that the continuum was always divisible into divisibles were Roger Bacon, Albert the Great, Thomas Aquinas, Siger of Brabant, Giles of Rome, Richard of Mediavilla, John Duns Scotus, Walter Burley, William of Ockham, and John Buridan and 'his school', i.e., Nicole Oresme, Albert of Saxony, and Marsilius of Inghen. Like Duhem, Maier believed that in the later fourteenth century some authors came to conceptions of infinity that were surprisingly close to those of modern set theory.[194]

In 'John Buridan on Infinity', John Murdoch and Hans Thijssen claim that in questions III.14–19, Buridan 'presents the most central aspect of his treatment of infinity, which is most characteristic about his approach in natural philosophy and metaphysics in general, namely, the further development of Aristotelian notions and doctrines in terms of medieval logic and semantics.'[195] Later in the same article, they write:

193 Duhem, *Medieval Cosmology*, 131. At the time Duhem wrote, Oresme's questions on the *Physics* had not been identified. Guy Beaujouan later identified the work in the Biblioteca Capitular y Colombina in Seville.

194 Maier, *Die Vorläufer Galileis*, 170.

195 J.E. Murdoch & J.M.M.H. Thijssen, 'John Buridan on Infinity', in: J.M.M.H. Thijssen & J. Zupko (eds), *The Metaphysics and Natural Philosophy of John Buridan*, Leiden [etc.] 2001 (Medieval and early modern science, 2), 127–149, at 127–128.

> During the fourteenth century, another important distinction [beyond Aristotle's distinction between the potential and the actual infinites] was introduced, namely that between the categorematic and syncategorematic uses of the term 'infinite' or 'infinitely many' in different propositions. Anneliese Maier believed that it was merely a matter of terminology that a categorematic infinite corresponded to an actual infinite, whereas a syncategorematic infinite was equivalent to a potential infinity. Nowadays, the generally accepted view is that the distinction between the categorematic and the syncategorematic infinite is at the heart of the logic of the infinite. It is part of a new approach in natural philosophy which focuses on propositional analysis. Until now, the examples used to illustrate this approach have been mainly taken from the works of William Ockham or the Mertonians and have largely neglected Buridan. The latter, however, was one of the most consistent practitioners of this logico-semantic approach towards natural philosophy.[196]

Studies of Ockham's writings on the logic or semantics of 'infinity' help to provide a background for understanding what Buridan has to say on the categorematic and syncategorematic infinites.[197]

Question III.14—Whether there is some sensible body infinite in act (*Utrum sit aliquod corpus sensibile actu infinitum*).[198] Aristotle, *Physics*, III, 4, 204a34–206a7; Averroes, *In Physicam*, III, comm. 40–55.

1. S: *Utrum magnitudo possit crescere in infinitum* (III.36) *Utrum elementum, si sit infinitum, corrumperet aliud* (III.41)[199]
2. G[1]: *Utrum quod non sit ponere infinitum in actu sit per naturam magnitudinis simpliciter aut magnitudinis naturalis* (III.18)

---

196 Murdoch & Thijssen, 'John Buridan on Infinity,' 129.

197 See J.E. Murdoch, 'William of Ockham and the Logic of Infinity and Continuity,' in: N. Kretzmann (ed.), *Infinity and Continuity in Ancient and Medieval Thought*, Ithaca [etc.] 1982, 165–206. See also J.E. Murdoch, 'Infinity and Continuity,' in: N. Kretzmann, A. Kenny & J. Pinborg (eds), *The Cambridge History of Later Medieval Philosophy. From the Rediscovery of Aristotle to the Disintegration of Scholasticism, 1100–1600*, Cambridge 1982, 564–591.

198 The table of questions adds: 'Quod saepe maius corpus ceteris paribus non agit intensius vel fortius in corpus sibi impositum quam minus' (5^{26-27}).

199 Trifogli, *Liber Tertius* Physicorum, 108–119, lists questions III.21 through III.42 in this work as involved with infinity.

3. Boethius of Dacia: *Utrum corpori secundum quod corpus repugnet infinitum* (III.28)
4. Geoffrey of Aspall: *Utrum infinitum in actu repugnet magnitudini unde magnitudo sive magnitudini unde naturale est* (III.15)[200]
5. Radulphus Brito: *Utrum sit aliquod corpus sensibile infinitum in actu* (III.12) *Utrum de ratione corporis sit quod sit superficie determinatum* (III.13)
6. Bartholomew of Bruges: *Utrum possibile sit esse magnitudinem actu infinitam* (III.13)
7. John of Jandun: *Utrum possit esse corpus actu infinitum* (III.10)
8. William of Ockham (*Expositio*): III, cap. 10–12 (tt. 40–55, 204b4–206a2)[201]
9. Walter Burley (*Expositio*): *Utrum magnitudini secundum quod est in materia repugnat crescere in infinitum* (III.D.17)[202]
10. Nicole Oresme: *Utrum sit aliquod corpus actu infinitum* (III.11)
11. Hugolinus of Orvieto: *Utrum sit possibile esse aliquod corpus infinitum extensive* (21 = III.6)[203]
12. Albert of Saxony: *Utrum sit aliquod corpus sensibile actu infinitum* (III.11)
13. Marsilius of Inghen: *Utrum in maiori corpori sit maior virtus* (III.5.1)[204]
14. Johannes Marsilii (?): *Utrum de facto sit aliquod corpus naturale actu infinitum* (III.10)
15. Lawrence of Lindores: *Utrum sit aliquod corpus sensibile actu infinitum* (III.13)
16. Benedictus Hesse: *Utrum sit aliquod corpus sensibile actu infinitum* (III.31)
17. John of Celaya: *An naturaliter dabile sit infinitum* (III.9.1)[205]

---

200 According to Zimmermann, *Verzeichnis*, 161, this is question III.13.

201 Ockham, *Expositio*, III, cap. 10–12, 519–539.

202 Burley, *Expositio in Physicam* (1501), 73va–75ra: 'Nullum corpus sensibile est actu infinitum ... Nullum corpus sensibile est infinitum ex hoc quod aliquod elementorum ex quibus componitur, est actu infinitum ... Non est dare corpus sensibile infinitum ex hoc quod unumquodque elementorum ex quibus componitur est infinitum ... Impossibile est corpus sensibile consimilium partium similiter esse actu infinitum.'

203 This question corresponds to Buridan's questions III.14 and III.15. Hugolinus, *Quaestiones Physicorum*, 21 = III.6, 27: 'Primus articulus. In quorum primo videbitur de quaesito an sit possibile esse aliquod corpus actu infinitum extensive, et an de facto aliquod tale sit in natura ... Tertia ⟨conclusio⟩ est quod non est demonstratum de facto nullum esse corpus infinitum extensive.'

204 Marsilius of Inghen, *Abbreviationes Physicorum*, 12rb: 'Quinto notandum quod Philosophus probat in isto tractatu hanc conclusionem: nullum corpus sensibile sensu tactus est actu infinitum.'

205 Celaya, *Expositio Physicorum*, III.9.1, f. 123va–124vb: 'Pro huius questionis enodatione est

18. Domingo de Soto: *Utrum infinitum sit naturaliter possibile* (III.3)
19. Conimbricenses: *Possitne viribus naturae dare in rebus infinitum, an non* (III.8.1)

At the beginning of the question, Buridan writes: 'I now intend to treat this question only textually, narrating what Aristotle says on this subject, and narrating his arguments, so that their value will be clear' (124^{9-11}). He will be concerned in this question only with the extension of bodies taken in the categorematic sense, dealing only with terrestrial and not with celestial bodies, leaving the latter for questions on *De caelo*. Agreeing with Aristotle, Buridan claims that there is no sensible body infinite in act (125^{1-3}). In support of this claim, Buridan repeats Aristotle's arguments and objections against them. To Aristotle's argument that an infinite body would have an infinite force and would thus corrupt another body instantaneously, Buridan replies that the argument is not demonstrative, because the force might be greater extensively, but not intensively (127^{1-11}).

---

notandum quod ille terminus "infinitum" biffariam solet capi: uno modo sincathegoreumatice, et de ista acceptione visum est in exponibilibus nostris. Alio modo solet capi cathegoreumatice, et isto modo capitur in proposito. Et adhuc capitur dupliciter. Uno modo ut attribuitur multitudini, et solet sic describi: infinitum multitudine est multitudo cuius consequenter numerando non est dabilis ultima unitas. Ex ista diffinitione sequitur quod in quolibet continuo est infinita multitudo, quoniam in quolibet continuo quantumcumque parvo sunt infinite partes proportionales. Infinitum magnitudine diffinitur a Philosopho sic: est cuius semper est aliquid extra accipere, id est, infinitum magnitudine est ens extensum sine termino ... Pro solutione huius argumenti est advertendum quod, si loqueremur secundum viam realium, necesse est ponere lineam girativam infinite longam, ut lucet de illa que girat omnes partes proportionales alicuius corporis pedaliter profundi. Sed loquendo secundum viam nominalium non est necessarium ponere aliquam lineam infinite longam, saltem tenendo quod in omnibus difformibus oportet reducere difformitates ad uniformitatem. Nam diceretur quod longitudo linee girative non est infinita; ideo quia oportet reducere difformitates ad uniformitatem, quibus reductis liquidum evadet quod illa linea non erit infinite longa ... Ulterius est advertendum quod Monachus et Buridanus tenent (*in marg.*: Monachus et Buridanus in 17 q., 3 phisi.) quod nulla linea girat omnes partes proportionales continui, quia nec illa que girat primam partem proportionalem, nec illa que girat secundam, nec illa que girat tertiam, et sic consequenter in infinitum. Ideo dicunt quod nulla est linea girativa infinite longa. Sed in hoc non videtur apparentia ...' There are many individuals named in Celaya's questions on Book III of the *Physics*, but this is the only mention I noticed of Buridan.

Aristotle's second argument to prove that there is not an infinite simple body, is that, if it existed, it would follow that its natural place would be infinite, which is impossible (127^{12-14}). There are several arguments to support this conclusion. Then also, if there were an infinite body, it would follow that the infinite simple body would not have a place, neither a natural place nor a violent place (128^{9-11}). Furthermore, the infinite simple body would be totally heavy or totally light. If it were totally heavy, there would be nothing light, because the heavy body would fill all imaginable space. Or, alternatively, it would be totally light (128^{17-19}).

Then Buridan turns to Aristotle's arguments that the infinite body could not be a body composed of simple bodies. If there were one infinite body, there would be no place for another infinite body. Against this argument, it might be argued that there are said to be two separate infinite times, one to the past, and one to the future (129^{27-28}). To this argument, Buridan responds:

> It should be said that this is a mathematical imagination, not a natural one, because there is no natural reason why a body would be infinite in one direction in this way and not in the other direction (130^{3-6}).

In reply to the fifth principal argument, Buridan then writes that it is not necessary to believe that whatever can be imagined or thought of must also exist in reality. Just because an infinite can be imagined does not imply that an infinite exists or could exist:

> When it is said that the intellect and imagination must be moved by things, I say that this is true. Therefore to a simple concept it is necessary that some thing correspond either in the present or in the past. But in compounding simple concepts, falsity or fiction can arise. I say 'falsity' if in a proposition we affirmatively compound terms which do not supposit for the same thing or if we negatively compound terms which supposit for the same thing. Even if this is not in the way of a proposition, but as a determination or something determinable we can compound simple concepts in an affirmative or negative way, as 'white man' or 'not-white man'. And if simple terms have supposition for the same thing, the negatively compound term (*terminus negative compositus*) will be fictitious, i.e., it will have supposition for nothing, as in 'a man not able to laugh' (*homo non risibilis*), and if they do not have supposition for the same, the affirmatively compound term (*terminus affirmative compositus*) will be fictitious, as in 'a man able to whinny'. And thus it is in the present case, because from reality (*a rebus*) we have a concept of a body and we

> have a concept of finite or terminate, and we compound these negatively, forming the complex term 'non-finite body', or 'infinite body', which now is a fictitious term and supposits for nothing (131^{7-22}).

It is true per se that every body is finite, but it is not so manifest that every accident of a body must be finite (132^{9-11}). After this relatively straightforward report of Aristotle's arguments, albeit phrasing them in the terminology of the *logica moderna*, Buridan turns to a less physical and more logical analysis of the question whether there can be an infinite magnitude, setting aside arguments that arise from Aristotelian physics.

Question III.15—Whether there is some infinite magnitude (*Utrum sit aliqua magnitudo infinita*).[206]

1. **Er349(2)**: *Utrum possibile sit aliquam magnitudinem esse infinitam* (III.10) *Utrum contingat excellere omnem magnitudinem determinatam per additionem* (III.16)
2. **Ka11**: *Utrum sit possibile magnitudinem esse infinitam* (III.12) *Utrum contingat excellere magnitudinem quamcumque datam per appositionem* (III.16)
3. Giles of Rome: *De quadam consequentia: si extra caelum esset spatium infinitum, de necessitate esset ibi corpus infinitum* (III.D.26)
4. Thomas Wylton: *Utrum sit ponere infinitum secundum magnitudinem in actu* (III.12)
5. Walter Burley (*Expositio et quaestiones*): *Utrum possibile sit aliquid esse infinitum* (III.6)[207]
6. John of Jandun: *Utrum possit esse corpus actu infinitum* (III.10)
7. William of Ockham (*Expositio*): III, cap. 7 (*Quod ad naturalem scientiam pertinet determinare de infinito*) (tt. 24–30, 202b30–203b15)[208]

---

206 The table of questions adds: 'Quod Deus non potest creare magnitudinem actu infinitam, licet omni creata vel creabili posset creare maiorem. Quod extra hunc mundum non est spatium. Quod annihilato omni eo quod est infra orbem lunae, non esset inter eius latera vacuum neque distantia. Quod in ita parvo corpore vel loco, sicut est granum milii, posset creari maius corpus quam sit mundus et moveri velociter motu recto etc. Quod possibile est corpora neque tangere se neque distare' (6^{2-8}).

207 In the *Quaestiones*, this is question 32.

208 Cf. Ockham, *Expositio*, III, cap. 7, §1, 499: 'Notandum quod quamvis nihil sit actu infinitum ita quod sit actu extensum in infinitum, habens partes in actu infinitas eiusdem quantitatis, aliquid tamen est divisibile in infinitum. Propter quod habet naturalis considerare

8. Walter Burley (*Expositio*): *Utrum magnitudini secundum quod est in materia repugnat crescere in infinitum* (III.D.17)[209]
9. Francesc Marbres: *Utrum effectus actualiter infinitus sit possibilis a Deo produci vel si eius productio includat repugnantiam ex terminis* (III.3)[210]
Utrum infinitum secundum quod infinitum sit ab intellectu humano naturaliter cognoscibile (III.4)[211]

---

quomodo est aliquid infinitum et quomodo non ... Secundo notandum quod Philosophus dicit passionem, punctum et instans forte non esse finita nec infinita ... Tertio notandum quod non intelligunt Philosophus et Commentator quod punctus sit quaedam res distincta a linea, quae nec sit finita nec infinita, quia nulla res est in istis inferioribus inanimatis quin sit extensa, sed loquuntur secundum opinionem famosam quae ponit tales res indivisibiles.'

209 Cf. *Utrum naturalis et mathematicus considerent de eisdem dimensionibus* (IV.D.5). See also question IV.1, *infra*, where this question is also cited as relevant to the ontological status of quantity.

210 Francesc Marbres, *Quaestiones Physicorum*, III.3, 38^{ra-rb}: 'Respondeo ad primum. Ubi sunt multi modi solvendi. Primus est modus Francisci de Mayronis, qui dicit quod plura actu divisa sunt simpliciter plura. Tamen plura actu indivisa ut abstrahunt a divisione plura sunt secundum quid, sed sunt simpliciter unum propter indivisionem. Ponere autem infinitum secundum quid nulla est repugnantia, ut dicit. Partes ergo in continuo sunt infinite secundum quid, et ideo constituunt infinitum secundum quid, quod in rerum natura non repugnat. Aliter respondet Thomas Anglicus ponens istam distinctionem: quod in continuo sunt duplices partes. Quedam actuales, quedam potentiales. Partes actuales dicuntur partes eiusdem quantitatis, et ex numero et ex quantitate talium consurgit magnitudo totius. Partes vero potentiales dicuntur partes eiusdem proportionis, et secundum numerum illarum partium non variatur corpus secundum magnitudinem et parvitatem, quia tot sunt partes tales in grano milii vel in millesima parte eius quot in toto universo, quia utrobique sunt infinite ... Tertio est alius modus solvendi Aureoli quod in continuo non sunt partes infinite, sed finite in infinitum, ita quod nunquam est devenire ad ultimum. Ratio enim, ut dicit, est forma successiva ... Aliter dicit quidam doctor quod in continuo non sunt partes finite nec infinite numero actu, quia nullum numerum constituunt. Divisio enim continui numerum causat ... Hic dicit quidam alter ... Tertia conclusio est quod Deus non potest producere aliquam formam infinitam secundum perfectionem ...'

211 Francesc Marbres, *Quaestiones Physicorum*, III.4, 39^{ra-rb}: 'Respondeo. Ubi sunt duo modi dicendi. Primus est cuiusdam doctoris qui ponit quod de infinito possumus loqui dupliciter: uno modo quantum ad quid nominis, alio modo quantum ad quid rei (*in marg.*: Opinio sancti Thome) ... Ideo dicitur aliter secundum doctorem Scotum quod de infinito possumus loqui dupliciter: vel quantum ad rationem infinitatis in se, vel quantum ad illud quod subest infinitati, vel quantum ad rem subiectam infinitati. Primo modo loquendo de infinito in se, sive potentiali sive actuali, sic ipsum intelligimus non solum quantum ad quid nominis, sed etiam quantum ad quid rei. Loquendo autem de infinito quantum ad

10. Nicole Oresme: *Utrum sit aliquid infinitum* (III.10)[212]
11. Hugolinus of Orvieto: *Utrum sit possibile esse aliquod corpus infinitum extensive* (21 = III.6)[213]
12. Albert of Saxony: *Utrum aliqua dimensio sit infinita* (III.12) *Utrum possibile sit esse aliquam magnitudinem actu infinitam et etiam an possibile sit aliquod esse spatium actu infinitum* (III.13)
13. Marsilius of Inghen: *Utrum sit aliqua magnitudo actu infinita* (III.5.2) *Utrum per aliquam potentiam contingat esse aliquam magnitudinem infinitam* (III.6.3)
14. Johannes Marsilii (?): *Utrum possibile sit esse aliquam magnitudinem actu infinitam* (III.9)
15. Lawrence of Lindores: *Utrum sit aliqua magnitudo actu infinita* (III.14)
16. Benedictus Hesse: *Utrum aliqua sit magnitudo actu infinita* (III.33)[214]
17. John Mair: *An implicet contradictionem dare infinitum actu, hoc est querere an Deus de sua potentia absoluta possit producere infinitum magnitudine*

---

illud quod subest infinitati realiter, vel quantum ad rem subiectam infinitati, sic dicitur quod neutrum infinitum intelligitur vel cognoscitur.'

212 Oresme, *Questiones super Physicam*, III.10, 360: 'Quantum ad secundum solvenda est questio secundum opinionem que videtur Aristotelis. Ideo sit prima conclusio quod in permanentibus per se existentibus, que sunt actu supposita, infinitum cum verbo de presenti vel propositio indefinita cum verbo de presenti omni modo est neganda, categorematice et syncategorematice.'

213 Hugolinus, *Quaestiones Physicorum*, 21 = III.6, 27–28: 'Prima conclusio est quod est admittendum tamquam imaginabile esse aliquod corpus infinitum. Secunda est quod non est impossibile esse aliquod corpus infinitum extensive ... In secundo articulo videbitur an corpus infinitum extensive, si esset, posset moveri aliquo motu. Prima conclusio est quod corpus infinitum omniquaque et secundum omnem dimensionem, si esset, posset moveri motu alterationis. Secunda est quod huiusmodi corpus posset moveri motu generationis et corruptionis. Tertia est quod idem corpus posset moveri motu rarefactionis et condensationis secundum aliquid sui. Quarta est quod huiusmodi corpus infinitum non posset esse motum aliquo motu recto. Quinta est quod videtur difficile quod ipsum posset moveri circulariter; quod tamen non dico pro nunc fore impossibile.'

214 Benedictus Hesse, *Quaestiones Physicorum*, III.33, 376: 'Quamvis praesens quaestio posset includi in ista quaestione "Utrum sit aliquod corpus sensibile" etc., et etiam determinatio ipsius posset haberi ex determinatione illius quaestionis, tamen praesens quaestio non superflue resumitur, quia Buridanus movet quaestionem specialiter, propter movere aliqua dubia de infinito et solutionibus eorum ... Ex isto sequitur quod quaestio prior loquitur de infinito secundum potentiam naturalem, sed praesens quaestio secundum potentiam supernaturalem.'

vel intensione, vel infinitam multitudinem rerum non constituentium aliquod unum (2)[215]

18. Domingo de Soto: *Utrum de potentia Dei absoluta possit fieri supranaturaliter infinitum in actu* (III.4)
19. Conimbricenses: *Possitne infinitum actu divina virtute produci, an non* (III.8.2) *Utrum quacunque specie possit alia perfectior a Deo in infinitum creari* (III.8.3) *Num quodlibet infinitum a Deo cognoscatur* (III.8.4)[216] *Possitne intellectus creatus propriis viribus infinita cognoscere* (III.8.5)[217] *An intellectus creatus ad infinita simul cognoscenda divinitus excitari queat* (III.8.6)

Given that there is no infinite sensible or moveable body, might there be an infinite body or magnitude of any kind? It is argued that there can be, because

---

215 John Mair, *Propositum de infinito*, 140–142: 'Septimo arguitur: implicat in qualibet parte proportionali unius hore creare pedale et addere posterius priori; ergo probatio tertia nulla est ... Sed quicquid sit, erunt infiniti lapides totales. Istud argumentum Albertus de Saxonia dicit in scriptis phisicorum, generat ei fidem, in me non generat opinionem. Buridanus etiam eiusdem tempestatis regens hoc argumento utitur. Ex isto nullo pacto habetur quod aliqua pars proportionalis hore sit ultima, quemadmodum non est ultimus lapis creatus, sed post quemlibet sunt infiniti.'

216 Collegium Conimbricense, *Commentarii Physicorum*, III.8.4, 393–394: 'Articulus 1. Omne infinitum cognosci a Deo, nullum tamen successive posse cognosci ... Sit 2. conclusio: Deus non solum intelligit infinita cognitione simplicis intelligentiae, qua concipit res secundum se; sed etiam notitia intuitiva, qua eas sub esse existentiae comprehendit ... Hanc conclusionem fusius explanat D. Thomas 1 part., quaest. 14, artic. 12 & 13. Nec sibi repugnat sanctus doctor, dum aliis in locis ut 1 *Contra gent.*, ca. 69 asserit Deum intuitiva cognitione non intelligere infinita. Loquitur enim tunc de rebus per se existentibus, ut interpretatur Capr. in 3, distinctione 14, quaest. 2, artic. 1, Caiet. 3 p., quae. 10, art. 3, & Ferr. 1 *Contra gentes* c. 69 ...'

217 Collegium Conimbricense, *Commentarii Physicorum*, III.8.5, 396: 'Non defuere, qui arbitrarentur infinitum perfectione, hoc est, Deum Opt. Max., posse a nobis in hac vita solius naturae facultate clare ac perspicue videri nisi mentes nostrae concreta vitiorum caligine offusae tanti luminis splendore perstringerentur; proinde si quis animum excoleret, atque omnino repurgaret, nullo praeterea auxilio ad tam excellentem cognitionem in qua nostra beatitudo consistit, indigere. Hunc errorem sacrae paginae testimoniis iam antea confutatum, damnavit Concilium Viennense sub Clemente 5, quod decretum habetur in Clementina, Ad nostrum, de haereticis. Quare de huiusmodi infiniti cognitione nulla quaestio relinquitur. De aliis tamen, hoc est, de infinito extensione et multitudine, ad quae cetera infinita reducuntur, disceptatio est, cui nos hac conclusione faciemus satis. Non potest intellectus creatus naturali sua virtute percipere infinita, sive simul, id est non intelligendo unum post aliud, sive successive tempore finito.'

if outside the heaven there is some space, it is infinite, because there is no reason why there would be space of some certain quantity there rather than greater, unless it were supposed (*positum*) to be infinite (133^{8-10}). According to the argument, it is a fabrication (*fictitium*) to say there is a finite space outside the heavens unless this is inferred from the Sacred Scripture, as when Scripture says there are waters above the heavens (Genesis 1, 6) (133^{13-16}). Buridan's response to this question is referred to more than once in Book IV, in the section on the vacuum, so it seems to be significant. The essential point is whether, if there were a vacuum, it would have intrinsic dimensions. The starting assumption is that to have three dimensions is a characteristic of bodies (*corpora*)—a tight linkage that René Descartes would later make a principle of his physics. Aristotle had asserted that the cosmos is a plenum and that, as it was famously expressed, nature abhors a vacuum, meaning that any motions or processes that might seem likely to cause a vacuum to exist would immediately be prevented or undone. So in the cosmos as Aristotle understood it there is no opportunity to know what would be the case if there were any space not occupied by a body.

The ancient atomists, on the other hand, assumed that the world consists of atoms and empty space, and the atomists' empty space did seem to have three dimensions. Aristotle understood, then, that the ancient atomists might have identified place as three-dimensional empty space. But he argued against this view that if empty space were understood to have three dimensions, then when a body was put into it, there would be two overlapping sets of dimensions in the same volume—this was understood to be equivalent to saying that there would be a penetration of dimensions or two bodies in the same place, something that never occurs naturally. Well before Buridan (as early as Philoponus in the sixth century) some commentators on Aristotle had considered the arguments and decided, against Aristotle, to identify place with three-dimensional empty space, however this might seem to others to be a self-contradiction.[218] How could an empty space or vacuum, either inside or outside the cosmos, being nothing, have properties such as dimensions? If it should be believed by faith that God could form and create space outside the existing world, how might this occur? According to Buridan, God could create magnitude or extension outside the existing cosmos by creating a body, even something as small as a bean, and God could extend that magnitude or extension by moving the bean.

218 A fourteenth-century advocate of the view of place as three-dimensional space is Gerard of Odo. See P.J.J.M. Bakker & S.W. de Boer, '*Locus est spatium*. On Gerald Odonis' *Quaestio de loco*,' *Vivarium*, 47 (2009), 295–330.

(The list of condemned theses published by the Bishop of Paris in 1277 famously included the thesis that God could not move the whole cosmos for the reason that it would leave behind a vacuum.)

God could create outside the cosmos other spheres and other worlds and all finite magnitudes whatsoever, as many as God wishes, such that given all finite created things, he could create double, ten times, or a hundred times as great, in every other proportion of one finite to another. It is not, however, necessary to believe that God could create an infinite magnitude, because if this occurred, God could not create a greater (it is contradictory [*repugnat*] for something to be greater than something infinite in act) and it is incongruous that God could make a creature proportionate to his power such that he could not make and create something greater and more perfect (136^{1-5}).

To the principal argument that outside the cosmos there is infinite space because, if outside the heaven there is some space, it must be infinite because there is no reason why there would be space of some certain finite quantity rather than a greater quantity, unless it were posited to be infinite (133^{1-3}), Buridan replies that this does not necessarily follow because of God's supernatural power. God can create just as much space as pleases him—one needs no other explanation of the quantity than the simple will of God (140^{18-21}). In Buridan's opinion, however, there is no space beyond the bodies apparent to us or beyond the bodies we believe exist on the basis of Sacred Scriptures (133^{13-15}, 140^{21-23}). Buridan writes:

> Although it cannot be demonstrated that outside the world there is no space and magnitude, because God could make there both magnitude and space, nevertheless I am of the opinion that there is not space or magnitude or another world there. And to this end Aristotle adduces natural reasons in Book I of *De caelo*, and they will be dealt with there. Therefore I only propose on this matter such a persuasion that it is not likely (*verisimile*) that God would make there another world or other worlds, because if he had wanted to make more worldly creatures than he did, it would not have been necessary for him to make other worlds, because he could have made this world twice as large or a hundred times as large. And if God did not make there another world or other worlds, there seems to be no reason why he would make space there, because he made his world from nothing (*quia illud de nihilo deserviret ad istum mundum*) and it would appear to be needless (*frustra*) (136^{20}–137^{3}).
>
> And again, if there were space there, God could annihilate it, with this world remaining as it is, and this having been done, everything which appears would be saved. And if there were space there, while it remained

> finite, still there would remain all the difficulties for arguing that beyond this there would be more space, as there were for arguing that there would be space beyond the heaven. Therefore that space would do nothing to save the appearances or even to avoid the apparent difficulties. Such, however, should not be posited unless it follows from the words of Sacred Scripture (137^{4-10}).

Another way to approach this problem is to consider what might happen if God annihilated everything inside the orb of the moon. Buridan writes:

> Then it seems that it should be believed, and I believe, that God could annihilate everything that is below the orb of the moon or within the concavity of the orb of the moon with the orb remaining, and even with the whole heaven remaining in the same magnitude and figure that it now has. This case having been posited, it should be seen what would follow.
>
> I say that if this case is posited, nothing would be in or within the concavity of the orb of the moon, because the whole has been posited to be annihilated. And thus there would not be any space and also there would not be a vacuum within this concavity, because it implies a contradiction that nothing is inside the concavity and that a vacuum or space is within the concavity, because since this proposition 'vacuum is within the concavity' or 'space is within the concavity' is affirmative, it is necessary, if it is true, that the term 'space' or 'vacuum' has supposition for something. Therefore this follows: 'vacuum is within the concavity, therefore something is within the concavity'; and this is contradictory to the first proposition, which said that nothing was within the concavity.
>
> From this it follows that the arguments that are made to show that outside of the heaven there is space are invalid, because it could be argued similarly concerning what would happen within the heaven, assuming that the whole of what is inside had been annihilated, and nevertheless the conclusion would not be true, so the argument would not be valid (137^{11}–138^{2}).

But what about the distance between the sides of the evacuated lunar orb, and what about the motion of a stone approaching or moving away from the pole of the orb? Buridan writes:

> I say that in this it is difficult to satisfy the imagination because it always appears to the imagination that there is space there, as it always appears to sense that the sun is not larger than a horse and that it is much smaller

> than the earth. Nevertheless, in such things the intellect ought to correct such appearances of sense or imagination. I say, therefore, because we are not talking of cases that are naturally possible, but miraculously possible, that God could form there a very big body, that is bigger than this world. And it is true that that body would not be in that small body or place circumscriptively or commensurably. This should be believed, because within the small quantity of the Host and in its small place there is the body of Christ as large as it was at the last supper (*cena*) and as large as it is in paradise, and configured in the same way—indeed in any quantitative part of the Host, however small, there is the whole large body of Christ, optimally configured. But this magnitude of the body of Christ is not in the Host in a way commensurable to the magnitude of the Host. And no less could God make a bigger body in the place and with the magnitude of a millet seed. On account of this I also conclude that within the concavity of the orb of the moon God could also make a body a hundred times larger than the world, not changing the magnitude and figure of the orb of the moon. Indeed if there were there a body in a circumscriptive way measurable by the magnitude of the orb, it could not be larger than it now is, because it would be necessary for its diameter to be the third part approximately of the circular line drawn in the concavity of the orb of the moon (138^{6-28}).

On this basis, Buridan goes on to argue about distances and motions within the evacuated sphere of the moon: there a motion no matter how fast would not move a body closer to one side or the other. Buridan writes:

> Afterwards I also say that in a small space, as in the space of ten feet, God could move a stone or a very big body for a whole year very fast in straight motion, and nevertheless this stone would not exit from that space at rest—indeed it would not approach a corner of that space or move away from another corner. This is made clear as follows: if God should move his hand from his head to his feet, the motion will be as long as the body of Christ is long, and nevertheless that hand will be no closer to any corner of the Host nor farther from another corner, because the hand would not be moved in comparison to the position or the magnitude of the Host, but according to the position and in comparison to the magnitude of the body of Christ (138^{29}–139^{9}).

While Buridan continues to speculate about what might be possible for God, he repeats in various ways that he leaves the determination of such questions to the Masters of Theology and wants to acquiesce to their determination, decree, or ordination. Or he says he has not spoken assertively, but in the manner of disputation, raising doubts so that he may be taught the truth by others (141^{17-18}). Clearly, the doctrine of the presence of the body of Christ in the Eucharist influences his speculation:

> And I do not assert all of this, but in asserting this, or any of this, or not asserting this, I submit myself totally to the decree and ordination of the holy Church and the doctors of Sacred Scripture (139^{9-11}).

Question III.16—Whether some gyrative line is infinite (*Utrum linea aliqua gyrativa sit infinita*).[219]

I have not found uses of the gyrative line in questions on the *Physics* before Buridan. Richard Kilvington, however, developed cases involving the gyrative line in his questions on *De generatione et corruptione*, said to have been written in the 1320s.[220] In his questions on the *Sentences*, likely from the later 1330s, Roger Roseth makes use of the gyrative line without explaining its construction, meaning that it was commonly known. Buridan had already referred to a gyrative line in question I.12 of his *Quaestiones Physicorum*.[221] Buridan's question I.12 was *Utrum omnia entia naturalia sint determinata ad maximum*. Since the gyrative line could be taken as mathematical or physical, one might look to find a gyrative line considered in questions about nature, mathematics, or theology. Hence, I list questions in such authors as Radulphus Brito and Thomas Wylton, without knowing if they have any mathematical examples.

---

219 The table of questions adds: 'Quod non est ultima medietas proportionalis; et vide ibi de hoc multas conclusiones. De isto verbo "incipit". Quod non est neque fuit infinitum tempus, etiam concessa opinione Aristotelis de aeternitate mundi, licet syncategorematice perpetuum fuerit tempus' (6^{10-13}).

220 E. Jung & R. Podkoński, 'Richard Kilvington on Continuity', in: C. Grellard & A. Robert (eds), *Atomism in Late Medieval Philosophy and Theology*, Leiden [etc.] 2009 (Medieval and early modern science, 9), 65–84. Jung and Podkoński refer also to Roger Roseth, *Lectura super Sententias*, q. 5, a. 2 (*Utrum aliqua creatura possit esse infinita*), ed. O. Hallamaa, Helsinki 2005 (Helsingin Yliopiston systemaattisen teologian laitoksen julkaisuja, 18), 262–285 (e.g. 266), and John Mair, *Propositum de infinito*, 12–52.

221 Buridan, *Quaestiones Physicorum*, I.12, ed. Streijger & Bakker, 126^{1-2}: 'Non enim est longissima linea gyrativa, sicut ostendetur in tertio libro.'

1. Radulphus Brito: *Utrum sit infinitum in potentia* (III.17)
2. Thomas Wylton: *Utrum contingat esse magnitudinem tantam in actu quanta est in potentia* (III.15)
3. Walter Burley (*Expositio*): *Utrum naturalis et mathematicus considerent de eisdem dimensionibus* (IV.D.5)
4. Nicole Oresme: *Utrum infinite partes proportionales sint in continuo* (III.15)[222]
5. Albert of Saxony: *Utrum aliqua dimensio sit infinita* (III.12)[223] *Utrum possibile sit esse aliquam magnitudinem actu infinitam et etiam an possibile sit aliquod esse spatium actu infinitum* (III.13)
6. Marsilius of Inghen: *Utrum sit aliqua longitudo infinita* (III.5.3)[224]
7. Johannes Marsilii (?): *Utrum possibile sit esse aliquam magnitudinem actu infinitam* (III.9)[225]
8. Lawrence of Lindores: *Utrum sit aliqua linea gyrativa infinita* (III.25)
9. Benedictus Hesse: *Utrum aliqua linea gyrativa sit infinita, accipiendo 'infinitum' categorematice* (III.37)

---

222 Oresme, *Questiones super Physicam*, III.15, 406: 'Tertia propositio est quod procederetur in talibus in infinitum secundum quamlibet proportionalitatem maioris inequalitatis continuam, verbi gratia quod prima pars sit dupla ad secundam (duplam *ed.*), et secunda ad tertiam, etc.' I list this question because proportional parts play an important role in Buridan's arguments about the gyrative line.

223 [Albert of Saxony], *Quaestiones Physicorum*, III.12, 2: 557 and 563: 'Et arguitur primo quod sic: aliqua linea est infinita; ergo aliqua dimensio est infinita. Consequentia tenet, ex eo quod omnis linea est dimensio. Antecedens probatur nam: infinitae sunt lineae gyrativae in ista columna, quarum quaelibet est maior una certa data ... Ad rationes. Ad primam patet ex iam dictis: unde, quamvis infinita sit linea gyrativa, nulla tamen linea gyrativa est infinita, nec aliqua est linea gyrativa composita ex omnibus gyris.'

224 Marsilius of Inghen, *Abbreviationes Physicorum*, 12^{rb-vb}: 'Et arguitur quod sic quia: linea gyrativa est infinita; igitur. Tenet consequentia. Antecedens probatur quia: componitur ex infinitis gyris, quarum quelibet est pedalis et tales constituunt infinitam longitudinem ... Pro tertia dubitatione est notandum diligenter quod linea gyrativa dicitur linea gyrans circa corpus. Tunc imaginemur corpus columnare cuius diameter sit pedalis. Secundo quod circa eius partes proportionales transeat linea gyrativa, sic quod in qualibet parte proportionali fiat una gyra; vocantur autem partes proportionales consequenter se habentes, partes sic se habentes quod qualis est proportio prime ad secundam, talis est secunde ad tertiam, tertie ad quartam, et sic deinceps ...' Marsilius argues for fifteen conclusions 'ex quibus sequi videtur conclusio responsalis quod non est aliqua linea curva infinita' (13rb).

225 Johannes Marsilii (?), *Quaestiones Physicorum*, III.9, 43va: 'Et arguitur quod sic: primo de linea girativa partium proportionalium continua.'

10. John Mair: *an sit aliquid infinitum extensive vel intensive* (1)[226]
11. Franciscus Suárez (1548–1615), *Disputationes Metaphysicae*, disp. 40 (*De quantitate continua*), section 5 (*Utrum in quantitate continua sint puncta,*

---

226 John Mair, *Propositum de infinito*, 16–18: 'Quinta propositio: aliqua est linea infinite longa. Probatur: gira girans omnes partes proportionales illius linee recte est infinite longa; ergo propositum. Patet: omnis linea continens infinitas partes equales huic pedalitati, demonstrando pedalitatem meam, non communicantes, est infinite longa; sed gira que girat omnes partes est huiusmodi; igitur. Consequentia tenet in darii, huiusmodi a parte predicati in minore dat intelligere subiectum maioris. Nec valet dicere sicut dicunt Monachus et Buridanus: nulla girat omnes, quia nec prime partis proportionalis gira girat omnes partes proportionales, nec secunde, et sic de singulis; una enim singularis est falsa demonstrando totalem giram que girat omnes; si illam non demonstres, ascensus nihil valet.' In the margins of the 1510 and 1519 editions, we read: 'Buridanus tertio physicorum. Monachus. Albertus saxo.' Elie, *Le traité*, 18, n. 1, suggests that the 'Monachus' referred to by Mair is the Franciscan John of Rodington, who, in his commentary on Book I of the *Sentences* discusses whether God could produce an infinite outside himself. Elsewhere, according to Elie, Mair refers to Rodington as 'Monachus anglicus'. John Mair, *Propositum de infinito*, 20–22: 'Sed dubitabis de linee gyrative infinitate, et merito. Et sic argumentor ... Respondetur tenendo lineam esse accidens latitudinis et profunditatis immune (prout recitavimus superius in vigesima quarta distinctione); mihi persuasum est non solum quod infinita est linea gyrativa, quia hoc Joannes ⟨Buridanus⟩ concedit, sed quod linea gyrativa est infinita; dicit aliqua transit per tres, aliqua per centum, per omnes est aliqua protensa, sed nulla transit per omnes. Sed illam categoricam de copulato extremo dicit se nullatenus concedere: aliqua est protensa per decem, centum et mille, et semper stat in ostio sue spelunce: nulle sunt omnes, quod eum movet; et dicit hoc (ut patet in decima sextu [!] questione illius libri) si non teneatur linea more realium sive opposito modo, non refert; sed illam de copulato predicato quam negat inconcusse concedo tenendo positionem realium de linea. Sed tenendo oppositum (quod videtur rationabilius) augetur difficultas, et secundum illum conceditur quod infinita est linea gyrativa capiendo "infinitum" syncategorematice ... (44) Ex his habes methodum aptam ad investigandum an ex quocunque corpore quantolibet parvo sine rarefactione Deus potest facere actu infinitum. Hoc argumentum factum (ut mea fert opinio), si ponderetur, probatur tenendo lineam more mathematicorum et realium esse latitudinis atque profunditatis expertem. Non ignoro quid Joannes Buridanus dicat et Albertus Saxo. Ipsi enim dicunt quod nulle sunt omnes partes continui et infinite longa est linea gyrativa syncategoreumatice. Verum nulla linea gyrativa est infinite longa, sed certum est quod nulle sunt omnes partes continui, capiendo "omnes" distributive, verum de "omnes" collective refragor, et illorum placitum miror cum teneant constanter totum esse suas partes simul sumptas. Marsilius in hac distinctione insequitur sicque probat quod nulle sint omnes quando de aliqua multitudine dicitur: est omnia simul accepta respectu alicuius termini, tunc requiritur quod ille terminus pro nullo supponit in propositione de presenti quando illud sit unitas illius multitudinis. Exemplificat quod, si duodecim sunt omnes apostoli Dei, requiritur quod nullus sit apostolus Dei quando erit

lineae et superficies quae sint verae res, inter se et a corpore quanto realiter distinctae)[227]

Buridan opens the question by saying that he will not make an issue here (*non facio vim ad praesens*) whether there are lines indivisible according to width and depth, or not, because if it is conceded, then the question plainly proceeds, and if it is not conceded, then by 'gyrative line' he will understand a part of the columnar body proceeding in the surface of the column around the column, not from the point from which it would be imagined to begin, but from another point further on, imagined with a small latitude and depth (142^{5-11}). The gyrative line would go around the column once in its first half, around again in the next quarter, around again in the next eighth, and so forth according to proportional parts (142^{11-17}). If the line in the first proportional part has a certain width, certainly it could not have that width in all the succeeding parts, because some would be less wide, so the width would have to decrease proportionally. Since this would make discussing the question more cumbersome, he will continue writing as if there were indivisible points and lines, knowing that the conversion to the other way of talking, without indivisibles, is always possible (143^{4-7}). He supposes that the column is of such a

aliqua unitas illorum duodecim ...' In a footnote (50–51, n. 1), Elie translates a passage from Mair's commentary on the *Sentences*, Book III, q. 3, d. 13, which says in part: 'Ce que disent Buridan et Marsile est sans fondement, à savoir qu'il n'existe pas de spirale qui embrasse toutes les parties proportionelles, car, disent-ils, il n'y a pas de parties qui soient toutes les parties proportionelles, étant donné qui ni deux, ni trois, ni quatre parties ne sont toutes les parties proportionnelles, et ainsi de suite. Je réponds que chacune de ces propositions particulières est fausse concernant toutes les parties, même si l'on soutient (ce que l'on juge plus exact) que toute longueur est aussi largeur et profondeur, et que la longueur d'un corps difforme doit être mesurée par la ligne supposée la plus longue: cela n'y change rien.' In an appendix (220–240), Elie translates passages from Louis Coronel, *Physicae perscrutationes*, to which, he claims, Mair is responding.

227 See the following passage (cited according to the digital edition by S. Castellote: http://www.catedraldevalencia.es/castellote/d40.htm): 'Quinto, haec eadem difficultas tacta in punctis locum habet in lineis existentibus in magnitudine finita; idemque est de superficiebus. Et praeterea occurrit specialis difficultas, nam sequitur in corpore pedali, verbi gratia, esse lineam infinitae longitudinis simpliciter, et inter duo extrema puncta clausam, quae est aperta repugnantia ... Et simile argumentum vulgare est de linea gyrativa, quae circuit omnes partes continui, proportionales quidem in longitudine, aequales autem in crassitie, secundum quam omnes partes illius lineae habent aequalitatem respectu alicuius certae longitudinis; et tamen sunt infinitae, sicut et partes proportionales.' Suárez' response is at paragraph 22.

width that every turn of the line around it is at least a foot long. Then he states four arguments that the line will be infinitely long (143^{8}–144^{17}). On the other side is Aristotle, who seems to deny entirely that something is infinite in act. Also it's hard to comprehend that something infinite can be within determinate limits (144$^{18-21}$).

The question is difficult for Buridan, he admits, and he will have to make suppositions that he will not support until Book VI, for instance that every line and every part of a line is divisible, and that no continuum is composed of indivisibles (144$^{22-24}$). On this basis, he immediately formulates six conclusions:

1. Beginning from one end of column B, and proceeding towards the other end by proportional halves one after the other, none is the last proportional half (144^{26}–145^{1}).
2. Let one end of column B be A, and the other end C, and let the proportional halves begin from A going towards C: then according to this process no proportional half attains end C, nor is any closer to C than any other, because (if so) that would be the last, and it was said that none is the last (145$^{10-15}$).
3. Suppose that some mobile begins to move on column B starting from the end C and proceeding (on the gyrative line) towards A: then that mobile reaches no proportional half, taken in the aforesaid way, before any of the others. I.e., there is no first proportional half reached: if there were a half reached before any of the others, that would be the last in the preceding order, which was said to be impossible (145$^{19-23}$).
4. A mobile beginning to move over column B in this way begins to reach no two or more proportional halves equally fast, but always one before another, because no two are equally near to point C. Indeed, always according to the infinite division the prior is more remote from point C and the later is closer to it. And nevertheless, the nearer and the farther do not begin to be reached together, but the nearer earlier than the farther.[228] Therefore none (*nullas plures*) at once begin to be reached (145$^{24-30}$).
5. If something traverses a line through all the proportional halves of the line B in the gyrative way described, it is infinite in length, because it is composed of infinite lengths a foot or more long, as was argued; and it cannot be determined that it is terminated at some terminus (146$^{1-5}$).
6. If there is a gyrative line infinite on one side, it is necessary to concede that there is some line infinite on both sides, because in column B we

228 Here Buridan seems to be taking the prior to be that which is closer to A, where the proportional parts begin, even when the motion begins from C.

> can designate a middle point—call it D—, and from that point designate proportional halves going in infinitum towards A and other proportional halves proceeding toward C, which are joined together and unified in point D. Thus this total line will be infinite on either side (146[6–11]).

These conclusions Buridan proposes as true or as what he believes to be true, demonstrated or demonstrable. The fourth conclusion is categoric, while the fifth and sixth are conditional. So there is a 'great doubt' (*magna dubitatio*) for some whether a stone that touched the column at end C begins to touch the proportional parts ordered from point A, or when it begins to touch some one or more of them (146[14–19]). Buridan writes:

> To this it is necessary to respond—since we do not posit in time indivisible instants—that it is necessary to expound the verb 'to begin' (*incipit*) either negatively, by the time in which it does not touch them, or affirmatively, by the time in which it touches them. These two times are immediate to each other and continuous, for the time of quiet is immediate and continuous with the time of motion, and in the whole time of quiet this stone touches none of the proportional parts, but in the whole time of motion it touches several of them (146[20–26]).

This leads to a series of further conclusions:

7. In the last part of the time in which the stone rests it begins to touch these proportional parts, expounding 'to begin' negatively, because in that whole time it does not touch and in all time following immediately it touches (146[27–30]).
8. Expounding 'to begin' affirmatively, the stone begins to touch these proportional halves in the first part of the time in which it moves, because in every time in which it moves over this column, it touches some one or more than one of these proportional halves; and nevertheless whatever first part of the time is taken, it is true that immediately before it did not touch some of these proportional halves (147[1–6]).
9. In infinitely many diverse times, this stone begins to touch these proportional halves, because infinite are the last parts of the time of rest, and infinite also are the first parts of the time of motion, as was said of the termini of the line in question III.8. And nevertheless if 'to begin' (*incipit*) were expounded negatively, then in every last part of the time of rest, it begins to touch, and if it were expounded affirmatively, then in every first part of the time of motion it begins to touch.

Therefore, in infinite parts of time, of which any one is time, it begins to touch (147^{7-14}).

10. Whenever this stone begins to touch one of those proportional halves, it begins to touch several. This is clear by the third conclusion, because to no proportional half does it reach before any other, and because it does not begin to touch in an indivisible instant (because there are no such instants), but in divisible time it begins to touch, and every time is divisible in an earlier part and a later part, and when in an earlier part it touches some part, in a later part it touches another.[229] And thus it is clear that this tenth conclusion is not in opposition to the fourth, because although in the same time the stone begins to touch several parts, nevertheless it does not at once or equally quickly touch, but one earlier and another later, because in every time there is earlier and later (147^{15-25}).

These conclusions, Buridan writes, he posits as not doubtful to him, whether they are true or false (147^{26-27}). He admits, however, that there might be a doubt about the negative exposition of the verb 'to begin', which he states and replies to. He still does not know the answer to the main question, however, whether the gyrative line is infinitely long. He is not in doubt whether it is infinite syncategorematically, and will immediately make this his eleventh conclusion.

11. The gyrative line is infinite in the syncategorematic sense (148^{9-15}).

But to clarify the categorematic sense, he continues with a further set of conclusions:

12. No straight line drawn (*protracta*) in column B from end A toward end C is infinite, because if it reaches terminus C, it will be terminated at that end, and it if does not reach it, it will be less, and nothing that is terminated at something less is infinite. Every infinite, if it existed, would be larger than any finite (148^{17-21}).
13. To every gyrative line extended through proportional halves in this way corresponds a straight line drawn through the same proportional halves (148^{22-24}).
14. If a gyrative line is drawn through all the proportional halves of column B in the aforesaid way, and not beyond all of them, so it is necessary that

229 Here I use the English 'to touch' to translate both 'tangere' and 'attingere'.

there be a straight line drawn through all these proportional halves and not beyond all of them (149^{3-6}).

15. There is no single straight line drawn through these halves unless it is drawn beyond all of them, because if it were drawn through the whole column up to the end C, such that it touched the body extrinsic to the column that the column touches, this would be drawn beyond all, since none of the proportional halves touches the end C, as was said in the second conclusion. If, however, it were not drawn up to the end C, then, since it is not infinite, as was said in the twelfth conclusion, it follows that it is terminated at some other terminus within C; but beyond all termini within C are some proportional halves, because still between that terminus and C there are some parts of the column divisible in halves. Therefore this line would not be drawn through all the proportional halves. Therefore there is no straight line drawn through all halves which is not drawn beyond all. And if there were some such, it would be infinite, which is against the twelfth conclusion (149^{16}–150^{1}).
16. From the two preceding conclusions it follows that there is no single (*una*) gyrative line drawn in such a way through all the proportional halves of column B, because no such line is drawn beyond all, as is apparent by the case, and nevertheless it is not drawn through all unless it is drawn beyond all, as is clear from the two preceding conclusions. Therefore, no gyrative line is drawn through all proportional halves. But some one is drawn through two, some through a hundred, some through a thousand, and so forth for any number (150^{3-9}).

This analysis allows Buridan to formulate his principal conclusion (*conclusio principalis*): no gyrative line drawn through the proportional halves of the column is infinite in length, because none would be put forth as infinite unless it was drawn through all proportional halves, and none is drawn through all, therefore no such line is infinite in length (150^{10-13}).

Beyond this, however, Buridan concedes that 'through all the proportional halves is drawn a line' (*per omnes protracta est una linea*). He concedes this proposition, he says, because in this proposition 'line' does not have confused supposition, but rather determinate (because it is 'una linea') (150^{13-16}). Although Buridan's grammatical point is understandable, I am not sure what this line is, unless it would be the one that goes beyond all the proportional parts.

As a corollary, Buridan argues that similarly no time is perpetual or infinite, even if we concede with Aristotle that always and eternally there was one continuous time, such that we would concede that eternal or infinite is time, insofar as the terms 'eternal' and 'infinite' are taken syncategorematically. However, we could not concede that time is infinite categorematically (150^{17-22}). The situation is made more complex because Buridan is trying to handle the time at which a proposition is true without assuming indivisible instants.

In his replies to the principal arguments, Buridan makes some points clearer: there is no single gyrative line, nor a single straight line that goes through all the proportional halves of the column, but there are infinitely many such lines, none of which is the last:

> To the first principal argument he replies: there is no (*nulla est*) such gyrative line going through all the proportional halves (151^{2-3}).
>
> The fourth principal argument argues well that, if there were some line drawn through all the proportional halves, it would be infinite, and terminated at no terminus. But there is no such line (152^{3-4}).

As Buridan himself admits, question III.16 is difficult. He had to work harder on it than on some other questions. As in other questions, he used primarily the tools of the *logica moderna* to distinguish between different possible positions. I have not discovered the history of the gyrative line before Buridan.

Question III.17—Whether for every number there is a greater number (*Utrum omni numero sit numerus maior*).[230]

1. Roger Bacon: *Queritur an sit ponere numerum infinitum infinitum in actu* (p. 147) *Queritur utrum continuum sit divisibile in infinitum* (p. 159) *Queritur utrum magnitudo naturalis sit divisibilis in infinitum* (p. 160)[231] *Queritur quomodo est infinitas in numeris: queritur an appositio numeri sit in infinitum* (p. 164) *Queritur an divisio magnitudinis sit in infinitum* (p. 165)
2. **Er349(2)**: *Utrum magnitudo sit in infinitum divisibilis* (III.21)

---

230 The table of questions adds: 'Quod in continuis nulla unitas est indivisibilis. Omnis binarius est centenarius. An numerus par est numerus impar. Non sunt plures partes vel pauciores in linea *b* quam in eius medietate. Quod nullus numerus est alio maior secundum multitudinem. Quomodo igitur salvantur principia mathematica' (6^{15-19}).

231 Bacon, *Questiones supra libros octo Physicorum*, 160–161: 'Dubitatur de magnitudine naturali, utrum sit divisibilis in infinitum. Quod non: quia naturalis resolvit usque ad minimam materiam, quia forma naturalis appropriat sibi materiam in qua minori operari non potest ... Oppositum videtur: Aristoteles docet dividere magnitudinem in infinitum; et loquitur de naturali; ergo etc. ... Quod concedo, quod magnitudo naturalis est a parte rei divisibilis in infinitum in quantum naturalis ... tamen a parte sensus no vadit in infinitum ...'

3. S: *Utrum magnitudo inquantum huiusmodi sit divisibilis in infinitum* (III.34) *Utrum magnitudo secundum quod naturalis est possit dividi in infinitum* (III.35)
4. William of Clifford: *An magnitudo sit divisibilis in infinitum* (III.25)
5. Boethius of Dacia: *Utrum in discretis sit infinitum in actu* (III.31) *Utrum numerus possit esse infinitus* (III.32)
6. Geoffrey of Aspall: *An sit infinitum in discretis* (III.11)
7. Giles of Rome: *Utrum hic numerus qui vadit in infinitum separari possit a decisione continui* (III.46) *Utrum sit aliquis numerus separabilis a decisione continui* (III.48)
8. Radulphus Brito: *Utrum numerus separetur a continuo* (III.20) *Utrum numerus sit multiplicabilis in infinitum* (III.22)
9. Thomas Wylton: *Utrum infinitum possit esse multitudine vel etiam possit esse* (III.13) *Utrum numerus creetur ex divisione continui* (III.16)
10. Walter Burley (*Expositio et quaestiones*): *Utrum omne continuum sit divisibile in semper divisibilia* (90 = VI.5)
11. Bartholomew of Bruges: *Utrum sit dare multitudinem actu infinitam* (III.15)
12. John of Jandun: *Utrum per appositionem contingat excellere quantamcunque magnitudinem* (III.13)[232]
13. William of Ockham (*Quaestiones*): *Utrum possit evidenter probari quod numerus sit alia res a rebus numeratis* (107) *Utrum secundum intentionem Philosophi numerus sit alia res a rebus numeratis* (108) *Utrum secundum intentionem Philosophi numerus sit actu infinitus* (109) *Utrum de facto numerus sit alia res a rebus numeratis* (110) *Utrum numerus binarius et ternarius distinguantur specie* (111)
14. Walter Burley (*Expositio*): *Utrum omnis numerus possit esse actu numeratus* (III.D.14) *Utrum magnitudo possit dividi in infinitum* (III.D.18)
15. Francesc Marbres: *Utrum numerus sit aliqua entitas realis absoluta distincta essentialiter a rebus numeratis* (IV.7)[233]

---

232 Jandun, *Quaestiones Physicorum*, III.13, 50ra: 'Tunc dico ad quaestionem tria. Primo dico quod per divisionem contingit excellere quantamcunque parvitatem. Secundo dico quod per appositionem contingit excedere omnem multitudinem praeexistentem finitam et determinatam. Tertio quod per appositionem quantitatis non contingit etc.'

233 Francesc Marbres, *Quaestiones Physicorum*, IV.7, 47ra–48va: 'Respondeo. Ubi sic procedam: primo enim inquiretur si numerus est ens absolutum distinctum a rebus numeratis; secundo utrum habeat esse circumscripto omni actu intellectus; tertio movebuntur quedam dubia de numero. Quantum ad primum est opinio duorum doctorum quorum quilibet ponit unam propositionem. Prima propositio est fratris Occam qui ponit quod

16. Nicole Oresme: *Utrum sit aliqua multitudo actu infinita* (III.16)
17. Hugolinus of Orvieto: *Utrum sit dare aliquam multitudinem actu infinitam* (22 = III.7)[234]
18. Albert of Saxony: *Utrum omni magnitudine data contingat dare maiorem et omni numero dato contingat dare maiorem* (III.15)

---

numerus non distinguitur realiter a rebus numeratis, nec essentialiter, sic quod numerus ternarius trium lapidum non distinguitur essentialiter a tribus lapidibus ... Secunda propositio est cuiusdam alterius doctoris qui ponit quod numerus est forma respectiva ... Ex quo quidem dicto concludit Franciscus de Marchia aliud dictum quod cum ratio totius univeralis pro qua provenit numerus uniformiter salvetur in quolibet genere quod numerus causatus ex divisione continua est eiusdem rationis cum numero cuiuscumque generis alterius. Hoc autem dimitto tanquam dubium. Hoc premisso ponit istam conclusionem: quod numerus non habet esse formaliter circumscripta anima vel eius actu .... Tenet enim Landulphus quod numerus non est aliquod unum ens ut recitabitur in sequentibus ... Item ad idem arguit Franciscus de Marchia: nulla forma una secundum speciem specialissimam absoluta fundatur supra rem cuiuslibet generis; unde omnes ille forme differunt genere quarum subiecta immediata differunt genere ... Secunda dubitatio est ista: a qua unitate dicitur numerus unus formaliter. Respondeo. Ubi sciendum est quod sunt hic multi modi dicendi. Unus modus est fratris Landulphi super primum *Sententiarum*, qui ponit istam conclusionem: quod numerus formaliter et realiter est quedam multitudo et non aliqua unitas nec aliquod unum. Ideo dicit quod fatuum est dicere et querere propter quid numerus habet unitatem suam, cum oppositum insit sibi.'

234 Hugolinus, *Quaestiones Physicorum*, 22 = III.7, 28–29: 'In quarum prima ⟨decisione⟩ videbitur an de facto sit aliqua multitudo entium infinita. Et dato quod sit, quae et qualis sit illa multitudo. Prima conclusio est quod in omni continuo est dare multitudinem actu infinitam partium realiter distinctarum. Secunda est quod in aliquo continuo infinitae sunt partes aequales secundum diversas dimensiones, puta secundum longitudinem et latitudinem, quarum quaelibet secundum se totam est extra aliam. Tertia est quod nullae sunt partes in aliquo continuo sic infinitae et distinctae quod quaelibet sit extra aliam, ita quod inter eas sit aliqua simpliciter prima et aliqua secunda et aliqua tertia et sic deinceps. Quarta conclusio est quod infinitae sunt partes in continuo aequales secundum omnem dimensionem, puta secundum longitudinem et profunditatem, quae tamen partes sunt se mutuo includentes. Quinta est quod nullae sunt partes aequales secundum aliquam dimensionem infinitae, quae sint se penitus excludentes. Sexta conclusio est quod infinitae sunt partes aequales secundum omnem dimensionem in aliquo continuo actu, quae etiam sunt se penitus excludentes. Inter has conclusiones secundum logicam nulla est repugnantia, quamvis hoc prima facie videatur ... In secunda vero ⟨decisione⟩ videbitur de divisione continui quomodo ipsum sit divisibile in infinitum. Prima conclusio est quod de virtute sermonis loquendo nullum continuum est divisibile in infinitum. Secunda est quod in infinitum omne continuum est divisibile. Tertia est quod continuum sic est divisibile quod sine contradictione potest esse divisum actu secundum quamlibet sui partem.'

19. Marsilius of Inghen: *Utrum omni numero dato contingat dare maiorem et est aliquis numerus maior* (III.6.1)[235]
20. Lawrence of Lindores: *Utrum omni numero sit numerus maior* (III.16)
21. Benedictus Hesse: *Utrum omni numero dato est dare numerum maiorem supposito quod nullum continuum sit compositum ex indivisibilibus sed quod omne divisibile sit divisibile in partes quarum quaelibet ultra est divisibilis* (III.38)

Buridan's treatment of this question is shaped by his assumption that numbers count things in the external world, especially the parts into which a continuum, while remaining uncut, could theoretically be divided. If numbers are produced when a person counts things, does this mean that there are possible numbers that await counting? Or should it be said that numbers exist if there are so many countable things in the world even if they have yet to be counted? If the same magnitude is the basis for 2, if it is divided in two parts, and the basis for 3, if it is divided in three parts, does this mean that the same physical thing is the basis for an even and an odd number, so that everything that is odd is also even?

Buridan writes that in this question he wishes to treat the difficulties that arise concerning numbers connected to parts of a continuous magnitude (155^{11-17}). He will only consider more or less as related to the size of magnitudes or of multitudes (155^{24-26}). The conclusions that he can formulate are the following:

1. There is no indivisible unity: every unity is divisible and any part of it is divisible. Unity is the same as one thing, as multitude and number are many and numbered things. Therefore every unity is a magnitude and every magnitude is divisible (155^{28}–156^{4}).
2. Every double is a triple and a hundredfold, because since a whole is its parts, it follows that any unity is fifty unities; therefore a double is a hundred unities, and is thus a hundredfold. However, this is not always true, as for instance, three humans are not two humans (156^{5-13}).

---

235 Marsilius of Inghen, *Abbreviationes Physicorum*, 13^{va-vb}: 'Hic queri solet utrum omni numero dato contingat dare maiorem et est aliquis numerus maior ... Pro prima dubitatione sciendum est quod duplex est numerus, scilicet numerus numerans et numerus numeratus. Numerus numerans dicitur conceptus mentalis quo anima numerat vel etiam ipsamet anima, et de hoc non intelligitur questio. Potest tamen dici quod aliquis est numerus numerans quo nullus est maior. Patet, quia omnes anime intellective et omnes conceptus quibus ipsamet numerat sunt in certo numero ...'

3. Every unity is greater than a binary and a hundredfold, because every unity is a magnitude. So a binary can be larger than a triple, etc. (156^{14-20}).

If the basis of numbers in magnitudes is accepted, how can the distinction between even and odd numbers be maintained, as well as other common mathematical statements? Buridan decides in favor of a series of further claims:

4. No number is odd. According to Buridan, this seems to be true *de virtute sermonis* (157^{13-20}).
5. No equal is unequal and no equal to something is unequal to something, but every thing equal to something is unequal to something and vice versa (157^{23-25}).
6. There are not more or fewer parts in line B than in its half or vice versa. Take a small circle around the pole of the last sphere, say a circle with a diameter of a foot, and take the equinoctial circle. There are not more parts in the equinoctial circle than in the small circle (158^{6-12}).
7. No three things are more than two (159^{1}).
8. No number is greater than another number in multitude. This follows from the preceding claim, because each number corresponds to divisions of a whole, and each whole can have various divisions (159^{5-8}).

Of course, given that these conclusions are counter to accepted truths of mathematics, Buridan has to explain how they can be reconciled. For example, arithmetic assumes that unity is indivisible and that there is a smallest possible number. Also arithmetic distinguishes between odd and even numbers (159^{11}–160^{4}). Buridan explains why he has set forth his conclusions:

> Because of this it should be diligently noted that the term 'number' has supposition for several things distinct from each other, even though it is taken with a demonstrative pronoun, as 'this number'. Otherwise 'number' does not have supposition for anything. And it connotes that these many things are numbered or numerable, i.e., that they are known or it is knowable how many they are by reason or by a mental concept treating them discretely (*per rationem sive per conceptum animae discretivum*), i.e., by which the soul understands and can understand these many things discretely and divided from each other. And therefore Aristotle says well in Book IV of this work that, if there cannot be a numberer (*numerans*), namely the soul, there could not be number, since it belongs to the definition (*ratio*) of number that it is numbered or numerable. If there could not be a soul numbering, there could not be the *ratio* which the word 'num-

ber' connotes, and if the connotation of a term is lacking, the term has supposition for nothing; therefore nothing would be a number (160[5–17]).

The same thing can be understood under very diverse concepts of its parts, as we can think of two parts or three parts of the same line (162[18–21]).

How does one profit from such different ways of speaking of things according to different concepts? To this it should be said that the science of numbering, i.e., arithmetic, was invented principally and finally for measuring motion, and magnitudes, and times. Whence in the natural order geometry presupposes (*supponit*) arithmetic, according to Aristotle. For it cannot be known how great a large continuum is except by distinguishing its parts separately and knowing their quantity, such as feet or ells, and by counting (*numerando*) them (164[1–7]).

By what has just been said, many usual ways of talking in arithmetic can be understood, such as that every double is equal to another double and less than a triple. All such things mathematicians and natural scientists understand of the numbers of things of the same kind, because other things are not properly comparable to each other, i.e., according to a certain proportion (164[10–16]).

Having seen these things, it can well be conceded with Aristotle that for every number there is a greater number according to multitude, in this sense that, however many unities the distinguishing reason (*ratio discretiva*) can distinguish, the distinguishing reason can distinguish more. And thus also would be conceded that infinite is number, taking 'number' syncategorematically, because in the aforesaid sense there is not so much but that there may be more. But it does not follow hence that number is infinite, just as it did not follow that if infinite is a gyrative line that a gyrative line is infinite, etc. Here again the distinction between syncategorematic and categorematic senses is Buridan's fundamental tool (165[1–8]).

Question III.18—Whether in any continuum infinite are the parts (*Utrum in quolibet continuo infinitae sint partes*).[236]

1. Roger Bacon: *Queritur quid est infinitum, queritur de diffinitione infiniti, que est: infinitum est cujus nichil est extra, an sit bona* (p. 158) *Queritur de*

---

236 The table of questions adds: 'Quomodo exponitur "infinitum" tam categorematice quam syncategorematice sumptum. Quod implicat contradictionem esse magnitudinem infinitam vel multitudinem infinitam capiendo "infinitum" categorematice. Quod valde differenter est dicendum secundum diversas expositiones "infiniti" syncategorematice sumpti' (6[21–25]).

alia diffinitione infiniti, que est: infinitum est cujus quantitatem accipientibus semper est aliquid sumere extra (p. 159) *Queritur utrum continuum sit divisibile in infinitum* (p. 159)

2. Boethius of Dacia: *Utrum continuum aliquod possit esse infinitum secundum divisionem* (III.35)
3. Geoffrey of Aspall: *An sit infinitum in quantitate continua* (III.12)
4. Giles of Rome: *Utrum sit idem infinitum secundum divisionem vel secundum appositionem* (III.38) *Unde est quod in divisione magnitudo non habet statum* (III.40)
5. Radulphus Brito: *Utrum infinitum sit eodem modo in divisione magnitudinis et in appositione ad magnitudinem* (III.18)
6. John of Jandun: *Utrum magnitudo sit divisibilis in infinitum* (III.12)[237]
7. William of Ockham (*Expositio*): III, cap. 9 (t. 34, 204a2–7)[238]
8. Walter Burley (*Expositio*): *Utrum magnitudo possit dividi in infinitum* (III.D.18) *Utrum divisio magnitudinis sit ratione materiae* (III.D.19) *Utrum subiectum recipiens infinitatem sit materia* (III.D.20) *Utrum continuum sit divisibile in semper divisibilia* (VI.D.5)[239]

---

237 Jandun, *Quaestiones Physicorum*, III.12, 49rb: 'Ad quaestionem dico breviter quod magnitudo est divisibilis in infinitum, quia quod est divisibile in semper divisibilia est divisibile in infinitum, quia in tali semper remaneret aliquid dividendum; sed magnitudo est divisibilis in semper divisibilia, quia aut est divisibilis in semper divisibilia, aut in indivisibilia simpliciter; sed non est dicendum quod sit divisibilis in indivisibilia, quia unumquodque est compositum ex illis in quae est divisibile ... Quare relinquitur quod magnitudo non est divisibilis in aliqua indivisibilia, immo in divisibilia, et sic est divisibilis in infinitum.'

238 Ockham, *Expositio*, III, cap. 9, § 2, 513: 'Quintus modus est quo aliquid dicitur "infinitum" secundum appositionem vel divisionem vel utroque modo. Et est talis modus secundum Commentatorem quod semper actus "servat potentiam", hoc est quando unum dividitur actualiter, adhuc potest aliud dividi et non dividitur; similiter quando unum apponitur, adhuc aliud quod non apponitur potest apponi ... Nam bene potest procedi in infinitum dividendo et apponendo, et tamen nihil est actu infinitum non habens finem, sed oportet necessario quod sint aliqua infinita multitudine, quae tamen omnia simul iuncta faciunt unum finitum, immo unum valde parvum.' The issue appears in many other places in Ockham's works, as can be seen from the indices to the critical edition of Ockham's philosophical and theological works.

239 In book VI for Burley the issue is not infinity taken in different senses, but the fact that some of his contemporaries, viz., Gerard of Odo, had argued that supposed continua are composed of atoms or indivisibles. Burley writes (*Expositio in Physicam* [1501], 177rb): 'Quero utrum continuum sit divisibile in semper divisibilia. Et videtur quod non, quia in quibuscumque est minimum, in illis non procedit divisio in infinitum; sed in rebus naturalibus continuis est minimum; ergo in illis non procedit divisio in infinitum ...'

9. Nicole Oresme: *Utrum infinite partes proportionales sint in continuo* (III.15) *Utrum sit aliqua multitudo actu infinita* (III.16)
10. Albert of Saxony: *Utrum infinitae sint partes in continuo* (III.14)[240]
11. Marsilius of Inghen: *Utrum omni dato contingat dari maiorem* (III.6.1) *Utrum in continuo sint infinite partes* (III.6.2)
12. Lawrence of Lindores: *Utrum in quolibet continuo sint infinitae partes* (III.17) *Utrum possibile sit esse magnitudinem actu infinitam et continuum in omnes suas partes esse divisum* (III.18)
13. Benedictus Hesse: *Utrum possibile sit infinitam esse magnitudinem et in infinitas partes lineam esse divisam* (III.40)[241]

Buridan begins his determination by writing:

> This question contains many difficulties:
>
> The first is that, since 'infinite' may be taken categorematically and syncategorematically, how should it be expounded in either way? And since words are as one pleases (*ad placitum*), many pose expositions as it pleases them, and according to the requirements of their expositions, it is necessary to speak consequently, because the meaning of a term is the principle of every discipline, as Aristotle says.
>
> And it seems to me that taking 'infinite' categorematically, Aristotle would expound it for magnitudes as being extended without terminus or non-terminated extension. And thus it makes no difference to say 'infinite

---

Burley's reply to this question continues to the end of f. 178rb. On Burley's position vis-à-vis Gerard of Odo, see A. Maier, 'Handschriftliches zu Wilhelm Ockham und Walter Burley—Addenda,' in: Ead., *Ausgehendes Mittelalter. Gesammelte Aufsätze zur Geistesgeschichte des 14. Jahrhunderts*, 1, Roma 1964, 469–479, at 477. On Gerald of Odo's view of the composition of continua, see S.W. de Boer, 'Gerard of Odo on the Atomistic Structure of Continua. A Discussion and Edition of a Tract Found in Ms. Madrid, Biblioteca Nacional 4229,' *Documenti e studi sulla tradizione filosofica medievale*, 23 (2012), 387–427.

240 [Albert of Saxony], *Quaestiones Physicorum*, III.14, 2: 588, 'Quarta conclusio: infinitae sunt partes proportionales in continuo ... Quinta conclusio: nullae partes continui sunt infinitae.' Cf. question III.10, 2: 540: 'Sciendum est quod aliquis terminus, qui aliquando solet teneri syncategorematice et aliquando categorematice, quando ponitur circa subiectum alicuius propositionis, tunc solemus eo uti solum syncategorematice; quando vero ponitur a parte praedicati, solemus eo uti solum categorematice.'

241 Benedictus Hesse, *Quaestiones Physicorum*, III.40, 395: 'Nota de primo: responsio prima intelligitur quod non est possibile infinitam esse magnitudinem, capiendo "infinitum" categorematice; sed capiendo syncategorematice, tunc bene est possibile.'

> magnitude' and 'magnitude infinite', as it makes no difference to say 'man white' or 'white man'. And in each case the term 'man' has determinate supposition without any confusion if it has supposition for something, unless there is something else in the proposition causing confusion (i.e., confused supposition). And it was said previously that a magnitude is not infinite nor infinite is magnitude, nor a gyrative line is infinite in length, nor infinite is a gyrative line, etc. Nor also is time infinite nor infinite is time (169^{4-19}).[242]

Buridan then makes an analogous claim about infinite multitudes. There is no infinite multitude categorematically understood and no determinate number is infinite (169^{20}–170^{8}). The properties of 'infinite' taken categorematically if it has supposition for something, are:

> The term 'infinite' is privative, and opposite to 'finite'. Nothing is both finite and infinite in magnitude, multitude, length, duration, etc. (170^{24}–171^{4}).
>
> If something is infinite according to magnitude, nothing is greater than it. An infinite cannot be reduced by subtracting from it. If past time is infinite, no past days are greater (in multitude) than past years. No days are all the past days, as no proportional halves of line B are all its proportional halves (171^{5}–172^{2}).[243]
>
> It is impossible for an infinite magnitude to exist or for an infinite time to exist. Nor should the infinite perfection of God be denied on account of this (172^{3-10}).
>
> Another property of the infinite, if it existed, is that it could not be assigned determinately and properly how much it would be, nor, concerning a multitude, how many it would be (172^{11-13}).
>
> In addition, it seems that it is not possible for there to be an infinite magnitude, because it would follow that the whole would not be bigger than its part, the opposite of which is especially supposed for a quantitative whole (172^{20-22}).

Buridan concludes his discussion of the categorematic infinite by raising and replying to a doubt about the eternity of the world: even if eternally was the

242 See Murdoch & Thijssen, 'John Buridan on Infinity,' 130–134.

243 Here Buridan writes of 'halves' (*medietates*) where normally one would expect him to say 'proportional parts', because the proportional parts used in his examples happen to be halves.

world, and time, and motion, nevertheless there would be no infinite time, and thus every past time was finite. If this is the case, it is asked whether there would be more days than years in past time (173[5–8]). Buridan responds by conceding that there were more days than years, and that the days were more than the years, because in the relevant propositions no word precedes the word 'were' (*fuerunt*) which would distribute its connotation. Therefore the word 'were' stands for past time indefinitely, so that, if the proposition is verified for some past time, it is verified absolutely. And it is, for instance, verified for the time that has passed since the birth of Christ, although it would not be true for infinite past time (173[20]–174[12]).

Buridan then continues on for the syncategorematic sense of 'infinite'. There are diverse ways of expounding propositions including 'infinite' in a syncategorematic sense, for instance, in magnitude: 'so great and not so great but that more', or in multitudes: 'so many and not so many but that more.' To these expositions, however, Buridan prefers a simpler exposition, which he claims is equivalent, namely to say that 'infinite is B in magnitude' means that 'for every B there is a B greater', and 'infinite is B in length' means that 'for every B there is a longer B', and so forth for infinite in velocity, or slowness, or smallness, etc. Buridan means the same thing by 'infinite according to length' as by 'infinitely long' and by '*in infinitum* long' (174[15–24]). On the basis of these expositions of 'infinite' in the syncategorematic sense, Buridan concludes:

1. Infinite is the gyrative line in length, because given any gyrative line, there is another longer, and there is no gyrative line but that there is another longer (175[1–3]).
2. This proposition is false: 'infinite is body', because not for every body is there a larger body. Indeed, even if there were some infinite body, taking 'infinite' categorematically, still 'infinite is body' would be false taking 'infinite' syncategorematically, because there would be some body, namely the categorematically infinite body, than which no body would be greater. On the other hand, if there were a categorematically infinite body, the proposition 'infinite is finite body' would be true (syncategorematically), because than for every finite body there would be a greater body (175[7–16]).
3. This proposition is false: 'infinite in length is this gyrative line', and similarly this one: 'infinitely fast is this motion', and so forth for other similar propositions. These are false, because the word 'this' (*haec, iste*) means that the same thing would have to be included (twice) in the exposition, as in 'this gyrative line is longer than this gyrative line' etc. Other propositions in which the word at issue supposits determinately are likewise false (175[17]–176[2]).

The remaining two conclusions concerning multitudes are more difficult, because they concern parts in a continuum divided in reason, but not separated, and their multitudes or numbers:

4. This proposition is false: 'infinite according to multitude are parts in a continuum.' From this it follows that this proposition is also false: 'infinite is number according to multitude' (176^{3-8}).
5. Between the infinite parts of B there is a discrete numerical ratio (*ratio numeralis discretiva*) (176^{9-16}).

The fourth conclusion echoes a claim made in question III.17, according to which the halves of a continuum already contain many smaller parts.[244] These claims may raise the question what is meant by the previous authorities, who say 'infinite are the parts in a continuum' and 'for every number there is a greater number'.

Up to this point, only about half of Buridan's treatment of this question has been surveyed.[245] After the five conclusions just listed, Buridan raises 'strong doubts' (*fortes dubitationes*) about his fifth conclusion on syncategorematic infinites (176^{17}–181^{6}). The answer to his fifth doubt is noteworthy. He asks:

> How does the term 'B' supposit in propositions such as: 'Infinite is B' or 'infinitely long is B', and so forth for other propositions (180^{24-26})?

He replies:

> Response, not only concerning this proposition but concerning all others which require an exposition or expositions: if some term is taken only once in the proposition expounded and it requires being taken several times in the exposition or expositions, and if in those several acceptations it supposits with diverse suppositions, in such a case it seems to me it should be said that that term in the proposition to be expounded does not supposit with a single (kind of) supposition, but with those several suppositions. And I say that in this proposition 'infinitely long is B' the term 'B' supposits with distributive supposition and with merely confused supposition, because, when I say 'than every B is B longer', the first

244 See conclusion 6 in question III.17 (158^{6-7}): 'There are not more or fewer parts in line B than in its half, or conversely …'

245 Further discussion may be found in Murdoch & Thijssen, 'John Buridan on Infinity.'

> 'B' has distributive supposition and the second has merely confused supposition (180^{27}–181^{5}).

At this point, Buridan suggests an alternative exposition of 'infinite' in the syncategorematic sense. According to multitude 'infinite' can be expounded as: 'infinite is B signifies that 2 are B, and 3 are B, and 100 are B, and 1000 are B, and so forth without end' (*sic sine statu*). In the case of magnitudes, the increase may be by multiplication as: 'infinitely long is B signifies that, give the length of some B, for example of a foot, then there is B of two feet, and B of three feet, and B of 100 feet, and so forth without end' (18$^{17-12}$). In accordance with this exposition, Buridan proposes the following set of conclusions:

1. Infinite are the parts of a continuum according to multitude, because 2, 3, 100, and so forth without end; indeed this line is of infinite parts, because of 2, 3, 100, etc. (181$^{20-22}$).
2. Infinite is a gyrative line in length because, if there is some line of a foot, so there is one of two feet, and another of a hundred feet, and so forth without end (182$^{15-17}$).
3. This proposition is false: 'a gyrative line is infinite in length', because here the term 'line' supposits determinately, because no term confounding it (i.e., causing it to have confused supposition) precedes (182$^{18-20}$).
4. This proposition is false: 'Infinitely fast is motion' (183^{1}). The reasoning for this claim seems simply to be factual: there is a maximum velocity in the cosmos. A similar reasoning leads Buridan to claim that this proposition is false: 'infinitely long is a body according to a straight line' (183$^{1-5}$).
5. This proposition is true: 'infinitely small is a magnitude', and also this one: 'infinitely slow is a motion', and similar propositions (183$^{6-7}$).

On this basis, what are the properties of the term 'infinite' taken syncategorematically? Buridan mentions the following:

> The first property is that the terms 'finite' and 'infinite' are not opposed to each other, as the terms 'human' and 'every' are not opposed. This can be seen in true affirmative predications such as: 'infinite are the finite parts of this continuum' and 'infinitely slow is motion finitely slow.'
>
> Another property is that, if something is infinite in length, than it there is a longer, and if infinite were a body, than it there would be greater.
>
> Infinite in length could be cut by taking away a finite many times (183$^{16-25}$).

On this basis, Buridan can reply fairly easily to the principal arguments. For instance, to the third principal argument he replies:

> In this sense Aristotle denied an actual infinite or infinites and conceded an infinite or infinites in potency, because it is not true that infinite is body, but it is true that infinite can be body, because it is not true that for every body there is a greater body, but for every body there can be a greater body (184^{16-20}).

Although there is insufficient space here to explain Buridan's reasoning in every case, it seems clear that he is more interested in applying the tools of logic to solving problems of the infinite (and to demonstrating to his students or readers how such tools are to be applied) than he is concerned to advocate the one and only right answer to every question.[246]

Question III.19—Whether it is possible that infinite is a magnitude and whether it is possible that in infinite parts a line may be divided (*Utrum possibile sit infinitam esse magnitudinem et in infinitas partes lineam esse divisam*).[247]

1. S: *Utrum magnitudo possit crescere in infinitum* (III.36)
2. Thomas Wylton: *Utrum contingat esse magnitudinem tantam in actu quanta est in potentia* (III.15)
3. Bartholomew of Bruges: *Utrum contingat tantam esse magnitudinem actu quanta est in potentia* (III.16)
4. John of Jandun: *Utrum magnitudo sit divisibilis in infinitum* (III.12)
5. William of Ockham (*Quaestiones*): *Utrum haec sit concedenda de virtute sermonis 'continuum potest dividi in infinitum'* (66) *Utrum omni magnitudine finita possit esse aliqua magnitudo maior* (67) *Utrum quaelibet pars continui sit in continuo actu vel potentia tantum* (68) *Utrum partes sint in actu in continuo actualitate totius tantum* (69) *Utrum in continuo sint*

246 Cf. H. Hugonnard-Roche, 'Le possible et l'imaginaire dans la physique d'Albert de Saxe,' in: J. Biard (ed.), *Itinéraires d'Albert de Saxe. Paris-Vienne au XIVe siècle*. Actes du Colloque organisé le 19–22 juin 1990 dans le cadre des activités de l'URA 1085 du CNRS à l'occasion du 600e anniversaire de la mort d'Albert de Saxe, Paris 1991 (Etudes de philosophie médiévale, 69), 161–173.

247 The table of questions continues: 'Quod Deus in qualibet medietate proportionali huius diei potest creare unum lapidem pedalem, sed non est possibile ipsum creare in qualibet unum lapidem pedalem. Quod aliqua universalis est impossibilis, cuius omnes singulares sunt possibiles et compossibiles' (6^{27}–7^{3}).

infinitae partes actu (70) *Utrum in continuo sint infinitae partes totaliter distinctae inter se* (71)

6. Walter Burley (*Expositio*): *Utrum magnitudo possit dividi in infinitum* (III.D.18) *Utrum divisio magnitudinis sit ratione materiae* (III.D.19) *Utrum subiectum recipiens infinitatem sit materia* (III.D.20)[248] *Utrum continuum sit divisibile in semper divisibilia* (VI.D.5)
7. Nicole Oresme: *Utrum contingat magnitudinem tantam esse in actu quantam contingat esse eam in potentia* (III.9)
8. Hugolinus of Orvieto: *Utrum sit dare aliquam multitudinem actu infinitam* (22 = III.7)
9. Albert of Saxony: *Utrum infinitae sint partes in continuo* (III.14) *Utrum omni magnitudine data contingat dare maiorem et omni numero dato contingat dare maiorem* (III.15)
10. Marsilius of Inghen: *Utrum omni dato contingat dari maiorem* (III.6.1) *Utrum in continuo sint infinite partes* (III.6.2)
11. Lawrence of Lindores: *Utrum in quolibet continuo sint infinitae partes* (III.17) *Utrum possibile sit esse magnitudinem actu infinitam et continuum in omnes suas partes esse divisum* (III.18)
12. Benedictus Hesse: *Utrum possibile sit infinitam esse magnitudinem et in infinitas partes lineam esse divisam* (III.40)

Like question III.18, question III.19 is more about logical methods than about infinity in particular. In question III.15, Buridan had already discussed whether God could create an infinite magnitude. In question III.18, the infinite divisibility of the continuum was also an issue. In question III.19, there are six principal arguments to the effect that it is possible for there to be an infinite magnitude and that in infinite parts a line may be divided (186^{7}–188^{16}). It is argued on the contrary that Aristotle denied the infinite division of a continuum and that it has been said earlier that it implies a contradiction for a magnitude to be infinite (188^{18-24}).

After saying that the question appears difficult to him (189^{1}), Buridan presents extensive arguments in favor of the following conclusions:

248 Burley, *Expositio in Physicam* (1501), 78rb: 'In prima igitur parte huius particule Philosophus concludit istam conclusionem quod infinitum est in potentia, et hec est 30ª conclusio huius libri … (80rb) Et potest hic poni conclusio 34ª huius libri sub hac forma, scilicet quod non contingit magnitudinem esse in potentia maiorem omni magnitudine finita … (85ra) Et quod nulla magnitudo possit crescere in infinitum probatur, et hec est 42ª conclusio huius libri.'

1. This proposition is impossible: 'line B is divided in all its proportional halves', and similarly this one is impossible: 'line B will today be divided in all its proportional halves', where by 'division' real separation is meant (189^{3-6}). This conclusion is proved by considering a stone touching the end of the line and asking whether any proportional half of the line touches the stone if the division and separation has been completed. It is not possible that more than one proportional half touches the stone, nor is it possible that the last proportional half touches the stone, because there is no last proportional half (189^{7}–190^{2}). If none of the proportional halves touches the stone, then there will be part of the line that has not been divided, against the hypothesis (190^{3-6}).
2. This proposition is impossible: 'in any proportional half of this day, God creates a stone of a foot size', or this one: 'in any proportional half of this day God will create a foot-size stone'. Here by 'create' is not meant conserve in existence, but newly create (190^{16-23}). Among the previous arguments that imply this conclusion are that that is impossible to which follows that a magnitude is infinite; but if God created a stone in each proportional part of a day, an infinite body would result. Creating stones in this way is not more possible than adding to a body gnomons in proportional parts, since that would always result in a body with a given figure; but an infinite body can have no figure (191^{8-20}). Buridan continues with various different scenarios for what might happen in such cases, all of which have impossible consequences.
3. Whether it is true or false, this claim seems to follow, whether it is probable or sophistical: it is not possible by any power for there to be an infinite magnitude. This is proved first because a strong argument for proving that it is possible, namely that God could create a stone of a foot size in every proportional half of an hour or a day, has been destroyed (190^{15-20}).

Then an alternative is proposed. If the world is eternal, suppose that God created a stone of a foot size in every past day. It is possible that God created the world from eternity (194^{1-6}).[249] Buridan replies that in cases like this something can be possible in a divided sense, but impossible in a compounded sense. It is not necessary to concede that this is possible in a compounded sense: God in any day created a stone of a foot size conserving them always (194^{7-11}). But why isn't this possible? If the world were eternal, it could rain every day: why could God not create a stone in every day?

249 In his short work *On the Eternity of the World* (*De aeternitate mundi*), Thomas Aquinas had argued for the possibility of creation from eternity.

One response to this would be to say that there is no potentiality for the past. It is not possible now, even by the power of God, to do something in the past that was not done in the past (194[18]–195[2]). But what if (per impossibile) God had done it, would there now be infinite stones? Buridan replies that, if God had done it, there would indeed now be infinite stones, but of this consequence both the antecedent and the consequent are impossible, and if the world were eternal, they would always have been impossible (195[3–13]).

Buridan goes on by formulating the following conclusions:

5. In every proportional half of this day God could make a foot-size stone conserving it ever afterwards (195[23–24]). This claim is about the present, so there is no problem about potentiality for the past, and there is currently no end to time, so there is no problem about completing an infinite process of creation.
6. An infinite magnitude can exist, taking 'infinite' syncategorematically, because there cannot be so great a finite that there could not be a greater, even twice as great, a hundred times as great, and so forth without end. And since there cannot be an infinite magnitude, it follows, absolutely speaking (*simpliciter loquendo*), that than every possible magnitude there can be a larger, double, 100 times as great, etc., by divine power (196[3–8]).

What then about the supposition that God creates a stone in every proportional half of a day? Let us assume that this has been done. Buridan replies again that a universal proposition concerning possibility (*de possibili*) in a divided sense does not imply a corresponding univeral proposition about existence (*de inesse*), but it suffices that any singular proposition about existence be possible (196[12–14]). Here Buridan again deploys the tools of logical analysis: just because all singular propositions can be true together in a divided sense does not imply that they can all be true together in a compounded sense. Take the following proposition: 'all stars I can see' (without a miracle). Given that some stars are visible at different times of the year, or from different locations, it does not follow that this proposition is possible: 'all stars I see' (196[14–16]). In the case of bodies that might be created, larger and larger bodies are possible, but this does not imply that all infinitely many stones can be in existence at the same time.

Another question that arises is whether God could separate and conserve separately in a divided sense all the parts of line B. Certainly he could not do it in a collective sense (*collective*), because there are no parts that are all the parts (*nec omnes sunt aliquae nec aliquae sunt omnes*) (197[5–10]). But could God do it in a divided sense? Some say that, if a predicate appellates a form or formality, then it should be conceded that, if a proposition about possibility (*de possibili*) is true, then the corresponding proposition about existence (*de inesse*) should be conceded with the complete predicate such that (*ita quod*) nothing impossible follows from it (197^{11-20}). Although this opinion might plausibly be sustained, Buridan's opinion inclines to the opposite, namely that this proposition is true: 'I can see every star', and that God can separate and conserve separately all the parts of line B, because everyone concedes that it is not necessary to posit in existence what is possible, retaining the whole predicate. If it is true that this white thing can be black, it does not follow that it is possible that 'this white thing is black.' What does follow is that it is possible that 'this is black', pointing to the thing that previously was white (197^{21}–198^{3}). This can be compared to the case of saying that this proposition is true: 'I will drink tomorrow', although this proposition never will be true: 'I drink tomorrow.' What will be true tomorrow is that 'I drink', omitting 'tomorrow' from the proposition. And Buridan gives other similar examples in which the whole predicate is not retained in the statement of what is true in the future.

In the case at hand concerning God's separating and keeping separate all the proportional halves of a line, the statement that it will be possible does not necessarily imply that it will be possible all at once. The proposition does not connote simultaneity of time, but more likely succession over time. Some of the individual possibilities may be true at the same time, but not all of them (198^{24}–199^{1}).

In reply to the principal arguments, Buridan argues that God does not apprehend the parts of a continuum successively. God knows of some parts how many they are, but he does not know how many all are, because there are no parts that are all of the parts (200^{1-10}).

4 The Questions on Book IV

The main subjects of Book IV of Aristotle's *Physics* are place, vacuum, and time. Aristotle is trying to discover the most reasonable definitions '*quid rei*' of these physical concepts. Although Aristotle knew that Leucippus and Democritus had said that the world consists of atoms and empty space, he did not believe that in the real world there is anything that might qualify as empty space. In the section on the vacuum he argues this point. So the Aristotelian cosmos is a spherical plenum outside of which there is nothing, not even empty space. Within the cosmos, there are five basic elements. In the sublunar region there are two heavy elements, earth and water, which move naturally toward the center of the cosmos, thus producing a roughly spherical earth, much but not all of which is covered by water. The heavens consist of thick, contiguous spherical shells made of the fifth element, aether. The outermost shell, containing the fixed stars, rotates rapidly east to west once a day, carrying around other concentric aether spheres, each containing one planet or luminary, each moving west to east at its own rate. From the outside in, these planets are Saturn, Jupiter, Mars, Sun, Venus, Mercury, and the Moon. In between the watery earth at the center of the cosmos and the aether shells at the outside, there are rough spheres, or regions, of two light elements, air and fire, which naturally move as close to the orb of the moon as possible, the fire outside the air. For Aristotle, place has to explain why heavy bodies naturally fall down, or towards the center of the cosmos, and why light bodies naturally move up, or away from the center.

4.1 *Buridan's Questions on Place: Questions IV.1–6*

In Book IV, chapters 1–5 (208a27–213a11), Aristotle discusses place. He concludes that the 'place' (*locus*) of a body is the innermost unmoving surface of the body surrounding it. Thus the place of a bird is the innermost surface of air surrounding it, and the place of a fish in a lake is the innermost surface of the water surrounding it. There are immediate problems with this definition, however. What is the place of the entire cosmos, if there is nothing outside it? If a boat is anchored in a moving river, what is its place, as the water surrounding it continuously changes? Should the unmoving banks of the river be considered the place of the ship rather than the moving water in the river? What is the place of a tree, whose roots are in the ground, and about whose branches the wind blows? What about wine in a barrel on a moving ship? The innermost surface of the barrel touching the wine is not moving relative to the wine, and the barrel may not be moving relative to the ship, but the whole ship may be moving. What is the relevant place of the wine? By the fourteenth century, an alternate definition of place ('*ubi*' rather than '*locus*') made it depend on the distances to the center of the cosmos and/or to points on the heavens (with a conception something like that of a GPS system).[250]

In Buridan's time, there was another issue concerning Aristotle's definition of place as the innermost unmoving surface of the surrounding body,

250 For an overview of the medieval discussions about place (both '*locus*' and '*ubi*'), see E. Grant, 'The Medieval Doctrine of Place: some Fundamental Problems and Solutions,' in: A. Maierù & A. Paravicini Bagliani (eds), *Studi sul XIV secolo in memoria di Anneliese Maier*, Roma 1981 (Storia e letteratura, 151), 57–79, and Id., 'The Concept of *ubi* in Medieval and Renaissance Discussions of Place,' *Manuscripta*, 20 (1976), 71–80.

namely that nominalistic practitioners of the *logica moderna* (most famously William of Ockham), analyzing the personal supposition (*suppositio personalis*) of terms in propositions about place (what these terms refer to in the external world), had accepted the view that the sorts of indivisibles found in Euclid's geometry, such as points, lines, and surfaces (and more broadly instants and indivisibles of motion or '*mutata esse*') do not really exist in the external world. On this view, there is no single thing in the external world for which the word 'surface' (*superficies*) can supposit or have supposition. Instead, the term 'surface' was said to have 'confused supposition' for ever thinner layers of the medium or body surrounding what is in place, such as the closest tenth, or the closest one-hundredth, or the closest one-thousandth, and so forth. But the program to provide expositions of all propositions containing the word 'surface' into one or more other propositions meaning the same thing (but avoiding the use of the word 'surface' or other words that name an indivisible) was cumbrous to carry out. Consequently, nearly everyone agreed that often it was best to continue to make use of the traditional terms for indivisibles, but to assume that all such talk could, if desired, be translated into propositions avoiding these terms. One finds, therefore, authors frequently saying 'if place is an (indivisible) surface, then such-and-such will follow, but if place is a body, then something else will follow.' It was commonly said that the word 'place' supposits for a body or bodies, but connotes, or has appellation, for the terminus or limit (*ultimum*) of the body, so that the definition of the word 'place' should include the word 'surface' rather than the word 'body'. Aristotle, it was said, frequently spoke inexactly when his writing seemed to presuppose the existence of indivisible surfaces.

The first four questions of Book IV of Buridan's *Quaestiones Physicorum* are related to Aristotle's definition of place. Before choosing between the alternative definitions proposed by earlier authors, Aristotle lists the characteristics that are commonly associated with place. One such characteristic is that a place is thought to be equal to what is placed in it. This is the topic of the first question on Book IV.

Question IV.1—Whether every place is equal to what is placed in it (*Utrum omnis locus sit aequalis locato suo*).[251] Aristotle, *Physics*, IV, 3, 211a2 (cf. *Auctoritates Aristotelis*, 2: 127, '*locus est aequalis locato*'); Averroes, *In Physicam*, IV, comm. 30.

251 The table of questions adds: 'Quod eiusdem locati sunt infinita loca propria. Idem locatum habet locum proprium maiorem se et locum proprium minorem se' (201[3–5]).

1. **Er349(2)**: *Utrum locus sit aequalis locato* (IV.11)
2. **Ka11**: *Utrum locus sit aequalis locato* (IV.13)
3. William of Clifford: *An locus sit aequalis locato* (IV.23)
4. Radulphus Brito: *Utrum locus sit aequalis locato* (IV.10)
5. Giles of Rome: *Utrum locus differat a vacuo per hoc quod est equalis locato* (IV.25)
6. John of Jandun: *Utrum locus sit aequalis locato* (IV.3)[252]
7. William of Ockham (*Expositio*): IV, cap. 5 (t. 30, 210b32–211a7)[253]
8. Walter Burley (*Expositio*): *Utrum naturalis et mathematicus considerent de eisdem dimensionibus* (IV.D.5)[254]
9. Nicole Oresme: *Utrum locus sit equalis locato* (IV.2)

---

252 Jandun, *Quaestiones Physicorum*, IV.3, 51va: 'Tunc ad quaestionem dico quod locus communis non est aequalis locato speciali, quia quod continet aliquid et adhuc aliud, non est ei aequale—hoc est manifestum; sed locus communis continet locatum speciale et adhuc aliud, ut coelum continet homines et multa alia, et domus similiter; quare etc. Secundo dico quod locus proprius non est aequalis locato secundum omnes dimensiones ... Tertio dico quod locus proprius bene est aequalis locato aequalitate continentiae ... Ex his igitur patet quod locus proprius est aequalis corpori locato secundum continentiam, sic intelligendo quod non est aliquod corpus continens quin sit aliquod contentum replens ipsum totum neque est aliquod corpus contentum et inclusum qui⟨n⟩ sit aliquod continens ipsum totum includens, continens et circumscribens; et sic utique intelligendum puto aequalitatem loci et locati.'

253 Ockham, *Expositio*, IV, cap. 5, § 2, 43: 'Sciendum est quod non est universaliter verum quod locus est aequalis locato, quia tunc esset tanta profunditas loci quanta locati et e converso, quod non est verum. Et ideo debet intelligi quod est aequalis in longitudine et latitudine, hoc est, locus est ita longus et ita latus sicut locatum et e converso. Et ideo ad salvandum istam proprietatem loci non oportet ponere unam superficiem distinctam a corpore cuius est, quae circumdet locatum, et unam aliam superficiem in locato, quae sint simul, sicut intellexerunt antiqui ante tempora Aristotelis, sicut etiam tenent aliqui moderni recedentes a philosophia Aristotelis, quamvis credant se habere intentionem ipsius' (the editors identify '*aliqui moderni*' as Giles of Rome).

254 Burley, *Expositio in Physicam* (1501), 88va: 'Intelligendum quod, quamvis res mathematice et res naturales proprie loquendo differant, tamen quantitas naturalis et quantitas mathematica nullo modo differunt realiter. Immo una est linea naturalis et mathematica, et superficies naturalis et superficies mathematica, et corporeitas que est quantitas eadem manens et est naturalis et mathematica. Et ita est de omnibus aliis de quibus utraque scientia considerat. Nam eandem lineam quam naturalis considerat prout est in materia sensibili, eadem considerat mathematicus prout abstrabitur secundum intellectum a materia sensibili, et eodem modo est de superficie et corpore quod est quantitas, et etiam de figuris et de omnibus aliis que possunt considerari prout sunt in materia et prout abstrahuntur a materia.'

10. Albert of Saxony: *Utrum locus sit aequalis locato* (IV.2)
11. Marsilius of Inghen: *Utrum locus sit equalis locato* (IV.1.1)
12. Johannes Marsilii (?): *Utrum locus sit equalis locato* (IV.4)
13. Lawrence of Lindores: *Utrum omnis locus sit aequalis locato* (IV.1)
14. Benedictus Hesse: *Utrum omnis locus sit aequalis suo locato* (IV.10)

This was originally an exegetical question, arising from Aristotle's text. It became more than this when, as described above, Ockham and other nominalists denied the real existence of a key concept in Aristotle's definition of place, namely the concept of 'surface' (*superficies*). Even before that, however, there was a problem lying behind this question. According to Aristotle, place is the innermost surface of the surrounding body, something supposed to be two-dimensional, whereas the body in place is assumed to be three-dimensional. Thus the first principal argument in question IV.1 begins as follows:

> What is located is a body and place is a surface; but no surface is equal to a body ... (204^{7-8}).

Whoever compiled the table of questions for Book IV chose as notable results of Buridan's question IV.1 that of the same located body there are infinite proper places and that the same located body has a proper place larger than it and a proper place smaller than it. These results follow if the innermost limit of the surrounding body is not indivisible, but rather could be the inner half, or the innermost one-hundredth part, or the innermost one-thousandth part, and so forth—there is no last part immediate to the contained body. Buridan admits in passing that Aristotle uses many words that are not technically exact, so that it is difficult to talk properly about place (205^{13-14}).

Buridan argues first that the same thing is a proper or a common place with respect to diverse contents: the inner side of the orb of the moon is the common place of the element earth and of other contained things, but it is the proper place of the entire contents taken together, all the four elements, and all the mixed bodies (206^{1-3}). His second conclusion is that, according to the accepted exposition of the term 'place', the highest sphere of the world is not in place, because nothing contains it. And there are other exceptions: fire in its sphere has no proper place because nothing contains it that does not also contain what is inside the fire, such as air, water, and earth. A human does not have a proper place that does not contain anything else, because anything containing the human also contains the contents of the stomach, the air in the lungs, and so forth. It is true, Buridan admits, that if place were defined not as Aristotle defined it but as a separate space (*spatium separatum*) coinciding with the body, then every body would have a proper place. But he himself supposes that the definition of place as a separate space is not correct (206^{4-17}). Here it might be noted that Nicole Oresme in his questions on the *Physics* writes that the definition of place as a separate space would be the most plausible or probable definition if it were not uncommon (*inconsueta*).[255]

Buridan's third conclusion is that there are infinite proper places of a given thing (this being understood in the syncategorematic sense of infinite: no matter how many places are posited, there are still more). Thus every spherical (or orbicular) part of the orb of the moon touching the sphere of fire is a proper place of the terrestrial realm, and there are (syncategorematically) infinite such last parts of the orb of the moon. If we imagine the orb of the moon divided into more or fewer spherical shells, there will always be an innermost shell of the given division. This follows, Buridan explains, because he does not posit any surface distinct from a body (*non ponimus superficiem distinctam a corpore nec terminum corporis distinctum a corpore*) (206^{17}–207^{1}).

The claim mentioned in the table of questions for Book IV, that every body will have a proper place larger and smaller than itself, is substantiated in conclusions 4 and 5 from other ways of determining the size of a place in terms of length, breadth, and depth. A piece of wax may not increase in size (i.e., volume) if, starting as a sphere, it is stretched into something long and narrow.

Question IV.2—Whether place is the terminus of the containing body (*Utrum locus sit terminus corporis continentis*).[256] Aristotle, *Physics*, IV, 4, 212a20–21; Averroes, *In Physicam*, IV, comm. 33.

1. **Er349(1)**: *Utrum locus sit ultimum continentis* (IV.8)
2. **Er349(2)**: *Utrum locus sit superficies* (IV.17)
3. **Ka11**: *Utrum locus sit spatium infra latera continentis sive dimensiones separatae a corporibus sensibilibus* (IV.15) *Utrum locus sit superficies* (IV.16)
4. **L1386(1)**: *Utrum locus sit ultimum continentis* (III.5)

255 Oresme, *Questiones super Physicam*, IV.6, 464. On Oresme's view of place, see S. Kirschner, 'Oresme's Concepts of Place, Space, and Time in his Commentary on Aristotle's *Physics*,' *Oriens–Occidens. Cahiers du Centre d'histoire des sciences et de philosophies arabes et médiévales*, 3 (2000), 145–179, esp. 146–164. One of the few medieval authors who defended the view of place as separate, three-dimensional space is Gerald of Odo. See Bakker & De Boer, '*Locus est spatium*'.

256 The table of questions continues: 'De proprietatibus loci. Quare locus dicitur esse superficies et non corpus, cum omnis superficies sit corpus' (201^{6-8}).

5. William of Clifford: *An locus sit superficies* (IV.20)
6. Thomas Wylton: *Utrum locus sit ultimum corporis continentis* (IV.5)
7. John of Jandun: *Utrum locus sit ultimum continentis* (IV.4)[257]
8. Walter Burley (*Expositio et quaestiones*): *Utrum locus sit ultimum corporis continentis immobilis* (IV.2)[258]
9. William of Ockham (*Expositio*): IV, cap. 6 (t. 33, 211a23–29)[259]
10. William of Ockham (*De successivis*): II: *Tractatus de loco*[260]
11. William of Ockham (*Quaestiones*): *Utrum superficies sit res distincta a corpore* (65) *Utrum locus sit aliqua res absoluta distincta a corpore locante* (72) *Utrum Philosophus posuerit locum distinctum a corpore locante et locato* (73) *Utrum locus sit corpus continens* (75)
12. Walter Burley (*Expositio*): *Utrum locus sit ultimum corporis continentis* (IV.D.12) *Utrum locus sit sola superficies* (IV.D.13)
13. Francesc Marbres: *Utrum locus sit aliqua entitas absoluta eadem essentialiter cum superficie* (IV.1)[261]

---

257 Jandun, *Quaestiones Physicorum*, IV.4, 52rb: 'Post hoc dicendum ad quaestionem quod locus est ultimum corporis continentis. Et hoc probatur quia: aut locus est materia, aut forma, vel spatium positum vel interceptum inter latera continentis, vel est ultimum ipsius corporis continentis ... Sed locus non est forma, nec materia, nec spatium imaginatum separatum ... Ergo relinquitur quod locus est ultimum corporis continentis.'

258 In the *Quaestiones*, this is question 36.

259 Ockham, *Expositio*, IV, cap. 6, § 1, 51: 'Secundo sciendum est quod non est imaginandum quod in corpore locante sit aliqua superficies vel aliquis finis seu terminus distinctus realiter a corpore locante, quo modo albedo distinguitur a subiecto secundum se totam, et quod illa superficies vel finis seu terminus est locus aequalis locato.'

260 Ockham, *Tractatus de successivis*, 69–70: 'Consequenter videndum est de loco, quem Philosophus quarto *Physicorum* sic definit: "Locus est ultimum corporis continentis contigui immobile primum." Ad evidentiam cuius est advertendum quod non est imaginandum quod in corpore locante sit aliqua superficies vel aliquis situs distinctus realiter a corpore locante, quo modo albedo distinguitur secundum se totam a subiecto, et quod illa superficies, quae est finis vel terminus, est locus aequalis locato ... Primo igitur probatur quod locus non est talis alia res a corpore locante quia: superficies non est talis alia res a corpore distincta secundum se totam; igitur nec locus. Consequentia patet, quia superficies et locus non distinguuntur realiter. Quod patet per hoc, quia Philosophus ponit quod locus est ultimum corporis continentis, et secundum Commentatorem commento 33 est finis et terminus; sed nulla alia res a superficie est finis corporis.'

261 Francesc Marbres, *Quaestiones Physicorum*, IV.1, 39vb–40vb: 'Respondeo. Ubi sic procedam: primo enim recitabo unum modum dicendi de loci entitate et quidditate; secundo aliter dicam; tertio videbitur de loci immobilitate. Quantum ad primum est una opinio Aurioli supra secundum *Sententiarum*, qui ponit quod locus nihil aliud est quam determinata positio hic vel ibi ... Sed est dubitatio iuxta predicta in quo genere sit ipse locus. Respon-

14. Nicole Oresme: *Utrum locus sit superficies* (IV.1)
15. Hugolinus of Orvieto: *Utrum locus sit substantia vel accidens* (23 = IV.1)[262]
16. Albert of Saxony: *Utrum locus sit superficies* (IV.1)
17. Marsilius of Inghen: *Utrum locus sit superficies corporis continentis* (IV.1.2)

---

deo: illi qui dicunt quod locus non est nisi ipsa circumscriptio activa dicunt quod locus non est nisi de genere *ubi*. Distinguuntur enim duplex *ubi*, scilicet activum et passivum ... Alii dicunt quod in nullo genere est, sed reducitur ad genus quantitatis. Unde Scotus in hac materia procedit multum dubie. Non enim dicit in quo genere est. Verumtamen dicit quod non est in genere quantitatis. Potest tamen probabiliter dici quod non tantum propter rationes me cogentes, sed eo quia maior videtur probabilitas ab hac parte quam ab alia, quod ipse locus est de genere quantitatis. Et hoc videtur probare prima auctoritas Aristotelis in *Predicamentis*, qui ponit locum esse distinctam speciem quantitatis a superficie. Ergo sequitur quod sit quantitas ... Respondeo: quantum ad immobilitatem istam loci sunt multi modi dicendi. Unus modus dicendi est cuiusdam doctoris qui ponit quod in loco est aliquid materiale et aliquid formale. Materiale in loco est ultimum, id est ultima superficies corporis idem continentis et ambientis. Formale vero est respectus determinatus ad polos universi: illud quod est materiale, ut dicit, est mobile, si consideretur secundum se. Illud tamen ultimum, scilicet formale, ut stat sub ordine tali ad polos vel ad centrum respectu universi, est immobile ... Hec opinio non placet mihi. Nam variato subiecto variatur accidens necessario ... Alia est opinio Thome Anglici, qui premittit quod de ratione loci proprie dicti sunt tria vel quatuor que ponuntur in eius ratione. Primum est quod sit ultimum corporis, non corpus, ut posuerunt ponentes locum esse spatium. Secundum est quod sit ultimum corporis ambientis et continentis. Per hoc excluditur superficies secundum eius rationem quidditativam, quamvis materialiter sit idem. Tertium quod sit immobilis, per quod distinguitur a vase, quod est transmutabile a loco ad locum. Quarto ponitur primum, per quod excluditur locus communis ... Istum modum dicendi adhuc non bene capio ... Unde dicit Landulphus quod sancti non stabunt supra convexum celi empyrei sic quod totum corpus stet in vacuo, sed toti erunt infra celum empyreum. Quid autem sit de hoc vere cum formidine loquendum est.'

262 Hugolinus, *Quaestiones Physicorum*, 23 = IV.1, 29–30: 'Primus articulus. In quorum primo articulo perscrutabitur de natura loci quid sit in se. Et ex hoc patebit an sit substantia vel accidens. Prima conclusio est quod locus non est superficies distincta realiter a subiecto nec aliquod accidens inhaerens. Secunda est quod nullum corpus secundum quod corpus est locus. Tertia est quod omne corpus secundum quod ambiens vel continens est locus. Et sumo hic ly "secundum quod" non reduplicative, sed specificative. Quarta est quod omnis locus est realiter et formaliter corpus ... Tertius articulus. In tertio articulo videbitur an locus sit per se et proprie in aliquo uno genere vel in pluribus generibus reponatur ... Quinta ⟨conclusio⟩ est quod accipiendo "esse in praedicamento" primo modo large, tam iste terminus "locus" quam ista res quae est locus sunt in pluribus praedicamentis.'

18. Johannes Marsilii (?): *Utrum locus sit spacium separatum* (IV.1) *Utrum locus sit spacium contentum* (IV.2) *Utrum locus sit ultima superficies corporis continentis* (IV.3)
19. Lawrence of Lindores: *Utrum locus sit terminus corporis continentis* (IV.2)
20. Benedictus Hesse: *Utrum locus sit terminus corporis continentis* (IV.12)

Buridan's second question deals more directly with the semantic problems involving use of a term like 'surface' that appears to refer to indivisibles even though in the real outside world there are no indivisibles. If the term 'surface' in talking about place always has supposition for bodies, why should people continue to use the word 'surface'? The table of questions for Book IV follows question 2 with: 'on the properties of place', and the question 'why is place said to be a surface and not a body, since every surface is a body?' (201^{6-8}).

The principal arguments in question IV.2 largely have to do with the relation of place to the categories. In the *Categories*, Aristotle writes that place is a species of quantity distinct from surfaces. Elsewhere it seems that place does not belong to the category of quantity, but rather to the category of *ubi* (where). Places are contrary to each other, but surfaces are not. Up and down apply to place but not to surface. Place is thought to be immobile, but surfaces such as the innermost surface of the orb of the moon move (211^{6}–212^{3}). Buridan begins his determination of the question by saying that he wants to proceed 'textually' (*textualiter*), explaining Aristotle's arguments (212^{17-18}). He does not agree with all of the properties that Aristotle ascribes to place. According to Aristotle, the ancients had four main opinions about place. One said that place was the matter of the located body; the second said that place is the form of the body. The third opinion was that place is the separate space equal in every dimension to the located body. We imagine such a space to receive the body, just as we imagine an infinite space beyond the heaven (214^{11}–215^{8}). With respect to this third opinion, Buridan comments: 'Without doubt, if there were such a space, it would be reasonable to say that it would be place. And thus, because everyone can imagine that there is such a space, it was the opinion of many that such a space is place' (215^{8-11}). The insensibility of the air encourages many uneducated people to imagine this view, as well as to hold that there is a vacuum or a separate space in a vase that contains only air (215^{11-14}). The fourth opinion on place is that of Aristotle, namely that it is the innermost unmoving surface of the surrounding body (215^{15-17}).

Buridan argues that place is neither matter nor form, nor an accident inhering in the body in place. Place is not a separate space, because there is no such separate space as is sometimes imagined (215^{20}–216^{16}). Moreover such a separate space would not help explain the properties of place and Aristotle's persuasions used to support it are not effective. Buridan calls these added arguments 'persuasions', he says, because if there were a separate space, the persuasions could easily be refuted, while the arguments against the existence of a separate space are more demonstrative (217^{26-28}). With the three alternative definitions of place rejected, there remains only Aristotle's definition as the correct one. Nevertheless, surfaces are the termini of bodies, as lines are the termini of surfaces, and points the termini of lines. So why prefer to say that place is the surface of the containing body rather than saying that place is the containing body (218^{26-28})? Buridan admits that in truth place is the body containing what is located, but place is said to be proper for the same reason (*ratio*) that it is said to be surface—because it is divisible in two dimensions, not considering the third dimension (219^{1-9}). Bodies cannot interpenetrate, so their only contact is in two dimensions, i.e., according to their surfaces. This is what Aristotle means by his definition (219^{13}–220^{5}). Thus Buridan ends on a conciliatory note, supporting the view that indivisibles like surfaces are not separately existing things in the external world and yet justifying Aristotle's definition of place in terms of surface.

Question IV.3—Whether place is immobile (*Utrum locus sit immobilis*).[263] Aristotle, *Physics*, IV, 4, 212a14–21; Averroes, *In Physicam*, IV, comm. 41.

1. **Er349(1)**: *Utrum locus sit immobilis* (IV.9)
2. **Ka11**: *Utrum locus sit immobilis* (IV.17)
3. **L1386(1)**: *Utrum locus sit immobilis* (IV.6)
4. Geoffrey of Aspall: *An locus sit immobilis* (IV.4)
5. William of Clifford: *An locus sit mobilis per accidens, immobilis tamen per se* (IV.28) *An locus sit immobilis per naturam infusam ab octava sphaera* (IV.29) *An locus sit immobilis* (IV.31)
6. Radulphus Brito: *Utrum locus sit immobilis* (IV.14)
7. Giles of Rome: *Utrum locus sit quid immobile* (IV.32)
8. Thomas Wylton: *Utrum locus sit immobile* (IV.6)[264]

263 The table of questions continues: 'De alia acceptione "loci" ab ista quam definit Aristoteles. Quare dicit Aristoteles locum esse immobilem, cum sit mobilis sicut locatum' (201^{9-11}).

264 See C. Trifogli, 'Thomas Wylton on the Immobility of Place,' *Recherches de Théologie et Philosophie médiévales*, 65 (1998), 1–39 (with an edition of Wylton's question IV.6).

9. John of Jandun: *Utrum locus sit immobilis* (IV.6)[265]
10. Walter Burley (*Expositio et quaestiones*): *Utrum locus sit mobilis* (IV.3)[266]
11. William of Ockham (*Expositio*): IV, cap. 7 (tt. 41–42, 212a14–30)[267]
12. William of Ockham (*Quaestiones*): *Utrum sit idem locus numero corporis continue quiescentis quando corpus circumstans continue movetur circa illud* (77) *Utrum secundum veritatem locus sit immobilis* (78)
13. Walter Burley (*Expositio*): *Utrum locus sit immobilis* (IV.D.14)
14. Hugolinus of Orvieto: *Utrum locus sit substantia vel accidens* (23 = IV.1)[268]

---

265 Jandun, *Quaestiones Physicorum*, IV.6, 54va: 'Et ideo dico aliter ad quaestionem. Primo quod locus est immobilis per se, saltem primo et principaliter. Secundo dico quod locus sic non est immobilis quod nullo modo moveatur, nec per se nec per accidens. Tertio videbitur quo modo vel quibus modis locus est immobilis.'

266 In the *Quaestiones*, this is question 37.

267 Ockham, *Expositio*, IV, cap. 7, 79–93. Cf. the following passages (80–86): 'Circa istam immobilitatem loci sciendum est quod diversi diversimode nituntur salvare eam. Dicunt enim aliqui (ed.: Giles of Rome) quod in loco est duo considerare, scilicet illud quod est materiale in loco, ut est superficies corporis continentis, et illud quod est ibi formale, ut ordo ad universum. Ordo autem ad universum semper manet immobilis ... Sed isti in nullo declarant immobilitatem loci, nec etiam unitatem loci, quando aliquo corpore quiescente corpus circumdans ipsum movetur ... Confirmatur, quia sicut accidens absolutum non potest idem numero esse in diversis subiectis distinctis, ita accidens quod ipsi ponunt respectivum non poterit manere idem in diversis subiectis ... Ideo dicunt alii (ed.: John Duns Scotus) magis appropinquando veritati, quod quando aliquod tale corpus movetur circa aliud quiescens, non manet idem locus numero, sed est alius et alius numero, est tamen idem locus numero per aequivalentiam ... Ideo dico quod de virtute sermonis debet concedi quod locus est mobilis, quia nihil est in istis inferioribus generabilibus et corruptibilibus quin sit mobile motu locali per se vel per accidens ... Et ita cum locus in istis inferioribus non possit esse nisi substantia vel accidens, oportet quod locus sit mobilis vel per se vel per accidens ... Verumtamen pro intentione Philosophi sciendum est quod Philosophus intendit dicere quod locus est immobilis per aequivalentiam, hoc est quod tantum valet locus ad salvandum omnia quae ponuntur de loco, sicut si realiter esset immobilis ... Et hoc idem intendit Commentator, quando dicit in diversis locis quod locus est immobilis essentialiter, quod vocat locum esse immobilem per aequivalentiam.' See also Ockham, *Tractatus de successivis*, 86–87.

268 Hugolinus, *Quaestiones Physicorum*, 23 = IV.1, 29–30: 'Secundus articulus. In secundo articulo videbitur de immobilitate loci, quae ponitur in eius definitione, utrum locus sit simpliciter immobilis vel non. Prima conclusio quod omnis entitas quae est locus, est mobilis ... Tertia conclusio quod de virtute sermonis accepto loco pro aliquo significato partiali vel principali, puta pro illo pro quo supponit, omnis locus est mobilis. Quarta est quod accepto loco pro eius significato totali, locus est immobilis.'

15. Albert of Saxony: *Utrum locus sit immobilis* (IV.3)
16. Marsilius of Inghen: *Utrum locus sit immobilis* (IV.1.3)
17. Benedictus Hesse: *Utrum locus sit immobilis* (IV.13)

How can place be immobile, as Aristotle's definition asserts, if places are surfaces, all surfaces are in bodies, and all bodies move? Aristotle includes 'immobile' in his definition of place, Buridan explains, because he wants to differentiate place from something like a vase, which also contains the body within it, but which normally moves when the body contained moves—when wine is moved, the vase or barrel or whatever else contains it, also moves—otherwise the wine would not stay together. Place, however, stays behind. When a feather floats with the wind, the air surrounding it may stay the same, and thus its place stays the same according to Aristotle's definition of place, while the normal conception is that it moves (222^{15}–223^{4}).

Averroes has three theses on this subject.[269] First he claims that every place is mobile, because every place is the surface of the containing body, and every body is mobile (223^{6-8}). Second he argues that place is not mobile alone and by itself, but is moved only with the body of which it is the surface (223^{9-11}). Third he maintains that place is not necessarily moved when the body in it is moved, because place is extrinsic to the contained body (223^{13-14}). According to Averroes, Aristotle argues that place is immobile, not because place does not sometimes move with what it contains, but because it is not necessary for place to be moved with what it contains (223^{17-19}).

Buridan does not deny that Averroes' claims are true, but he thinks they are not sufficient to provide criteria to know when something is moved locally and when it remains at rest. To move is to change places and to rest is to stay in the same place. How can it be that the towers of Notre Dame in Paris are said to have been in the same place ever since they were built when the air containing them constantly changes (223^{20}–224^{3})? Many expositors of Aristotle, Buridan reports, have distinguished what is material in place from what is formal.[270] What is material is the innermost surface of the surrounding body, whereas place in the formal sense depends on the distance or nearness to the heaven or earth or other bodies that are at rest (224^{4-16}). There is still a problem even with this distinction, however, because the distances to heaven

---

269 Averroes, *In Physicam*, IV, comm. 41, 139L–140B.

270 Giles of Rome is sometimes credited with this idea. See E.D. Sylla, 'Transmission of the New Physics of the Fourteenth Century from England to the Continent,' in: S. Caroti & P. Souffrin (eds), *La nouvelle physique du XIVe siècle*, Firenze 1997 (Biblioteca di Nuncius. Studi e testi, 24), 65–110, at 77.

or to resting bodies are measured on bodies which themselves may vary over time.[271] Why is distance supposed to be formal and surface not, rather than the reverse?

This opinion is close to the truth, Buridan admits, but it is poorly expressed and lacks clarity (225^{1-2}). He therefore proposes eleven conclusions of his own, which he takes to be both true and properly expressed:

1. Every place is mobile, because every place is body and body is mobile (225^{4-5}).
2. Every place, at least if it is some one thing, is mobile per se, i.e., by itself (*solitarie*), without something other (*aliquid aliud*) moving with it, or it with the other (*alio*). I say 'other' because every surface of some body could at least by the power of God be separated and conserved separately from other parts, and then God could move it separately (225^{10-15}).

It remains to be seen how this conclusion should be interpreted. At face value it seems to imply that God could conserve a surface not part of any body, but Buridan accepts Ockham's view that in reality there are not indivisible surfaces, but only (infinitely many in the syncategorematic sense) thinner and thinner layers of the surrounding body.

3. It is not possible naturally for a body in place to exit its place without motion of the place or some part of it, because the place surrounds the located body on every side. Thus the located body cannot exit its place on any side until it penetrates the dimensions of place on that side, which is impossible, or unless some part of the containing body cedes and moves out of the way (225^{16-21}).
4. It is equally possible for place to be moved with the located body resting as for the body to move with the place resting. If fire rested in its sphere, nonetheless the orb of the moon would be moved. Likewise with the towers of Notre Dame resting, the air surrounding them is moved by the wind (225^{22-25}).
5. The same body may remain in the same proper place continually through some whole time, while nevertheless it is continually moved through that whole time with a motion that we customarily call local motion. In this way the sun and moon and all the stars move and a feather carried by the wind is moved (225^{26}–226^{4}).

271 Note that Buridan here says the same thing Ockham had said, as reported above.

In all of these cases the sun, moon, stars, and feather are supposed to remain within the same surrounding surface, while the larger celestial orb or mass of air containing them is moved.

6. It is possible that some body rests while its place continually changes, as in the case of the towers of Notre Dame or in the case of a piece of wood fixed on the shore as the tides come in and out (226^{5-9}).
7. Likewise there can be motion that is commonly called local according to its intrinsic properties without any change of place—even without any place (226^{10-12}).[272]
8. It is impossible for us to perceive by sense that something is moved in local motion unless we perceive that it changes its relation to some body or something else distant from it. In this way those who are in the hold of a ship moving rapidly, not looking outside, do not perceive whether the ship is moved or rests (226^{14-19}).
9. We judge that local motion occurs when we perceive diverse bodies continually to be situated differently relative to each other (226^{20-22}).

Actually, Buridan remarks, this should not be considered a conclusion but a principle (*communis animi conceptio*), because everyone judges this way, although they cannot judge with certainty which bodies are moved or which are moved faster, unless they know which of them are not moved or unless they have other similar information (226^{22-28}).

10. The tenth conclusion—or perhaps proposition known in itself by definition—is that everything that is continually moved locally through some total time continually relates differently to place (*se habet aliter et aliter secundum locum vel ad locum*), because to be moved or changed is to be positioned differently before and after, by the definition of the term (227^{1-5}).
11. We often use the word 'place' in a different sense than the one previously stated. This is so when we say that the stars are moved locally and that the highest sphere is moved (227^{8-20}).

Everyone concedes that the question 'where?' (*ubi*) is seeking to know the place of something, but we may answer with something that does not fit the definition of place. Thus when it is asked 'where is the sun?' the answer may

272 This could apply to the rotation of the outermost sphere.

be 'in the east'. Or the answer to the question 'Where is Saint-Denis?' may be 'ten miles to the north of Paris' (227^{21}–228^{12}). So Buridan says that with 'place' just as with words like 'healthy' or 'time' and almost all other words, we should recognize that there are many senses of which one may be primary. The primary sense of the word 'time' refers to the motion of the heavens, but we also say 'time is dear' or 'the time is serene'. The primary sense of 'place' should be related to what is changed when a body is moved or what is not changed when a body is at rest (228^{13-22}). When Aristotle says that place is immobile, he is trying to differentiate what is true of a container like a vase, which moves with its contents, from what is true of place in general. In his long answer to this question, only partly summarized here, Buridan appears to be trying to explain in the best possible way the Aristotelian conception of place. Although there are points that may benefit from further thought, Buridan does not appear dissatisfied with what Aristotle has to say about place.

Question IV.4—Whether the definition of place Aristotle assigns is good, in which it is said 'place is the first immobile terminus of the containing body' (*Utrum definitio loci quam assignat Aristoteles sit bona, qua dicitur 'locus est terminus corporis continentis immobilis primum'*). Aristotle, *Physics*, IV, 4, 211a23–29, 212a20–21; Averroes, *In Physicam*, IV, comm. 33.

1. Thomas Wylton: *Utrum locus sit ultimum corporis continentis* (IV.5)
2. William of Ockham (*Expositio*): IV, cap. 6 (tt. 33–40, 211a23–212a14)[273]
3. William of Ockham (*Quaestiones*): *Utrum Philosophus posuerit locum distinctum a corpore locante et locato* (73)
4. Walter Burley (*Expositio*): *Utrum locus sit ultimum corporis continentis* (IV.D.12)[274]

273 Ockham, *Expositio*, IV, cap. 6, 50–79. Cf. the following passages (55–56): 'Propter istas ergo rationes et alias quas alibi feci, et etiam alias quas faciam alibi tam in isto libro quam in aliis, dicendum est quod locus non est alia res a superficie, nec superficies est alia res secundum se totam distincta a corpore. Nec aliquam talem rem secundum se totam distinctam a corpore includit, quia in corpore nulla res est quin sit pars corporis, vel accidens eius, vel locatum contentum in illo ... Et ideo pro intentione Aristotelis dicendum est quod non intendit quod locus sit alia res a corpore locante et sit una alia res terminans ipsum corpus, sed ipsum corpus se ipso, ex hoc ipso quod non habet partes extensas ulterius, terminatur ... Sed intentio Philosophi est dicere quod locus est corpus continens aliud ubique sibi contiguum, ita quod non est aliqua pars corporis contenti exterior, quae scilicet tangit aliud corpus, quin contiguetur alicui parti loci, nec e converso. Nec aliud intendit Philosophus per ultimum corporis continentis.'

274 Burley, *Expositio in Physicam* (1501), 98[vb]: 'Ad hanc igitur questionem quidam de novo

5. Nicole Oresme: *Utrum locus sit ultimum corporis continentis* (IV.3)
6. Albert of Saxony: *Utrum definitio loci sit bona in qua dicitur: locus est terminus corporis continentis immobilis primum* (IV.4)
7. Marsilius of Inghen: *Utrum diffinitio loci sit bona* (IV.1.4)
8. Johannes Marsilii (?): *Utrum locus sit ultima superficies corporis continentis* (IV.3)
9. Lawrence of Lindores: *Utrum definitio loci sit bona, in qua dicitur: locus est terminus corporis continentis immobilis primum* (IV.3)
10. Benedictus Hesse: *Utrum definitio loci sit bene posita* (IV.14)
11. Domingo de Soto: *Utrum diffinitio loci eiusque conditiones recte sint a Philosopho constitutae* (IV.1)
12. Conimbricenses: *Rectene Aristoteles loci naturam explicarit* (IV.1)

Asking whether Aristotle's definition of place is good is a very standard kind of question to be included in a commentary on Aristotle's *Physics*. Buridan replies briefly that Aristotle's definition as it stands is incomplete and requires several supplements (234[12–13]). The place of a body should not be continuous (*continuum*) with the body in place, but it should be immediate (*immediatum*) to it. Aristotle makes these additions explicitly, although he does not present the definition with the additions included. With the additions the definition of place is very good (*valde bona*) (235[12]). When Aristotle says that place is the first immobile terminus of the containing body, rather than using the word 'surface', one may understand that 'place' adds connotations to the term 'surface', so that 'place' is a qualification (*passio*) of the subject 'surface' (235[17–21]). Although he said in his reply to the first question that 'we do not posit (*non ponimus*) surface distinct from body, nor a terminus of body distinct from body' (206[26]–207[1]), Buridan continues to use the word 'surface' in explaining, 'although place is surface and body, nevertheless it is described better as surface than as body,

---

incipientes philosophari dicunt quod locus non est ultimum corporis continentis. Unde de virtute sermonis debet dici quod locus est corpus continens et circumdans aliud corpus undique ita quod quelibet pars corporis continentis que est ultima versus locatum tangit aliquam partem locati et e converso quelibet pars locati que est ultima versus locum tangit aliquam partem loci. Unde quelibet pars corporis locantis immediate tangens locatum secundum omnem partem ultimam locati dicitur locus illius locati. Unde breviter dicitur quod locus nihil aliud est quam corpus locans circundans locatum undique. Et quod ultimum corporis locati non sit locus probatur per hoc quia: non est dare aliquod tale ultimum, quoniam non est dare aliquam superficiem habentem longitudinem et latitudinem carentem profunditate, ut imaginantur ponentes puncta, lineas et superficies esse res distinctas.'

because it agrees more strongly that something is called the proper place for the reason that it is called surface than for the reason that it is called body' ($237^{8–11}$).

Question IV.5—Whether the earth is in water or in the surface of water as in its proper and natural place (*Utrum terra sit in aqua sive in superficie aquae tamquam in loco suo proprio et naturali*).[275]

1. **Er349(1)**: *Utrum locus terrae sit ultimum aquae* (IV.11)
2. John of Jandun: *Utrum terra sit in superficie aquae sicut in suo loco naturali* (IV.7)[276]
3. William of Ockham (*Expositio*): IV, cap. 8 (t. 46, 212b13–22)[277]
4. Walter Burley (*Expositio*): *Utrum locus naturalis terre sit superficies concava totius aque, et sic de aliis elementis* (IV.D.18)
5. Albert of Saxony: *Utrum terra sit in aqua sive in superficie concava ipsius aquae tamquam in loco sibi naturali et proprio* (IV.5) *Utrum concavum orbis lunae sit locus naturalis ipsius ignis* (IV.6)
6. Marsilius of Inghen: *Utrum terra sit in aqua tanquam in loco proprio et naturali etc.* (IV.1.5)
7. Johannes Marsilii (?): *Utrum aqua sit locus naturalis terre* (IV.5)
8. Lawrence of Lindores: *Utrum terra sit in aqua tamquam in loco suo naturali et proprio vel in superficie aquae* (IV.4)
9. Benedictus Hesse: *Utrum locus proprius terrae sit concava superficies aquae* (IV.15)
10. Domingo de Soto: *Utrum omne corpus locum sibi vindicet naturalem atque adeo omne ens necessario sit in loco uno* (IV.2)

---

275 The table of questions continues: 'Quod multa sunt corpora quae non habent loca propria, quia saepe unius corporis locus proprius non est una res, sed multae et valde diversae. De diversitate locorum naturalium diversorum corporum. Unde locus dicatur naturalis vel violentus. Utrum, quando de profundo terrae aufertur aliqua pars terrae, aer descendit naturaliter vel violenter ad replendum' ($201^{15–20}$).

276 Jandun, *Quaestiones Physicorum*, IV.7, 55rb: 'Et ideo aliter dicatur secundum Aristotelem quod superficies aquae informata virtute naturali conservativa ipsius terrae immediate se habens ad terram est locus terrae.'

277 Ockham, *Expositio*, IV, cap. 8, § 4, 101: 'Ex quo concludit Philosophus quod propter hoc terra est in aqua sicut in loco, aqua autem est in aere sicut in loco, et aer est in aethere sicut in loco, et aether in caelo. Caelum autem secundum se totum, comprehendendo omnia corpora caelestia, non est in loco.'

This question concerns the body of the earth as a whole, in some places covered with water and in other places not, in order to investigate in detail how Aristotle's definition of place must be fleshed out to deal with the actual world, in which very few bodies in place are homogeneous and in which the surroundings of bodies in place are almost always composed of heterogeneous parts. The answer is divided into two main parts: first concerning proper places, which contain all of a given thing and nothing else, and second concerning natural places (239[19–22]). It is notable how much empirical detail, and not only general Aristotelian theory, enters into Buridan's reply. Thus, how does one explain the amount of water under the earth—rain sinking into earth and springs of water emerging from earth (238[18–24])? Sometimes heavy bodies naturally move upward and light bodies downward in order to prevent the emergence of a vacuum (250[24–26]). This is because of the common properties of bodies which maintain continuity. When water sinks into earth, it is likely replacing air trapped in cavities underground rather than replacing earth, thus a heavier body moves under a lighter one.

Is the natural place of earth the point at the center of the cosmos or the innermost surfaces of the water and air, not to mention fires that have been set on the ground, ants, and human feet surrounding it (242[1–13])? Buridan replies that the center of the world is the whole earth and this is not its own place. If the center were taken for an indivisible point according to mathematical imagination, that point would not be place, because it contains nothing, and earth would not move to it after it was already next to other earth under the other elements (251[14–19]).

Buridan's long answer to this question includes nine conclusions concerning proper place and eight conclusions concerning natural place. His seventh conclusion concerning natural place is:

> No place is said to be natural or violent for some body because of the precise essence of place, other things being set aside, because the precise essence of place is dimension, which would not be of a different character (*ratio*) in the heaven than in the earth, setting aside the substances subject to those dimensions and the active and passive qualities. Indeed, as Aristotle says in Book IV of the *Physics*, the dimension of a vacuum, if it existed, would not be of a different character (*ratio*) from the dimension of a plenum, to the extent that it is the precise character and essence of dimension ... Therefore, place is not said to be natural or violent because of the simple character of dimension, but because of the character of the nature subject to dimension that is place or the character of the nature or of the natural qualities with that dimension, which

is the place of the things existing in the subject of that dimension which is place (246^{19}–247^{6}).

Beyond the character of the bodies whose dimensions provide place, Buridan mentions also influences from the heavens on the various places below the moon as these influences vary according to distance (248^{2-6}). Unlike his answers to most of the other questions on the *Physics*, Buridan's long reply to this question is mainly directed to giving the best possible presentation of Aristotelian physics, rather than, say, to deploying the tools of the *logica moderna*.

Question IV.6—Whether the last sphere, i.e., the highest sphere, is in place (*Utrum ultima sphaera, scilicet suprema, sit in loco*).[278]

1. Roger Bacon: *Queritur an celum habeat locum* (p. 216) *Queritur utrum celum habeat locum per se vel per accidens* (p. 216) *Queritur quid est locus celi* (p. 217) *Queritur utrum celum alico modo habeat locum in quo* (p. 220) *Queritur de toto universo utrum habeat locum* (p. 221) *Queritur de loco orbium planetarum* (p. 221)
2. **Er349(1)**: *Utrum ultima sphaera sit in loco per se* (IV.18)
3. **Er349(2)**: *Utrum ultima sphaera sit in loco per se* (IV.24) *Utrum ultima sphaera sit in loco per suas partes* (IV.25) *Utrum ultima sphaera sit in loco per centrum naturale* (IV.26)
4. **Ka11**: *De ultima sphaera utrum sit in loco* (IV.18)
5. **L1386(1)**: *Utrum octava sphaera per se sit in loco* (IV.13)
6. Geoffrey of Aspall: *Qualiter caelum est in loco* (IV.5)
7. William of Clifford: *An caelum sit in loco* (IV.40) *An caelum sit in loco per se vel per accidens* (IV.41)
8. Radulphus Brito: *Utrum ultima sphaera sit in loco* (IV.15)
9. Giles of Rome: *Ubi dictum fuit totum celum non esse in loco* (IV.33) *Philosophus videtur sibi contradicere quod celum sit in loco per accidens* (IV.34) *Utrum celum sit aliquo modo in loco* (IV.36) *Utrum celum sit in loco per partes* (IV.37) *Utrum celum sit in loco propter ultimam superficiem continentem ipsum* (IV.38) *Utrum celum sit in loco per centrum* (IV.39) *Utrum celum sit in loco per se* (IV.40) *Utrum celum moveatur in loco per se* (IV.41) *Utrum motus celi debeat dici solum motus in situ* (IV.42)
10. Thomas Wylton: *Utrum celum sit in loco* (IV.9)

278 The table of questions adds: 'An ultima sphaera movetur secundum locum vel secundum situm. De opinionibus Commentatoris et sancti Thomae' (201^{21-23}).

11. Walter Burley (*Quaestiones*): *Utrum celum sit in loco* (III.38)
12. John of Jandun: *Utrum ultima sphaera sit in loco* (IV.9)[279]
13. William of Ockham (*Expositio*): IV, cap. 8 (t. 43, 212a31–212b3)[280]
14. William of Ockham (*Quaestiones*): *Utrum octava sphaera sit in loco per se aut per accidens* (79) *Utrum octava sphaera moveatur per se* (80)
15. Walter Burley (*Expositio*): *Utrum suprema spera sit in loco vel non* (IV.D.15) *Utrum suprema spera sit in loco per accidens* (IV.D.16) *Utrum suprema spera sit in loco per centrum* (IV.D.17)
16. Francesc Marbres: *Utrum ultima sphera sit in loco per se* (IV.2)[281]
17. Hugolinus of Orvieto: *Utrum omnia entia sint in loco* (24 = IV.2)[282]

---

279 Jandun, *Quaestiones Physicorum*, IV.9, 57va: 'Circa hoc ergo dicuntur quatuor secundum ordinem. Primo quod ultima sphaera, sive sit sphaera stellarum fixarum sive alia, est in loco aliquo modo. Secundo quod non est in loco per se. Tertio dico quod est in loco per accidens. Quarto quod est in loco per centrum.'

280 Ockham, *Expositio*, IV, cap. 8, §1, 94–99. Cf. Ockham, *Tractatus de successivis*, 94–95: 'Item nota ultimo secundum opinionem Commentatoris quod caelum est in loco per accidens, ita quod aequivoce dicitur caelum esse in loco et alia corpora. Verumtamen definitio quam Aristoteles dat de loco, non competit caelo nec per se nec per accidens ... Item nota quod Commentator non vocat in isto commento centrum aliquod indivisibile immobile existens in medio mundi, sicut aliqui imaginantur, sed vocat centrum corpus quiescens quod secundum omnes partes suas ultimas aequaliter distat a caelo.'

281 Francesc Marbres, *Quaestiones Physicorum*, IV.2, 41^{ra-rb}: 'Respondeo. Ubi pono tres modos dicendi ... Primus modus est Aurioli, qui est: motus localis non est ad ubi, sed est ad situm, ita quod celum per suum motum continuum non acquirit aliud et aliud ubi, sed tantum alium et alium situm ... Ideo est alius modus dicendi, qui videtur Egidii, qui ponit quod aliquid dicitur esse in loco dupliciter: uno modo secundum se totum, alio modo secundum partes suas. Tunc dicit quod celum non est in loco secundum se totum, sed secundum partes suas, hoc est, secundum spheras inferiores que comparantur ad ipsum sicut partes ad totum vel, secundum alium modum exponendi, secundum partes ipsum integrantes ... Ideo est alius modus dicendi Commentatoris, qui ponit quod primum mobile non est per se in loco, sed solum per accidens ratione centri, quod est per se in loco, circa quod ipsum celum movetur, ita quod motus eius non dicitur localis eo quod sit ad locum, sed eo quod sit vel est circa locum ... Et ideo dico aliter. Pro cuius evidentia sciendum quod ista questio de primo mobili habet difficultatem quantum ad philosophos, non autem quantum ad theologos. Secundum enim fidem primum mobile continetur a celo empyreo, et ideo est in loco. Secundum autem philosophos primum mobile a nullo continetur, sed continet omnia ... Respondeo ad primam difficultatem de immobilitate loci, que est difficultas cuiuscumque opinionis communis, licet sit satis dictum in precedenti questione. Verumtamen dicit Franciscus de Marchia quod locus est immobilis immobilitate opposita motui locali quo movetur corpus locatum ad ipsum locum.'

282 Hugolinus, *Quaestiones Physicorum*, 24 = IV.2, 30–31: 'Primus articulus. Primo: an ipsa sim-

18. Albert of Saxony: *Utrum omne ens sit in loco* (IV.7)
19. Marsilius of Inghen: *Utrum celum sit in loco* (IV.1.6)
20. Johannes Marsilii (?): *Utrum omne ens sit in loco* (IV.7)
21. Benedictus Hesse: *Utrum ultima sphaera sit in loco* (IV.16)
22. Conimbricenses: *Utrum supremum coelum in loco sit* (IV.2)

This last question on place is one that had very frequently been asked by earlier commentators since it involves the failure of Aristotle's definition of place to provide a place for something having nothing outside it. The question is easy to answer, Buridan writes, once it is recognized that the highest sphere is not in place taking 'place' in its proper sense, but it is in place in a looser sense, insofar as we perceive the motion of the highest sphere (on account of the motion of the fixed stars) in relation to the terrestrial realm (254^{1-8}).

Interesting in Buridan's answer is that he quotes previous commentators. Averroes' opinion that the highest sphere is in place accidentally by virtue of its center (namely the earth) is 'entirely absurd' (*omnino absurdum*), Buridan argues, because terrestrial bodies are in place by virtue of the heavens and not vice versa. If God moved the heaven in a straight line, it would be necessary for the center of the cosmos to move or for there to be continually different centers (255^{3-10}). Avicenna, Buridan continues, argues that the last sphere is not moved according to place but according to position (*secundum situm*), i.e., by changing distances to other bodies (255^{11-15}). Buridan is amazed by the arguments of Averroes and Thomas Aquinas against Avicenna's view. Averroes only bases himself on the authority of Aristotle. And Aquinas argues that motion with respect to position is relative, whereas Aristotle holds there is no motion in the category of relation. Moreover, according to Aquinas, position is indivisible, and there is no motion with regard to indivisibles (255^{16-24}). All this can be dispensed with when it is recognized that Aristotle was not using 'place' in its strict or proper sense when he said that the highest heaven is in

---

pliciter indivisibilia sint in loco et quomodo sint in loco ... Secundus articulus. Secundo: an omnia corpora infra ultimam sphaeram contenta sint in loco. Prima conclusio est quod non omnia corpora infra ultimam sphaeram contenta sunt in loco circumscriptive. Secunda est quod non omnia corpora talia sunt in loco per coexistentiam nec videlicet simul coexistunt alicui quod per se est in loco, et illud posset dici esse in loco coexistentiae, si licitum esset talia fingere ... Tertius articulus. Et tertio videbitur an ipsa sphaera ultima sit in loco. Prima conclusio est quod coelum et ultima sphaera non est in loco circumscriptive. Secunda est quod non est in loco secundum coexistentiam, ut ipsa accidentia et formae sunt in loco. Tertia est quod ultima sphaera est per accidens in loco. Quarta quod ultima sphaera est in loco per suum centrum.'

place because it moves—i.e., we perceive the heaven to move by its changing relations to other bodies. Position no more belongs to the category of relation than place. Furthermore, Aquinas' opinion that position is indivisible makes no sense, if you consider the example of the motion of a person from sitting to standing over time (256^{1-23}). In reply to the principal arguments, then, Buridan concludes that it is most true (*verissimae*) that the highest sphere is not in place in the proper sense (256^{25-26}).

4.2 *Buridan's Questions on the Vacuum: Questions IV.7–11*

Questions IV.7–11 are related to the definition and existence of empty places or vacuums as discussed by Aristotle in Book IV, chapters 6–9 (213a11–217b20); Averroes, *In Physicam*, IV, comm. 50–86.

Question IV.7—Whether it is possible for a vacuum to exist (*Utrum possibile sit vacuum esse*).[283] Cf. Aristotle, *Physics*, IV, 6, 213a32–b2; Averroes, *In Physicam*, IV, comm. 50–62.

1. S: *Utrum vacuum secundum quod ponitur dimensio separata sit* (III.35)
2. **Er349(1)**: *Utrum vacuum sit in rerum natura* (III.24)
3. **Er349(2)**: *Utrum vacuum sit aut possit esse in rerum natura* (IV.33)
4. William of Clifford: *Utrum sit vacuum extra caelum* (IV.49) *Utrum sit vacuum in ipsis corporibus* (IV.6)
5. Radulphus Brito: *Utrum vacuum sit* (IV.18)
6. Thomas Wylton: *Utrum vacuum possit esse in natura* (IV.10)
7. Walter Burley (*Expositio et quaestiones*): *Utrum vacuum posset esse in natura* (IV.5, incomplete)[284]
8. John of Jandun: *Utrum necessarium sit esse vacuum* (IV.10)[285]
9. William of Ockham (*Expositio*): IV, cap. 11 (t. 60, 214a16–20)[286]
10. Francesc Marbres: *Utrum vacuum possit esse in natura* (IV.4)
11. Nicole Oresme: *Utrum naturaliter possit esse vacuum in hoc mundo* (IV.7)

---

283 The table of questions adds: 'Ostenditur quod non naturaliter' (201^{24-25}).

284 In the *Quaestiones*, this is question 39.

285 Jandun, *Quaestiones Physicorum*, IV.10, 59[va]: 'Et ideo dicendum aliter breviter ad quaestionem quod non est possibile esse vacuum simpliciter.'

286 Cf. Ockham, *Expositio*, IV, cap. 11, § 4, 125: 'Sciendum est quod Commentator, commento 60, ponit rationem per quam probat quod non est talis dimensio per se separata, dicens: "illud quod invenitur per se est corpus; dimensiones vero inveniuntur in corporibus. Et quia vacuum est dimensio exsistens per se, sequitur ut non sit, quoniam si esset, tunc praedicamentum quantitatis esset separatum, et tunc quantitas esset substantia".'

12. Hugolinus of Orvieto: *An sit possibile esse vacuum in natura* (25 = IV.3.1)
13. Albert of Saxony: *Utrum vacuum esse sit possibile* (IV.8)
14. Marsilius of Inghen: *Utrum vacuum possit esse* (IV.2.1)
15. Johannes Marsilii (?): *Utrum possibile sit esse vacuum* (IV.13)
16. Lawrence of Lindores: *Utrum possibile sit vacuum esse* (IV.5)
17. Benedictus Hesse: *Utrum possibile sit esse vacuum naturaliter* (IV.17)
18. Conimbricenses: *Utrumne vacuum in natura detur an non* (IV.9.1)

In this question, Buridan considers only whether a vacuum is naturally possible (261^{1-2}), whereas in question IV.8, he will consider whether a vacuum is possible supernaturally. He defines a vacuum as a 'place not filled' (*locus non plenus*) (261^{12}). If it were possible for a vacuum to exist, then 'plenum' and 'vacuum' would be passions of 'place'. In other words 'plenum' and 'vacuum' in propositions would have supposition for place and 'plenum' would connote that there was a body contained in the place, while 'vacuum' would connote that there was no body contained in the place. 'Vacuum' would be privative in opposition to 'plenum' (261^{12-23}).

According to Buridan, place can be imagined in two ways:

> If there were space without the magnitudes of natural bodies, in which would be received natural bodies without the space ceding, then that space could well be posited to be place (261^{25}–262^{1}).[287]

In this case, a vacuum would have a corporeal dimension as great in length, breadth, and depth, as the natural body that would fill it, if it were put in it. Aristotle speaks of place in this way in the *Categories*. Such a vacuum does not and cannot exist naturally, although it could be be produced by divine power (262^{13-18}).

In the *Physics*, by contrast, Aristotle declares that place is the surface of the body containing what is in place. In that case, if the contents of a place were supposed to be annihilated, then what was left where the body had formerly been would have no dimensions even though the container, say the sphere of the moon, kept its shape rigidly as before. The only real entity for which 'vacuum' could supposit would be what had been the container, for example the sphere of the moon. 'A vacuum would be a mobile thing, because it would be the orb of the moon or a part of it, namely the concave surface of it' (263^{5-6}).

287 Trifogli, *Oxford Physics in the Thirteenth Century*, 141–164, labels this the 'immersive' notion of place.

What Buridan means by this is that the thing in the outside world for which the word 'vacuum', or better 'evacuated', can have supposition is the orb of the moon (i.e., the crystalline aether sphere that carries the moon around in its monthly motion).

In nature no things are distant from each other between which nothing mediates. Moreover it is not natural for two bodies to have differing distances from each other, while those bodies always retain the same magnitude and figure, and every part remains immediate to the parts it was previously immediate to. This is known by experimental induction (*experimentalis inductio*):

> Every universal proposition in natural science that can be proved by experimental induction such that in many singular cases of it, it is found clearly to be true, and in no case is there an exception, should be conceded as a principle. Thus Aristotle says well that it is necessary (*oportet*) for many principles to be accepted and known by sense, memory, and experience—indeed otherwise we could not know that all fire is hot. But by such experimental induction it is apparent to us that no place is void, because everywhere we find some natural body, such as air or water or something else. And we experience that we cannot separate one body from another without another body intervening (264^{3-12}).

Buridan goes on to mention common examples of nature avoiding a vacuum such as what occurs with a drinking straw. All in all, Buridan's discussion of the natural possibility of a vacuum here is pretty standard and not original. If there is anything to be noted, it is his emphasis that natural science has its foundation in experience.

Question IV.8—Whether the existence of a vacuum is possible by any power (*Utrum possibile sit esse vacuum per aliquam potentiam*). Cf. Aristotle, *Physics*, IV, 7, 213b30–214b11.

1. S: *Utrum vacuum secundum quod ponitur dimensio separata possit esse* (IV.36) *Utrum vacuum possit imaginari* (IV.37) *De comparatione vacui ad alia impossibilia, utrum scilicet magis sit impossibile vel minus vacuum esse quam grave quiescere in loco non proprio, ut sursum* (IV.38)
2. G[1]: *Cum vacuum dicatur uno modo dimensiones mathematicae separatae, alio modo dicitur vacuum locus privatus corpore, ideo primo quaeratur an hoc modo possibile sit vacuum esse extra imaginationem* (IV.21) *An vacuum sit imaginabile* (IV.22)

3. M3: *An vacuum possit esse extra imaginationem* (IV.21) *De imaginatione vacui* (VI.22)
4. William of Clifford: *An contingat aliquo modo ponere vacuum in rerum natura* (IV.47)[288] *An vacuum sit imaginabile* (IV.48)[289] *An vacuum sit vel esse possit* (IV.49)[290]
5. Radulphus Brito: *Utrum si totum spatium quod est infra latera caeli esset vacuum, latera caeli concurrerent* (IV.20.)
6. Thomas Wylton: *Utrum posito vacuo ipsum sit receptivum corporis* (III.11)
7. Walter Burley (*Quaestiones*): *Utrum vacuum possit esse* (III.39)
8. John of Jandun: *Utrum necessarium sit esse vacuum* (IV.10)
9. Nicole Oresme: *Utrum ad vacuum esse sequatur contradictio* (IV.8)
10. Lawrence of Lindores: *Utrum possibile sit vacuum esse* (IV.5)
11. Benedictus Hesse: *Utrum possibile sit per potentiam supernaturalem esse vacuum* (IV.18)
12. Conimbricenses: *Possitne virtute divina dari vacuum* (IV.9.2) *Utrum virtus angelica possit vacuum in naturam inducere* (IV.9.3)

This very short question contains Buridan's well-known statement explaining why he is considering what God might do to cause a vacuum to exist, even though Masters of Arts had taken an oath not to dispute any purely theological question and, if they disputed a question that was both theological and philosophical, to determine it in accordance with faith (268^{5-19}). The practice of considering what would follow if God caused something to be the case that does not and cannot occur in nature was most probably one of the most important factors in undermining various parts of the Aristotelian world picture, such as the assertion that there are no vacuums in the cosmos, often captured in the saying that 'nature abhors a vacuum.'[291]

288 According to Zimmermann, *Verzeichnis*, 177, this is question IV.47. In Trifogli, *Liber Quartus* Physicorum, 239, this question is IV.51, *Utrum absolute loquendo possit vacuum aliquo modo poni in natura.*

289 According to Zimmermann, *Verzeichnis*, 177, this is question IV.48. In Trifogli, *Liber Quartus* Physicorum, 240, this question is IV.52, *Utrum vacuum possit intelligi vel imaginari.*

290 According to Zimmermann, *Verzeichnis*, 177. Interestingly, Trifogli, *Liber Quartus* Physicorum, 240–242, also reports four experiments *de vacuo*. The first concerns two flat tablets that are pulled apart. The second concerns a vase filled with water inverted without the water coming out. The third concerns a sealed rigid metal vase within which water shrinks without letting anything else enter. The fourth involves a clepsydra.

291 On this saying and its medieval interpretations, see E. Grant, 'Medieval Explanations and Interpretations of the Dictum that "Nature Abhors a Vacuum",' *Traditio*, 29 (1973), 327–355.

The arguments that even God could not cause a vacuum to exist mostly consist in saying that it would imply a contradiction for a vacuum to exist, and even God cannot cause what is self-contradictory (267^{6-22}). Buridan concludes that God could cause a vacuum to exist by annihilating everything inside the orb of the moon in either of two ways: either such that within the evacuated orb there would be three dimensions, or such that there would be no dimensions or extension. The first possibility corresponds to the conception of place as having dimensions while the second possibility corresponds to Aristotle's definition of place as the innermost surface of the surrounding body, the latter modified so as to refer to ever thinner parts of the orb of the moon, rather than to its inner surface (269^{8-19}). Buridan says that he believes that God could cause a vacuum to exist, and does not prove it by natural arguments, but only says what appears possible to him (269^{1-2}). Thus the first way of imagining a vacuum with dimensions would imply that penetration of dimensions is possible, something possible for God if not naturally possible (269^{3-7}).

There is a rather frivolous claim that results from his conclusion, namely that it explains how one could see or hear through a vacuum. This happens because the word 'vacuus' (as in 'aer esset vacuus') referring to a spherical region of air within which everything else had been evacuated (water and earth, etc.), would have supposition for the spherical region of air. Two men within the thickness of that shell of air could see and talk to each other (269^{10}–270^{1})!

The first principal argument to this question had argued: it follows that, if there is a vacuum, it is a place, as follows from the nominal definition of vacuum. But it also follows, if a vacuum exists, that it is *not* a place, because it does not agree with the description of place, according to which place is the surface of the containing body, because it would contain nothing: so this is a contradiction (267^{6-11}). Buridan replies that Aristotle's description of place, as the innermost surface of the surrounding body, is not totally satisfactory (*simpliciter bona*). This is so because a universal proposition should not be falsified on account of some possible case in which the definition is affirmed of what is defined. But if there were a vacuum, not every place would contain what is in that place (*si esset vacuum, non omnis locus esset continens locatum*). Aristotle gave his description of place because he believed that there could not be an empty place, Buridan argues, and Aristotle's description of place is very good for the whole expression 'proper place of body.' Every proper place of some body is the surface of the body containing that body immediate to it and separate from it, etc. Since Aristotle did not intend his description to be valid except for natural cases, it follows that he did well according to the requirements of his intention (270^{7-21}).

Question IV.9—Whether in the motions of heavy and light bodies to their natural places the whole successiveness arises from the resistance of the medium (*Utrum in motibus gravium et levium ad sua loca naturalia tota successio proveniat ex resistentia medii*).[292] Aristotle, *Physics*, IV, 8, 214b13–215a24. Cf. Averroes, *In Physicam*, IV, comm. 64–70.

1. **Er349(1)**: *Utrum, remota densitate medii aut impedimento facto per densitatem et raritatem, maneat motus localis in medio* (IV.25)
2. William of Clifford: *De vacuo per comparationem ad motum* (IV.50)[293] *An vacui ad plenum sit proportio* (IV.54)[294]
3. Radulphus Brito: *Si grave descenderet per vacuum deorsum, utrum in tali descensu esset aliqua successio* (IV.22)
4. Giles of Rome: *Si descenderet grave in vacuo, utrum in illo descensu essent plura mutata esse* (IV.125) *Utrum talis descensus mensuretur uno instanti* (IV.126) *Utrum talia plura instantia mensurantia plura mutata esse in vacuo haberent continuitatem* (IV.127) *Utrum illa plura instantia differant re vel ratione solum* (IV.128) *Utrum in illis pluribus instantibus sit successio* (IV.129)
5. Thomas Wylton: *Queritur, supposito vacuo et cum hoc quod grave poneretur in vacuo, et cum hoc posito quod grave descendat, utrum ille descensus erit subitus vel successivus* (IV.13)
6. Walter Burley (*Quaestiones*): *Utrum resistentia medii sit tota causa successionis in motu gravis et levis* (IV.41)
7. John of Jandun: *Utrum proportio velocitatis unius motus ad velocitatem alterius motus sit secundum proportionem medii ad medium in spissitudine et tenuitate, sive in subtilitate et grossitie* (IV.12)[295] *Utrum succes-*

292 The table of questions continues: 'Quid est successio quae est principaliter a motore. Quid est resistens in motu simpliciter gravis deorsum. An partes resistunt sibi invicem tendentes quaelibet ad centrum. Quod una pars aquae respectu alterius non inclinat sursum nec deorsum, sed ratione gravioris vel levioris coniuncti, et sic de terra etc. Quod gravitas vel levitas aquae vel aeris non est composita ex partibus diversarum rationum, sicut esset tepiditas. De resistentia in gravibus mixtis. Unde provenit successio in motu caeli' (201^{2-9}).

293 According to Zimmermann, *Verzeichnis*, 177, this is question IV.50. In Trifogli, *Liber Quartus* Physicorum, 243, this question is IV.58: *De vacuo ut pure nihil per comparationem ad motum*.

294 According to Zimmermann, *Verzeichnis*, 177, this is question IV.54. In Trifogli, *Liber Quartus* Physicorum, 248, this question is IV.64.

295 Jandun, *Quaestiones Physicorum*, IV.12, 62ra: 'Dicamus igitur ad quaestionem quod in

sio in motu gravium et levium proveniat vel causetur ex resistentia medii (IV.13)[296]

8. William of Ockham (*Quaestiones*): *Utrum successio in motu locali recto sit praecise propter resistentiam medii ad mobile seu ad motorem* (87)
9. William of Ockham (*Brevis summa*): IV, cap. 4[297]
10. Walter Burley (final *Expositio*): Tract. 2, cap. 4[298]

---

istis gravibus et levibus inanimatis proportio velocitatis motus naturalis ad velocitatem eiusdem mobilis est secundum proportionem mediorum ad invicem in subtilitate et grossitie.'

296 Jandun, *Quaestiones Physicorum*, IV.13, 62va: 'Aliqui voluerunt, sicut Avempace et sui sequaces, quod non provenit successio in motu gravis aut levis ex resistentia medii propter rationes praemissas. Provenit enim successio huius motus ex situ opposito, ut dicunt, et etiam ex priori et posteriori magnitudinis. Sed dico secundum Aristotelem et Commentatorem quod successio motus gravis et levis provenit proprie ex resistentia medii pleni, per quod natus est fieri motus talis, ita scilicet quod sine tali resistentia non esset successio.'

297 Ockham, *Brevis summa*, IV, cap. 4, 70–71: 'Tertio notandum pro quaestione quam movet Commentator, an scilicet motus sit in tempore propter resistentiam medii, quod sola distantia terminorum, omni alio circumscripto, sufficit ad hoc quod motus sit continuus. Quod probatur sic: omne quod prius tangit medium quam extremum movetur successive; sed ubi est sola distantia terminorum, sine omni resistentia medii, mobile prius accedit ad medium, ut patet in motu caeli cui nihil resistit; igitur etc. Sed contra illud arguitur primo quia: Philosophus videtur dicere quod, si motus esset in vacuo, tunc esset in instanti, quia ibi non est medium resistens. Secundo sic quia: in hoc improbat Commentator Avempace qui ponit motum esse continuum propter duo; sed haec positio ponit sic; igitur etc. Tertio sic: Commentator dicit frequenter quod resistentia medii est causa successionis in motu. Ad primum dico quod Philosophus non dicit istam conclusionem quod tunc motus fieret in instanti, sed potius quod motus in vacuo nec esset in tempore nec in instanti, ut dictum est. Ad secundum dico quod Commentator non improbat eum in hoc, nec etiam tale recitat de eo; sed improbat eum quoad hoc quod dicit velocitatem et tarditatem esse res distinctas a motu, ut patet in commento. Ad tertium dico quod resistentia est dupliciter: quaedam positiva, quae est rei exsistentis in medio quae impedit motum. Alia est privativa, quae vocatur sola distantia terminorum, qualis tantummodo est in caelo; et ista distantia necessario requiritur ad omnem successionem.'

298 Cf. Burley, *Expositio in Physicam* (1501), 113va–114rb: 'Capitulum quartum tractatus secundi, in quo ex parte velocitatis et tarditatis in motu, et etiam ex parte nature vacui, probatur vacuum non esse … In prima igitur particula huius disgressionis Commentator ponit opinionem Avempace circa istam propositionem et rationem probantem quod illa propositio non est vera: qualis est proportio medii ad medium in spissitudine et subtilitate, talis est proportio motus ad motum in velocitate et tarditate. Et quod illud non sit verum probavit Avempace per unam rationem, que est talis …' On ff. 113va–119rb Burley offers an extended discussion of the natural motions of elemental bodies in various media, and possibly in a vacuum, related to Averroes' comments 71–78 on Book IV.

11. Richard Kilvington: *Utrum aliquod corpus simplex possit moveri aeque velociter in vacuo et in pleno* (3)[299]
12. Nicole Oresme: *Utrum grave simplex in movendo habeat resistentiam in se ipso* (IV.9) *Utrum in motu locali gravium et levium simplicium requiratur medium, et hoc propter successionem* (IV.10) *Utrum grave simplex aut leve moveretur in vacuo localiter successive* (IV.11) *Utrum mixtum in vacuo localiter moveretur successive* (IV.12)
13. Hugolinus of Orvieto: *Utrum resistentia medii sit praecise causa successionis in motu gravium et levium* (25 = IV.3)[300]
14. Albert of Saxony: *Utrum grave simplex habeat resistentiam intrinsecam quantum ad motum eius deorsum et consimili modo leve simplex quantum ad eius motum sursum* (IV.9)
15. Marsilius of Inghen: *Utrum in motu gravium et levium tota successio sit ex resistentia medii* (IV.2.2)
16. Johannes Marsilii (?): *Utrum grave simplex habeat resistentiam intrinsecam per quam potest fieri successio* (IV.8) *Utrum in motibus gravium et levium simplicium necessario requiratur medium extrinsecum* (IV.9) *Utrum in motibus corporum simplicium talis sit proportio motus ad motum in velocitate et tarditate qualis est proportio medii ad medium in raritate et densitate* (IV.10) *Utrum velocitas motus sequatur proportionem potentie moventis ad suam resistentiam* (IV.11)
17. Lawrence of Lindores: *Utrum in motibus gravium et levium ad sua loca naturalia tota successio proveniat ex resistentia medii* (IV.6)
18. Benedictus Hesse: *Utrum in motibus gravium et levium ad sua loca naturalia tota successio proveniat ex resistentia medii* (IV.25)

Buridan's response to this question is very long, and he himself suggests at the end that it might be divided into three separate questions, one concerning bodies that are purely heavy or light, a second concerning heavy and light bodies

---

299 Cf. E. Jung, 'Motion in a Vacuum and in a Plenum in Richard Kilvington's Question: *Utrum aliquod corpus simplex posset moveri aeque velociter in vacuo et in pleno* from the Commentary on the *Physics*,' in: J.A. Aertsen & A. Speer (eds), *Raum und Raumvorstellungen im Mittelalter*, Berlin [etc.] 1998 (Miscellanea mediaevalia, 25), 179–193.

300 Hugolinus, *Quaestiones Physicorum*, 25 = IV.3, 31–32: 'Secundus articulus. Et dato quod sit, videbitur in secundo articulo an in vacuo vel in aliquo medio, si esset, quod nullam haberet resistentiam, grave descenderet in ipso subito vel successive ... Secunda conclusio est ... quod resistentia medii posita non est praecise causa quare gravia et levia moventur successive. Tertia conclusio est quod distantia terminorum non est causa successionis in motu gravium et levium.'

that are mixed, and a third concerning the heaven (291^{7-10}). He excludes voluntary motions or those that are the result of free will (273^{3-9}). The question occurs because Aristotle had argued that the velocity of a naturally falling or rising body is inversely proportional to the resistance of the medium (or its tendency to an opposite motion), so that, if the resistance were zero, the velocity would be infinite, a self-contradiction. A vacuum would have no resistance, so the possible existence of a vacuum implies a self-contradiction. Averroes (in comment 71 of Book IV) had argued that, if the resistance of the medium were not the only cause of succession in natural motion of purely heavy or light bodies, then Aristotle's argument that in a vacuum a purely heavy body would move instantaneously would be invalid.

Buridan first clarifies the meaning of some words, such as 'succession' (*successio*), which indicates that a motion will not be all at once but take time. From this it immedately follows that succession does not only arise from the resistance of the medium, but more principally derives from the motor (272^{9-16}).[301] Resistance, Buridan argues, may be called an inclination of the mobile to the disposition opposite to that intended by the motor. Motion will not occur unless the power of the mover exceeds the power of the resistance. The greater the proportion of the mover to the resistance, the faster the motion will be (272^{17-22}). If there is no resistance at all, then there will be instantaneous mutation if the mover is applied instantly to the mobile and not successively (272^{22-24}). Buridan's further conclusions are:

2. It is impossible that, a motor having been sufficiently applied to the mobile, there is motion without resistance, distinguishing motion from instantaneous mutation (273^{10-14}).
3. It is necessary that in all natural motion of a heavy body downwards there be resistance to the mover (273^{18-19}).
4. In such motions, prime matter does not resist the mover, because prime matter either is inclined to no place and to no disposition, or it is indifferent to all inclinations (273^{21-25}).
5. In something purely and simply heavy, there is nothing that resists the mover moving this heavy body downward naturally (274^{3-4}).
6. In natural motion of something simply heavy downwards, the medium through which it is moved resists the mover (274^{11-12}). This conclusion is followed by an extended discussion of the general subject. In the course of the

301 This claim is not explicitly listed as a conclusion, but since the conclusions in most of the manuscripts start with conclusion 2, this might be considered the first conclusion.

discussion, Buridan indicates that the center of the world should not be considered to be an indivisible point (276^{1-2}).
7. Sometimes in local motion of a heavy body downwards, there is resistance other than that from the medium through which it moves (277^{13-15}).

Buridan then turns to the second part of the question, that concerning the motion of bodies other than those that are purely heavy and purely light. This includes water and air. He bases his discussion on material taken from *De caelo et mundo* and supposes that the very same quality moves water down if it is in air, up if it is in the earth, and resists being moved up or down if it is in its natural place. He doesn't prove this but assumes it from Book IV of *De caelo* (278^{1-16}). He also supposes from Book IV of *De caelo* and from *De generatione et corruptione*, that mixed bodies participate in some way in the natural qualities of the elements. They participate in heat by reason of fire or air, and in cold by reason of earth or water. They also participate in lightness and heaviness. The qualities moving mixed bodies are not simple as were the qualities of the elements, but composed of parts and degrees (278^{17-24}). Then he states the following conclusions:

1. When air existing in water ascends and air existing in fire descends, the air in that motion has no intrinsic resistance (279^{1-3}).
2. A mixed heavy body, if it were in fire, would descend naturally and would not have intrinsic resistance (279^{12-13}).
3. It seems it should be posited that a mixed heavy or light body, when it is moved in air or water, up or down, will have intrinsic resistance, supposing that it is mixed of the four elements (280^{18-21}). To doubts about this conclusion Buridan responds that in a tepid body the degrees of hot and cold do not act on each other, but the tepid body acts on a hotter body by reason of its coldness or resists it, and similarly it acts by reason of its heat in what is colder or resists it (281^{8-11}).

In all of this, Buridan seems to be reporting the sorts of arguments and evidence that might have been common in the university in the preceding decades.

Then Buridan turns to a more difficult issue, that of the motion of the heavens, where the causes of motion were thought to be different from those in the terrestrial realm (281^{18-19}). His conclusions are the following:

1. The *primum mobile* has no intrinsic resistance to its motion or to its mover, because the natural perfection of the *primum mobile* is to be moved contin-

ually and not to rest in some terminus or place or site (*in aliquo termino vel loco vel situ*) (281^{20-22}).
2. Any celestial sphere is moved without intrinsic resistance (282^{1-2}).
3. The third conclusion is difficult, namely that no celestial sphere in its motion or motions has any resistance (282^{3-4}). There is no extrinsic resistance on the part of God or the intelligences, because they are not opposed to each other, nothing can resist divine power because of its infinity, and the spheres are not continuous with each other or connected (*colligata*), so they do not resist each other or drag each other (282^{4-10}).

Against this last conclusion there are seven difficult arguments (*difficiles rationes*), to which Buridan replies. The fifth of these arguments, which is ascribed to Thomas Aquinas, is that in all local motion there is a resistance caused by the incompossibility of the termini (*incompossibilitas terminorum*). Because a body cannot be in two places at the same time, it takes time to get from one place to another irrespective of other resistance (283^{5-11}). In his reply, Buridan argues that there is no *incompossibilitas terminorum* in the heavens (290^{5-7}).

But Buridan has a far more complex reply to the second argument against the conclusion that there is no resistance in the heavens, which runs as follows:

> It would follow that a fly, or at least an intelligence that did not have greater power (*fortioris potentiae*) than a fly, could move the heaven with a faster motion than that with which it now moves with the diurnal motion, because to move faster a greater power is not required except to overcome a larger resistance (282^{14-18}).

Buridan writes he does not see how this argument could be resolved, unless by conceding resistance in the heaven or by turning to the opinion of Avempace, called the most subtile philosopher (*subtilissimus philosophus*) by Buridan:

> It was his opinion, as I believe, that setting aside all resistance, whether in causing motion or something else, the termination of the effect comes from the termination of the moving power (*ex determinatione potentiae moventis provenit determinatio effectus*) and so great a power could not produce a greater or more intense effect ... Then what Aristotle and the Commentator say must be denied ... Then the resistance of the medium would add other degrees of slowness to the slowness already resulting from the termination of the power moving without any resistance (287^{1-18}).

According to this, Buridan continues, it is necessary to correct what was said before, namely that to the extent that there is a greater ratio of motive force to resistance, there is a greater velocity, and when the ratio is less, the velocity is less, and when there is no resistance, there is no succession (287^{19-22}). What was said before is true only of the added slowness. Buridan concludes:

> And it does not appear to me that this imagination of Avempace could be demonstratively disproved (288^{5-6}).

Moreover, if Avempace's imagination were not conceded, there is another imagination which also could not be demonstratively disproved, although it is not in accord with the opinion of Aristotle, namely that not any active power can act in any passive, but only a determinate power in a determinate. Heat acts on an opaque body touching it (*approximatum*), but light cannot act on the body, and the sun does not heat the celestial bodies. Gravity and levity, heat and cold do not move the heaven (288^{7-15}). So Aristotle's ideas about the proportions of forces, resistances, and velocities need not apply in the heavens.

It is at this point that Buridan mentions that he also does not know how to disprove demonstratively the imagination that it is impetus put into the heavens by God at creation that causes their motion and which never wears away because of resistance (288^{22}–289^{2}). If this were true, one would not have to posit intelligences moving the heavens (289^{3-4}). This reply has been picked up by historians interested in ideas that seem to be pointing in the direction of the law of inertia, but it is just one in a list of possible imaginations. Buridan concludes his reply to the second argument as follows:

> Through these three imaginations the second argument, which concerned the fly, is solved. It is entirely absurd to say that an infinitely small power could move the heavens (289^{17-19}).

Buridan's reply to this question, then, might be classified as primarily natural philosophical. He does not make extended use of the tools of logic, although he does mention more than once that he intends to speak of a syncategorematic infinite and not a categorematic infinite. There is also some mathematics related to the relation of forces, resistances, and velocities, but nothing that would indicate that Buridan had studied Thomas Bradwardine's *De proportionibus velocitatum in motibus*, although Anneliese Maier found evidence in Buridan's *Physics* that he knew Bradwardine's work.[302] In this question, Bur-

---

302 Maier, *Die Vorläufer Galileis*, 98: 'Die dritte Gruppe schliesslich bilden die grossen Pariser

idan did not directly address problems of dynamics, but rather his attention was focused on providing alternative answers to the question at hand.

Question IV.10—Whether, if there were a vacuum, a heavy body would move in it (*Utrum, si vacuum esset, grave moveretur in eo*).[303] Cf. esp. Aristotle, *Physics*, IV, 8, 215a24–31; Averroes, *In Physicam*, IV, comm. 71.

1. S: *Posito vacuo cum distantia quaeritur utrum grave possit moveri deorsum* (IV.40)
2. Giles of Rome: *Utrum tota causa quare requiritur tempus in motu sit impedimentum medii* (IV.65) *Utrum quantitas spatii faciat ut requiratur tempus in motu* (IV.66) *Utrum in vacuo sit motus in instanti* (IV.67) *Utrum sit illa proportio rari ad rarum que est temporis ad tempus* (IV.69) *Utrum in vacuo animalia moverentur et utrum moverentur in non tempore* (IV.70)
3. Radulphus Brito: *Utrum gravia et levia possint moveri in vacuo* (IV.21) *Utrum animalia possint moveri in vacuo* (IV.24)
4. Thomas Wylton: *Utrum posito vacuo possit fieri transitus in vacuo* (IV.12)
5. John of Jandun: *Utrum, si vacuum esset, motus localis posset fieri in ipso* (IV.11)[304] *Utrum, si vacuum esset, motus localis animalis posset fieri in ipso* (IV.14)[305]
6. William of Ockham (*Expositio*): IV, cap. 14 (tt. 71–75, 215a24–216a26)[306]

Naturphilosophen, mit denen die Bradwardinesche Funktion Eingang in die eigentlichen physikalischen Theorien findet. Denn auch Johannes Buridan und seine Schüler haben die Bradwardinesche Lehre ohne weiteres akzeptiert und sie als wesentlichen Bestandteil in ihre eigenen physikalischen Systeme aufgenommen.' As evidence, Maier refers to Buridan's questions VII.7–8.

303 The table of questions adds: 'An in vacuo vel ultra supremam sphaeram posset homo extendere bracchium suum' (202^{10-12}).

304 Jandun, *Quaestiones Physicorum*, IV.11, 61ra: 'Propter quod dicendum est ad quaestionem secundum Aristotelem et Commentatorem quod, si esset vacuum, nullus motus localis posset fieri in ipso.'

305 Jandun, *Quaestiones Physicorum*, IV.14, 63va: 'Et ideo potest dici aliter, scilicet quod, si esset vacuum et animal posset ibi esse per aliquod tempus, tamen non posset in ipso moveri.'

306 Ockham, *Expositio*, IV, cap. 14 (*Utrum gravia et levia in vacuo moverentur*), 140–162. Cf. the following passages (140): 'In ista parte ponit Philosophus alias rationes ad probandum motum non esse si vacuum est. Verumtamen sciendum est quod non probat universaliter nullum motum esse si vacuum est, sed specialiter probat motum gravium et levium non esse si vacuum est ... (146) Tertio notandum quod Commentator hic, commento 71, movet quaestionem an motus sit in tempore praecise propter impediens ... (151) Sed contra ista videtur esse Commentator, primo quia Commentator reprobat Avempace propter

7. Walter Burley (*Expositio*): *Utrum corpora simplicia gravia possint moveri in vacuo* (IV.D.21) *Utrum corpus mixtum possit moveri in vacuo* (IV.D.23) *Utrum animal possit moveri in vacuo motu progressivo* (IV.D.24)
8. Nicole Oresme: *Utrum grave simplex aut leve moveretur in vacuo localiter successive* (IV.11)
9. Hugolinus of Orvieto: *Utrum resistentia medii sit praecise causa successionis in motu gravium et levium* (25 = IV.3)
10. Albert of Saxony: *Utrum in omni motu gravium et levium requiratur medium resistens* (IV.10) *Utrum, si vacuum esset, grave moveretur in ipso* (IV.11) *Utrum, si vacuum esset, aliquid posset moveri in ipso velocitate finita sive motu locali sive alterationis* (IV.12)
11. Marsilius of Inghen: *Utrum, si vacuum esset, grave simplex moveretur in eo* (IV.2.3) *Utrum in vacuo aliquid posset moveri successive* (IV.2.4)
12. Johannes Marsilii (?): *Utrum, si vacuum esset, possit aliquid in eo moveri* (IV.12)
13. Lawrence of Lindores: *Utrum, si vacuum esset, grave simplex moveretur in eo* (IV.7)
14. Benedictus Hesse: *Utrum, si vacuum esset, grave moveretur in eo* (IV.26)
15. Domingo de Soto: *Utrum, si quid moveretur per vacuum, moveretur in instanti* (IV.3)
16. Conimbricenses: *Utrum corpus in vacuo moveri possit an non* (IV.9.4) *Utrum posito vacuo corpus in instanti an tempore cieretur* (IV.9.5)[307]

---

hoc quod ponit duplex tempus requisitum ad motum, scilicet unum naturale propter distantiam terminorum et aliud propter resistentiam medii ... (157) Sed in motu corporum caelestium est resistentia moti ad motorem, quae resistentia non est renisus et reactio vel contrainclinatio—tunc enim motus ille esset violentus—sed est talis resistentia quod unum corpus caeleste non est natum moveri nisi a determinata intelligentia et non ab alia, secundum intentionem Commentatoris.'

307 Collegium Conimbricense, *Commentarii Physicorum*, IV.9.5, 73–74: 'Articulus I. *Non nisi tempore cieri posse*. Magnus Albertus hoc loco tract. 2, capite 6 & septimo, Gregorius in 2, distinct. 6, quaestione 3, art. 2, Averroes hoc in libro, comm. 71, Aegidius, Burlaeus, Iandunus & Saxonia putant motum in vacuo non tempore, sed momento peractum iri. Idemque censuisse autumant Aristotelem 8 ca. huius libri. Verum contraria opinio aientium talem motum necessario debere tempus consumere, etsi a M. Alberto dicatur in intellectum cadere non posse, videtur tamen & intellectu facilis & veritati consentanea. Eam vero secutus est post Avempace & Avicennam, D. Thomas in 4, d. 44, q. 2, art. 3, quaestiuncula 2, Scotus in 2, d. 2, quae. 9, aliique complures ... Articulus II. *Explicatio argumentorum quibus Aristoteles probare visus est motum in vacuo momentaneum fore*.'

This question occurs here because Aristotle argued that, if there were a vacuum, a heavy body would fall instantaneously in it (assuming velocity of fall is inversely proportional to the resistance of the medium and that a vacuum would have no resistance). But instantaneous motion is self-contradictory, and so a vacuum must be impossible. As can be seen above, this question is very common in questions commentaries on the *Physics*, but what authors include in their answers may be quite different. A common point is that Aristotle's argument is not compelling, because simple distance may be a reason why motion takes time irrespective of the resistance of a medium. Hence Aristotle's argument that velocity of a simple body in a vacuum will take no time does not hold.

In reply to the question, Buridan says that it is a conditional equivalent to a consequence, namely the consequence: 'a vacuum exists, therefore a heavy body is moved in it.' The question merely asks whether this consequence is valid (292^{9-12}). Buridan turns immediately to his conclusions:

1. Aristotle would have conceded this consequence as valid, and similarly that 'a vacuum exists, therefore no heavy body is moved in it', and that there is motion in the vacuum in an instant, and that there would not be motion in it in an instant, because he believed that it was absolutely (*simpliciter*) impossible for a vacuum to exist, and from the impossible follows anything. All the consequences that Aristotle states are true and not contradictory, because the antecedents are impossible (292^{13-20}).

But, Buridan writes, we concede that a vacuum is possible, i.e., by divine power, so we cannot argue as Aristotle does (293^{5-7}). This leads to the second conclusion:

2. This is not a good consequence: 'a vacuum exists, therefore a heavy body moves in it', because, posited that a vacuum existed (since this is possible), nevertheless perhaps no heavy body would be in it, or even if there were a heavy body in it, perhaps it would be at rest, either by God's power or otherwise (293^{8-11}).
3. In addition, this is not a good consequence: 'a vacuum exists, therefore a heavy body does not move in it', because it is possible that it would move in it, at least by divine power (293^{12-14}).
4. It is possible for a heavy body to move in a vacuum, i.e., by divine power. This is no less possible than for the whole world to move in a straight line, which was discussed in question III.15 (293^{17-20}).

But staying closer to Aristotle's argument and that of inquirers, Buridan posits the case that there is a vacuum, for instance that with the air surrounding the spheres of water and earth remaining in its quantity and orbicular figure as it now is, everything within the air is annihilated such that the air would be evacuated according to this imagination, and a stone would be placed in that evacuated air. The question then arises whether the stone would move naturally descending (293^{21-26}). Buridan replies that this can be imagined in two ways. In one way, the stone would be within the concavity of that air as the earth and water now are. In the other way, the stone would be between the concave and the convex surface of the air, as if it were in the middle of the region of air (293^{27-30}). This imaginary case leads to a further set of conclusions:

5. If the stone were in the evacuated air in the second way, such that it had air under it, it would move down naturally by its gravity until it was below the air and did not have air under itself, because the inclinations of heavy and light bodies to moving upwards and downwards are according to the exigency of the bodies next to them. For example, wood existing in water would ascend to be above the water, if the water were in a vase fastened in the highest place, just as if it were in the bottom of a well (294^{1-7}).
6. The sixth conclusion concerns the first way of imagining that the stone was in a vacuum, namely if that stone were entirely under the air touching the air on one side of the concave surface of the air. Then that stone would not move, nor would it descend further through its gravity. This conclusion follows, because the stone would not have anything lighter under it, and therefore it would have no inclination to be under something else rather than under that which it was already under (294^{16-21}).

Someone could say that the vacuum will be imagined as a simple dimension without natural substance or natural qualities, existing commensurably where the earth and water now are, i.e., within the sides of that air. In that case, Buridan argues:

7. Still that stone would not move naturally by its gravity, because it would not be below something heavier, nor above something lighter, nor receding from the air would it have some body above or below next to it, heavy or light, heavier or lighter. Therefore, as Aristotle says well, there would be no reason why it ought to be more inclined upwards or downwards, to one side or to the other (295^{9-14}). Nor does an argument about moving towards or away from the heaven have any value. Also, if the vacuum were supposed to be a simple

dimension separate and immobile, it would not be naturally penetrable, and therefore the stone could not move through it (295^{15-24}).

8. If a simple heavy body moved in a vacuum, it would move in an instant and equally fast in a plenum as in a vacuum on account of having no resistance (295^{25-27}).

Aristotle composes two arguments for these implications in his text. And Buridan notes (as was found in earlier authors) that Aristotle assumes that the opinion of Avempace is not true. But Buridan himself does not know how to disprove it, and is more inclined to it than to the opposite. If the opinion of Avempace is true, then the eighth conclusion should not be conceded, and the arguments of Aristotle would not be valid (296^{27}–297^{4}).

Here Buridan inserts the (Stoic) argument about what would happen if a man were at the lower boundary of the air by divine power, whether the man could move his arms below the air, given that there would be no space there (*cum tamen illic nullum sit spatium*). A similar question would arise whether, if a man were above the last sphere, he could move his arm further beyond it (297^{5-10}). This argument leads Buridan to formulate the last conclusion of this question:

9. A man could move his limb beyond the sphere, because nothing extrinsic would resist him. If he did this, where his limb was there would be space and the dimension of his body, although before he raised his limb beyond the sphere, there was no space there (297^{11-17}).

Question IV.11—Whether rarefaction and condensation are possible, or whether it is possible for something to be rarefied or to be condensed (*Utrum rarefactio et condensatio sint possibiles vel utrum possibile sit aliquid rarefieri vel condensari*).[308] Aristotle, *Physics*, IV, 8–9, 216b12–217b28; Averroes, *In Physicam*, IV, comm. 79–86.

1. William of Clifford: *An necesse sit ponere vacuum propter condensationem et rarefactionem* (IV.57)
2. Radulphus Brito: *Utrum rarefactio et condensatio sit possibilis* (IV.25)
3. Thomas Wylton: *Utrum rarefactio et condensatio sint motus ad quantitatem vel ad qualitatem* (IV.14) *Utrum rarefactio et condensatio sint possibiles* (IV.15)

308 The table of questions continues: 'Quod sic et per compressionem etc. Quod omnis rarefactio est per generationem magnitudinis et condensatio per corruptionem' (202^{13-15}).

4. William of Ockham (*Expositio*): IV, cap. 16–17 (tt. 79–86, 216b22–217b28)[309]
5. William of Ockham (*Quaestiones*): *Utrum rarefactio vel condensatio sit possibilis* (92) *Utrum raritas et densitas differant realiter a substantia rara vel densa* (93) *Utrum secundum intentionem Philosophi et Commentatoris raritas et densitas distinguantur realiter a corpore raro et denso* (94)[310]
6. William of Ockham (*Brevis summa*): IV, cap. 3[311]
7. Walter Burley (*Expositio*): *Utrum rarefactio terminetur ad maioritatem sive extensionem maiorem* (IV.D.26) *Utrum rarefactio et condensatio sint possibiles* (IV.D.27) *Utrum rarefactio sit motus ad quantitatem* (IV.D.28)
8. Nicole Oresme: *Utrum aliquid possit condensari* (IV.14) *Utrum in rarefactione acquiratur nova quantitas et similiter in condensatione deperdatur precedens* (IV.15)
9. Hugolinus of Orvieto: *Utrum in rarefactione aliqua nova realitas acquiratur* (26 = IV.4)
10. Albert of Saxony: *Utrum condensatio et rarefactio sint possibiles* (IV.13)
11. Marsilius of Inghen: *Utrum rarefactio et condensatio sint possibiles* (IV.2.5)

309 Ockham, *Expositio*, IV, cap. 16–17, 167–194. Cf. the following passages (167): 'Dicit ergo primo quod sunt quidam qui propter rarum et densum opinantur vacuum esse. Arguunt enim sic: si vacuum non est, non est raritas nec densitas. Sed si non est raritas neque densitas, vel oportet quod omnino non sit motus, supple generationis, nec motus localis, vel quod unum corpus non possit moveri localiter nisi omnia moveantur localiter ... (171) Secunda consequentia, quam fecerunt antiqui, vera est, ista scilicet: si raritas et densitas non sunt, ergo vel motus non est, vel moto uno turbabitur caelum, vel generatio ex elementis erit aequalis, ita quod si in aliquo tempore ex aqua generetur aer, oportet quod in eodem tempore ex aere generetur aqua aequalis ... (184) Sic ergo patet quod secundum intentionem Philosophi non oportet ponere talem rem mediam inter substantiam et qualitates. Quia potest salvari quod substantia sit quanta et una maior quam alia et quod sit longa, lata et profunda per hoc solum quod habet partes substantiales distinctas realiter quae non sunt simul, et quarum aliquae minus approximantur quam aliae, et quod inter eas non est aliquod medium. Et per istum modum faciliter salvatur rarefactio et condensatio ...'

310 Ockham, *Quaestiones*, 94, 652–653: 'Et ideo dico quod sicut curvitas et rectitudo non differunt realiter a recto et curvo, ita raritas et densitas non differunt realiter a raro et denso. Et illam conclusionem reputo veram.'

311 Ockham, *Brevis summa*, IV, cap. 3, 68: 'Tertio notandum quod rarum et densum non sunt res distinctae ab aliis per se ... quia ista possunt successive competere eidem propter solum motum localem. Unde raritas est ipsum corpus rarum; immo omnino idem significatur per concretum et abstractum. Et sic dicendum est de denso.'

12. Johannes Marsilii (?): *Utrum possibile sit aliquid rarefieri* (IV.14)
13. Lawrence of Lindores: *Utrum rarefactio et condensatio sint possibiles, id est, utrum possibile sit aliquid rarefieri vel condensari* (IV.8)
14. Benedictus Hesse: *Utrum rarefactio et condensatio sint possibiles* (IV. 27)

Rarefaction properly speaking does not occur when a rarer substance enters pores in a denser substance, but when a body that previously was smaller becomes larger with no extrinsic body entering between its parts, and similarly condensation occurs when a body becomes smaller with nothing leaving it (300^{5-8}). Buridan's first conclusion is that rarefaction and condensation are possible through heating and cooling. This obviously happens when one element is generated from another, and Buridan gives other cases from experience (300^{15}–301^{4}).

The second conclusion is that condensation is possible by compression without alteration (301^{5-7}). The third conclusion is that, analogously, rarefaction must also be possible without alteration. Air especially may be rarefied and condensed. If a wine barrel is tightly sealed, it is difficult to draw off wine unless air is allowed into the container, but once half of the wine has been drawn off, more will flow out because the air in the container will expand without more entering it (301^{13}–302^{11}). In such a case, if what is exerting pressure on the air is removed, it will naturally revert to its original volume (302^{12-20}). This is similar to what happens with a bow which, after it is bent, springs back to its original shape.

Buridan's fourth conclusion, which he considers probable (*probabilis*), is that in every rarefaction magnitude or dimension is generated and in every condensation magnitude or dimension is corrupted—similarly to what he argued in question I.8 (303^{1-4}). Condensation does not occur solely by local motion, because, if it were, it would be easier to achieve without alteration than is observed to be the case. Just as more heat makes something hotter, and more light makes the subject brighter, so more dimension makes a body more extended (303^{26}–304^{2}).

Other plausible speculations (*persuasiones*) might be added. There are two ways in which dimensions can be together. In one way one dimension can be the subject of another, just as a substance is the subject of an accident. Everyone who believes that every extended thing is magnitude or dimension concedes that dimensions can be together. In another way the dimensions are thought to be together in position (*situs*). This opinion was disproved in question I.8. Buridan himself thinks it is impossible that several dimensions can be together and not make the subject more extensive and larger than did one one of them (304^{8-24}).

4.3 *Buridan's Questions on Time: Questions IV.12–16*

In questions IV.12–16, Buridan deals with time, following chapters 10–14 of Book IV of Aristotle's *Physics* (217b29–224a16) and Averroes' comments 87–134.[312] He begins with questions relating to Aristotle's definition of time, i.e., that time is the measure of motion with respect to before and after. Buridan elaborates Aristotle's view, rather than opposing it, although a slight divergence from Aristotle comes in question IV.16, on whether time would exist although no intellective soul existed, where Aristotle argues that there would be no time if there were no intellective soul to do the numbering, and Buridan argues that for time to exist it is sufficient that there be motions that are numberable—perhaps by someone who would be created the next day.

For both Aristotle and Buridan, the units or measures of time are days, months, years, and so forth, all based on the motions of the heavenly bodies. The most proper measures of time, Buridan argues, are based on the daily rotation of the first moved or *primum mobile*. Astronomers take this motion as a basic measure of a day (this would be twenty-three hours and 56 minutes, as we measure), whereas common people (*vulgares*) determine the length of a day by the motion of the sun, which adds the proper motion of the sun around the ecliptic in a year to the motion of the *primum mobile*, resulting in a day of 24 hours.

Buridan says nothing here about other much slower apparent motions of the stars, now attributed to precession of the equinoxes and then attributed to a very slow rotation of what was called the eighth sphere, commonly estimated at one degree in a century. Along the way Buridan comments on the possibilities of precision:

> It should also be noted that we cannot measure natural motions altogether precisely and punctually (*praecise et punctualiter*), i.e., according to the mode of mathematical consideration. For we cannot with a scale know precisely that a pound of wax is equal to a pound of lead, for there may be an excess of such a small quantity that we would not perceive the excess. But measuring approximately (*ad prope*) is often sufficient,

312 Buridan's views on time are discussed by D.-J. Dekker, 'Buridan's Concept of Time. Time, Motion and the Soul in John Buridan's Questions on Aristotle's *Physics*,' in: J.M.M.H. Thijssen & J. Zupko (eds), *The Metaphysics and Natural Philosophy of John Buridan*, Leiden [etc.] 2001 (Medieval and early modern science, 2), 151–163, and Id., *De tijdfilosofie van Johannes Buridanus († ca. 1360). Een historisch-wijsgerige studie met editie van Buridanus'* Quaestiones super octo libros *Physicorum* Aristotelis (secundum ultimam lecturam), *IV, 12–16*, Ph.D. dissertation, Radboud University, Nijmegen 2003, 31–90.

> according to the saying that one should not sweat the details (*de modico non est curandum*) (332^{3-8}).

In his reply to question IV.13, Buridan explains how number, which is discrete, appears in the definition of time, even though the magnitudes traversed in the motions underlying time are continuous (314^{22}–315^{3}, 315^{19-23}). Along the way, Buridan remarks that time and motion, like distance, are continuous, as are the intensive distances gained in alteration (316^{20-24}).

Is time the measure of all motions (question IV.14)? It is argued that it is not, because there is some sempiternal motion and infinites are not measurable (320^{5}). Within this question, there is a detailed description of the ways in which measuring may occur starting with counting, the measurement of geometric extension by superposition of a measuring stick, and by use of mathematics, including multiplication of the sides of a rectangle for a measure of area, and proportionalities, as in measuring the circumference of the earth by determining the length on the surface of the earth corresponding to a given angle at the center and then using the proportion of that angle to 360 degrees (as Eratosthenes had done) (323^{10}–324^{19}).

In her *Oxford Physics in the Thirteenth Century*, Cecilia Trifogli reports that the early English commentators argue, against Aristotle, for the extra-mental reality of both number and time.[313] Time, however, depends on motion, and so the characteristics of time, for instance, continuity, depend upon the characteristics of motions, which, in turn, are related to what is gained or lost in motion. Later scholastic discussions about time and these related issues are studied by, among others, Pierre Duhem and Anneliese Maier.[314]

Question IV.12—Whether time is motion (*Utrum tempus sit motus*).[315] Aristotle, *Physics*, IV, 10–11, 217b29–220a26; Averroes, *In Physicam*, IV, comm. 87–108.

---

313 Trifogli, *Oxford Physics in the Thirteenth Century*, 20.

314 See Duhem, *Le système du monde*, VII, 363–461 (English translation: Duhem, *Medieval Cosmology*, 295–363) and A. Maier, *Metaphysische Hintergründe der Spätscholastischen Naturphilosophie* (Studien zur Naturphilosophie der Spätscholastik, 4), Roma 1955 (Storia e letteratura, 52), 47–137. See also the essays collected in P. Porro (ed.), *The Medieval Concept of Time. Studies on the Scholastic Debate and its Reception in Early Modern Philosophy*, Leiden [etc.] 2001 (Studien und Texte zur Geistesgeschichte des Mittelalters, 75).

315 The table of questions adds: 'Quod tempus est successivum. Quod dicitur diversis intentionibus in tantum quod aliquando vinum aut panis dicitur tempus' (202^{16-18}).

1. **Er349(1)**: *Utrum ⟨tempus⟩ sit idem quod motus* (IV.39)
2. **L1386(1)**: *Utrum tempus sit motus* (IV.28)
3. Radulphus Brito: *Utrum tempus sit* (IV.28) *Utrum tempus sit motus* (IV.29)
4. Giles of Rome: *Utrum tempus sit de numero entium* (IV.79) *Quomodo differt motus a tempore per hoc quod tempus est similiter et ubique et apud omnia* (IV.85) *Utrum tempus sit idem quod motus* (IV.86) *Utrum tempus possit esse sine motu* (IV.87) *Utrum tempus sit passio motus* (IV.131) *Utrum tempus sit mensura primi motus* (IV.137)
5. Thomas Wylton: *Utrum tempus sit essentialiter motus primus* (IV.17)
6. Walter Burley (*Expositio et quaestiones*): *An tempus sit* (IV.6)[316]
7. John of Jandun: *Utrum tempus sit essentialiter motus primus* (IV.18)[317]
8. William of Ockham (*Expositio*): IV, cap. 18–20 (tt. 87–98, 217b29–219a10)[318]
9. William of Ockham (*Quaestiones*): *Utrum tempus sit motus secundum rei veritatem* (40) *Utrum secundum intentionem Philosophi et Commen-*

316 In the *Quaestiones*, this is question 42.

317 Jandun, *Quaestiones Physicorum*, IV.18, 65[ra]: 'Ad quaestionem possunt dici duo. Primo quod tempus non est motus absolute. Secundo quod tempus non est motus primus, scilicet prima circulatio.'

318 Ockham, *Expositio*, IV, cap. 18–20, 194–218. Cf. the following passages (195–196): 'Intelligendum est hic quod istae rationes non sunt simpliciter sophisticae, sed aliquo modo concludunt. Probant enim sufficienter quod tempus non est aliqua res per se una, secundum se totam distincta ab omnibus rebus permanentibus et ab omni re permanente. Pro quo sciendum est quod, sicut dictum est de motu in tertio libro, quod non est aliquid distinctum ab omnibus entibus absolutis et permanentibus, sicut homines communiter imaginantur, sed hoc nomen "motus" importat multas res permanentes, videlicet mobile et illud quod adquiritur mobili, et quod unum adquiritur post aliud, quod potest fieri ex hoc quod primo est unum et non aliud, et postea est illud aliud, et sic de aliis … ita est imaginandum de tempore, scilicet quod tempus non est aliquid unum secundum se totum distinctum a rebus permanentibus, sed hoc nomen "tempus" importat illa eadem quae importat hoc nomen "motus". Et propter hoc dicit Commentator, commento 88, quod "talia nondum habent esse completum, sed esse eorum componitur ex actione animae in eo quod est in eis extra animam; et entia completa sunt illa in quorum esse nihil facit anima, ut post declarabitur de tempore" … Sed esse eorum componitur ex actione animae in eo quod in eis est extra animam, hoc est aliqua importata per talia non habent esse extra animam, quamvis possint cognosci ab anima, et aliquid importatum ab eis est extra animam … (204–205) Propter quod non est intelligendum quod Philosophus intendat probare quod tempus sit aliqua res distincta realiter secundum se totam a circulatione caeli, sicut multi ponunt, quia talis conclusio nullo modo sequitur ex premissis quas accipit. Sed intendit probare quod circulatio caeli et tempus non sunt convertibilia … (208) Sciendum est, sicut prius, quod ista ratio non probat quod haec sit falsa "tempus est motus".'

tatoris haec sit vera 'tempus est motus' (41) *Utrum secundum intentionem Philosophi haec sit vera 'motus caeli est tempus'* (42)

10. Walter Burley (*Expositio*): *Utrum sequatur: plures sunt celi, ergo plura sunt tempora* (IV.D.31) *Utrum tempus sit motus celi* (IV.D.33)[319]
11. Francesc Marbres: *Utrum tempus sit mensura cuiuslibet durantis plus quam per instans* (IV.8)[320]
12. Hugolinus of Orvieto: *Utrum tempus habeat aliquod esse extra animam praeter motum* (27 = IV.5)[321]
13. Albert of Saxony: *Utrum tempus sit motus caeli* (IV.14)
14. Marsilius of Inghen: *Utrum tempus sit et quid sit* (IV.3.1)
15. Johannes Marsilii (?): *Utrum tempus sit motus* (IV.15)
16. Lawrence of Lindores: *Utrum tempus sit motus* (IV.9)
17. Benedictus Hesse: *Utrum tempus sit motus* (IV.30)
18. Conimbricenses: *Utrumne tempus a motu distinguatur* (IV.14.2)

Buridan notes that everyone concedes that time is that by which we number things in time (*temporalia*), in relation to (*quantum ad*) their durations. We ask how great is the natural life of humans, and we say 70 years. We ask how far the way is from Paris to Avignon, and we say 12 days. Whatever moves uniformly over a given space traverses equal spaces in equal times, and longer spaces in

---

319 Cf. also the following passage (Burley, *Expositio in Physicam* [1501], 124[rb]): 'Sed sic occurrunt duo dubia. Primum est an tempus habeat esse ... Ad primum dubium arguitur primo per rationem Philosophi ... ut dicit Commentator: tempus est de numero entium quorum esse completur per anima; sed nullum tale habet esse extra animam ... In oppositum huius est Philosophus ... Circa primam questionem dico duo. Primo quod tempus habet esse ... Quoad primum dico quod tempus est, et hoc est manifestum, ut dicit Commentator ...'

320 Francesc Marbres, *Quaestiones Physicorum*, IV.8, 50[rb]: 'Sed est dubitatio qua mensura mensurantur operationes angeli. Thomiste dicunt quod mensurantur tempore discreto. Putant enim et dicunt quod tempus discretum est quoddam ens de genere quantitatis constans ex instantibus ad invicem succedentibus, ita quod Aristoteles non fecit mentionem de isto tempore, quia posuit substantiam angeli non distingui ab eius operatione; et ideo credebant quod eadem mensura mensuretur substantia angeli et eius operatio.'

321 Hugolinus, *Quaestiones Physicorum*, 27 = IV.5, 33: 'Primus articulus. In quorum primo videbitur an tempus sit aliquod ens successivum distinctum realiter ab omnibus permanentibus ... Secundus articulus. In secundo articulo videbitur an tempus sit ens formaliter continuum vel discretum (distinctum *ed.*) ... Tertius articulus. In tertio videbitur de quaesito principali. Prima conclusio est quod ad hoc quod tempus sit non necessario requiritur motus extra. Secunda est quod ad hoc quod tempus sit necessario requiritur apprehensio alicuius motus. Tertia conclusio ... omnis motus apprehensus potest esse mensura alicuius.'

longer times, shorter spaces in shorter times. From this it follows that time is a successive thing (*res successiva*) having earlier parts and later parts which do not exist at the same time (307^{22}–308^{7}).

Buridan's conclusions are the following:

1. Time is a successive thing, of which the parts are not together (*simul*), but one earlier and another later, and when the earlier exists the later is not yet, and when the later exists the earlier is no longer (308^{25}–309^{2}).
2. Time is motion, because every successive continuum is motion or mutation, because such a successive continuum is continually otherwise and otherwise, which is to be changed (309^{3-5}).
3. Time in its strictest sense (*propriissime acceptum*) is the prime motion (of the outermost sphere), because it is of the nature (*ratio*) of motion that it be the measure of motions; therefore time in its strictest sense ought to be that motion which is most properly said to be the measure of others, because in any genus it is more reasonable that the first be the measure of the others, than vice versa (309^{13-17}). Also the prime motion is the most regular motion, since it does not move faster one day and slower another day (309^{21-22}).
4. Among ordinary people (*vulgares*) the motion of the sun composed of the daily motion and the sun's own motion, is taken as time more than some other motion, because, beyond 'motion', 'time' connotes that it is the measure of other motions (310^{3-6}). This motion is the most obvious (*notissimus*), because most obvious to sense, and the measure ought to be more known than what is measured (310^{9-13}).
5. Mechanical operators (*operatores mechanici*) often use their own operations as a measure of time, because from habit their operation is well known to them. In this way they know when it is the third hour, or time to eat, when they do not see the sun. Sometimes also the church clock is used for time, although properly speaking it is not time (310^{14-21}).
6. In a more remote way, the word 'time' sometimes has supposition for temporal things, because their existence is measured by time. Thus we say the time is serene, or rainy, or dear, or causing illness, or peaceful, or bellicose, or cold or hot, etc. In this way time is air, or bread or wine, or humans, etc. (310^{22-27}).[322]

In his reply to the principal arguments, Buridan writes that astronomers take time to be determined by the prime motion, because they use intellectual

322 I.e., the word has personal supposition for these things, while connoting other things implied by it.

reasoning to compute the positions of the stars even when they do not see them, whereas ordinary people use sensible things or imagination to judge time (312^{1-7}). In this question Buridan is taking a mainstream or middle of the road approach, neither emphasizing exegesis of Aristotle's text nor using very frequently the tools of the *logica moderna*.[323]

Question IV.13—Whether the definition of time is good in which it is said 'time is the number of motion with regard to before and after' (*Utrum definitio temporis in qua dicitur 'tempus est numerus motus secundum prius et posterius' sit bona*).[324] Aristotle, *Physics*, IV, 11, 219a10–b9, 220a24–26 (*Auctoritates Aristotelis* 2: 137); Averroes, *In Physicam*, IV, comm. 99–102.

1. **L1386(1)**: *Utrum tempus sit numerus motus per se* (IV.32)
2. **L1386(2)**: *Utrum tempus sit numerus* (IV.29)
3. Radulphus Brito: *Utrum tempus bene diffiniatur dicendo 'tempus est numerus motus'* (IV.35)
4. Giles of Rome: *Utrum tempus sit numerus* (IV.92) *Utrum tempus sit numerus numeratus* (IV.93) *Videtur quod tempus prius et posterius non differant nisi secundum rationem* (IV.100) *Videtur quod etiam numero sit idem tempus prius et posterius* (IV.101)
5. Thomas Wylton: *Utrum tempus sit numerus motus* (IV.19) *Utrum tempus sit numerus motus secundum prius et posterius in motu* (IV.20)
6. Walter Burley (*Expositio et quaestiones*): *Utrum diffinitio temporis sit bene data* (IV.12)[325]
7. John of Jandun: *An diffinitio temporis sit convenienter assignata* (IV.19)[326]
8. William of Ockham (*Expositio*): IV, cap. 21 (tt. 99–102, 219a10–b9)[327]

---

323 Cf. Dekker, 'Buridan's Concept of Time.'

324 The table of questions adds: 'De proprietatibus temporis' (202^{20}).

325 In the *Quaestiones*, this is question 45.

326 Jandun, *Quaestiones Physicorum*, IV.19, 66[ra]: 'Ad quaestionem dicendum quod ista diffinitio est bona.'

327 Ockham, *Expositio*, IV, cap. 21, 219–229, esp. 222–223: 'Intelligendum est hic primo quod semper ad hoc quod tempus sit, requiritur motus medius inter nunc. Quantumcumque enim aliquis imaginatus fuerit prius et posterius, nisi imaginetur motum medium, non imaginabitur tempus. Unde si aliquis imaginetur mutationem subitam, bene imaginatur prius et posterius, non tamen imaginatur tempus, quia non imaginatur motum sed mutationem tantum. Secundo notandum est quod Philosophus per duo nunc, et Commentator per duo instantia, non intelligunt aliquas duas res realiter distinctas secundum se totas quarum una corrumpatur per adventum alterius, et inter quas sit motus tamquam res dis-

9. William of Ockham (*Quaestiones*): *Utrum tempus sit numerus motus secundum prius et posterius* (47)
10. Walter Burley (final *Expositio*): *Utrum tempus sit numerus motus* (IV.D.35)[328]
11. Nicole Oresme: *Utrum tempus sit numerus motus secundum prius et posterius* (IV.18)
12. Albert of Saxony: *Utrum definitio temporis sit bona qua dicitur: tempus est numerus motus secundum prius et posterius* (IV.15)
13. Marsilius of Inghen: *Utrum diffinitio temporis sit bona* (IV.3.2)
14. Johannes Marsilii (?): *Utrum tempus sit numerus motus secundum prius et posterius* (IV.17)
15. Lawrence of Lindores: *Utrum definitio temporis sit bona in qua dicitur: tempus est numerus motus secundum prius et posterius* (IV.10)
16. Benedictus Hesse: *Utrum definitio temporis sit bene posita* (IV. 31)
17. Domingo de Soto: *Utrum diffinitio temporis recte fuerit a Philosopho constituta* (IV.4)
18. Conimbricenses: *Rectene tempus ab Aristotele definitum fuerit an non* (IV.14.1)

How can it be that the measures of motion are numbers, which are discrete, whereas motion itself is said to be continuous? After the principal arguments, and after some suppositions, Buridan concludes:

1. It is of the nature (*ratio*) of motion, or of the term 'motion', to have distinctions (*discretio*) between the parts of motion that is time. Thus we divide the motion of the heaven into years, days, hours, and minutes, etc. (315^{9-10}).
2. It is reasonable to say that time is number, because it is a measure, as has been said, and not without distinction between the parts of what is time. A measure, however, with distinction between parts is called number. This is of the essence (*ratio*) of number, that it is a discrete measure according to parts (315^{19-23}).

---

tincta, ad modum quo unum lignum est medium inter duos lapides. Ponere enim tales res est impossibile. Sed volunt dicere quod nunquam percipimus tempus nisi quando dicimus in anima nostra primo "hoc nunc est ibi vel ibi" et poterius dicimus "hoc hunc est ibi vel ibi", et quod postquam primo fuit verum dicere "hoc est in *a*" et antequam verum fuit dicere "hoc nunc est in *b*", verum fuit dicere "hoc movetur".'

328 Burley, *Expositio in Physicam* (1501), 130ra: 'Circa hoc capitulum quero tres questiones. Quarum prima est an tempus sit motus. Secunda an tempus sit consequens motum. Tertia an tempus sit numerus motus secundum prius et posterius.'

3. Time is the measure of motion, because it is of the nature of time that it be a discrete measure, as has been said; but it is not a measure of permanent things as permanent, because it has been said that time is a successive thing, and a successive thing is not a measure of a permanent thing in the same way as a permanent thing (315^{24}–316^{3}).
4. To express the sense of 'time' it is not enough to say that time is the number of motion, but there must be added 'according to before and after' because motion is divisible in two ways. In one way it is divisible according to the division of the mobile ... And in this way, time is not the measure of motion ... In another way motion is divisible in earlier and later parts, as according to the division of the space of which one part is traversed earlier and others follow, or according to the division of the degrees of quality that are acquired. And according to this division and quantity of motion, time measures motion. In twice the time, the motion is double if it is equally fast, or a quality twice as intense is acquired (316$^{10-27}$).
5. Aristotle's definition is a good description of time, because it explains the meaning (*significatio*) and all the connotations of the term 'time', and it is convertible with what is defined (317$^{4-6}$).

In reply to the third principal argument Buridan writes that infinite is number and infinite is time, taking 'infinite' syncategorematically, as was said in question III.16. Nevertheless there is no infinite number and no infinite time, taking 'infinite' categorematically (318$^{7-10}$). Again in this question, Buridan seems to take a moderate middle of the road approach (setting aside possible questions about the eternity of the world mentioned at the end), and making moderate use of the analysis of propositions using the *logica moderna*.

Question IV.14—Whether of any motion one chooses, time is the measure (*Utrum cuiuslibet motus tempus sit mensura*).[329] Aristotle, *Physics*, IV, 14, 223a29–223b1; Averroes, *In Physicam*, IV, comm. 132.

1. Radulphus Brito: *Utrum motus habeat esse in tempore* (IV.41)
2. Giles of Rome: *Utrum sit unum tempus omnium temporalium* (IV.135) *Utrum tempus diversorum sit unum pro numero numerato* (IV.136)

329 The table of questions continues: 'Quid est mensurari aliqua mensura. De quinque diversis modis mensurandi. De duplici magnitudine motus et quae res sit utraque. De mensuratione parvi per magnum' (202$^{21-24}$).

3. Thomas Wylton: *Utrum tempus sit essentialiter motus primus* (IV.17) *Utrum secundum unitatem* (melius: *veritatem?*) *hec consequentia valeat: si sint plures celi, erunt plures tempora* (IV.18) *Quis sit ille motus cuius numerus secundum prius et posterius est tempus* (IV.21)
4. Walter Burley (*Quaestiones*): *Utrum tempora multiplicentur secundum multiplicationem motuum* (IV.47)
5. John of Jandun: *Utrum tempus mensuret motum secundum quod quantus est* (IV.24)[330]
6. William of Ockham (*Expositio*): IV, cap. 27 (t. 132, 223a29–b12)[331]
7. William of Ockham (*Quaestiones*): *Utrum tempus sit mensura cuiuslibet motus* (48)
8. Walter Burley (*Expositio*): *Utrum sequatur: plures sunt celi, ergo plura sunt tempora* (IV.D.31) *Utrum quodlibet tempus habeat suam propriam durationem* (IV.D.37) *Utrum tempus sit quantitas inherens immediate ipsi motui* (IV.D.38) *Utrum tempus sit idem ubique* (IV.D.50)
9. Nicole Oresme: *Utrum omnia sint in tempore* (IV.19)
10. Hugolinus of Orvieto: *Utrum tempus habeat aliquod esse extra animam praeter motum* (27 = IV.5)
11. Albert of Saxony: *Utrum omnis motus sit mensurabilis tempore* (IV.17)
12. Marsilius of Inghen: *Utrum cuiuslibet motus tempus sit mensura* (IV.3.4)
13. Johannes Marsilii (?): *Utrum omne ens sit in tempore* (IV.18)
14. Lawrence of Lindores: *Utrum cuiuslibet motus tempus sit mensura* (IV.11)
15. Benedictus Hesse: *Utrum cuiuslibet motus tempus sit mensura* (IV.36)
16. Conimbricenses: *Num praeter tempus aliae aliarum rerum durationes constituendae sint* (IV.14.3)

Buridan's reply to this question is very long. He writes that in his judgement the question is very difficult and implies further difficulties (323^{3-4}). What does it mean to measure? And what are the ways of measuring? There may be intrinsic and extrinsic measures, some using measuring instruments, and others using mathematics, such as multiplying length and width to get area. In this question Buridan has in mind a method like that of Eratosthenes, calculating the circumference of the earth by measuring a distance on the surface that corresponds to a measured fraction of a full circle of 360 degrees

330 Jandun, *Quaestiones Physicorum*, IV.24, 71va: 'Ad quaestionem breviter dico quod tempus mensurat motum secundum quod quantus est.'

331 Ockham, *Expositio*, IV, cap. 27, § 5, 296: 'Dicit ergo primo quod aliquis poterit dubitare qualis motus tempus sit numerus, utrum scilicet tempus sit numerus cuiuslibet motus vel non, et utrum sit numerus omnium motuum aeque primo vel non.'

(325^{1-11}). Or a clock might be used to measure the motion of the sun (325^{17-19}). Buridan states and elaborates eight conclusions:

1. Sometimes we measure by simple enumeration, as in counting days, we conclude that they are 1 or 10 (326^{2-4}).
2. Sometimes we count days or months, and because we know how much time is in a day and how many days in a month, we conclude that a month is so many times the time in a day (326^{5-8}).
3./4. We do not measure time by time, or motion by motion, using superposition, because the parts of time are fleeting and not permanent. Therefore we cannot put one over the other (326^{9-12}).
5. We measure other motions, whether local motions or alterations, by the time established by the motion of the highest sphere, using proportional divisions of the time and of the measured motion (326^{21-25}). Buridan elaborates this in detail for local motions and alterations, taking account of the dimensions of the mobile in space as well as time.
6. Not all motions are actually measured by time, because for every time that measures motion there is another larger time. Some motions cannot be measured, because they are imperceptible to us, such as the motions of fish at the bottom of the sea (331^{7-15}).
7. There is no specific or general type of motion that is in principle immeasurable by us (331^{16-17}).
8. It is probable (*probabilis*) that there is no motion that on account of its great or small magnitude is incapable of being measured by time, and, likewise there is probably no time that on account of its large or small magnitude is incapable of measuring motion (332^{9-12}). Sometimes we can measure using proportionality, for instance, if it takes Saturn 30 years to complete its oblique circle, and if we assume its motion is regular, we can determine how far it would move in a year.

There is detailed discussion of measurement in relation to each of these conclusions.

Question IV.15—Whether rest is measured by time (*Utrum quies mensuretur tempore*).[332] Aristotle, *Physics*, IV, 12, 221b7–16; Averroes, *In Physicam*, IV, comm. 118.

---

332 The table of questions adds: 'Quod res quae non mutatur non mensuratur tempore; ideo nec quies. Quomodo esse rei non mutatae mensuratur. Quomodo sine motu et tempore posset esse maior duratio et minor duratio et quomodo esset prius et posterius' (202^{25-28}).

1. **Er349(1)**: *Utrum quies primus sit in tempore* (IV.47)
2. **Ka11**: *Utrum quiescens sit in tempore* (IV.52)
3. **L1386(1)**: *Utrum quies sit in tempore* (IV.34)
4. **L1386(2)**: *Utrum ipsum quiescens sit in tempore* (IV.39) *Utrum omnis quies sit in tempore* (IV.40)
5. Boethius of Dacia: *Utrum quies sit actus* (III.16)
6. Giles of Rome: *Unde sequitur: tempus est mensura motus per se, ergo quietis per accidens* (IV.110) *Utrum mobile vel quiescens secundum quod est aliquid quantum mensuretur tempore* (IV.112)
7. Thomas Wylton: *Utrum tempus sit mensura quietis* (IV.28)
8. John of Jandun: *Utrum tempus sit mensura quietis* (IV.25)[333]
9. William of Ockham (*Expositio*): IV, cap. 25 (t. 118, 221b7–16)[334]
10. Nicole Oresme: *Utrum omnia sint in tempore* (IV.19)
11. Albert of Saxony: *Utrum quies mensuretur tempore* (IV.18)
12. Marsilius of Inghen: *Utrum tempus sit mensura quietis* (IV.3.5)
13. Lawrence of Lindores: *Utrum quies mensuretur tempore* (IV.12)
14. Benedictus Hesse: *Utrum quies mensuretur tempore* (IV.37)

Buridan concludes that:

1. An absolutely permanent thing, i.e., one that is not changed at all (*quae nulla mutatione mutatur*) is not measured by time (337^{22-23}).
2. Rest is not measured by time, because rest is nothing else over and above the thing resting, and if it is resting entirely, it is a permanent thing undergoing no change, and as such it is not measurable by time (338^{7-10}).
3. Every thing that does not always exist is measured by time with respect to its existence (338^{11-12}).
   In what sense is this true? It should be said that properly speaking neither a human nor the existence of a human is measured by time, unless this is by virtue of the motion or mutation of something else than the human or the

333 Jandun, *Quaestiones Physicorum*, IV.25, 71vb: 'Dicendum quod quies mensuratur tempore. Quod patet ratione Aristotelis in littera, nam quies est privatio motus in subiecto apto nato moveri. Non enim omne immobile quiescit, sed privatum motu, aptum autem natum moveri. Et sic patet quod quies est privatio.'

334 Ockham, *Expositio*, IV, cap. 25, § 4, 269: 'Sciendum est quod non est intelligendum quod quies mensuretur tempore tamquam res distincta tam a tempore quam a motu et requiescente. Sed per istam propositionem "quies mensuratur tempore" intelligunt Philosophus et Commentator istam propositionem "per tempus potest sciri quamdiu aliquod corpus quiescit".'

existence of the human. The verbs 'to be' (*esse*), 'to be in the future' (*fore*), and 'to have been' (*fuisse*) consignify times. The existence of a human is not measured by time in the sense that what 'the existence of a human' has supposition for is measured by time, but in the sense that the connotations of 'a human is', 'a human will be', and 'a human was' are measured by time (339^{1-8}).

There remains a great doubt (*fortis dubitatio*), because it seems that even if there were no motion, something that has no quantity except duration could still be measured with respect to duration, which would imply that time would still exist. Suppose that before the world was created and, consequently, before there was any motion, God created three angels, A, B, and C, and that he first created A, and later B and C together. In such a case it would be necessary (*oportet*) to concede that A endured longer than angels B and C, and that angels B and C endured for an equal length of time. A's duration would even be in a ratio to the durations of B and C, because A's duration is no different than if the motion of the heavens were co-existent with the angels (339^{27}–340^{16}).

If this were conceded, another great incongruity would follow, because the duration of angel A, or of B and C, cannot be anything else but the angels themselves or God, because in the case posited nothing else is supposed to exist, and yet there is no way to compare the angels except by their perfections, and the angel supposed to last for a longer time may be less perfect than the angels existing a shorter time (340^{17-27}). Buridan comments:

> This is difficult because neither sense nor imagination applies to (*cadit super*) such a case, but only the intellect. And perhaps not even the intellect can be convinced by arguments deduced from sense that the given cases are possible. But this should simply be believed. Therefore the human intellect must so to speak beg (341^{1-5}).[335]

In this case, Buridan's begging or borrowing seems to involve adopting additional conditions or suppositions:

> And nevertheless it appears to me that angel A was prior to angel B, not because it was in a prior time, but because it existed when B did not exist. And it can be said that that priority was temporal in a conditional sense,

335 See Sylla, *'Ideo quasi mendicare oportet intellectum humanum'*, 228–230.

> because if time had coexisted with those angels, as time coexists with those things that are now coexisting with time, then angel A would have been prior in time to the succeeding time in which angel B or angel C existed. And thus similarly, in the aforesaid conditional sense and from an external denomination, we might say that angel A endured longer than angel B endured, and even in a certain ratio (*proportio*), as twice or three times as long, because, that is, angel A would have coexisted with two or three times as long a time as the time with which angel B coexisted, if there were coexisting times ... In no other way could we perceive or express how much earlier or how much longer angel A endured than angel B ... So too we cannot say that we perceive how great the motion we call local motion is except by the relation of the mobile to something else, which however has no effect on that motion (341^{10}–342^{2}).[336]

Here Buridan reasons hypothetically and he makes use of assumptions that were common at the time about the characteristics of angels, but he makes no mention of condemnations or of God's absolute power.

Question IV.16—Whether time would exist even though there was no intellective soul (*Utrum tempus esset, quamvis non esset aliqua anima intellectiva*).[337] Aristotle, *Physics*, IV, 14, 223a16–224a17; Averroes, *In Physicam*, IV, comm. 130–134.

1. **Er349(1)**: *Utrum tempus habeat esse praeter animam* (IV.52)
2. **Ka11**: *Utrum tempus potest esse praeter animam* (IV.56)
3. **L1386(2)**: *Utrum, si impossibile sit esse numerantem, impossibile* ⟨*sit*⟩ *esse numerabile* (IV.42)
4. Giles of Rome: *Utrum tempus sit solum ens apud animam* (IV.133) *Ad quid oportet prius et posterius numerari in motu ab anima* (IV.134)
5. Thomas Wylton: *Utrum tempus habeat esse extra animam ita quod tempus possit habere esse perfectum quod sibi debetur absque consideratione anime replicantis vel numerantis partes motus ad invicem* (IV 30)[338]

336 See Sylla, *'Ideo quasi mendicare oportet intellectum humanum'*, 230–231.

337 The table of questions continues: 'Quid iste terminus "numerus" connotat super illum terminum "multitudo". Utrum esset numerus, licet non esset numerans vel licet non posset esse numerans' (203^{2-4}).

338 For an edition of Wylton's question, see Trifogli, 'Il problema dello statuto ontologico del tempo,' 515–527.

6. John of Jandun: *Utrum tempus sit extra animam humanam* (IV.27)[339]
7. William of Ockham (*Expositio*): IV, cap. 18–20 (tt. 87–98, 217b29–219a10)
8. William of Ockham (*De successivis*): III: *Tractatus de tempore*[340]
9. William of Ockham (*Quaestiones*): *Utrum tempus possit esse sine anima* (49)[341]
10. Walter Burley (*Expositio*): Book IV, Tractatus III[342]

---

339 For an edition of Jandun's question, see Trifogli, 'Il problema dello statuto ontologico del tempo,' 528–548. According to the Venice 1551 edition, Jandun replies as follows (72vb): 'Tunc dico ad quaestionem tria. Primo quod tempus quantum ad illud quod significatur hoc nomine "tempus" sicut materiale est actu praeter animam. Secundo dico quod ipsum tempus quod est formale significatum huius nominis "tempus" non est actu praeter animam, sed ab anima. Tertio dico quod istud idem significatum est extra animam in potentia.'

340 Ockham, *Tractatus de successivis*, 96–97: 'Philosophus quarto *Physicorum* capitulo de tempore investigans utrum tempus sit, adducit quasdam rationes probantes tempus non esse, ut patet intuenti processum eius. Quae rationes omnino non sunt sophisticae, sed aliquo modo concludunt. Probant enim sufficienter quod tempus non est aliqua res distincta secundum se totam ab omni re permanente et ab omnibus rebus permanentibus ... Ita consequenter imaginandum est de tempore, quod tempus non est aliquid secundum se totum distinctum a rebus permanentibus; sed hoc nomen "tempus" importat illa eadem quae hoc nomen "motus", quorum aliqua sunt extra animam, aliqua non sunt extra animam vel non coexistunt extra animam, quae possunt cognosci ab anima.'

341 Ockham, *Quaestiones*, 49, 529: 'Quia solutio istius quaestionis multum dependet ex logica, ideo primo videndum est quae consequentiae et propositiones sunt concedendae in ista materia et quae negandae ... Circa primum sciendum quod prima consequentia concedenda in ista materia est ista "tempus est, igitur anima est", quia secundum Philosophum tempus est prius et posterius in motu, numerata ab anima numerante ista "prius" et "posterius", et ideo quando haec est vera "prius et posterius in motu numerantur ab anima", haec est vera "tempus est", et quando non, non est vera. Et propter hoc sicut ista consequentia est formalis "prius et posterius in motu sunt numerata ab anima numerante, igitur anima est", ita est ista consequentia formalis "tempus est, igitur anima est".'

342 Burley, *Expositio in Physicam* (1501), 146vb–147vb: 'Istud est octavum et ultimum capitulum huius tractatus, in quo determinatur de tempore in comparatione ad suam causam, scilicet in comparatione ad animam ... Intelligendum quod sicut hec vox "lapis intellectus" significat quoddam copulatum esse ex lapide et ex intellectione, et tamen vere dicitur de lapide—nam hec est vera "lapis est lapis intellectus"—ita hoc nomen "tempus" significat unum copulatum ex quantitate successiva motus et ex actione anime, et tamen hec vox "tempus" vere dicitur de quantitate successiva motus, quoniam sola quantitas successiva motus est tempus, sicut solus lapis ⟨est⟩ lapis intellectus. Et qualitercumque accipiatur "tempus" sive pro eo quod denominat sive pro eo quod significat, semper est illa consequentia bona "tempus est, ergo anima est", et hec similiter "nulla anima est, ergo nullum tempus est", si hec consequentia est bona "lapis intellectus est, ergo anima

11. Francesc Marbres: *Utrum tempus sit aliqua entitas preter animam existens* (IV.5)[343]
12. Nicole Oresme: *Utrum tempus sit ab anima* (IV.17)
13. Albert of Saxony: *Utrum tempus sit ab anima* (IV.16)[344]
14. Marsilius of Inghen: *Utrum tempus sit ab anima* (IV.3.3)

---

est". Si tamen aliquod nomen imponatur precise ad significandum quantitatem successivam motus, ut si A signifcaret precise quantitatem successivam motus, hec consequentia non valeret "A est, igitur anima est", sicut illa consequentia non valeret "lapis est, ergo anima est", quamvis sequatur "lapis intellectus est, ergo anima est".' Cf. C. Trifogli, 'Motion and Time' in: A.D. Conti (ed.), *A Companion to Walter Burley: Late Medieval Logician and Metaphysician*, Leiden [etc.] 2013 (Brill's Companions to the Christian Tradition, 41), 267–299, esp. 296.

343 Francesc Marbres, *Quaestiones Physicorum*, IV.5, 43va–45ra: 'Et ideo sciendum quod est opinio Aurioli qui ponit quod tempus nihil aliud est quam successio motus sive omnium succedentium. Unde est mora vel successio motus et omnium mutabilium ... Et hoc videtur intentio Augustini II *Confessionum*, capitulo secundo: tempus non est aliud quam discretio successivorum ... Item Avicenna in secundo libro suorum *Physicorum* dicit quod tempus non est nisi continuitas motus, non continuitas permansiva, ergo successiva; hoc autem nihil est aliud quam successio; ergo etc. Aliter describit tempus Landulphus dicens quod tempus est quantitas continua successiva et mora motus distincta ab eo in quo infingitur prius et posterius ad mensurandum motum. Aliter dico: ubi sciendum quod in tempore est aliquid materiale et aliquid formale ... Respondeo. Ubi sciendum quod est una opinio Aurioli, que etiam videtur Philosophi and Commentatoris, quod omnium temporalium est unicum tempus. Et hoc probatur sic a dicto doctore quia: tempus constituitur per fluxum ipsius nunc; sed impossibile est esse nisi unum nunc; ergo impossibile est esse nisi unicum fluxum, et per consequens nisi unicum tempus ... Respondeo. Ubi sciendum quod est una opinio Aurioli qui ponit duas propositiones. Prima est quod tempus acceptum per modum quantitatis indeterminate est quantitas continua et non discreta. Secunda propositio est quod tempus acceptum per modum quantitatis determinate et mensurate est quantitas composita ex continua et discreta ... Ergo illa ratio Aurioli non probat nisi quod discretio insit sibi tanquam accidens, ut si dicerem lineam trium palmorum, dico ipsam lineam substratam ternalitati, que est accidens eius ... Quarta dubitatio est ista: utrum tempus sit passio vel sequela motus. Respondeo. Ubi sciendum quod est una opinio Gerardi Odonis ponens duas conclusiones. Prima est quod tempus habuit esse ante mundi initium et inceptionem cuiuslibet creature sic quod fuit ab eterno ... Secunda conclusio quam ponit est ista: quod tempus non est passio nec sequela motus ... Per te ab eterno fuit tempus; ergo ab eterno habuit aliquod subiectum; et istud non fuit Deus, cum Deus non sit subiectum accidentis; ergo ab eterno fuit aliqua creatura que fuit subiectum temporis; sed hoc est contra veritatem et istum doctorem et fidem; ergo etc. Nec videtur salva sui reverentia secundum veritatem esse dictum quod tempus nihil habet pro subiecto, ut dicit, sed est quoddam consequens ipsum Deum vel sequela.'

344 This question is missing in the 1516 Lokert edition.

15. Johannes Marsilii (?): *Utrum tempus sit motus celi* (IV.15)
16. Benedictus Hesse: *Utrum si nulla esset anima intellectiva, nullum esset tempus* (IV.40)

Buridan recognizes that Aristotle and Averroes argued that the existence of time in a complete sense requires a soul numbering motion. He believes that numerability is sufficient. His conclusions are the following:

1. If there were no intellect, there would be no time, nor could there be—actually nothing would exist, because God is intellect, and if God did not exist, nothing could exist. From the impossible anything follows, especially from the assumption that God does not exist (344^{13-16}).
2. God does not measure anything in the sense that we measure, because we measure one quantity that was previously unknown by another quantity that we know. Thus to measure is to know by measuring a measurabile quantity previously unknown. Nothing, however, is unknown or doubtful or imperfectly known to God. God knows evidently how great any quantity is and how great it would be, if there were no other quantity by which to measure it (344^{17-23}).

Nevertheless it seems to Buridan that the term 'number' does not connote actual numeration, but numerability (345^{5-6}). From this follows:

3. If every knower except God were annihilated, with only things without intellect remaining, things would still be numbers because, even though they would not be counted, nevertheless they would be numerable and could be counted, because God could immediately create humans and souls which could count those things, and this suffices for these things to be numbers (345^{14-19}).

Buridan raises the following question: if the word 'number' had been imposed to mean actual numeration—which is possible, since the meanings of terms are open to choice (*ad placitum*)—, what should be said if there were no knower except God? One could understand God's knowing of all things separately as actual counting. Alternately, if one does not allow God's counting, then even if there were two stones, they would not be counted if there were no (human) knowers (345^{23}–346^{3}). Buridan comments: 'These conclusions could be counted with the other conclusions, but it makes no difference, because I believe that "number" in fact connotes not actual numeration but numerability' (346^{10-12}). Hence he continues:

4. If there could not be a numberer, then there could not be a number, as Aristotle well argues. But, Buridan admits, this is true because the antecedent is impossible, and from the impossible anything follows (346^{13-18}).
5. If there could not be an intellect, there could not be a number (346^{19-20}). This also follows immediately from the rule that from the impossible, anything follows. It also is proved because no soul except the intellective soul is able to count. A persuasion for this is that a hen knows it has many chicks, but it does not realize it if you take away two or three from ten in the nest. Similar experiments (*experimentum*) can be made with sows and piglets (346^{21}–347^{4}).

In sum, Buridan believes that God and the world must exist for time to exist, but time would exist if there were no human intellects, because 'time' connotes not actual, but potential numeration. It may be, however, Buridan writes, that Averroes thought that the word 'time' connoted actual and not only potential numeration.

5 Conclusion (and Warning to the Reader)

This Guide to the Text of Books III and IV of Buridan's *Quaestiones Physicorum* has been finished before starting to prepare the Guide for Books V–VIII. It is possible that what appears in the Guide to the later books will raise doubts about what has been included in the Guides to Books I–II and III–IV. If this happens, it will be reported in the third volume of the edition of Buridan's *Quaestiones*. It is hoped that the information provided here is for the most part accurate, but it is incomplete. First of all, many medieval commentaries on the *Physics* have been neglected, and even those that are included—for instance the works of William of Ockham—are only selectively reported. There are many other medieval commentaries on the works of Aristotle (and not only on the *Physics*) or on Peter Lombard's *Sentences* that would easily have enabled a finer-tuned interpretation of Buridan's *Quaestiones Physicorum* surveyed here. For the majority of listings of question titles from works of other authors, I have not studied the actual content of the question, which may be more or less than I imagined. I apologize to the authors of the many secondary works about the material reported here whose work would have corrected or enriched what is reported here, if only I had had time to read and remember it in detail. Let me close by recommending again that anyone intending to use this Guide for their own future research of Buridan's views make use also of Jürgen Sarnowsky's 1989 *Die Aristotelisch-Scholastische Theorie der Bewegung*.

Studien zum Kommentar Alberts von Sachsen zur Physik des Aristoteles. At the very least the indexes to Sarnowsky's book can be used, for example, as an index to Albert of Saxony's *Quaestiones Physicorum*, which, indirectly can serve as an *introitus* to Buridan's *Quaestiones*. The indices to Ockham's philosophical and theological works in the St. Bonaventure editions are very useful for the same purpose.

Bibliography

This bibliography contains all publications cited in the Introduction and the Guide to the Text. For the sources referred to in the edition, see above XIX.

Primary Sources

[Albert of Saxony], *Expositio et quaestiones in Aristotelis Physicam ad Albertum de Saxonia attributae*, ed. B. Patar, 3 vols, Louvain-la-Neuve, Leuven 1999 (Philosophes médiévaux, 39–41).

Albert of Saxony (?), *Quaestiones super libros Physicorum*, manuscript London, Wellcome Institute for the History of Medicine, cod. L.15.

Albert of Saxony, *Quaestiones eximii Doctoris Magistri Alberti de Saxonia in octo libros Physicorum Aristotelis*, in: G. Lokert (ed.), *Quaestiones et decisiones physicales insignium virorum*, Paris 1516 (URL: http://www.mdz-nbn-resolving.de/urn/resolver.pl?urn=urn:nbn:de:bvb:12-bsb10195484-7).

Aristotle, *Physica, translatio vetus*, ed. F. Bossier & J. Brams, Leiden [etc.] 1990 (Aristoteles Latinus, VII/1).

Aristotle, *Physics*, tr. R.P. Hardie & R.K. Gaye, in: J. Barnes (ed.), *The Complete Works of Aristotle*, The Revised Oxford Translation, Princeton 1984 (Bollingen Series 71/2), 1: 315–446.

Averroes, *Expositio media super libros Physicorum Aristotelis*, in: *Aristotelis opera cum Averrois commentariis*, 4, Venezia 1562–1574 (repr. Frankfurt am Main 1962).

Averroes, *In Physicam*, in: *Aristotelis opera cum Averrois commentariis*, 4, Venezia 1562–1574 (repr. Frankfurt am Main 1962).

Benedictus Hesse, *Quaestiones super octo libros Physicorum Aristotelis*, ed. S. Wielgus, Wrocław [etc.] 1984.

Boethius of Dacia, *Quaestiones super libros Physicorum*, ed. G. Sajó, København 1974 (Corpus philosophorum Danicorum medii aevi, 5/2).

Collegium Conimbricense, *Commentarii in octo libros Physicorum Aristotelis*, Lyon 1594 (repr. Hildesheim 1984).

Domingo de Soto, *Super octo libros Physicorum Aristotelis subtilissimae quaestiones*, Venezia 1582 (URL: http://www.mdz-nbn-resolving.de/urn/resolver.pl?urn=urn:nbn:de:bvb:12-bsb10150499-5).

Francesc Marbres (*alias* John the Canon), *Quaestiones super octo libros Physicorum*, Venezia 1520.

Geoffrey of Aspall, *Quaestiones in Physicam*, manuscript Oxford, Merton College Library, cod. 272, 88ra–118vb.

Giles of Rome, *Commentaria in octo libros Phisicorum Aristotelis*, Venezia 1502 (repr. Frankfurt am Main 1968).

Hugolinus of Orvieto, *Commentarius in quattuor libros Sententiarum*, I, ed. W. Eckermann, Würzburg 1980 (Cassiciacum, Supplement, 8).

Hugolinus of Orvieto, *Quaestiones super quattuor libros Physicorum*, ed. W. Eckermann, *Der Physikkommentar Hugolins von Orvieto OESA. Ein Beitrag zur Erkenntnislehre des spätmittelalterlichen Augustinismus*, Berlin [etc.] 1972 (Spätmittelalter und Reformation, 5).

Johannes Marsilii (?), *Quaestiones subtilissimae Johannis Marcilii Inguen super octo libros Physicorum secundum nominalium viam*, Lyon 1518 (repr. Frankfurt am Main 1964).

John Buridan, *Quaestiones super octo libros Physicorum Aristotelis (secundum ultimam lecturam), Libri I–II*, ed. M. Streijger & P.J.J.M. Bakker, Leiden [etc.] 2015 (Medieval and early modern science, 25).

John Buridan, *Summulae de Dialectica*, ed. G. Klima, New Haven (CT) and London 2001 (Yale library of medieval philosophy).

John Mair, *Propositum de infinito*, in: H. Elie, *Le traité* De l'infini *de Jean Mair*, Paris 1938.

John of Celaya, *Expositio in octo libros Physicorum Aristotelis cum quaestionibus eiusdem secundum triplicem viam beati Thomae, realium et nominalium*, Paris 1517.

John of Jandun, *Quaestiones super 8 libros Physicorum Aristotelis*, Venezia 1551 (repr. Frankfurt am Main 1969).

John the Canon, see Francesc Marbres.

Liber decanorum Facultatis Philosophicae Universitatis Pragensis, ab anno Christi 1367 usque ad annum 1585, Pars I, Praha 1830 (Monumenta historica Universitatis Carolo-Ferdinandeae Pragensis, 1/1).

Marsilius of Inghen, *Abbreviationes super octo libros Physicorum Aristotelis*, Venezia 1521 (URL: http://daten.digitale-sammlungen.de/bsb00090905/image_1).

Nicole Oresme, *Questiones super Physicam (Books I–VII)*, ed. S. Caroti, J. Celeyrette, S. Kirschner & E. Mazet, Leiden [etc.] 2013 (Studien und Texte zur Geistesgeschichte des Mittelalters, 112).

Radulphus Brito, *Quaestiones in Physicam I–VIII*, manuscript Paris, Bibliothèque Nationale de France, cod. lat. 18160, 3ra–79vb.

Richard Kilvington, *Quaestiones quattuor super Physicam*, manuscript Venezia, Biblioteca Nazionale Marciana, cod. lat. VI.72, 81ra–112rb, 168ra–169va.

Roger Bacon, *Questiones supra libros octo Physicorum*, ed. F.M. Delorme & R. Steele, Oxford 1935 (Opera hactenus inedita, 13).

Roger Bacon, *Questiones supra libros quatuor Physicorum*, ed. F.M. Delorme & R. Steele, Oxford 1928 (Opera hactenus inedita, 8).

Roger Roseth, *Lectura super Sententias*, ed. O. Hallamaa, Helsinki 2005 (Helsingin Yliopiston systemaattisen teologian laitoksen julkaisuja, 18).

Thomas Aquinas, *Commentary on Aristotle's Physics*, tr. R.J. Blackwell, R.J. Spath & W.E. Thirkel, revised edition, Notre Dame (IN) 1999 (first ed. New Haven [CT] 1963).

Thomas Aquinas, *In octo libros Physicorum Aristotelis expositio*, ed. M. Maggiòlo, Torino, Roma 1965.

Thomas Wylton, *Quaestiones in Physicam*, manuscript Cesena, Biblioteca Malatestiana, cod. S.VIII.2, 2^{r}–141^{v}.

Walter Burley, *Expositio et quaestiones librorum Physicorum*, manuscript Cambridge, Gonville and Caius College Library, cod. 448 (409), 172–543.

Walter Burley, *Expositio in libros octo De phisico auditu*, Venezia 1501 (repr. Hildesheim 1972).

Walter Burley, *Quaestiones super libros Physicorum*, manuscript Basel, Universitätsbibliothek, cod. F.V.12, 108^{r}–171^{v}.

Walter Burley, *Tractatus secundus de intensione et remissione formarum*, Venezia 1496.

William of Clifford, *Compilationes super librum Physicorum Aristotelis*, manuscript Cambridge, Peterhouse Library, cod. 157, I, 43ra–104va.

William of Clifford, *Quaestiones in Physicam I–V, VII*, manuscript Siena, Biblioteca Comunale degli Intronati, cod. L.III.21, 39vb–46vb, 46vb–64ra.

William of Ockham, *Brevis summa libri Physicorum*, ed. S. Brown, St. Bonaventure (NY) 1984 (Opera philosophica, 6).

William of Ockham, *Expositio in libros Physicorum Aristotelis*, Libri IV–VIII, ed. R. Wood, R. Green, G. Gál, J. Giermek, F. Kelley, G. Leibold & G. Etzkorn, St. Bonaventure (NY) 1985 (Opera philosophica, 5).

William of Ockham, *Expositio in libros Physicorum Aristotelis*. Prologus et libri I–III, ed. V. Richter & G. Leibold, St. Bonaventure (NY) 1985 (Opera philosophica, 4).

William of Ockham, *Philosophia naturalis Guilielmi Occham*, Roma 1637 (repr. Vaduz-Liechtenstein 1963).

William of Ockham, *Quaestiones in libros Physicorum Aristotelis*, ed. S. Brown, St. Bonaventure (NY) 1984 (Opera philosophica, 6).

William of Ockham, *Quaestiones in librum secundum Sententiarum* (*Reportatio*), ed. G. Gál & R. Wood, St. Bonaventure (NY) 1981 (Opera theologica, 5).

William of Ockham, *Scriptum in librum primum Sententiarum—Ordinatio*, ed. G.I. Etzkorn, St. Bonaventure (NY) 1977 (Opera theologica, 3).

William of Ockham, *Summa logicae*, ed. Ph. Boehner, G. Gál & S. Brown, St. Bonaventure (NY) 1974 (Opera philosophica, 1).

William of Ockham, *Summula philosophiae naturalis*, ed. S. Brown, St. Bonaventure (NY) 1984 (Opera philosophica, 6).

William of Ockham, *Tractatus de successivis*, ed. Ph. Boehner, *The Tractatus de successivis attributed to William Ockham*, St. Bonaventure (NY) 1944 (Franciscan Institute Publications. Philosophy Series, 1).

Secondary Literature

Antony, H., (ed.), *Mittellateinisches Wörterbuch bis zum ausgehenden 13. Jahrhundert*, III, München 2007.

Bakker, P.J.J.M., 'Syncatégorèmes, concepts, équivocité. Deux questions anonymes, conservées dans le ms. Paris, B.N., lat. 16.401, liées à la sémantique de Pierre d'Ailly (c. 1350–1420),' *Vivarium*, 34 (1996), 76–131.

Bakker, P.J.J.M., & D.-J. Dekker, 'Antoine Andrée ou Jean le Chanoine? A propos de l'authenticité du commentaire de la *Physique* conservé dans le Ms. Cambridge, Gonville et Caius College, 368 (590),' *Bulletin de philosophie médiévale*, 42 (2000), 101–131.

Bakker, P.J.J.M., & S.W. de Boer, '*Locus est spatium*. On Gerald Odonis' *Quaestio de loco*,' *Vivarium*, 47 (2009), 295–330.

Biard, J., 'La logique de l'infini chez Jean Mair,' *Les études philosophiques*, 3 (1986), 329–348.

Boer, S.W. de, 'Gerard of Odo on the Atomistic Structure of Continua. A Discussion and Edition of a Tract Found in Ms. Madrid, Biblioteca Nacional 4229,' *Documenti e studi sulla tradizione filosofica medievale*, 23 (2012), 387–427.

Büttner, J., P. Damerow & J. Renn, 'Galileo's Unpublished Treatises. A Case Study on the Role of Shared Knowledge in the Emergence and Dissemination of an Early Modern "new science",' in: C.R. Palmerino & J.M.M.H. Thijssen (eds), *The Reception of the Galilean Science of Motion in Seventeenth-Century Europe*, Dordrecht 2004 (Boston studies in the philosophy of science, 239), 99–117.

Caroti, S., 'Some Remarks on Buridan's Discussion on Intension and Remission,' *Vivarium*, 42 (2004), 58–85.

Courtenay, W.J., 'Radulphus Brito, Master of Arts and Theology,' *Cahiers de l'Institut du Moyen-Age Grec et Latin*, 76 (2005), 131–158.

Cross, R., *The Physics of Duns Scotus. The Scientific Context of a Theological Vision*, Oxford 1998.

Dekker, D.-J., 'Buridan's Concept of Time. Time, Motion and the Soul in John Buridan's Questions on Aristotle's *Physics*,' in: J.M.M.H. Thijssen & J. Zupko (eds), *The Metaphysics and Natural Philosophy of John Buridan*, Leiden [etc.] 2001 (Medieval and early modern science, 2), 151–163.

Dekker, D.-J., *De tijdfilosofie van Johannes Buridanus († ca. 1360). Een historisch-wijsgerige studie met editie van Buridanus'* Quaestiones super octo libros *Physicorum* Aristotelis (secundum ultimam lecturam), *IV, 12–16*, Ph.D. dissertation, Radboud University, Nijmegen 2003.

Dewender, Th., *Das Problem des Unendlichen im ausgehenden 14. Jahrhundert. Eine Studie mit Textedition zum Physikkommentar des Lorenz von Lindores*, Amsterdam 2002 (Bochumer Studien zur Philosophie, 36).

Donati, S., 'Commenti parigini alla Fisica degli anni 1270–1300 ca.,' in: A. Speer (ed.), *Die Bibliotheca Amploniana. Ihre Bedeutung im Spannungsfeld von Aristotelismus, Nominalismus und Humanismus*, Berlin [etc.] 1995 (Miscellanea mediaevalia, 23), 136–256.

Duhem, P., *Le système du monde. Histoire des doctrines cosmologiques de Platon à Copernic*, 10 vols, Paris 1913–1959.

Duhem, P., *Medieval Cosmology. Theories of Infinity, Place, Time, Void, and the Plurality of Worlds*, ed. and tr. R. Ariew, Chicago [etc.] 1985.

Ebbesen, S., & J. Pinborg, 'Bartholomew of Bruges and his Sophisma on the Nature of Logic,' *Cahiers de l'Institut du Moyen-Age Grec et Latin*, 39 (1981), iii–xxvi, 1–76.

Fuchs, J.W., O. Weijers & M. Gumbert-Hepp, *Lexicon Latinitatis Nederlandicae medii aevi. Woordenboek van het middeleeuws Latijn van de noordelijke Nederlanden*, III, Leiden 1986.

Geréby, G., '*Ignorato motu, ignoratur natura*. Logic and Physics in *Sophisma* II.110 of Albert of Saxony,' in: J. Biard (ed.), *Itinéraires d'Albert de Saxe. Paris-Vienne au XIVe siècle*. Actes du Colloque organisé le 19–22 juin 1990 dans le cadre des activités de l'URA 1085 du CNRS à l'occasion du 600e anniversaire de la mort d'Albert de Saxe, Paris 1991 (Etudes de philosophie médiévale, 69), 175–189.

Grant, E., 'Medieval Explanations and Interpretations of the Dictum that "Nature Abhors a Vacuum",' *Traditio*, 29 (1973), 327–355.

Grant, E., 'The Concept of *ubi* in Medieval and Renaissance Discussions of Place,' *Manuscripta*, 20 (1976), 71–80.

Grant, E., 'The Medieval Doctrine of Place: some Fundamental Problems and Solutions,' in: A. Maierù & A. Paravicini Bagliani (eds), *Studi sul XIV secolo in memoria di Anneliese Maier*, Roma 1981 (Storia e letteratura, 151), 57–79.

Hugonnard-Roche, H., 'Le possible et l'imaginaire dans la physique d'Albert de Saxe,' in: J. Biard (ed.), *Itinéraires d'Albert de Saxe. Paris-Vienne au XIVe siècle*. Actes du Colloque organisé le 19–22 juin 1990 dans le cadre des activités de l'URA 1085 du CNRS à l'occasion du 600e anniversaire de la mort d'Albert de Saxe, Paris 1991 (Etudes de philosophie médiévale, 69), 161–173.

Jung, E., 'Motion in a Vacuum and in a Plenum in Richard Kilvington's Question: *Utrum aliquod corpus simplex posset moveri aeque velociter in vacuo et in pleno* from the Commentary on the *Physics*,' in: J.A. Aertsen & A. Speer (eds), *Raum und Raumvorstellungen im Mittelalter*, Berlin [etc.] 1998 (Miscellanea mediaevalia, 25), 179–193.

Jung, E., 'Works by Richard Kilvington,' *Archives d'histoire doctrinale et littéraire du Moyen* Age, 67 (2000), 181–223.

Jung, E., & R. Podkoński, 'Richard Kilvington on Continuity,' in: C. Grellard & A. Robert (eds), *Atomism in Late Medieval Philosophy and Theology*, Leiden [etc.] 2009 (Medieval and early modern science, 9), 65–84.

Jung, E., & R. Podkoński, 'The Transmission of English Ideas in the Fourteenth Century. The Case of Richard Kilvington,' *Mediaevalia Philosophica Polonorum*, 37/3 (2008), 59–69.

Kaluza, Z., *Nicolas d'Autrécourt. Ami de la vérité*, Paris 1995 (Histoire littéraire de la France, 42/1).

Kaye, J., *A History of Balance, 1250–1375. The Emergence of a New Model of Equilibrium and its Impact on Thought*, Cambridge 2014.

Kirschner, S., 'Oresme on Intension and Remission of Qualities in His Commentary on Aristotle's *Physics*,' *Vivarium*, 38 (2000), 255–274.

Kirschner, S., 'Oresme's Concepts of Place, Space, and Time in his Commentary on Aristotle's *Physics*,' *Oriens–Occidens. Cahiers du Centre d'histoire des sciences et de philosophies arabes et médiévales*, 3 (2000), 145–179.

Klima, G., 'Nominalist Semantics,' in: R. Pasnau & C. van Dyke (eds), *The Cambridge History of Medieval Philosophy*, 1, Cambridge 2010, 159–172.

Latham, R.E., *Dictionary of Medieval Latin from British Sources*, I, Oxford 1997.

Maier, A., 'Die naturphilosophische Bedeutung der scholastischen Impetustheorie,' in: Ead., *Ausgehendes Mittelalter. Gesammelte Aufsätze zur Geistesgeschichte des 14. Jahrhunderts*, 1, Roma 1964, 353–379.

Maier, A., 'Handschriftliches zu Wilhelm Ockham und Walter Burley—Addenda,' in: Ead., *Ausgehendes Mittelalter. Gesammelte Aufsätze zur Geistesgeschichte des 14. Jahrhunderts*, 1, Roma 1964, 469–479.

Maier, A., 'Zu Walter Burleys Traktat *De intensione et remissione formarum*,' in: Ead., *Zwei Grundprobleme der scholastischen Naturphilosophie* (Studien zur Naturphilosophie der Spätscholastik, 2), Roma 1968 (Storia e letteratura, 37), Nachträge, 315–352.

Maier, A., *Die Vorläufer Galileis im 14. Jahrhundert* (Studien zur Naturphilosophie der Spätscholastik, 1), Roma 1949 (Storia e letteratura, 22).

Maier, A., *Metaphysische Hintergründe der Spätscholastischen Naturphilosophie* (Studien zur Naturphilosophie der Spätscholastik, 4), Roma 1955 (Storia e letteratura, 52).

Maier, A., *On the Threshold of Exact Science. Selected Writings of Anneliese Maier on Late Medieval Natural Philosophy*, ed. and tr. S.D. Sargent, Philadelphia 1982.

Maier, A., *Zwei Grundprobleme der scholastischen Naturphilosophie* (Studien zur Naturphilosophie der Spätscholastik, 2), Roma 1968 (Storia e letteratura, 37).

Maier, A., *Zwischen Philosophie und Mechanik* (Studien zur Naturphilosophie der Spätscholastik, 5), Roma 1958 (Storia e letteratura, 69).

Murdoch, J.E., 'Infinity and Continuity,' in: N. Kretzmann, A. Kenny & J. Pinborg (eds), *The Cambridge History of Later Medieval Philosophy. From the Rediscovery of Aristotle to the Disintegration of Scholasticism, 1100–1600*, Cambridge 1982, 564–591.

Murdoch, J.E., 'William of Ockham and the Logic of Infinity and Continuity,' in: N. Kretzmann (ed.), *Infinity and Continuity in Ancient and Medieval Thought*, Ithaca [etc.] 1982, 165–206.

Murdoch, J.E., & J.M.M.H. Thijssen, 'John Buridan on Infinity,' in: J.M.M.H. Thijssen & J. Zupko (eds), *The Metaphysics and Natural Philosophy of John Buridan*, Leiden [etc.] 2001 (Medieval and early modern science, 2), 127–149.

Patar, B., *La Physique de Bruges de Buridan et le Traité du Ciel d'Albert de Saxe*, 2 vols, Longueuil 2001.

Porro, P., (ed.), *The Medieval Concept of Time. Studies on the Scholastic Debate and its Reception in Early Modern Philosophy*, Leiden [etc.] 2001 (Studien und Texte zur Geistesgeschichte des Mittelalters, 75).

Sarnowsky, J., *Die Aristotelisch-Scholastische Theorie der Bewegung. Studien zum Kommentar Alberts von Sachsen zur Physik des Aristoteles*, Münster 1989 (Beiträge zur Geschichte der Philosophie und Theologie des Mittelalters, n. F., 32).

Schabel, C., 'Francesc Marbres, a.k.a. *Iohannes Canonicus*,' *Bulletin de philosophie médiévale*, 56 (2014), 195–200.

Sylla, E.D., '*Disputationes Collativae*: Walter Burley's *Tractatus Primus* and Gregory of Rimini's *Lectura super primum et secundum Sententiarum*,' *Documenti e studi sulla tradizione filosofica medievale*, 22 (2011), 383–464.

Sylla, E.D., 'Guide to the Text,' in: John Buridan, *Quaestiones super octo libros Physicorum Aristotelis (secundum ultimam lecturam). Libri I–II*, ed. M. Streijger & P.J.J.M. Bakker, Leiden 2015 (Medieval and early modern science, 25), XLIII–CLXXV.

Sylla, E.D., '*Ideo quasi mendicare oportet intellectum humanum*: The Role of Theology in John Buridan's Natural Philosophy,' in: J.M.M.H. Thijssen & J. Zupko (eds), *The Metaphysics and Natural Philosophy of John Buridan*, Leiden [etc.] 2001 (Medieval and early modern science, 2), 221–245.

Sylla, E.D., 'Transmission of the New Physics of the Fourteenth Century from England to the Continent,' in: S. Caroti & P. Souffrin (eds), *La nouvelle physique du XIVe siècle*, Firenze 1997 (Biblioteca di Nuncius. Studi e testi, 24), 65–110.

Sylla, E.D., 'Walter Burley's Practice as a Commentator on Aristotle's *Physics*,' *Medioevo*, 27 (2002), 301–372.

Sylla, E.D., *The Oxford Calculators and the Mathematics of Motion. Physics and Measurement by Latitudes*, New York [etc.] 1991.

Thijssen, J.M.M.H., 'Introduction,' in: John Buridan, *Quaestiones super octo libros Physicorum Aristotelis (secundum ultimam lecturam). Libri I–II*, ed. M. Streijger & P.J.J.M. Bakker, Leiden 2015 (Medieval and early modern science, 25), XIII–XLII.

Thijssen, J.M.M.H., *Johannes Buridanus over het oneindige. Een onderzoek naar zijn theorie over het oneindige in het kader van zijn wetenschaps- en natuurfilosofie*, Ph.D. dissertation, Radboud University, 2 vols, Nijmegen 1988.

Thijssen, J.M.M.H., *John Buridan's Tractatus de infinito. Quaestiones super libros Physicorum secundum ultimam lecturam, Liber III, quaestiones 14–19*, Nijmegen 1991 (Artistarium. Supplementa, 6).

Thijssen, J.M.M.H., 'The Buridan School Reassessed. John Buridan and Albert of Saxony,' *Vivarium*, 42 (2004), 18–42.

Trifogli, C., 'Due questioni sul movimento nel commento alla *Physica* di Thomas Wylton,' *Medioevo*, 21 (1995), 31–73.

Trifogli, C., 'Il problema dello statuto ontologico del tempo nelle *Quaestiones super Physicam* di Thomas Wylton e di Giovanni di Jandun,' *Documenti e studi sulla tradizione filosofica medievale*, 1 (1990), 491–548.

Trifogli, C., 'Motion and Time' in: A.D. Conti (ed.), *A Companion to Walter Burley: Late Medieval Logician and Metaphysician*, Leiden [etc.] 2013 (Brill's Companions to the Christian Tradition, 41), 267–299.

Trifogli, C., 'The Reception of Averroes' View on Motion in the Latin West. The Case of Walter Burley,' in: P.J.J.M. Bakker (ed.), *Averroes' Natural Philosophy and its Reception in the Latin West*, Leuven 2015 (Ancient and Medieval Philosophy, Series I, 50), 127–139.

Trifogli, C., 'Thomas Wylton on Motion,' *Archiv für Geschichte der Philosophie*, 77 (1995), 135–154.

Trifogli, C., 'Thomas Wylton on the Immobility of Place,' *Recherches de Théologie et Philosophie médiévales*, 65 (1998), 1–39.

Trifogli, C., *Liber Quartus* Physicorum *Aristotelis: Repertorio delle Questioni. Commenti inglesi ca. 1250–1270*, Firenze 2007 (Corpus philosophorum medii aevi. Subsidia, 16).

Trifogli, C., *Liber Tertius* Physicorum *Aristotelis: Repertorio delle Questioni. Commenti inglesi ca. 1250–1270*, Firenze 2004 (Corpus philosophorum medii aevi. Subsidia, 13).

Trifogli, C., *Oxford Physics in the Thirteenth Century (ca. 1250–1270): Motion, Infinity, Place and Time*, Leiden [etc.] 2000 (Studien und Texte zur Geistesgeschichte des Mittelalters, 72).

Wood, R., 'Walter Burley's *Physics* Commentaries,' *Franciscan Studies*, 44 (1984), 275–327.

Zimmermann, A., *Verzeichnis ungedruckter Kommentare zur Metaphysik und Physik des Aristoteles aus der Zeit von etwa 1250–1350*, Leiden [etc.] 1971 (Studien und Texte zur Geistesgeschichte des Mittelalters, 9).

Iohannis Buridani

Quaestiones super octo libros Physicorum Aristotelis (secundum ultimam lecturam)

Libri III–IV

Conspectus siglorum et compendiorum

C	København, Kongelige Bibliotek, cod. Ny kgl. Samling 1801 fol.
G	Kraków, Biblioteka Jagiellońska, cod. 1771
I	Liège, Bibliothèque de l'Université, cod. 114 C
P	Vaticano (Città del), Biblioteca Apostolica Vaticana, cod. Vat. lat. 2163
p	Iohannes Buridanus, *Subtilissimae Quaestiones super octo Physicorum libros Aristotelis*. Paris: Petrus le Dru impensis Dionysii Roce, 1509.

Codices qui sequuntur rarius adhibiti sunt

A	Carpentras, Bibliothèque Inguimbertine, cod. 293 (L. 289)
B	Frankfurt am Main, Stadt- und Universitätsbibliothek, cod. Praed. 52
Br	Bratislava, Archív mesta Bratislavy, cod. E.L.5
D	Kraków, Biblioteka Jagiellońska, cod. 659
E	Kraków, Biblioteka Jagiellońska, cod. 660
Er	Erfurt, Universitätsbibliothek, cod. CA F. 300
F	Kraków, Biblioteka Jagiellońska, cod. 661
H	Kremsmünster, Bibliothek des Benediktinerstiftes, cod. CC 169
J	Buenos Aires, Biblioteca Nacional, cod. 342R
K	Stralsund, Stadtarchiv der Hansestadt Stralsund, cod. 1050
L	Paris, Bibliothèque Nationale de France, cod. lat. 14723
La	Lambach, Bibliothek des Benediktinerstiftes, cod. Ccl. 175
M	Salzburg, Stiftsbibliothek St. Peter (Erzabtei), cod. b.IX.24
N	Salzburg, Universitätsbibliothek, cod. M.II.311
O	Torino, Biblioteca Nazionale Universitaria, cod. G.IV.10
Ox	Oxford, Balliol College Library, cod. 97
Pb	Paderborn, Erzbischöfliche Akademische Bibliothek, cod. VVa 12
Q	Vaticano (Città del), Biblioteca Apostolica Vaticana, cod. Vat. lat. 2164
R	Wien, Bibliothek des Dominikanerkonvents, cod. 107/73
S	Wien, Österreichische Nationalbibliothek, cod. 5112
T	Wien, Österreichische Nationalbibliothek, cod. 5332
U	Wien, Österreichische Nationalbibliothek, cod. 5338
V	Wien, Österreichische Nationalbibliothek, cod. 5367
W	Wien, Österreichische Nationalbibliothek, cod. 5424
X	Wien, Österreichische Nationalbibliothek, cod. 5458
Y	Wien, Österreichische Nationalbibliothek, cod. 5481
Z	Zaragoza, Biblioteca Capitular de la Seo, cod. 15–61

 | DOI: 10.1163/9789004322356_002

add.	*addidit*
al. m.	*alia manu*
codd.	*codices*
corr.	*correxit*
del.	*delevit*
hom.	*homoeoteleuton*
inf.	*infra/inferiore*
inv.	*invertit*
lin.	*lineam*
marg.	*margine*
om.	*omisit*
praem.	*praemisit*
ras.	*rasura*
rep.	*repetivit*
seq.	*sequitur*
spat.	*spatium*
sup.	*supra/superiore*
transp.	*transposuit*
†...†	*verba resecta cum margine exteriore*

AA	*Auctoritates Aristotelis*

⟨Tabula quaestionum tertii libri *Physicorum*⟩

60va C 41ra p Tabula quaestionum tertii libri *Physicorum*.

⟨1⟩ Prima quaestio est ista: utrum necesse est ignorato motu ignorare naturam. De distinctione modalium inter sensum compositum et sensum divisum. Quomodo differt 'ignoro motum' et 'motum ignoro' tam ex parte negationis implicitae in nomine privativo quam de appellatione termini sequentis hoc verbum 'cognosco'.

⟨2⟩ Secunda quaestio est utrum ad alterationem requiritur fluxus distinctus ab alterabili et a qualitate secundum quam est alteratio.

⟨3⟩ Tertia quaestio est utrum qualitates contrariae possunt se compati simul in eodem subiecto secundum aliquos gradus ipsarum. De triplici contrarietate, scilicet terminorum, propositionum et rerum aliarum. Quae caliditas cui frigiditati sit contraria. Quomodo contraria maxime distant.

51va P ⟨4⟩ Quarta quaestio utrum qualitas secundum quam est alteratio | per se
50va G et proprie dicta, continua et temporalis, | acquiritur tota simul vel pars post partem. Quomodo qualitas est divisibilis quantitative et gradualiter. Quid debet intelligi per 'gradum qualitatis intensibilis'.

⟨5⟩ Quinta quaestio est utrum in alteratione pars qualitatis quae prius acquiritur manet cum parte quae posterius acquiritur. Utrum caliditas dicatur intensior. Quid intenditur.

⟨6⟩ Sexta quaestio est utrum motus localis est vel utrum haec est vera 'motus localis est'. Quod per 'praesens' non debet intelligi instans indivisibile, sed tempus divisibile. Quod oportet exponere 'mutari' per aliter et aliter se habere prius et posterius, scilicet utrobique de praesenti. Quod haec est vera 'Socrates est sedens et Socrates est non sedens'. Quod ad affirmativam de praedicato infinito non sequitur negativa de finito. Quomodo praesens simpliciter dicitur respective praeteritum vel futurum. Quanto tempore

---

2 physicorum] *om.* p 3 ista] *om.* GPp 3–4 necesse ... naturam] ignoto motu necesse est naturam ignorare C 4 distinctione] determinatione C ‖ modalium] motabilium P 4–5 sensum[1] ... divisum] sensum compositum et divisum Pp : subiectum divisum et subiectum compositum C 5 differt] differunt G 6 appellatione] ampliatione p 8 quaestio est] *om.* GPp 9 quam] qua G 10 quaestio est] *om.* GPp 11 ipsarum] ipsorum C 11–12 contrarietate] qualitate C 14 quaestio] *om.* GPp 15 et proprie] est propria C ‖ pars] *add.* vel C 16 quantitative] qualitative G 18 quaestio est] *om.* GPp 19 quae] *add.* in alteratione P ‖ acquiritur[2]] *add.* et P 21 quaestio est] *om.* GPp 22 praesens] *praem.* tempus P : instans C 23 quod] quia P ‖ exponere] *add.* per P 24 praesenti] *add.* et G 26 de[2]] *add.* praedicato P 26–27 praesens simpliciter] *inv.* GPp 27 respective] *add.* ad p

liceat uti pro praesenti. De istis terminis ‘cras’, ‘hodie’, ‘heri’. Quod est bona consequentia ‘in omni praesenti tempore Socrates movetur, igitur in omni tempore Socrates movetur’.

⟨7⟩ Septima quaestio est utrum motus localis est res distincta a loco et ab eo quod localiter movetur. Quod ultima sphaera potest moveri motu quo movetur sine loco. Quod motus ultimae sphaerae est res pure successiva.

⟨8⟩ Octava quaestio est ista: utrum de | necessitate motus localis sit 60vb C habere terminos positivos praeter fluxum. De duplici divisione et magnitudine motus. Quod omnis continui finiti termini sunt prima et ultima partes eius. Quod eiusdem lineae infiniti sunt termini et in infinitum termini est terminus. Quod saepe in motu recto intenduntur termini extrinseci.

⟨9⟩ Nona quaestio est utrum motus sit de essentia termini ad quem est motus. Hoc declaratur de motu locali, de alteratione, de augmentatione, diminutione, generatione, corruptione et instantanea mutatione.

⟨10⟩ Decima quaestio est utrum omnis motus est actus entis in potentia. Quod est actus moventis et eius quod movetur et entis in actu et entis in potentia et imperfectus actus.

⟨11⟩ Undecima quaestio est utrum definitio motus sit bona in qua dicitur etc. Non convenit mutationi instantaneae.

⟨12⟩ Duodecima quaestio est utrum omnis motus est subiective in mobili vel in movente vel in utroque.

⟨13⟩ Tredecima quaestio est utrum omnis actio est passio et omnis passio est actio. De quo praedicamento est ille terminus ‘motus’. An simplex corruptio est actio.

⟨14⟩ Quarta decima quaestio est utrum est aliquod corpus sensibile actu infinitum. Quod saepe maius corpus ceteris paribus non agit intensius vel fortius in corpus sibi impositum quam minus.

---

1 liceat] licet P ‖ cras hodie] *inv.* GPp 2 praesenti tempore] *inv.* P 4 quaestio est] *om.* GPp ‖ a loco et] *om.* P 5 potest] posset GPp 6 ultimae] ultimo C ‖ res] *add.* distincta C 7 quaestio est ista] *om.* GPp 9 omnis] omnes p ‖ termini] *post* sunt G : *om.* P ‖ partes] pars G 10 quod] et P 11 quod] quomodo GPp ‖ recto] *om.* C 12 quaestio est] *om.* GPp 12–13 est motus] est P : *om.* Gp 13 de^{3}] et P ‖ augmentatione] *add.* et GP : *add.* de p 14 diminutione] *add.* de GPp ‖ generatione] *add.* et GP : *add.* de p ‖ et] de GP 15 quaestio est] *om.* GPp ‖ potentia] *add.* et C 16 quod^{1}] quid P ‖ et^{2}] *add.* quod C 17 et] est C ‖ imperfectus actus] *inv.* p : actus imperfectus et secundus G : actus perfectus et secundus P 18 quaestio est] *om.* GPp ‖ in] *om.* C 19 etc.] quod p : motus est qui P 20 quaestio est] *om.* GPp ‖ est^{2}] *om.* C 21 utroque] utrobique G 22 quaestio est] *om.* GPp 22–23 et … actio] et e converso G : vel e converso P 23 est ille terminus] sit P ‖ simplex] simpliciter C 24 est] sit P 25 quaestio est] *om.* GPp 26 saepe] sic G 27 impositum] appositum p

⟨15⟩ Quinta decima quaestio est utrum est aliqua magnitudo infinita. Quod Deus non potest creare magnitudinem actu infinitam, licet omni creata vel creabili posset creare maiorem. Quod extra hunc mundum non (41rb p) est spatium. Quod annihilato omni eo quod est infra orbem lunae, non | esset inter eius latera vacuum neque distantia. Quod in ita parvo corpore vel loco, sicut est granum milii, posset creari maius corpus quam sit mundus et (51vb P) moveri velociter motu recto etc. Quod possibile est corpora neque | tangere se neque distare.

⟨16⟩ Sexta decima quaestio est utrum sit aliqua linea gyrativa infinita. Quod non est ultima medietas proportionalis; et vide ibi de hoc multas (50vb G) conclusiones. De isto verbo 'incipit'. Quod non est | neque fuit infinitum tempus, etiam concessa opinione Aristotelis de aeternitate mundi, licet syncategorematice perpetuum fuerit tempus.

(61ra C) ⟨17⟩ Septima decima quaestio | est utrum omni numero est numerus maior. Quod in continuis nulla unitas est indivisibilis. Omnis binarius est centenarius. An numerus par est numerus impar. Non sunt plures partes vel pauciores in linea *b* quam in eius medietate. Quod nullus numerus est alio maior secundum multitudinem. Quomodo igitur salvantur principia mathematica.

⟨18⟩ Octava decima quaestio est ista: utrum in quolibet continuo infinitae sunt partes. Quomodo exponitur 'infinitum' tam categorematice quam syncategorematice sumptum. Quod implicat contradictionem esse magnitudinem infinitam vel multitudinem infinitam capiendo 'infinitum' categorematice. Quod valde differenter est dicendum secundum diversas expositiones 'infiniti' syncategorematice sumpti.

⟨19⟩ Nona decima quaestio est utrum possibile est infinitam esse magnitudinem et in infinitas partes lineam esse divisam. Quod Deus in qualibet medietate proportionali huius diei potest creare unum lapidem pedalem,

---

1 quaestio est] est P : *om.* Gp 2 potest] posset GPp 3 creata] creato G : creatura p ‖ posset] possit P 4 infra] intra p 5 neque] nec Pp : vel G ‖ ita] uno P ‖ corpore] *corr. ex* spatio C : tempore G 6 posset creari] possit creari G : possit deus creare P 7 moveri] movere C ‖ etc.] etiam G : *om.* P ‖ neque] nec GPp 8 se] *om.* p ‖ neque] nec GPp 9 quaestio est] *om.* GPp 10 ultima] ulla P ‖ proportionalis] proportionabilis p 11 neque] nec GPp 11–12 infinitum tempus] *inv.* GPp 12 etiam] etc. C : etc. etiam p 13 perpetuum] propositum C ‖ fuerit] fuit p 14 quaestio est] *om.* GPp 15 est[2]] *add.* numerus P 16 est] sit P 17 quod] *om.* GPp 18 alio maior] *inv.* P 19 mathematica] mathematicalia P 20 quaestio est ista] *om.* GPp ‖ utrum] *om.* C 20–21 infinitae sunt partes] sunt partes infinitae P 24 expositiones] opiniones G 26 quaestio est] *om.* GPp 27 in[1]] *om.* C 28 medietate] parte G ‖ huius] unius G

sed non est possibile ipsum creare in qualibet unum lapidem pedalem. Quod aliqua universalis est impossibilis, cuius omnes singulares sunt possibiles et compossibiles.

Et sic est finis tabulae.

---

1 ipsum ... qualibet] ipsum in qualibet parte creare GP : illum in qualibet creare p 2 impossibilis] possibilis G 3 compossibiles] *add.* etc. Gp 4 sic ... tabulae] sic finitur tabula quaestionum tertii libri physicorum incipiunt consequenter quaestiones eiusdem G : sic est finis tertii libri physicorum incipiunt quaestiones tertii libri physicorum p : haec est tabula tertii libri physicorum etc. P

⟨III.1⟩

⟨Utrum necesse sit ignorato motu ignorare naturam⟩

Quaeritur primo circa tertium librum *Physicorum* utrum necesse est ignorato motu ignorare naturam.

Arguitur quod non quia:

⟨1⟩ Quod est per se notum non indiget quod per aliud cognoscatur; sed Aristoteles dicit quod ridiculum est demonstrare naturam esse eo quod hoc est per se notum; igitur etc.

⟨2⟩ Item secundo sine illo potest cognosci natura, sine quo potest esse; sed ipsa potest esse sine motu, ut natura terrae.

⟨3⟩ Item tertio, sicut natura est principium motus, ita est principium quietis; ideo sicut potest cognosci per motum, ita potest cognosci per quietem sine motu.

⟨4⟩ Item quarto substantia est prior accidente notitia, tempore et definitione, ut dicitur septimo *Metaphysicae*; igitur potest cognosci sine cognitione accidentium. Sed natura est substantia et motus est accidens. Igitur etc.

⟨5⟩ Item quinto, sicut res se habet ad esse, ita ad veritatem sive ad cognitionem. Hoc patet secundo *Metaphysicae*. Cum igitur accidens non sit causa substantiae in essendo, ipsum etiam non erit causa eius in cognoscendo.

⟨6⟩ Item sexto effectus innatus est cognosci per causam; ideo causa est
61rb C notior; et notius | non dependet in cognoscendo ex minus | notis; igitur cum
51ra G natura sit causa motus, ipsa non dependet ex motu in cognoscendo.

---

4 quaeritur … physicorum] quaeritur circa tertium librum primo P : circa tertium librum quaeritur primo p 6 arguitur] *praem.* et G : et arguitur primo P 7 quod[1]] *praem.* illud P ‖ indiget] *add.* illo P ‖ cognoscatur] cognoscitur P 9 igitur etc.] *om.* P 10 secundo] *om.* GPp 11 sed] et P ‖ ut natura terrae] *add.* igitur etc G : igitur P 12 tertio] *om.* GPp 13 sicut] *add.* natura P 14 motu] *add.* igitur P 15 quarto] *om.* GPp 15–16 tempore et definitione] definitione et tempore Pp : definitione etc. G 18 etc.] *om.* P 19 quinto] *om.* GPp ‖ habet] habent Gp 20 cum igitur] *inv.* GPp ‖ causa] *add.* subiecti sive G 21 eius] *sup. lin.* C : *om.* P 22 sexto] *om.* GPp ‖ innatus est] *inv.* GPp 23 notis] noto GPp 24 ex] a G

8–9 Cf. Aristoteles, *Physica*, II, 1, 193a3–4 16 Cf. Aristoteles, *Metaphysica*, VII, 1, 1028a32–33 20 Cf. Aristoteles, *Metaphysica*, II, 1, 993b30–31

 | DOI: 10.1163/9789004322356_004

⟨7⟩ Item septimo quod est simpliciter contingens, numquam fiet necessarium propter positionem alicuius possibilis in esse; sed est simpliciter contingens quod ignoretur natura, et possibile est quod ignoratur motus; igitur quantumcumque motus ignoretur, non est necesse | naturam ignorari etc. 52ra P

Oppositum dicit Aristoteles in principio huius tertii libri.

In ista quaestione contingunt multae difficultates, et logicae et naturales. Sed statim ab | ista quaestione ego excipio cognitionem qua Deus et intelli- 41va p
gentiae cognoscunt alia entia, et solum in quaestione loquar de cognitione nostra humana et naturali modo habita.

Tunc igitur distinguo quod multum refert dicere 'necesse est ignorato motu ignorare naturam' et dicere 'ignorato motu necesse est ignorare naturam', quoniam prima solet vocari composita, et est sensus eius proprius quod haec est necessaria 'ignorato motu ignoratur natura'; secunda autem propositio solet vocari divisa.

Tunc ponam istam conclusionem primam, scilicet quod ista propositio quae vocatur divisa est simpliciter falsa de virtute seu proprietate sermonis, sicut arguebat ultima ratio, scilicet quod propter quodcumque possibile, si ponatur in esse, non esset necessarium illud quod secundum se est contingens.

Deinde notandum est quod hoc verbum 'ignoro' est privative oppositum huic verbo 'cognosco'. Et ita in hoc verbo implicatur negatio quae potest distribuere terminum sequentem, sed non praecedentem. Et ideo refert dicere 'motum ignoro' et 'ignoro motum', sicut refert dicere 'motum non cognosco' et 'non cognosco motum', nam in prima haec dictio 'motum' non

---

1 septimo] *om.* GPp ‖ fiet] fieret GPp **2** propter positionem] per potentiam P **3** ignoratur] ignoretur p **4** naturam ignorari] ignorare naturam P ‖ etc.] *om.* Gp **5** libri] *om.* G **6** et[1]] *om.* GP **7** ab ista] a G ‖ ego] *sup. lin.* C : *om.* GPp **8** et] sed P ‖ in quaestione loquar] loquor GPp **10** igitur] ego P **12** quoniam] quia p ‖ et] vel P **15** tunc … primam] et tunc pono conclusionem primam P : et tunc statim pono primam conclusionem Gp **16** seu proprietate sermonis] sermonis sive proprietate P **17** arguebat] *add.* illa GPp **18** esset] est P ‖ illud] id p ‖ contingens] *add.* etc. p **19** deinde notandum est] deinde notandum p : deinde nota G : notandum P ‖ hoc verbum] *om.* P **20** et ita] ita quod P ‖ verbo[2]] *add.* ignoro Pp **21** praecedentem] *add.* ipsum P : praecedente p ‖ et] *om.* P **21–22** refert … sicut] refert dicere m†…† ignoro et igno†…†tum sicut *in marg.* C **23** haec dictio] iste terminus P

---

5 Cf. Aristoteles, *Physica*, III, 1, 200b14–15

distribuitur et in secunda distribuitur. Propter quod non sequitur 'motum ignoro, igitur ignoro motum', quia si aliquis videat Socratem currere et non habeat notitiam de fluxu et refluxu maris, ipse motum ignorat et tamen non ignorat motum. Unde illae duae stant simul 'motum ignoro' et 'motum non ignoro', et istae duae contradicunt 'ignoro motum' et 'non ignoro motum'. Et simili modo est de 'ignorare naturam' et de 'naturam ignorare'.

Pono igitur secundam conclusionem, scilicet quod possibile est me simul motum ignorare et naturam non ignorare, ut si materiam vel aliquam formam cognosco et aliquem motum, ut fluxum maris, non cognosco. Immo adhuc probabiliter et vere, ut puto, ista conclusio potest sic augmentari
61va C quod possibile est me simul omnem naturam cognoscere et nullum | motum cognoscere. Et ponitur talis casus quod ego habeo conceptum communem omnis substantiae et non habeo conceptum alicuius accidentis. Iste casus videtur esse possibilis, quia licet forte a principio non possumus cognoscere substantiam sine cognitione accidentis aut praevia aut simul existente, tamen intellectus postea potest abstrahere conceptum substantiae a conceptu accidentis et perseverare in conceptu substantiae dimittendo conceptum accidentis, quia si per ista sensibilia venimus in cognitionem Dei et quod sine praecognitione sensibilium non possumus Deum
51rb G intelligere, | tamen postea possibile est multa syllogizare et speculari de Deo non considerando actualiter de istis sensibilibus. Et sic etiam aliquando possumus considerare de quidditatibus substantiarum nihil considerando de earum accidentibus. Si igitur ille casus concedatur esse possibilis et
52rb P nos etiam capiamus 'cognoscere' pro actuali cognitione, | non pro habituali, et etiam capiamus 'ignorare' pro privatione actualis apprehensionis, non pro privatione habitualis, tunc omnem naturam cognoscimus et nul-

1 et] sed P 2 videat] videt p ‖ non] nullam GPp 5 et[1]] *om.* P ‖ et[3]] *om.* P 6 de[2]] *om.* GPp 7 igitur] *om.* P ‖ scilicet] *om.* P 8 motum ignorare] *inv.* P ‖ ut] quia P ‖ materiam] naturam P 9 motum ut] *om.* P 10 adhuc … vere] adhuc probatur etiam vere P : adhuc probabitur et vere p : probabiliter adhuc ut videtur et G ‖ ista conclusio potest] ista conclusio posset G : potest enim P ‖ augmentari] argumentari p : argui iterum P 11 simul] *om.* P 12 et … casus] et ponitur casus talis G : et ponatur casus p : pono casum P ‖ ego] *om.* P 13 alicuius] *praem.* aliquem Gp : aliquem P 14 videtur esse] est P ‖ a] in P 17 conceptu[2]] conceptum G 17–18 dimittendo] omittendo G 18 venimus] veniremus G 19 et quod] ABCGHLMP : et Up : et quamvis T ‖ praecognitione] cognitione P ‖ possumus] possimus GP 20 syllogizare et speculari] syllogizari G 21 actualiter] *corr. sup. lin. ex* accidentia C : accidentaliter p : accidentia P 22 considerando] considerantes GPp 23 concedatur] conceditur G 24 cognitione] *add.* et G 25 et] *om.* P ‖ apprehensionis] *add.* et Pp 26 habitualis] habituali P

lum motum cognoscimus et per consequens nullam naturam ignoramus et omnem motum ignoramus. Manifestum enim est per casum quod nullum motum actu cognoscimus. Sed probo quod omnem naturam cognoscimus quia: isto conceptu communi omni substantiae omnem substantiam indifferenter cognoscimus, et tunc arguam sic: omnem substantiam cognoscimus; omnis natura est substantia; igitur omnem naturam cognoscimus.

Recolligendo igitur apparet mihi quod in ista principali secunda conclusione contineri possunt quattuor conclusiones partiales.

Prima conclusio est quod possibile est me simul motum ignorare et naturam non ignorare.

Se|cunda conclusio est quod possibile est me simul omnem naturam 41vb p cognoscere et nullum motum cognoscere loquendo de actuali apprehensione.

Tertia conclusio est quod possibile est me simul omnem motum ignorare et nullam naturam ignorare loquendo de ignorantia opposita solum actuali apprehensioni.

Quarta conclusio est quod non est necesse quod omnem motum ignorans sit naturam ignorans.

Et tunc ad praedictos sensus ista negaretur 'necesse est ignorato motu ignorare naturam'.

Sed tamen si vellemus loqui de notitia habituali et de ignorantia sibi opposita, tunc esset mihi probabile quod naturaliter non sit possibile aliquem habere omnis substantiae notitiam sine habituali notitia motuum et multorum accidentium, quia a principio prius cognoscimus accidentia quam substantias vel saltem simul uno conceptu | communi et confuso; et 61vb C huiusmodi primae notitiae ita sunt faciles et saepe revertentes ad actum quod numquam naturaliter potest auferri a nobis habitus ex eis acquisitus. Et haec sunt satis difficilia et indigent alibi speciali consideratione.

---

1 consequens] *add.* etiam Gp 2 enim est] *inv.* Gp 3 probo] probatio P 5 et] *om.* P ‖ arguam sic] arguo sic Gp : sic arguitur P 6 omnem naturam cognoscimus] etc. G 7 recolligendo] intelligendo p 7–8 in ... possunt] in ista secunda principali conclusione contineri possunt Gp : ista secunda conclusio principalis potest continere P 8 partiales] principales C 9 conclusio] *om.* GPp ‖ simul] *om.* P 11 conclusio] *om.* P 14 conclusio] *om.* P 15 solum] *om.* P 17 quarta conclusio est] ideo (inde P) quarto concluditur GPp ‖ omnem] omne G 19 et] *om.* P ‖ ista negaretur] isti negarentur G 21 vellemus] velimus G 24–25 accidentia quam substantias] substantias quam accidentia P 25 simul] *inf. lin.* C : *add.* sub p : *om.* P 26 huiusmodi] huius P ‖ primae] *om.* p 27 potest ... acquisitus] possunt auferri a nobis habitus ex eis acquisiti G

Notandum est, sicut alias dixi saepe, quod haec verba 'appeto', 'cognosco' etc. faciunt terminos sequentes, cum quibus construuntur, appellare rationes secundum quas significant ea quae significant. Propter quod possibile est me simul venientem cognoscere et non cognoscere venientem, quia quando dico quod cognosco venientem, sensus est de proprietate sermonis quod cognosco illum secundum illam rationem secundum quam ipse dicitur veniens, et forte non est ita, immo ego cognosco sub alia ratione, videlicet secundum quam dicitur Socrates vel homo.

Notandum est etiam, ut alias dictum est, quod hoc nomen 'natura' non significat substantiam secundum conceptum simplicem et substantialem, sed secundum conceptum relativum ad motum. Et ideo per modum connotationis conceptus a quo sumitur nomen 'naturae' implicat in se conceptum motus et non potest esse sine illo.

Et tunc pro tertia conclusione principali vel etiam pro tertio articulo possumus inferre conclusiones partiales.

Prima est quod non sequitur 'naturam cognosco, igitur ego cognosco
52va P naturam' vel | 'motum cognosco, ergo cognosco motum', quia res illas quae
51va G sunt natura et motus possum cognoscere | sub aliis rationibus et non sub illis
secundum quas dicuntur natura et motus.

Secunda conclusio est quod etiam non sequitur 'ignoro naturam, igitur naturam ignoro' et sic de motu, quia quod non cognosco ignoro et e converso, sed a destructione consequentis sequitur ex priori conclusione quod non valet consequentia 'non cognosco naturam, igitur naturam non cognosco'.

Tertia conclusio est quod impossibile est me simul cognoscere naturam et non cognoscere motum, quia me cognoscere naturam significat quod cognosco eam secundum eam rationem secundum quam dicitur natura;

---

1 notandum est] nota G ‖ sicut] *om.* P 1–2 appeto cognosco etc.] cognosco appeto etc. Gp : cognosco appeto et similia P 2 terminos] *add. sup. lin.* se C 4 quia] et G 7 cognosco] *add.* eum P 9 notandum est] nota G ‖ etiam … est^2] *om.* P 11–12 connotationis] *add.* iste G 12 nomen naturae] hoc nomen natura p 14 tertia] secunda G : illa P ‖ principali] *om.* G ‖ etiam] *om.* GPp 14–15 possumus] possimus G 15 inferre] *add.* plures GPp ‖ partiales] particulares p : *om.* P 16 quod] *om.* P ‖ ego] *om.* GPp 17 vel … motum^2] *om.* C 18 possum] possumus C 19 natura] naturae p 20 etiam] *om.* P 22 a destructione] ad destructionem G ‖ ex] cum p 27 eam^2] illam GPp

---

1 Cf. Iohannes Buridanus, *Summulae, De suppositionibus* (ed. Van der Lecq, 4.5.3, 83–87); cf. Iohannes Buridanus, *Sophismata*, cap. 4, soph. 9–15 (ed. Scott, 73–89, ed. Pironet, 79–99).
9 Cf. Iohannes Buridanus, *Quaestiones super libros Physicorum*, II, q. 4 (ed. Streijger, Bakker, 265–266)

quam rationem non possum habere sine ratione secundum quam motus dicitur motus; quam si habeo, cognosco motum; igitur etc.

Quarta conclusio est quod impossibile est me simul ignorare motum et non ignorare naturam. Et haec conclusio sequitur ex praecedenti.

Et tunc ad istos sensus concedi debet propositio Aristotelis de qua quaerebatur, scilicet quod necesse est ignorato motu ignorare naturam, hoc est dictu quod illa est necessaria 'ignorans motum est ignorans naturam', si aliquis est ignorans motum.

Tunc ergo ad rationes. | 62ra C

⟨1⟩ Ad primam quod naturam esse non sic est per se notum, quin prius et magis sit notum motum esse, sed est per se, id est faciliter, notum. Ideo non dicit Aristoteles quod demonstrare naturam esse sit ridiculum, sed dicit quod temptare demonstrare naturam esse est ridiculum, quia 'temptare' notat difficul|tatem et inconveniens est facilia per difficilia demonstrare, 42ra p
quia esset petitio principii.

⟨2⟩ Ad aliam dico quod natura potest cognosci sine cognitione motus, sed non est possibile cognoscere naturam, quin cognoscatur motus.

⟨3⟩ Ad aliam dico quod non est possibile me cognoscere quietem et non cognoscere motum, eo quod in ratione quietis implicatur ratio motus, et omnino in ratione privationis implicatur ratio habitus. Tamen possibile est me simul quietem cognoscere et nullum motum cognoscere, ut si rem quiescentem viderem et numquam cognoscerem motum, ego quietem cognoscerem, quia res quiescens est quies, et tamen nec cognoscerem quietem nec motum.

⟨4⟩ Alia ratio solvenda est ex septimo *Metaphysicae*, ubi videri debet quod illa auctoritas 'substantia est prior accidente secundum notitiam' debet intelligi solum ad istum sensum quod res prius, id est perfectius, cognoscitur per praedicata substantialia, id est quidditativa, quam per praedicata denominativa. Magis enim et perfectius cognosco equum vel bovem, si cognosco

---

2 etc.] *om.* P 3 est[1]] sequitur Gp : sequitur scilicet P 4 praecedenti] praecedente Gp 6 est[1]] *om.* P 7 dictu] dictum P 8 motum] naturam CGp 9 tunc ... rationes] tunc ad rationes dicendum est p : tunc ad rationes dicendum G : ad rationes dicendum est P 10 sic est] *inv.* P 13 esse] *om.* G 16 dico] dicitur GPp ‖ cognitione motus] *inv.* GPp 17 non est possibile] est impossibile P 18 dico] dicitur P 19 eo quod] quia P ‖ et] etiam P 20 implicatur ratio habitus] ratio habitus implicatur G 22 cognoscerem] cognovissem Pp : vidissem G 25 ex septimo] septimo Pp : undecimo G ‖ ubi] ibi G 26 notitiam] *add.* etc. G 27 quod] quia Gp ‖ cognoscitur] cognoscatur P 28 id est] scilicet p : *om.* C

---

13 Cf. Aristoteles, *Physica*, II, 1, 193a3–4

quid ipse est, quam si cognoscerem qualis aut quantus aut ubi ipse est. Sic enim exponit se ipsum Aristoteles in dicto libro.

⟨5⟩ Alia auctoritas est intelligenda quod quantum ad scientiam propter quid causae priores in esse in cognitione sunt priores, sed quantum ad scientiam quia est, quae praecedit, non oportet ita esse.

⟨6⟩ Similiter dicitur ad rationem sequentem.

⟨7⟩ Alia ratio est pro prima conclusione.

Haec de quaestione.

---

1 ipse^{2}] *om.* GPp 2 libro] loco GPp 4 quid] quod p ‖ esse … priores^{2}] *corr. ex* cognitione sunt priores in esse C ‖ in^{2} … priores^{2}] sunt priores in cognitione GPp 5 praecedit] *add. in marg.* ab effectibus G 7 pro prima] proprium P 8 haec de quaestione] *om.* GPp

---

2 Cf. Aristoteles, *Metaphysica*, VII, 1, 1028a36–b2

⟨III.2⟩

⟨Utrum ad alterationem requiratur fluxus distinctus ab alterabili et a qualitate secundum quam est alteratio⟩

Nunc videatur | de definitione et quidditate motus et motuum quaerendo 51vb G utrum ad alterationem requiritur fluxus distinctus ab alterabili et a qualitate secundum quam est alteratio.

Arguitur quod sic:

⟨1⟩ Auctoritate Commentatoris tertio *huius* et quinto, qui | distinguit de 52vb P motu pro forma fluente et pro fluxu formae, vel etiam pro perfectione secundum quam est motus et pro via ad illam perfectionem. Modo per 'formam fluentem' vel per 'perfectionem secundum quam est motus' intelligit in alteratione qualitatem, | ut caliditatem vel frigiditatem, et per 'fluxum' vel 'viam' 62rb C non potest intelligere nisi aliquam successionem distinctam ab huiusmodi qualitate. Aliter nihil valent istae distinctiones.

⟨2⟩ Item actio et passio distinguuntur a qualitate et a substantia non solum numero vel specie, immo genere etiam generalissimo. Sed alteratio pro fluxu et via est de praedicamento actionis vel passionis, ut dicit Commentator; et hoc apparet, quia aliter alterare et alterari non essent agere et pati. Alterabile autem est substantia, ut homo vel aqua qui calefiunt. Igitur in alteratione fluxus vel via differt ab alterabili et ab illa qualitate secundum quam fit alteratio.

⟨3⟩ Item alteratio est vere motus. Tunc arguitur sic: omnis motus est naturae successivae et non permanentis, scilicet capiendo 'motum' pro fluxu; sed alterabile et qualitas sunt naturae permanentis; igitur differunt.

---

5 nunc ... quaerendo] deinde secundo ut videatur de definitione et quidditate motus et motuum quaerendum est G : quaeritur secundo de definitione et quidditate motus et motuum quaerendum est p : quaeritur secundo P 6 requiritur] requiratur Gp ‖ a] *om.* p 8 arguitur] *praem.* et G 9 huius] *post* quinto P 10 vel] et G 11 pro via] *add. in marg.* alias praevia C 12 per] *om.* G 15 qualitate] *add.* quia G ‖ valent] valerent p 16 a^{2}] *om.* GPp 17 etiam] *om.* P 19 aliter] *om.* G 21 ab^{2}] *om.* p 22 fit] est G

---

9 Cf. Averroes, *In Physicam*, III, comm. 4, f. 87C–D; V, comm. 9, f. 215A–C 18–19 Cf. Averroes, *In Physicam*, III, comm. 4, f. 87D

 DOI: 10.1163/9789004322356_005

⟨4⟩ Item cum omnis motus mensuretur tempore, oportet in alteratione aliquid esse quod mensuretur tempore distinctum ab eo quod non mensuratur tempore. Sed alterabile non mensuratur tempore, quia possibile est magis et minus alterari in eodem tempore. Nec qualitas quae acquiritur mensuratur tempore, quia possibile est quod eodem tempore acquiratur qualitas maior et qualitas minor tam intensive quam extensive. Igitur ibi concurrit aliud successivum quod mensuratur tempore.

⟨5⟩ Iterum calefactibile et caliditas erunt, quando non amplius erit calefactio; igitur calefactio differt ab utroque illorum.

42rb p ⟨6⟩ Iterum terminus | differt ab eo cuius est terminus; sed caliditas est terminus ad quem calefactionis; igitur calefactio differt a caliditate. Et credendum est quod adhuc magis differt a calefactibili. Igitur etc.

⟨7⟩ Iterum quinto *huius* dicitur quod motus non est motus neque generationis generatio, neque tamquam subiecti neque tamquam termini; sed quando aqua calefit, calefactio est generatio caliditatis; igitur ipsa non est caliditas. Et per syllogismum expositorium arguitur sic: haec generatio non est generationis; haec generatio est caliditatis huius; igitur haec caliditas non est generatio. Igitur alteratio quae est generatio caliditatis est distincta a caliditate.

Oppositum arguitur quia:

⟨1⟩ Quaereretur utrum ille fluxus esset prior illa qualitate vel posterior; et neutrum potest dici convenienter. Non enim esset prior, quia si est fluxus additus qualitati, ille fundatur in illa qualitate, et fundatum est prius. Nec
52ra G potest dici quod sit posterior, quia per illum fluxum | acquireretur illa qualitas et ille fluxus poneretur tamquam via ad illam qualitatem et tamquam generatio ipsius; via autem non est posterior termino ad quem itur nec generatio posterior generato.

---

1 cum ... tempore] cum omnis motus mensuratur tempore p : omnis motus mensuratur tempore ergo P 2 mensuretur] mensuratur P 4 magis] maius Gp ‖ alterari ... tempore] alterari eodem tempore Gp : eodem tempore alterari P 5 quod] *add.* in p ‖ acquiratur] acquiruntur G 9 calefactio] *om.* G ‖ differt] *corr. in marg. ex* dicitur C : differet P 12 etc.] *om.* P 16 et] sed G 21 quaereretur] quaeretur G ‖ qualitate] caliditate p 22 convenienter] *om.* P ‖ est] esset GP 23 fundatur] fundaretur P ‖ fundatum] fundamentum Pp 24 per] post P ‖ acquireretur] acquiretur P 25 illam] *om.* G ‖ qualitatem] caliditatem p 26 itur] igitur p 27 generato] *add.* igitur etc. G

---

13 Aristoteles, *Physica*, V, 2, 225b15

⟨2⟩ Iterum si illa qualitas generetur generatione | [62va C] distincta a se, pari ratione oportet quod illa generatio generetur, cum habeat esse post non esse. Et si dicatur generari se ipsa sine alia ulteriori generatione, ita poterat hoc dici de illa qualitate. | [53ra P] Et si iterum generatur alia generatione, procederetur sic in infinitum, quod est inconveniens.

⟨3⟩ Iterum Commentator concedit quod illa forma fluens, scilicet qualitas quae acquiritur pars post partem, est motus secundum veriorem acceptionem motus.Et hoc etiam apparet verum. Sibi enim convenit definitio motus, nam quamdiu aqua calefit, aliquid est acquisitum de illa caliditate et aliquid acquirendum. Et omnino illa caliditas est actus ipsius alterabilis quantum ad acquisitum, existentis adhuc in potentia ad illud quod restat acquirendum. Modo si concessum sit quod illa qualitas vere est motus, tunc non est aliqua ratio quare ad motum salvandum oportet ponere aliquid ultra additum.

⟨4⟩ Iterum si esset talis fluxus additus, ille esset actio vel passio, sicut dicit Commentator. Sed frustra poneremus in alteratione actionem vel passionem additam qualitati quae acquiritur, quia sine tali additione possumus totum salvare dicendo quod illa qualitas secundum diversas rationes dicitur actio vel passio, scilicet dicitur actio secundum quod producitur ab agente et passio secundum quod recipitur in passo. Sed nihil debet poni frustra in natura. Igitur etc.

⟨5⟩ Iterum non poneretur talis fluxus additus nisi ad salvandam successionem; sed sine tali addito salvatur successio per hoc quod continue est una pars gradualis illius caliditatis prius et alia sequitur in esse posterius; ideo omnino frustra poneretur talis additio. Et apparebit quod frustra poneretur per hoc quod rationes quae talem fluxum arguere videntur solventur.

⟨6⟩ Iterum generatio instantanea salvaretur ex eo solo quod forma aliqua esset in materia in qua non erat ante, et non prius una pars eius quam alia, sine aliquo alio addito. Igitur pari ratione generatio continua caliditatis,

---

1 generetur generatione distincta] generaretur generatione distincta G : generaretur distincta generatione P 2 oportet] oporteret P 3 poterat hoc] *inv.* GPp 4 iterum generatur] *inv.* GPp ‖ procederetur] produceretur *sed add. sup. lin.* procederetur C : procedetur P : procederet G 5 sic in infinitum] in infinitum tali modo P 10 omnino] ideo GPp 11 restat] *add.* ad P 13–14 additum] *om.* G 16 poneremus in alteratione] in alteratione poneremus GPp 17 possumus] possimus G 21 igitur] *om.* p ‖ etc.] *om.* P 22 iterum] tunc P ‖ salvandam] salvandum Pp 27–28 aliqua esset] *inv.* C

---

6 Cf. Averroes, *In Physicam*, III, comm. 4, f. 87D 15–16 Cf. Averroes, *In Physicam*, III, comm. 4, f. 87D

quae est alteratio, salvaretur, si illa caliditas esset in subiecto continue pars post partem, postquam non erat in eo, absque hoc quod oporteret ponere aliud additum.

Ad istam quaestionem respondeo quod in vera alteratione, ut in calefactione, non est aliquis fluxus alius quam illa caliditas quae continue acquiri-
42va p tur pars post partem, et etiam frigiditas quae continue abicitur e converso | pars post partem. Et ad hoc probandum sufficiunt mihi rationes quae immediate ad hoc adductae fuerunt.

Tunc respondendum est ad rationes ante oppositum.

62vb C ⟨1⟩ Ad | auctoritatem Commentatoris dico quod Commentator non posuit illam distinctionem nisi ad ponendum differentiam secundum rationem inter haec nomina 'qualitas' et 'alteratio', 'caliditas' et 'calefactio', 'magnitudo' et 'augmentatio' et huiusmodi. Haec enim nomina referendo singula singu-
52rb G lis supponunt pro eisdem rebus, | sed ista nomina 'qualitas' et 'caliditas' et 'magnitudo' non connotant fluxum, id est non connotant successionem in essendo unam partem prius quam aliam; et ista nomina 'alteratio', 'calefactio' et 'augmentatio' hanc connotant. Et si aliud intenderet Commentator, negaretur.

53rb P ⟨2⟩ Ad aliam | dicitur quod eadem res est qualitas, actio et passio. Sed nomina illa differunt secundum diversas rationes secundum quas imposita fuerunt ad significandum eandem rem.

⟨3⟩ Ad aliam dicitur quod caliditas, quando acquiritur pars post partem, est res successiva ad illum sensum quod continue una pars succedit alteri in essendo, ita quod incipit esse posterius una pars quam alia; et haec successio sufficit ad motum alterationis. Hoc tamen non obstante illae partes bene permanent simul, quando factae sunt.

⟨4⟩ Ad aliam dico quod qualitas secundum intensionem suam gradualem mensuratur tempore non ita, quin in eodem tempore posset acquiri maior caliditas in uno subiecto et minor in alio, sed sic quia, si fuerint alterationes aeque veloces secundum intensionem qualitatis, si tempora alterationum

4 quaestionem] *add.* ego Gp 5 aliquis fluxus alius] alius fluxus P 6 pars post partem] *om.* G ‖ continue ... converso] e converso abicitur P 9 tunc] ideo GPp ‖ ante] in GPp 13 et huiusmodi] *om.* P 14 et^1] *om.* GPp ‖ et^2] *om.* G 16 partem] *om.* p 17 hanc] *add. sup. lin.* seu haec C : haec GPp ‖ commentator] *add.* ipse G 23 quod] quia GPp 24 pars] *om.* Pp 25 tamen] *om.* P 25–26 bene permanent] manent P 27 dico] dicitur G 28 in ... posset] posset eodem tempore GP : possit eodem tempore p 30 alterationum] alterationis P

sint aequalia, qualitates acquisitae erunt aequalis intensionis. Et si tempus unius fuerit duplum tempori alterius, una qualitas erit duplicis intensionis ad aliam. Et de modo mensurandi motum per tempus dicetur magis in quarto libro.

⟨5⟩ Ad aliam conceditur quod calefactio erit, quando non erit calefactio. Et talia fuerunt saepe dicta.

⟨6⟩ Ad aliam dico quod calefactio haec et terminus eius differunt vel different sicut pars et totum. Quamdiu enim est calefactio, adhuc restat de termino aliquid acquirendum; ideo calefactio non est nisi pars eius quod erit terminus ipsius. Terminus enim erit caliditas totalis quae tempore alterationis acquirebatur pars post partem. Sed pars caliditatis est caliditas. Et de huiusmodi terminis motuum dicetur magis in sequentibus.

⟨7⟩ Ad aliam potest dupliciter responderi. Uno modo quod Aristoteles non vult | negare veritatem illius propositionis 'generationis est generatio', sed 63ra C vult dicere quod non est propria locutio, sed est propria locutio 'caliditatis est generatio', sicut non est propria locutio dicere quod privationis est privatio vel quod materiae est materia, sed est propria locutio dicere quod materiae est privatio, quamvis privatio non est alia a materia. Nec est propria locutio 'deitatis est deitas' vel 'Dei est Deus', sed 'Dei est deitas', licet nihil aliud sit Deus et deitas. Similiter non est propria locutio 'generatio generatur', sed 'caliditas generatur', sicut non est propria locutio 'privatio est privata', sed 'materia est privata'. Non enim est propria locutio, licet sit vera, si concretum abstractum suum denominet, sed debet denominare alterum terminum. Similiter non est propria locutio, si genitivus construatur cum nominativo, quod illi sint eidem termini, sed debent esse diversi, ita quod supponant pro diversis rebus vel saltem differant secundum concretum et abstractum.

---

1 erunt] erant C ‖ aequalis intensionis] *inv.* GPp 2 fuerit] quaerit G ‖ tempori] de tempore G 3 dicetur] videtur P 4 quarto] quinto G 5 conceditur] *om.* G ‖ calefactio[1]] *omnes codd.* (*desunt* BPbZ) ‖ quando] *add.* ipsa Gp 6 fuerunt saepe] *inv.* Pp : et consimilia sunt saepe prius G 9 termino] *add.* ad quem p 10 erit[1]] erat C ‖ ipsius] *add. in marg.* caliditatis totalis seu C 11 et] *om.* P 12 huiusmodi terminis motuum] huius termino motus P ‖ sequentibus] sequenti P 13 potest dupliciter responderi] respondetur dupliciter P 14 illius] huius Gp : huiusmodi P 15 est propria[2]] impropria P 17 est[1]] sit Gp 17–18 materiae] materia P 18 est[2]] sit GPp ‖ est[3]] sit C 21 caliditas] qualitas p 22 non enim] ideo non P 23 abstractum suum denominet] denominet suum abstractum G : denominat suum abstractum p : denominet abstractum P ‖ alterum] alium GPp 24 construatur] construitur Pp : construeretur G 26 saltem] *add.* quod GPp ‖ concretum et abstractum] abstractum et concretum P

3–4 Cf. inf., IV, q. 14 6 Cf. Iohannes Buridanus, *Quaestiones super libros Physicorum*, II, q. 3 (ed. Streijger, Bakker, 261[13–14]); I, q. 18 (ed. Streijger, Bakker, 186[8–10]) 12 Cf. inf., III, qq. 8–9

Aliter potest dici quod caliditas quae generatur non est generatio qua ipsa
42vb p 52va G genera|tur, quia illa caliditas quae generatur nondum est | et illa generatio
est. Unde totalis caliditas quae erit generatione completa, generatur, et illa
53va P non est, sed aliqua caliditas est, quae erit pars | illius, et illa est generatio. Et
de hoc dicetur post.

Haec de quaestione.

---

4 caliditas] qualitas p ‖ est^3] erit p ‖ et^2] *om.* P 5 de] *om.* C **6** haec de quaestione] etc. GP : sequitur quaestio tertia p

---

5 Cf. inf., III, q. 9; cf. Iohannes Buridanus, *Quaestiones super libros Physicorum*, V, q. 9, concl. 3–4 (ed. Parisiis 1509, f. 91va)

⟨III.3⟩

⟨Utrum qualitates contrariae, ut albedo et nigredo, caliditas et frigiditas, possint se compati simul in eodem subiecto secundum aliquos gradus ipsarum⟩

Quaeritur tertio utrum qualitates contrariae, ut albedo et nigredo, caliditas et frigiditas, possint se compati simul in eodem subiecto secundum aliquos gradus ipsarum.

⟨1⟩ Arguitur primo quod non. Sequeretur enim quod illae non essent contrariae nec repugnantes, quod implicat contradictionem.

⟨2⟩ Item idem esset simul calidum et frigidum, quod est impossibile. Consequentia patet, quia omnis caliditas dat esse calidum et omnis frigiditas dat esse frigidum; et tamen omnis gradus caliditatis est caliditas et frigiditatis frigiditas; igitur etc.

⟨3⟩ Item oppositio contrarietatis attenditur secundum rationem speciei, non secundum rationem individuorum. Hoc ego suppono. Et hanc suppositionem probo quia: decimo *Metaphysicae* dicitur quod unum est uni tantum contrarium, quod non esset verum de unitate numerali. Secundo quia: in eodem decimo videtur Aristoteles intendere quod contrarietas sit differentia specifica et quod debeat attendi secundum differentiam specificam et formalem. Ex hoc autem supposito videtur sequi quod omnis caliditas ab omni frigiditate differat specie; et per consequens etiam omnis gradus calidita-

---

6 quaeritur tertio] *add.* cum qualitates secundum quas est alteratio sunt (sint p) contrariae Pp : cum qualitates secundum quas est alteratio sint contrariae G 6–7 ut … frigiditas] ut albedo nigredo caliditas frigiditas *sup. lin.* C : ut albedo nigredo caliditas et frigiditas P 7 possint] possunt P 9 arguitur primo] *praem.* et G : arguitur P ‖ sequeretur enim] quia sequeretur G : quia sequitur Pp 11 simul] *om.* C ‖ calidum et fridigum] caliditas et frigiditas *sed add. sup. lin.* calidum et frigidum C 12 dat[2]] debet p 13 tamen] *om.* p 14 etc.] *om.* P 15 speciei] *add.* et GP 17 tantum] *om.* C 19 sit differentia] indifferens est P 20 et[1]] *om.* P 21 autem] *om.* G 22 frigiditate differat] frigiditati differt p 22–22.1 et … specie] *in marg. inf.* C : *om.* (*hom.*) G

---

17 Aristoteles, *Metaphysica*, X, 5, 1055b30; cf. *AA*, 1: 245 19 Cf. Aristoteles, *Metaphysica*, X, 8, 1057b35–1058a28

 DOI: 10.1163/9789004322356_006

tis ab omni gradu frigiditatis differat specie; et per consequens non possint simul stare in eodem. Sequitur igitur quod nullus gradus caliditatis possit
63rb C stare simul | cum aliquo gradu frigiditatis.

Iterum hoc confirmatur quia: plus distat distantia formali et specifica omnis albedo ab omni nigredine quam aliqua rubedo ab aliqua albedine, et sic magis contrariantur et repugnant et minus possunt esse simul; sed aliqua est albedo, scilicet perfecta et intensa, quae non potest simul stare cum rubedine; igitur nulla albedo potest stare simul cum aliqua nigredine.

⟨4⟩ Item sequeretur quod motus contrarii possunt esse simul in eodem, ut calefactio et frigefactio, quod est impossibile saltem secundum eandem partem quantitativam subiecti mobilis. Sed consequentia probatur quia: motus non dicuntur contrarii neque repugnantes nisi propter contrarietatem et repugnantiam terminorum, verbi gratia alterationes non dicuntur contrariae, ut calefactio et frigefactio, nisi propter contrarietatem qualitatum, scilicet caliditatis et frigiditatis, ut habetur quinto *huius*; igitur si caliditas et frigiditas possunt esse in eodem simul secundum aliquos gradus, sequeretur etiam quod calefactio et frigefactio possunt esse simul et sic idem simul frigefieret et calefieret, quod est impossibile.

⟨5⟩ Item sequeretur quod idem agens simul calefaceret et frigefaceret idem passum, ut aqua tepida. Est calefactibilis et frigefactibilis. Sit igitur activum approximatum in quo simul sunt caliditas et frigiditas. Quia igitur caliditatis est calefacere calefactibile sibi approximatum et frigiditatis est frigefacere frigefactibile sibi approximatum, sequitur quod illud activum per caliditatem suam calefaciet et per frigiditatem frigefaciet passum.

⟨6⟩ Item omnes sic concedentes gradus caliditatis stare cum gradibus
52vb G 53vb P frigiditatis dicunt quod omnino in calefactione, | quantum | de caliditate acquiritur, tantum de frigiditate corrumpitur et e converso. Et ideo, si potest

---

1 differat] differt p ‖ consequens] *add.* cum p ‖ possint] possunt GP 2 simul stare] *inv.* p ‖ sequitur igitur] sequitur enim G : et sequitur P : sequitur p 4 distat] distant GP 5 albedine] nigredine C 7 simul] *post* stare GP : *om.* p 9 possunt] possent GPp 10 et] *om.* G ‖ quod est impossibile] *om.* P 11 sed] *om.* P 12 neque] nec P 14 et] *om.* GPp 16 eodem] *add.* subiecto G 16–17 sequeretur] sequitur Gp 17 possunt] possent G ‖ esse simul] *inv.* p 18 frigefieret et calefieret] frigefit et calefit G : frigefiet et calefiet P : calefiet et frigefiet p 19 sequeretur] sequitur Gp 20 passum] passivum GPp ‖ calefactibilis et frigefactibilis] calefactibile et frigefactibile GPp 21 simul] *post* sunt GP : *om.* p 22–23 et … approximatum] *om.* (*hom.*) P 24 passum] passivum GPp 25 stare] *om.* P 27 corrumpitur] amittitur P ‖ et[2]] *om.* P

---

15 Cf. Aristoteles, *Physica*, V, 5, 229b21–22; 229a32–b1

ostendi quod caliditas posset remitti sine generatione alicuius frigiditatis, illa positio et opinio debet ex toto auferri.

Tunc igitur aliqui nituntur probare quod caliditas potest remitti sine aliqua generatione frigiditatis. Et ad hoc ponunt suppositiones. Quarum prima est quod caliditas et frigiditas sunt res naturae permanentis. Secunda est quod est dare pri|mum instans rei permanentis in esse et quod non 43ra p est dare ultimum instans non esse ipsius; unde octavo *Physicorum* dicitur quod instans debet attribui posteriori passioni. Tertia suppositio est quod instantia non sunt sibi invicem proxima sive immediata in tempore. Et hoc apparere debet sexto *huius*; ibi enim ostendetur quod inter quaecumque duo | instantia est tempus medium. 63va C

Tunc igitur ponitur casus quod *a* sit calidissimum sine aliquo gradu frigiditatis et approximetur agens frigidum corrumpens vel remittens illam caliditatem. Et sic illa caliditas remittetur et generabitur frigiditas secundum sic opinantes; igitur oportet quod sit primum instans in quo erit frigiditas. Et istud instans non erit instans a quo incipit motus, quia in illo instanti nihil adhuc motum est; ideo nulla est ibi frigiditas, cum ante non esset. Nec in instanti alio immediato erit frigiditas, quia instantia non sunt ad invicem immediata. Igitur in alio instanti distante erit primo frigiditas. Et tunc sequitur, cum inter istud instans et primum a quo motus incipiebat esset tempus medium et alteratio qua saltem caliditas remittebatur, quod in illo tempore erat remissio caliditatis sine aliqua generatione frigiditatis.

Iterum hoc confirmatur quia: non immediate per motum continuum fit mobile de termino a quo ad teminum ad quem; sed calefactio est de frigiditate in caliditatem tamquam de termino a quo ad terminum ad quem; igitur non immediate venit alterabile de frigiditate in caliditatem. Sed immediate veniret, nisi prius remitteretur frigiditas quam esset caliditas. Igitur prius est remissio frigiditatis quam aliqua generatio caliditatis.

---

1 posset] possit Pp 2 positio et opinio] opinio seu positio P 3 igitur aliqui] *inv.* P 4 et] *om.* G 5 secunda] *add.* suppositio P 7 non] in Pp ‖ ipsius] eiusdem P ‖ unde] ut G 7–8 unde … passioni] *om.* P 8 tertia suppositio est] tertia suppositio p : tertio supponitur GP 9 in] etiam P 10 apparere debet] apparebit P ‖ ibi enim] ubi enim p : ubi P 12 calidissimum] calefactivum p 13 vel] et p 16 incipit] *add. in marg.* seu incepit C 18 instanti alio immediato] alio instanti immediato p : instanti immediate alio P 19 alio] aliquo p ‖ et] *om.* P 20 cum] quod C ‖ esset] esse P 23 fit] venit G 24–25 sed … quem] *om.* (*hom.*) P 28 frigiditatis] caliditatis P ‖ aliqua] *om.* P

---

7 Cf. Aristoteles, *Physica*, VIII, 8, 263b9–26 10 Cf. Iohannes Buridanus, *Quaestiones super libros Physicorum*, VI, q. 1 (ed. Parisiis 1509, ff. 93vb–94va)

⟨7⟩ Item sequeretur quod aqua frigida fieret statim frigidissima et congelata, quod apparet falsum. Consequentia patet, quia in illa aqua esset multa frigiditas et parva caliditas; ideo frigiditas multum excedens et obtinens corrumperet statim illam caliditatem, quae propter sui parvitatem esset valde parvae resistentiae.

⟨8⟩ Iterum ex quo dicis quod per unum gradum caliditatis qui acquiritur removeretur unus gradus frigiditatis, sequitur quod illi habent ad invicem incompossibilitatem et repugnantiam. Sed si unus gradus caliditatis habeat incompossibilitatem vel repugnantiam cum aliquo gradu frigiditatis, sequeretur pari ratione quod quilibet gradus caliditatis ad quemlibet gradum frigiditatis habeat incompossibilitatem et repugnantiam et quod non possint simul existere. Et patet ex hoc consequentia, quia omnes gradus caliditatis sunt ad invicem eiusdem rationis; ideo si unus gradus caliditatis | [63vb C] uni gradui frigiditatis est incompossibilis, sequitur quod quilibet cuilibet est incompossibilis.

⟨9⟩ Iterum cum visus non decipiatur | [54ra P] circa obiectum | [53ra G] proprium, sequitur quod visus cognosceret albedinem et nigredinem in rubedine vel in viriditate, quod est falsum.

⟨10⟩ Iterum sequeretur quod calefactio esset duplex motus, quod videtur falsum. Consequentia patet, quia esset unus motus secundum caliditatem quae acquiritur et alius motus secundum frigiditatem quae abiceretur.

⟨11⟩ Iterum sequeretur quod in motu de albo in rubeum prius deveniretur in nigrum, quia antequam esset rubedo, oportet albedinem remitti, quod non esset sine generatione nigredinis. Et illud apparet falsum quinto *huius*, ubi dicitur quod medium est in quod prius mutat mutans quam in extremum.

⟨12⟩ Iterum sequeretur quod eiusdem speciei essent motus de albedine in rubedinem et de albedine in viriditatem, quia non acquireretur nisi nigredo, sicut non corrumperetur nisi albedo; | [43rb p] et illi dicunt quod, quantum de uno contrario corrumpitur, tantum de alio generatur et e converso.

---

1 sequeretur] sequitur p ‖ statim] *ante* fieret Gp : *om.* P 2 quia] quod p ‖ illa] *om.* G ‖ multa] magna p 3 ideo frigiditas] *om.* G 4 sui] suam GPp 6 quo] *add.* tu Gp 7 removeretur] removetur Gp : remittitur P ‖ illi] *add.* gradus G 9–10 sequeretur] sequitur GPp 11 possint] possunt Pp 12 existere] *add.* in eodem G ‖ consequentia] consequens G 14 quilibet] *om.* CG 16 decipiatur] decipitur G 17 in^2] *om.* GPp 19 sequeretur] sequitur p ‖ calefactio] in calefactione P 21 motus] *om.* GPp ‖ abiceretur] abicitur Gp 22 sequeretur] sequitur p ‖ in^2] ad P 23 oportet] oporteret GPp 27 sequeretur] sequitur p ‖ essent] esset P

---

24 Cf. Aristoteles, *Physica*, V, 3, 226b23–25

⟨13⟩ Iterum cum ex calidissimo fiat frigidissimum, remittitur caliditas in duplo et iterum residuum in duplo et sic in infinitum. Ad quod sequitur quod frigiditas in fine esset actu infinita, quod est falsum.

⟨14⟩ Iterum positio illa videtur ponere quod non possit forma remitti nisi per mixtionem contrarii. Et hoc videtur manifeste falsum de lumine, quod remittitur et non habet contrarium.

Oppositum arguitur quia:

⟨1⟩ Ex terminis contrariis potest intellectus formare unam propositionem categoricam, et etiam ex propositionibus contrariis potest intellectus formare unam hypotheticam copulativam vel disiunctivam. Quod non esset possibile, nisi termini contrarii et propositiones contrariae possent esse simul in intellectu.

⟨2⟩ Iterum differentiae divisivae alicuius generis sunt contrariae, ut habetur decimo *Metaphysicae*; et tamen illae sunt termini possibiles existere simul in eodem intellectu, immo etiam possibiles esse simul quantum ad res significatas et supponere pro eisdem rebus. Eadem enim res est magnitudo et numerus, linea et superficies et corpus, prout postea dicetur.

⟨3⟩ Iterum decimo *Metaphysicae* dicitur quod corruptibile et incorruptibile sunt contraria; et tamen simul stant, immo unum est subiectum alterius, ut materia et forma in composito.

⟨4⟩ Iterum opiniones contradictoriorum sunt contrariae, ut habetur quarto | *Metaphysicae* et in *Peri hermeneias*; et sunt tamen simul in eodem 64[ra] C
intellectu, si vera sit opinio Commentatoris quod idem est intellectus in numero in omnibus hominibus.

⟨5⟩ Iterum simul in aere sunt gravitas et levitas, quia descenderet naturaliter, si esset in sphaera ignis (et non nisi per gravitatem), et ascenderet naturaliter, si esset in aqua (per levitatem).

---

1 fiat] fit *sed add. sup. lin.* fiat C 5 mixtionem] admixtionem G ‖ et] *om.* P 9 categoricam] *sup. lin.* C : *om.* GPp 10 esset] est P 11 contrariae possent] possunt P 13 alicuius] *om.* P 14 et] *om.* P 15 possibiles esse simul] possibile est esse simul G : possibile est simul esse Pp 16 res] *om.* Pp 17 et[2]] *om.* GPp ‖ et corpus] *sup. lin.* C : corpus P 19 et] *om.* P 21 opiniones] oppositiones P 22 quarto] decimo P ‖ in[1]] *add.* primo C : *om.* P 23–24 in ... hominibus] in omnibus hominibus in numero P 26–27 sphaera ... in] *om.* (*hom.*) P

---

14 Cf. Aristoteles, *Metaphysica*, X, 7, 1057b5–8 17 Cf. Iohannes Buridanus, *Quaestiones super libros Physicorum*, VI, q. 3 (ed. Parisiis 1509, ff. 95[va]–96[rb]) 18 Aristoteles, *Metaphysica*, X, 10, 1058b26–27 22 Cf. Aristoteles, *Metaphysica*, IV, 3, 1005b28–29; cf. Aristoteles, *De interpretatione*, 7, 17b3–5 23–24 Cf. Averroes, *In De anima*, III, comm. 4–5 (ed. Crawford, 383–413)

⟨6⟩ Iterum sexto *huius* dicitur quod illud quod movetur habet partim de termino a quo et partim de termino ad quem, cum tamen illi termini sint contrarii.

⟨7⟩ Iterum tertio *Topicorum* dicitur quod albius est quod est nigro impermixtius.

⟨8⟩ Et similiter quinto *huius* dicitur quod magis et minus dicitur cui magis aut minus contrarii insint.

54rb P ⟨9⟩ Et ibidem etiam dicitur quod medium est ultimum quodam modo; | unde fuscum quodam modo est nigrum ad album et e converso.

⟨10⟩ Et etiam in *De sensu et sensato* dicitur quod colores medii sunt ex contrariis; et sicut colores, ita sapores et odores.

Notandum est quod contrarietas est oppositio quaedam et repugnantia. Et principalioribus modis dicitur tripliciter contrarietas: una terminorum
53rb G significativorum, alia propositionum, tertia rerum | quae non sunt propositiones neque termini significativi, ut caliditas, frigiditas.

Termini proprie dicuntur contrarii, si uterque sit positivus et non possint supponere simul pro eodem, sed successive bene, ut isti termini 'album' et 'nigrum', vel si supponant pro formis contrariis, ut isti termini 'albedo' et 'nigredo', vel etiam largius si sint termini privative oppositi, ut 'caecus' et 'videns', 'caecitas' et 'visus', 'corruptibile' et 'incorruptibile', 'corporeum' et 'incorporeum' et huiusmodi. Et adhuc multis aliis modis dicuntur termini contrarii secundum attributionem ad istos modos praedictos.

Propositiones dicuntur contrariae vel quia sunt de contrariis praedicatis, ut 'Socrates est albus', 'Socrates est niger', vel quia sunt de contrario

---

1 partim] partem GP 2 partim ... quem] partem de termino ad quem G : de termino ad quem etiam partem P 6 et[1]] *om.* P ‖ dicitur[2]] dicuntur G 7 aut] et G ‖ contrarii insint] contrarii insunt GP : contraria insunt p 8 est ... modo] quodam modo est ultimum GPp 9 quodam ... nigrum] est nigrum quodam modo Pp ‖ e converso] album ad nigrum GPp 10 et[1]] *add.* omnino GPp ‖ ex] *add.* extremis P 11 sicut ... odores] T : ita colores sapores odores C, *sed add. in marg.* sicut colores ita sapores et odores : ita colores sapores odores (*add.* etc. Lp) LPp : ita colores et sapores et odores A : ita sapores odores etc. HM : ita sapores et odores etc. B : similiter sapores et odores U ‖ colores ... odores] qualitates mediae sunt ex contrariis ut colores sapores etc. G 12 notandum est] notandum P : nota G ‖ oppositio quaedam] *inv.* Pp : quaedam dispositio G 14 significativorum] significatorum G 15 caliditas] *add.* et GP 16 possint] possunt GP 17 et] *om.* p 18 et] *om.* P 21 huiusmodi] de huius P 22 modos] terminos C 23 quia] *om.* p 24 albus] *add.* et p

---

1 Cf. Aristoteles, *Physica*, VI, 4, 234b15–16; cf. *AA*, 2: 174 4 Cf. Aristoteles, *Topica*, III, 5, 119a27–28 6 Aristoteles, *Physica*, V, 2, 226b7–8 8 Aristoteles, *Physica*, V, 1, 224b32, 34–35; cf. *AA*, 2: 148, 149 10 Cf. Aristoteles, *De sensu et sensato*, 4, 442a12–13; 442b27

modo enuntiandi, ut universalis affirmativa et universalis negativa de eodem subiecto et eodem praedicato.

De contrarietate autem secundum quam dicimus propositiones aut terminos contrarios, statim dicendum est quod possunt esse simul | in eodem 43va p subiecto, scilicet in eodem intellectu formante ex terminis contrariis propositionem unam categoricam et ex propositionibus contrariis unam hypotheticam, sicut prius arguebatur. Et loquor de propositionibus et terminis mentalibus. Et si aliquis dicere velit de vocalibus et scriptis, dicat sicut voluerit. Dico igitur quod dictorum terminorum et propositionum contrarietas et repugnantia non attenditur | in essendo simul, sed propositionum contra- 64rb C rietas attenditur in essendo simul veras et terminorum contrariorum repugnantia attenditur in praedicando simul affirmative et vere de eodem termino discreto sive singulari. Et nihil plus est dearticulandum de istis nunc, quia quaestio non intelligebatur de istis.

Res autem quae non sunt termini significativi neque propositiones, si dicantur contrariae, debent esse repugnantes in essendo simul in eodem subiecto, ita scilicet quod impossibile sit eas vel eis similes esse simul in eodem subiecto. Sed tamen cum hac condicione principali requiruntur bene aliae condiciones ad proprie dictam contrarietatem rerum. Secunda enim proprietas est quod illae res vel eis similes possint esse in eodem subiecto successive, ita quod habeant ad invicem transmutationem circa idem subiectum. Tertio etiam requiritur quod maxime distent ab invicem; et non oportet quod sit distantia proprie dicta, scilicet quantitativa, sed graduali, prout intensissima caliditas diceretur maxime distare ab intensissima frigiditate | eo quod longioris temporis esset transmutatio ab intensissima ad 54va P intensissimam quam a remissa ad remissam. Et si aliae condiciones requirantur ad propriissime dictam contrarietatem, tamen istae sunt principaliores.

---

2 eodem] *praem.* de p : *om.* P **3** autem] igitur GPp **6** unam²] *add.* propositionem P **7** sicut prius arguebatur] ut dicebatur prius P ‖ propositionibus et terminis] terminis et propositionibus GPp **8** et¹] *om.* P ‖ et²] aut GPp ‖ sicut] si P **9** et] *om.* G **10–11** et … contrarietas] *om.* (*hom.*) P **10** attenditur] intenditur *sed add. inf. lin.* attenditur C ‖ essendo] eodem *sed add. sup. lin.* essendo C **11** simul] *add.* esse P ‖ et] sed etiam p ‖ contrariorum] contrarietas seu G **13** sive singulari] sive de singulari P : *om.* p ‖ dearticulandum] articulandum G **14** non … istis] de istis non est G : *add.* in eodem subiecto P **16** dicantur] dicuntur G **16–17** in² … scilicet] eo P **17** simul] *om.* G **20** illae] *om.* p ‖ eis similes possint] eis consimiles possint P : eis consimiles possunt G : sibi consimiles possint p **21** subiecto] *om.* p **22** etiam] *om.* G ‖ distent] distant P **23** quod] *add.* haec GP : *add.* hoc p **23–24** graduali] gradualis P **25** esset] est G **26–27** condiciones requirantur] *inv.* GPp

Verum est quod aliquando etiam media dicuntur contraria ad invicem quodam modo et medium extremo, scilicet remissa caliditas remissae frigiditati vel tepiditas caliditati perfectae, sed haec non est proprie dicta neque perfecta contrarietas.

Et iuxta dicta solvetur una dubitatio quia: videtur quod nulla caliditas sit frigiditati contraria, quia sit caliditas intensissima in subiecto *a*. Si igitur illi sit frigiditas contraria, tunc arguitur: vel illa frigiditas contraria est vel non est. Non debet dici quod non est, quia quod nihil est nulli est contrarium plus 53va G quam chimaera. Nec etiam frigiditas quae est est illi caliditati contraria, | quia omnis frigiditas quae est est in aliquo subiecto; et numquam potest esse in subiecto *a*, cum accidens non possit transire de subiecto in subiectum. Igitur nulla est sibi contraria.

Dico igitur quod illi caliditati contraria est omnis frigiditas intensissima quae est, quia quae se invicem corrumpunt, si approximentur, habent ad invicem contrarietatem vel perfectam vel imperfectam; modo si aliqua talis frigiditas quae est approximetur tali caliditati, illae corrumperent se invicem vel saltem remitterent.

64va C Ad rationem igitur quae fiebat dicendum est quod non dixi quod formae | contrariae possunt esse in eodem subiecto successive, sed dixi quod ipsae vel sibi similes. Haec enim caliditas numquam potest esse naturaliter in subiecto frigiditatis huius, sed sibi consimilis potest esse in eo, et hoc sufficit.

Tunc igitur loquendo de contrarietate proprie dicta formarum quae non sunt propositiones neque termini significativi, ponendae sunt conclusiones.

Prima conclusio est quod impossibile est contraria simul esse in eodem subiecto, quia dictum est quod repugnant quantum ad esse simul in eodem subiecto; ideo implicat contradictionem quod sint simul in eodem subiecto 43vb p et quod sint contraria. | Et hoc consonat sensui, quia numquam vidimus

---

1 quod aliquando etiam] quod etiam aliquando P : etiam quod aliquando G ‖ contraria ad invicem] ad invicem contraria GPp 2 scilicet] ut GPp 3 dicta] *add.* contrarietas G ‖ neque] nec p 5 et] *om.* P ‖ solvetur] solvitur GPp ‖ nulla] una G 6 illi] *add.* caliditati P 7 tunc arguitur] *om.* GPp ‖ contraria^{2}] *om.* GPp 10 est^{2}] *om.* G 11 possit] posset P 13 contraria est] *inv.* P 14 quia] item C ‖ quae^{2}] *om.* P ‖ approximentur] approximantur p : *add.* quia P 16 approximetur] approximaretur GPp 18 rationem … fiebat] rationes quae fiebant P 19 possunt] possent G 20 similes] consimiles GPp ‖ potest] *rep.* p 21 frigiditatis huius] *inv.* GP : huiusmodi frigiditatis p ‖ consimilis] similis Pp ‖ eo] eodem G 24 conclusio] *om.* G ‖ simul esse] *inv.* Gp 26 sint] sunt P ‖ subiecto^{2}] *om.* p 27 sint] sunt P

in eodem esse simul caliditatem et frigiditatem intensissimas vel etiam albedinem et nigredinem intensissimas et sic de aliis. Sed tamen nolo negare nec etiam ad praesens affirmo, quin Deus miraculose posset facere simul in eodem subiecto caliditatem et frigiditatem intensissimas. Sed hoc non est possibile fieri naturaliter. Et si hoc est possibile per potentiam divinam, tunc oportet dicere quod contraria non repugnant simpliciter in essendo simul in eodem subiecto, sed repugnant in essendo simul per potentiam naturalem.

Secunda conclusio est auferendo hanc dictionem 'simul', scilicet quod impossibile est contraria | 54vb P esse in eodem subiecto (dico et secundum eandem partem illius subiecti), quia propositio est de possibili in sensu composito et est sensus quod haec est impossibilis 'contraria sunt in eodem subiecto', et hoc est verum (dico tamen naturaliter).

Et aliqui ponunt istam tertiam conclusionem in sensu diviso quod contraria possunt esse in eodem subiecto, licet non simul, quia haec dictio 'possunt' ampliat bene suppositionem ad possibilia quae nondum sunt; modo summa caliditas potest esse in isto subiecto, et summa etiam frigiditas potest esse in eodem subiecto, et istae sunt contrariae.

Sed ego dimitto istam conclusionem, quia non apparet mihi vera, quia licet bene concederem quod summa caliditas et summa frigiditas possunt esse in isto subiecto, tamen non concederem quod istae sunt contrariae, quia vel utraque earum non est vel saltem altera earum non est, et quod nihil est nulli est contrarium. Immo impossibile est ipsas esse contrarias, quia impossibile est ipsas esse. Si enim una vocetur *b* et alia *a*, haec est impossibilis naturaliter, quod *a* et *b* sunt. Tamen si aliquis vellet dicere quod summa caliditas possibilis et non existens sit contraria summae frigiditati existenti, ipse haberet concedere quod contraria possunt esse in eodem subiecto. Unde etiam concessi quod contraria vel eis similia possunt esse in eodem subiecto, quia si haec caliditas non potest esse in subiecto huius frigiditatis, | 64vb C tamen simile sibi potest esse in eo, quia haec dictio 'simile' ampliatur ad supponendum pro possibilibus.

---

1 eodem] *add.* subiecto G ‖ intensissimas] intensissimam G **1–2** vel … intensissimas] *om.* (*hom.*) P **2** sed] et GP **3** etiam ad] *om.* P ‖ posset] possit p : potest P **4** sed] et C **7** subiecto] *om.* Pp **8** scilicet quod] *om.* p **9** et] *ante* dico p : quod G **10** possibili] impossibili Pp **12** hoc] *add.* non P **13** et] quod P : *om.* Gp **14** eodem subiecto] *inv.* P **16** etiam] *om.* P **17** contrariae] *add.* ergo etc. G : *add.* igitur P : *add.* etc. p **18** ego dimitto] *corr. sup. lin. ex* licet C : ego dimittam P **18–19** quia[1] … licet] †…†et mihi vera quia licet *in marg.* C **19** licet] *add.* ego Gp ‖ summa[2]] *om.* P **21** vel[1]] *om.* P ‖ earum[1]] *om.* Gp ‖ non[1]] *om.* C **23** alia] altera p **24** quod[1]] *om.* GPp **27** subiecto] *om.* p ‖ etiam] *add.* ego GPp **28** subiecto[1]] *om.* GPp ‖ in[2]] *add.* eodem p ‖ huius] *om.* P **29** simile sibi] similis G ‖ eo] eodem P

53vb G Pono ergo tertiam conclusionem quod possibile est | esse simul aliquos gradus caliditatis cum aliquibus gradibus frigiditatis in eodem subiecto. Verbi gratia in tepiditate sunt aliqui gradus frigiditatis et aliqui gradus caliditatis simul. Hoc probatur primo quia: si in minus calido non esset aliqua frigiditas, immo solum caliditas, sequeretur quod nulla esset resistentia calefacienti innato intensius calefacere; consequens est falsum; igitur etc. Falsitas consequentis manifesta est, quia si non esset resistentia, illud statim fieret calidissimum vel saltem ita calidum, sicut illud innatum esset calefacere; quod apparet falsum, immo continue et temporaliter fit caliditas intensior, etiam postquam approximatum est calefaciens. Sed consequentia principalis patet, quia simplex caliditas non resisteret caliditati vel calefacienti, sicut videmus quod lumen remissum, quod non habet contrarium admixtum, non resistit illuminationi vel illuminanti fortiori; ideo statim approximato lucido est tanta illuminatio quanta fieri potest ab illo lucido.

Item sequeretur quod minus calidum approximatum magis calido non remitteret illud magis calidum, immo potius intenderet, quod apparet fal-
55ra P sum. Consequentia manifesta | est sicut de lumine; lumen enim remissum unius candelae non remitteret de lumine solis.

Item si fiat motus de calidissimo ad frigidissimum, hoc ego suppono et probabitur post, quod non simul corrumpitur tota caliditas intensissima quae erat ante initium motus, sed corrumpitur pars post partem. Diceret igitur adversarius quod nihil generatur de frigiditate in subiecto illo, donec tota caliditas est corrupta. Et tunc sequeretur manifestum inconveniens,
44ra p quia ego pono | quod tepiditas sit medium quasi per aequidistantiam inter summam caliditatem et summam frigiditatem in illo motu. Tunc ergo in tempore illius tepiditatis erit caliditas remissa usque ad non gradum vel prope et non erit adhuc aliqua frigiditas. Ergo tunc vel nulla erit caliditas

---

1 tertiam conclusionem] aliam conclusionem tertiam talem GPp ‖ esse simul] *inv.* GPp 3 verbi gratia] unde C 3–4 frigiditatis ... caliditatis] caliditatis et aliqui gradus frigiditatis GPp 4 hoc] *om.* P ‖ primo] *om.* Gp ‖ in minus] in nimis G : maius P 5 solum] *corr. ex* totum C : sola P 5–6 esset ... innato] resistentia calefacienti innata esset G : resistentia calefacienti innato P 6 calefacere] calefaceret P ‖ consequens] quod p 7 illud] *om.* G 8 esset] est p 9 quod ... intensior] *om.* C 10 est] *rep.* P 11 resisteret] resistere videtur P ‖ sicut] *om.* P 12 quod[2]] *sup. lin.* C : quia GP ‖ admixtum] aliquod medium *sed add. in marg.* admixtum C 15 sequeretur] sequitur p 16 intenderet] intenderent p 18 remitteret] remitteretur G 19 hoc] *om.* GPp 21 ante] *om.* C 23 caliditas] *add.* intensissima P ‖ est] esset G ‖ et] *om.* P ‖ sequeretur] sequitur Pp 24 tepiditas] *add.* b GPp 27 non] *om.* G ‖ nulla] *om.* G

---

20 Cf. inf., III, q. 4

secundum quam sit ille motus, vel erit ita debilis quod non sentiretur vel quod ipsa nullam haberet actionem | notabilem; et hoc est manifeste falsum, 65ra C immo illud sic tepidum notabiliter remitteret de intensa frigiditate vel etiam de intensa caliditate et refrigeraret ferrum ignitum et calefaceret manus tuas, si essent nimis frigidae. Et illa ratio facit mihi magnam fidem.

Quarta conclusio est quod impossibile est aliquos gradus formarum contrariarum esse simul in eodem subiecto, quia quamvis sint simul in tepido aliqui gradus caliditatis et aliqui gradus frigiditatis, tamen nulla sunt contraria quorum illi gradus sunt gradus vel partes, immo haec est impossibilis 'gradus simul existentes sunt gradus vel partes contrariorum'; illi enim gradus non sunt gradus vel partes intensissimae caliditatis vel intensissimae frigiditatis, quia illae non sunt et eius quod nihil est nihil est pars vel gradus.

Quinta conclusio quod possibile est simul esse aliquos gradus caliditatis qui fuerunt gradus vel partes summae caliditatis, cum aliquibus gradibus qui erunt gradus sive partes summae frigiditatis. Hoc enim esset in tepido, si fieret | motus de calidissimo ad frigidissimum, quia in illo tepido remanerent 54ra G aliqui gradus caliditatis qui fuerunt partes et gradus caliditatis intensissimae, antequam inciperet motus, et cum hoc essent ibi generati aliqui gradus frigiditatis qui erunt partes frigiditatis intensissimae perfecto motu. Hoc enim est verum, si concedamus gradus caliditatis prius acquisitos manere cum illis qui posterius acquiruntur, quod postea declarabitur esse verum.

Ex istis dictis sequitur sexta conclusio, scilicet quod contrarietas formarum non attenditur ex simplicibus | rationibus formarum, sed ex quantitate 55rb P graduum. Sic enim debet attendi contrarietas, sicut apparet earum incompossibilitas. Sed frigiditatis ad caliditatem non est incompossibilitas, cum possint esse simul. Sed frigiditatis tantorum graduum ad caliditatem tantorum est incompossibilitas, ut intensissimae ad intensissimam vel multum intensae ad multum intensam, sed intensae ad remissam vel medio modo remissarum ad invicem nulla est incompossibilitas. Verbi gratia ponamus

---

1 sit] fit G ‖ sentiretur] sentietur Pp : sensirent G 3–4 frigiditate … caliditate] caliditate vel etiam de intensa frigiditate p 5 nimis] minus P 6 quarta conclusio est] tunc ergo (ego G) pono quartam conclusionem scilicet (*om.* P) GPp 7 quia] quod G ‖ quamvis] quamquam P 8 gradus²] *om.* P 9 sunt] sint p 12 non] nihil P 13 conclusio] *add.* est GPp 16 ad] in P ‖ quia] *om.* G 17 fuerunt] fierent P ‖ et] vel P 17–18 caliditatis intensissimae] *inv.* G 19 intensissimae] *add.* in G 20 caliditatis] qualitatis GP 21 qui] quae C 22 ex istis dictis] ex dictis iam G : et ex dictis p : et igitur ex concessis P 24 contrarietas] *in marg.* C : *om.* GP 26 possint] possit p ‖ tantorum] totorum P 27 ut] et C 29 ad invicem] *post* est P

---

21 Cf. inf., III, q. 5

quod perfecta caliditas sit graduum decem et etiam frigiditas perfecta sit 65rb C decem graduum. Tunc quinque gradus caliditatis et quinque | gradus frigiditatis simul existentes constituunt unam qualitatem mediam et perfectam, quae etiam est decem graduum. Nec illi quinque gradus ad istos quinque habent aliquam oppositionem vel repugnantiam nec agunt ad invicem, quia non habent oppositionem. Similiter etiam octo gradus caliditatis et duo gradus frigiditatis constituunt unam formam existentem perfectam mediam, quae etiam est decem graduum et quae est propinquior caliditati perfectae quam frigiditati perfectae. Et illi gradus nullam habent ad invicem oppositionem nec actionem vel passionem. Nec umquam illa forma media vel proportio illorum graduum ad invicem, quae est proportio octo ad duo sive quattuor ad unum, corrumperetur, si non esset aliud agens consistens in alia 44rb p proportione huiusmodi qualitatum formaliter vel | virtualiter.

Et ita patet quomodo contraria maxime distant, scilicet quia calidissimum a frigidissimo distat per viginti gradus, quia antequam sit perventum a calidissimo ad frigidissimum, oportet decem gradus caliditatis corrumpi et decem frigiditatis generari. Et tepidum medium medie distat a calidissimo et frigidissimo, quia ab utroque per decem gradus; si enim de tepido fiat calidissimum, oportet corrumpi quinque gradus frigiditatis et generari quinque caliditatis. Et sic proportionaliter in aliis. Ubi non est contrarium, ut in lumine, pura privatio luminis non distaret a lumine intensissimo nisi per decem gradus, quia nihil oporteret corrumpi ad hoc quod esset perfectum lumen, sed solum fieri decem gradus luminis. Tepiditas autem a tepiditate non distat, sed sunt similes.

Tunc igitur solvendae sunt rationes.

⟨1⟩ Ad primam concedendum est quod contraria non possunt se compati in eodem subiecto nec aliqui gradus contrariorum, nisi hoc exponatur ad sensum quintae conclusionis praedictae.

---

1 graduum decem] *inv.* GPp ‖ etiam] *om.* P ‖ frigiditas perfecta] *inv.* GPp 2 gradus²] *om.* P 4 quinque²] *add.* non P 5 oppositionem vel repugnantiam] repugnantiam vel oppositionem G 6 etiam] *om.* GP 7 frigiditatis] *add.* simul existentes GPp ‖ existentem perfectam mediam] mediam perfectam GPp 8 propinquior] *add.* etiam C 10 umquam] *om.* G 12 unum] sex p : *add.* non G ‖ consistens] coassistens p 13 huiusmodi] huius P ‖ qualitatum] *add.* vel GPp 14 et] *om.* P ‖ distant scilicet quia] distent quia scilicet P 15 a frigidissimo] ad frigidissimum P ‖ distat] distant G ‖ viginti] decem C 15–16 perventum a] perventus de GP : proventum de p 16 corrumpi] corrumpere GP 18 tepido] *add.* medio GPp 19 corrumpi] corrumpere Gp 20 et] *om.* P ‖ proportionaliter] proportionabiliter p ‖ in] ab P ‖ est] esset GPp 21 pura] *om.* p 25 solvendae sunt] respondendum est ad p : solvendum est ad P ‖ rationes] *add.* principales GPp

⟨2⟩ Ad secundam dicendum est quod totalis caliditas | denominat simpli- 55va P citer esse tale et | non pars eius; ideo etiam lignum trium pedum non dicitur 54rb G bipedale, sed tripedale. Et cum hoc medium, ut tepidum, ab utraque qualitate denominatur non simpliciter, sed secundum quid et respective. Unde quinto *huius* dicitur quod fuscum dicitur album ad nigrum et nigrum ad album.

⟨3⟩ Ad aliam dictum est quod contrarietas caliditatis ad frigiditatem non attenditur secundum rationes simplices caliditatis et frigiditatis, sed secundum rationes intensissimae caliditatis et intensissimae frigiditatis. Et ita | 65va C debet intelligi quod non sit nisi unum uni contrarium, prout unitas attenditur non solum ex unitate specifica formarum, sed etiam ex unitate specifica quantitatis graduum. Quando etiam dicitur quod contrarietas est differentia specifica, hoc est concedendum, non solum specifica secundum rationes simplices formarum, sed etiam secundum rationes quantitativas graduum; ideo non sequitur illud quod dicebatur sequi. Et ita etiam nihil valet illud quod dicebatur, scilicet quod quaelibet albedo a qualibet nigredine plus distat distantia formali quam a rubedine.

⟨4⟩ Ad aliam credo esse dicendum quod proprie loquendo calefactio et frigefactio non sunt res contrariae, immo sunt res mediae inter terminum a quo est motus et inter terminum ad quem est motus; qui si essent simul, essent contrarii. Et de contrarietate motuum dicetur in quinto *huius*.

⟨5⟩ Ad aliam dicendum est quod calidum vel frigidum non calefacit vel frigefacit aliud in gradu simili calidum vel frigidum, quia nullam habet ad ipsum contrarietatem.

⟨6⟩ Ad aliam rationem longam dicendum est quod ipsa deficit vel in supponendo quod instantia sunt res indivisibiles in tempore, vel etiam supponendo quod instans medium inter generationem et corruptionem debeat

---

1 dicendum est] dicitur P ‖ totalis] tota G **3** hoc] habet p **10** prout] quod P **11** etiam] *om.* P **13** concedendum] *add.* sed GPp **15** et] *om.* P **16** scilicet] *om.* P ‖ a qualibet] cum quaelibet P ‖ plus] *post* distat (17) p : *post* formali (17) P : *om.* G **17** rubedine] *add.* etc. Gp **20** inter] *om.* Gp **21** motuum] motus P ‖ huius] libro GP **22** dicendum est] dicitur P **23** aliud] *ante* vel^2 (22) P ‖ in … frigidum] (in simul sive *del.*) in (*add. sed del.* tali) gradu simili (calidum vel frigidum *corr. sup. lin. ex* caliditatis et frigiditatis) C : in simili gradu calidum vel frigidum p : calidum sive frigidum in simili gradu caliditatis vel frigiditatis G **23–24** habet ad ipsum] habent ad invicem p **25** in] *om.* GPp **26** sunt] sint Gp ‖ vel] et C ‖ etiam] *om.* P **27** et] vel P ‖ debeat] debet P

---

5 Aristoteles, *Physica*, V, 5, 229b15–20 **21** Cf. Iohannes Buridanus, *Quaestiones super libros Physicorum*, V, q. 4 (ed. Parisiis 1509, ff. 86rb–88ra)

magis attribui formae vel passioni praecedenti quam sequenti vel e converso. De quibus dicetur in sexto *huius* et in octavo.

Ad confirmationem dicendum est quod huiusmodi calefactionis terminus non dicitur caliditas secundum illam rationem secundum quam dicitur caliditas simpliciter, sed secundum rationem secundum quam dicitur vel diceretur caliditas tanti gradus.

⟨7⟩ Ad aliam dictum est quod ista frigiditas intensa et illa caliditas remissa simul existentes non agunt nec contrariantur ad invicem.

44va p ⟨8⟩ Ad aliam | dictum est etiam quod non oportet quod unus gradus frigiditatis habeat repugnantiam vel contrarietatem ad unum gradum caliditatis nec quinque ad quinque nec octo ad duos, sed decem ad decem et octo ad
55vb P octo et octo ad sex vel octo ad quattuor haberent repugnantiam, quia | repugnantia et oppositio est ex quantitate graduum, ut dictum est.

⟨9⟩ Ad aliam dicendum est quod qualitates simul indistincte secundum situm existentes vel etiam simul secundum eandem lineam species suas multiplicantes visus percipit indistincte et confuse. Ideo sol videtur rubeus in mane; et si lumen multiplicatur per vitrum croceum et per vitrum blavium sibi invicem supposita, totum apparebit viride.

⟨10⟩ Ad aliam dicitur quod non est ibi duplex motus, sed ibi est duplex mutatio, scilicet generatio formae unius et corruptio alterius. Unde non est inconveniens, cum motus debeat esse de affirmato in affirmatum, ut dicitur
54va G quinto *huius*, quod ipse | sit compositus ex generatione et corruptione.

⟨11⟩ Ad aliam dicendum est quod non omne illud dicitur nigrum simpli-
65vb C citer in quo | est aliquis gradus nigredinis, sed oportet quod totalis qualitas sit nigredo sine aliqua albedine vel cum pauca. Et omnino prius venimus

---

2 et] vel p **3** ad] *praem.* et G : *add.* aliam P ‖ est] *om.* p ‖ huiusmodi] huius Pp **5** caliditas ... dicitur] *om.* (*hom.*) G ‖ simpliciter] *ante* dicitur^{2} (4) Pp **7** dictum] dicendum p ‖ illa caliditas] alia C **8** agunt nec contrariantur] contrariantur nec agunt GPp **9** etiam] *ante* dictum Gp : *om.* P **10** habeat] habet P ‖ unum] *om.* P ‖ gradum] graduum C **11–12** sed ... octo^{1}] *om.* P **12–13** quia repugnantia] *om.* p **16** confuse] *add.* et p **18** sibi] *add.* ad GPp ‖ supposita] superposita G **19** ibi^{1}] *ante* non P **19–20** sed ... mutatio] *om.* P **20** formae unius] *inv.* GPp **21** inconveniens] *add.* quod p ‖ affirmatum] affirmato p ‖ dicitur] *add.* in G : dicetur in Pp

---

2 Non invenimus in octavo libro. Cf. Iohannes Buridanus, *Quaestiones super libros Physicorum*, VI, q. 6 (ed. Parisiis 1509, f. 99ra–99vb), ubi legimus: 'In octavo huius dicit Aristoteles quod instans transmutationis debeat attribui posteriori passioni et hoc videtur esse intelligendum de transmutatione instantanea. Quamvis ergo de hoc quaerere pertinet et directe ad octavum librum, tamen ... quaeremus modo de hoc et erit quaestio sexta utrum instans transmutationis debeat attribui posteriori passioni, id est termino ad quem'. **22** Cf. Aristoteles, *Physica*, V, 1, 225b1–5

in qualitatem totalem quae dicitur rubedo, quam in illam quae simpliciter dicitur nigredo. Non enim totalis qualitas dicitur nigredo statim quando est aliquis gradus nigredinis.

⟨12⟩ Ad aliam potest dici quod ad generationem rubedinis vel viriditatis non solum concurrunt gradus albedinis vel nigredinis, quia per commixtiones iretur continue de albissimo in nigerrimum sine hoc quod transiretur per dictas qualitates. Ideo ad eas cum albedine et nigredine concurrunt vel lux vel diaphaneitas vel opacitas vel aliquid tale per quod illi colores medii sic specifice diversificantur. Sed de hoc debet esse specialis perscrutatio in *De sensu et sensato*.

⟨13⟩ Ad aliam dico quod est in proposito, sicut est de appositionibus et divisionibus magnitudinum. Nam si illud quod resecatur ab una apponatur alteri et ab una resecatur medietas, ita quod fiat in duplo minor, secundum talem modum loquendi non oportet quod magnitudo cui resecatum apponitur fiat in duplo maior quam ante. Unde quantum resecatur ab hac, tantum apponitur illi, tamen non oportet, si hoc diminuatur per dimidietatem, quod alterum duplicetur.

⟨14⟩ Ad aliam dicendum est quod intensio et remissio qualitatis non est intrinsece ex mixtione contrarii ad gradum eiusdem rationis et speciei vel ex ablatione gradus a gradu. Tamen in quibus qualitatibus est per se et proprie motus alterationis, haec consequuntur, scilicet additio gradus unius qualitatis et remotio alterius et e converso.

Item etiam consequenter videndum est quomodo rationes in oppositum procedunt.

⟨1–2⟩ Dicendum est igitur quod termini contrarii et propositiones | con- 56ra P
trariae bene stant simul in eodem subiecto, ut dictum est. Differentiae autem generum sunt termini significativi sicut et ipsa genera. Sed de illis in speciali est discutiendum in decimo *Metaphysicae*.

1–2 simpliciter dicitur] *inv.* P 2 quando] quod Gp 5 gradus ... nigredinis] albedines vel nigredines P 5–6 commixtiones] mixtiones GP 6 iretur continue] *inv.* P 8 aliquid] aliquod p ‖ colores medii] *inv.* p 11 est^2] *om.* GPp 12 illud] *om.* GPp 14–15 apponitur] apponatur p 16 illi] *add.* et GPp ‖ diminuatur per dimidietatem] dimidiatur Pp 17 duplicetur] *add. sup. lin.* dupletur C : dupletur Pp 18 dicendum est] dicitur P 20 ablatione] alteratione C ‖ a gradu] ad gradum G 21–22 additio ... remotio] addito gradu unius qualitatis et remoto C 27 illis] aliis P 28 est] *add.* dicendum seu P

10 Cf. Iohannes Buridanus, *Quaestiones super De sensu et sensato*, q. 13 (ed. Parisiis 1516, ff. 35rb–36ra) 28 Cf. Iohannes Buridanus, *Quaestiones super libros Metaphysicorum*, X, q. 2 (ed. Parisiis 1518, ff. 61vb–62rb)

⟨3⟩ Corruptibile autem et incorruptibile non sunt contraria. Sunt enim isti termini 'corruptibile' et 'incorruptibile' contrarii, id est privative oppositi.

⟨4⟩ Opinio Commentatoris de intellectu humano quod sit idem in omnibus hominibus est neganda.

⟨5–10⟩ Omnes aliae rationes et auctoritates arguunt quod cum aliquibus gradibus frigiditatis possunt stare aliqui gradus caliditatis. Et hoc est concessum. Et ad istum sensum debent exponi auctoritates. De gravitate tamen et levitate aeris facienda est specialis pertractatio in quarto *Caeli et mundi*.

66[ra] C 44[vb] p Et sic patet quaestio praesens et etc. |

1–2 sunt[2] ... incorruptibile] sed isti termini corruptibile et incorruptibile sunt Pp **3–4** omnibus hominibus] *inv.* p **6** possunt] possint p ‖ gradus] *om.* p **7** et[1]] *om.* P ‖ tamen] autem P **8** aeris] *add.* et aquae GP : *add.* et aquae frigidae p ‖ pertractatio] perscrutatio Pp **9** et[1] ... etc.] et sic est finis P : etc. et sic est finis quaestionis p : *om.* G

---

8 Cf. Iohannes Buridanus, *Quaestiones super libros De caelo et mundo*, IV, q. 7 (ed. Moody, 264–269)

⟨III.4⟩

⟨Utrum qualitas secundum quam est alteratio per se et proprie dicta, continua et temporalis, acquiratur tota simul vel pars post partem⟩

Quaeritur quarto utrum qualitas secundum quam est alteratio per se et proprie dicta, continua et temporalis, acquiritur tota simul vel pars post partem, verbi gratia caliditas quae apud calefactionem acquiritur, utrum acquiritur tota simul vel pars post partem.

⟨1⟩ Arguitur quod tota simul per hoc quod omnis forma est indivisibilis; ideo statim nihil acquiritur vel totum. Sed quod omnis forma sit indivisibilis et simplex patet per auctorem *Sex principiorum*, | qui dicit definiendo formam 54vb G quod forma est simplici et invariabili essentia consistens; et definitio debet convenire omni contento sub definito.

⟨2⟩ Item sicut se habet motus localis ad locum vel ubi, ita alteratio ad qualitatem; sed locus ad quem corpus naturale movetur naturaliter, non acquiritur continue per motum localem pars post partem, sed in fine totus simul; igitur etc. Maior patet per simile. Minor declaratur imaginando quod lapis existens sursum in aere descendat in aerem et iterum per aquam, donec sit in fundo aquae, scilicet in eius infima superficie. Constat quod, quamdiu movetur per aerem, adhuc nihil habet de illo loco infimo, scilicet de illa superficie aquae, immo nec etiam quamdiu descendit per aquam, sed solum in fine habet illum locum, id est continetur ab illa superficie. Et ita per illum motum non acquirit continue illum locum partem post partem.

---

5 et] *om.* p 7–8 verbi ... partem] *in marg. sup.* C 8 acquiritur tota simul] illa caliditas tota simul acquiritur p 9 arguitur] *praem.* et G 11 auctorem] auctoritatem P : auctoritatem auctoris G 12 consistens] existens C 13 omni] cuilibet P 14 locum] locus C : motum P ‖ ubi] A : ad ubi T : ubi (*sup. lin.*) ibi U : *add.* vel ibi P : *add.* ad ibi BCGL : *add.* ad ubi HMp 15 qualitatem] quantitatem G 17 etc.] *om.* P 18 descendat] descendit Pp 19 sit] *om.* P ‖ fundo] profundo P 20 infimo] inferiori C ‖ illa] *om.* p 21 descendit] dependit G 22 locum] locus C ‖ id est] et P 23 locum] *add.* per GPp

---

11 Anonymus, *Liber sex principiorum*, I, 1 (ed. Minio-Paluello, 35); cf. *AA*, 33: 1

 | DOI: 10.1163/9789004322356_007

⟨3⟩ Item quaereretur utrum pars primo acquisita remanet cum parte quae posterius acquiritur vel non. Et horum utrumque posset argui esse impossibile, sicut videbitur in alia quaestione.

⟨1⟩ Oppositum arguitur auctoritate Commentatoris tertio *huius* describentis motum, dicentis quod motus est generatio partis post partem illius perfectionis ad quam continue tendit mobile.

⟨2⟩ Item in sexto *huius* dicit Aristoteles quod necesse est esse easdem sive proportionales divisiones temporis et motus et ipsius moveri et quod movetur et in quo movetur; et per hoc quod dicit 'in quo movetur' intelligit 56[rb] P formam illam | quae per motum acquiritur vel qualitatem vel magnitudinem vel spatium quod pertransitur. Modo qualitas illa non esset proportionaliter divisibilis tempori, nisi acquireretur temporaliter pars post partem. Et ob hoc etiam dicit Aristoteles in eodem sexto quod omne quod movetur est partim in termino a quo et partim in termino ad quem.

Notandum est quod qualitas dupliciter imaginatur divisibilis. Uno modo 66[rb] C secundum divisionem | quantitativam subiecti. Et secundum illum modum quaerere utrum qualitas acquireretur pars post partem est idem quod quaerere utrum alterabile prius alteratur secundum unam eius partem quam secundum aliam, et posterius secundum illam aliam. Ad hoc potest faciliter responderi quod, cum alteratio fiat per contactum agentis extrinseci ad passum, ut si ignis calefaciat aquam in potto supra ignem, alterabile prius alteratur secundum partem propinquiorem agenti quam secundum remotiorem. Unde sic intendit Aristoteles in sexto *huius* quod alterabile dividitur proportionaliter tempori. Tamen postquam perventum est ad hoc quod ultima pars incipit alterari, tunc bene alterabile secundum omnes eius partes alteratur

---

2 posset] potest GPp ‖ argui] *add.* posse C 2–3 esse impossibile] *inv.* P 3 alia] ista p 4 auctoritate ... describentis] per commentatorem tertio huius et aristotelem describentes P 5 dicentis] dicens G : *om.* Pp 6 tendit mobile] *inv.* P : etc. Gp 7 in] *om.* G ‖ huius] *om.* Pp 8 proportionales] proportionabiles p ‖ divisiones] dimensiones G 10 qualitatem] caliditatem p 11 proportionaliter] proportionabiliter p 12–13 ob hoc etiam] hoc P 15 notandum est] nota G 17 quaerere] quaereretur C ‖ acquireretur] acquiritur GPp ‖ est idem quod] et est idem C 19 et] aut C ‖ illam] aliam et CG ‖ ad] *praem.* et p 20 responderi] respondere P ‖ extrinseci] *post* passum (20–21) P 21 potto] poculo G 23 in] *om.* Gp 23–24 proportionaliter tempori] proportionabiliter tempore p 24 est] esset P

---

3 Cf. inf., III, q. 5 4 Averroes, *In Physicam*, III, comm. 4, f. 78C–D 7 Aristoteles, *Physica*, VI, 4, 235a15–17 13 Cf. Aristoteles, *Physica*, VI, 4, 234b15–16 23 Aristoteles, *Physica*, VI, 4, 235a15–17

simul, ita quod in qualibet parte quantitativa qualitas intenditur, licet sit semper qualitas intensior usque ad cessationem alterationis in parte propinquiori alteranti quam in parte remotiori, nisi forte ex alio casu ali|ter 45ra p contingeret.

Alio modo qualitas imaginatur divisibilis in partes graduales simul existentes in eodem subiecto sine differentia situs, secundum quas subiectum dicitur magis tale et minus tale, et quod huiusmodi partium una prius acquiritur, deinde alia et alia, et sic continue efficitur illud subiectum magis tale et minus tale.

Multi igitur dubitaverunt | utrum caliditas est sic divisibilis et utrum sic pars 55ra G eius post partem acquiratur. Et quidam de hoc posuerunt duas conclusiones: prima fuit quod qualitas acquiritur tota simul essentialiter; secunda fuit quod non acquiritur tota simul gradualiter.

Primam conclusionem nituntur probare multis rationibus quia:

⟨1⟩ Vel illae partes qualitatis essent eiusdem rationis ad invicem vel diversarum; sed neutro modo potest dici. Probatio: primo non potest dici quod diversarum, quia sic non ponuntur formae esse compositae, sed simplices. Immo etiam hoc nomen 'caliditas' est una species specialissima vel hoc nomen 'albedo'; et quaelibet talis pars caliditatis est caliditas essentialiter; ideo omnes illae partes essent eiusdem speciei secundum suas essentias. Et sicut aliqua sunt eiusdem speciei, ita sunt eiusdem rationis. Igitur non posset dici quod illae partes essent ad invicem diversarum rationum. Sed etiam probatur quod non essent eiusdem rationis quia: sint quattuor partes *a*, *b*, *c*, *d* existentes in eodem subiecto et secundum eandem partem eius. Tunc alterans per consequens est omnino aequaliter et consimiliter approximatum uni earum sicut alteri. Igitur non est ratio aliqua quare illud agens prius corrumperet *a* quam *b* vel *c* | aut e converso. Igitur vel corrumpit omnes 56va P istas simul vel nullam corrumpit. Tunc igitur, si calefaciens | nullam partem 66va C frigiditatis corrumpat, sequitur pari ratione quod nullam partem caliditatis

2 usque ad] post p ‖ cessationem] cessionem C 3 forte] *add.* est G ‖ aliter] similiter C 4 contingeret] contingit G : contigerit Pp 5 graduales] *add.* scilicet GPp 6 quas] quos Gp : quod C ‖ subiectum] solum G 7 huiusmodi] huius P 8 et alia] *om.* P ‖ sic] *add.* tandem et P ‖ magis] *om.* C 10 caliditas] qualitas p ‖ et utrum] ut C 11 acquiratur] acquiritur Pp : *om.* G 13 quod] *add.* qualitas p 14 rationibus] *add.* primo GPp 15 vel[1]] *post* qualitatis Pp : *om.* G 16 primo] *add.* quod P 17 diversarum] *add.* rationum GPp 19 talis pars] *inv.* P ‖ est] esset GPp 20 ideo omnes illae] igitur illae omnes p 21 aliqua] aliquae P 21–22 posset] potest P 22 ad invicem] *ante* essent P : *post* rationum p 23 partes] *add.* scilicet p 26 aliqua] *om.* P 27 quam b vel] vel b quam P 28 istas simul] *inv.* Gp 29 corrumpat] corrumpit GPp

generat. Et si corrumpit omnes illas partes frigiditatis simul, tunc etiam pari ratione generat omnes illas partes caliditatis simul. Et ita sequitur quod totalis caliditas acquiritur simul, non pars gradualis post aliam. Et ista ratio est bene difficilis.

⟨2⟩ Item caliditas non manet eadem quantum ad gradus per totum motum; ergo si non maneat una et eadem essentialiter, ipsa nullo modo manet una. Et tamen non maneret eadem essentialiter, si pars essentialis post aliam partem essentialem generaretur, cum nec totum sit idem parti nec partes eaedem inter se. Sed si illa caliditas non maneat una nec essentialiter nec gradualiter, sequitur quod calefactio non poterit dici unus motus, quod est falsum. Et patet consequentia, quia dicitur in quinto *huius* quod ad unitatem motus requiritur unitas formae vel dispositionis secundum quam est motus.

⟨3⟩ Item sicut manet idem nomen significans formam, sic manet forma eadem. Sed si caliditas continue intendatur, nomen significans gradualiter bene mutatur, ut quia prius est minus calidum, postea magis calidum, sed nomen significans simpliciter essentiam non mutatur; est enim calidum ante et post. Unde calidum non fit calidum, sed fit calidius.

⟨4⟩ Iterum hoc illi confirmant auctoritate Aristotelis quinto *Metaphysicae* dicentis quod diversa sunt specie quaecumque in eadem substantia entia habent differentiam. Modo illae partes caliditatis haberent ad invicem differentiam, quia haec non esset illa, et essent omnino in eadem substantia, scilicet in eodem subiecto, et secundum eandem partem eius. Ergo istae partes essent ad invicem differentes specie, quod reputatur falsum.

Deinde per hanc primam conclusionem probatur secunda conclusio quia:

⟨1⟩ In omni motu necesse est esse aliquam successionem alicuius ad alte-
45[rb] p rum; sed in calefactione nulla esset talis successio, quando idem sub|iectum,

---

1 si … simul] A : si corrumpet omnes illas partes scilicet frigiditatis simul HM : si corrumpit omnes illas partes simul LU : si omnes partes illas caliditatis corrumpat simul T : si omnes (*add. sed del.* alias) illas partes caliditatis corrumpat simul C : si corrumpat omnes illas partes caliditatis simul P : si corrumpit omnes illas partes caliditatis simul p : sic corrumpit omnes illas partes caliditatis simul G : sic corrumpit omnes illas partes simul scilicet caliditatis B 2 caliditatis] frigiditatis C 3 simul] *add.* et G 5 eadem] *om.* P 6 maneat] manet GP ‖ nullo modo] C ?, *corr. ex* uno modo 8 parti] toti G ‖ partes] *add.* sint G 9 maneat] manet P 10 poterit] potest P 11 dicitur] *post* huius P 13 sicut] si C 14 si] cum p 15 est minus] *inv.* P 17 fit[2]] *om.* P 18 confirmant] confirmat G 21 essent] esset G ‖ in] entia p 24 primam] *om.* P 27 talis] totalis P

---

11 Cf. Aristoteles, *Physica*, V, 4, 227b27–28 18 Cf. Aristoteles, *Metaphysica*, V, 10, 1018b1–7

cum sit calidum, fit calidius, nisi acquiratur caliditas pars post partem vel secundum distinctionem essentialem vel secundum | (55rb G) gradualem; sed dictum est quod non secundum essentialem; ergo dicendum est quod secundum gradualem.

⟨2⟩ Et confirmatur ista conclusio quia: sicut mutatur nomen significans formam, ita proportionaliter debet concedi quod mutetur forma; sed licet non mutetur nomen significans formam simpliciter, scilicet essentiam formae, ut 'calidum' vel 'caliditas', tamen mutatur nomen significans formam gradualiter, ut 'minus calidum' et 'magis calidum'.

Ista opinio non videtur mihi vera, nisi exponatur ad sensum improprium. | (66vb C) Videtur enim imaginari quod sint gradus distincti ab essentia caliditatis et quod continue sit alius et alius | (56vb P) gradus et non sit alia et alia essentia. Et hoc non credo esse verum; et ideo pono aliquas conclusiones.

Prima est quod continue in calefactione aliquid caliditatis acquiritur prius et aliquid caliditatis acquiritur posterius in eodem subiecto et secundum eandem partem eiusdem, quia necesse est quod, si *a* calefit continue, quod continue fit minus calidum et magis calidum. Sed ipsum sic aliter et aliter se habere non posset salvari nisi per aliquam dispositionem posterius existentem quae non erat prius vel e converso, cum non possit salvari per habitudinem ipsius *a* ad aliquod extrinsecum, quia omni extrinseco per intellectum circumscripto, retento solum quod continue calefiat, adhuc sic aliter et aliter se haberet. Et cum hoc salvari non posset per diversam habitudinem vel situm partium ipsius ad invicem. Nam in quaestione de distinctione figurae a figurato visum fuit quod non potest fieri nisi altero dictorum modorum quod aliquid aliter se habeat prius et posterius. Sed concesso quod

---

1 cum sit] quod fit p ‖ acquiratur] acquireretur Gp ‖ partem] *add.* quia C 3 est quod[2]] *inv.* G 5 quia sicut] sicut enim p 6 proportionaliter] proportionabiliter G ‖ sed] scilicet G 7 scilicet] secundum Cp 8 caliditas] calidius P 9 gradualiter] gradualem CG ‖ magis] maius P ‖ calidum[2]] *add.* etc. G 10 videtur] apparet Pp ‖ sensum improprium] bonum sensum G 11 imaginari] imaginandum G 13 esse verum] *inv.* P ‖ pono] ponam P 14 prima] *add.* conclusio p 15 caliditatis acquiritur] *om.* P 16 eiusdem] eius Pp 17 quod] et G ‖ fit] sit GPp ‖ minus calidum] *corr. sup. lin. ex* magis C : minus P ‖ ipsum] ipsam p 17–18 et aliter] *sup. lin.* C : *om.* GPp 18 posset] possit G 18–19 posterius] *sup. lin.* C : *om.* GP 19 cum] tamen (alias cum *in marg.*) P 20 ad] *add.* b G ‖ aliquod] aliquid P 21 sic] *om.* P 22 hoc] *add.* etiam Gp ‖ posset] possit GPp 24 a] et P ‖ fieri] salvari (*post* modorum (25)) GPp ‖ altero] *add.* prius p 25 aliquid] aliquis p ‖ aliter] *add.* et aliter Gp ‖ prius et] ad p

---

23–24 Cf. Iohannes Buridanus, *Quaestiones super libros Physicorum*, II, q. 3 (ed. Streijger, Bakker, 261^{27}–262^{25})

necesse est, si *a* sit calidius quam ante, aliquid esse quod ante non erat vel e converso, non solum extrinsece, sed in ipso *a*, omnes concederent quod illud aliud sit caliditas vel aliquid caliditatis, scilicet pars aut gradus aut huiusmodi.

Secunda conclusio est quod gradus caliditatis non est alia res a caliditate, scilicet si ponatur quod in calefactione illud quod prius acquiritur et quod posterius acquiritur vocetur gradus caliditatis. Tunc probatur conclusio quia: ponamus quod iam *a* sit calidum et quod continue per totam diem fit calidius; igitur continue acquiritur alius gradus post alium; sed numquam acquiritur caliditas quae iam erat; ergo cum omnis illa alteratio sit calefactio, sequitur quod calefactio non est motus ad caliditatem tamquam ad terminum ad quem, quod est inconveniens. Patet consequentia, quia ad illud quod iam est totum acquisitum non est motus tamquam ad terminum ad quem.

Item omnino illi gradus additi ponerentur frustra, quia quidquid salvaretur ponendo illos, salvaretur sine illis, scilicet successio, ut alias dicebatur, et continuitas, quia ita possumus dicere quod continue acquiritur pars caliditatis post partem sicut gradus post gradum, et unitas motus, quia ad unitatem etiam numeralem motus non oportet aliud quam subiectum manere 67ra C totaliter idem, immo de forma vel dispositione secundum quam est motus | sufficit quod sit unitas continuitatis in succedendo partem post partem, quod salvatur ponendo quod caliditas continue acquiritur pars post partem sine gradibus aliis. Ideo superflue ponuntur.

55va G Item quaecumque rationes alicuius apparentiae videntur arguere | quod caliditas non sit partibilis in partes quae acquiruntur una prius, alia posterius in eodem subiecto, illae similiter arguerent de illis gradibus, quoniam 45va p quicumque gradus signaretur, ille esset una forma ac|cidentalis, sicut cali- 57ra P ditas, et esset divisibilis in partem quae | prius acquiritur et in partem quae

---

1 sit] fit G ‖ ante[2]] *sup. lin.* C : *om.* P 2 sed] *add.* scilicet P 3 aut[2]] et CP 6 scilicet] *om.* G ‖ ponatur] ponamus Gp 7 quod] *om.* G ‖ vocetur] vocentur Gp 7–8 probatur conclusio] ergo conclusio probatur Gp 8 iam] *post* a G : *om.* p ‖ per] *add.* unam Gp 9 fit] fiat Gp ‖ calidius] calidus G ‖ igitur] *corr. sup. lin. ex* tunc C : *praem.* tunc GPp ‖ alius] *post* gradus Gp : unus P ‖ numquam] non Gp 10 quae] quia G 12 inconveniens] *add.* et Gp 13 iam] non G 15 quia] *om.* P 15–16 salvaretur] salvatur p 17 possumus] possemus G 18 et] *om.* G 19 numeralem] naturalem p 21 sit] sicut p ‖ in] *om.* G ‖ partem[1]] pars G 24 apparentiae] apparentis p 25 caliditas] *add. sup. lin.* alias qualitas C : qualitas GPp ‖ non] quae sic G : *om.* p ‖ partibilis] *add.* sic P ‖ prius] *add.* et Gp 26 illae] illi p ‖ illis] aliis G 27 quicumque] qui P ‖ gradus] *add.* significaretur vel P ‖ signaretur ille esset] signarentur illi essent C 28 et[1]] *om.* C

---

16 Cf. sup., III, q. 2, 17[22–26]

posterius acquiritur. Et tunc quaereretur utrum illae partes essent eiusdem rationis ad invicem vel diversarum, sicut quaerebatur de caliditate. Et sic nihil prodest ponere tales gradus additos.

Item hoc esset ponere in alteratione fluxum additum qualitati secundum quam est alteratio, quod prius fuit improbatum.

Tertia conclusio est quod qualitas secundum quam est alteratio continue acquiritur una pars eius post aliam etiam in eodem subiecto et secundum eandem partem subiecti illius. Et haec conclusio sequitur ex duabus praecedentibus. In prima enim conclusione dictum est quod continue acquiritur aliquid prius et aliud posterius; et haec non sunt gradus sive res distinctae a caliditate sive a partibus caliditatis, ut dicit secunda conclusio.

Tunc ad rationes alterius opinionis dicendum est.

⟨1⟩ Ad primam quod illae partes sunt eiusdem rationis et speciei specialissimae de praedicamento qualitatis. Ad hoc autem quod contra hoc arguebatur respondebitur in alia quaestione.

⟨2⟩ Ad aliam dicitur quod caliditas per totam alterationem est una numero etiam essentia caliditatis, non unitate indivisibilitatis, sed unitate continuitatis in succedendo partem post partem. Immo est una verius quam Sequana, quae tamen per centum annos dicitur idem fluvius numero, scilicet unus. Dico enim quod caliditas manet verius una, inquantum pars prius acquisita manet cum parte posteriori sive quae posterius acquiritur, ut dicetur in alia quaestione, et fit ex eis una totalis caliditas.

⟨3⟩ Ad aliam respondetur similiter. Essentia enim caliditatis quae non est aliud quam caliditas manet eadem modo praedicto, sed tamen habet diversas partes quarum quaelibet est caliditas.

---

1 quaereretur] quaeretur P 2 et] *om.* P 4 item] et iterum Gp ‖ in alteratione] *rep.* P 4–5 secundum … alteratio] secundum quam est (*sup. lin.*) alteratio C : *om.* P 5 prius fuit] *inv.* GPp 7 eius] *sup. lin.* C : *om.* P ‖ etiam] et G 8 subiecti illius et] illius subiecti GPp ‖ duabus] *add.* conclusionibus p 10 aliquid] unus G : aliud P ‖ aliud] alius G ‖ et[2]] *om.* P ‖ res] ita G 11 secunda] *om.* p ‖ conclusio] *add.* ergo etc. GPp 12 tunc] *add.* ergo Gp 12–13 alterius … et] *om.* G 13–14 specialissimae] *add.* eiusdem p 14 praedicamento] prima specie p ‖ hoc[2]] *sup. lin.* C : *om.* P 14–15 arguebatur] *corr. ex* arguitur (?) C : arguitur GPp 16 totam] totalem GPp ‖ est] esset G 17 etiam] *praem.* et G : et p 18 immo] *add.* etiam G ‖ una verius] *inv.* p ‖ quam] *add.* sit G 19 quae] qui Pp ‖ centum] decem P ‖ dicitur idem] dicitur unus Gp : dicebatur ille P 19–20 scilicet unus] *om.* Gp 20 manet verius una] HUp : verius manet una T : verius est una M : manet verius in a ACGLP : manet verius in illa B 21 posteriori sive] *om.* Gp 23 similiter] consimiliter GPp 24 habet] habens GPp

---

5 Cf. sup., III, q. 2 15 Cf. inf., III, q. 5 22 Cf. inf., III, q. 5

67[rb] C ⟨4⟩ Ad auctoritatem Aristotelis dico quod per | 'habere differentiam' intendebat idem quod esse diversarum rationum secundum suos conceptus absolutos et essentiales, scilicet qui fiunt sine connotatione aliena. Sic enim non differunt quae sunt eiusdem speciei.

⟨1–2⟩ Ad alias, quae arguunt quod sit diversitas inter caliditatem et gradus caliditatis, dicitur quod illi gradus sunt verae partes caliditatis et verae partes ex quibus remanentibus constituta est una totalis caliditas.

Ad rationes principales.

⟨1⟩ Ad primam dicendum est quod omnis forma proprie dicta sic est indivisibilis quia non est divisibilis in partes diversarum rationum essentialium, sed bene est divisibilis in partes eiusdem rationis.

⟨2⟩ Ad aliam quae arguit de motu locali dicetur in quaestione quae de hoc fiet in speciali.

⟨3⟩ De tertia ratione fiet quaestio sequens etc.

1–2 intendebat] intelligit G : *praem.* ipse p **2** conceptus] effectus p **3** essentiales] essentiam P ‖ scilicet] *om.* G ‖ qui] quae GP ‖ connotatione] contrarietate G ‖ sic] si G ‖ non] *post* quae (4) p **6** dicitur] dico G ‖ sunt … caliditatis²] *rep.* G ‖ verae²] substantiae P **9** ad primam] ad primum p : *om.* P **10** rationum] *add.* scilicet p **13** in speciali] specialis GPp **14** etc.] et sic est P : *add.* sequitur quinta p

---

12–13 Cf. inf., III, q. 8

⟨III.5⟩

⟨Utrum in alteratione pars qualitatis quae prius acquiritur maneat cum parte quae posterius acquiritur⟩

Quaeritur quinto utrum in alteratione pars qualitatis quae prius acquiritur manet cum parte quae posterius acquiritur.

⟨1⟩ Arguitur primo quod non quia: sicut est de motu locali, ita est de alteratione, | [57rb P] per simile; sed in motu locali mobile non retinet locum prius acquisitum cum loco quem posterius acquirit, sed exit ab eo; igitur etc.

⟨2⟩ Item vel istae qualitates prius vel | [55vb G] posterius acquisitae remanerent distinctae ab invicem vel indistinctae; sed neutrum potest dici; igitur etc. Primum non potest dici, scilicet quod indistinctae, quia prima pars est, quando nondum est secunda; igitur prima pars non est illa secunda; et ipsa numquam erit illud quod ipsa non est; igitur numquam erit illa secunda; ideo erit distincta ab ea. Sed probatur quod non essent distinctae quia: non secundum speciem, ut dicebatur in alia quaestione; nec secundum numerum, quia earum materia esset una, scilicet subiectum, et tamen | [45vb p] dicitur quinto *Metaphysicae* quod illa sunt unum numero quorum materia est una.

⟨3⟩ Iterum omnes concedentes quod pars prius acquisita manet cum parte quae posterius acquiritur, dicunt quod per appositionem partis cum parte fit qualitas intensior et per abiectionem partis a parte fit remissior, ita quod idem subiectum fit calidius quam ante per hoc quod cum caliditate praecedente advenit alia caliditas in eodem subiecto. Igitur si hoc possit improbari, debebit concedi quod non sic manet. Ideo descendo ad hoc improbandum.

---

5 quaeritur] *praem.* deinceps G : *praem.* ideo consequenter p **7** primo] *om.* P ‖ est[1]] *om.* Gp ‖ est[2]] *om.* Gp **8** retinet] retineret C **9** quem posterius acquirit] qui posterius acquiritur G ‖ sed] *add.* sic G **10** remanerent] remanent G **11** igitur etc.] *om.* Gp **12** primum ... quod] primo non potest dici quod Pp : primo non potest dici G **13** igitur] *om.* P **15** ea] eo p ‖ essent distinctae] est distincta C **16** dicebatur] dicitur p **17** earum] eorum CP **18** dicitur quinto] arguitur primo G ‖ unum] *add.* in Gp **22** fit[2]] fieret P **23** idem] illud p **24** hoc possit] *inv.* P

---

16 Cf. sup., III, q. 4 **18** Aristoteles, *Metaphysica*, V, 6, 1016b32–33; cf. *AA*, 1: 129

 | DOI: 10.1163/9789004322356_008

⟨3.1⟩ Et arguo primo sic quia: sequeretur quod lumen esset infinite intensum in hac domo, quod est absurdum. Consequentia probatur quia: si continue per unam horam approximetur lucidum huic domui, tunc continue lumen fit intensius, ita quod in quolibet instanti esset lumen intensius quam esset ante illud instans, ita quod in quolibet instanti illius horae generaretur 67va C lumen. Tunc igitur, cum sint ibi infinita instantia, tunc signetur una pars | unius instantis. Infinitae tantae partes generabuntur in aliis instantibus infinitis et manebunt simul. Sed quocumque gradu signato quantumcumque parvo, infinitum esset illud quod esset ex infinitis tantis aggregatum.

⟨3.2⟩ Item secundo ad idem arguitur sic: sequeretur quod albedo esset infinitae perfectionis, quod est falsum. Probatur consequentia supponendo primo quod in ordine essentialis perfectionis entium una species excedit aliam secundum omnia eius supposita, ita quod omnis equus est omni asino perfectior essentialiter, licet potest esse e converso de aliquibus perfectionibus accidentalibus. Ita erit concedendum quod omnis albedo est omni nigredine perfectior perfectione essentiali. Secundo etiam supponitur quod albedo una tanto est alia albedine perfectior, quanto est ea albedine intensior, si sit intensio per additionem gradus ad gradum, quia sic dupliciter intensa ad aliam esset ad illam dupla in essentia et per consequens in perfectione essentiali. Tunc arguitur sic: haec albedo est in duplo perfectior sua 57va P medietate | et in triplo perfectior sua parte quae est sua tertia pars et sic in infinitum. Et sic est aliqua albedo in infinitum perfectior. Et tamen omnis albedo omni nigredine est perfectior. Igitur haec albedo est in infinitum perfectior nigredine, quod non est possibile, nisi sit infinitae perfectionis.

Vel arguitur sic: in infinitum est minus albedo perfecta quam sit haec albedo signata, quia quanto est minor gradualis, tanto est minus perfecta;

---

1 sequeretur] sequitur p ‖ lumen] aliqua P 3 approximetur lucidum] *inv.* GPp ‖ domui] domi p 5 esset … instanti] *om.* (*hom.*) C 7 unius] illius G 7–8 instantibus infinitis] *inv.* GPp 8–9 quantumcumque parvo] quamcumque parvus G 9 infinitum] *praem.* in G ‖ tantis] causatis P ‖ aggregatum] congregatum Gp 10 item] *om.* Gp ‖ sic] *add.* quia G ‖ sequeretur] sequitur p 12 essentialis] HM : essentiali ABCGLPTUp ‖ excedit] exedat C 14 potest] possit G : posset p 14–15 perfectionibus accidentalibus] perfectionibus accidentalibus (*corr. sup. lin. ex* essentialibus) C : *add.* et p : actionibus et G 15 concedendum] concedendo G 16 secundo] ita G 17 tanto est] *inv.* p ‖ ea] *corr. sup. lin. ex* alia C : illa G ‖ albedine[2]] *om.* GPp 18 additionem] albedinem G ‖ sic dupliciter] sicut dicitur p 19 aliam] *add.* ita p ‖ illam] aliam C 21 medietate] medie p ‖ parte … pars] tertia medietate *sed add. in marg.* sua parte quae est sua †…† pars quae pars sit in †…† imperfectior toto C 22 in[1]] *om.* p 23–24 perfectior] *add.* omni P 24 nigredine] nigredini G 25 albedo perfecta] *inv.* Gp 26 est[1]] sit P ‖ minor] minus (*del.*) *sed add. sup. lin.* alias minor C : minus P

et | tamen est in infinitum minor gradualis. Sed omni albedine haec nigredo 56ra G est minus perfecta. Igitur in infinitum haec nigredo est hac albedine minus perfecta. Sed hoc impossibile esset, nisi haec albedo esset in infinitum perfecta.

⟨3.3⟩ Item tertio arguitur sic: sit *a* calefaciens *b* continue per totam diem usque ad summam caliditatem; quod quidem *b* erat frigidissimum. In medietate diei corrumperet medietatem frigiditatis inducendo aliquantam caliditatem. Igitur cum *b* non resistat calefacienti nisi per suam frigiditatem, sequitur quod post illam medietatem primam diei *b* erit in subduplo minoris resistentiae. Ideo alteratio erit velocior in duplo. Propter quod sequitur quod in secunda medietate proportionali diei tantum de caliditate acquiretur, quantum acquirebatur in prima medietate; et pari ratione ita erit de qualibet alia medietate proportionali diei. Igitur cum infinitae sint huiusmodi medietates proportionales, sequitur quod infinitae | erunt partes calidita- 67vb C tis acquisitae, quarum quaelibet erit tanta quanta erat illa quae in prima medietate acquirebatur. Igitur si manent simul, caliditas in fine diei erit infinita. Consequentia patet, quia quacumque quantitate data sive graduum sive aliorum, compositum ex infinitis tantis esset infinitum.

⟨3.4⟩ Item quarto arguitur quia: per | dictum modum non posset fieri qua- 46ra p litas intensior quam ante, quia vel prior pars esset intensior quam ante vel posterior vel totum ex eis aggregatum; sed quodlibet illorum est impossibile. Probatio: primo non potest dici quod pars prior sit intensior quam ante, quia oportet imaginari sicut de magnitudine; modo si lineae pedali alia linea pedalis addatur, non propter hoc illa linea prima pedalis erit maior quam ante, immo solum esset pedalis, licet congregatum esset bipedale. Ita igitur prima pars caliditatis propter additionem secundae non esset intensior quam ante, licet congregatum esset intensius. Sed etiam non potest dici de hoc congregato, quia quod intenditur vel fit intensius quam esset ante, necesse est quod prius sit minus intensum et posterius magis intensum; sed

---

1 et] *rep.* G : *om.* P ‖ tamen] *corr. sup. lin. ex* cum C : tunc G 3 hoc] *sup. lin.* C : *om.* GPp ‖ esset[1]] *corr. ex* est C : est GPp 5 item] *om.* Gp ‖ sic] *praem.* ad idem G : ad idem si p ‖ totam] totum istum G 6 b] *om.* P ‖ erat] esset p 7 diei] *add. sup. lin.* et C ‖ inducendo] indicendum p 8 resistat] resistit G : resiat p ‖ calefacienti] *add. sup. lin.* alias caliditati C : caliditati Gp 9 primam] *om.* G ‖ b] *om.* p 10 erit] est G 11 medietate] parte P ‖ tantum … acquiretur] tantundem caliditatis acquireretur (acquiritur p) GPp 13 sint] sunt P 19 item] *om.* Gp ‖ quia] *corr. sup. lin. ex* quod C : quod Pp ‖ dictum] praedictum p ‖ posset] possit GPp 19–20 qualitas] caliditas p 23 lineae] in linea G 24 linea prima] *inv.* GPp 25 congregatum] aggregatum G 27 congregatum] aggregatum G ‖ sed] *om.* p 28 vel] cum p ‖ esset] *om.* G 29 necesse est] oportet P ‖ sit] fit G ‖ magis] maius G

congregatum minus intensum prius non erat, quia non erat prius. Et pari
57vb P ratione nec pars posterior intenditur, quia non erat ante. Igitur sic | non fit intensio formae.

⟨3.5⟩ Item quinto adhuc sequeretur quod calidum remissum producere posset in corpore frigido caliditatem valde intensam et corrumpere totam suam frigiditatem, quod apparet falsum. Consequentia probatur quia: cum omnes gradus caliditatis sint eiusdem rationis, si posset producere unum, pari ratione et centum et mille et sic deinceps, sicut Aristoteles arguit in quarto *huius* quod, si unum corpus potest penetrare alterum, pari ratione centum et mille, et quaelibet possent penetrare se. Etiam si remisse calidum posset corrumpere unum gradum frigiditatis huius corporis frigidi, pari ratione, immo fortiori, poterit corrumpere alterum et sic consequenter, quia erit minor resistentia, quando plus erit corruptum de frigiditate.

⟨3.6⟩ Item sexto adhuc oporteret dicere quod totalis calor esset intensior
56rb G sua medietate | graduali, et sic idem haberet caliditatem intensam et frigiditatem intensam; ideo esset simul intense calidum et remisse calidum, quod est impossibile.

⟨3.7⟩ Septimo arguitur quia: motus gravis descendentis deorsum fit con-
68ra C tinue velocior; et tamen non potest | dici quod intendatur gravitas per additionem gradus posterioris ad gradum priorem, quia in motu locali pars posterior non manet cum priori, ut dicetur post.

⟨3.8⟩ Et eodem modo apparet hoc de lumine quod in mane intenditur continue in hac domo. Tamen non remanet lumen prius factum cum posteriori posito quod aer movetur continue per ventum. Sic cum radius non moveatur cum aere, sed semper remaneat eodem situ ad solem, et cum etiam non moveatur de subiecto in subiectum, oportet quod continue generetur novus radius et quod non maneat genitus prius cum genito posterius.

---

1 minus ... prius[2]] non erat prius minus intensum quia non erat prius Pp : non erat minus intensum G 4 item] *om.* Gp ‖ adhuc sequeretur] ad hoc sequitur p 4–5 producere posset] *inv.* Gp 5 valde intensam] *inv.* P ‖ et] vel Gp 6 probatur] patet G 7 posset] potest GP : possunt p 8 aristoteles arguit in] arguit aristoteles GPp 9 potest] posset Gp ‖ alterum pari] aliud eadem P ‖ ratione] *add.* et G 10 quaelibet possent] quodlibet posset CP : per consequens mille possent p ‖ se] *add.* et G : *add.* quod est falsum et p 11 huius] huiusmodi p 13 quando] quanto GPp 14 item] *om.* Gp ‖ adhuc] secundum hoc p 15 idem] *add.* simul GPp 18 septimo] *praem.* item P ‖ quia] quod P 22–27 et ... posterius] *om.* Cp 23 tamen] *praem.* et P 24 movetur continue] *inv.* P ‖ sic] *add.* enim P 25 remaneat] remanet in P

---

9 Cf. Aristoteles, *Physica*, IV, 6, 213b7–8 21 Cf. inf., III, q. 6, 62–63

⟨3.9⟩ Deinde etiam arguitur quod non est possibile qualitatem sic remitti, scilicet per corruptionem aliquorum graduum aliis remanentibus quia: omnes gradus sunt eiusdem rationis ad invicem, et omnes etiam sunt uniformiter approximati corrumpenti, cum non sunt situaliter extra invicem; ideo nulla esset ratio quare aliqui plus corrumperentur quam alii vel aliqui prius quam alii; ideo multi simul corrumperentur vel omnes.

Oppositum arguitur quia:

⟨1⟩ Lumina duarum candelarum in camera illuminatarum diffunduntur simul per totum aerem camerae et manent simul per totum aerem. Quod enim remaneant distincta apparet per distinctas umbras quas facit corpus opacum in medio camerae positum, et etiam per hoc quod videmus unam candelam non videndo aliam, ex eo scilicet quod illa lumina essent simul in medio camerae, tamen unum multiplicatur ultra ad oculum meum et non alterum; quod non esset ita, si essent omnino idem in medio in quo sunt simul. | Et manifestum est propter hoc quod illa lumina sunt simul, quod tale 58ra P lumen est intensius. Aer enim intensius est illuminatus in quo illa lumina sunt simul, quam in quo propter obstaculum est unum illorum tantum. Et omnino manifestum est quod per speculum concavum congregando plures radios ad eundem locum, nos facimus lumen ita intensum quod comburit combustibile in loco congregationis radiorum positum. Ideo manifestum est quod per congrega|tionem qualitatis in eodem subiecto fit subiectum 46rb p intensius tale vel magis tale.

⟨2⟩ Iterum aliter non remaneret qualitas symbola eadem in generato et corrupto, ut caliditas, quando ex aere fit ignis; cuius oppositum dicitur secundo *De generatione*. Consequentia patet, quia tunc caliditas intenditur.

⟨3⟩ Iterum sicut est in augmentatione quantum ad extensionem, ita in alteratione quantum ad intensionem; sed in augmentatione non corrumpi-

1 deinde etiam] inde G ‖ qualitatem] caliditatem p 4 sunt] sint Gp 5 aliqui plus] *inv.* P 5–6 vel ... alii] *om.* (*hom.*) C 5 aliqui[2]] quare Pp 7 arguitur] *add.* quod secundus gradus maneat cum primo et tertius cum primo p 9 camerae ... aerem[2]] *add.* distincta ab invicem p : *om.* (*hom.*) G 10 apparet] patet P ‖ facit] faceret GPp 11 et] *om.* P ‖ videmus] video GPp 12 eo scilicet] eis G ‖ quod] *add.* si P 13 ultra] *om.* G 14 ita] *om.* p ‖ omnino idem] *inv.* p 15 est] *add.* quod Pp ‖ quod[1]] *om.* G ‖ quod[2]] *om.* p 16 intensius est] *inv.* p 17 simul] *om.* p 19 ita intensum] *inv.* P 21 qualitatis] plurium partium caliditatis p 24 et] *add.* in P ‖ ut caliditas] *om.* G 27 est] *corr. sup. lin. ex* sicut C : *om.* GPp

25 Cf. Aristoteles, *De generatione et corruptione*, II, 4, 331b21–24

tur quantitas prior adveniente posteriori per quam totum redditur extensius; igitur etc.

68rb C ⟨4⟩ Iterum sequeretur quod habitus morales essent de | facili corruptibiles, quod est falsum. Et patet consequentia, quia faciliter per similes operationes intenduntur, ut patet secundo *Ethicorum*.

56va G Dico breviter quod in vera et proprie dicta | alteratione, ut in calefactione, pars qualitatis quae prius acquiritur manet cum illa quae posterius acquiritur, quia aliter sequeretur quod qualitas in quiete signata fuisset instanter acquisita, ita quod nihil erat de ea, antequam esset tota; et ad talem non est motus per se, temporalis et continuus; et tamen omnes concedunt quod ad illam qualitatem, ut ad caliditatem, erat per se alteratio quae erat continua et temporalis.

Iterum sequeretur quod secundum illam caliditatem signatam in quiete non magis fuit motus vel mutatio temporalis quam secundum animam intellectivam; modo consequens est absurdum. Et consequentia probatur quia: sicut nihil erat de anima, antequam ipsa esset tota, sed erat mutatio temporalis secundum res alias praevias, ita nihil erat de illa caliditate, antequam ipsa esset tota, sed erat praecedens mutatio temporalis secundum alias res.

Iterum sequeretur quod in alteratione continua nulla qualitas durante illa alteratione duraret nisi per instans indivisibile. Consequens est falsum; igitur et antecedens. Falsitas consequentis patet primo, quia supponimus quod instans indivisibile nihil est, sicut dicetur in sexto libro; ideo si aliquid esset per solum instans indivisibile, ipsum nihil esset.

Secundo etiam probatur falsitas consequentis quia: si essent instantia indivisibilia, tamen ipsa non essent continua ad invicem neque proxima in
58rb P tempore; ideo etiam illa non possunt | ad invicem continuari neque fieri

---

1 quam] quem G ‖ extensius] intensius p 3 sequeretur] sequitur p 4 patet consequentia] *inv.* G 5 ut] et hoc totum GPp 6 et] *add.* in p 7 qualitatis] caliditatis C ‖ illa] parte G 8 sequeretur] sequitur p ‖ signata] assignata G ‖ instanter] instantanee p 9 ad talem] ad talem (*corr. sup. lin. ex* simile) C : ad tale p : tale G 10 motus] *post* se GPp 11 quae] quia p 13 sequeretur] sequitur scilicet p ‖ illam] *add.* qualitatem vel P 14 magis] maius G ‖ fuit] fit p ‖ secundum animam] secundam P 15 et] *om.* P 16 sicut] *om.* C 17 praevias] primas p ‖ erat] *om.* p 18 praecedens] *post* temporalis P 19 sequeretur] sequitur p ‖ nulla] una G 21 supponimus] supponitur p 22 sicut] ut P ‖ aliquid] aliquis p 23 nihil] numquam GPp 24 si] *corr. sup. lin. ex* licet C : licet GPp 25 continua] *post* invicem GPp 26 etiam] et G ‖ illa non] alia numquam P ‖ possunt] possent Gp

---

5 Cf. Aristoteles, *Ethica Nicomachea*, II, 1, 1103a14–b25 22 Cf. Iohannes Buridanus, *Quaestiones super libros Physicorum*, VI, q. 4 (ed. Parisiis 1509, ff. 96rb–98va)

proxima quae essent per instantia, scilicet quorum quodlibet duraret per solum instans; et per consequens, si omnis qualitas duraret in alteratione per solum instans, illae qualitates non possent se habere continue in alteratione; ideo alteratio non esset continua, quod est falsum, quia omnis motus est continuus.

Sed tunc consequentia principalis probatur ponendo quod fiat motus de frigidissimo ad calidissimum. Constat quod in qualibet caliditate vel in parte caliditatis acquisita vel acquisibili per talem motum posset quiescere. Posset enim quiescere in prima medietate, in prima decima et in prima centesima et sic ubique, ita quod nullum esset signabile etiam per imaginationem aut intellectum, quin ibi posset alteratio quietari saltem per potentiam divinam. Tunc igitur arguitur sic de caliditatibus in illo motu existentibus: omnis caliditas in qua mobile | posset quiescere duraret per solum instans; et in 68va C
omni caliditate quae esset in illo motu mobile posset quiescere; igitur omnis caliditas quae esset in illo motu esset per solum instans. Minor declarata est. Et maior probatur quia: sit caliditas *b* in qua mobile poterat quiescere. Tunc si quievisset, oportet imaginari divisionem temporis motus a tempore quietis; et imaginetur modo mathematico instans indivisibile intermedium, quod vocetur *c*. Tunc constat quod caliditas *b* fuit in tempore quietis. Et nihil eius fuit in tempore motus, scilicet praecedente instans *c*, quia semper erat qualitas remissior. Et tunc, si mobile non quievisset, constat quod etiam caliditas *b* non fuisset in tempore sequente instans *c* nec | aliquid eius, 46va p
quia semper fuisset intensior. Igitur vel numquam fuisset nec aliquid eius vel fuisset solum per istud instans indivisibile. Nec est simile sicut de re pure successiva, sicut post dicemus esse motum localem, quia talis res,

1 quae] sed p ‖ per instantia] per (*in marg.*) instantia G : instantia *sed add. in marg.* alias instantanea C : instantia Pp 2–3 et … instans] *om.* (*hom.*) p 2 omnis] *om.* P 3 non … alteratione] *om.* G ‖ se] *om.* p 4 esset] est P 7 frigidissimo ad calidissimum] calidissimo ad frigidissimum C ‖ quod] *add.* tunc P ‖ caliditate] *add. sup. lin.* alias qualitas (?) C ‖ in[2]] *om.* GPp 8 motum] *add.* mobile p ‖ posset[1]] possit G ‖ posset[2]] possit G 9 medietate] *add.* et GPp 10 ubique] *om.* G ‖ aut] ad P 12 sic] *om.* P ‖ caliditatibus] qualitatibus C 14 caliditate] qualitate C 15 caliditas] *add.* in qua P ‖ esset[1]] est p 16 poterat] potest P 17 quievisset] quiescit p ‖ a] cum P 18 imaginetur] imaginamur P ‖ modo] *add.* eo p 19 vocetur] vocatur Pp 20 scilicet] *add.* tempore G ‖ praecedente] praecedenti G 21 etiam] *add.* quod p 22 sequente] sequenti P 23–24 quia … vel] ergo G 23 vel] nihil P 24 vel] *om.* p ‖ solum per istud] per solum istud GP : per illud solum p ‖ per] *rep.* C ‖ sicut] *om.* p 25–52.1 sicut … successiva] *om.* (*hom.*) G ‖ res scilicet pure] res (*sup. lin.*) scilicet pure C : res pure P : esset proprie p

25 Cf. inf., III, q. 6, 62–63

scilicet pure successiva nec potest esse in instanti indivisibili, si ponatur, 56[vb] G nec in tempore quietis, sed est pars post | partem in tempore in quo mobile movetur. Unde nihil eius potest signari quod esset mobili quiescente. Et si aliquis diceret quod talis res, scilicet pure successiva, est caliditas illa quae est tempore calefactionis, tunc oporteret dicere quod nulla caliditas quae est tempore calefactionis maneret, si mobile quiesceret, quod videtur absurdum, quia magna caliditas inveniretur remanere.

Iterum arguitur ad principale quia: si non manet prius acquisitum cum posterius acquisito, tunc etiam pari ratione oportet dicere quod, quando 58[va] P remittatur caliditas nihil remanet de caliditate | priori quae erat intensior. 68[vb] C Sed | ego probo quod hoc sit impossibile quia: sequeretur quod intensissima caliditas corrumperetur instantanee, quia tota simul et quia esset in toto tempore praecedente motum et nihil post. Sed hoc est impossibile dicere, quia sequeretur quod ipsa non haberet aliquam resistentiam frigefacienti et per consequens nec caliditas sequens haberet resistentiam, quia esset minoris virtutis; ideo statim totum esset frigidum, quod est falsum.

Item sequeretur quod frigidissimum generaret per se caliditatem valde intensam, quod est falsum. Consequentia patet, quia frigidissimum remitteret per se caliditatem intensissimam et per consequens corrumperet eam totam; et tamen statim post esset adhuc caliditas valde intensa, quae non erat ante nec aliquid eius; ideo esset de novo generata, et non ab alio generante quam ab illo frigidissimo, quia non ponitur ibi aliud agens.

Aliqui respondent quod frigidissimum corrumpit per se illam caliditatem summam, sed non generat per se caliditatem aliam, sed per accidens et per modum sequelae provenit illa caliditas sequens ad corruptionem summae caliditatis.

Sed haec responsio non valet, quia illa summa caliditas ante corrumpitur quam alia generatur, ex quo non possunt simul stare haec et illa nec partes earum; et quod ante corrumpitur non generat illud quod post generatur; ideo illa summa caliditas non generat caliditatem sequentem, sed frigiditas

---

2 nec] ut G 3 nihil] vel G ‖ mobili] mobile P 4 res] *om.* p ‖ scilicet] *om.* GP ‖ illa] *add.* scilicet P 5 est tempore calefactionis] tempore calefactionis est P 6 est] *add.* in G 6–7 videtur absurdum] est falsum et absurdum videtur P 7 inveniretur] videtur G 8–9 acquisitum … acquisito] acquisita cum posterius acquisita P 10 remittatur] remittitur GPp ‖ quae erat intensior] *om.* P 13 nihil] numquam G ‖ post] plus C ‖ sed] et P 14 quia] *add.* sic G ‖ sequeretur] sequitur p ‖ frigefacienti] *add. in marg.* alias frigiditati C 17 sequeretur] sequitur p 20 totam] *add.* et sequitur P 22 ibi] *om.* P 24 sed[2]] immo GPp 25 modum] motum C 27 ante] prius G 28 generatur] generetur GPp

solum generaret eam. Et tamen oportet quod habeat per se generans et per se movens, cum secundum eam fit per se motus et alteratio, quae est quaedam generatio.

Et iterum, quod provenit ab aliquo per modum sequelae, hoc est, quia ab illo provenit aliud primo ad quod | naturaliter illud consequitur, verbi gra- 69ra C
tia si a calefactione provenit per se et primo caliditas et consequenter per modum sequelae raritas et levitas, quia consequuntur naturaliter caliditatem in materia ad hoc disposita. Sed si aqua sit calidissima et remittendo corrumpatur illa tota caliditas, nihil remanet vel generatur ad quod debeat consequi naturaliter caliditas, cum forma aquae sibi magis determinat naturaliter frigiditatem quam caliditatem. Igitur nihil est dicere quod generetur caliditas per modum sequelae ad aliud et non per se.

Nec valet etiam quod aliqui dicunt, quod quando corrumpitur sic summa caliditas, non potest statim induci frigiditas, sed minor caliditas, quia subiectum debet esse dispositum ad illud quod | recipit; sed calidissimum non est 57ra G
dispositum ad recipiendum frigiditatem; ideo non recipit frigiditatem, sed remissam caliditatem.

Contra hoc arguitur quia: antequam generetur | alia caliditas, corrumpitur 58vb P
illa prima caliditas. Ideo illa prima, cum nihil sit, nihil valet ad generationem eius quod sequitur, propter quod | materia in generatione alterius caliditatis 46vb p
nihil plus est disposita naturaliter ad illam caliditatem quam ad frigiditatem, nisi aliquid de illa prima caliditate remaneat.

Iterum ad principale arguitur sic: sequitur quod caliditas remissa eadem specie generaret caliditatem multo intensiorem se, quod videtur falsum et inconveniens. Et consequentia patet sicut prius, quia remitteret de summa caliditate et sic generaretur statim caliditas valde intensa.

Item omnes sic dicentes dicunt quod, sicut non manet gradus caliditatis prior cum posterius adveniente, ita multum minus manet gradus frigiditatis

---

1 generaret] generat p ‖ et[1]] *om.* P 2 movens] *add.* et G ‖ fit] sit GPp 4 iterum] *add.* non valet dicere p 6 si] sicut P ‖ calefactione] calefaciente Gp : caliditate P 7 consequuntur naturaliter] *inv.* GPp 9 tota] *om.* G ‖ vel generatur] *add. sed del.* quam C : nihil generatur aliud p : naturaliter G 9–10 debeat consequi naturaliter] debeat consequi (*in marg.*) naturaliter C : naturaliter debeat consequi Gp : naturaliter debet consequi P 10 sibi] si G ‖ determinat] determinet Gp 13 sic] *post* caliditas[1] (14) *sed corr.* C : *om.* G 14 non … caliditas[2]] *om.* (*hom.*) P ‖ statim] sic G 15 illud] aliud p 18 generetur] *post* caliditas p 19 ideo] immo P ‖ nihil[2]] non p 20 materia] *om.* G 21 ad[2]] *om.* C 22 de] *om.* P ‖ remaneat] remanet P 23 sequitur] sequeretur P ‖ remissa] *add.* in GPp 24 falsum et] *in marg.* C : *om.* GPp 26 generaretur] generetur G ‖ caliditas] *add.* adhuc GPp 27 dicentes] concedentes G 28 multum] multo GPp

cum gradu caliditatis; et tamen prius determinatum est quod manent simul; igitur etc.

Item suppono quod illud quod nihil est nihil generat. Omne enim agens agit secundum quod est in actu, ut dicitur tertio *huius*. Sint igitur summa caliditas et summa frigiditas approximatae. Certum est quod remitterent se invicem; ideo totaliter corrumperent se invicem secundum adversarios. Et prius corrumpuntur quam alia caliditas vel alia frigiditas generetur. Quid igitur generaret caliditatem sequentem et frigiditatem? Constat enim quod nullum posset assignari generans, nisi recurreremus ad universale generans, 69rb C scilicet ad Deum; quod in proposito non esset naturalis recursus. |

Tunc igitur propter solutiones rationum et propter maiorem huius materiae explicationem notanda sunt aliqua.

Primo notandum est quod magnitudo bipedalis non est maior quam pedalis ex eo quod habeat plures partes, quia utraque infinitas habet partes; ideo una non plures quam alia. Sed ex eo dicitur maior, quia potest aliqua pars signari, quod bipedalis magnitudo habet plures partes tantas quam pedalis. Vel secundum propriam descriptionem bipedalis dicitur maior pedali, quia continet tantam quanta est pedalis et amplius. Ita etiam caliditas *a* non dicitur intensior caliditate *b*, quia habeat plures partes graduales, quia utraque infinitas habet, sed quia habet plures tantas quanta est signata vel signabilis, aut quia tantundem et amplius, sicut dicebatur. Licet enim non sit proprie quantum nec aliquantum nisi in magnitudine vel numero, tamen modo proportionali et secundum attributionem est quantum et aliquantum vel etiam magnitudo in perfectionibus rerum vel in potentiis, licet etiam istae sint indivisibiles, ut in potentiis intelligentiarum, et etiam magis in intensionibus et remissionibus qualitatum et in multis aliis.

---

1 manent] manet C ‖ simul] prius G 4 sint] si p 5 approximatae] approximatur p 6 invicem^{2}] *add.* sed p 8 enim] *om.* GPp 9 recurreremus] recurrens p 10 quod] qui C ‖ esset] est p 11 solutiones rationum] *add.* ad oppositum adductarum Pp : rationes solutionum ad oppositum adductarum G 12 notanda] neganda G 13 notandum est] notum est G : *om.* P 14 utraque] utrumque C 15 una non plures] non una plures GP : non magis una p 16 signari] *add.* sic P ‖ partes] *om.* p 16–17 tantas] *om.* P 18 amplius] maius G ‖ etiam] *om.* P 19 quia] cum G ‖ habeat] habet P 21 est] *add.* una GPp 22 enim … proprie] proprie non sit P ‖ quantum] *corr. in marg. ex* generatum C : generatum p ‖ nec] vel P 23 attributionem] attributione p 24 vel^{2}] *sup. lin.* C : *om.* GPp 25 etiam] *om.* P 26 et^{1}] *om.* P ‖ intensionibus] intentionibus p

---

1 Cf. sup., III, q. 4 4 Cf. Aristoteles, *Physica*, III, 3, 202a16–17

Tunc igitur imaginandum est quod, si in aliquo subiecto sit aliquanta pars caliditatis per intensionem et iterum cum ea ad|veniret [59ra P] alia tanta pars, tunc totalis caliditas esset in duplo intensior quam erat pars prima, et si decem tantae apponerentur, esset in decuplo intensior; et esset aeque intensa caliditas, si istae decem fierent successive | [57rb G] una post aliam.

Notandum est etiam quod, licet omnes partes vel gradus caliditatis sint eiusdem rationis et speciei quantum ad nomen significans simpliciter qualitatem sine relatione aut mensura, tamen bene sunt diversarum rationum et specierum quantum ad nomina significantia relative vel quantitative, ut 'magis' vel 'minus', 'intensum' vel 'remissum', 'duplum' vel 'quadruplum', 'duo' vel 'quattuor' etc. Nec oportet omnes gradus ad invicem esse aequales secundum intensionem, sicut nec omnes partes magnitudinis sunt ad invicem aequales secundum extensionem. Immo magni gradus dividuntur in minores et illi iterum in minores et sic in infinitum, sicut est de magnitudine extensionis. Sed in hoc est differentia, quia partes magnitudinis possunt ab invicem separari cum permanentia earum etiam per | [48ra p] potentias naturales; gradus autem qualitatis non possunt sic separari, sed solum sic quod unus potest corrumpi altero manente. Concedendum est tamen quod possunt separari cum permanentia eorum per potentiam divinam. | [69va C]

Tunc est respondendum ad rationes.

⟨1⟩ Ad primam, quae est de motu locali, non dicetur, donec de illo quaeretur.

⟨2⟩ Ad secundam dicendum est quod illae partes sunt distinctae ab invicem, licet etiam sint unum totum, et sunt ad invicem unum secundum speciem et plura secundum numerum, licet cum hoc sint unum totum in numero. Et ista auctoritas quinti *Metaphysicae* debet exponi quod illa nomina dicuntur unum numero, id est de ipsis vere praedicatur hoc praedicatum 'unum numero', quorum materia est una, non loquendo de mate-

1 sit] fit p 2 intensionem] intentionem p || ea] alia P || tunc] *om.* p 3 erat pars prima] esset prima pars P || et] *add.* etiam G || decem] decimo p 4 tantae] tantum P || decuplo] duplo Cp || esset[2]] esse P 6 etiam quod] *om.* P 9 et] aut G || vel] aut P 10 vel[1]] aut P 11 etc.] etiam G : et p : *om.* P || esse] *ante* ad Pp 13 immo] *om.* G 14 et sic] *om.* Gp 15 ab] ad C 16 permanentia] permanentiis G 17 qualitatis] caliditatis Gp || quod unus] quia unus GP : quia unius p 18–19 possunt separari] possent separari p : possint separare G 19 eorum] earum p : *om.* P 20 tunc est respondendum] tunc respondendum est p : respondendum est G : *om.* P 23 dicendum est] dicitur P 24 sunt] sint p 25 licet] *add.* non p || sint] sunt G 27 unum] *add.* in GPp || id est] et P 28 unum] *add.* in GPp

21–22 Cf. inf., III, q. 6

ria proprie, sed capiendo 'materiam' pro illa re pro qua nomen supponit, hoc est dictu quod, si res est eadem pro qua uterque terminorum supponit, tunc de illis terminis significative sumptis verificatur hoc praedicatum 'unum numero'.

⟨3⟩ Ad aliam concedo quod bene dicto modo fiunt intensiones et remissiones qualitatum.

Tunc ad rationes quae contra hoc arguunt.

⟨3.1⟩ Ad primam dicendum est quod illa ratio imaginatur instantia indivisibilia; quae non sunt. Sed concedi debet quod in qualibet parte illius temporis generatur aliquis gradus, et in maiori tempore maior gradus, et in minori minor. Et si signetur unus gradus, non erunt infiniti tanti, quia etiam si signetur una pars horae, non erunt infinitae tantae.

59rb P ⟨3.2⟩ Ad aliam dicendum est | quod illa quae sunt diversarum rationum non sunt comparabilia proprie, scilicet secundum aliquam determinatam proportionem numeralem, ut habetur in septimo *huius*. Unde homo excedit asinum in perfectione essentiali sine aliqua proportione numerali, et tamen non est infinitae perfectionis, cum sit aliud perfectius. Ita etiam, licet albedo esset perfectior nigredine sine aliqua determinata proportione et mensura, tamen non esset infinitae perfectionis, quia est aliud perfectius. Et aliqui ponunt exemplum bene apparens quia: simili argumento probaretur quod iste angulus rectilineus esset infinitus, quia est maior in duplo quam angulus rectilineus qui est sua medietas, et est maior in centuplo quam angulus rectilineus qui est sua centesima, et sic in infinitum; et tamen omnis 57va G angulus rectilineus est omni angulo contingentiae maior; | ergo iste angulus rectilineus est infinite maior isto angulo contingentiae, quod non esset, nisi esset infinitus. Et etiam manifestum est quod una intelligentia est perfectionis maioris quam esset ista albedo quantumcumque multiplicata, et

---

1 proprie] *ante* de (54.28) P 2 quod] *om.* P ‖ est^{2}] sit GPp ‖ terminorum] *add.* illorum p : illorum P 3 praedicatum] nomen G 4 unum] *add.* in GPp 6 qualitatum] *om.* P 7 tunc ad rationes] ad rationes tunc p : quantum ad rationes tunc P 10 minori] *add.* tempore G 11 gradus] *om.* G 11–12 infiniti ... erunt] *om.* (*hom.*) G 11 etiam si] *inv.* P 13 dicendum est] dicitur P ‖ rationum] specierum GPp 14 scilicet] *om.* p ‖ aliquam] *om.* G 15 numeralem] naturalem CGp 16 perfectione] *add.* scilicet G ‖ essentiali] naturali P ‖ numerali] naturali CPp ‖ et] *sup. lin.* C : *om.* P 17 cum ... perfectius] *om.* P ‖ sit aliud] est (*del.*) aliud sit (*in marg.*) C : *inv.* p 19 et] *om.* P 20 argumento] *add.* bene P 23 et^{2}] *om.* P 24 contingentiae] incidentiae P 24–25 ergo ... contingentiae] †...† angulus rectilineus est †...† maior alio angulo †...† *in marg.* C 25 est infinite] est infinitum P : in infinitum est p ‖ contingentiae] incidentiae P 26 et] *om.* P 26–27 perfectionis maioris] *inv.* GPp

---

15 Cf. Aristoteles, *Physica*, VII, 4, 249a3–8

non oportet propter hoc concludere quod sit infinitae perfectionis simpliciter, cum aliquid sit perfectius. Unde quantum ad gradus specialissimarum distinctionum quattuor partes de inferiori specie non essent propinquiores gradui speciei superioris quam una illarum partium. | Et iterum non esset 69vb C mihi notum de formis extensis et materialibus quod quaelibet pars formae de superiori specie sit perfectior forma de inferiori specie. Verbi gratia non oportet dicere quod quaelibet pars quantitativa formae equi sit perfectior totali forma asini, licet bene concedatur quod omnis forma equi perfecta in sua specie sit perfectior forma asini etiam perfecta in sua specie.

⟨3.3⟩ Ad aliam dico quod, si *a* calefaciens sit calidissimum et agendo repatiatur, manifestum est quod ratio non valet. Ideo ponatur quod non repatiatur, sed semper remaneat calidum. Tunc concedo quod in medietate horae remota est medietas frigiditatis, scilicet remoti sunt quinque gradus frigiditatis, posito quod ipsa erat decem graduum, et generati sunt quinque gradus caliditatis. Sic conce|do quod remota est medietas resistentiae, sed 48rb p etiam remota est medietas activitatis, quia cum *b* factum est tepidum, *a* non est sibi contrarium nec corruptivum ipsius nisi ratione graduum in quibus *a* superat caliditatem ipsius *b* genitam. Modo non superat eam nisi | in 59va P quinque gradibus; unde quantum ad quinque gradus esset sibi similis. Sic igitur *a* agebat a principio ratione decem graduum et nunc ratione quinque graduum solum. Ideo manifestum est quod motus debet esse aeque velox sicut prius, et non velocior. Certum est enim quod extremum ad extremum habet in duplo plus de distantia graduali quam ad medium.

⟨3.4⟩ Ad aliam respondetur probabiliter quod loquendo proprie qualitas non intenditur nec fit intensior, sed subiectum intenditur, id est fit intensius tale. Ad talem sensum et non ad alium conceditur quod qualitas intenditur. Vel potest dici quod qualitas non sic intenditur quod totaliter eadem esset prius minus intensa et post magis intensa, sed ad illum sensum dicitur

1 oportet] apparet P 2 quantum] quantumcumque C ‖ specialissimarum] specificatarum P : specificarum p 4 et iterum] ita etiam P ‖ esset] est GPp 5 de] in Pp ‖ extensis et materialibus] materialibus et extensis G 6 sit] fit p ‖ specie^{2}] *om.* P 7 quantitativa] *post* equi p 8 concedatur] conceditur G 10 aliam] secundam C ‖ calidissimum et] *add.* in P : calidum in G 11 ponatur] ponamus GP 12 calidum] calidissimum Pp 13 remoti] remotae G 14 posito] *praem.* et p : et P 16 factum est] factum sit Gp : sit factum sit P ‖ a] *om.* P 17 ipsius] ipsi G 18 a] *om.* GPp 19 gradus] *om.* GP 20 nunc] non C 21 motus ... velox] debet esse aeque velox motus p : debet esse aeque velox scilicet motus P 22 est enim] *inv.* GPp 23 quam] quantum C 25 intenditur^{1}] *add.* proprie P ‖ fit^{1}] sit G 26 tale] *add.* et GPp 27 non sic] *inv.* P ‖ quod^{2}] quia p 28 magis] maius G ‖ intensa^{2} ... sensum] ad istum sensum enim P

intendi, quia continue qualitas quae est posterior est intensior et plurium graduum tantorum quam esset qualitas quae esset prius. Et alii etiam dicunt quod nec pars prior nec pars posterior intenditur, sed congregatum, ad illum 70ra C sensum quod qualitas congregata | incipit esse intensior non quam ipsa erat ante, sed intensior quam esset qualitas quae erat ante. Et manifestum est quod haec solutio et praecedens revertuntur in idem quantum est ex parte rei, et sunt verae.

⟨3.5⟩ Ad aliam dictum fuit iam quod remisse calidum, cum reduxerit frigidum ad gradum sibi similem, amplius non aget nec habebit sibi contrarietatem. Ideo nec amplius potest corrumpere aliquem gradum frigiditatis nec generare aliquem gradum caliditatis.

⟨3.6⟩ Ad aliam potest dici quod subiectum non denominatur simpliciter nisi a totali qualitate sua, non a partiali, sicut homo non dicitur parvus ex eo quod habet unam parvam partem.

⟨3.7–3.8⟩ Ad alias duas, quae arguunt de velocitate motus et de lumine, dictum fuerit prius quod ita bene est forma intensa, si plures magni gradus 57vb G generentur simul, | sicut si successive. Modo simul generantur in casibus praedictis de motu et de lumine propter hoc quod virtus agentis sit fortior. Et de his dicetur magis post.

⟨3.9⟩ Ad aliam dicunt aliqui quod ordo provenit ex Dei scientia et voluntate. Qui ad omne quod fit coagit, immo agit principaliter; illi igitur gradus prius auferuntur, quos Deus scit et vult esse prius auferendos.

Alii dicunt quod illi gradus prius et posterius generantur, et qui posterius generantur, illi prius corrumpuntur. Et ad hoc ponitur quaedam persuasio, 59vb P videlicet quod materia appetit formas | futuras sive generandas ratione privationis, ut apparet primo *Physicorum*, id est ea ratione qua caret eis. Quanto igitur magis caret caliditate, tanto magis appetit caliditatem; ideo

---

1 posterior] posterius p ‖ et] ad P 2 esset prius] erat prius P : prius erat G ‖ et alii etiam] alii P 3 pars^2] *om.* G 4 quod] quia Pp ‖ quam] quod p 5 esset] *om.* p ‖ et manifestum] manifestum enim P 8 iam] *om.* P 9 aget] *corr. ex* agit C : ageret G 10 potest] poterit GPp 11 gradum] *om.* Gp 12 potest dici] dicitur P 13 non a partiali] *om.* P ‖ sicut] *add.* enim C : *add.* etiam p 14 partem] *ante* parvam p : personam P 15 ad] *praem.* et P 16 fuerit] fuit p : est G 17 generentur] generantur GP 17–18 casibus praedictis] *inv.* P 18 de^2] *om.* Pp ‖ lumine] *add.* et G 19 et] *om.* P 20 aliqui] *om.* P 22 scit et] sic P ‖ esse prius] *inv.* Pp 23 prius] post p 24 generantur] generentur P ‖ prius] post P ‖ quaedam] *om.* P 26 physicorum] huius p ‖ id est] scilicet P ‖ ea] *om.* p ‖ eis] *add.* in G

---

16 Cf. sup., III, q. 5, 55$^{1-5}$ 19 Cf. inf., III, q. 7, ratio 3 et ad 3 26 Cf. Aristoteles, *Physica*, I, 9, 192a22

etiam tanto caliditas sibi adveniens magis adhaeret sibi; igitur cum magis careat caliditate in principio calefactionis, gradus prius advenientes fortius adhaerent.

Sed aliqui obiciunt ponendo casum quod Deus in subiecto *b* faciat simul instanter caliditatem intensam et ultra permittat procedere naturam modo naturali: quomodo tunc corrumpentur illi gradus a frigido superveniente? Utrum simul omnes vel aliqui prius et alii posterius?

Respondetur quod, licet Deus generaret omnes gradus simul, tamen posset in eis ponere ordinem quantum ad fortiorem adhaerentiam illi subiecto, ac si successive et modo naturali essent geniti; | et iterum etiam taliter pos- 70rb C
set eos generare et conservare quod numquam corrumperentur. Etiam ita est quod, cum Deus facit unum miraculum ponendo res extra modum naturalem, ipse facit aliud miraculum reducendo, sicut in passione Christi fecit modo | supernaturali moveri lunam ad eclipsandum solem, ut aliqui dicunt, 48va p
et postea etiam modo supernaturali reposuit eam ad locum suum etc.

---

1 tanto caliditas sibi] sibi tanta caliditas P **2** caliditate] calefaciente G ‖ gradus prius advenientes] prius agentes P **6** tunc] ergo G ‖ corrumpentur] corrumperentur p **7** simul omnes] *inv.* P ‖ et alii] alii Gp : aliqui P **10** geniti] generati GPp **10–11** posset] potest P **12** quod] *om.* P **13** ipse] ipsa P ‖ miraculum] *om.* G ‖ fecit] facit p **15** postea] post p ‖ ad locum suum] in loco suo P ‖ etc.] et sic est finis quaestionis etc. G

⟨III.6⟩

⟨Utrum motus localis sit vel utrum haec sit vera 'motus localis est'⟩

Nunc venio ad quaerendum de motu locali. Et erit sexta quaestio utrum motus localis est vel utrum haec est vera 'motus localis est'.

Et primo suppono quod motus localis nec sit locus nec sit illud quod movetur localiter, prout hoc ostendetur in septima quaestione.

Arguitur quod non quia:

⟨1⟩ Dictam suppositionem concedentes ponunt quod motus ita sit res pure successiva quod pars prior non manet cum parte posteriori. Tunc igitur arguitur, sicut arguebat Aristoteles quarto *huius* de tempore, quia: partium motus una est tota praeterita, et sic amplius non est, et alia est tota futura, et sic nondum est.

⟨2⟩ Vel arguitur sic: cuiuscumque divisibilis necesse est, si ipsum est, omnes partes eius esse vel saltem aliquam; sed motus est divisibilis in partem priorem et posteriorem, quarum neutra est, ut dictum est; igitur etc.

⟨3⟩ Item illa propositio non est concedenda ad quam sequuntur contradictoriae; sed ad istam 'motus localis est' sequuntur contradictoriae. Probatio quia: ille motus, si est, habet partem praeteritam, quae sit *a*, et partem futuram, quae sit *b*. Tunc sequitur quod *a* est et quod *a* non est, quia *a* iam praeteriit, igitur *a* non est; et tamen *a* est pars huius motus, igitur *a* est. Et ita etiam sequitur quod *b* non est, quia est futurum, et quod *b* est, quia est pars. Et sequitur quod ille motus qui ponitur esse non est, quia *a* et *b* non sunt; et tamen illa sunt motus et motus non est aliud quam *a* et *b*, quia totum est
60ra P suae partes, si ipsum est; | igitur ille motus non est.

---

4 *Inde ab hac quaestione usque ad III, q. 17 pro G nimis laeso codicem I adhibuimus* ‖ nunc … quaestio] nunc vero est quaerendum consequenter de motu locali et erit sexta quaestio I : quaeritur sexto P 5 est[1]] sit P 7 septima] alia IPp 8 arguitur] *add.* primo Pp 9 ponunt] dicunt P 9–10 ita … pure] sit ita pure res P 10 manet] maneat Ip ‖ igitur] *om.* P 11 arguebat aristoteles] *inv.* P 12 tota praeterita] *inv.* P 16 et] *add.* in partem p 21 tamen] cum I 22 est[3]] *add.* patet p ‖ pars] *add.* a P

---

11 Cf. Aristoteles, *Physica*, IV, 10, 218a5–6

 | DOI: 10.1163/9789004322356_009

⟨4⟩ Item sequeretur quod eadem res, scilicet ille motus, simul fieret et corrumperetur, quod est impossibile. Impossibilitas consequentis patet, quia eiusdem generatio et corruptio sunt contrariae, ut dicitur quinto *huius*. Et etiam, quia generatio est ad esse et corruptio ad non esse; ideo in termino ad quem idem simul esset et non esset, si simul generaretur et corrumperetur.

Sed consequentia principalis probatur quia: motus ille fieret, quando ipse esset, et corrumperetur etiam, quando ipse esset, et sic simul. Ostendo quod ipse fit, quando ipse est, primo auctoritate Aristotelis | in tertio *huius*, 70va C qui dicit quod esse eius consistit in fieri aliud post aliud. Secundo idem patet, quia si *b* movetur per unam totam horam praecise, ille motus non fit per se totus ante illam horam nec in primo instanti illius horae, quia tunc statim esset totus factus; igitur fit in toto tempore illius horae, et in illo est. Similiter probatur quod ipse corrumpitur, quando est, quia: ille motus non corrumpitur ante illam horam, quia nondum est vel fuit et res non corrumpitur, antequam fit; nec corrumpitur post illam horam, immo iam totus corruptus est; igitur in tempore illius horae corrumpitur, et tunc ipse est.

⟨5⟩ Iterum sequitur quod idem simul fieret et esset factum, immo non fieret nisi quando esset factum. Consequens est falsum; igitur antecedens. Falsitas consequentis probatur per Aristotelem sexto *huius* dicentem quod impossibile est quod mutatum est, quando mutatum est, mutari in illud in quod iam mutatum est. Sed consequentia manifesta est, quia motus non fit nisi quando est, ut dicebatur; et tamen quando est, factus est, cum esse sit terminus ad quem factionis et generationis.

⟨6⟩ Item in sexto *huius* ponit Aristoteles pro regula quod omne quod factum est, quando primo factum est, est, propter hoc quod terminus ad quem factionis est esse. Sed totalis motus qui fit per unam horam factus est primo in ultimo instanti illius horae et non ante. Igitur in illo instanti est. Sed hoc est falsum, si sit res successiva modo praedicto. | 48vb p

---

1 sequeretur] sequitur p **3** et[2]] *om.* P **7** motus ille] *inv.* P **8** ostendo] *add.* igitur IPp **10** esse eius] omne ens P ‖ in] *om.* P ‖ idem] etiam P **11** per] secundum P **12** totus] totum P **13** esset] *om.* P **16** fit] sit p **17** est] *ante* totus IPp **19** sequitur] sequeretur IP **20** antecedens] etc. Ip : *om.* P **21** aristotelem] *add.* in p ‖ dicentem] qui dicit P **23** sed] *om.* P ‖ manifesta est] patet P **26** in] *om.* p **30** praedicto] *add.* igitur P : *add.* etc. I : *add.* ergo etc. p

---

3 Cf. Aristoteles, *Physica*, V, 1, 225a33–34 **9** Cf. Aristoteles, *Physica*, III, 6, 206a21–22, 31–33 **21** Cf. Aristoteles, *Physica*, VI, 5, 235b25–26 **26** Cf. Aristoteles, *Physica*, VI, 5, 235b6–7, 14–15

Oppositum arguitur quia:

⟨1⟩ Licet secundum veritatem motus sit res distincta a mobili et loco, ut in alia quaestione dicetur, tamen sequitur quod ipse est, quia nihil est idem vel diversum, nisi sit ens, ut dicitur decimo *Metaphysicae*.

⟨2⟩ Et Aristoteles in primo *huius* dicit quod negantes motum esse vel pluralitatem entium negant principia doctrinae scientiae naturalis.

⟨3⟩ Et oportet concedere tempus praesens esse; et tamen tempus, si est,
47rb I est motus localis, ut dicetur in quarto *huius*. |

Ego suppono ex sequenti quaestione quod motus localis nec est locus nec illud quod movetur localiter, et tamen quod motus localis est. Nec oportet hoc probare, quia apparet ad sensum et est supponendum in scientia naturali tamquam principium doctrinae.

Et quia per 'praesens' non debemus intelligere rem indivisibilem in tem-
60rb P pore, quia declarabitur in sexto libro quod nulla talis res indivisibilis | sit in continuo sive in magnitudine sive in motu sive in tempore, igitur per 'praesens' oportet intelligere tempus divisibile. Et tamen tempus est, quia nec grammaticus nec logicus ponit differentiam inter praesens, praeteritum et futurum nisi per esse, fuisse et fore. Igitur tempus est et per consequens
70vb C motus est. |

Iterum aliter non posset dici quod aliqua propositio vocalis esset; et sic nulla esset vera aut falsa, quod est absurdum dicere. Et patet consequentia, quia sicut tu arguis de partibus motus quae non sunt simul, ita quando subiectum propositionis vocalis est, praedicatum nondum est. Et propositio non est, si eius praedicatum non est. Et quando praedicatum est, subiectum iam amplius non est, sed praeteriit. Et tamen propositio non est, si eius subiectum non est. Igitur propositio vocalis numquam esset, si talia argumenta valerent.

Solum igitur restat videre quomodo talia successiva dicuntur esse. Et Aristoteles tertio et quarto *huius* statim determinat quod esse talium consistit

1 quia] quod p   3 in alia quaestione] *om.* I   4 dicitur] dicetur p ‖ decimo] quarto CI 5 dicit] dixit p ‖ negantes] negans p   8 localis] *om.* ABHILMPTUp ‖ huius] *add.* ergo etc. Ip   9 ego] hic p ‖ nec[1]] non I ‖ nec[2]] *add.* est Pp   16 tamen] *corr. sup. lin. in* quod C 20 vocalis] *om.* p   23 nondum] tunc non p ‖ est[2]] *om.* I   25 tamen] cum C

3 Cf. inf., III, q. 7   4 Cf. Aristoteles, *Metaphysica*, X, 3, 1054b18–19   5 Cf. Aristoteles, *Physica*, I, 2, 184b25–185a5   8 Cf. inf., IV, q. 12   14 Cf. Iohannes Buridanus, *Quaestiones super libros Physicorum*, VI, q. 4 (ed. Parisiis 1509, ff. 96rb–98va)   29 Cf. Aristoteles, *Physica*, III, 6, 206a18–b3; IV, 10, 217b33–218a2

non in essendo simul partes eorum, sed in essendo unam partem post aliam, scilicet in semper aliud et aliud fieri. Sic enim sunt dies et anni, ut dicit. Et expresse dicit ibidem non stare simul quod prius est cum eo quod posterius est. Dicit enim successionem infinitam in divisionibus magnitudinum esse permanente accepto, in hominibus autem et tempore corruptis, ita tamen ut non sit deficere. Et idem determinat in quarto libro.

Ut igitur videamus manifestius quomodo tempus sit et motus localis, notandum est, quod semper supponendum est usque ad sextum librum, quod in linea nullum est punctum quod sit res indivisibilis, et quod in tempore nullum est instans quod sit res durationis indivisibilis, et ita de momento in motu.

Ex quo statim sequitur falsitas cuiusdam opinionis dicentis quod successivum, ut tempus vel motus, ex eo dicitur esse secundum quendam modum attributionis, quia quoddam indivisibile est, scilicet instans, ad quod copulantur partes eius.

Secundo etiam ex hoc concludendum est quod per 'praesens' oportet intelligere tempus divisibile cuius una pars est prius et alia pars est posterius. Et illud tempus, cum sit praesens, est et quaelibet eius pars est et est praesens, licet una pars prius et alia posterius. Et haec etiam conclusio probatur quia: impossibile est aliter bene describere moveri vel mutari. Nam si dicas 'hoc aliqualiter se habet et aliter se habuit', non sequitur nisi quod mutatum est. Et si dicas 'hoc aliqualiter se habet et aliter se habebit', non sequitur nisi quod mutabitur. Etiam si dicas 'hoc aliqualiter se habet et aliter se habuit et aliter se habebit', non sequitur nisi quod mutatum est et mutabitur. Et ex nullo istorum sequitur quod mutatur. Igitur ad exponendum hoc verbum 'mutatur' per verbum de praesenti | oportet dicere quod hoc se habet | aliqualiter prius et aliter posterius, vel saltem si non sit motus | de affirmato in affirmatum, sed sit generatio vel corruptio, quae est de affirmato in negatum vel econtra, oportet dicere quod hoc in priori parte huius temporis praesentis non est et in posteriori est vel e converso.

60va P 71ra C 49ra p

1 in[1]] *om.* I ‖ eorum] earum p 3 est] *corr. sup. lin. ex* sed C : *om.* I 5 permanente] permanentem p ‖ corruptis] corruptionis CI 7 et] vel P 8 notandum est] *inv.* P ‖ quod] et IPp ‖ est[2]] *om.* IPp 10 et] *om.* P 11 momento] moto C 12 cuiusdam] *post* opinionis p : eiusdem P 13 tempus vel motus] motus vel tempus Pp ‖ ex] *om.* p 14 quoddam] quod p ‖ est] *om.* I 16 ex] ad p ‖ concludendum est] concludimus p 17 pars est[2]] *om.* P 18 est[1]] *om.* p ‖ et[2]] tamen etiam P ‖ eius pars] *inv.* Pp 19 pars] *om.* IPp ‖ etiam] *om.* IP 20 quia] quod I ‖ bene describere] *inv.* P 23–24 mutabitur ... quod] *om.* (*hom.*) C 23 etiam] et Ip 24 et[2]] *add.* quod P

Verum est quod aliqui volunt aliter describere mutari, quia aliqualiter se habet et aliter immediate se habebat et aliter etiam immediate se habebit. Sed hoc nihil valet, quia quando tu dicis quod aliqualiter se habet, vel tu intelligis quod sic se habet in instanti indivisibili, et hoc iam remotum est; vel tu intelligis quod sic se habet in tempore divisibili et praesente, ut in hac hora, et tunc ante hanc horam immediate se haberet aliter et immediate etiam post hanc horam se habebit aliter, tamen possibile est quod in hac tota hora se habet consimiliter; ideo non mutatur. Ideo numquam ex tali expositione sequitur quod ipsum mutatur.

Sed contra hoc est duplex obiectio et una magna dubitatio.

Prima obiectio est quia: tunc esset verum dicere 'Socrates currit et Socrates sedet'; et quicumque sedet, non currit; igitur esset verum dicere 'Socrates currit et non currit'. Et sic copulativa ex contradictoriis composita esset vera. Et hoc confirmatur fortius quia: esset verum dicere 'Socrates est currens et Socrates est non currens'; sed ad affirmativam de praedicato infinito sequitur negativa de praedicato finito; igitur esset verum dicere 'Socrates est currens et Socrates non est currens'.

Secunda obiectio est quod Aristoteles in sexto *huius* determinat quod
47[va] I omne quod movetur movebatur prius. Et hoc esset falsum ponendo casum | quod tempore *a* utamur tamquam praesente. Tunc sequitur quod quaelibet pars eius est praesens; et nullum praesens est praeteritum vel futurum; ideo nulla pars temporis *a* est praeterita. Et ponamus quod *b* movetur praecise et adaequate per tempus *a*. Tunc est verum dicere quod *b* movetur et numquam prius movebatur, quia hoc verbum 'movebatur' est praeteriti temporis et in nullo tempore praeterito movebatur secundum casum positum.

Sed tunc etiam sequitur fortis dubitatio, scilicet si debeamus uti aliquo tempore tamquam praesente, quantum debet esse istud tempus?

---

1 mutari] *add.* scilicet Pp ‖ aliqualiter] BHLMTUp : aliter ACIP 2 aliter[1]] aliqualiter P ‖ immediate[1] ... aliter[2]] *om.* (*hom.*) C ‖ aliter[2] ... se[2]] immediate se aliter P 3 aliqualiter] aliter p 5 et] *add.* in p ‖ praesente ut] praesente et I : praesenti et P 6 tunc] *praem.* licet I : *add.* licet p : licet P 7 etiam] *om.* Pp ‖ hac] *om.* p 8 habet] haberet I 10 duplex] *add.* dubitatio I ‖ una] *add.* est p 11 esset] est P 13 et[1]] *add.* socrates P ‖ et[2]] si P 14 et[1]] *om.* P 15 currens] *add.* quia prius est (*om.* p) currens et posterius non currens Ip 18 quod[1]] quia IPp ‖ in] *om.* P ‖ sexto] secundo C 21 pars eius] *inv.* p ‖ ideo] ergo P 24 quia] et P 25 tempore praeterito] *inv.* I 26–27 uti aliquo tempore] uti aliquo toto tempore p : aliquo toto tempore uti P

---

18 Cf. Aristoteles, *Physica*, VI, 6, 236b32–237a17

Ad primam obiectionem respondeo quod possibile est hanc esse veram 'Socrates currit et Socrates sedet' et similiter istam 'Socrates est sedens et Socrates est non sedens'. Socrates enim non mutatur, nisi aliter et aliter se habeat. Sed dico hanc esse impossibilem 'Socrates currit et Socrates non currit' et similiter istam 'Socrates est currens et Socrates non est currens'. Et causa huius est, quia cum dico 'Socrates est currens', haec est vera ratione cuiusdam praesentis temporis quod est pars totalis praesentis; et ista 'Socrates est | non currens' etiam est vera ratione alterius temporis praesentis quod 71rb C est pars totalis praesentis. Sed cum dico 'Socrates non est currens' negatio praecedit hoc verbum | 'est'; ideo negat pro omni tempore distributive. Sen- 60vb P sus enim est quod Socrates in nullo tempore praesente est currens, quod est falsum.

Ex hoc corollarie concluditur quod non est bona consequentia formalis ab affirmativa de praedicato infinito ad negativam de praedicato finito. Cuius oppositum videtur determinare Aristoteles in libro *Peri hermeneias*. Sed dicendum est quod regula Aristotelis cum aliis circumstantiis debet intelligi, si utamur tamquam praesente illo solo tempore pro quo toto illa affirmativa est vera; si tamen utamur tempore maiori pro praesente, regula non valet.

Pro secunda obiectione notandum est | quod istis nominibus 'praesens', 49rb p 'praeteritum' et 'futurum' possumus uti absolute. Et tunc nullum praesens est praeteritum vel futurum. Et cum omnis pars praesentis sit praesens, nulla pars praesentis est praeterita vel futura. Immo si hac totali die uteremur tamquam praesente, hora prima esset praesens et hora meridiei esset praesens et hora completorii esset praesens; et quaelibet horarum esset praesens, licet haec prius et alia posterius. Et nulla istarum horarum est praeterita vel futura. Et esset verum dicere quod hodie pulsatur ad primam et ad vesperas. Et ita conceditur, sicut ratio arguebat, quod non omne quod movetur movebatur, quia si *b* movetur per hanc totam diem et numquam ante

4 sed dico] dico tamen P 5 et[1]] *om.* P 6 et ... currens] *om.* (*hom.*) P 7 praesentis temporis] *inv.* IPp 8 est vera] *inv.* p ‖ ratione] *post* praesentis p 9 currens] sedens C 11 praesente] praesenti P ‖ est[2]] *ante* in P 11–12 quod est falsum] *om.* P 13 ex] *praem.* et I ‖ bona consequentia] consequentia bona et P 14 ab] ex IPp ‖ praedicato[2]] *om.* Ip 15 determinare aristoteles] *inv.* Pp 20 obiectione] *ante* secunda P : dubitatione C ‖ notandum est] *inv.* P 21 et[1]] vel p 25–26 praesens] *om.* Pp 26 et alia] et illa P : illa Ip ‖ est] esset IPp 28 et] *om.* P

15 Cf. Aristoteles, *De interpretatione*, 10, 20a23–26

initium huius diei movebatur, verum est simpliciter dicere quod numquam movebatur.

Alio modo utimur dictis nominibus respective et est usus valde consuetus et valde communis, scilicet prout temporis praesentis diceremus partem priorem esse praeteritam respectu posterioris et partem posteriorem esse futuram respectu prioris. Et ita, si temporis praesentis pars prior sit *a* et pars posterior sit *b*, diceremus quod *a* est praeteritum et *b* est futurum. Et ita omne tempus praesens est compositum ex praeterito et futuro. Et esset verum dicere quod praesens est praeteritum et quod praeteritum est praesens et quod praesens est futurum et quod futurum est praesens. Et non sequitur '*a* est praeteritum, igitur non est', '*b* est futurum, igitur non est'. Et est ista respectiva acceptio temporis, sicut dicit Aristoteles quinto *huius* quod fuscum est album ad nigrum et nigrum ad album. Sic enim idem tempus est praeteritum et futurum respectu diversorum, ut si totali die uteremur tamquam praesente, hora meridiei esset praeterita respectu horae vesperarum et futura respectu horae primae. Et secundum istum usum verum est quod omne quod movetur movebatur prius, omne quod movetur movebitur posterius et omne quod est fuit prius et erit posterius, 71va C si nihil est per | solum instans indivisibile ('si' pro 'quia', cum nihil sit instans indivisibile).

Sed tunc respondendum est ad dubitationem quae sequebatur. De qua sunt valde diversae opiniones.

Prima opinio est quod de virtute sermonis illo tempore debemus uti 61ra P tamquam praesente, quando loquimur; quod quidem tempus | adaequate coexistit propositioni quam proferimus. Unde si dico vel dixi 'Socrates currit', dico vel dixi verum, si tunc, quando hoc dicebatur, Socrates currebat, et si non, non.

---

4 valde communis] *inv.* P 5 respectu posterioris] *praem.* et p : *om.* C 7 pars] *om.* P 8 ita] ideo C 11 igitur[1]] *add.* a P ‖ est[2]] *add.* et P ‖ igitur[2]] *add.* b P 12 est ista] *inv.* p : ita est P ‖ temporis] quia IP : *om.* p 14 et] vel I ‖ ut] et P ‖ si totali] cito tali I 15 uteremur] utamur P 16 et[2]] *om.* P 17 prius] *add.* et P 18 movebitur] movebatur P ‖ et[1]] *om.* Ip 19 si pro quia] ABBrCEErGHIKLLaMNPRSTXYZp : sed pro quia Q : scilicet (*del. ?*) quia W : scilicet DFJ : *om.* OOxUV : *deest* Pb ‖ cum] *om.* C ‖ sit] significat P 21 respondendum est] *inv.* Pp 23 opinio] *om.* P ‖ de] *om.* I 25 vel dixi] *om.* P 26 hoc dicebatur] hoc dicebam IP : hanc dicebam p

---

12–13 Aristoteles, *Physica*, V, 1, 224b34–35; cf. *AA*, 2:149

Sed haec opinio non videtur mihi vera propter multa. Primo, quia sequeretur quod copulativa ex duabus contradictoriis composita esset vera, ut ista 'Socrates currit et Socrates non currit'; et hoc non debet concedi. Consequentia patet, quia utraque categoricarum esset vera. Nam si tempore praecise quo haec proponitur 'Socrates currit' Socrates curreret, sic haec esset vera, et si tempore in quo praecise haec propositio | 'Socrates non currit' proponetur [47vb I] Socrates non curreret, iam haec esset vera, quia non negatur aliud tempus quam praesens, quando dicitur 'Socrates non currit', et non est praesens nisi quod sibi coexistit, in quo non currit.

Secundo etiam sequeretur quod ego disputans tecum non possem tibi contradicere, et sic periret principalis meta quam intendunt disputantes, quod est falsum. Consequentia probatur quia: si tu dicis primo quod Socrates currit et ego post quod Socrates non currit, uterque forte dicit verum, si uterque utatur pro praesente illo tempore quod praecise coexistit propositioni suae; ideo non contradicunt. Et si illi non proponant propositiones suas praedictas in temporibus extra invicem existentibus, tamen possibile est quod ego non possum ita velociter proponere sicut tu. Ponamus igitur quod breviori tempore tu proponis quod Socra|tes non currit quam ego pos- [49va p] sim proponere quod Socrates currit. Numquam tibi contradico, quia tu dicis verum et ego etiam, si toto tempore tuae propositionis Socrates non currit et tamen immediate post ipse currat. Nam propositio mea extendit se ultra et ita ego pro principio habeo aliquod tempus in quo currit et quod sufficit ad veritatem propositionis meae, quia non distribuo 'tempus praesens', sed indefinite sumo.

Et ex interemptione huius opinionis ego ultra concludo quod nos etiam de virtute sermonis, si volumus invicem disputare et conferre, debemus aliquando uti tamquam praesente in loquendo illo tempore quod non coexistit propositioni nostrae nec aliqua pars eius, quia aliter sequeretur quod nullus disputans cum alio sciret sibi contradicere vel etiam suae proposi-

---

1 quia] *om.* P 1–2 sequeretur] sequitur p 2 duabus] *post* contradictoriis P : duobus p ‖ composita] *om.* IPp ‖ ista] *om.* P 3 non[2]] *om.* P 4 si] *om.* P ‖ praecise] praeterito CP 5 currit] *add.* et P ‖ sic haec] et sic IPp 6 si tempore in] tempore etiam IPp ‖ propositio ... proponetur] proponitur socrates non currit IPp 7 iam] ideo Ip 8 quando ... praesens[2]] †...†citur Sortes non currit et †...† praesens *in marg.* C 10 sequeretur] sequitur p 12 quod[2]] *om.* P 14 uterque] *add.* solum Pp 15–16 propositiones suas] *inv.* IPp 17 proponere] loqui p : *om.* P 18 tu] *ante* breviori P 18–19 possim] possum P : possem p 20 currit] currat I 21 currat] curret p 22 et ita ego] ego ita P 24 indefinite] indeterminate *sed add. sup. lin.* seu indefinite C 25 interemptione] interpretatione p 28 sequeretur] sequitur p 29 etiam] *om.* P

tioni aequipollentem dare, posito quod propositiones essent de praesenti et de inesse; et hoc est inconveniens dicere. Consequentia probatur quia: si dico 'Socrates currit', tu non potes scire mihi contradicere, donec protuli 71vb C orationem meam, quia ante nescis quid debeam dicere; sed post tu | non contradiceres dicendo 'Socrates non currit', nisi utaris tamquam praesente illo tempore eodem quo ego utebar, quod non coexistit propositioni tuae; igitur etc.

Nec valet evasio dicentium quod tu contradiceres mihi dicendo 'Domine, Socrates non currebat, quando vos protulistis illam orationem "Socrates currit"'. Constat enim quod oratio mea et oratio tua non essent de eodem subiecto et eodem praedicato; quod tamen requiritur ad contradictionem.

Si ergo dicendo 'Socrates currit' ego utor tamquam praesente illo tempore 61rb P quod coexistit propositioni meae, oportet, si tu vis mihi contradicere, | dicere quod Socrates non currit, et quantum ad hoc verbum 'currit' oportet te uti tamquam praesente illo eodem tempore quo ego utebar, licet non coexistat propositioni tuae.

Et iterum syllogizando oportet uti eodem tempore tamquam praesente in maiore, minore et conclusione. Aliter non valeret syllogismus. Nam si omne *b* est *a* praecise, quando ego profero maiorem, et omne *c* est *b*, quando profero minorem, non sequitur quod umquam omne *c* sit *a*.

Alii dicunt quod, si esset nunc tempus indivisibile, nos deberemus eo uti tamquam praesente, sed quia non est, ideo debemus de virtute sermonis tamquam praesente uti tempore minimo secundum sensum.

Sed iterum ista opinio non valet, quia sicut dicit Aristoteles in primo *De generatione* de mixtione ad sensum quod aliquid mihi esset mixtum et tibi non mixtum, si haberes visum meliorem quam ego, ita tempus posset tibi esse praesens quod numquam mihi posset esse praesens. Non enim esset idem minimum habenti meliorem sensum et debiliorem.

Item sequeretur quod impossibile esset propositionem vocalem esse, quia non est possibile aliquam esse in tempore minimo secundum sensum.

---

4 ante] antea P ‖ quid debeam] quid debeo P : quod debeo p **5** contradiceres] potes contradicere P **6** eodem] *add.* modo P **8** dicendo] *om.* I **9** vos] *om.* p **11** et] *add.* de p **12** si] sic p **15–16** coexistat propositioni tuae] coexistat tuae propositioni Ip : coexistit tuae propositioni P **19** ego] *om.* P ‖ omne[2]] omnis P **20** umquam] numquam C ‖ omne] omnis P **21** tempus indivisibile] indivisibile IP : divisibile p ‖ eo uti] *inv.* p **23** uti] *ante* tamquam Pp **24** in] *om.* IPp **25** mihi esset] *inv.* IPp **26** tempus posset tibi] tibi tempus potest P **27** mihi posset] *inv.* p : possit mihi I : potest mihi P

24–25 Cf. Aristoteles, *De generatione et corruptione*, I, 10, 328a12–15

Item tu non perciperes Socratem ambulare, quia tu non percipis ipsum ambulare in tempore minimo secundum sensum, et tu non percipis ipsum ambulare nisi isto tempore quod est praesens, quando tu hoc percipis.

Videtur igitur mihi esse dicendum quod aliquando utimur tamquam praesente maiore tempore, aliquando minore. Et nobis licet uti tamquam praesente quantocumque volumus, ita tamen quod ad idem tempus debent se referre disputantes simul, ut non sit ratione diversi usus aequivocatio in suis sermonibus. Quod autem longo tempore utamur aliquando tamquam praesente apparet, quia non solum dicimus 'in hac | [72ra C] praesente die', sed etiam 'in hoc praesente anno', et loquendo per verbum de praesenti dicimus annum continere duodecim menses. | [49vb p] Immo aliquando tempore perpetuo utimur tamquam praesente dicentes quod tempus est perpetuum et quod caelum semper movetur. Sed hoc non assero ad praesens.

Et est sciendum quod, si hoc totali anno utamur tamquam praesente, verum est simpliciter et absolute dicere quod primo est Ianuarius et secundo Februarius et sic deinceps. Et ita duodecim mensium quilibet mensis est. Et de omni eo quod est in aliquo illorum duodecim mensium, verum est dicere quod ipsum est, licet unum prius | [48ra I] et alterum posterius. Et tamen hoc non obstante concedimus secundum usum respectivum illum mensem fuisse et alium fore.

Sed aliqui contra hoc obiciunt quia:

⟨1⟩ Tunc sequeretur quod haec esset vera 'Antichristus est' et etiam ista 'Aristoteles est' et etiam ista 'Antichristus et Aristoteles sunt', quod videtur falsum.

⟨2⟩ Item sequeretur quod | [61va P] ego currerem, quando ego sederem, quia quae eodem tempore sunt, unum est, quando alterum est; et tamen eodem tempore, scilicet isto anno, ego currerem et sederem. Et sic etiam una contradictoriarum esset vera, quando altera esset vera.

---

1 tu[1]] *om.* P ‖ perciperes] percipies p ‖ tu[2]] *om.* P ‖ ipsum] illum p **4** esse] *om.* P **5** maiore] maiori IPp ‖ minore] minori IPp **7** simul] similiter p ‖ diversi usus] *inv.* I **9** quia] quod P ‖ praesente] praesenti Pp ‖ sed] vel P **10** praesente] praesenti p ‖ loquendo] *add.* etiam p **14** et] sed P **15** secundo] *add.* est p **16** et ita] ita etiam P **17** aliquo] *add.* eorum p : *rep.* P **18** et[2]] *om.* P **19** obstante] absolute C ‖ et] sed I **20** alium] illum Pp **22** esset] est p **22–23** antichristus ... aristoteles[1]] aristoteles est et ista etiam antichristus p : aristoteles IP **23** etiam ista] *inv.* IPp ‖ antichristus et aristoteles] aristoteles et antichristus Pp **24** falsum] absurdum IPp **26** et] *om.* P **27** et sederem] *om.* p **28** vera[1]] *om.* P ‖ vera[2]] falsa p

⟨3⟩ Item sequeretur quod esset verum dicere 'dies crastina est', 'dies hodierna est', et sic esset verum dicere 'ego curro cras', 'ego curro heri'. Et haec sunt absurda.

⟨1⟩ Ad primam concedo quod non est inconveniens nec absurdum concedere quod Aristoteles et Antichristus sunt, iste prius per longum tempus et ille posterius, quando volumus uti tempore aeterno vel tempore decem mille annorum tamquam praesente, sicut de parvo tempore non est inconveniens dicere quod Socrates est sedens et currens prius et posterius. Tamen hoc videtur absurdum aliquibus de isto longo tempore, quia raro consuevimus uti tanto tempore tamquam praesente, sed magis consuevimus uti minore, scilicet nobis vel sermoni nostro coexistente; secundum quem usum non est verum dicere quod Antichristus vel Aristoteles est, vel etiam quod Augustus vel Aprilis sunt, et sic de aliis.

⟨2⟩ Ad aliam concedo quod haec est vera 'Socrates currit, quando ipse sedet', sed non est verum quod, quandocumque ipse currit, ipse sedet. Ipse enim in quodam tempore currit in quo sedet, sed non in omni currit in quo sedet. Et concedo etiam quod ista numquam est concedenda 'Socrates est currens, quando ipse sedet', si utamur tamquam praesente illo solo tempore quod praecise coexistit cursui eius; tunc enim non esset verum dicere quod
72rb C Socrates sedet. Sed etiam dictum est prius quod, quamvis est | verum dicere quod currit et sedet vel quod est currens et non currens, tamen non est verum dicere quod currit et non currit.

⟨3⟩ Ad aliam dicendum est quod iste terminus 'hodie' est terminus discretus. Significat idem enim quod 'hac die' et est terminus discretus sicut 'hic homo'. Et propter pronomen demonstrativum requiritur quod demonstretur certa dies, si iste terminus 'hodie' vel 'hac die' supponat pro aliquo; et non demonstretur proprie nisi praesens. Ideo oportet, si vere dicimus 'hodie

---

1 sequeretur] sequitur p ‖ esset] est p 2 est] *om.* C 3 haec] *sup. lin.* C : ista p ‖ absurda] *add.* etc. I 4–5 concedere] credere p : *praem.* dicere simul P 5 et[2]] *om.* P 6 decem] *om.* p 7 sicut] sed P 8 tamen] *praem.* licet P 10 sed] licet p : licet liceat P : licet et liceat I ‖ minore] breviore Ip : breviori P 11 sermoni] sermone IPp 12 antichristus vel aristoteles] aristoteles vel antichristus IPp ‖ etiam] *om.* P 13 vel] et I 15 quod] *om.* IPp ‖ quod ... sedet[2]] *in marg.* (*corr. ex* quando) C ‖ currit ... ipse[3]] *om.* (*hom.*) IPp 16 enim] *sup. lin.* C ‖ quo[1]] *add.* non P ‖ sed] se I 17 et concedo etiam] ergo concedendo P 17–18 est currens] currit IPp 19 coexistit] coexisteret p 20 socrates] *om.* IPp ‖ est verum] sit verum p : verum sit P 21 et[2]] *add.* est Pp : *add.* quod est I 22 dicere] *om.* IPp ‖ et] *add.* quod p 23 dicendum est] dicitur P 24 idem enim] *inv.* IPp 25 requiritur] *om.* P 25–26 demonstretur] demonstratur P 26 die] *om.* P 27 demonstretur] demonstratur IP

currit', quod utamur hac die pro praesente. Sed iterum isti termini 'cras', 'heri' significant connotando futurum et praeteritum. Et cum hoc connotant immediationem ad diem praesentem demonstratam. Et ideo, si utamur tempore continente tres dies tamquam praesente, nulla earum proprie et absolute potest dici 'cras' vel 'heri', sed quamcumque earum demonstraremus, possemus de illa dicere 'hodie'. Et si illa die Socrates curreret, tunc illa demonstrata vere diceremus 'hodie Socrates currit'. Sed si sola media harum trium utamur pro praesente, tunc de prima illarum vere dicimus 'heri' et de tertia 'cras'. Et non est sic verum quod heri est vel quod cras est vel quod heri currit | vel quod cras currit, sed heri cucurrit et cras curret. Et ideo concedo 61vb P
quod numquam est bene dictum 'cras currit', 'heri currit'.

Et | per hoc solvitur sophisticatio qua arguitur quod haec non est vera 50ra p
'semper caelum movetur', quia per locum a toto in tempore sequitur ista conclusio 'cras caelum movetur', quod est inconveniens.

Solutio: dico quod ille terminus 'cras' non continetur nisi sub isto termino 'tempore' supponente pro futuro; et iste terminus 'tempus' pro nullo futuro supponit loquendo absolute, quando omni tempore utimur tamquam praesente, sicut est in hac propositione 'semper caelum movetur'; ideo non valet iste descensus. Sed utendo hac sola die tamquam praesente valeret descensus sic: 'semper movebatur, igitur heri movebatur'.

Ultimo ex dictis inferendum est corollarie quod est bona consequentia 'omni tempore praesente Socrates movetur, igitur omni tempore movetur', quia aequivalet 'omni tempore praesente' et 'omni tempore', cum non sit ibi ampliatio suppositionis ad alia quam ad praesentia. Et ideo etiam ulterius, si quis reputat aequivalere 'semper' et 'omni tempore', manifestum est quod ipse habet concedere, si hac sola die utamur pro praesente, quod bene sequi-

1 currit] *om.* p ‖ praesente] praesenti p ‖ cras] *add.* et IPp 2 futurum et praeteritum] praeteritum et futurum p 3 et] *om.* P 5–6 demonstraremus] demonstremus P 7 vere diceremus] *post* currit p ‖ media] *om.* I ‖ harum] horarum P 9 tertia] secunda p ‖ est vel quod[1]] et P 10 concedo] concedendo P 12 est] sit IPp 13 quia] et P ‖ sequitur] sequeretur P 14 conclusio] *add.* ergo Ip : *add.* quod P ‖ est inconveniens] non est conveniens P 15 solutio] *om.* p ‖ continetur] ponitur p 16 tempore supponente] tempus supponente P : tempus supposito I 19 valeret] valet P 20 semper] *add.* caelum p 22 tempore[2]] *add.* socrates IPp 23 aequivalet] aequivalent IPp ‖ tempore praesente] tempore praesenti P : praesenti tempore p ‖ non sit ibi] ibi non sit P 24 et] *om.* IPp 25 quis] aliquis IPp ‖ reputat] *add.* haec p : *add.* haec valere vel P 26 ipse] ille p ‖ sola die] *inv.* IPp

tur 'Socrates hac tota die movetur, igitur Socrates semper movetur', quia manifeste sequitur quod ipse omni tempore movetur. Sed non determino quod aequivaleant.

Tunc ad rationes principales statim respondendum est.

⟨1–3⟩ Ad tres primas rationes quod praesentis temporis nulla pars est
72va C praeterita vel futura loquendo absolute. | Sed si aliqua dicatur praeterita vel futura loquendo respective, non sequitur quod ipsa non est, sicut ante dictum est.

⟨4⟩ Ad aliam dicendum est quod pure successivum non fit neque corrumpitur, nisi quando est, et suum esse est ipsum fieri et ipsum corrumpi.
48rb I Aristoteles | etiam dicit quod esse eius consistit in semper fieri aliud post aliud et quod talia semper sunt in generatione et corruptione. Nec talis generatio et corruptio sunt contrariae, immo sunt simul et sunt idem, et utrumque nomen exponitur per 'esse'. Tempus enim aliquod vel motum aliquem fieri est ipsum esse et ante non fuisse et ipsum corrumpi est ipsum esse et postea non fore. Et quando ultra arguitur 'si simul fieret et corrumperetur, sequeretur quod in termino mutationis idem simul esset et non esset', nego consequentiam loquendo de pure successivis. Sed quomodo et qui sunt termini in motu sive locali sive alterationis, dicetur post.

⟨5⟩ Ad aliam dicitur quod motus non est factus, quando est, sed fit; econtrario modo in permanentibus.

⟨6⟩ Ad ultimam dico quod regula est vera de permanentibus, scilicet quorum esse est in simultate partium, sed falsa est de pure successivis etc.

Et per consequens debet esse finis quaestionis.

---

1 tota die] *inv.* P 4 statim] *post* est Pp 5 rationes] *add.* dicitur P ‖ pars est] *inv.* P 6 sed] sicut P 7 non[1]] *add.* tamen P 9 dicendum est] dicitur P ‖ neque] nec P 11 esse eius] *inv.* IPp ‖ in] *om.* P 12 quod] *om.* P 12–13 generatio] *praem.* rei P : regeneratio p 13 sunt[3]] *om.* P 15 et[1] … esse[2]] †…† fuisse et ipsum †…† est ipsum esse *in marg.* C 16 postea] post IPp ‖ arguitur] *add.* quod IPp 17 sequeretur] sequitur Pp 18 pure] puris C ‖ sunt] sint p 19 post] *om.* I 20 est[1] … est[2]] factus est P 21 modo] vero I : *om.* Pp 22 permanentibus] manentibus IPp 23 pure] puris P ‖ etc.] *om.* I 24 et … quaestionis] et sic finitur quaestio I : *om.* Pp

---

11–12 Cf. Aristoteles, *Physica*, III, 6, 206a21–22, 31–33 19 Cf. inf., III, q. 9

⟨III.7⟩

⟨Utrum motus localis sit res distincta a loco et ab eo quod localiter movetur⟩

Quaeritur septimo utrum | motus localis est res distincta a loco et ab eo quod localiter movetur. 62ra P

Arguitur quod non quia:

⟨1⟩ Si totum potest salvari sine re addita mobili et loco, frustra illa poneretur, et hoc est inconveniens; sed omnia sine hoc possunt salvari. Probatio quia: esset motus localis, si continue *a* mobile esset supra aliam et aliam partem spatii *b*, licet non poneretur aliud esse; et salvaretur successio et prioritas et posterioritas per diversas partes spatii quae secundum situm habent ordinem et positionem circumscripto alio addito.

⟨2⟩ Item sequeretur quod Deus potest separare et separatim conservare motum sine mobili et loco, immo ipsis annihilatis; quod videtur inconveniens, quia tunc esset motus et nihil moveretur.

⟨3⟩ Item quomodo intenderetur velocitas motus, | cum non remaneat pars prior cum posteriore? Et tamen sic debet fieri intensio, scilicet per additionem partis ad partem remanentem in eodem subiecto. 50rb p

⟨4⟩ Item Commentator et alii ponunt quod motus est de essentia termini ad quem; et in motu locali terminus ad quem non est nisi locus qui acquiritur; igitur motus localis est de essentia loci.

⟨5⟩ Item potest argui quod motus non esset. Sed de hoc dictum est in alia quaestione; ideo dimitto.

Oppositum arguitur quia:

⟨1⟩ Nec esse loci nec esse mobilis consistit in fieri, immo utrumque est perfecte factum, nisi sit aeternum; sed esse motus localis vel temporis con-

4 quaeritur septimo] consequenter quaeritur I ‖ utrum] *rep.* P ‖ et] vel P 7–8 poneretur] ponerentur CPp 8 omnia] *om.* C ‖ probatio] probo I 9 continue] *post* mobile P ‖ supra] super p 10 partem spatii] *inv.* p ‖ poneretur] ponerentur p 11 posterioritas] posteritas P 13 sequeretur] sequitur Pp ‖ potest] posset IPp ‖ separare] salvare P 17 posteriore] posteriori Pp 20 in] ita in P : ita I ‖ non] nihil P 22 potest] posset IPp ‖ de] *om.* C

19–20 Cf. Averroes, *In Physicam*, V, comm. 48, f. 237A **22–23** Cf. sup., III, q. 6

 | DOI: 10.1163/9789004322356_010

72vb C sistit in fieri aliud post aliud; igitur esse motus non | est esse loci vel eius quod movetur. Igitur non est de essentia alicuius eorum nec per consequens est aliquod eorum. Consequentia patet, quia idem est esse hominis, essentia hominis et homo.

⟨2⟩ Item tam mobile quam locus est naturae permanentis; et motus non, sed naturae successivae.

⟨3⟩ Item cum tempus sit motus, prout debet videri in quarto libro, si tempus non est mobile nec locus, sequitur etiam quod neque motus. Sed ego probo quod tempus non est mobile nec locus quia: dicit Aristoteles quarto *huius* quod nullae partes diversae temporis simul sunt, nisi una includat aliam; sed mobilis et loci omnes partes simul sunt; igitur etc.

⟨4⟩ Item in quolibet instanti temporis sempiterni est sphaera lunae et locus eius, cum numquam desinant; et in nullo instanti est motus, prout habetur in sexto *huius*; igitur etc.

Item arguitur quod motus non sit locus quia:

⟨5⟩ Motus est subiective in eo quod movetur, ut dicitur tertio *huius*; et non est locus in eo quod movetur, sed in locante forte quiescente.

⟨6⟩ Etiam ultimae sphaerae est per se motus, immo primo; et eius non est
62rb P per se locus. | Immo vel ultimae sphaerae non est locus vel ille secundum Commentatorem est locus centri, et ille non est motus ultimae sphaerae.

Item arguitur quod motus non sit ipsum mobile, scilicet quod movetur:

⟨7⟩ Motus enim est actus ipsius mobilis, ut dicitur tertio *huius*; et idem non est actus sui ipsius; igitur etc.

⟨8⟩ Item motus est in mobili subiective, ut dicitur tertio *huius*; et idem non est subiective in se.

---

1 esse[1]] ille C ‖ est] *om.* I ‖ vel] *add.* esse p 2 non] nec P ‖ de] *om.* P 4 homo] *add.* prout manifestari debet quarto metaphysicae IPp 5 et] sed P 7 debet videri] videbitur P 8 sequitur etiam] sequeretur P 8–9 ego probo quod] *om.* I 10 diversae] diversi P 11 etc.] *om.* P 12 temporis sempiterni] temporis sempiternum I : tempus sempiternum C 13 desinant] desinat P ‖ prout] ut IPp 14 sexto] quarto C ‖ etc.] *om.* P 16–17 ut … movetur] *in marg.* C 16 dicitur] *add.* in Pp 18 est[1]] *post* se P 19 locus[1]] *ante* per P 21 movetur] *add.* et (etiam P) arguitur sic Pp 22 enim] *om.* IPp 23 igitur etc.] *om.* IPp 24 tertio huius] in isto tertio Pp

---

7 Cf. inf., IV, q. 12 9–10 Cf. Aristoteles, *Physica*, IV, 10, 218a11–14 14 Cf. Aristoteles, *Physica*, VI, 3, 234a31 16 Cf. Aristoteles, *Physica*, III, 2, 202a13–14 20 Cf. Averroes, *In Physicam*, IV, comm. 43, f. 142G–H 22 Aristoteles, *Physica*, III, 2, 202a7–8 24 Cf. Aristoteles, *Physica*, III, 2, 202a13–14

⟨9⟩ Item cum locus qui acquiritur sit terminus ad quem et locatum sit mobile, manifestum est quod mobile non est de essentia termini ad quem; et tamen motus est de essentia termini ad quem secundum Commentatorem.

⟨10⟩ Item si terra moveretur sursum, non est verum quod ille motus sit illa terra, quia ille motus est innaturalis et violentus ipsi terrae, terra autem non est sibi ipsi innaturalis et violenta.

⟨11⟩ Item sequeretur quod materia esset motus, si Deus eam solitarie moveret; et sic ipsa esset actus, quod est inconveniens.

Antiquiores non dubitaverunt de ista | [48va I] quaestione, sed concorditer concesserunt motum localem esse aliam rem a mobili et loco. Sed iam posteriores moderni propter rationes praedictas posuerunt quod motus non sit alia res a mobili.

Sed ad videndum de hoc oportet supponere quid nominis, quia sine hoc non potest esse disputatio, ut patet quarto *Metaphysicae* et in libro *Posteriorum* et etiam in libro *De sensu*, ubi dicitur quod quid nominis est principium doctrinae. Omnes igitur concedunt quod motus est mutatio quaedam et moveri mutari. Et in quinto *huius* dicit Aristoteles, et est per se notum, quod mutari est aliter et aliter se habere prius et posterius, vel saltem est prius aliqualiter se habere et posterius taliter non se habere aut e converso. Unde Aristoteles dicit sic: 'quoniam autem omnis | [50va p] mutatio est a quodam in | [73ra C] quoddam (manifestat utique nomen; post aliud enim aliquid et aliud quidem prius aliud autem monstrat posterius)'. Et Commentator dicit sic: 'hoc est per se manifestum, quoniam dum res fuerit in eadem dispositione, tunc illic non erit transmutatio'.

3 commentatorem] *add.* ergo etc. Ip : *add.* igitur P 4–6 item ... violenta] *in marg. inf.* C 4 moveretur] movetur IP 6 est] *add.* sic p 7 sequeretur] sequitur p 8 sic] *rep.* I ‖ esset] est I 9 concorditer] *om.* P 13 sed] et IPp ‖ de hoc] *om.* C 14 patet] apparet IPp 14–15 et ... posteriorum] †...†o posteriorum *in marg.* I 15 et] *om.* P ‖ etiam] *om.* p ‖ sensu] *add.* et sensato P 16 igitur] *add.* alii I ‖ motus] *add.* localis p 17 moveri] *add.* est IPp ‖ et[2] ... aristoteles] sed in quinto dicit aristoteles p : sed dicit aristoteles in quinto huius P 19 taliter] *post* non I : *post* habere[2] Pp 21 quoddam] quiddam I ‖ aliquid] aliud I 22 et aliud] et aliquid P : *om.* C ‖ quidem] quod p ‖ monstrat] manifestat p : *om.* P 24 tunc] termini C

3 Cf. Averroes, *In Physicam*, V, comm. 48, f. 237A 14–15 Cf. Aristoteles, *Metaphysica*, IV, 7, 1012a21–22; cf. Aristoteles, *Analytica posteriora*, I, 1, 71a12–16; cf. Aristoteles, *De sensu et sensato*, 1, 437a12–15 (?) 20–22 Aristoteles, *Physica*, V, 1, 224b35–225a2 22–24 Averroes, *In Physicam*, V, comm. 7, f. 211B–C

Tunc igitur ego pono conclusiones.

Prima est quod possibile est ultimam sphaeram moveri motu quo movetur sine loco. Probatur sic quia: si ultima sphaera et alia fierent unum continuum per potentiam divinam, ita quod totus mundus esset unum corpus continuum, tunc nullus esset locus secundum Aristotelem, quia nulla esset 62va P superficies corporis continentis divisi et tangentis. Unde Aristoteles | ponit totum mundum non habere locum nisi ratione partium, quarum una locat aliam, quia continet eam et est divisa ab ea et tangens ipsam. Hoc enim requiritur ad hoc quod sit locus. Unde si Deus omnia corpora annihilaret praeter istum lapidem, ipse lapis non amplius esset in loco. Et tamen illo casu posito adhuc esset possibile quod Deus moveret simul circulariter totum mundum. Hoc probo per quendam articulum Parisius condemnatum, in quo dicitur quod Deus non potest movere simul totum mundum motu recto—error; et non est ratio quare magis posset movere ipsum motu recto quam motu circulari.

Item sicut motu diurno movet omnes sphaeras caelestes simul cum ultima sphaera, ita potest omnia alia, scilicet inferiora, volvere simul. Et si ipse possit omnia volvere simul, cum modo sint ad invicem discontinuata, non minus hoc potest, si essent facta unum continuum. Igitur potest totum mundum movere, licet non esset locus.

Item oporteret concedere, si totus mundus esset unum continuum, quod Deus extra posset formare unum granum milii tangens illum mundum, et quod illo grano milii formato quiescente Deus potest sic volvere illum mundum quod continue alia et alia pars eius tangeret illud granum milii; et tamen illo posito ille mundus non haberet aliquem locum. Et etiam, si Deus illo grano formato posset sic volvere illum totum mundum, ita posset sine illo.

---

1 igitur ego] ego I : ergo p : tibi P 2 prima] *add.* conclusio Ip || est[2]] esset IPp 3 sic] *om.* P || fierent] fuerint p 8 enim] non *sed add. in marg.* alias enim P 10 ipse] iste Pp || amplius esset] *inv.* I 13 potest] possit IPp 14 error] *in marg.* C || posset] possit I : potest P || movere ipsum] *inv.* Pp 16 item] et iterum IPp 17 potest] posset IPp 17–18 et … simul] *in marg.* C || si ipse] si I : se ipso p 18 possit] potest IPp || modo] non I || discontinuata] discontinua p : distantia P 19 potest[1]] posset IPp || potest[2]] posset IPp 20 movere] *sup. lin.* C : *om.* I || esset] est p : esse I 21 oporteret] oportet IPp 22 tangens] contangens P 23 quod] *add.* in p || potest sic] posset sic Ip : sic posset P 24 milii] *om.* IPp 25 et[1]] *om.* P || locum] locus C || et[2]] *om.* P 26 volvere] solvere I || illum totum] *inv.* Ip || posset[2]] possit I : potest P 27 illo] *add.* grano Pp : *add.* loco I

6–8 Cf. Aristoteles, *Physica*, IV, 5, 212b8–11 12 Cf. *Chartularium Universitatis Parisiensis* (ed. Denifle, Chatelain, I, no. 473, 546)

Secunda conclusio est quod ultima sphaera non solum ex eo movetur quod se habet continue aliter et aliter ad ipsam terram vel ad aliquod aliud corpus. Probatio quia: non minus moveretur, si omnia alia volverentur cum illa sine alio motu eorum; et tamen tunc non se haberet per talem motum aliter et aliter | ad aliquod aliud corpus. 73rb C

Iterum non minus oportet se habere aliter et aliter quod movetur motu recto, quam quod movetur circulariter. Et tamen ad moveri recte non oportet se habere aliter et aliter ad aliud corpus, quia si a Deo totus mundus moveretur simul motu recto, non propter hoc se haberet aliter et aliter ad terram, licet quiesceret, sicut nunc habet, scilicet si terra volveretur. Igitur ex aliter se habere ad terram non sequitur quod moveatur. Ideo non solum ex eo movetur quod aliter se habet ad terram vel ad aliud corpus, quoniam ita diceremus de alio corpore, sicut diximus de terra.

Item sequeretur quod ex motu difformi terrae vel partium eius ultima sphaera difformiter et irregulariter moveretur, quod est falsum. Consequentia patet, quia ex tali motu irregulari | terrae vel partium eius ultima sphaera 62vb P difformiter et irregulariter se haberet ad terram aliter et aliter.

Tertia conclusio est quod ultimam sphaeram moveri est eam intrinsece aliter et aliter se habere prius et posterius. Probo quia: per quid | nominis 48vb I moveri est aliter et aliter se habere prius et posterius; et tamen moveretur, licet | non se haberet aliter et aliter prius et posterius ad aliquod extrinse- 50vb p cum, ut apparuit per conclusiones praecedentes; igitur etc.

Sed aliqui respondent quod moveri est aliter et aliter se habere ad aliquod quiescens aut simpliciter, si aliquid quiescit, aut sub condicione, quia si aliquid quiesceret, se haberet ad illud aliter et aliter.

Sed ista evasio nihil valet, quia possibile est quod ultima sphaera moveretur de facto, licet nihil de facto quiesceret; igitur ipsa nullo modo se haberet de facto aliter et aliter ad aliquod quiescens nec ad aliquod extrinsecum. Igitur si non se haberet aliter et aliter intrinsece, ipsa nullo modo se haberet aliter et aliter de facto. Ideo nullo modo mutaretur de facto. Nam ad mutari requiritur aliter et aliter se habere simpliciter de facto et non solum sub condicione.

---

1 est] *om.* I 2 terram] *om.* CI ‖ ad²] *om.* P 4 illa] ea IPp ‖ eorum] earum C ‖ se] *ante* tunc P 6 iterum] *praem.* et IPp 6–8 quod ... aliter²] *om.* (*hom.*) p 7 circulariter] motu circulari P ‖ et] *om.* P 10 scilicet] sicut C ‖ aliter] *add.* et aliter p 14 sequeretur] sequitur p 15 difformiter] difformitur p ‖ moveretur] movetur P 17 se haberet] *post* aliter² p 18 tertia conclusio est] tertia conclusio I : tunc sit tertia conclusio scilicet p 19–20 probo ... est] probatio per quid nominis quia non movetur propter p ‖ probo ... posterius] *om.* (*hom.*) P 20 posterius] *add.* ad aliquod extrinsecum p 22 igitur etc.] *om.* P 25 illud] aliud p 27 ipsa] ista p 31 simpliciter] *add.* et P

Item numquam habitus debet describi per privationem sibi oppositam, immo oportet quod fiat e converso; sed iste terminus 'quiescere' est privatio opposita huic termino 'moveri'; igitur mala est descriptio dicens et declarans quid nominis 'moveri', quod moveri sit aliter et aliter se habere ad aliquod quiescens. Nam loco huius termini 'quiescens' pono eius descriptionem. Tunc idem omnino describeretur per se ipsum, scilicet sic: moveri est aliter 73[va] C et aliter se habere ad illud quod est aptum natum moveri et | non movetur; et hoc est manifestum inconveniens.

Quarta conclusio est quod motus ultimae sphaerae non est sphaera illa nec locus eius. Primo manifestum est quod non est locus eius, quia possibile est quod moveretur, licet non haberet locum, ut dictum est; et quia, si habet locum, tamen ille est divisus ab ea, motus autem eius non est divisus ab ea, cum dictum sit quod ipsa intrinsece aliter et aliter se habet. Etiam nec ille motus est illa sphaera, quia ut dictum fuit in quaestione de distinctione figurae a figurato, non est imaginabile vel possibile quod aliquid se habeat aliter quam se habebat ante, nisi hoc sit ad aliquod extrinsecum, vel nisi hoc sit propter aliquid esse quod ante non erat, aut non esse quod ante erat; sed duo primi modi non habent locum in motu ultimae sphaerae, ut patet ex dictis; igitur oportet concedere tertium modum. Et tamen quantum ad substantiam ultimae sphaerae nihil est quod non esset ante, et nihil erat ante quod non sit modo; igitur aliud a sphaera est quod ante non erat vel e converso; et hoc non est nisi motus vel partes eius; igitur etc.

63[ra] P Item motus ultimae sphaerae | non est mutatio eius substantialis, nec in ordine ad aliquod extrinsecum, nisi hoc acciderit ad aliter se habere intrinsece, ut praedictum est. Igitur est mutatio secundum dispositionem aliam a substantia sphaerae et sibi inhaerentem.

Item aliter et aliter se habet intrinsece; igitur est alteritas alicuius ab aliquo intrinsece; et non substantiae sphaerae ad se ipsam circumscripto omni alio; igitur est alia dispositio, et illa est motus eius.

---

1 item] et iterum IPp 2 immo] sed p : ergo non P 4 et aliter] *om.* IPp ‖ aliquod] *om.* IPp 5 nam] *add.* si P ‖ pono eius] ponemus p 6 idem omnino] *inv.* I : omnino idem terminus Pp ‖ describeretur] describetur p 7 et[2]] *rep.* C 8 manifestum] manifeste IPp 9 est[1]] *om.* I ‖ sphaera illa] *inv.* IPp 11 est[2]] *add.* prius p 13 se habet] *ante* aliter[1] Ip ‖ etiam] *praem.* sed Ip 14 ut] sicut Ip 15 a figurato] ad figuratum I ‖ aliquid] aliquod p 17 aliquid] aliquod C ‖ ante erat] *inv.* P 19 tertium modum] aeternum motum P 20 est … nihil[2]] *in marg.* C : *om.* (*hom.*) P 21 ante[2]] *sup. lin.* C : *om.* P 23 eius] *om.* P 24 ad[2]] *add.* aliud C 25 praedictum] dictum IPp 27 habet] *praem.* habere I : habere p ‖ igitur] *om.* p 28 substantiae sphaerae] *inv.* P 29 omni] *om.* IPp

---

14–15 Cf. Iohannes Buridanus, *Quaestiones super libros Physicorum*, II, q. 3, ad 7. (ed. Streijger, Bakker, 261[27]–262[25])

Quinta conclusio est quod motus ultimae sphaerae est distinctus ab ultima sphaera et a loco eius, si habeat locum, quia est et non est hoc nec illud; igitur etc.

Sexta conclusio est quod motus ultimae sphaerae est res pure successiva, cuius scilicet est pars prior et pars posterior non manentes simul, quia si esset res permanentis naturae, tunc secundum illam ultima sphaera non se haberet aliter et aliter prius et posterius plus quam secundum eius magnitudinem vel figuram vel alia eius accidentia permanentia, quod est falsum.

Tunc igitur respondendum est ad rationes.

⟨1⟩ Ad primam manifestum est quod sine dispositione superaddita non potest | salvari quod ultima sphaera se habeat aliter et aliter intrinsece, sicut 51ra p se habet.

⟨2⟩ Ad aliam dico quod non plus reputarem impossibile quod esset motus et nihil moveretur vel mutaretur, quam quod esset albedo et nihil esset album. Neutrum est possibile naturaliter et utrumque est possibile supernaturaliter. Sed de hoc quod dicitur, quod implicat contradictionem esse motum localem et non esse | locum, ego dico quod motus ultimae sphae- 73vb C rae vel navis in fluvio non dicitur localis quia necesse sit quod secundum illum mutetur locus, sed quia secundum communem cursum naturae omne quod movetur illo motu variat de facto habitudinem localem vel situalem ad aliquod aliud. Et omnino ille motus quem vocamus localem potest non esse localis, quia nullus mutaretur locus nec situs ad aliquam aliam rem; sed tunc non possem illum percipere. Non ergo vocatur motus localis quia ad ipsum sit locus necessarius, sed quia percipi non potest, nisi appareret mutatio loci vel situs rei ad aliam rem. Unde existentes in navibus in mari velociter et simul motis non percipiunt quod illae moveantur.

⟨3⟩ Ad aliam dictum fuit prius quod forma ita bene redderetur intensa, si plures gradus | vel maiores gradus generarentur simul, sicut si generarentur 49ra I

2 a] *om.* P ‖ si habeat locum] si habeat locus C : *om.* P ‖ non] nec P 3 igitur etc.] *om.* p 8 vel^{1}] et p 9 tunc … est] *om.* P 10 superaddita] addita Pp : habita I 11 potest] possit P : posset Ip 13 aliam] secundam IPp ‖ impossibile] inconveniens Pp : *om.* I 15 est^{1}] enim esset P 17 motus] *add. sup. lin.* alias locus C 18 non] *add.* non debet esse localis vel non P 19 quia] *om.* P 20 motu] loco CI 21 aliquod] aliquid Pp ‖ omnino] omnis Pp ‖ potest] possit P : posset Ip 23 possem] possemus Pp ‖ non ergo] *inv.* P ‖ motus] *corr. ex* locus C : *om.* Pp 24 potest] posset IPp ‖ nisi appareret] quia apparet P 25 rem] *om.* p ‖ mari] *add.* aeque p 27 quod] quia I ‖ forma] *post* bene Ip : *post* redderetur P 28 maiores gradus generarentur] maior gradus generaretur P

27 Cf. sup., III, q. 5, 55^{1-5}

unus prius et alter posterius cum permanentia eorum. Modo a fortiori lucido et fortiore motore ceteris paribus generarentur plures gradus simul luminis 63rb P vel velocitatis | et maiores quam a debilibus; ideo esset intensius lumen et intensior velocitas.

⟨4⟩ Ad aliam dicitur quod termini intrinseci qui sunt de necessitate motuum quos vocamus locales, non sunt loca, sed sunt partes extremae illorum motuum, sicut partes extremae lineae sunt termini lineae.

⟨5⟩ Ad aliam rationem responsum fuit in alia quaestione.

Haec de quaestione.

---

2 fortiore] fortiori P : a fortiore I : a fortiori p 8 aliam] ultimam IPp 9 haec de quaestione] et sic est finis quaestionis I : etc. Pp

---

8 Cf. sup., III, q. 6

⟨III.8⟩

⟨Utrum de necessitate motus localis sit habere terminos positivos praeter fluxum, scilicet terminum a quo et terminum ad quem⟩

Quia statim tactum fuit de terminis motus localis, potest quaeri utrum de necessitate motus localis sit habere terminos positivos praeter fluxum, scilicet terminum a quo et terminum ad quem.

Arguitur quod sic quia:

⟨1⟩ Oportet motum localem esse de uno loco ad alium locum; aliter non esset localis. Et illa loca sunt terminus a quo et terminus ad quem.

⟨2⟩ Item mobile, antequam moveatur, debet esse in termino a quo, et quando motum est, debet esse in termino ad quem; sed antequam moveatur et postquam motum est, nihil est de fluxu; igitur oportet esse alios terminos a fluxu.

⟨3⟩ Item cum motus aliquis sit finitus et terminatus a parte ante et a parte post, necesse est quod habeat terminos. Sed quod illi termini sint | distincti 74ra C
a fluxu patet, quia fluxus est idem quod ille motus; et tamen illi termini sunt distincti ab illo motu, tum quia sunt plures et diversi motu existente uno et eodem, tum etiam quia idem non est sui ipsius terminus.

⟨4⟩ Item omnis motus debet esse de affirmato in affirmatum sive de subiecto in subiectum, ut habetur quinto *huius*; modo talia affirmata in motu locali non sunt momenta nec partes illius motus; igitur praeter haec omnia oportet ponere alios terminos positivos. Quod autem illi termini non sunt momenta probatur quia: momenta et instantia sunt privativa sicut et puncta. Sed etiam probatur quod illi termini non sunt partes illius motus quia: sequeretur quod terminus iterum indigeret habere terminum et sic in infinitum, quod est inconveniens. Et probatur consequentia quia: pars

---

5 quia] *praem.* octavo Ip ‖ quia … quaeri] quaeritur octavo P ‖ tactum fuit] *inv.* I 8 arguitur] *praem.* et IP 9 alium] alterum P ‖ locum] locus C 13 alios terminos] *inv.* Pp 16 quod[1]] ut p ‖ termini] *om.* Pp ‖ sint] sunt I 17 et] sed P 18 tum] *om.* C 19 etiam] *sup. lin.* C : *om.* IPp 24 quia] quod I ‖ et[2]] haec I 25 sunt] *add.* etiam Ip 26 sequeretur] sequitur p ‖ iterum indigeret] *inv.* IPp 27 in] *om.* p

---

21 Cf. Aristoteles, *Physica*, V, 1, 225b1–5

motus est motus; et non est motus infinitus; igitur est motus finitus et ter-
51rb p minatus; et per consequens oportet quod ha|beat terminos; igitur oporteret terminum semper habere terminos.

⟨5⟩ Item Commentator distinguit in motu inter viam sive fluxum et perfectionem ad quam vadit, quae est terminus.

⟨6⟩ Item grave motum non solum intendit fluxum, quia tunc praeter intentionem quiesceret deorsum.

⟨7⟩ Item illa requiruntur ad motus locales elementorum, quae si non haberent diversas virtutes, non fierent motus illorum elementorum; sed si
63va P non essent loca diversarum virtutum, scilicet sursum | et deorsum, non fierent motus locales naturales elementorum, quia non esset ratio quare hoc magis moveretur deorsum quam illud, et illud magis sursum quam illud, sicut bene dicit Aristoteles; igitur ad illos motus requiruntur loca diversarum virtutum. Et non apparet quod requirantur nisi tamquam termini. Igitur etc.

Oppositum arguitur quia: non apparet quod in motu locali praeter fluxum qui est motus debeant poni termini nisi loca vel momenta indivisibilia, sicut diceretur de punctis in linea; sed loca non requiruntur, ut dictum est; nec momenta indivisibilia, quia non concedimus illa esse sicut nec puncta indivisibilia in linea, immo etiam haec nomina 'punctum', 'instans' et 'momentum' sunt privativa, ut habetur tertio *De anima*; igitur etc.

Item notandum est quod dupliciter imaginatur motus habere magnitudinem et divisibilitatem et per consequens finitatem et infinitatem, ut apparet sexto *huius*. Uno modo secundum extensionem et divisionem mobilis, quia in alia parte mobilis est alia pars motus. Et sic omnes partes motus sunt

1 est motus[3]] est p : *om.* P 2 et per consequens] ergo P ‖ oporteret] oportet I 3 semper] *ante* terminum Ip : *om.* P 6 fluxum] motum p 7 intentionem] intensionem P 9 haberent] habeant p ‖ fierent] fieret P ‖ si] *om.* P 10–11 non[2] … elementorum] motus locales naturales illorum non fierent p 11–12 hoc magis] *inv.* P 14 requirantur] requiruntur p ‖ etc.] *om.* P 16 debeant] debent P ‖ termini] *om.* CI ‖ nisi] nec I : *om.* P 17 diceretur] dicetur C ‖ in] et CIp 18 nec[1]] *om.* p 19 etiam] et P 19–20 punctum instans et] punctus linea P 20 etc.] *om.* P 21 item] *om.* IPp ‖ est] *om.* P 21–22 magnitudinem] *add.* continuitatem Pp 22 finitatem et infinitatem] infinitatem et finitatem P : finitam et infinitam etc. I 23 uno modo] *om.* I 24 est] et I 24–83.1 sic … simul] simul sunt omnes partes motus P

4–5 Cf. Averroes, *In Physicam*, III, comm. 4, f. 87C–D; V, comm. 9, f. 215A–C 13 Cf. Aristoteles, *Physica*, IV, 8, 215a1–14 18 Cf. sup., III, q. 7 20 Cf. Aristoteles, *De anima*, III, 6, 430b20–21 23 Cf. Aristoteles, *Physica*, VI, 4, 234b21–23 simul secundum durationem et tempus, non una pars prius quam alia, quia mobile | (74rb C) simul per omnes partes movetur et simul incipit moveri et simul etiam desinit moveri, nisi fiat discontinuatio partium vel plicatio. Et sic ego puto quod est eadem magnitudo qua mobile extenditur et qua motus extenditur, sicut dicerem quod est eadem magnitudo qua lapis et albedo et durities et alia eius accidentia | (49rb I) extenduntur. Sic igitur est tantus motus quantum est mobile, sicut dicitur in *Praedicamentis* quod tanta est albedo, quanta est superficies subiecta ei. Et sic motus caeli, licet sit aeternus, est terminatus sicut et caelum per terminos quantitatis caeli. Ideo si sciatur qui sunt per se termini magnitudinis caeli, dicetur quod illi sunt per accidens termini substantiae caeli et motus eius. Sed de illa magnitudine vel extensione motus non intelligebatur principaliter haec quaestio.

Alio modo motus habet magnitudinem, extensionem et divisibilitatem secundum partes temporis ad invicem priores et posteriores in eodem mobili et secundum eandem partem eius sibi succedentes. Et sic motum caeli dicimus infinitum. Et potest haec magnitudo motus vocari magnitudo durationis. Sic igitur motus diceretur infinitus, qui semper esset vel qui esset perpetuus. Et diceretur finitus a parte ante, qui non semper est, sed ante aliquando nihil eius erat; et a parte post, cuius aliquando nihil erit.

Tunc ponuntur conclusiones.

Prima est quod motus | (63vb P) perpetuus, si est aliquis perpetuus, non habet terminos ipsius, scilicet quibus terminetur secundum durationem, quia est infinitus secundum durationem; et infinitum, si est aliquid, est interminatum et non habens terminos. Tamen non est inconveniens quod huiusmodi motus infiniti sint termini suarum partium.

Secunda conclusio est quod partium eius essent termini, quia possent eius partes | (51va p) signari finitae, ut una revolutio diurna vel duae; et omnis magni-

---

1 et tempus] mobilis p ‖ pars] *om.* IPp 3 etiam] *om.* P ‖ vel plicatio] *add.* partium P : *om.* p 4 et] ex P 5 sicut] *add.* ego P : sic ego p ‖ et[1]] est C : *om.* I 7 sicut] ut P ‖ est[3]] *om.* p 9 et] *om.* P 12–13 non ... motus] *om.* (*hom.*) P 14 temporis] tempore I ‖ et] vel p 15 partem eius] *inv.* p : eius partes P 17 sic] si P ‖ esset[1]] erat P 18 et] *add.* motus IPp ‖ finitus] infinitus I 19 et] *om.* I ‖ cuius aliquando nihil] aliquando nihil eius C 21 est aliquis] *inv.* IPp 22 scilicet] *om.* p 23–24 aliquid ... tamen] aliquid est interminatum †...† habens terminos tamen *in marg.* C 24 huiusmodi] huius P 25 sint] sunt I 26 secunda ... partium] secunda †...† p†...† *in marg.* I 26–27 possent eius partes] possent partes eius Ip : possunt partes eius P

---

7 Cf. Aristoteles, *Praedicamenta*, 6, 5b7–8

tudinis, immo omnis continui finiti, necesse est esse terminum ex omni latere ad quod est finitum.

Tunc igitur sumatur aliqua pars motus ultimae sphaerae finita secundum durationem, quae vocetur *a*, et sit tanta quanta est una revolutio diurna solis. Erit tertia conclusio quod motus *a* habet terminum a parte ante et terminum a parte post, quia utrobique est finitus.

Quarta conclusio est quod illi termini non sunt extrinseci a sphaera illa et sic de aliis motibus localibus, quia non minus esset ille motus finitus et terminatus, si non esset aliquid extrinsecum, vel etiam non minus esset terminatus, si omnia alia essent illi sphaerae continua et moverentur cum illa; et tamen tunc non potest assignari quid illorum extrinsecorum esset terminus vel a parte ante vel a parte post.

Item quod maneret idem ante et post et continue per totum istum motum, illud non deberet dici terminus nec a parte ante nec a parte post, quia
74va C debet differre terminus a parte ante et a parte post | et per totum medium; et tamen possibile est quod mota tota sphaera extrinseca maneant continue ante et post et in medio.

Item licet ultima sphaera in una eius revolutione se habeat continue aliter et aliter ad terram vel ad corpus aliud, tamen non intendit sic se habere vel sic tamquam ad terminum, quia tunc ibi deberet quiescere.

Item sexto *huius* patet quod sphaera non mutat locum realiter, licet secundum rationem; igitur a parte rei non potest ex loco vel terra vel alio tali assignari terminus differens ab altero termino vel a medio.

Tunc igitur sit quinta conclusio quod etiam motus finiti ultimae sphaerae, puta unius revolutionis, nec caelum nec magnitudo caeli vel aliqua pars caeli aut magnitudinis est terminus sive a parte ante sive a parte post, quia quodlibet istorum ita est ante sicut post et sicut in medio.

---

5 erit] *praem.* et IPp ‖ a] *om.* P ‖ terminum[2]] *om.* P 6 finitus] *add.* quaere p 8 motibus localibus] *inv.* p : mobilibus localibus I 8–9 finitus et] *om.* p 9 aliquid] aliquod p 10 illi] omni I 10–11 cum illa] *om.* I 11 potest] posset p : possit I ‖ quid] quin CI 13 maneret] manet I ‖ continue] *post* motum (13–14) P 15 et[2]] *om.* I 16 mota tota sphaera] mota tota (*sup. lin.*) sphaera C : mota sphaera Ip : sphaera mota P ‖ extrinseca maneant continue] manerent extrinseca continue Pp : extrinseca manerent I 19 corpus aliud] *inv.* IPp ‖ intendit] intenditur IPp ‖ se] *om.* I 20 ad terminum] terminus IPp 21 patet] dicitur I : dicit aristoteles P ‖ locum] locus C 22 igitur] *om.* I ‖ potest] *add.* sic p : *add.* dici I 23 altero] alio IPp 24 tunc … conclusio] quinta conclusio est IPp ‖ etiam] *om.* P 26 aut] ac I 27 ita] *om.* P

---

21 Cf. Aristoteles, *Physica*, VI, 9, 240a29–b7

Sed tunc quaeretur rationabiliter qui sunt termini quibus illa revolutio est terminata.

Et ego dico quod de hoc est dicendum, sicut esset dicendum de terminis lineae vel magnitudinis proportionaliter. Ponamus igitur quod esset virga decem pedum longitudinis | et nullum aliud esset corpus in mundo. Illa 64ra P esset finita et terminata secundum suam longitudinem; et non per aliquod extrinsecum, quia non esset aliquod extrinsecum. Nec essent termini eius puncta indivisibilia vel superficies indivisibiles secundum profunditatem, quia supponimus talia non esse. Nec totalis linea sive virga esset terminus eius, quia propter resecationem medietatis ex una parte non resecaretur aliquid de termino eius ex alia parte. Et haec omnia concessa sunt per se. | Ex his concluditur quod termini lineae vel virgae sunt eius partes 49va I quantitativae, scilicet prima pars et ultima pars, quia vel oportet ponere illas esse terminos illius lineae vel aliqua praedictorum, quod est improbatum. Et ita dicemus motus finiti terminos esse primam partem eius et ultimam.

Sed tunc oriuntur dubitationes plures.

⟨1⟩ Una est quod quaelibet pars lineae finitae est linea finita et per consequens habens terminos. Ideo si terminus lineae esset pars lineae, sequeretur quod terminus lineae haberet terminum et ille terminus alium terminum et sic in infinitum.

⟨2⟩ Item secundum Aristotelem terminus est indivisibilis; et nulla pars lineae vel magnitudinis est indivisibilis.

⟨3⟩ Item partium ad invicem continuarum idem est terminus, per definitionem continuorum; sed nulla est eadem pars linearum extra invicem situaliter existentium; igitur terminus earum non est aliqua pars earum nec alicuius earum.

---

1 sed tunc quaeretur] tunc ergo quaeretur I : tunc ergo quaeritur Pp ‖ sunt] sint p 3 sicut] sic P 4 proportionaliter] proportionabiliter I 9 talia] tales p ‖ sive] vel P 9–10 terminus eius] *inv.* P 11 aliquid] aliquod Ip 13 ponere] *om.* IPp 15 terminos] *add.* illius lineae (vel aliqua *del.*) praedictorum C ‖ esse … eius] eius esse primam partem p : eius esse primam partem eius IP 17 tunc] *om.* IPp ‖ plures] *om.* P 18 quod] quia IPp 19 ideo si terminus] *om.* p 20 terminus[1]] terminis p 21 in] *om.* p 23 vel magnitudinis] *post* indivisibilis Pp 24 terminus] *add.* et P 25 continuorum] continuarum p ‖ eadem] earum P 26 earum[1]] eorum P ‖ earum[2]] eorum P 27 earum] eorum P

---

22 Cf. Aristoteles, *Physica*, VI, 1, 231a25–26 24–25 Cf. Aristoteles, *Physica*, V, 3, 227a11–13

⟨4⟩ Item terminus lineae est infra quem est tota linea; sed infra nullam 51vb p 74vb C partem | eius est ipsa | tota, quia sequeretur quod illa pars esset infra semet ipsam, quod est impossibile.

⟨5⟩ Item totum terminatum est infra suos terminos; et tamen nulla tota linea est inter aliquas partes eius, quia tunc illae partes essent inter se ipsas.

⟨6⟩ Item si lineae pedalis terminus sit aliqua pars eius, oportet primo quaerere quanta sit illa pars, scilicet utrum medietas illius lineae vel quarta pars vel decima vel centesima. Et non poterit dici rationabiliter quae vel quanta sit illa.

⟨7⟩ Item tu dicis quod ipsius motus termini sunt prima pars eius et ultima; et hoc est contra Aristotelem in sexto *huius*, qui determinat quod in motu vel in tempore quantumcumque finito non est dare primam partem.

Et istae rationes sunt sophisticae et faciles ad solvendum, quia non est difficultas ex parte rerum, sed verborum; de quibus exprimendi sunt mentis conceptus.

⟨1⟩ Dico igitur ad primam quod, cum linea sit in infinitum divisibilis et quod in infinitum pars habeat partem, non debet reputari inconveniens quod in infinitum terminus habeat terminum. Nam linea pedalis habet partem primam inter suas centesimas, et illa prima centesima iterum habet partem primam inter suas centesimas, et sic in infinitum. Si igitur cuiuslibet lineae dicamus terminum esse suam primam centesimam, nos dicemus in infinitum termini esse terminum.

⟨2⟩ Ad aliam rationem dicendum est quod simpliciter et de virtute sermonis terminus lineae est indivisibilis in partes quarum quaelibet sit terminus illius lineae. Unde si ultima centesima lineae *b* ponatur esse terminus eius 64rb P et illa ultima centesima dividatur in duas medietates, | prima illarum medietatum non est terminus illius totalis lineae *b*, quia ultra est aliquid eius accipere.

⟨3⟩ Ad aliam dico quod partium ad invicem continuarum utraque habet suos terminos proprios, ita quod nullus terminus unius est terminus alterius.

---

1 linea] *add.* illa Pp 2 eius] illius p ‖ sequeretur] sequitur p ‖ semet] se IPp 6 primo] prius p : *om.* P 7–8 scilicet ... pars] *om.* (*hom.*) P 7 utrum] *add.* sit p 12 in] *om.* p ‖ quantumcumque] quomodocumque p 13 sophisticae] solutae C 16 sit] sunt I 17 debet] debeat P 19 centesimas] *corr. in marg. ex* decimas C : centesima I 19–20 et ... centesimas] *om.* (*hom.*) P 21 primam] partem p 23 rationem] *om.* P 24 indivisibilis] divisibilis p 25 eius] *om.* I 27 est[1]] erit Pp

---

11 Cf. Aristoteles, *Physica*, VI, 5, 236a7 sqq.

Sed tamen terminus unius et terminus alterius sunt quoddam unum, scilicet una linea (quod non est ita in discontinuis), ita quod si linea *a* et linea *b* sunt ad invicem continuae, erit una linea continua, cuius una pars erit terminus proprius lineae *a* et alia pars erit terminus proprius lineae *b*. Sic igitur intelligitur quod continuorum ultima sunt unum, licet ultimum unius non sit ultimum alterius.

⟨4–5⟩ Tunc ad alias rationes dicendum est quod totum terminatum est infra terminum vel inter terminos ad talem sensum negativum quod nihil eius est | ultra terminum vel extra terminos. Et non oportet ad alium sensum 75ra C concedere.

⟨6⟩ Et tunc ad aliam dico quod, quantacumque est pars lineae quae est prima vel ultima, illa est terminus lineae, sive sit prima sive ultima medietas, sive prima vel ultima tertia vel quarta vel centesima. Ideo concludo quod lineae *b* infiniti sunt termini. Sed tamen bene dicitur quod unius lineae non sunt nisi duo termini ad istum sensum quod non possunt unius lineae signari plures termini quam duo, quin unus participet cum alio, sicut pars cum toto etc.

⟨7⟩ Ad ultimam dicendum est quod nulla est pars prima lineae vel motus sic quod ipsa sit omni alia parte prior, sed bene est sic prima pars quia nulla alia est prior ea. Et etiam Aristoteles in sexto *huius* intendit quod temporis finiti nulla est pars prima pertransita sive praeterita, quia prius medietas est praeterita quam ipsa; et sic de motu. Et ita etiam de magnitudine spatii nulla est prima pertransita a mobili, cum tamen infinitae sint primae sic quod infinitae simul incipiunt vel simul incipiunt pertransiri, sed illae sunt communicantes sic quod una est pars alterius.

Tunc istis visis de magnitudine dicendum est simili|ter de quocumque motu 52ra p locali finito quod termini eius intrinseci sunt extremae partes eius, scilicet prima et ultima. Et dicantur omnia sicut dictum est de linea.

---

1 quoddam] quodam modo p 4 et] *om.* P 5 igitur] *om.* Pp 7 tunc] *praem.* et Ip : *om.* P ‖ rationes] *praem.* duas I : duas p : duas quod totum P 9 ad] *om.* Ip 11 et tunc] *om.* IPp ‖ est pars lineae] sit pars lineae I : pars lineae sit Pp 12 vel] *corr. in marg. ex* et C : *add.* centesima p ‖ illa] ille P ‖ sive[2]] vel IPp 13 tertia ... vel[3]] vel decima vel p : tertia sive decimalis P ‖ concludo] *add.* corollarie IPp 15 quod] quia Pp 17 etc.] *om.* p 18 ultimam] aliam p ‖ pars prima] *inv.* Pp 19 omni] omnia p 20 alia ... ea] est prior alia P ‖ et] *om.* Pp 21 sive] vel I ‖ medietas] *add.* eius Pp 22 ita] ideo C 23 prima] primo p : *om.* P ‖ cum tamen] cum I : tamen Pp ‖ sint] sunt IPp 24 vel ... pertransiri] vel simul inceperunt pertransiri I : pertransire P 26 tunc] *om.* IPp 27 eius[1]] *om.* I

Tunc specialiter dicendum est de motu recto.

Et est prima conclusio quod motus rectus non indiget terminis ea ratione (49vb I) qua rectus, | quia sicut Deus potest movere caelum motu infinito circulari, ita potest ipsum movere motu infinito recto; et motus infinitus non haberet terminum.

Alia conclusio est quod secundum cursum naturalem et potentias naturales necesse est omnem motum rectum habere terminos, quia necesse est ipsum esse finitum et non perpetuum, ut dicetur in octavo libro. Non enim potest esse motus rectus naturaliter nisi infra orbem lunae, et ibi non est spatium rectum nisi finitum; et tamen sexto *huius* dicitur quod non est motus (64va P) infinitus rectus in spatio finito. Et de hoc | est illic dicendum.

Alia conclusio est quod omne grave vel leve motum naturaliter per gravitatem vel levitatem intendit terminum extrinsecum, scilicet locum sibi naturalem. Aliter enim non esset ratio quare grave magis moveretur deorsum quam sursum, vel etiam quare magis quiesceret deorsum quam sursum, et e converso de levi.

Et est sciendum quod saepe animalia et homines per suos motus locales intendunt sibi aliquid extrinseci acquirere, ut hirundo aerem temperatum, (75rb C) et homo, cum velit calefieri, quaerit ignem et equus herbam | vel aquam. Sed tamen homo aliquando movetur non intendens terminum, sed solum intendens motum vel aliquid quod non fit per esse in termino motus, sed per moveri habetur et conservatur, ut si quis ambulat hinc inde ad campos propter conservationem sanitatis vel quia ambulando vel ludendo delectatur.

Tunc ad rationes principales.

⟨1⟩ Ad primam concedendum est quod motus ad hoc quod sit localis indiget loco, tamen motus qui est localis vel sibi similis secundum intrinsecam essentiam, potest esse sine loco et esset non localis. Concedendum est etiam

---

1 dicendum est] *inv.* P ‖ dicendum … recto] *sup. lin.* C 2 terminis] termino Pp 3 infinito] finito P 4 potest ipsum] posset ipsum IP : posset ipse p 6 quod] *om.* p 9 est] potest esse P 10 tamen] *add.* in Pp ‖ est] potest esse Pp : *om.* I 11 rectus] *om.* CI 12 vel leve] *om.* C 13 vel] aut Pp ‖ scilicet] secundum p 14 enim] *om.* IPp ‖ grave magis moveretur] grave moveretur magis p : moveretur magis P 15 quare] *ante* etiam p : quod P 17 est] *post* sciendum IP : *om.* p 18 intendunt] intendant C 19 velit] vult I 20 homo aliquando] *inv.* Pp 21 fit per] est per I : potest P ‖ motus] motum P 25 concedendum est] *corr. ex* dicendum est C : *inv.* P 26 vel] et P 27 potest] posset IPp ‖ esset] esse P ‖ etiam] *om.* I

---

8 Cf. Iohannes Buridanus, *Quaestiones super libros Physicorum*, VIII, q. 7 (ed. Parisiis 1509, ff. 115rb–116rb) 10 Cf. Aristoteles, *Physica*, VI, 2, 233a32–34 11 Cf. Iohannes Buridanus, *Quaestiones super libros Physicorum*, VIII, q. 7 (ed. Parisiis 1509, ff. 115rb–116rb)

quod de facto omnis motus rectus est de uno loco ad alterum locum intelligendo cum moderatione quae dicitur in quarto *huius*; nec est possibile esse aliter naturaliter. Sed non est ita de motu circulari.

⟨2⟩ Ad aliam dicendum est ut prius quod illud quod movetur non indiget ad hoc quod moveatur quod sit in aliquo loco vel termino extrinseco sive ante motum sive post motum sive quando movetur. Sed si motus sit finitus, necesse est quod in principio primus terminus motus sit in eo quod movetur et in fine ultimus terminus. Et per totum tempus et per totam durationem motus oportet quod totus ille motus sit in illo mobili referendo singula singulis, scilicet quod secundum primam partem durationis prima pars motus sit in mobili et secundum secundam secunda et sic deinceps. Et de facto verum est hoc in motu recto, quod mobile ante motum est in uno termino, scilicet loco extrinseco, et post in alio.

⟨3⟩ Ad aliam concedo quod quilibet terminus est distinctus ab illo cuius est terminus; et ideo terminus motus est distinctus ab illo motu. Et non sequitur 'est distinctus ab illo motu, igitur est distinctus a motu'.

⟨4⟩ Ad aliam dicendum est quod non oportet omnem motum esse de affirmato in affirmatum tamquam de termino eius in terminum eius, si motus sit infinitus. Nec est de necessitate motus quod sit aliqua affirmatio vel negatio. Tamen omnis motus finiti vel etiam mutationis finitae termini sunt verae res quae sunt partes eius. Sed ad illum sensum dicimus motum esse de affirmato in affirmatum, et non generationem vel corruptionem, quia in exponen|do quid nominis generationis vel corruptionis nobis apparentis | utimur affirmatione et negatione dicendo 'hoc prius est et posterius non est' vel e converso; sed in motu, prout res apparet nobis moveri, utimur duplici affirmatione, ut quod prius est album et post nigrum; similiter 'hoc prius est parvum et post magnum' vel 'prius pedale et post tripedale' vel e converso, vel 'hoc prius est hic et posterius illic' vel saltem 'prius se habet taliter secundum situm ad aliquod determinatum et post aliter'. Quamvis enim hoc non sit de necessitate motus, tamen hoc requiritur ad hoc quod percipiatur motus.

52rb p
64vb P

2 dicitur] dicetur Pp ‖ huius] *om.* IPp 3 naturaliter] *ante* esse (2) Pp 4 dicendum est] dicitur P 5 extrinseco] *add.* extra p 8 ultimus] ulterius C : ultime P ‖ et per[2]] vel IPp 12–13 scilicet loco] scilicet in loco P : scilicet termino I : *om.* p 15 et[2]] sed Pp 17 est] *om.* P 18 in[2]] ad P ‖ motus sit] aliquis est Pp 20 tamen] cum Ip ‖ etiam] *om.* P 23 vel corruptionis] *post* apparentis P : *om.* p 24 vel] et P 25 prout] *add.* est C : ut p 26 prius est[1]] *praem.* hoc Pp : *inv.* I ‖ similiter] *om.* Pp 27 pedale] bipedale Pp ‖ vel[2]] et p ‖ e converso vel] *om.* P 28 prius[2]] *sup. lin.* C : *om.* I 29 ad] et p ‖ determinatum] demonstratum IP

2 Cf. Aristoteles, *Physica*, IV, 5, 212a31–b22

De hoc autem quod dicitur, quod terminus haberet terminum in infinitum, dictum est prius.

75va C ⟨5⟩ Ad aliam. Quando | Commentator dicit de via et fluxu etc., de hoc satis dictum fuit in quaestionibus de alteratione.

⟨6–7⟩ Ad aliam concessum est quod grave intendit locum deorsum. Et simili modo etiam dicendum est ad aliam, scilicet ad ultimam. De istis enim duabus rationibus dictum est satis in positione.

Haec de quaestione.

---

1 quod dicitur] obicitur P ‖ in] *om.* IPp 3 quando] quod Pp ‖ et] *add.* de p ‖ etc.] *om.* P 3–4 de^{2} ... fuit] dictum fuit satis IPp 5 est] *om.* p ‖ locum deorsum] locus (*sup. lin.*) deorsum *corr. ex* deorsum locus C 6 etiam] *post* est I : *om.* Pp ‖ aliam scilicet ad] *om.* IPp 8 haec de quaestione] et sic finitur octava quaestio I : etc. p : *om.* P

---

4 Cf. sup., III, q. 2

⟨III.9⟩

⟨Utrum motus sit de essentia termini ad quem est⟩

Continuando sermonem de terminis motus quaeritur utrum motus sit de essentia termini ad quem est.

⟨1⟩ Et arguitur quod non, sicut de linea. Nam linea non est de essentia puncti terminantis eam vel etiam non est de essentia | ultimae partis eius, sed e 50ra I
converso pars est de essentia totius.

⟨2⟩ Item illa stant simul quorum unum est de essentia alterius; sed motus non est, quando est terminus ad quem, nec e converso. Habitibus enim praesentibus in materia cessat motus, ut habetur in libro *De generatione.*

⟨3⟩ Item motus est successivus, ut calefactio, et terminus eius est naturae permanentis, ut caliditas. Sic autem oppositionem habentia non sunt unum de essentia alterius.

⟨4⟩ Item motus est medius inter terminos; medium autem non est de essentia extremi, sed potius e converso.

⟨5⟩ Item motus aliquando plus habet de termino a quo quam de termino ad quem; igitur magis deberet dici esse de essentia termini a quo quam ad quem.

⟨6⟩ Item eiusdem essentiae essent motus contrarii, quod est falsum. Consequentia patet, quia ad eundem terminum possunt terminari, ut calefactio frigidi et frigefactio calidi ad tepiditatem, et motus ex aqua sursum et motus ex igne deorsum ad locum aeris.

Item specialiter arguitur de motu locali quia:

---

4 quaeritur] *sup. lin.* C : *add.* nono IPp 5 est] *sup. lin.* C : *om.* p 6 et] *om.* I 8 est] *om.* IPp 9 stant simul] *inv.* P 10 est^{2}] *add.* ibi P 11 praesentibus] existentibus *sed add. sup. lin.* alias praesentibus C ‖ libro] *om.* P 12 et] est p 12–13 eius ... permanentis] eius est permanentis naturae p : eius naturae permanentis I : permanentis naturae P 15 motus] *om.* p ‖ medium autem] et medium Pp : medium I 17 item] tunc P ‖ motus aliquando] *inv.* IPp 18 magis deberet] *inv.* p ‖ quam] *add.* termini IP 20 essentiae] *om.* C 22 aqua] qua P ‖ motus^{2}] *om.* P 23 locum] locus C : motum P

---

11 Cf. Aristoteles, *De generatione et corruptione*, I, 7, 324b16–17; cf. *AA*, 4: 16

 | DOI: 10.1163/9789004322356_012

⟨7⟩ Si sit motus lapidis de sphaera ignis usque ad terram, quando est in aere, nihil habet de terminis, scilicet de loco ignis aut terrae, vel etiam nihil habet de partibus extremis motus totalis.

⟨8⟩ Item diversas et contrarias habent operationes, ut forte quia terminus motus, scilicet deorsum, est frigefaciens et motus naturaliter calefacit.

⟨1⟩ Oppositum allegatur ab Aristotele et Commentatore tertio *huius*. Dicit enim Aristoteles quod motus non est praeter res ad quas est motus.

⟨2⟩ Et Commentator dicit quod motus non differt a perfectione ad quam tendit nisi per magis et minus.

⟨3⟩ Et etiam in secundo *huius* dicit quod motus est de specie eius ad quod 65ra P venit, sed differt secundum magis | et minus.

⟨4⟩ Et etiam ad istam intentionem dicit Aristoteles quod illud quod movetur habet partim de termino a quo et partim de termino ad quem.

⟨5⟩ Item dictum est prius quod in alteratione non est fluxus additus qualitati secundum quam est alteratio; ideo oportet quod alteratio sit de essentia illius qualitatis.

Notandum est quod Aristoteles et Commentator non dixerunt ista verba, scilicet quod motus est de essentia termini ad quem. Sed expositores ita locuti 75vb C sunt propter auctoritates | Aristotelis et Commentatoris prius allegatas. Et ideo non solum est videndum de veritate illius propositionis, sed omnino 52va p viden|dum est quomodo motus se habeat ad suos terminos.

Primo videndum est de motu locali, cuius aliqui termini sunt intrinseci, qui sunt partes eius prima et ultima, et aliqui termini extrinseci, qui sunt

1 lapidis] *corr. sup. lin. ex* localis C : lapis I 2 terminis] termino C 3 totalis] localis *sed add. sup. lin.* alias totalis C : localis P 4 habent] *codd.* (habet BrL, *deest* Pb) : *exspectes* haberet 5 frigefaciens] frigefactivus IPp ‖ naturaliter] naturalis *sed add. sup. lin.* alias naturaliter C 8 et] sed p 10 et] vel P ‖ de] *om.* I 11 venit] vadit *sed add. in marg.* venit alias C 12 et etiam] et (*sup. lin.*) etiam C : et P 13 partim[1]] partem I ‖ partim[2]] partem I 14 prius] *om.* P 16 essentia] *add.* termini ad quem p ‖ qualitatis] *add.* etc. I 17 notandum] *praem.* et I 17–18 notandum ... quem] *om.* p 17 et] vel IP 19 auctoritates ... allegatas] auctoritatem aristotelis et commentatoris prius allegatam C ‖ et[2]] *om.* P 20 est videndum] est p : oportet videre P 21 suos terminos] *inv.* IPp 22 aliqui termini sunt] aliqui (*sup. lin.*) termini sunt C : aliquando sunt termini P : termini sunt I : sunt termini p 23 qui[1]] quod P ‖ aliqui] aliquando IPp ‖ qui[2]] quae P

7 Aristoteles, *Physica*, III, 1, 200b32–33 8 Cf. Averroes, *In Physicam*, III, comm. 4, f. 87C 10 Cf. Averroes, *In Physicam*, II, comm. 14, f. 53E 12 Cf. Aristoteles, *Physica*, VI, 4, 234b15–16; cf. *AA*, 2: 174 15 Cf. sup., III, q. 2

loca diversa, scilicet terminus a quo: locus in quo mobile erat immediate, antequam moveretur; et terminus ad quem: locus in quo est, quando desinit moveri.

De terminis igitur intrinsecis dictum est prius quod illi sunt partes motus cuius sunt termini; pars autem est de essentia totius et non totum de essentia partis loquendo proprie; ideo sic motus non est de essentia sui termini, sed e converso. Potest etiam dici quod sunt eiusdem speciei, quia quaelibet pars motus est motus. In talibus enim pars et totum sunt eiusdem speciei, quia eodem nomine specifico nominantur.

Sed quomodo de istis terminis esset verum quod mobile, dum movetur, habet partim de termino a quo et partim de termino ad quem?

Respondeo quod, si toto tempore quod adaequate coexistit illi motui utamur tamquam praesente, mobile non solum habet partim de termino a quo et partim de termino ad quem, immo habet utrumque terminum totum, quia habet totum illum motum unam partem post aliam.

Si vero loquamur de terminis extrinsecis, scilicet de locis, tunc statim dicendum est quod motus non est de essentia termini nec e converso; immo motus est subiective in mobili, locus autem non. Sed tamen mobile, dum movetur, dicitur esse partim in termino a quo et partim in termino ad quem, quia si per 'terminum a quo' intelligimus locum praecisum in quo mobile est, antequam incipiat moveri, et per 'terminum ad quem' intelligimus locum alium praedicto loco aequalem et sibi immediatum, quem immediate intrat mobile exeundo a primo loco, tunc manifestum est quod, antequam moveatur, est totaliter in primo loco, et postquam perfecte motum est, est totaliter in secundo loco; et in isto toto tempore motus nec est totaliter in primo loco nec totaliter in secundo loco, sed semper habet aliquam partem eius in primo loco et aliam in secundo, et sic est partim in termino a quo et partim in termino ad quem. Sed si motus sit longus, ita quod inter locum in quo est, antequam moveatur, et locum | in quo est, quando cessat moveri, 50rb I sit magnum spatium intermedium (et tunc primum locum | vocemus termi- 65rb P num a quo et ultimum locum terminum ad quem), manifestum est quod, quando movetur in spatio intermedio, tunc nihil habet de illo termino a quo nec etiam aliquid habet de illo termino ad quem, sed est totaliter extra eos.

1 diversa] diversi I || mobile] *om.* IPp 4 igitur] ibi p 7 potest] posset IPp 12 quod[2]] *add.* non P 13 mobile] *om.* P || partim] partem Pp 14 partim] partem p : *om.* P 20 per terminum] in termino P || intelligimus] intelligamus I 21 intelligimus] intelligamus IPp 22 loco] modo P 24 motum est] *inv.* p 25 isto] *om.* IPp 26 loco[1]] *om.* P || totaliter] *praem.* est p : *om.* P || loco[2]] *om.* Ip 26–27 loco[2] ... secundo] *om.* (*hom.*) C 28 inter] *add.* motum I 31 locum] *om.* P 32 illo] *om.* P 33 aliquid habet] *inv.* P || illo] *om.* P

76ra C Nunc venio ad motum alterationis. De quo sciendum est quod in | alteratione proprie dicta continue una qualitas corrumpitur pars post partem et alia generatur, ut in calefactione frigiditas corrumpitur et caliditas generatur. Modo ita est quod aliquando totalem qualitatem quae abicitur vocamus terminum a quo et totalem qualitatem quae acquiritur vocamus terminum ad quem. Verbi gratia, si sit motus de frigidissimo ad calidissimum, perfecta frigiditas esset terminus a quo et perfecta caliditas esset terminus ad quem.

Tunc ergo posito quod tamquam praesente uteremur illo tempore in quo adaequate mobile movetur, ego dico primo quod illo tempore nec est terminus a quo nec est terminus ad quem, sed erat terminus a quo, antequam moveretur, et erit terminus ad quem, quando perfectus erit motus.

Secundo ex hoc sequitur quod nec termini a quo est aliqua pars nec etiam termini ad quem, quia eius quod nihil est nulla est pars. Ideo quod movetur nec habet partem termini a quo nec habet partem termini ad quem.

Dico tamen tertio quod mobile, quamdiu sic movetur, habet aliquid quod 52vb p fuit pars termini a quo et habet aliquid quod erit | pars termini ad quem, quia nec tota frigiditas quae erat terminus a quo est corrupta nec tota caliditas quae erit terminus ad quem est generata. Et ad illum sensum debet intelligi quod mobile, dum movetur, habet partem de termino a quo et partem de termino ad quem, vel etiam quod ipsum est partim in termino a quo et partim in termino ad quem.

Quarto etiam dico quod nihil est in motu de essentia termini ad quem, sed bene est in motu aliquid quod erit de essentia termini ad quem, et aliquid quod fuit de essentia termini a quo. Tamen nec motus fuit de essentia termini a quo nec erit de essentia termini ad quem, quia calefactio est composita ex caliditate et frigiditate, et tale compositum numquam fuit de essentia summae frigiditatis nec umquam erit de essentia summae caliditatis.

Dico igitur finaliter quod ad istum sensum et non ad alium debet intelligi quod motus sit de essentia termini ad quem, quia aliquid eius erit de essentia termini ad quem. Et sic etiam est de essentia termini a quo, quia aliquid eius fuit de essentia termini a quo.

Sed tunc potest quaeri quare magis solet dici quod sit de essentia termini ad quem quam de essentia termini a quo.

---

7 esset[2]] est P **9** mobile] *om.* IPp ‖ quod] *add.* in p **10** est] *om.* IP **11** moveretur] movetur P **14** habet partem termini[2]] *om.* P **18** erit] AHLMTU : erat CIPp : est B ‖ sensum] *om.* I **19** partem[1]] partim Pp ‖ partem[2]] partim p : *om.* P **22** etiam] *om.* P **22–23** sed ... quem] *om.* (*hom.*) P **24–25** tamen ... quo] *om.* (*hom.*) P **29** sit] est P

Respondeo quod hoc est, quia agens intendit terminum ad quem et non intendit terminum a quo. Ideo non est semper verum simpliciter loquendo, sed est verum quod motus alterationis, quantum ad illud quod agens intendit, est vel erit de essentia termini ad quem, et quantum ad hoc numquam est vel fuit vel erit de essentia | termini a quo. 65va P

Consequenter dicendum est de generatione et corruptione. | 76rb C De qua sciendum est quod aliqua solet dici instantanea, quia quod generatur generatur totum simul, ita quod non prius una pars eius quam alia; alia autem est generatio successiva et continua, sicut esset generatio caliditatis in calefactione et sicut esset generatio aeris ex aqua, si continue pars aquae post partem resolveretur in aerem. Et ita proportionali modo dicendum est de corruptione.

Nunc igitur primo dicendum est de generatione temporali et continua, ut si consideremus in calefactione generationem caliditatis nihil attendendo ad corruptionem frigiditatis. Et ponamus casum quod illud quod nihil habet de caliditate alteretur, donec sit calidissimum. Tunc igitur est quaestio qui sunt termini illius generationis caliditatis. Et ego dicam quod dupliciter possunt illius generationis assignari termini. Uno modo realiter intrinseci, et illi sunt pars quae primo generatur et pars quae ultimo generatur, sicut de motu locali dicebatur. Alio modo assignantur termini secundum denominationes affirmativas vel negativas praedicati supponentis pro eo quod generatur, ut quia dicimus prius 'non erat albedo' et postea 'est albedo' vel prius 'nihil erat de albedine', scilicet ante motum, et postea 'est tanta albedo' cessante motu. Et sine dubio de istis terminis sic acceptis loquebatur Aristoteles quinto *huius*, quando dixit quod motus est de affirmato in affirmatum. | 50va I Hoc enim non erat intelligendum quod motus inciperet ab aliquo affirmato et desineret ad aliquod affirmatum, quia nihil ad motum proficit affirmatio vel negatio, sed hoc erat intelligendum quod ante motum poteramus de eo quod movetur bene dicere 'hoc est sic dispositum vel sic se habens ad

---

1 respondeo] respondetur IPp 2 intendit] *om.* Pp 3 ad illud] *om.* P 5 vel[1]] *om.* IPp ‖ quo] *add.* etc. I 6 dicendum est] *om.* I 7 est] *om.* p ‖ generatur[2]] *om.* p 8 eius] *om.* P 10 et] *om.* CI ‖ pars aquae] aliqua pars IPp 11 aerem] aliam I 13 igitur primo] *om.* P ‖ temporali et continua] continua et temporali P 14 nihil] *post* attendendo P : non p 16 sit] fiat P 17 sunt] sint Ip ‖ dicam] dico IPp 18 intrinseci] intrinsece p 20 de] in IPp 21 vel] et p ‖ praedicati supponentis] puta supponentes C 22 et] *om.* IP 24 istis] *om.* IPp 25 dixit] dicit p 29 bene] vere IPp

---

25 Cf. Aristoteles, *Physica*, V, 1, 225b1–5

hoc vel ad illud' et post motum etiam verum est dicere quod hoc est aliter dispositum vel aliter se habens ad hoc vel ad illud. Et quando etiam ibidem dicebatur quod generatio est de negato in affirmatum, hoc non est intelligendum quod ad generationem requiritur aliqua affirmatio vel negatio, sed hoc sic intelligitur, quia de termino sic supponente pro eo quod generatur verum
53ra p erat prius ne|gative loquendo dicere quod ipsum non est, et post affirmative quod ipsum est. Et e converso proportionali modo dicendum est de corruptione. Quaeri igitur potest utrum generatio est de essentia talis termini ad quem. Et statim apparet quod non. Generatio enim vel corruptio caliditatis non est de essentia orationum nostrarum.

Et tunc dicendum est de generatione instantanea. De illa manifestum est quod ipsa intrinsece non habet aliquid prius et aliud posterius; ideo
65vb P sic dicitur indivisibilis, scilicet indivisibilitate opposita | divisioni in partes
76va C quarum una sit prius quam alia. Et sicut aliquid dicitur indivisibile, | sic non dicitur habere terminum vel terminos; ideo tali generationi vel corruptioni nulli debent assignari termini secundum durationem vel successionem.

Verum est tamen quod tali generationi bene assignamus terminos secundum denominationes affirmativas et negativas, scilicet quia dicimus ipsam prius non esse et posterius esse. Hoc enim possemus dicere de creatione animae. Nos igitur in generatione animae Socratis utimur hac oratione 'anima Socratis non est' tamquam termino a quo, et non oportet quod aliquid sibi correspondeat; et utimur hac oratione 'anima Socratis est' tamquam termino ad quem. Sic igitur dicimus generationem vel creationem esse mutationem de non esse ad esse tamquam de termino a quo ad terminum ad quem. Et veritati istarum orationum correspondent in re diversa tempora. Et quia Aristoteles credidit quod in omni generatione formae necesse sit materiam praesupponi carentem illa forma, et materia forma carens dicitur privatio, ideo terminum a quo dixit esse privationem. Et haec non est propria locutio, quia privatio non est nisi materia, quae nec est terminus a quo nec terminus

---

1 verum est] *inv.* I 2 ad^2] *om.* P ‖ quando] quia p ‖ ibidem] idem p : illud P 4 aliqua] *post* negatio P ‖ hoc] *om.* P 5 quia] quod P ‖ sic^2] *corr. ex* pro C : *om.* Pp 6 est] *add.* prius p 7 proportionali modo] proportionabili modo I : proportionaliter P 11 instantanea] *add.* et Pp 12 ipsa] illa p ‖ intrinsece] *post* habet P ‖ aliud] aliquid P 14 dicitur] *add.* prius P 15 vel corruptioni] *om.* P 16 successionem] *add.* et ita proportionaliter est dicendum de corruptione tali P 18 denominationes] durationes p ‖ et] vel P 19 possemus] possumus p 27 illa forma] *inv.* Pp ‖ forma carens] *inv.* IP : illa carens forma p 28 dixit] dicit p 29 quae] quia P

---

28 Cf. Aristoteles, *Physica*, I, 7, 189b30–191a22

ad quem, cum remaneat tam ante quam post, et tam post quam ante. Sed illa assignatio terminorum est solum secundum locutionem affirmativam et negativam, ut quia materia prius non habet illam formam. Ideo bene dicit Aristoteles primo *huius* quod in generatione terminus a quo non multiplicat contra materiam et formam, sed secundum rationem et locutionem multiplicat.

Si vero aliquis quaerat de augmentatione et diminutione, ego dicam primo quod in ea bene assignantur termini affirmativi secundum locutionem, ut quia hoc prius erat minus et post maius vel e converso.

Sed si quis quaerat de terminis realibus, dicetur quod dupliciter fit augmentatio: uno modo per rarefactionem, alio modo per nutritionem.

Et illa quae fit per rarefactionem imaginatur ab aliquibus fieri aliquo trium modorum. Uno modo sine alicuius generatione et corruptione per modum motus localis partium solum, scilicet recedentium ab invicem vel accedentium ad invicem; et tunc dicatur sicut de motu locali. Sed alio modo per generationem unius qualitatis quae vocatur raritas, et corruptionem alterius quae vocatur densitas, sicut est in vera alteratione vel sicut est de generatione et corruptione. Tertio modo per generationem magnitudinis in eodem subiecto cum magnitudine praeexistente, sicut generaretur caliditas cum caliditate praeexistente, quando aliquid fit calidius; et tunc dicetur sicut de generatione huiusmodi caliditatis.

De augmentatione autem quae fit per nutritionem | pertinet ad librum 76vb C
De | *generatione*, ubi dictum est quod ibi concurrunt omnia alia genera 66ra P
motuum, scilicet motus localis nutrimenti ad membra, alteratio ipsius et tandem generatio et corruptio substantialis, scilicet conversio nutrimenti in substantiam | aliti. Et praeter huiusmodi motus et mutationes non est ibi 50vb I
alius motus simplex, sed ex additione quantitatis nutrimenti ad quantitatem praeexistentem resultat totale corpus esse maius quam esset | praeexistens. 53rb p
Et de omnibus illis modis motuum et mutationum dictum est prius. Vel

1 tam[1]] terminus *et add. sup. lin.* tam C 3 materia] *add.* prima p 4 huius] physicorum Pp ‖ generatione] augmentatione P 8 assignantur] affirmantur I 9 prius erat] *inv.* IPp 10 sed] si C ‖ de terminis] *om.* P 14 partium] *om.* CI 15 accedentium] accidentium p ‖ ad invicem] *om.* CI ‖ et] *om.* P ‖ sed] *om.* IPp 17 in] de P ‖ est[2]] *om.* Pp 19 praeexistente] *ante* cum C 20 dicetur] diceretur p : diceret P 21 huiusmodi] huius P 22 augmentatione] magnitudine C ‖ librum] *om.* P 25 et corruptio] *om.* P 26 aliti] nutriti I : *om.* C ‖ ibi] *om.* p

4 Cf. Aristoteles, *Physica*, I, 7, 190b24–191a1 23 Cf. Iohannes Buridanus, *Quaestiones super libros De generatione et corruptione*, I, q. 16 (ed. Streijger, Bakker, Thijssen, 126–128)

etiam aliqui dicunt quod quantitate praeexistente possumus uti tamquam termino a quo et quantitate nutrimenti adveniente tamquam termino ad quem.

Tunc ad rationes dicendum est.

⟨1⟩ Ad primam dictum fuit satis quomodo termini lineae se habent ad lineam.

⟨2⟩ Ad aliam dictum est quod bene, quando motus est, termini eius sunt unus prius et alter posterius, quia sunt partes eius. Et de alteratione concessum est quod non stat simul cum termino eius a quo vel ad quem, sed est bene cum aliquo quod fuerit termini a quo, et cum aliquo quod erit ipsius termini ad quem, immo ex illis integratur alteratio.

⟨3⟩ Ad aliam dictum est quod tam motus localis quam eius termini intrinseci sunt res pure successivae. De alteratione dictum est quod forma existens in quiete et permanentia bene successive generata fuit, una pars prius et alia pars posterius.

⟨4⟩ Ad aliam dictum est quod motus totalis non est medius per exclusionem terminorum et eorum quae fuerunt vel erunt de essentia terminorum.

⟨5⟩ Ad aliam, concessum est de termino a quo sicut de termino ad quem. Sed dictum est quod magis consuevimus dicere quod motus sit de essentia termini ad quem, quia ille intenditur ab agente.

⟨6⟩ Ad aliam concessum est quod loca non sunt de essentia motuum localium nec e converso. Et si calefactio et frigefactio finiuntur ad tepiditatem, tamen secundum aliud et aliud tepiditas est terminus ad quem calefactionis et frigefactionis, quia est terminus frigefactionis quantum ad frigiditatem acquisitam, et calefactionis quantum ad caliditatem acquisitam.

⟨7–8⟩ Duae ultimae rationes arguunt de motibus localibus et de terminis extrinsecis eorum etc.

---

1 etiam aliqui] *inv.* IPp 2 adveniente] *om.* P 4 tunc] *om.* P 5 habent] habeant P 6 lineam] lineas P : invicem C 8 prius … quia] post et alter prius quae P 8–9 concessum est] *inv.* p 9–10 est bene] *inv.* IPp 10 aliquo[1]] alio C ‖ fuerit] fuit IPp ‖ termini] *praem.* ipsius p : terminus CP ‖ et] vel C 10–11 ipsius termini] termini p : terminus P 12 quam] quod I 13 dictum] dicendum IPp 15 pars] *om.* IPp 16 dictum] manifestum IP ‖ totalis] localis *sed add. sup. lin.* alias totalis C ‖ medius] medium p 17 fuerunt vel erunt] erunt vel fuerunt Pp 18 a … termino[2]] *om.* (*hom.*) p 19 dictum] dicendum p ‖ magis] *om.* P 22 et si] et sic CI : ut si P 26 de[2]] *om.* P 27 etc.] *om.* p : et sic finitur nona quaestio I

⟨III.10⟩

⟨Utrum omnis motus sit actus entis in potentia⟩

Quaeritur decimo accedendo ad definitionem motus utrum omnis motus est actus entis in potentia etc.

Arguitur quod non quia:

⟨1⟩ Videtur implicare contradictionem. De ratione enim | actus est quod 77ra C ipso subiectum dicitur in actu, ut quod albedine subiectum dicitur actu album et nigredine nigrum. Et ita motu subiectum dicitur actu moveri; ideo debet concedi quod omnis motus | est actus entis in actu, licet ante fuisset 66rb P in potentia.

⟨2⟩ Item si motus esset actus, sequeretur quod mobile et movens essent idem, quod est falsum. Consequentia patet, quia res distinctae distinguuntur per suos actus, prout allegatur ex septimo *Metaphysicae*, et etiam secundo *De anima* habetur quod potentiae distinguuntur per actus; ideo illa debent dici idem, quorum idem est actus; et tamen moventis et mobilis esset idem actus, si motus esset actus, quia ipse esset actus tam moventis quam mobilis, ut patet in illo tertio.

⟨3⟩ Item ipse esset actus imperfectus, ut dicit Aristoteles. Sed hoc est impossibile, quia 'actus' significat idem quod perfectio, et perfectio non debet dici imperfecta, sicut nec albedo nigra.

⟨4⟩ Item vel esset actus primus vel secundus. Non primus, quia non est substantia, et substantia debet esse prior accidente, ut dicitur septimo *Metaphysicae*. Nec est actus secundus, quia praecedit substantiam et generatio-

---

3 decimo] *om.* I ‖ accedendo … motus[1]] *ante* quaeritur I : *om.* P ‖ omnis] *om.* I 4 etc.] *om.* Pp 7 dicitur[1]] dicatur Pp ‖ albedine] albedinis P 8 nigredine] nigredinis P ‖ motu] *praem.* de C : motus P ‖ moveri] movere P 9 actu] potentia P 11 mobile et movens] movens et mobile IPp 13 et] *om.* P 14 habetur] *om.* C 15 dici] concedi P ‖ idem est] *inv.* Ip ‖ tamen] *corr. sup. lin. ex* cum C : cum I 18 actus] actu P 19 actus] actu P 22 esse prior] *inv.* p ‖ accidente] accident I 23 et] ad P

---

13 Cf. Aristoteles, *Metaphysica*, VII, 13, 1039a7; cf. *AA*, 1: 187 14 Cf. Aristoteles, *De anima*, II, 4, 415a18–20 17 Cf. Aristoteles, *Physica*, III, 3, 202a15–20 18 Cf. Aristoteles, *Physica*, III, 2, 201b31–32 22–23 Aristoteles, *Metaphysica*, VII, 1, 1028a32–33

 | DOI: 10.1163/9789004322356_013

nem substantialem, immo motus caeli est primus omnium mutationum, ut habetur in octavo *huius*.

⟨5⟩ Item reducendo ad quaestiones praecedentes potest argui quod motus non est et per consequens quod non est actus.

⟨1⟩ Oppositum dicit Aristoteles.

⟨2⟩ Item motus non est privatio, immo quies sibi opponitur sicut privatio habitui; nec est potentia pura nec compositum, quia est in subiecto; igitur est
53va p actus. Et est entis in potentia, quia eius quod movetur; et illud quod move|tur potest moveri, aliter non moveretur; et posse moveri est esse in potentia ad moveri.

Notandum est quod omnium praedicamentorum praedicata dicuntur aliquando de suis subiectis mediante hoc verbo 'est' expresso vel implicito, et aliquando mediante hoc verbo 'potest', ut quod 'ille est homo' et 'Antichristus potest esse homo', et 'ille est albus' et 'ille potest esse albus', et 'ille agit vel patitur' et 'ille potest agere vel pati'. Modo per 'esse' intelligimus esse in actu, si non sit praedicatum ampliativum ad praeteritum vel futurum, et per 'posse esse' intelligimus esse in potentia.

Secundo notandum est quod capiendo communiter et large 'potentiam' bene ad esse sequitur posse esse et per consequens ad esse in actu sequitur esse in potentia. Sed tamen proprie loquendo posse et potentia restringun-
52ra I tur | ad ea quae nondum sunt in actu. Sic enim intelligitur quod actus et potentia sive esse in actu et esse in potentia opponuntur et non conveniunt
77rb C eidem simul respectu eiusdem. |

Tertio notandum est quod etiam proprie loquendo et quantum spectat ad propositum, illud vocatur actus, quo adveniente aliquid dicitur esse in actu, cum ante diceretur in potentia et non in actu. Sic enim anima humana dicitur actus hominis, quia qui ante erat homo in potentia dicitur iam esse

---

1 mutationum] *add. sup. lin.* alias motuum C 2 huius] *add.* ergo etc. I : *add.* igitur P : *add.* etc. p 3 potest] posset IP 4 consequens quod] *add.* motus p : accidens P 5 dicit] determinat IPp 6 motus] *om.* IP ‖ opponitur] opposita C 8 quia eius] *sup. lin.* C, *add. sup. lin.* alias est quia oportet C ‖ illud] idem P 9 moveri[1]] *add.* quod P 11 est] *om.* Pp 12 expresso vel implicito] explicite vel implicite P 16 et] ut P 18 secundo notandum est] notandum secundo P ‖ potentiam] *add.* et posse Pp 19 sequitur[2]] *om.* IPp 21–110.25 ad … realem] *transp. post* infinitae (122.14) I 22 esse[2]] *om.* P 23 eidem simul] *inv.* P : simul eidem et Ip 24 tertio] *om.* P ‖ est] *om.* P ‖ etiam] *om.* P ‖ et] *om.* P 25 aliquid] aliud P ‖ esse] *om.* IPp 27 ante erat] *inv.* p

---

2 Cf. Aristoteles, *Physica*, VIII, 7, 260b29–30 3 Cf. sup., III, q. 6, 60^{8}–61^{30} 5 Cf. Aristoteles, *Physica*, III, 1, 201a10–11, 201b4–5

homo per adventum animae | humanae; et sic albedo dicitur actus albi, quia albedine est album quod ante poterat esse album. 66va P

Et ex istis iam sequitur prima conclusio, scilicet quod motus est actus eius quod movetur et moventis, quia quod poterat ante moveri et non movebatur movetur ipso motu praesente, et etiam quod poterat movere movet ipso motu. Et etiam agens actione agit, cum ante posset agere, et patiens patitur, cum ante posset pati; ideo dicimus quod actio est actus agentis et passio patientis.

Sed adhuc circa hoc notandum est quod, licet caelum semper movetur et sic numquam dicatur proprie in potentia ad moveri, tamen caelum non semper movebatur revolutione hodierna, sed poterat moveri ea et nunc movetur ea. Ideo adhuc manifestum est quod motus caeli est actus ipsius, scilicet quo actu movetur; et ita etiam est actus Dei moventis, quia Deus prius poterat movere illa revolutione qua movet modo.

Secunda conclusio est quod motus est actus entis in actu, quia est actus eius quod movet et eius quod movetur; sed ista verba 'movere' et 'moveri' significant esse in actu secundum talem dispositionem, sicut 'posse movere' et 'posse moveri' significant esse in potentia; igitur etc. Et universaliter per quid nominis apparet quod omnis actus quo aliquid dicitur in actu, cum ante diceretur in potentia, est actus entis in actu.

Tertia conclusio est quod omnis motus est actus entis in potentia. Et hoc patet primo de motu perpetuo, quia quando caelum movetur una revolutione, adhuc est in potentia ad hoc quod moveatur alia revolutione. Sed etiam hoc patet de omni motu non perpetuo, quia omne quod tali motu movetur adhuc non est perfecte motum, sed potest esse perfecte motum; igitur adhuc est in potentia ad esse perfecte motum. Et iterum, quando aliquid movetur, ipsum prius movetur secundum priorem partem motus et posterius secundum posteriorem; et quando movetur secundum priorem, non movetur secundum posteriorem, sed potest moveri secundum eam; igitur adhuc est in potentia, non solum ad esse perfecte motum, sed etiam ad moveri secundum aliquam partem motus.

---

1 per] propter P 3 et] *om.* p ‖ iam sequitur] *inv.* IPp 5 ipso[1]] ipsum P ‖ praesente et] praesenti P 6 etiam] *praem.* sic Ip : sic P ‖ posset] potest C 9 adhuc] *om.* Pp ‖ caelum semper movetur] caelum semper moveatur IP : semper moveatur caelum p 10 dicatur proprie] *inv.* IPp ‖ moveri] motum *sed add. sup. lin.* alias moveri C 13 movetur] moveretur I ‖ quia] quod P 14 movet modo] *inv.* IP 15 est[1]] ponitur IPp 16 movere] movetur P 19 dicitur] *add.* esse P 22 patet primo] *inv.* Ip 24 etiam hoc] *inv.* IPp ‖ patet] apparet p ‖ non] omnino P 27 movetur[2]] movebatur C ‖ priorem partem] *inv.* IPp

Quarta conclusio est quod omnis motus est actus imperfectus, quamdiu est motus, quia in his quae innata sunt fieri vel facta esse, nihil dicitur perfectum, nisi sit totum factum, quoniam in hac dictione 'perfectum' haec dictio 'per' designat complementum; sed quandocumque motus est totus factus, ipse non amplius est mo| [53vb p] tus, quoniam omnis motus consistit in fieri aliud post aliud. Igitur bene contingit motum esse perfectum et completum, sed tunc non est motus, immo fuit motus. | [77va C]

Quinta conclusio est quod motus est actus secundus et non actus primus. | [66vb P] Ut appareat quomodo hoc intelligitur, notandum est quod non proprie capitur 'primum' et 'secundum', quando distinguimus actum in actum primum et secundum. Non enim dicitur actus primus quia non sit actus prior. Aristoteles enim vocat scientiam actum primum et considerare actum secundum, licet anima sit actus prior ipsa scientia. Sed cum nomina sint ad placitum, philosophi vocaverunt actum primum omnem formam vel habitum a quo procedit vel etiam innata est procedere operatio, et actum secundum vocaverunt operationem respectu formae vel habitus a qua vel a quo procedit. Sic enim scientiam vocamus actum primum respectu considerationis, et considerare actum secundum. Sic animam etiam vocamus actum primum respectu operationum vitalium, et istas operationes vocamus actus secundos. Sic etiam gravitatem et levitatem vocamus actus primos respectu motuum localium elementorum, et illos motus actus secundos. Omnis autem motus est operatio procedens ab aliqua forma vel ab aliquo actu agente; ideo omnem motum dicimus esse actum secundum respectu actus agentis a quo procedit.

Tunc ad rationes dicendum est.

⟨1⟩ Ad primam quod motus est actus entis in actu et cum hoc entis in potentia. Idem enim est bene in actu et in potentia respectu diversorum.

⟨2⟩ Ad aliam dico quod non est inconveniens diversorum esse eundem actum diversimode, scilicet unius in quo recipitur et alterius a quo producitur. | [52rb I] Et sic est de motu respectu moventis et mobilis.

---

3 quoniam] quia p **4** dictio] praepositio I ‖ motus est] *inv.* P **5** omnis] esse IPp **6** igitur ... completum] *om.* P **8** est[1]] *om.* I ‖ et] *om.* IPp ‖ actus[2]] *om.* P **9** ut] *praem.* et IPp ‖ est] *om.* P **12–13** actum secundum] *inv.* Cp **15** etiam innata est] innata est P : est innata Ip **18** animam etiam] *inv.* IPp **19** et] *om.* P **19–20** vocamus] *in marg.* C : *om.* IPp **20** secundos] secundus P ‖ primos] primus P **21** elementorum] *om.* p ‖ secundos] secundus P **22** ab[1]] ad I **23** esse] *om.* IPp **25** tunc ad rationes] ad rationes IP : ad rationes ergo p **27** enim] *sup. lin.* C : *om.* I ‖ est bene] *inv.* P : est esse p ‖ in[2]] *om.* P **29** et] *om.* p

---

12 Cf. Aristoteles, *De anima*, II, 1, 412a22–23

⟨3⟩ Ad aliam dico quod perfectum vel perfectio potest dici uno modo quod a nullo exceditur in nobilitate et entitate vel simpliciter vel in suo genere. Secundo modo dicitur aliquid perfectum vel perfectio, quia est totum factum. Et istis duobus modis non est verum quod omnis actus sit perfectus vel perfectio. Tertio modo perfectio vocatur dispositio inhaerens subiecto vel proveniens ab agente, conveniens illi subiecto et illi agenti. Et sic motus bene est perfectio, sed talis perfectio bene dicitur imperfecta imperfectione opposita duobus primis modis perfectionis.

⟨4⟩ Ad aliam dictum est quomodo et quare motus dicatur actus secundus. Unde licet aliquis motus praecedat formam generandam, tamen non praecedit actum sui moventis respectu cuius dicitur actus secundus.

Haec de quaestione ista decima.

---

2 vel^2] *om.* I   3 dicitur aliquid] *inv.* I   3–4 factum] perfectum *corr. ex* factum C   4–5 perfectus vel perfectio] perfectio vel perfectus P   5 perfectio vocatur dispositio] vocatur perfectio P   6 proveniens] procedens P ‖ illi^2] *om.* P   7 bene est] *inv.* P   9 dictum] dicendum I ‖ dicatur] vocatur P   10 praecedat] praecedit P   11 sui] suae P   12 haec … decima] et sic est finis quaestionis I : est sic est finis P : etc. p

⟨III.11⟩

⟨Utrum definitio motus sit bona in qua dicitur quod motus est actus entis in potentia secundum quod in potentia⟩

Quaeritur undecimo utrum definitio motus sit bona in qua dicitur quod motus est actus entis in potentia secundum quod in potentia.

Arguitur quod non per argumenta quaestionis praecedentis in qua arguebatur quod motus non sit actus entis in potentia. Sed illa dimitto et arguo aliter sic:

67ra P ⟨1⟩ Motus est | accidens; et accidentis non est definitio, ut patet ex septimo *Metaphysicae*; et hoc etiam probatur secundo *Posteriorum*.

77vb C ⟨2⟩ Item idem non debet definiri per opposita; et tamen | actus et potentia sunt opposita, ut habetur nono *Metaphysicae* et prooemio *De anima* a Commentatore.

⟨3⟩ Item si esset bona, sequeretur quod omnis albedo vel anima esset motus, quod est falsum. Consequentia patet, quia albedo est actus entis in potentia ad nigredinem. Et anima etiam est actus entis in potentia, scilicet corporis potentia vitam habentis, ut dicitur secundo *De anima*; et secundum quod in potentia, quia anima non est actus corporis nisi secundum quod corpus est in potentia subiectiva respectu animae; igitur sibi conveniret definitio motus.

⟨4⟩ Item sequitur 'entis in potentia secundum quod in potentia, igitur
54ra p entis in omni poten|tia', quod est impossibile. Et patet consequentia, quia illa reduplicatio 'secundum quod' debet reduplicare de omni.

---

5 quaeritur undecimo] consequenter quaeritur I ‖ quod] *om.* I 8 sit] *rep.* P ‖ actus] *add.* et sic non est actus p ‖ arguo aliter] *inv.* p 10 accidentis ... definitio] accidentia non debent definiri p ‖ ut patet ex] ut accipitur ex I : patet P : ex p 11 et hoc] *om.* P 12 debet] potest P ‖ et tamen] sed P 17 entis] *om.* P 18 potentia vitam] *inv.* P 20 in] *om.* Pp 22 sequitur entis] sequeretur entis P : sequitur p 23 et] *om.* P ‖ patet consequentia] *inv.* I 24 reduplicare] reduplicari Pp

---

7 Cf. sup., III, q. 10, 99^5–100^4 10–11 Cf. Aristoteles, *Metaphysica*, VII, 4–5, 1030a14–1031a14; cf. Aristoteles, *Analytica posteriora*, II, 7, 92b28–29 13 Cf. Aristoteles, *Metaphysica*, IX, 3, 1047a17–20; cf. Averroes, *In De anima*, I, comm. 6 (ed. Crawford, 10) 18 Cf. Aristoteles, *De anima*, II, 1, 412a27–28

 | DOI: 10.1163/9789004322356_014

⟨5⟩ Item haec definitio convenit aliis a motu; igitur non est bona. Antecedens probo quia: generatio et corruptio non sunt motus, ut dicitur quinto *huius*; et tamen eis convenit ista definitio. Convenit etiam doctrinationi, ut vult Aristoteles; et tamen doctrinatio non est motus, quia ad primam speciem qualitatis non est motus, ut dicitur septimo *huius*.

⟨6⟩ Item ista definitio convenit tempori et infinito; quae tamen non sunt motus, immo sunt passiones entium naturalium diversae a motu.

⟨7⟩ Item cum dictum sit in alia quaestione quod motus est actus entis in actu et entis in potentia, videtur quod ita bene deberet poni in definitione motus quod sit actus entis in actu, sicut quod sit actus entis in potentia; et tamen hoc non ponitur; igitur male etc.

Oppositum arguitur per Aristotelem, qui ponit istam definitionem tamquam bonam.

Notandum est quod ad motum requiritur mobile et forma vel dispositio secundum quam mobile se habet aliter et aliter prius et posterius. Et illud mobile tripliciter potest se habere ad illam formam vel dispositionem. Uno modo in pura potentia, quia adhuc nihil habet de ea; et tunc nondum movetur secundum illam. Alio modo, quia perfecte habet vel habuit illam; et sic non amplius movetur secundum illam, sed motum est. Tertio modo, quia aliquid illius habet vel acquisivit, sed non acquisivit illam totam; et isto modo necesse est mobile se habere ad illam dispositionem, quando movetur secundum eam.

Deinde etiam notandum est quod hoc nomen 'motus' non supponit pro mobili, ut dictum fuit, sed pro dispositione secundum quam mobile se habet aliter et aliter prius et posterius.

Ex istis concludendum est quod motus, scilicet illa dispositio secundum quam est motus, est actus ipsius mobilis quantum ad illud quod mobile habet de illa | dispositione; et illud mobile est adhuc in potentia ad illud 67rb P
quod ultra acquirendum est; ideo motus est actus entis in potentia.

1 haec] ista P 3 doctrinationi] *add.* et P 4–5 primam … motus] speciem quantitatis non est motus P : primam qualitatis I 6 convenit] conveniret IPp 10 quod[1]] quid P ‖ quod[2]] quid P ‖ et] *om.* P 11 etc.] *om.* IPp 12 arguitur] patet P ‖ qui] quoniam p 14 est] *om.* P 19 non amplius] *inv.* P ‖ modo] *om.* C 20 et] *om.* P 23 etiam] *om.* IP ‖ hoc nomen] *om.* P 26 ex] *praem.* et I 27 mobile] *add.* se I

2–3 Cf. Aristoteles, *Physica*, V, 1, 225a26–27, 32 4 Cf. Aristoteles, *Physica*, III, 1, 201a18–19 5 Cf. Aristoteles, *Physica*, VII, 3, 245b8–9 8 Cf. sup., III, q. 10, 101[15–31] 12 Cf. Aristoteles, *Physica*, III, 1, 201a10–11, 201b4–5 24 Cf. sup., III, q. 7

Sed iterum notandum est quod, cum mobile fuerit partim in actu et partim in potentia | [78ra C] secundum illam dispositionem secundum quam est vel esse potest motus, hoc potest adhuc esse dupliciter: uno modo quod sit in permanentia, scilicet sine actuali tendentia ad illud de illa dispositione ad quod adhuc est in potentia; alio modo quod sit in actuali tendentia ad istud. Et si fuerit primo modo, tunc non movetur secundum illam dispositionem. Verbi gratia, si aqua fuerit calefacta usque ad tepiditatem et ibi permaneat sine processu ad ulteriorem caliditatem, tunc amplius non movetur secundum caliditatem. Sed si adhuc sit processus ad ulteriorem caliditatem, tunc movetur adhuc secundum caliditatem. Igitur ad excludendum illam permanentiam in dispositione media, | [52va I] oportuit apponere in definitione motus aliam clausulam. Et ob hoc dicit Aristoteles 'secundum quod in potentia'. Igitur bene dictum est quod motus est actus entis in potentia secundum quod in potentia.

Sed merito dubitatur quare illa clausula 'secundum quod in potentia' removet talem permanentiam in dispositione media et quare designat processum sive tendentiam ad illud quod restat habendum.

Respondeo quod, si aqua sit remisse calida et in illa remissa caliditate quiescens non esset in potentia ad caliditatem ampliorem, non minus illa caliditas aquae esset actus aquae quam modo est. Et ideo illa caliditas aquae, quamdiu est in permanentia, non dicitur actus aquae sub illa ratione sub qua aqua est ulterius in potentia, sed simpliciter dicitur actus eius sine connotatione quod sit vel non sit in potentia ad aliquid ulterius. Sed non possit esse nec intelligi tendentia ad ulte|[54rb p]rius, quam connotat hoc nomen 'calefactio', quin intelligeretur esse in potentia ad aliquod ulterius. Ideo caliditati convenit hoc nomen 'calefactio' vel 'motus' secundum illam rationem secundum quam est in potentia ad aliquid ulterius; ideo bene illa actualis tendentia designatur per illam clausulam 'secundum quod in potentia'.

Et tunc dicendum est quod haec est bona definitio motus, quia declarat explicite et convertibiliter significationem et omnem connotationem huius termini 'motus'. Ibi enim ponitur 'actus' tamquam genus, et distinguitur per hoc motus a pura potentia vel privatione. Et ponitur ibi etiam 'entis in

---

3 esse potest] *inv.* I ‖ adhuc] *om.* P 5 quod[1]] quam C 9 ulteriorem caliditatem] *inv.* P 11 apponere … motus] in definitione motus apponere Ip : in dispositione motus apponere P 12 aliam] illam P ‖ dicit] dixit IP 16 designat] desinat C 17 sive tendentiam] sine tendentia Cp 18 aqua sit] aqua Ip : aliqua P 20–21 aquae quamdiu] aquae quae diu C : quae quamdiu P 23 vel] *add.* quod p ‖ possit] posset Pp 25 quin] quae p ‖ aliquod] aliquid Pp 27 aliquid] *om.* IPp ‖ bene illa] *inv.* P 28 quod] quam P 30 connotationem] *add. sup. lin.* alias cognitionem C 31 distinguitur] differt IPp 32 etiam] *om.* IPp

potentia' ad exprimendum mobile cuius motus est actus ad differentiam actus perfecti, quo habito mobile amplius non movetur secundum illam dispositionem. Et ponitur 'secundum quod in potentia' ad excludendum quietem in dispositione media et ad connotandum | tendentiam etc. Et haec 78rb C requiruntur et sufficiunt ad hoc quod sit motus.

Tunc ergo solvendae sunt rationes.

⟨1⟩ Ad primam dico quod accidentis, id est termini connotativi, non potest esse definitio | simpliciter quidditativa propter connotationem quam opor- 67va P tet explicare et quae modo quidditativo explicari non potest. Et hoc voluit Aristoteles septimo *Metaphysicae*. Sed tamen accidentis bene est descriptio et definitio causalis per praedicata vel alia designantia connotationes vel causas accidentium et terminorum connotativorum. Sic est hic.

De argumento autem Aristotelis secundo *Posteriorum* dicendum est quod, cuius est demonstratio tamquam conclusionis scitae per demonstrationem, eius non est definitio. Conclusio ista enim, ut quod triangulus habet tres etc., non definitur, licet hoc nomen 'conclusio' definiatur. Quando autem dicit Aristoteles quod accidentis est demonstratio, sensus debet esse quod conclusionis in qua accidens, id est passio, praedicatur de subiecto est demonstratio; et iam dictum est quod eius non est definitio.

⟨2⟩ Ad aliam conceditur quod haec nomina 'actus' et 'potentia' proprie dicta opponuntur, si sumantur respectu eiusdem et secundum idem et eodem modo. Sed respectu diversorum non opponuntur, et sic est hic.

⟨3⟩ Ad aliam dicitur quod, si anima et albedo sint in quiete et in permanentia, non convenit eis haec definitio, quia non convenit eis haec clausula 'secundum quod in potentia' ad illum sensum ad quem ponitur in dicta definitione, sicut dictum est.

⟨4⟩ Ad aliam dicitur quod haec dictio 'secundum quod' non tenetur hic reduplicative, sed specificative ad specificandum rationem secundum quam motus dicitur actus entis in potentia, scilicet quod hoc est secundum illam

2 actus] *add. in marg.* alias habitus C 4 connotandum] *add. in marg.* seu cognoscendum C 9 explicari] explicare P ‖ voluit] vult IPp 11 et] *praem.* bona p : bona vel P 12 connotativorum] *add.* et IPp 13 autem] *om.* P 15 conclusio ... quod] conclusio enim quod P : conclusio enim ut p : concludo enim quod ut quod I 16 non] modo p ‖ conclusio] *codd.* (concludo P, *om.* ErHKMR, *deest* Pb) : *exspectes* triangulus 18 de] *add.* suo p 20 conceditur] dicitur P ‖ haec nomina] *om.* P 21 eiusdem] *add.* et hoc concedo P 23 ad] *praem.* sed P ‖ in^{2}] *om.* IPp 27 tenetur] capitur p 29 quod] quia IPp

17 Aristoteles, *Analytica posteriora*, II, 7, 92b12–13

rationem secundum quam mobile intelligitur tendere secundum actualem processum ad illud ad quod dicitur esse in potentia.

⟨5⟩ Propter aliam rationem sciendum est quod hoc nomen 'motus' invenitur capi quattuor modis. Uno modo valde communiter, prout intelligitur se extendere ad omnem mutationem sive temporalem sive instantaneam, etiam si sit creatio indivisibilis. Et ita non definitur hic, quia quod vocatur mutatio instantanea sive indivisibilis non est proprie mutatio neque motus, sicut in sexto libro videbitur.

Secundo modo 'motus' capitur proprie, scilicet prout se extendit ad omnem mutationem divisibilem et continuam secundum successionem, sed non tamen ad hoc quod vocamus mutationem instantaneam vel simplicem creationem. Et sic definitur hic motus. Ideo bene sic aliqua generatio est motus, et aliqua corruptio et doctrinatio et assimilatio et figuratio, si sint divisibiles et continuae secundum successionem.

Tertio modo capitur 'motus' magis proprie, et sic distinguitur contra 'generationem' et 'corruptionem', per hoc quod dicimus motum esse de affir-
52vb I mato in affirmatum, generationem | autem et corruptionem dicimus esse de
54va p negato in affirmatum et | de affirmato in negatum. Et ita non definitur hic
78va C motus, sed largius, ut dicebatur. |

Adhuc quarto modo dicitur 'motus' maxime proprie. Et sic ad motum cum praedictis condicionibus requiritur quod dispositio secundum quam dicitur
67vb P esse motus proveniat directe et immediate a movente et | non per modum sequelae ad aliam dispositionem secundum quam per se et immediatius dicitur esse motus. Et sic dicitur in quinto *huius* et in septimo quod non est motus in ad aliquid nec secundum primam vel etiam secundum quartam speciem qualitatis. Sed etiam manifestum est quod sic non definitur hic motus, sed largius, scilicet in secundo modo, quia cum illo secundo modo convertitur ista definitio.

---

1 tendere] intendere P **2** illud] aliud C ‖ esse] *om.* P **3** aliam] illam P ‖ est] *om.* p **6** creatio] *add. in marg.* alias mutatio C **8** sexto] octavo I **9** capitur proprie] *inv.* P **11** non tamen] *inv.* IPp ‖ mutationem] *om.* P **11–12** simplicem] simpliciter p **13** et[2]] est C ‖ figuratio si] significatio sed P **16** dicimus motum] *inv.* IPp **18** de ... hic] e converso non ita hic definitur P **20** adhuc] *om.* P ‖ dicitur] debent C **22** movente] *add. in marg.* seu motore C : *om.* I **24** dicitur[1]] dicatur IPp ‖ esse] *om.* C **25** ad] *om.* p ‖ nec] ut C ‖ vel etiam secundum] nec etiam secundum I : nec secundum P : nec etiam p **26** speciem] species p ‖ etiam] *om.* IPp

---

8 Cf. Iohannes Buridanus, *Quaestiones super libros Physicorum*, VI, q. 5 (ed. Parisiis 1509, ff. 98va–99ra) **24** Cf. Aristoteles, *Physica*, V, 2, 225b11; VII, 3, 245b8–9

⟨6⟩ Ad aliam concedo quod ista definitio convenit tempori, quia tempus est motus. De infinito autem dico quod, cum ipsum sit in potentia, tamen non est motus, nisi continue sit tendentia et processus ad illud ad quod est in potentia; et si sic, tunc est motus et sibi convenit definitio motus. Nec obstat, si illi termini 'motus' et 'tempus' sint diversae passiones huius termini 'corpus' vel 'corpus naturale' vel 'ens naturale', quia diversi termini bene supponunt pro eadem re.

⟨7⟩ Ad ultimam dicendum est quod per istam clausulam 'motus est actus', si est actus alicuius, statim sequitur quod illud ipso est in actu; ideo non oportebat amplius exprimere quod esset actus entis in actu. Sed per hoc non erat significatum quod illud esset in potentia; ideo oportebat dicere 'entis in potentia'.

Haec etc.

---

2 tamen] cum I 3 illud] id p 4 est[1]] *praem.* quid C : *om.* I ‖ si] *om.* P 5 et] *om.* I ‖ sint] sunt Ip 9 si est actus] *om.* p ‖ ipso] ipse Pp 11 significatum quod] significatum quin C : significat quod I 13 haec etc.] et sic sit finis quaestionis I : et sic est finis P : etc. sequitur alia quaestio p

⟨III.12⟩

⟨Utrum omnis motus sit subiective in mobili vel movente vel in utroque⟩

Quaeritur duodecimo utrum omnis motus est subiective in mobili vel movente vel in utroque.

⟨1⟩ Arguitur primo quod sit in movente quia: actus est in eo cuius est actus; et motus est actus moventis, ut dicit Aristoteles; igitur etc.

⟨2⟩ Iterum forma vel dispositio denominans aliquod subiectum est in illo subiecto; et tamen ab illo motu denominatur movens movere; igitur etc.

⟨3⟩ Item dispositio secundum quam aliquid se habet aliter et aliter prius et posterius est in eo quod sic se habet aliter et aliter. Nec mirum, quia tale movetur vel mutatur secundum illam dispositionem; et tamen oportet concedere quod dispositio secundum quam aliquid movetur vel mutatur est in illo quod sic movetur vel mutatur. Sed quod prius non est movens et post est movens se habet aliter et aliter prius et posterius, et non nisi secundum motum quo prius non movebat et posterius movet.

⟨4⟩ Item sicut se habet quies ad quiescentem, ita motus ad moventem, per simile. Sed quies est in quiescente; non enim posset assignari in quo esset quies lapidis, quando lapis quiescit, nisi diceretur esse in isto lapide. Igitur etc.

⟨5⟩ Item in quocumque est actio, in isto est motus; sed in omni movente vel agente est actio; igitur etc. Probo primo maiorem quia: omnis motus est
78vb C actio et omnis actio est motus, ut dicetur | in alia quaestione.

Deinde probo minorem:

⟨5.1⟩ Primo quia: sicut se habet passio ad passum, ita actio ad agens; sed omnis passio est in passo; igitur omnis actio est in agente.

---

4 quaeritur duodecimo] *add.* ut P : consequenter quaeritur I ‖ vel] *add.* in Pp 7 est] *om.* P ‖ igitur etc.] *om.* IPp 9 illo] ipso IPp ‖ movere] moveri p ‖ igitur etc.] *om.* IP 10 item] et iterum IPp 14 sed] secundum Cp 16 quo] ergo ut P ‖ movet] *add.* ergo etc. Ip : *add.* igitur P 17 moventem] *add.* et ultra C : *add.* ultra I 19 esse] *post* lapide Pp 19–20 igitur etc.] *om.* P 21 omni] *sup. lin.* C : *om.* IPp 22 primo] *ante* probo P : igitur C 23 actio[1]] actus p 25 se habet] *sup. lin.* C : *om.* IPp ‖ sed] et IPp

---

7 Cf. Aristoteles, *Physica*, III, 3, 202a15–17 23 Cf. inf., III, q. 13

 | DOI: 10.1163/9789004322356_015

⟨5.2⟩ Secundo quia: actio est actus proprius agentis.

⟨5.3⟩ Tertio quia: agens ipsa actione se habet aliter quam ante, quia agit et ante non agebat; ideo ipsa | [68ra P] actione mutatum est. Ideo actio est mutatio qua ipsum mutatur. Sed mutatio est in eo quod mutatur. Igitur etc.

⟨5.4⟩ Quarto quia: actio vel est in agente vel est in passo; sed non est in passo, quia tunc denominaret passum (omnis enim forma vel dispositio inhaerens alicui subiecto dat ei aliquod esse secundum quod est innata denominare ipsum); et tamen actio non denominat passum, sed passio tantum, quia passum non dicitur nisi pati.

⟨5.5⟩ Quinto quia: dicit Aristoteles quod actio et passio non sunt idem proprie; et hoc etiam arguetur in alia quaestione. Sed si dicamus eas non esse idem, non debemus | [54vb p] eas dicere esse in eodem subiecto, sed passionem in passo et actionem in agente.

Oppositum arguitur per Aristotelem et Commentatorem.

⟨1⟩ Dicit enim Philosophus: 'manifestum est quod motus est in mobili'.

⟨2⟩ Et arguit quod non in motore, saltem inquantum est motor, quia: si in motore, inquantum est motor, esset motus, sequeretur quod omne movens moveretur, quod est falsum de primo motore. Consequentia patet, quia omne in quo est motus subiective movetur.

⟨3⟩ Commentator etiam dicit quod actio motoris est eius et non in eo.

Notandum est quod in hac quaestione quantum ad motum localem oportet aliter et aliter loqui dicentes quod motus localis est ipsum mobile, et dicentes quod differt a mobili et loco.

Dicentes enim quod est idem quod mobile non possunt dicere quod motus localis sit subiective in mobili secundum realem | [51ra I] inhaerentiam, sed dicunt quod, quando Aristoteles et alii philosophi dixerunt motum esse

---

1 actio est] actione p ‖ actus proprius] *inv.* IPp 2 actione] actio est p 3 ideo[2]] et sic IPp 3–4 qua ipsum] qua ipsa P : secundum quam ipsum C 4 sed] vel P 5 est[2]] *om.* IPp 8 ipsum] TU : eam HM : *om.* ABCGILPp ‖ et tamen actio] et cum actus *sed add. sup. lin.* alias et tamen actio C 9 passum] *sup. lin.* C : passio P 11 dicamus] dicimus p 12 eas dicere] *inv.* IPp 13 agente] *add.* ergo etc. IPp 15 enim philosophus] enim aristoteles P : aristoteles p : enim I ‖ est[1]] esse IPp ‖ in] *om.* I 16 non in] *inv.* p ‖ est] *om.* IPp 17 in motore] motor C ‖ est] *om.* IPp 19 movetur] *add.* et IPp 20 et] *om.* Ip 22–23 dicentes] dicunt C 24 dicentes] dicunt C ‖ mobile] *add. sup. lin.* ideo (*corr. ex* ubi ubi) C 26 quando] cum IPp

---

10–11 Cf. Aristoteles, *Physica*, III, 3, 202b20–21; cf. inf., III, q. 13 15 Aristoteles, *Physica*, III, 3, 202a13–14 16 Aristoteles, *Physica*, III, 3, 202a28–31 20 Averroes, *In Physicam*, III, comm. 18, f. 92H

subiective in mobili, ipsi capiebant 'esse in' pro praedicatione denominativa, sicut consuevimus dicere quod talis passio vel tale praedicatum inest tali subiecto, si vere praedicetur de eo; et oportet capere terminos materialiter, ut si caelum movetur, motus caeli est in caelo, id est iste terminus 'motus' vere affirmatur de hoc nomine 'caelum' dicendo 'caelum est motus'. Sed de alteratione bene concederent quod est in alterabili tamquam in subiecto per realem inhaerentiam et capiendo terminos significative.

Nos autem dicimus primo quod omnis motus est subiective in mobili, scilicet in eo quod movetur, per realem inhaerentiam, sicut albedo esset in pariete, quia non apparet de hoc instantia, nisi ista esset de motu locali; et apparuit prius quod non est de hoc instantia; igitur etc.

79ra C Dico secundo | quod indefinite loquendo motus est in movente, quia contingit quod movens movetur, immo adhuc quod eodem motu movet et movetur, scilicet quia movet se, prout in septimo et octavo videbitur.

Tertio dico quod non in omni movente est motus, quia instantia est de primo motore. Non enim in eo est motus subiective, quia est immutabilis.

Quarto ex hoc concluditur quod, si in movente est motus, tamen non convenit moventi, secundum quod est movens, quod in eo sit motus; sed hoc convenit sibi ea ratione qua movetur. Sequitur enim 'moventi, secundum
68rb P quod | est movens, inest motus, igitur omni moventi'.

Ad rationes respondetur.

⟨1⟩ Ad primam dicendum est quod dupliciter potest esse actus alicuius. Uno modo tamquam eius a quo producitur, et tunc non oportet quod sit in eo subiective. Alio modo tamquam eius in quo recipitur, et sic est in eo. Sed primo modo motus est actus moventis.

---

1 ipsi capiebant] ipse capiebat p 2 talis passio vel] talis passio id est P : tale vel C 3 si] sive p ‖ eo] illo IPp 4 movetur] moveretur p 6 concederent] concederem P 7 et] *om.* P 10 de²] in P 11 hoc] eo P 12 movente] *add. sup. lin.* alias motore C 13 adhuc quod] *inv.* P 14 prout] ut Pp ‖ septimo et octavo] septimo et in octavo I : octavo et septimo P 15 tertio dico] *inv.* Pp ‖ quia] *om.* p 17 quod] quia I 18 hoc] *sup. lin.* C : haec p 19 enim] *add.* quod si P 20 est] *om.* IPp ‖ inest] inesset P ‖ motus] *add.* quod P ‖ igitur] *add.* in p ‖ moventi] *add.* inesset motus P 21 ad rationes respondetur] ad rationes I : tunc ad rationes Pp 22 dupliciter] duplex Ip ‖ alicuius] *ante* actus p : *om.* P 23 eius] *om.* IPp ‖ producitur] proceditur p

---

11 Cf. sup., III, q. 7 14 Cf. Iohannes Buridanus, *Quaestiones super libros Physicorum*, VII, q. 1 (ed. Parisiis 1509, ff. 103va–104ra); cf. Iohannes Buridanus, *Quaestiones super libros Physicorum*, VIII, q. 4 (ed. Parisiis 1509, ff. 112va–113va)

⟨2⟩ Ad aliam dicitur quod multae sunt denominationes ab extrinsecis, ut quod unus homo dicitur dives divitiis ab eo distantibus et honorificus ab honore existente in honorante, et corpus locari a loco sibi extrinseco. Ita agens denominatur agere vel movens movere motu et actione in aliis existentibus.

⟨3⟩ Ad aliam dico quod movens quod movet, cum ante non moveret, non se habet aliter et aliter intrinsece, nisi moveatur, sed se habet aliter et aliter ad mobile quod movetur ab eo, cum prius non movetur ab illo. Vel etiam extrinseca denominatione dicitur aliter et aliter se habere, quia in virtute ipsius motum ab eo aliter et aliter se habet.

⟨4⟩ Ad aliam dico quod est valde magna dissimilitudo, quia movere est agere et quiescere non. Et ideo grammatici bene dixerunt 'movere' esse verbum activi generis et non 'quiescere'. Et cum hoc motus est res distincta a movente, sed non quies a quiescente, quia privatio non est res distincta a privato sive a re privata.

⟨5⟩ Ad aliam dico quod omnis actio est motus, prout hic capimus 'actionem' sicut dicetur in alia quaestione. Et concedo quod in movente | vel in agente est actio, quando contingit quod agens movetur sive patitur, aliter non. 55ra p

⟨5.1⟩ Nec est verum quod actio se habet ad agens sicut passio ad passum, quia actio producitur ab agente et non passio a passo, immo ab agente.

⟨5.2⟩ Ad aliam dico quod tam actio quam passio est proprie motus agentis tamquam a quo producitur, et patientis tamquam in quo recipitur.

⟨5.3⟩ Ad aliam dicendum est, sicut dictum fuit de motu. Agens enim ipsa actione se habet aliter et aliter sicut movens motu, scilicet ad passum, non intrinsece.

⟨5.4⟩ Ad aliam dicendum est quod res extra non denominant | se invicem, sed termini significantes res. Modo iste terminus 'agere' significat actionem 79rb C

---

1 dicitur] dicendum est IPp ‖ extrinsecis] extrinseco P **2** unus] *om.* IPp ‖ honorificus] homo honoratus IP : homo honoratur p **3** locari] vocari p : locat P ‖ a loco sibi] sibi a loco P : a loco sive I **4** in] et P **6** quod movet] motu Pp : motum I **8** movetur[2]] moveretur IPp ‖ illo] eo IPp ‖ vel etiam] *inv.* P **10** motum] moventis *sed add. in marg* alias motus seu motum C ‖ ab eo] *om.* P **11** est[1]] *post* valde p : *post* magna IP ‖ quia] quod P **12** non et] non Ip : *om.* P ‖ esse] est p **13** verbum] *om.* C ‖ activi generis] *inv.* p : activum igitur P ‖ et[1]] sed Ip **15** privato … privata] privato Pp : re privata I **17** sicut] ut p : et P **17–18** movente … agente] movente vel agente Ip : agente vel movente P **19** non] *om.* P **20** ad[1]] *om.* I ‖ passum] patiens P **21** immo] *add.* etiam IPp **22** motus] actus p : actio P **23** in] a P **24** enim] in P **25** et aliter] *om.* IPp ‖ motu] movetur C

---

17 Cf. inf., III, q. 13

connotative, licet non supponat pro ea; et per consequens etiam significat passionem, cum sit idem actio et passio. Et sic etiam iste terminus 'pati' significat actionem et passionem. Ideo manifestum est quod terminus significans actionem et passionem denominat terminum supponentem pro agente dicendo quod *a* agit, et terminus significans actionem et passionem denominat terminum supponentem pro passo dicendo quod *b* patitur.

⟨5.5⟩ De ultima ratione dicetur in alia quaestione etc.

---

1 etiam] *om.* P 4 actionem et] *corr. ex* mutationem vel C : actionem vel P ‖ denominat] determinat p 5 et[1]] *add.* etiam IP 7 etc.] et sic finitur quaestio I : *om.* P

---

7 Cf. inf., III, q. 13

⟨III.13⟩

⟨Utrum omnis actio sit passio et econtra⟩

Quaeritur tredecimo utrum omnis actio est passio et econtra.

Arguitur quod non quia:

⟨1⟩ In libro *Posteriorum* dicitur quod essentiae praedicamentorum sunt impermixtae; et ideo etiam ibi dicitur | [68va P] quod propositiones negativae unius praedicamenti de alio sunt primae et immediatae. | [51rb I] Cum igitur haec nomina 'actio' et 'passio' sunt diversa praedicamenta, sequitur quod haec est falsa 'actio est passio' et quod haec est immediata et prima 'nulla actio est passio'. Et idem dicit Commentator decimo *Metaphysicae* quod quidditates praedicamentorum sunt diversae.

⟨2⟩ Item auctor *Sex principiorum* describit passionem quod passio est effectus illatioque actionis; idem autem non est effectus sui ipsius neque causa sui ipsius. Et si dicatur quod idem est causa unius et effectus alterius, tamen non procederetur in infinitum. Ideo erit passio ultimata quae erit effectus actionis et non erit actio, quia oporteret ultra esse passionem quae esset effectus ipsius. Et etiam ascendendo est dare actionem primam quae non est passio, quia oporteret esse aliam actionem priorem cuius illa esset effectus.

⟨3⟩ Item quinto *Metaphysicae* dicitur quod, quaecumque differunt genere, differunt specie; et quaecumque differunt specie, differunt numero; igitur quaecumque differunt genere, differunt numero. Sed actio et passio differunt genere, cum sint diversa generalissima. Igitur differunt numero. Et

---

3 quaeritur tredecimo] consequenter quaeritur I ‖ econtra] omnis passio est actio IPp **5** posteriorum] *corr. in marg. ex* praedicamentorum C : priorum I **7** alio] altero IPp **8** haec est falsa] hoc est falsum P **9** immediata et prima] prima et immediata IPp **10** et] *add.* ad P ‖ decimo] quarto P **13** illatioque … effectus²] illatioque actionis idem †…† est effectus *in marg.* C ‖ neque] nec IPp **15** tamen] cum I ‖ procederetur] procedetur IPp ‖ quae] *add. sed del.* et C : *add.* et I **16** quia] *om.* p **17** et] *om.* P ‖ ascendendo] ostendi quod P **18** priorem … esset] priorem cuius illa *corr. ex* cuius nec esset alius C ‖ illa] ille p **20** quaecumque] quicumque P **21** quaecumque] quicumque P **22** quaecumque] quicumque P **23** cum sint] quia sunt P

---

5 Cf. Aristoteles, *Analytica posteriora*, I, 15, 79b6–12 **6** Cf. Aristoteles, *Analytica posteriora*, I, 15, 79b12–20 **10** Averroes, *In Metaphysicam*, X, comm. 8, f. 257G **12** Anonymus, *Liber sex principiorum*, III, 29 (ed. Minio-Paluello, 41) **20** Aristoteles, *Metaphysica*, V, 9–10, 1017b27–1018b8

 | DOI: 10.1163/9789004322356_016

quaecumque differunt numero, verum est dicere quod hoc non est illud; igitur verum est dicere quod actio non est passio.

⟨4⟩ Item per locum a coniugatis sequitur 'actio est passio, igitur agere est pati et agens est patiens'; sed consequens non conceditur; igitur nec antecedens.

⟨5⟩ Item potentiarum specie diversarum debent esse diversi actus proprii, cum potentiae distinguantur per actus, secundo *De anima*, nono *Metaphysicae* et secundo *Ethicorum*; sed in genere potentiarum potentia activa et potentia passiva sunt diversae specie vel etiam genere; igitur non est idem actus proprius potentiae activae et potentiae passivae. Sed potentiae activae actus proprius est actio et passivae passio. Igitur non est idem actio et passio.

⟨6⟩ Item arguitur auctoritate Aristotelis dicentis sic: 'omnino autem dicere est: nec doctio cum doctrina nec actio cum passione idem est proprie, sed
79va C cui insunt haec, motus'. |

⟨1⟩ Oppositum patet, quia dicit Aristoteles quod idem est actus et non alius motivi et mobilis, activi et passivi; et tamen actus motivi et activi est actio et
55rb p actus | passivi et mobilis est passio; igitur etc.

⟨2⟩ Et ad idem dicitur secundo *De anima* quod idem est actus sensibilis et sensus, cum sensus patiatur a sensibili et sensibile agat in sensum. Unde eadem visione videtur lapis et videt oculus.

⟨3⟩ Item quinto *Metaphysicae* dicit Commentator quod inter agens et passum est dispositio media quae in respectu agentis dicitur actio et in respectu patientis dicitur passio.

⟨4⟩ Item tam actio quam passio est motus, cum utraque consistat in successione; et non est dicendum quod sint duo motus in hanc caliditatem apud calefactionem; igitur etc.

---

4 conceditur] est concedendum IPp **6** diversarum] *om.* P **7** actus] *add.* prout allegatur ex IPp ‖ anima] *add.* ex Ip : *add.* et P **8** et^{1}] *add.* ex IP **9** potentia] *om.* p ‖ diversae] diversa P ‖ est] *om.* I **12** aristotelis] *add.* in tertio huius Pp ‖ sic] si p **13** est proprie] *inv.* IPp **14** haec] *add. inf. lin.* alias hi C : hi IPp **16** et^{3}] *om.* P ‖ motivi et activi] activi et motivi IPp **18** et] *om.* P **20** videtur] videret I ‖ et] *add.* etiam P **21** dicit commentator] dicitur a commentatore I **22** est] *add.* quaedam P **22–23** et … dicitur] respectu patientis P

7–8 Cf. Aristoteles, *De anima*, II, 4, 415a18–20; cf. Aristoteles, *Metaphysica*, IX, 8–9, 1049b4–1051a33; cf. Aristoteles, *Ethica Nicomachea*, II, 1, 1103a26–b23 **12** Aristoteles, *Physica*, III, 3, 202b19–21 **15** Cf. Aristoteles, *Physica*, III, 3, 202a15–20 **18** Non invenimus in secundo libro; sed cf. Aristoteles, *De anima*, III, 2, 425b25–27 **21** Cf. Averroes, *In Metaphysicam*, V, comm. 25, f. 133G

Quia termini huius quaestionis dicuntur multipliciter, | ideo eos exponere et distinguere oportet. 68vb P

Dico igitur quod aliquando Commentator utitur hoc nomine 'actio' pro omni actu formali sive existente in permanentia et quiete seu in fluxu, ut quod anima diceretur actio corporis. Alio modo etiam vocatur actio actus purus, prout per modum actus secundi significatur, prout in Deo diceremus quod intelligere et velle vel etiam intellectio et volitio sunt actiones Dei immanentes. Alio modo etiam dicitur actio illud quod productum est ab agente, quamvis in quiete maneat, postquam agens desinit agere. Et istis modis non quaerimus de actione.

Alio modo dicitur actio operatio quae provenit et producitur ab agente et qua praesente agens ea dicitur agere. Et de ista intendit ista quaestio. Verum est quod iste modus aliquando restringitur magis, ut solum dicatur de actionibus electivis. Et sic dicitur in sexto *Metaphysicae* quod idem est agibile et eligibile. Et sic secundum strictam locutionem distinguitur in moralibus actio contra factionem et agibile contra factibile et habitus activus, qui est prudentia, contra habitum factivum, qui est ars. Sed non ita specialiter et stricte accipitur hic 'actio', immo sicut dictum est.

Deinde etiam de passione dicendum est quod valde multipliciter accipitur. Uno modo omne accidens dicitur passio eius cui inhaeret, ut albedo parietis, scientia animae, lux solis, magnitudo substantiae corporalis etc. Aliquando autem nomen passionis restringitur ad qualitates sensibiles; et sic ponitur tertia species qualitatis. Aliquando etiam appropriate dicitur passio de motibus appetitus sensitivi, sicut dicimus delectationem, tristitiam, | gaudium, iram, timorem et huiusmodi passiones; et sic communiter 51va I in moralibus utimur hoc nomine 'passio'. Aliquando etiam terminus connotativus respectu termini substantialis vel etiam terminus magis connotativus respectu termini minus connotativi dicitur passio, et alter terminus dicitur subiectum eius, si pro eodem supponunt; et dicimus passionem praedicari de subiecto denominative. Sic enim loquendo materialiter risibile est

---

4 permanentia] permanenti p ‖ seu] sive IPp 5 diceretur] dicitur p 6 prout[1] … significatur] †…† modum actus †…†tur *in marg.* C : prout actus per modum secundi significatur p ‖ prout[2]] sicut P 7 et[2]] vel p 8 etiam dicitur] etiam vocatur I : dicitur Pp 12 ea dicitur] dicitur eam C ‖ ista[1]] isto p 13 modus] motus P 14 et] *om.* P 16 activus qui] actus quae P 18 sicut] ut P 19 etiam] *om.* P ‖ dicendum est] dicitur P 20 ut] et sic P 22 autem] *om.* IP 23 etiam] autem p 24 sensitivi] *corr. ex* sensui (?) C, *et rescr. in marg.* 26 hoc] *om.* IPp ‖ passio] passionis IPp 29 supponunt] supponant Ip 29–30 praedicari] praedicare P 30 materialiter] naturaliter p ‖ est] dicitur IPp

---

14 Cf. Aristoteles, *Metaphysica*, VI, 1, 1025b24

passio hominis et par vel impar numeri et tempus motus etc. Aliquando 79vb C etiam | restringitur nomen passionis ad motus nocivos vel tristes, ut patet quinto *Metaphysicae*. Sed istis modis non volumus uti hoc nomine 'passio' in ista quaestione. Et ideo accipitur hic 'passio' pro operatione qua passum dicitur pati, sicut actione agens dicitur agere.

Et de huiusmodi actione et passione ponuntur conclusiones.

Prima conclusio est quod possibile est esse actionem sine passione, ut si Deus sine subiecto recipiente creat aliquam creaturam. Ista enim creatio est actio Dei ab eo procedens; sed haec non est passio, quia in nullo recipitur quod dicatur pati. Deus enim posset creare angelum, quod nihil aliud a Deo esset nisi iste angelus; et sic Deus ageret et non esset actio eius nisi ille 69ra P angelus qui crearetur. Ille enim angelus | non diceretur passio, quia passio debet recipi in passo, sicut actio debet provenire ab agente; et ille in nullo passo reciperetur.

Tamen si quis omnino vellet vocare passionem illud quod fit et produci- 55va p tur | in esse ab alio, sic isto modo angelus, quando crearetur, esset passio. Et sic omnis actio esset passio. Sed haec non est propria acceptio passionis, cum hoc nomen 'passio' dicatur a 'passo'.

Secunda conclusio est quod impossibile est esse passionem sine actione, quia passio est quaedam transmutatio passi vel non est sine transmutatione; et impossibile est quod aliquid fiat vel transmutetur nisi ab agente quod agat; et agens dicitur actione agens; igitur etc.

Tertia conclusio est quod impossibile est naturaliter actionem esse sine passione, quia nihil potest fieri naturaliter nisi ex subiecto praesupposito vel in subiecto praesupposito, ut dictum est in primo *huius*; ideo si agens aliquid producat, necesse est aliquid recipi in subiecto quod vocatur passum, et hoc erit passio vel non sine passione.

Quarta conclusio est quod omnis actio naturalis est passio et omnis etiam passio est actio, quia omnis actio est operatio quae producitur ab agente; et omnis talis recipitur in aliquo subiecto, et sic dicitur passio. Et ita omnis

---

1 etc.] *om.* P 4 accipitur hic] *inv.* IPp 7 conclusio] *om.* Ip 9 eo] ipso I || sed] *om.* C || recipitur] est realiter P 10 dicatur] diceretur p || posset] potest I 11 et sic] tunc I 12 non] *add.* proprie IPp || passio[1]] pati P 13 ille] ita P 14 passo] *om.* P 15 omnino] *om.* p 17 sic] *om.* P 18 hoc nomen] *om.* P || dicatur] dicitur P 19 esse passionem] *inv.* IP 22 dicitur] dicetur C 23 actionem esse] *inv.* Pp 25 in[2]] *om.* IPp || si] *om.* P 26–27 hoc erit] hoc vocatur I : haec erit p 28 est[1]] *om.* I

---

3 Cf. Aristoteles, *Metaphysica*, V, 21, 1022b19–20 25 Cf.

passio est operatio quae recipitur in passo; et omnis talis est sive producitur ab aliquo agente; ideo est actio.

Quinta conclusio est quod omnis motus est actio et passio, quia omnis motus est actus mobilis tamquam in quo recipitur, et sic est passio. Et omnis etiam motus est actus moventis tamquam a quo producitur, et sic est actio. Et haec dicta fuerunt prius, scilicet quod omnis motus est actus mobilis et moventis. Eadem igitur res est motus, actio et passio, sed nomina differunt secundum rationem, quia dicitur motus vel mutatio ratione successionis, et dicitur actio prout est ab agente, et | passio prout recipitur in passo. Et ita 80ra C
haec nomina 'actio' et 'passio' connotant diversa, scilicet agens et passum, et propter hoc dicta nomina praedicantur denominative de se invicem et non quidditative.

Sed tunc essent dubitationes in quo praedicamento essent isti termini 'motus' et 'mutatio', quia non apparent esse in aliquo praedicamento, nisi sint in praedicamento actionis et passionis; et tamen non sunt in illis praedicamentis, si haec nomina 'actio' et 'passio' praedicentur de eis denominative, quia genus debet praedicari de suis speciebus quidditative.

Item dubitatio est de corruptione quae esset sine generatione, utrum illa deberet dici actio vel passio aut non.

Ad primam dubitationem potest dici quod, cum praedicamenta ponantur ab Aristotele distingui penes diversos modos praedicandi de primis substantiis, scilicet de nominibus discretis significantibus substantias | per se 69rb P
subsistentes simpliciter et sine connotatione aliena nisi grammaticali, multi autem termini abstracti supponentes pro accidentibus non praedicantur vere de primis substantiis, ut 'albedo', 'caliditas' et 'magnitudo', ut credo, et 'motus' et 'calefactio' etc., ideo tales terminos abstractos oportet ponere in | 51vb I
praedicamentis secundum exigentiam terminorum suorum concretorum. Termini autem concreti motus et mutationis, calefactionis, alterationis etc. sunt 'movere' et 'moveri', 'alterare' et 'alterari' etc., de quibus iam vere et quidditative dicuntur 'agere' et 'pati'. Ideo 'motus' diceretur de praedicamento actionis prout eo movens dicitur movere, et de praedicamento passionis

2 ab] *om.* p ‖ est] dicitur I 3 est[1]] *om.* I 5 actus] *om.* p 6–7 mobilis et moventis] moventis et mobilis Pp 11 hoc] haec Pp 15 sint] *om.* P ‖ et[1]] vel IPp 16 eis] eo P 17 suis] *om.* I 20 dici] dicitur I 26 etc.] *om.* P ‖ tales terminos abstractos] talis terminus abstractus P 27 terminorum suorum] *inv.* IPp ‖ concretorum] *om.* P 28 mutationis] transmutationis I ‖ alterationis] actionis P 29 etc.] *sup. lin.* C : *om.* Pp

6 Cf. sup., III, q. 10 20–22 Cf. Iohannes Buridanus, *Quaestiones in Praedicamenta*, q. 3 (ed. Schneider, 17–18)

prout eo mobile dicitur moveri. Et possumus dicere quod, prout intelligimus quod motu mobile movetur, hoc nomen 'motus' connotat mobile et est quidditativa praedicatio 'motus est passio'; et prout intelligimus quod motu movens movet, hoc nomen 'motus' connotat movens et est quidditativa praedicatio 'motus est actio'.

Ad aliam dubitationem diceretur quod simplex corruptio non ita proprie 55vb p dicitur actio vel passio, sicut generatio, quia proprie loquendo ni|hil agitur et nihil in passo recipitur, et sic corrumpens nihil faceret. Tamen quia quodam modo ista nomina 'generatio' et 'corruptio' opponuntur quasi privative, et unum oppositorum debet sumi vel simpliciter vel reductive in eodem genere cum suo opposito, ideo large loquendo dicimus corruptionem actionem et corrumpere agere; et etiam quia ut in pluribus non est corruptio sine generatione.

Tunc ad rationes principales.

⟨1⟩ Ad primam dicitur quod essentiae vel quidditates praedicamentorum sic sunt impermixtae vel diversae quod numquam unum praedicamentum praedicatur de altero praedicamento quidditative nec aliquis terminus 80rb C unius praedicamenti | de altero termino alterius praedicamenti. Tamen bene praedicantur de se invicem denominative. Et sic etiam, quando Aristoteles dicit propositionem in qua unum praedicamentum negatur de altero esse primam et immediatam, non intendit quod omnis talis sit vera, sed per hoc vult dicere quod nulla est quidditativa in qua unum praedicamentum affirmaretur de altero vel etiam species unius praedicamenti de specie alterius. Et vult etiam dicere quod hoc manifestius est, quando unum praedicamentum dicitur de altero praedicamento, quam quando species dicitur de specie. Et hoc intelligit per 'primaevitatem' et 'immediationem'.

⟨2⟩ Ad auctoritatem auctoris *Sex principiorum* dico quod, ut mihi videtur, melius fuisset quod numquam illum librum fecisset.

⟨3⟩ Ad aliam videtur mihi quod ista auctoritas quinti *Metaphysicae* valde vera est in praedicamento substantiae. Sed de terminis connotativis, scili-

---

1 moveri] movere P ‖ quod] *om.* P 4 movens^{2}] mobile *sed add. sup. lin.* alias movens C 5 actio] passio P 8 quodam] quoddam p 9 ista nomina] *inv.* P ‖ opponuntur] ponuntur P 10 sumi] poni IPp 12 et^{2}] *sup. lin.* C : *om.* P 13 generatione] *add. sup. lin.* alias alteratione C 14 tunc] *add.* dicendum est P : *add.* respondendum est Ip 15 dicitur] dico p 16 numquam] *sup. lin.* C : *post* praedicamentum (16–17) p 18 praedicamenti^{1}] praedicant I ‖ altero] aliquo IPp 19–20 aristoteles dicit] *inv.* I 20 negatur] praedicatur I 25 praedicamento] *om.* I ‖ de^{2}] *om.* I 26 intelligit] intendit p 27 ut] *om.* C 28 numquam] *post* librum I 29 quinti] quintae C 30–121.1 scilicet] si C : et P

cet aliorum praedicamentorum, non est universaliter verum quod termini diversorum generum vel specierum non supponant pro eodem, et quod de ipsis significative sumptis non possit verificari hoc praedicatum 'idem in numero'. | Dicimus enim bene quod homo et musicus sunt idem in numero. 69va P

⟨4⟩ Ad aliam dicendum est quod locus a coniugatis non tenet nisi in praedicationibus quidditativis. Non enim sequitur, si album est dulce, quod albedo sit dulcedo. Sed si haec est vera et quidditativa 'albedo est color', sequitur quod album est coloratum. Unde iste locus valet directe ad terminandum problema de genere. Si enim actio est genus ad sectionem, sequitur quod agere est genus ad secare.

⟨5⟩ Ad aliam dicendum est quod non oportet universaliter potentiarum diversarum esse actus diversos, sed aliquando sufficit quod earum sit idem actus diversimode. Et sic est de agente et passo.

⟨6⟩ Ad aliam dicitur quod Aristoteles, cum dixit actionem non esse idem cum passione proprie, intendit solum negare praedicationem quidditativam esse unius illorum terminorum de reliquo. Et cum dicit hoc inesse eidem motui, non intendebat per 'inesse' realem inhaerentiam, sed veram praedicationem illorum terminorum de isto termino 'idem motus', ita quod haec sit vera 'idem motus est actio et passio'; ad quod sequitur quod actio est passio etc.

---

2 supponant] supponunt I ‖ et] ad P ‖ de] *om.* P 3 possit] potest P ‖ hoc praedicatum] *om.* I 4 musicus] musicum IPp 5 dicendum est] dicitur P 7 sit] est IPp ‖ quidditativa] *add.* quod p : *add.* et P 11 dicendum est] dicitur P 12 diversos] diversi P ‖ quod] quando P 14 dixit] dicit p 16 dicit] dixit IP ‖ eidem] *om.* Pp 19 et] *om.* p 20 etc.] *om.* P : et sic finitur quaestio I

⟨III.14⟩

⟨Utrum sit aliquod corpus sensibile actu infinitum⟩

Quaeritur consequenter circa tractatum de infinito. Et quaeritur quarto decimo utrum est aliquod corpus sensibile actu infinitum.

Arguitur quod sic rationibus quas tangit Aristoteles.

⟨1⟩ Primo quia: tempus et motus sunt infinita, cum sint perpetua sine principio et sine fine; igitur corpus est infinitum. Probatur consequentia primo quia: dicit Aristoteles sexto *huius* quod eiusdem rationis sunt tempus, motus et magnitudo secundum finitatem et infinitatem. Secundo quia: tempus et motus sunt passiones corporum naturalium; et passio non debet esse maioris entitatis quam suum subiectum.

80va C ⟨2⟩ Item corpus est divisibile in infinitum; et tale habet | infinitas partes. Modo infinitae partes reddunt infinitam extensionem. Probatur quia: cum quaelibet pars reddat aliquantam extensionem, plures partes reddunt maio- 53ra I 56ra p rem et infinitae | infi|nitam. Et hoc confirmatur quia: si magnitudo esset composita ex partibus indivisibilibus infinitis, ita quod duae earum redderent aliquantam extensionem, sequeretur quod illa esset infinita; igitur cum quaelibet pars magnitudinis sit maior et alteri addita reddat maiorem extensionem quam si esset indivisibilis, sequeretur quod, si sint infinitae, quod reddunt infinitam extensionem magis quam si essent indivisibiles.

⟨3⟩ Item si materia ex qua corpora generantur naturalia non esset infinita secundum magnitudinem, sequeretur quod aliquando cessarent generatio-

---

4–5 quaeritur[1] … decimo] consequenter (*om.* I) quaerendum est circa tractatum de infinito et quaeritur decimo quarto Ip : quaeritur quarto decimo circa tractatum de infinito est prima quaestio P 5 est] sit P 6 quod] *om.* p 7 primo] *om.* P ‖ tempus et motus] *corr. ex* tempus est motus C : motus et tempus P ‖ infinita] infinitum P 9 dicit] *post* aristoteles p : dixit I ‖ tempus motus] *inv.* P 11–12 maioris] HM : minoris ABCILPTUp 12 suum subiectum] *inv.* Ip 13 tale] *om.* P 15 reddat] reddit P ‖ partes] *sup. lin.* C : *om.* Pp 15–16 maiorem] *add.* extensionem I : *add. in marg.* †…† partes reddunt extensionem C 16 et[1]] *om.* C ‖ infinitae] *add.* partes reddunt I : indefinite C ‖ infinitam] *add.* extensionem I ‖ hoc] *add.* etiam p 17 earum] illarum IPp 18 sequeretur] sequitur p ‖ esset] essent p 20 sequeretur] sequitur Pp ‖ quod[2]] *om.* I 22 corpora generantur] *inv.* IPp ‖ esset] essent P 23 sequeretur] sequitur p

---

6 Cf. Aristoteles, *Physica*, III, 6, 206a9–12 9 Cf. Aristoteles, *Physica*, VI, 1, 231b18–20

 | DOI: 10.1163/9789004322356_017

nes entium naturalium in hoc mundo; cuius oppositum probatum est primo et secundo *De generatione*. Consequentia probatur quia: quodlibet corpus quod generatur capit sibi aliquam portionem materiae, cum | 69vb P non possit fieri ex nihilo; et sic tandem tota materia esset consumpta, cum omne finitum tandem consumatur per ablationem finiti multotiens factam; cum autem materia esset consumpta, non amplius esset generatio.

⟨4⟩ Item accepto aliquo corpore vel ipsum est infinitum, et habetur propositum; vel est finitum, et tunc finitur ad alterum corpus, quod iterum est finitum vel infinitum. Si infinitum, habetur propositum; si finitum, quaeritur ut prius procedendo in infinitum, et sic iterum revertitur in infinitum.

⟨5⟩ Item quantocumque corpore finito imaginato possumus imaginari aliud maius; igitur quantocumque corpore finito dato est aliud maius; ex quo sequitur quod est infinitum. Et probatur consequentia prima quia: imaginatio et sensus et intellectus sunt virtutes passivae ab obiectis suis, scilicet ab imaginabili, a sensibili et ab intelligibili; et virtus passiva non movetur nisi ab obiecto existente; igitur illud quod est sensibile, imaginabile vel intelligibile est, fuit vel potest esse aut potuit esse, quia propositio affirmativa non est vera, si subiectum non supponat pro aliquo praesente, praeterito vel saltem possibili; igitur cum infinitum sit imaginabile vel intelligibile, sequitur quod infinitum est vel saltem potest esse.

⟨6⟩ Item si non posset esse corpus infinitum, ista esset vera per se 'omne corpus est finitum'. Sed probatur quod non sit vera per se quia: ex bona definitione debent apparere omnia per se accidentia definiti, immo et causae illorum | 80vb C accidentium, ut dicitur in quarto *huius*; sed ex definitione corporis non apparet quod corpus debet esse finitum. Nam corpus describitur primo *Caeli* quod corpus est omniquaque extensum vel omniquaque divisibile, scilicet secundum longum, latum et profundum, vel quod corpus est

---

1 est] *om.* I 3 portionem] proportionem P 5 consumatur] consumitur P 7–8 infinitum ... est[1]] *om.* (*hom.*) I 8 est[1]] *om.* P ‖ quod iterum] et iterum vel illud p 9–10 habetur ... infinitum[1]] *om.* (*hom.*) p 10 in[2]] *om.* Pp 11 quantocumque] quantumcumque p ‖ finito] *om.* I 12 aliud[1]] *corr. sup. lin. ex* aliquod C : ad P ‖ dato] *om.* IPp 13 est] erit p : *om.* P 14 et[1]] *corr. sup. lin. ex* est C : *om.* P ‖ suis] *om.* I 15 a] et P ‖ ab] *om.* IPp 16–17 vel intelligibile] *om.* p 17 aut potuit esse] *om.* (*hom.*) I 18 praesente] praesenti P 21 posset esse] potest P ‖ vera] *om.* I 23 immo] *add.* etiam P 25 debet] debeat Ip 26 omniquaque[1]] undiquaque p : undique P 26–27 omniquaque divisibile] *corr. in marg. ex* aliquoque difficile C : undiquaque divisibile p : undique divisibile P

1–2 Cf. Aristoteles, *De generatione et corruptione*, I, 3, 317b33–318a27; II, 10, 336a15–b24; cf. Iohannes Buridanus, *Quaestiones super libros De generatione et corruptione*, II, q. 13 (ed. Streijger, Bakker, Thijssen, 257–258) 24 Cf. Aristoteles, *Physica*, IV, 4, 210b32–211a12 26 Cf. Aristoteles, *De caelo et mundo*, I, 1, 268a7

dimensiones, scilicet longitudo, latitudo et profunditas; et haec omnia non arguunt finitatem, quia convenirent corpori, licet esset infinitum.

⟨7⟩ Item posset argui quod extra caelum sit spatium infinitum. Sed hoc dicetur in alia quaestione.

Oppositum determinat Aristoteles hic et in primo *Caeli*. Et arguit hic Aristoteles sic: figura est de propriis passionibus corporis; igitur omne corpus est figuratum et per consequens finitum. Nam 'corpus esse figuratum' significat quod sit superficie vel superficiebus clausum, et tale est finitum.

Et istam quaestionem intendo nunc tractare textualiter solum, scilicet narrando quid de hoc dicit Aristoteles et narrando rationes eius, ut appareat quantum valent.

Licet igitur multis modis dicatur infinitum, ut secundum vigorem, secundum durationem, secundum divisionem etc., tamen hic solum intenditur de 70ra P infinito secundum magnitudinis extensionem, prout infinitum | diceretur corpus extensum sine terminis. Et non curamus hic de 'infinito' syncategorematice accipiendo, sed categorematice.

56rb p Sciendum est etiam quod duplicia | sunt corpora naturalia et sensibilia, scilicet quaedam mobilia motu recto, puta gravia et levia, quae sunt sensibilia sensu tactus et habentia ad invicem contrarietatem et actionem et passionem et sunt ex invicem generabilia et corruptibilia; alia sunt mobilia solum circulariter, quae non sunt gravia nec levia etc., scilicet corpora caelestia. Sed quia in isto libro nondum erat demonstratum quod illa corpora caelestia essent alterius naturae a gravibus et levibus et tactu sensibilibus, et cum in primo *Caeli* ostensum fuerit caelum esse alterius naturae, reiterabitur ibi tractatus de infinito quantum ad omnia corpora, et erit ibi determi-
53rb I natio perfectior quam | hic.

---

3 sit] est sic I ‖ sed] *add.* de IP 4 alia quaestione] quaestione sequente I 5 hic^{2}] *om.* p 6 figura] finita P 8 finitum] infinitum P 9 et] *om.* IPp ‖ intendo nunc] *inv.* P 12 licet … infinitum] licet ibi multis modis dicatur infinitum P : dicitur ergo infinitum multis modis p 13 divisionem] dimensionem p 15 curamus] quaeramus P 16 accipiendo] sumendo IPp 17 et] *om.* P 18 recto] ratione P 20 ex] extra P 21 non] nec p 21–22 caelestia] supercaelestia C 22 demonstratum] determinatum p 23 sensibilibus] sensibus P ‖ et^{3}] HT : *add.* ideo ACILMPp : *add.* igitur BU 24–25 reiterabitur] recitabitur p 25 ibi^{2}] *om.* P

---

4 Cf. inf., III, q. 15 5 Cf. Aristoteles, *Physica*, III, 5, 204b4–206a8; cf. Aristoteles, *De caelo et mundo*, I, 5–7, 271b1–276a17 6 Cf. Aristoteles, *Physica*, III, 5, 204b4–6 24 Cf. Iohannes Buridanus, *Quaestiones super libros De caelo et mundo*, I, q. 9 (ed. Moody, 42) 25 Iohannes Buridanus, *Quaestiones super libros De caelo et mundo*, I, q. 17 (ed. Moody, 77–82)

Et sic nunc de huiusmodi corporibus tactu sensibilibus ponamus cum Aristotele istam conclusionem quod nullum est corpus sensibile actu infinitum. Et est ad hoc prima ratio Aristotelis quia: si esset aliquod tale corpus infinitum, vel ipsum esset | simplex vel mixtum vel compositum ex simpli- 81ra C cibus; sed neutro modo est possibile; igitur etc.

Minor probatur. Primo quod non sit simplex, multis rationibus.

Prima ratio quia: sequeretur quod illud statim corrumperet omnia alia, quod apparet falsum. Consequentia probatur quia: talia corpora simplicia habent ad invicem contrarietatem et actionem et passionem; propter quod unum corrumperet reliquum, si multum excederet ipsum in virtute (et hoc supponitur). Deinde etiam supponitur quod in maiori corpore est maior virtus secundum proportionem magnitudinum, si cetera sint paria. Sed infinitum corpus in infinitum excederet finitum secundum magnitudinem; igitur et secundum virtutem. Igitur statim infinita velocitate corrumperet omne aliud corpus finitum.

Contra istam rationem Aristotelis sunt cavillationes. Prima est quod aliqui antiquorum per dictam rationem concesserunt, ut dicit Aristoteles, nullum quattuor vocatorum elementorum esse infinitum, scilicet nec ignem nec aerem nec aquam nec terram, quia propter contrarietatem istorum ad invicem infinitum corrumperet alia. Sed dixerunt infinitum esse quoddam aliud corpus medium, ex quo dicebant dicta quattuor elementa generari, quod non sic determinabat sibi tales contrarietates; ideo non corrumpebat alia.

Contra hanc cavillationem dicit Aristoteles quod inconveniens est in naturalibus concedere tale corpus, quia illud deberet nobis alicubi apparere, quoniam sicut ex eo generarentur quattuor vocata elementa, ita deberent

---

1 et sic] ergo IPp ‖ huiusmodi] huius P 3 ad hoc] *om.* P ‖ quia] quod I ‖ aliquod] *om.* IPp 4 vel[3]] sive IPp 5 neutro modo] neutrum p ‖ etc.] *om.* P 7 ratio quia] ratio est quod P : est quia p ‖ sequeretur] sequitur p ‖ omnia] *om.* P 8 consequentia probatur] *om.* P ‖ simplicia] sensibilia I 9 et[1]] *om.* P 10 reliquum] alterum IPp ‖ multum excederet] multum extenderet P : nullum excederet p ‖ et] *om.* IPp 11 etiam] *om.* P 12 proportionem] proportiones Pp ‖ magnitudinum] magnitudinis I ‖ sint] sunt I 13 excederet] extenderet P 14 et] etc. p : *om.* P ‖ corrumperet omne] corrumperet omne (*sup. lin.*) C : corrumperet quodlibet P : corraperet quodlibet p 16–17 aliqui antiquorum] antiqui P 17 per … concesserunt] propter dictam rationem concesserunt Ip : concesserunt propter dictam rationem P ‖ ut dicit aristoteles] ut dixit aristoteles I : *om.* P 18 scilicet nec] scilicet C : sed nec P 20 invicem] *add.* illud p 21 dicta] *post* elementa P 22 quod non] *praem.* et I : quoddam modo p 26 ex eo] ea P

---

3 Aristoteles, *Physica*, III, 5, 204b11 sqq. 17 Cf. Aristoteles, *Physica*, III, 5, 204b25–28 24 Aristoteles, *Physica*, III, 5, 204b29–34

aliquando in ipsum resolvi; nunc autem non apparet nobis illud, sed solum dicta quattuor elementa et mixta ex eis.

[70rb P] Item omne corpus per se subsistens, cuiusmodi ponebant | illud corpus, ex quo fit aliud corpus habet cum eo aliquam contrarietatem, quia generatio talis corporis ex tali corpore est secundum alterationem praecedentem, quam oportet esse de contrario in contrarium. Et tunc revertitur quod ipsum esset corruptivum aliorum et statim corrumperet alia.

Alia cavillatio est quia, licet sit verum quod in maiori corpore est maior virtus proportionaliter ceteris paribus, tamen hoc non est verum, si cetera non sint paria. Verbi gratia decem mensurae aeris vel forte mille non essent tantae virtutis nec tantae activitatis quantae esset una mensura ignis. Et ideo, si infinitum poneretur esse ignis, bene forte sequeretur quod corrumperet alia. Sed hoc non ita sequitur, si poneretur esse aer.

Contra istam responsionem replicat Aristoteles sic: si ignis ponatur finitus et aer infinitus, certum est quod iste ignis non esset infinitae virtutis, quia dicetur in octavo *huius* quod impossibile est in corpore finito esse virtutem infinitam. Tunc ultra de aere infinito poterimus accipere unam portionem [81rb C] aequalis quantitatis illi igni, | quae vocetur *a*. Tunc ista pars erit aliquan- [56va p] tae virtutis, licet parvae. Et oportebit aliquam esse | certam proportionem virtutis istius ignis ad virtutem ipsius *a*, quia necesse est proportionem determinatam finiti ad finitum esse. Sit igitur gratia exempli quod ille ignis sit virtutis maioris in centuplo quam *a*. Tunc in aere infinito poterimus accipere unam aliam partem in centuplo maiorem quam sit *a*, quae vocetur *b*. Tunc quia ceteris paribus in maiori corpore est maior virtus proportionaliter, sequitur quod *b* est in centuplo maioris virtutis quam sit *a*; igitur est virtutis aequalis illi igni. Et tamen aer infinitus, cum sit in infinitum maior quam *b*, est in infinitum maioris virtutis quam *b* et per consequens quam iste ignis. Ideo sequitur ut prius quod statim corrumperet illum ignem.

---

1 illud] *ante* nobis P : aliud p 7 esset] est P 8 quia] quod Pp 9 proportionaliter] proportionabiliter p 10 sint] sunt P 11 nec tantae activitatis] *om.* P ‖ quantae] quanta I : quanto C : quanti P ‖ et] *om.* P 12 esse ignis] esse (*sup. lin.*) ignis esse C ‖ bene forte sequeretur] bene formaliter sequitur p : forte sequeretur I 12–13 quod … sequitur] *in marg.* C 13 hoc] hic C : *praem.* de p ‖ sequitur] sequeretur P 15 esset] est IPp 16 huius] *om.* IPp 16–17 virtutem infinitam] *inv.* p 17 portionem] proportionem P 18 aequalis] aequalem P 19 oportebit] oportebat P 21 quod] *om.* P 21–22 sit virtutis maioris] maioris virtutis P 24 proportionaliter] proportionabiliter I 25 b] *add.* proportionaliter P ‖ sit] *om.* IPp 25–26 virtutis^2 … igni] aequalis virtutis ille ignis P 26 tamen … cum] *corr. ex* cum aer infinitus C : cum aer infinitus p ‖ b] *add.* sequitur quod p

---

14 Cf. Aristoteles, *Physica*, III, 5, 204b14–19 16 Cf. Iohannes Buridanus, *Quaestiones super libros Physicorum*, VIII, q. 10 (ed. Parisiis 1509, ff. 118ra–119ra)

Sed salvo nobiliori iudicio, quamvis istae rationes sunt subtiles, tamen non apparet mihi quod sunt demonstrativae, quia licet corpus, quanto est maius, tanto sit maioris virtutis extensive, si cetera sunt paria, tamen non oportet quod sit tanto maioris virtutis | [53va I] intensive. Verbi gratia totale mare non est notabiliter frigidius intensive quam pars eius quae replet dolium, et si ponas manum tuam in mari, illa non frigefiet notabiliter fortius quam si poneres eam in caccabo pleno aqua frigida. Unde non oportet semper quod maius corpus fortius agat quam minus corpus, quia verisimile | [70va P] est quod non agat ad omnem distantiam; ideo partes multum distantes non agunt. Quomodo enim parvus ignis candelae duraret in tanta multitudine aeris, non apparet. Valeat igitur ista prima ratio Aristotelis quantum valere poterit.

Secunda ratio Aristotelis ad probandum quod non sit corpus simplex infinitum est quia: sequeretur quod locus eius naturalis esset infinitus, cum locus debeat esse aequalis locato; sed hoc est impossibile. Quod probo multipliciter.

Primo quia: periret sursum et deorsum, nam in infinito non esset medium nec extremum; et tamen sursum et deorsum sunt extremum et medium.

Secundo sequeretur ex hoc quod non essent amplius diversi motus naturales diversorum elementorum, quia ex eo sunt diversi, quia unum naturaliter tendit sursum et aliud deorsum; et hoc non esset ita, si non esset sursum neque deorsum.

Tertio etiam quia: perirent dictae differentiae naturales locorum, per quas corpora naturalia diversa moventur ad loca diversa naturaliter.

Quarto quia: sequeretur quod lapis vel glaeba terrae, si esset in isto loco infinito, vel ubique moveretur vel ubique quiesceret, et sic de aliis corporibus; quod non apparet nobis verum, immo est dare loca ubi unum corpus moveretur sursum et aliud deorsum, et loca alia in quibus | [81va C] quiesceret. Con-

1 nobiliori iudicio] meliori iudicio Ip : iudicio meliori P ‖ sunt] sint p 2 sunt] sint IPp 3 sunt] sint IPp 4 tanto] in tantum I ‖ intensive] extensive P ‖ totale mare] totalis sequana P 5 frigidius intensive] frigidior extensive P ‖ replet] repleret Ip 6 si[1]] *add.* tu I ‖ mari] sequana P 7 eam] *om.* Ip ‖ pleno] plena P 8 corpus[2]] *sup. lin.* C : *om.* Pp 11 valeat] *post* aristotelis P ‖ poterit] potest IPp 13 sequeretur] sequitur p 14 debeat] debet Pp 17 nec] neque I ‖ et[3]] ad P 18 sequeretur] sequitur p ‖ ex hoc] *om.* P 19 sunt diversi] *inv.* I ‖ quia[2]] quod IPp 19–20 naturaliter tendit] intendit naturaliter P 20 sursum[1] ... deorsum] deorsum et aliud sursum Ip 22 dictae] *om.* IPp ‖ per] propter Ip ‖ quas] *add.* differentias p 24 sequeretur] sequitur p 27 moveretur] movetur CP ‖ quiesceret] quiescent P

12 Cf. Aristoteles, *Physica*, III, 5, 205a15–19

sequentia patet, quia totus locus naturalis corporis simplicis debet esse consimilis; ideo dicetur in quarto *huius* quod idem est locus totius et partis. Si autem locus infinitus esset totus consimilis naturae, non esset aliqua ratio quare aliquod corpus positum in eo magis moveretur vel quiesceret magis in una parte quam in alia; ideo vel ubique moveretur vel ubique quiesceret. Non esset ratio etiam aliqua quare moveretur magis ad unam partem quam ad aliam; ideo vel in omnem partem moveretur vel in nullam. Et haec omnia apparent impossibilia.

Tertia ratio ad conclusionem principalem est quia: sequeretur quod illud corpus infinitum non haberet locum, neque naturalem neque violentum; consequens est falsum; igitur et antecedens. Falsitas consequentis patet ex eo quod omne corpus sensibile et mobile motu recto, de quo hic loquimur, potest moveri motu recto aut naturaliter aut violenter; quod non posset esse nisi transeundo de uno loco in alterum. Et consequentia principalis
56vb p patet, quia locus | est extrinsece continens locatum; et impossibile est quod infinitum contineretur ab alio.

Quarta ratio est quia: illud corpus infinitum vel esset totum grave, et sic esset nihil leve, quia illud infinitum occuparet totum spatium etiam imaginabile, ideo non permitteret secum aliud corpus; vel esset totum leve, et sic periret grave. Et haec sunt inconvenientia. Vel una pars eius esset gravis et alia levis, et sic iam esset compositum et non simplex. Et etiam non posset dici ratio quomodo deberet partiri et quam proportionem deberet habere pars levis ad partem gravem.

70vb P Igitur concludendum | est ex dictis quod nullum est corpus simplex infinitum.

Sed postea probatur quod nullum sit compositum ex simplicibus infinitum quia: vel esset compositum ex simplicibus infinitis secundum multitudinem vel finitis secundum multitudinem; sed neutro modo est possibile; igitur

---

1 simplicis] finitus P 4 magis[1]] *post* moveretur IP : *om.* p 4–5 quiesceret ... vel[2]] *om.* (*hom.*) p 6 etiam] *ante* esset P : *ante* ratio Ip ‖ moveretur magis] *inv.* I 7 ad] *om.* p ‖ in[1]] ad *sed add. sup. lin.* in C : ad P ‖ et] moveretur P 9 conclusionem principalem] *inv.* P ‖ sequeretur] sequitur p 11–12 ex eo quod] ex hoc quod p : ex hoc quia I : quia P 14 nisi] non P ‖ uno] *om.* p ‖ alterum et] alium P 16 alio] *add.* etc. I 17 quia] quod IP 18 esset nihil] *inv.* Ip : periret P 19 ideo non] *om.* P 21 et[4]] *om.* P 22 quomodo] quare P 24 est[2]] *post* infinitum (24–25) P 26 quod] *add.* etiam Ip ‖ simplicibus] *add.* in p 27 vel] nihil sic P

---

2 Aristoteles, *Physica*, IV, 5, 212b1, 12–14

etc. Maior patet sufficienti divisione. Minor probatur. Primo quod non ex infinitis secundum multitudinem. Probatur multipliciter.

Primo quia: ista infinita essent principia illius compositi; et probatum est primo *huius* contra Anaxagoram quod non sunt infinita principia rerum naturalium.

Secundo quia: sicut ista simplicia essent diversarum naturarum, ita etiam eorum loca naturalia, quae etiam essent infinita; modo non est possibile loca sic esse infinita, cum motus naturales simplices sunt finiti et determinati, ut apparet primo *Caeli*. Et etiam, si essent loca infinita, non esset inter ea locus medius nec locus extremus, et sic non esset sursum nec | deorsum, et | 81vb C
argueretur sicut prius arguebatur. 53vb I

Deinde probatur quod non sit corpus infinitum compositum ex simplicibus finitis secundum multitudinem quia: hoc non posset esse, nisi unum istorum esset infinitum et alia finita vel quod plura eorum essent infinita, nam ex finitis secundum magnitudinem et secundum multitudinem non resultat nisi finitum; sed utrumque horum est impossibile, sicut probabitur. Primo non potest dici quod unum istorum simplicium sit infinitum et alia finita, quia corrumperet alia, sicut prius arguebatur. Satis enim fuit argutum prius quod non sit corpus simplex infinitum. Sed etiam arguo quod non possunt esse plura infinita quia: unum occuparet totum spatium imaginabile; ideo non compateretur aliud extra se. Et hoc etiam declarat Aristoteles per definitionem corporis infiniti, quae debet esse congregata ex definitione corporis et ex definitione infiniti; definitio autem corporis est 'quod est omniquaque distans' et definitio infiniti est 'quod est distans sine termino'; ideo corpus infinitum esset omniquaque distans sine termino, et sic nihil compateretur extra se. Ideo impossibile est esse duo corpora infinita.

Sed aliqui instant contra hanc rationem quia: sicut ponimus duo tempora infinita, scilicet praeteritum a parte ante et futurum a parte post, ita pos-

---

1 patet] apparet P 2 probatur] *add.* etiam P 3 primo] *om.* P ‖ principia] principalia P 4 non] *om.* P 6 etiam] et Ip 7 etiam] *add.* sic P ‖ possibile] impossibile P 8 sunt] sint Pp 9 et] *om.* P 13–15 multitudinem ... secundum[1]] *om.* (*hom.*) P 13 hoc] primo I 14 essent infinita] *inv.* p 16 sicut] ut P 17 primo] *add.* enim IPp ‖ istorum] eorum p 19 simplex infinitum] *inv.* P ‖ etiam] *om.* P 19–20 possunt] *praem.* sic P : possint I : poterunt p 21 etiam] *om.* P 23 ex] *om.* P 23–24 omniquaque] undiquaque p : undique P 24–25 et ... distans] *in marg.* C 25 omniquaque] undiquaque p : undique P 27 rationem] responsionem I

---

4 Aristoteles, *Physica*, I, 4, 187a16 sqq. 9 Cf. Aristoteles, *De caelo et mundo*, I, 7, 274b1–4 21 Cf. Aristoteles, *Physica*, III, 5, 204b18–21

sumus imaginari unum corpus a nobis extensum in infinitum ante nos et alterum retro.

Sed dicendum est quod ista est imaginatio mathematica, non naturalis, quia non esset ratio naturalis quare unum corpus sic ad unum latus protenderetur in infinitum et non ad aliud. Videmus enim corpora naturalia simplicia similiter undiquaque extendi et ob hoc se invicem continere.

Tunc ad rationes.

⟨1⟩ Ad primam dicitur quod Aristoteles concessisset infinitatem motus circularis et temporis. Nec intendit in sexto *huius* quod sit quantum ad hoc eadem ratio de magnitudine et de motu nisi forte quia magnitudo est infinita secundum durationem. Sed bene intendebat quod, si magnitudo est composita ex indivisibilibus, quod ita est de motu et de tempore, et si non, non. Et
57ra p etiam intendebat | quod super spatium finitum non potest fieri motus infi-
71ra P nitus nisi circulariter reiterando revolutiones. Nec etiam potest esse motus | finitus super totum spatium infinitum. Et haec videbuntur in sexto libro. Dicendum est etiam quod motus, licet infinitus sit, non est nobilior nec
82ra C maioris entitatis quam caelum, quod est | finitum, quia etiam caelum est infinitum secundum durationem; et est nobilius sic esse infinitum in permanentia quam in successione.

⟨2⟩ Ad aliam concedo quod magnitudo est in infinitum divisibilis et infinitas partes habet. Qualiter dicetur post. Et non sequitur inde extensio infinita. Et haec postea declarabuntur.

⟨3⟩ Ad aliam solvit Aristoteles primo *De generatione* per hoc quod generatio unius est corruptio alterius et non annihilatur materia in corruptione, sed quam habebat quod corrumpitur, illam accipit quod generatur; et sic numquam aliquid de materia corrumpitur.

1 in] *del.* C : *om.* I 3 est[1]] *om.* P ‖ est imaginatio] *inv.* P 4 quia] *post* naturalis p ‖ quare] dare P ‖ sic] si P 5 enim] *om.* p 5–6 naturalia ... undiquaque] simplicia naturalia undiquaque similiter I : simplicia similiter undique vel undique P 8 dicitur] dico IPp 10 de[2]] *om.* P ‖ nisi] non P ‖ est infinita] *in marg.* C : est in infinitum p 11 secundum durationem] *om.* P 12 de[2]] *om.* IP 14 nisi] *ante* motus (13) P ‖ circulariter] circularis etiam C 15 haec videbuntur] hoc videbitur p ‖ libro] huius p 16 licet infinitus sit] licet sit infinitus P : licet sit finitus I : qui sit infinitus p 20 aliam concedo] aliam conceditur I : aliam dicitur P : secundum conceditur p ‖ est] *post* infinitum Ip 20–21 infinitas ... qualiter] conceditur et qualiter infinitas partes habet P 23 aliam] tertiam p 25 sed] *add.* materiam p

15 Cf. Iohannes Buridanus, *Quaestiones super libros Physicorum*, VI, q. 9 (ed. Parisiis 1509, ff. 101rb–102rb) 22 Cf. inf., III, q. 18 23 Cf. Aristoteles, *De generatione et corruptione*, I, 3, 318a23–25; cf. *AA*, 4: 7

⟨4⟩ Ad aliam dicit Aristoteles quod differt esse finitum et tangere. Necesse est enim quod tangit tangere aliud, sed finitum non oportet esse finitum ad aliud. Immo corpus est finitum et terminatum per suas proprias et intrinsecas superficies.

⟨5⟩ Ad aliam dicit Aristoteles quod non oportet credere imaginationi vel intelligentiae, id est intellectioni, quia possumus imaginari te crescere ultra quantitatem quam habes et non oportet ita esse in re. Quando autem dicitur quod oportet intellectum et imaginationem moveri a re, dico quod verum est. Ideo conceptui simplici necesse est correspondere rem aliquam vel praesentem vel praeteritam. Sed in componendo conceptus simplices potest esse falsitas vel fictio. Dico 'falsitas', si in propositione componamus affirmative terminos non supponentes pro eodem vel si componamus negative terminos supponentes pro eodem. Sed etiam non per modum propositionis, sed per modum determinationis vel determinabilis possumus componere simplices conceptus modo affirmativo vel negativo, ut 'homo albus', 'homo non albus'. Et si termini simplices supponunt pro eodem, terminus negative compositus erit fictus, id est pro nullo supponens, ut 'homo non risibilis', et si non supponunt pro eodem, terminus affirmative compositus erit fictus, ut 'homo hinnibilis'. Et sic est in proposito, quia a rebus habemus conceptum corporis et habemus conceptum finiti vel terminati, et istos | negative com- 71rb P ponimus formando illum terminum complexum 'corpus non finitum' vel 'corpus infinitum', qui iam est terminus fictus et pro nullo | supponens. Sed 54ra I etiam, quando dicitur quod est bona consequentia '*a* est intelligibile, igitur *a* est, fuit vel potest esse', concedo. Et ideo dico quod omnes tales sunt falsae | de virtute sermonis et loquendo non materialiter, sed significative: 82rb C 'Deum non esse est intelligibile', 'chimaera est intelligibilis vel opinabilis', 'vacuum est imaginabile', 'corpus infinitum est imaginabile vel intelligibile'. Dico quod istae essent falsae et impossibiles, si impossibile est chimaeram esse, vacuum esse, corpus infinitum esse. Sed per tales terminos intelligimus veras res secundum conceptus complexos fictos. Et eodem modo diceretur quod isti termini 'chimaera', 'vacuum' et 'corpus infinitum' non significant

---

1 aliam dicit] aliam respondet IP : quartam respondet p ‖ finitum] infinitum I **2** non oportet] nihil habet p **5** aliam] quintam p **7** ita] ista P ‖ autem] ante p **8** a] in P ‖ quod[2]] *add.* maxime p **13** sed[2]] *add. in marg.* etiam C : *add.* etiam I **14–15** simplices conceptus] *inv.* IPp **15** affirmativo vel negativo] affirmative vel negative P **16** si] *add.* terminus vel p **17** id … supponens] *om.* p **17–19** et … hinnibilis] *om.* (*hom.*) p **19** a rebus] in rebus C : *om.* P **20** habemus] *om.* IPp ‖ istos] *add.* modos C **22–23** sed etiam] et P **24** est] *om.* P ‖ et] *om.* P **24–25** sunt falsae] *om.* p **25** et] *om.* P **28** essent] sunt P ‖ est] esset IPp **29** esse[2]] *add.* et IPp ‖ sed] semper P **30** veras res] *inv.* P ‖ et] *om.* P **31** significant] significarent P

chimaeram nec vacuum nec corpus infinitum (dico: semper, si impossibile sit talia esse), sed significant veras res secundum conceptus complexos fictos. Et bene verum est quod tales termini vocales bene significant conceptus realiter existentes in anima et ultra etiam significant res extra. Verbi gratia idem significat iste terminus 'vacuum' sicut haec oratio 'locus non repletus 57[rb] p corpore', et haec oratio significat et locum et plenum, quae sunt ve|rae res extra. Sed significat ea secundum talem complexionem conceptuum quod conceptus complexus pro nullo supponit.

⟨6⟩ Ad aliam concedo quod haec est vera per se 'omne corpus est finitum'. Sed non oportet quod omnia per se accidentia corporis vel alicuius alterius sint statim manifesta de isto per eius definitionem sine discursu. Immo multa indigent magno discursu et longo; et sic fit in proposito. Per processum enim scitur omne corpus esse finitum.

Et sic est finis quaestionis etc.

---

2 talia] tale (*corr. ex* tales) P 3 bene[2]] *om.* IPp 4 ultra etiam] *inv.* p 5–6 repletus corpore] plenus p : plenus corpore *sed del.* corpore I 6 locum] locus C 7 ea] eas IPp ‖ quod] cum C 9 aliam concedo] aliam dico I : sextam concedo p 14 et … etc.] *om.* IPp

⟨III.15⟩

⟨Utrum sit aliqua magnitudo infinita⟩

Licet fuerit concessum quod non sit corpus sensibile et mobile infinitum, sicut hic et in primo *Caeli* nititur ostendere Aristoteles, tamen adhuc restat dubitatio de corpore simpliciter vel de magnitudine. Ideo fiat nunc consequenter quinta decima quaestio utrum est aliqua magnitudo infinita.

Arguitur quod sic quia:

⟨1⟩ Si extra caelum est aliquod spatium, illud est infinitum, quia non est ratio quare esset ibi spatium alicuius quantitatis certae potius quam maioris, nisi poneretur infinitum. Et etiam, quia quocumque spatio finito extra caelum posito et concesso, remaneret quaestio et omnis difficultas ut prius, videlicet utrum ultra illud | finitum datum esset aliud spatium. 71va P Igitur ficticium est dicere quod ultra caelum sit spatium, nisi concedatur esse infinitum, nisi quantum ex Sacra Scriptura possemus inferre, videlicet quod sunt aquae super caelos. Unde dictum est in *Genesi:* 'divisit aquas ab aquis' et 'benedicite aquae omnes quae super caelos sunt Domino'. | 82va C Sed quomodo haec intelligenda sunt, non est praesentis speculationis nec tollit difficultatem quaestionis, quia quantumcumque poneretur ibi aqua vel aliud corpus finitum, diceremus ut prius quod, si ultra hoc esset spatium, ipsum deberet concedi infinitum, nisi aliquis omnino vellet ficte loqui et sine ratione.

Igitur si posset ostendi quod ibi sit spatium, debet concedi quod sit infinitum. Et illud spatium, quia haberet longitudinem, latitudinem et profunditatem, esset corpus. Ideo deberet concedi quod esset corpus infinitum.

Nunc igitur descendo ad arguendum quod ultra caelum vel aquas quae super caelos sunt etc. sit spatium.

3–6 licet … quaestio] quaeritur quinto decimo P **5** simpliciter] simplici p **5–6** fiat nunc consequenter] nunc fiat p **8** si] *om.* p ‖ spatium] *add.* et p **9** ibi] *om.* p ‖ quantitatis certae] *inv.* IPp **12** aliud spatium] *inv.* p **15** in genesi] *om.* IPp **17** sunt] sint Ip **19** diceremus] diceres I **20** ficte] stricte P **22** posset] possit IPp ‖ ibi] *add. sup. lin.* isto modo C : illo modo I : illic Pp **23** quia] quod p **24** concedi] dici I **25** ad arguendum] arguendo P ‖ caelum] *om.* P **26** etc.] *om.* p

---

4 Cf. Aristoteles, *Physica*, III, 5, 204b4–206a8; cf. Aristoteles, *De caelo et mundo*, I, 5–7, 271b1–276a17 **15** *Gn* 1, 6 **16** *Dn* 3, 60

 | DOI: 10.1163/9789004322356_018

⟨1.1⟩ Primo quia: supra caelum posset Deus creare unam fabam; quod non esset, si non esset supra; sed 'supra' significat locum vel spatium; igitur ibi est spatium, scilicet in quo crearetur illa faba.

⟨1.2⟩ Item si ibi crearetur faba, ipsa non esset in indivisibili, immo aliquod spatium divisibile occuparet; igitur ibi est spatium.

⟨1.3⟩ Item si Deus ibi crearet illam fabam nihil aliud creando, tamen posset istam movere motu recto elongando eam a caelo, et sic distaret a caelo plus et plus secundum quod plus et plus moveretur; sed non posset distare a caelo nisi per spatium intermedium; igitur ibi est spatium.

⟨1.4⟩ Iterum si extra caelum crearentur duae fabae, illae essent extra invicem et extra caelum; sed 'intra' et 'extra' significant locum vel spatium; igitur ibi est locus et spatium.

⟨1.5⟩ Item pono quod Deus extra hunc mundum formaret alios duos mundos et quod illi tres mundi contingerent se in punctis, sicut sphaerae imaginarentur se contingere. Constat quod inter istos tres mundos | 54rb I et puncta in quibus tangerent se esset spatium medium et distantia media. Aliter contingerent se non solum secundum illa puncta, sed secundum totas suas superficies; quod est impossibile.

⟨1.6⟩ Item non magis esset concedendum spatium intra orbem lunae, si omnia corpora intra existentia essent annihilata, quam debeat concedi extra caelum; sed tamen dicto ca|57va p su posito esset spatium inter latera orbis lunae. Probatio quia: adhuc latera | 71vb P orbis lunae distarent ab invicem sicut nunc, posito quod orbis lunae remaneret in sua quantitate et figura (non enim essent latera eius invicem proxima et tangentia); et tamen non distarent, nisi esset spatium medium per quod distarent.

⟨1.7⟩ Et iterum hoc confirmo quia: in casu posito oportet dicere quod inter latera illa esset vacuum; et cum propositio affirmativa non sit vera in qua terminus non supponit pro aliquo, oportet in | 82vb C dicto casu quod iste terminus 'vacuum' supponeret pro aliquo; et non posset pro alio supponere quam pro isto spatio; igitur ibi esset spatium.

---

1 primo] *om.* p 4 crearetur] *add.* illa I 6 illam fabam] *inv.* I : unam fabam P ‖ tamen] *om.* P 8 sed] *add.* tamen P ‖ posset] possit P 11 sed intra] intra enim P 12 et] vel I 13 alios duos] *inv.* I 14 illi] ibi P 14–15 imaginarentur] imaginantur P 16 tangerent] contingerent P ‖ esset] esse P 16–17 contingerent] contingeret CP 17 solum secundum] sicut Pp ‖ totas suas] *inv.* IPp 21 inter] intra I 22 probatio] probatur p ‖ latera orbis lunae] orbes I 24 invicem proxima] *inv.* P 26 et] *om.* p ‖ confirmo] confirmatur IPp ‖ oportet] oporteret IPp 27 latera illa] *inv.* IPp ‖ sit] esset Ip 28 supponit] supponat IP ‖ oportet] oporteret IPp 29 vacuum … aliquo] supponeret pro aliquo scilicet vacuum P ‖ alio] aliquo C

⟨1.8⟩ Et iterum ulterius videtur quod ibi esset spatium certae quantitatis, cuius diameter esset tertia pars lineae circularis imaginatae in concavo orbis lunae, quia tantae quantitatis erat corpus quod ibi fuit prius et quod ponitur esse annihilatum. Et etiam ibi potest iterum creari corpus tantae quantitatis et non maioris, posito quod semper remaneat orbis lunae in sua magnitudine et figura.

⟨1.9⟩ Iterum intra istud concavum orbis lunae posset Deus movere lapidem motu recto de uno polo ad alterum polum; et ille lapis continue elongaretur ab uno polo et appropinquaret ad alterum; et tamen non posset esse appropinquatio et elongatio sine distantia et spatio; igitur etc.

Oppositum arguitur quia:

⟨1⟩ Hoc esset ponere dimensionem sive accidens sine subiecto; quod reputatur impossibile naturaliter.

⟨2⟩ Item non magis debet concedi illic esse spatium quam concederetur intra orbem lunae, si totum quod est intra esset annihilatum, quia quaecumque rationes viderentur arguere quod illic esset spatium, illae similiter arguerent de intra orbem lunae. Et tamen in casu dictae annihilationis nihil esset intra orbem lunae, scilicet infra concavum eius, quia totum ponitur annihilatum; igitur nullum esset ibi spatium.

⟨3⟩ Item primo *Caeli* determinat Aristoteles quod extra caelum non est locus neque vacuum neque aliquod spatium vel aliqua dimensio. Et si hoc est verum, tunc nulla est magnitudo infinita.

Primo credendum est fide quod Deus posset ultra illum mundum formare et creare alias sphaeras et alios mundos et omnino alias magnitudines finitas, quantascumque vellet, ita quod omni finita creata posset creare maiorem in duplo, in decuplo, in centuplo, et sic de omni alia proportione finiti ad finitum.

---

1 esset] *add.* unum p 2 cuius] *add.* illa P 2–3 in … erat] circa I 3 fuit prius] *inv.* IPp || quod[2]] *om.* P 4 ibi] *post* iterum IPp 5 et non maioris] *om.* I 7 intra] inter C || posset] possit I || movere] *add.* unum p 9 ad alterum] *del.* ad C : alteri Ip || tamen] cum I : *om.* P 12 subiecto] substantia IP 14 magis … illic] debet magis illic concedi P 14–15 concederetur intra] concederetur inter P : concedere inter C 18 intra] inter P || infra] intra I || concavum eius] *inv.* IPp 21 neque[1]] nec p || neque aliquod] neque aliud C : nec aliquod Ip 23 primo] *add.* igitur p || fide] enim P || posset] *post* mundum P

---

20 Cf. Aristoteles, *De caelo et mundo*, I, 9, 279a11–18

Sed forte non oportet credere quod Deus posset creare magnitudinem actu infinitam, quia ista creata non posset creare maiorem (repugnat enim quod actu infinito sit aliud maius); et tamen inconveniens est quod Deus 72ra P posset facere creaturam potentiae suae | proportionatam, sic quod non posset maiorem et perfectiorem creare et facere. Et mihi videtur simile de 83ra C magnitudine | sicut de perfectione. Quamvis enim omni creatura formata et formabili Deus posset creare perfectiorem, tamen non posset creare aliquam infinitae perfectionis. Illa enim esset aeque perfecta vel non minus perfecta quam Deus; nam si esset minus perfecta, ipsa non esset infinitae perfectionis, sed posset creari perfectior. Et ita videtur de magnitudine et velocitate et intensione caliditatis et huiusmodi.

Item quantamcumque magnitudinem posset Deus formare, tantum ignem posset formare; igitur posset formare ignem actu infinitum. Et cum in maiori corpore sit maior virtus proportionaliter, ut vult Aristoteles, sequeretur quod ille ignis esset virtutis infinitae et sic non minoris virtutis quam Deus, quod omnino est impossibile.

57vb p Tamen de isto et de | pluribus, immo de omnibus quae dicantur in ista quaestione, ego dimitto determinationem dominis theologis et acquiescere volo determinationi eorum.

Sed ultra, quamvis non possit demonstrari quod extra mundum non sit spatium et magnitudo, quia Deus illic posset facere et magnitudinem et spatium, tamen ego opinor quod illic non sit spatium vel magnitudo vel 54va I alius mundus. | Et ad hoc adducit Aristoteles rationes naturales in primo *Caeli*, quae illic tractandae sunt. Ideo solum ad hoc pono talem persuasionem quia: non est verisimile quod Deus ibi fecerit alium mundum vel alios mundos, quia si plures creaturas mundanas voluisset fecisse quam fecerit, non oportebat facere alios mundos, quia potuisset istum mundum fecisse in

---

2 enim] *om.* P 4 posset] possit Ip 4–5 posset] possit IPp 5 maiorem ... facere] maiorem et perfectiorem facere Pp : facere maiorem et perfectiorem I 6 enim] *om.* P 7 posset[1]] *add.* formare sive p : possit I 9 esset[2]] est P 10 posset] possit P ‖ et[1]] *om.* P ‖ videtur] *praem.* mihi P : *add.* mihi p 11 huiusmodi] huius P 14 proportionaliter] *om.* I 14–15 sequeretur] sequitur p 16 omnino est impossibile] non est possibile p 17 dicantur] dicam Pp : dicta sunt I 21 posset] possit P ‖ et[2]] *om.* P 24 solum] *om.* C 24–25 talem persuasionem] *inv.* IPp 25 fecerit] faceret *sed add. in marg.* fecerit C : fecit p 26 fecerit] fecit IP

---

14 Cf. Aristoteles, *Physica*, VIII, 10, 266b8–11 23–24 Cf. Aristoteles, *De caelo et mundo*, I, 8–9, 276a18–278b8; 9, 279a11–18; cf. Iohannes Buridanus, *Quaestiones super libros De caelo et mundo*, I, qq. 19–20 (ed. Moody, 87–95)

duplo vel centuplo maiorem; et si non fecit ibi Deus alium mundum vel alios mundos, non apparet ratio quare fecisset illic spatium, quia illud de nihilo deserviret ad istum mundum et appareret esse frustra.

Et iterum, si esset illic spatium, Deus posset ipsum annihilare remanente isto mundo sicut est, et hoc facto omnia quae apparent salvarentur. Et si esset ibi spatium, dum tamen finitum, adhuc restarent omnes difficultates ad arguendum quod ultra adhuc esset spatium, quae fiebant ad arguendum quod esset spatium extra caelum. Ideo nihil proficeret illud spatium ad salvandum apparentia vel etiam ad evitandum difficultates apparentes. Tale autem non debet poni esse, nisi ex verbis Sacrae Scripturae sequatur.

Deinde mihi etiam videtur esse credendum fide et credo quod Deus posset annihilare omne quod est sub orbe lunae sive intra concavum orbis lunae remanente orbe lunae et etiam toto caelo remanente | secundum magnitudines (72rb P) et figuras quas nunc habent. Et isto casu posito videndum est quid sequeretur.

Dico igitur quod isto casu posito nihil esset infra sive intra concavum orbis lunae, | quia totum ponitur annihilatum. Et sic non esset aliquod spatium et (83rb C) etiam non esset vacuum intra vel infra huiusmodi concavum, quia implicat contradictionem quod nihil sit intra concavum et quod vacuum vel spatium sit intra concavum, quoniam cum haec propositio 'vacuum est intra concavum' vel 'spatium est intra concavum' sit affirmativa, oportet, si sit vera, quod iste terminus 'spatium' vel 'vacuum' pro aliquo supponat; ideo sequitur 'vacuum est intra concavum, igitur aliquid est intra concavum', et haec est contradictoria illius primae, quae dicebat quod nihil est intra concavum etc.

Et ex hoc patet quod istae rationes quae arguebant quod extra caelum sit spatium non valebant, quia similiter argueretur de intra caelum posito

---

1 vel[1]] *add.* in IP ‖ alios] *om.* P 3 ad] ultra p 4 iterum] *om.* I 5 salvarentur] *add.* sicut nunc P 7–8 ultra ... quod] *om.* (*hom.*) C 8 esset] sit IPp ‖ spatium[1]] *add.* et C 9 etiam] *om.* P ‖ evitandum difficultates apparentes] difficultates apparentes evitandas P 10 nisi] in se P ‖ scripturae] *om.* P 11 mihi etiam videtur] etiam mihi videtur Ip : videtur mihi P ‖ credendum] *add.* de P ‖ posset] possit I 12 orbe] orbi P 13 remanente orbe lunae] *sup. lin.* C : remanente orbi lunae P ‖ etiam] *om.* P 14 habent] habet P 15 sequeretur] sequitur Pp 16 isto] hoc IPp 17 aliquod] aliud CP ‖ et[2]] *om.* P 18 huiusmodi] huius P 22 spatium vel vacuum] vacuum vel spatium p 23 aliquid est] *inv.* p 23–25 et ... etc.] et haec est contradict†...† quae dicebat quod †...† est seu erat int†...† *in marg.* C 24 contradictoria] contradictionem P 25 etc.] *om.* IP 27 similiter argueretur] similiter arguetur P : consimiliter argueretur I

illo casu quod totum esset annihilatum; et tamen non concluderent verum; igitur non valent.

Sed tunc quaeritur quid esset de distantia laterum vel polorum orbis lunae ad invicem et quid esset de motu lapidis in appropinquando vel elongando ad polum.

Dico quod in hoc est difficile satisfacere imaginationi, quia semper apparet imaginationi quod ibi esset spatium, sicut semper sensui apparet quod sol non sit maior equo et quod sit valde minor terra. Tamen in talibus intellectus debet corrigere tales apparentias sensus et imaginationis. Dico igitur, quia non loquimur de casibus naturaliter possibilibus, sed miraculose possibilibus, quod in uno parvo spatio et loco, ut in loco grani milii vel sub eius magnitudine, Deus posset formare valde magnum corpus, scilicet maius quam sit mundus. Et verum est quod illud corpus non esset in illo parvo corpore vel loco circumscriptive vel commensurabiliter. Hoc credendum est, quia | [58ra p] sub parva quantitate hostiae et in eius parvo loco est corpus Christi ita magnum, sicut erat in cena et sicut est in paradiso, et ita figuratum; immo in qualibet parte quantitativa hostiae, quantumcumque parva, est totum corpus Christi magnum et optime figuratum. Sed haec magnitudo corporis Christi non se habet in hostia modo commensurabili ad magnitudinem hostiae. Et non minus posset Deus facere maius corpus in loco et cum magnitudine grani milii. Propter quod etiam concludo quod intra concavum orbis lunae posset etiam Deus facere maius corpus in centuplo quam sit mundus, non mutata magnitudine et figura orbis lunae. Sed | [72va P] non esset ibi circumscriptive nec modo mensurabili ad magnitudinem orbis. Immo si esset ibi corpus per modum circumscriptivum et mensurabilem ad magnitudinem orbis, illud non posset esse maius quam quod nunc est, quia oporteret eius diametrum | [54vb I] esse tertiam partem vel circiter lineae circularis protractae in concavo orbis lunae.

Postea ego etiam dico quod in parvo spatio, ut in spatio decem pedum, Deus posset | [83va C] movere unum lapidem vel unum corpus valde magnum per

---

1 totum] *corr. in marg. ex* nihil C : nihil Pp **3** esset] est P ‖ vel] *om.* p **5** ad] *sup. lin.* C : *om.* P **6** imaginationi] imaginationem P **7** esset] sit P ‖ sensui apparet] *inv.* P **9** tales] illas IPp ‖ imaginationis] imaginationes p **10** quia] quod CI ‖ casibus] talibus p **12** posset] possit I **14** vel[2]] et IPp **15** in] *om.* P **16** cena] *add. sup. lin.* alias terra C ‖ immo] imaginatio P **17** totum] tamen P **20** et cum] etiam in P **22** etiam] *om.* IPp **24** mensurabili] commensurabili P **25** modum] motum P **26** orbis] corporis I ‖ posset] possit I ‖ quod] *om.* IPp **29** ego etiam] etiam Ip : *om.* P ‖ ut ... pedum] puta spatium decem pedum P : *om.* I **30** posset] potest I

unum totum annum continue valde velociter motu recto, et tamen iste lapis non exiret ab illo spatio quiescente, immo nec alicui cono illius spatii approximaretur nec ab aliquo cono elongaretur.

Hoc declaratur quia: si Deus moveat manum de capite ipsius ad pedes, erit ita longus motus, quantum corpus Christi est longum; et tamen nulli cono hostiae fieret illa manus propinquior quam ante vel remotior, quia non moveretur manus secundum comparationem vel situm ad magnitudinem hostiae, sed secundum situm et comparationem ad magnitudinem corporis Christi. Et non assero haec omnia, sed in asserendo haec vel aliqua eorum aut non asserendo submitto me totaliter decreto et ordinationi sanctae ecclesiae et doctorum Scripturae Sacrae.

Deinde si haec essent vera et concessa, sequeretur quod lapidem infra orbem lunae Deus posset movere valde velociter motu recto et non recederet vel elongaretur ab uno latere orbis lunae nec accederet vel approximaretur ad alterum latus. Dico quod in praedicto casu annihilationis eorum quae sunt infra orbem lunae unus polus non tangeret alterum polum nec distaret ab altero polo secundum rectitudinem, quia non esset spatium rectum medium per quod distaret. Sed posset concedi distare secundum distantiam circularem vel curvam. Immo videtur mihi quod, si essent tres lapides pedales consequenter se habentes secundum rectitudinem et tangentes se et nihil esset plus in mundo et Deus annihilaret lapidem medium non approximando vel coniungendo lapides extremos ad invicem, illi lapides remanentes nec tangerent se invicem nec essent longe ab invicem nec prope, sicut etiam, si Deus separaret magnitudinem lapidis substantia eius remanente, lapis adhuc haberet partem aliam a parte, sed non haberet partem extra partem situaliter. Lapides igitur illi non distarent, quia nihil esset medium per quod distarent, et tamen necesse est quod per spatium medium sit et mensuretur | distantia distantium. Nec tamen oportet eos esse contiguos, 72vb P
quamvis nihil esset inter eos secundum rectitudinem, sicut nec poli orbis lunae tangerent se, quamvis nihil esset inter eos secundum rectitudinem.

---

1 valde] *om.* I 4 moveat] movet I 5 erit ita] *corr. sup. lin. ex* est C : erit IPp ‖ nulli] nullo Ip 6 fieret] fiet Ip ‖ propinquior] proximior IPp 7 moveretur] movetur IPp 12 sequeretur] sequitur p ‖ infra] intra I 14 vel approximaretur] vel approximarem C : nec approximaretur I 15 dico] *add.* ergo IPp 16 alterum] alium C 18 distaret] distant p ‖ posset concedi] potest concedi I : posset P 20 pedales] *om.* P 21 esset plus] *inv.* P 22 lapides[1] ... invicem] ad invicem lapides extremos (extremo P) IPp 23 tangerent] tangeret I 24 separaret] annihilaret IPp 28 distantium] *corr. in* ad distantiam C : *om.* I ‖ oportet] oporteret IPp 29 inter eos] interius P ‖ secundum rectitudinem] †...† rectitudinem *in marg.* C : *om.* Pp 29–30 sicut ... rectitudinem] *om.* (*hom.*) I

Et ita concluditur quod plus requiritur ad hoc quod duo corpora tangant se quam quod inter ea nihil sit. Et hoc forte quod plus requiritur est quod se 83vb C habeant | ad invicem commensurabiliter et secundum determinatos situs partium unius et partium alterius. Unde nec proprie magnitudo corporis Christi tangat magnitudinem hostiae, et illi lapides non sic haberent se ad invicem.

58rb p Sed tunc quaeritur utrum inter illos lapides absque motu eorum pos|set Deus iterum creare lapidem maiorem quam erat lapis annihilatus qui erat intermedius.

Et dico quod sic, scilicet lapis in centuplo maior et faciens magis distare illos lapides in centuplo quam distarent per lapidem annihilatum, antequam annihilaretur. Nec mirum, quia sine motu corporis *b* et sine distantia ipsius a se invicem posset Deus facere quod ipsum esset in diversis locis ab invicem distantibus. Sed talia non possunt fieri naturaliter nec imaginatio cadit super ea; ideo difficiliora sunt ad intelligendum.

Tunc igitur ad rationes.

⟨1⟩ Ad primam, quando dicitur 'si extra caelum sit spatium, ipsum esset infinitum', dico quod Aristoteles concessisset istam consequentiam. Sed tamen propter potentiam supernaturalem dico quod ipsa non est necessaria, quia Deus posset ibi creare spatium finitum, quantum placeret sibi; de cuius quantitate non esset quaerenda ratio nisi simplex voluntas Dei. Sed tamen opinor quod non sit ibi aliquod spatium, scilicet ultra corpora nobis apparentia et ea quae ex Sacra Scriptura tenemur credere.

⟨1.1⟩ Tunc ad rationes quae arguunt quod ibi sit spatium, quia supra vel ultra 55ra I posset Deus creare fabam, dico quod verum est. Et tunc extra esset spatium. | Sed illud spatium non esset nisi magnitudo fabae quae ante non erat.

⟨1.2⟩ Ad aliam dico quod illa faba nec crearetur in indivisibili nec in divisibili spatio, quia non esset in loco nec in aliquo spatio, sicut nec totalis

---

1 concluditur] BHILMPTUp : conceditur C : continetur A ‖ tangant] tangerent I 2 inter] praeter I ‖ hoc ... est] ad hoc forte plus requiritur I 3 habeant] habeat P 4 unius et partium] *om.* (*hom.*) p 5 tangat] tangit IPp ‖ haberent se] *inv.* Pp : se habent I 8 iterum creare] *inv.* I ‖ qui] cui I 10 dico] *corr. sup. lin. ex* iterum C : ego credo IPp ‖ lapis ... maior] lapidem in centuplo maiorem *et add. in marg.* alias lapis maior C 13 se invicem] *corr. ex* se C : se P : scilicet p 15 sunt] *post* intelligendum p 17 ad primam] *om.* Pp ‖ esset] est IPp 18 sed] *om.* P 20 posset] potest I 21 esset] est P 22 opinor] *add.* etiam p ‖ non sit ibi] ibi non sit p : ibi non est P 24 tunc] *add.* ergo Pp ‖ rationes quae arguunt] rationem quae arguit p 25 posset] potest I 26 sed] et p 28 divisibili spatio] *inv.* IPp

mundus est modo in loco vel in aliquo spatio. Sed collective loquendo omnes magnitudines partium mundi sunt omnia spatia.

⟨1.3⟩ Ad aliam concedo quod Deus posset istam fabam movere elongando eam continue a caelo. Sed hoc esset creando spatium in quo moveretur et per quod haberet certum situm ad caelum et ad partem caeli. Sed postquam Deus absolveret fabam habendo certum situm ad caelum, ipsa nec esset prope caelum | nec longe a caelo. 73ra P

⟨1.4⟩ Ad aliam | dicitur sicut ad primas rationes. 84ra C

⟨1.5⟩ Ad aliam diceretur de illis mundis, sicut diceretur de lateribus orbis lunae annihilatis contentis.

⟨1.6⟩ Ad aliam dictum est quomodo latera orbis lunae non distarent.

⟨1.7⟩ Ad aliam dictum est quod non esset vacuum intra concavum orbis lunae, sed illud concavum esset vacuum, quod modo est plenum. Idem enim est locus et plenum, et ille idem locus esset tunc vacuus; ideo vacuum esset una res magna et notabilis, quae esset caelum.

⟨1.8–1.9⟩ De aliis rationibus dictum fuit satis in positione.

Protestor ut prius quod praedicta non dico nec dixi assertive, sed disputative, movendo dubitationes, ut de his docear ab aliis veritatem.

Haec de quaestione etc.

---

1 in^{1}] *add.* aliquo P ‖ in aliquo] *om.* P 2 mundi] *om.* I 4 caelo] polo P 8 primas] priores IPp 11 latera orbis] orbes P 13 concavum] continuum IP 15 res magna] *inv.* I ‖ notabilis quae] notabilis quae (*corr. ex* quod) C : notabilis quia p : mobilis quia P 16 positione] *add.* quia P 17 quod] *om.* I ‖ assertive] affirmative *sed add. in marg.* seu assertive C 18 de his] eis P 19 haec de quaestione] et sic sit finis P : et sic est finis I : *om.* p

⟨III.16⟩

⟨Utrum linea aliqua gyrativa sit infinita⟩

Quaeritur sexto decimo utrum linea aliqua gyrativa sit infinita; et semper accipio 'infinitum' categorematice.

Et non facio vim ad praesens utrum sit dare lineas indivisibiles secundum latitudinem et profunditatem vel non, quia si concedantur, tunc quaestio plane procedit, et si non concedantur, tunc per 'lineam gyrativam' intelligemus partem corporis columnaris procedentem in superficie columnae circa columnam non ad punctum a quo imaginaretur incipere, sed ad alterum punctum ulterius imaginatum in directum columnae; quae quidem pars esset parvae latitudinis et parvae profunditatis. Et iterum linea gyrativa vel pars gyrativae, de qua est difficultas, imaginaretur procedere circa columnam per medietates proportionales columnae, ita quod prima gyra esset praecise circumeundo columnam incipiendo a principio columnae et terminando ad finem primae medietatis, et secunda gyra esset incipiendo continue a prima gyra terminando ad finem secundae medietatis proportio-
58va p nalis, et sic | deinceps procedendo per medietates proportionales. Et certum est quod istae medietates proportionales sunt sibi invicem inaequales in tantum quod prima medietas est dupla ad secundam et secunda dupla ad tertiam et sic deinceps. Et ideo, si non ponantur lineae indivisibiles secundum latitudinem et profunditatem, sed quod omnes sint aliquantae grossitiei, tunc oportebit dicere quod, si prima gyra sumatur certae grossitiei, non est possibile quod omnes aliae sint aequalis grossitiei, quia esset devenire tandem ad medietates proportionales quae non essent tantae grossitiei vel
73rb P latitudinis. Tamen si, quanto minorantur proportionales medietates, tanto | proportionaliter imaginetur minorari grossities illius partis gyrativae, tunc

3 quaeritur sexto decimo] consequenter quaeritur I ‖ linea aliqua] *inv.* IPp 5 et non] nec p 7 tunc] tamen IP 7–8 intelligemus] *corr. ex* intelligeremus C : intelligeremus I 8 superficie columnae] *add.* vel p : superficiem columnae P : superficie I 9 imaginaretur] imaginatur I 11 esset] erit I ‖ linea] *add.* esset p 12 gyrativae] gyrativa P ‖ imaginaretur] imaginarentur P 21 sed] *om.* P ‖ omnes sint] omnis sit IPp 23 devenire] evenire I 24–25 non … minorantur] *in marg. inf.* C 25 quanto] quantum I ‖ proportionales medietates] *inv.* IP 26 proportionaliter] *om.* I ‖ imaginetur] imaginentur P ‖ illius partis] *inv.* p

 | DOI: 10.1163/9789004322356_019

nihil obstabit, quin idem sit iudicium de finitate vel infinitate secundum longitudinem lineae gyrativae, sive ponatur indivisibilis secundum | 84rb C latitudinem et profunditatem sive non, sed ponatur proportionaliter diminui secundum grossitiem, sicut diminuuntur medietates proportionales. Ideo procedemus in proposito ac si essent puncta indivisibilia et lineae nullam habentes latitudinem et profunditatem, quia facilius est loqui et quoad propositum revertitur eadem sententia.

Tunc ego suppono quod columna sit ita grossa quod quaelibet gyra sit de longitudine pedali vel maiori. Quo supposito ego arguo quod circa illam columnam esset linea gyrativa infinita quia:

⟨1⟩ Linea composita ex infinitis pedalibus non participantibus est infinita; sed talis est linea gyrativa circa istam columnam procedens dicto modo per medietates proportionales; igitur etc. Maior videtur nota de se. Minor probatur quia: cum infinitae sint medietates proportionales columnae *b*, si esset linea transiens dicto | 55rb I modo per omnes medietates proportionales ipsius columnae, ipsa contineret infinitas gyras pedales vel maiores; igitur ipsa esset infinita; et tamen sic est una linea transiens gyrative per omnes istas medietates; igitur ista est infinita, scilicet secundum longitudinem. Et ego probo assumptum, scilicet quod sit una linea transiens per omnes medietates etc., quia: aliqua est quae transit non solum per duas medietates proportionales, sed per tres et quattuor et decem et centum et mille; et non est aliqua ratio, quin semper debeat esse una transiens per omnes.

⟨2⟩ Iterum idem probatur quia: linea *b* transit per duas medietates et ultra et per centum et per mille et sic sine statu, quia quemcumque statum tu dares, adhuc si essent ultra medietates proportionales, ista procederet per istas; igitur non est status, quin ipsa procedat per omnes.

⟨3⟩ Iterum cum sit linea continua et una procedens per primam medietatem et secundam et tertiam et usque ad centesimam et millesimam, quid obstaret, quin esset una procedens per omnes? Et ad quotam medietatem

---

1 quin] quid P ‖ idem sit] *inv.* IPp **3** proportionaliter] proportionabiliter I **4** grossitiem] grossitatem Pp **4–5** ideo procedemus] ideo procedamus P : iam procedemus p **5** ac] *om.* P ‖ puncta] *om.* P ‖ nullam] naturaliter P **8** tunc ego] tunc ergo ego p : item ergo P **9** quo] pro I ‖ ego] ex P : *om.* I **10** esset] est IPp **13** de] per Ip **14** cum] *om.* C ‖ sint] sunt P ‖ proportionales columnae] *inv.* P **15** medietates proportionales] *inv.* p **16** ipsius] illius IPp ‖ contineret] continet I **17** tamen] *om.* p **18** ista] ipsa I ‖ scilicet] *om.* p **19** sit una] *om.* P **20** transit] *add.* sic CP **21** et mille] *om.* p **22** semper] *om.* P **23** idem] *om.* P ‖ quia] quod C **23–24** ultra et] ultra etiam p : *om.* I **24** tu] *om.* P **25** ista] ipsa IPp ‖ per] secundum P **26** ipsa] illa IPp **28** et^{3}] *om.* P **29** esset una] *inv.* P

esset standum? Nulla potest de hoc assignari ratio. Igitur aliqua est linea una quae continue procedit per omnes.

Nec valet cavillatio qua dicunt aliqui quod bene est una quae per omnes procedit, sed nulla est quae procedat per omnes. Sufficit enim concedere quod est una quae per omnes procedit, quoniam ista esset infinita et ex infinitis pedalibus composita; non enim per omnes procederet, si solum per centum aut per mille procederet aut quantumcumque per alium numerum finitum.

73va P ⟨4⟩ Item ad nullum terminum est talis | linea terminata a parte post; igitur est infinita. Consequentia videtur de se nota. Antecedens probatur quia: non est terminata ad ultimum punctum columnae, quoniam ad illud punctum numquam possunt pertingere medietates proportionales consequenter se habentes (semper enim quaelibet relinquit ultra se aliam medietatem
84va C sibi aequalem); sed etiam ista linea non esset terminata ad aliquod | punc-
58vb p tum citra ultimum, quia quodcumque demonstretur, adhuc ultra pro|cedunt medietates proportionales per quas est una linea gyrativa a principio columnae incipiens et transiens per eas et omnes praecedentes.

Oppositum arguitur per Aristotelem, qui omnino videtur negare actu infinitum, et tamen secundum praedicta argumenta ista poneretur actu infinita.

Et etiam esset mirabile quod infra terminos certos comprehenderetur longitudo infinita.

Ista quaestio est mihi difficilis. Et est supponendum quod in sexto libro determinabitur, scilicet quod nullum continuum est compositum ex indivisibilibus; ideo omnis linea et omnis pars lineae est divisibilis.

Et tunc statim ponuntur conclusiones.

Una est quod incipiendo ab uno cono columnae *b*, procedendo versus alterum conum per medietates proportionales consequenter se habentes,

---

1 potest] *post* assignari P ‖ assignari] reddi p 2 omnes] *add.* medietates I 5 ex] *add.* illis p 6 si] sed C ‖ solum per] *om.* p 7 procederet] *ante* aut[1] I ‖ alium numerum] *inv.* p 11–12 punctum] *add.* columnae p 14 aliquod] aliquem P 15 demonstretur] *add. sup. lin.* seu daretur C : demonstratur P : demonstraretur I : detur p 17 et[1]] *sup. lin.* C : *om.* IPp 19 secundum] scilicet p ‖ infinita] infinitum p 20 terminos certos] *inv.* IPp 21 longitudo] multitudo p ‖ infinita] *add.* actu quare etc. p 22 mihi] valde P : *add.* valde Ip 25 statim ponuntur] statim (*om.* P) sequuntur IPp 26 una] prima p

---

18 Cf. Aristoteles, *Physica*, III, 6, 206a18–21, 206b12–13; 7, 207b10–15 22 Cf. Iohannes Buridanus, *Quaestiones super libros Physicorum*, VI, q. 4 (ed. Parisiis 1509, ff. 96rb–98va)

nulla est ultima medietas proportionalis, quia quaelibet medietas proportionalis relinquit ultra se aliam medietatem sibi aequalem, quae est divisibilis iterum in duas medietates, quarum prima est proportionalis cum praecedentibus. Sit igitur ultima medietas proportionalis in consequenter se habentibus, quae vocetur *a*. Constat quod ultra eam est alia medietas sibi aequalis quae contra ipsam fuit divisa, et ista est divisibilis in duas medietates, quae sunt *c* et *d*. Et iam medietas *c* est proportionalis praecedentibus consequenter se habens ad *a*; igitur *a* non erat ultima medietas proportionalis.

Ex hoc sequitur secunda conclusio: ponendo quod columnae *b* unus terminus sit *a* et alter sit *c* et quod incipiamus medietates proportionales ab *a* versus *c*, tunc est conclusio quod nulla est secundum illum processum medietas proportionalis attingens ad terminum *c* nec etiam quae sit propinquior ipsi *c* quam aliqua alia, quia illa esset ultima et dictum est quod nulla est ultima. Et quia aliqui cavillando volunt dicere quod est aliqua attingens vel terminata ad *c* exclusive et non inclusive, ideo dicamus quod nulla est attingens ad corpus extrinsecum quod columna tangit nec aliqua est illi propinquior quam alia. Et hoc sufficit nobis.

Tertia conclusio est: posito quod aliquod mobile inciperet moveri super columnam *b* incipiens a termino *c* procedendo versus *a*, tunc ad nullam | 55va I medietatem proportionalem secundum prius dictam divisionem illud mobile prius attingeret quam ad aliquam aliarum. | 73vb P Illa enim esset ultima in ordine praedicto, quod dicebatur esse impossibile.

Quarta conclusio est quod mobile sic incipiens moveri super columnam *b* nullas duas vel plures istarum medietatum proportionalium aeque cito incipit attingere, sed semper unam prius quam aliam, quia nullae duae sunt aeque propinquae puncto *c*, immo semper secundum infinitam divisionem prior est remotior a puncto *c* et | 84vb C posterior est sibi propinquior; et tamen propinquiorem et remotiorem non incipit simul attingere, sed prius propinquiorem quam remotiorem; igitur nullas plures simul incipit attingere.

---

1 ultima medietas] *inv.* P ‖ quia] et *sed add. sup. lin.* quia C **1–2** proportionalis relinquit] proportionalis reliquit p : relinquit P **4** medietas] *om.* C **5** eam] *add.* non P **6** ista est] *inv.* P **7** c[2]] *om.* p **8** habens] *add. in marg.* alias habentibus C : habentibus P ‖ erat] erit P **8–9** proportionalis] *om.* P **11** sit[2]] *om.* P **12** illum] *sup. lin.* C : *om.* IPp **15** est aliqua] *inv.* P **16** c] *om.* P **17** attingens] *add.* vel terminata P ‖ nec] *corr. sup. lin. in* quia C : quia P : *om.* p ‖ est[2]] *sup. lin.* C : *om.* Ip **19** est] *om.* I ‖ posito quod] quod si P **20** incipiens] incipiente P **22** illa enim] quia illa IPp **23** praedicto] prius dicto IPp ‖ impossibile] possibile C **26** incipit] inciperet P **28** remotior] *add. inf. lin.* divisioni C ‖ est[2]] et P **30** nullas] nullo modo P

Quinta conclusio est quod, si aliqua una linea per omnes medietates proportionales lineae *b* pertransit gyrative modo praedicto, ipsa est infinita secundum longitudinem, quia ex infinitis pedalibus vel maioribus est composita, sicut arguebatur; et non potest assignari quod ad aliquem terminum sit terminata.

Sexta conclusio est quod, si est una linea gyrativa sic infinita ex una parte, ita oportet concedere quod aliqua est infinita ex utraque parte, quia in columna *b* possumus signare punctum medium, quod sit *d*, et ab illo puncto versus *a* incipere medietates proportionales procedendo in infinitum versus *a*, et alia in infinitum versus *c*, quae copulantur et uniuntur in puncto *d*;
59ra p et sic ista linea totalis esset infinita ex utroque latere. Et ita | sequeretur quod corpus infinitum secundum longitudinem continebitur infra certos terminos, scilicet inter *a* et *c*, quae sunt termini columnae *b* finitae.

Istas conclusiones pono tamquam veras vel tamquam quas credo esse veras, demonstratas vel demonstrabiles. Sed quia in casu possibili posito in esse quarta conclusio posita est categorica, licet duae sequentes sint condicionales, ideo est magna dubitatio aliquibus quando lapis ille qui tangebat columnam in puncto *c* incipit tangere partes proportionales procedentes ordinate a puncto *a*, vel quando incipit tangere aliquam vel plures earum.

Et ad hoc oportet respondere, cum non ponamus in tempore instantia indivisibilia, quod oportet exponere hoc verbum 'incipit' vel negative per tempus in quo non tangit illas, vel affirmative per tempus in quo tangit illas. Haec autem duo tempora sunt immediata sibi invicem et continua. Tempus enim quietis est immediatum et continuum tempori motus; et in toto tempore quietis nihil tangit iste lapis de illis partibus proportionalibus, sed in toto tempore motus tangit plures earum.

Dicam igitur pro septima conclusione quod in ultima parte temporis in
74ra P quo quiescit incipit tangere istas partes proportionales exponendo | negative, quia in toto illo tempore non tangit et in omni tempore immediate sequente tangit.

---

1 medietates] partes p 5 terminata] terminatum I 6 est^{1}] *om.* I ‖ una^{1}] *om.* IPp 8 b] *om.* I ‖ punctum] *om.* I 11 totalis] *add. sup. lin.* a C : a totalis P ‖ esset] *add.* una Pp : *add.* una linea I ‖ sequeretur] sequitur p 12 continebitur] continuabitur p 13 scilicet] *add. in marg.* infra seu C ‖ c] b Ip ‖ quae] qui P ‖ finitae] *corr. in* infinitae C : infinitae P 14 pono] *post* veras^{1} p ‖ veras^{2}] *add.* et IPp 16 categorica] *add. in marg.* alias categorice (ca^{che}) C : categorice P 18 procedentes] praecedentes p 19 earum] alias P 23 illas] eas IPp 24 immediatum et continuum] continuum et immediatum p ‖ tempori] tempore P 25 tangit] tangeret P 27 septima] alia P

Sed dico pro octava conclusione quod exponendo affirmative incipit tangere istas medietates proportionales in prima parte temporis in quo illud movetur, quia in omni tempore in quo movetur super istam columnam | 85ra C tangit aliquam vel aliquas earum medietatum proportionalium; et tamen quacumque prima parte temporis illius accepta, verum est quod immediate ante non tangebat aliquam istarum medietatum proportionalium.

Nona conclusio est quod in infinitis temporibus diversis iste lapis incipit tangere istas medietates proportionales, quia infinitae sunt ultimae partes temporis quietis et infinitae etiam sunt primae partes temporis motus, sicut dicebatur de terminis lineae in octava quaestione huius tertii libri; et tamen, si 'incipit' exponatur negative, tunc in omni ultima parte temporis quietis incipit tangere, et si exponatur affirmative, tunc in omni parte prima temporis motus incipit tangere; igitur in infinitis partibus temporis, quarum quaelibet est tempus, incipit tangere.

Decima conclusio est quod, quandocumque iste lapis incipit tangere unam illarum medietatum proportionalium, tunc incipit tangere plures. Hoc patet per tertiam conclusionem, quia ad nullam prius attingit quam ad aliquam aliam, et quia non in instanti indivisibili incipit attingere, cum non sint talia instantia, sed in tempore divisibili | 55vb I incipit attingere; et omne tempus est divisibile in partem priorem et partem posteriorem; et cum in parte priori attingat aliquam partem, in parte posteriori attingit aliam. Et sic etiam patet quod haec decima conclusio non est contra quartam conclusionem, quia licet in eodem tempore plures incipiat attingere, tamen non simul nec aeque cito, sed unam prius et aliam posterius, quia in omni tempore est prius et posterius.

Has igitur conclusiones posui tamquam non mihi dubias, sive sint verae sive falsae.

Tamen dubium est de expositione negativa, quae videtur contradicere quartae conclusioni; nam in omni ultima parte temporis quietis incipit tangere non solum aliquam, sed plures, quia in omni tempore quietis non tangit et in omni tempore immediate sequente tangit plures; ideo simul

---

1 conclusione] *add.* scilicet p ‖ affirmative] *om.* P 2 istas medietates] *inv.* I 4 et] *om.* P 5 prima] *om.* p 7 est] sequitur IPp ‖ in] *om.* C 9 etiam sunt] *inv.* P ‖ sicut] ut P 10 quaestione] conclusione C 12 prima] *ante* parte Pp : *om.* I 13 motus] *om.* P 15 iste lapis] *inv.* P 16 incipit tangere] *om.* P 17 hoc] haec p ‖ attingit] tangit P 18 attingere] tangere I 19 sint] sunt P ‖ divisibili] indivisibili *sed del.* P 20 partem[2]] *om.* P 21 attingat] attingit P ‖ partem] *add.* et P ‖ attingit] attingat P 22 et sic etiam] sic P 25 et] *om.* p 28 quae] *corr. ex* quod C : quia IPp 30 aliquam] aliquem CP

---

10 Cf. sup., III, q. 8, 85–87

incipit tangere plures, quia in omni tempore ultimo quietis incipit tangere 59rb p plures, et non in aliqua parte illius temporis magis unam quam aliam. |

Respondeo primo quod non credo istam expositionem negativam esse propriam, sed semper debere exponi 'incipit' et 'desinit', scilicet per 'primo esse' et 'ultimo esse'. Et si concedatur exponi per 'non esse', tamen ista exponens, scilicet 'in tempore immediate erit', habebit veritatem secundum prius et posterius in proposito.

74rb P Sed adhuc est mihi ignotum de quaestione principali, videlicet utrum ista | dicta linea gyrativa sit infinita secundum longitudinem. Et non quaero utrum infinita est linea gyrativa capiendo 'infinitum' syncategorematice, quia sic statim ego ponerem pro undecima conclusione quod infinita est 85rb C linea gyrativa ad istum sensum quod, | quantacumque est linea gyrativa finita, alia est maior ea secundum longitudinem, quia si aliqua est decem gyrarum, alia est centum, et si aliqua est centum gyrarum, alia est mille, et sic sine statu; igitur non est dare tantam finitam, quin sit dare maiorem. Volo igitur inquirere de hac quaestione capiendo 'infinitum' categorematice.

Et pono istam duodecimam conclusionem quod nulla linea recta protracta in columna *b* de termino *a* versus terminum *c* est infinita, quia si attingat ad terminum *c*, erit isto termino terminata, et si non attingat ad ipsum, erit minor, et nulla quae est terminata aliqua minor est infinita; omnis enim infinita, si esset, esset quacumque finita maior.

Tertia decima conclusio est quod omni lineae gyrativae sic protensae per medietates proportionales correspondet linea recta protracta per easdem medietates proportionales. Hoc patet per inductionem, nam sicut per quamlibet medietatem proportionalem est vel potest imaginari una gyra, ita et una recta; et si duae gyrae protensae per duas medietates proportionales sunt una linea gyrativa, ita duae rectae sunt una recta (non enim minus unitae sunt duae rectae quam duae gyrativae), et si mille gyrae mille medietatum proportionalium sunt una gyrativa, ita mille rectae protractae per illas

---

2 parte] *om.* I 4 semper] *om.* P ‖ et desinit scilicet] *sup. lin.* C ‖ scilicet] *om.* Ip 5 concedatur] conceditur p ‖ ista] alia IPp 9 dicta] una I : *om.* Pp 10–12 capiendo ... gyrativa[1]] *om.* (*hom.*) p 12 quod quantacumque] quia quantacumque I : quia quantumcumque p 13 aliqua est] *inv.* P 15 sit] est P 16 de ... categorematice] capiendo infinitum categorematice de ista quaestione p 17 duodecimam] undecimam p 18 columna] columnam I ‖ de] *add.* isto I 18–19 attingat] attingit P 19 erit] erat p ‖ ad ipsum] *om.* P 20 terminata aliqua] *inv.* IPp 21 quacumque] quantacumque p 22 tertia decima] duodecima p ‖ est] *om.* Ip ‖ omni] omnes p 23 medietates] partes C 25 ita et] *inv.* P 26 gyrae] quae P 28 unitae] *post* rectae P ‖ gyrae] gyrativae *sed add. sup. lin.* alias gyrae C 29 illas] easdem P

mille partes sunt una recta, et sic in infinitum; et omnino, si infinitae sunt gyrae, ita infinitae sunt rectae correspondentes illis gyris.

Ex hoc sequitur quarta decima conclusio, scilicet quod si sit aliqua una linea gyrativa protracta per omnes medietates columnae *b* modo praedicto et non ultra omnes, ita oportet esse unam rectam protractam per omnes istas medietates proportionales et non ultra omnes, quia sicut tu capis primam gyram protendi per primam medietatem et non ultra, et secundam per secundam et non ultra, et sic deinceps, ita ego accipio rectas correspondentes, scilicet quarum prima transit praecise per primam medietatem et non ultra, et secunda per secundam et non ultra, et sic deinceps. Et sic manifestum est quod numquam tales rectae vel aliqua ex eis composita transeunt vel transit ultra omnes medietates proportionales. Et tamen, si aliqua gyrativa composita ex omnibus gyris transit per omnes, ita oportet quod una recta ex omnibus illis rectis composita transeat per omnes, quia oportet semper esse correspondentiam, ut dicit praecedens conclusio; igitur etc.

Quinta decima conclusio est quod nulla est linea recta una pro|tracta per 56ra I omnes istas medietates, nisi sit protracta | ultra omnes, quia si sit protracta 74va P per totam columnam usque ad terminum *c*, ita quod tangat corpus extrinsecum columnae quod columna tangit, ipsa est protracta ultra omnes, cum nullae sic attingunt ad terminum *c*, ut dictum fuit in secunda conclusione. Si vero sic non sit protracta usque ad terminum *c*, tunc cum non | sit infinita, 85va C ut dicebat duodecima conclusio, sequitur quod est ad aliquem alium terminum infra *c* sic terminata, ita quod inter ipsam et corpus extra quod columna tangit sit aliquid de columna; sed ultra omnem terminum infra *c* sunt aliquae medietates proportionales, cum adhuc inter illum terminum et *c* sit aliqua pars columnae divisibilis in medie|tates; igitur ista non esset protensa per 59va p omnes medietates proportionales. Igitur nulla est recta protracta per omnes quae non sit protracta ultra omnes. Et si esset aliqua talis, ipsa esset infinita;

---

1 et omnino] *om.* P 2 gyrae ... sunt] gyrae ita infinitae †...† *in marg.* C ‖ correspondentes] correspondens p 3 quarta decima] decima tertia p 4 protracta] pertracta p ‖ medietates] *add.* pertractas p 5 ita] illas p ‖ protractam] pertractam p 6 istas medietates] *inv.* I ‖ sicut] si C 6–7 primam] unam p 8 ego] *om.* I 10 et sic deinceps] *om.* P ‖ sic[2]] *om.* P 11 tales rectae] talis recta I 13 composita] *post* gyris IPp 14 rectis] *om.* p ‖ semper] *om.* P 15 igitur] *om.* I 16 quinta decima] decima quarta p ‖ recta] gyrativa C ‖ protracta] pertracta p 17 protracta[1]] pertracta p ‖ si sit] sicut P ‖ protracta[2]] pertracta p 19 protracta] pertracta p ‖ cum] quod P 20 nullae sic attingunt] mille sic attingant p : nulla sit I 22 dicebat duodecima conclusio] dici conclusio duodecima p 23 sic] *om.* I ‖ ita] sic P 24 sed] *add.* licet P ‖ c] se P 25 cum] *corr. sup. lin. in* tamen C : tamen P 26 divisibilis] divisibiles p ‖ esset] est p

quod est contra duodecimam conclusionem. Consequentia patet, quia nec est terminata ad terminum *c* nec ad alium terminum, sicut statim dicebatur.

Ex hac conclusione et praecedente sequitur sexta decima conclusio quod nulla est linea gyrativa una protracta dicto modo per omnes medietates proportionales columnae *b*, quia nulla talis est protracta ultra omnes, ut apparet per casum; et tamen non est protracta per omnes, nisi sit protracta ultra omnes, prout apparet ex duabus conclusionibus praecedentibus; igitur nulla gyrativa est protracta per omnes. Sed bene aliqua est protracta per duas, aliqua per centum, aliqua per mille, et sic de quantocumque numero.

Ex his infero principalem conclusionem, scilicet quod nulla linea gyrativa protracta per medietates proportionales columnae est infinita secundum longitudinem, quia nulla poneretur infinita, nisi esset protracta per omnes; et nulla est protracta per omnes; igitur etc. Concedo tamen quod per omnes protracta est una linea, quia ibi iste terminus 'linea' non supponit confuse, sed supponit determinate, cum dico quod linea est protracta per omnes vel etiam quod linea est per omnes protracta.

Et ego etiam corollarie concludo quod similiter nullum tempus est perpetuum vel infinitum, licet concederemus cum Aristotele semper et aeternaliter fuisse unum tempus continuum, ita quod bene concederemus aeternum vel infinitum esse tempus, prout hae dictiones 'aeternum' et 'infinitum' sumerentur syncategorematice. Sed nullum tempus vel motum concederem infinitum. Et ita intelligendum esset quod prius diximus de usu praesentis temporis, scilicet quod omni tempore et perpetuo et infinito possumus uti pro praesente, sed non tempore infinito vel perpetuo. Verba enim aliquando,
74vb P si proferuntur praepostera, debent intelligi ad sensum qui approbabitur, | quando de his fiet specialis perscrutatio, et ex intentione etc.

---

2 est] *om.* IPp 3 et] *om.* P ‖ conclusio] *add.* scilicet Ip 5–6 ultra ... protracta[1]] *om.* (*hom.*) P 9 aliqua[1]] *add.* protenditur p ‖ quantocumque numero] quocumque numero p : quantocumque igitur P 10 principalem conclusionem] *inv.* IPp ‖ linea gyrativa] *inv.* P 14 non] *sup. lin.* C : *om.* Ip 16 etiam quod] *inv.* P 17 etiam] *om.* P ‖ concludo] *corr. in marg. ex* concedo C : concedo P 18 licet concederemus] sed concedamus P 20 et] vel p 21 sumerentur] sumuntur p ‖ concederem] concederemus Ip 22 prius diximus] *inv.* IP : dicimus prius p ‖ usu] verbis P 23 infinito] *add.* tempore IPp 24 praesente] *add.* tempore P 24–25 aliquando si proferuntur] si aliquando proferuntur Ip : si proferantur aliquando P 25 si] *sup. lin.* C ‖ approbabitur] approbatur p : appropriatur I : proponitur P 26 fiet] fit Ip ‖ etc.] *om.* IPp

---

18 Cf. Aristoteles, *Physica*, VIII, 1, 250b10–252a5; cf. Aristoteles, *Metaphysica*, XII, 6, 1071b6–9

Tunc ergo respondeo ad rationes.

⟨1⟩ Et dico quod nulla est talis linea gyrativa transiens per omnes medietates proportionales. Et quando arguebatur quod immo, quia aliqua est una quae transit non solum per tres aut per quattuor, sed | per centum aut per 85vb C mille, concedo. Et quantumcumque numerum dixeris, aliqua est quae procedit per tot.

Sed quando tu dicis quod nulla est ratio, si aliqua procedat per tot, quare non esset aliqua quae procedit per omnes, dico quod immo est valde magna ratio, quia valde bene concederem secundum propositionem copulativam quod aliqua est protensa per tres et aliqua per decem et aliqua per centum vel per mille et sic in infinitum, sed non concedam istam categoricam de copulato extremo 'aliqua est protensa per tres et per decem et per centum et sic sine statu'. Similiter ego bene concedam quod per omnes est aliqua protensa, sed nulla est per omnes protensa. Et iterum, licet aliqua sit protensa per centum et aliqua per mille et sic de quocumque numero, tamen non sequitur quod aliqua est protensa per infinitas vel per omnes, quia nullae sunt infinitae et nullae sunt omnes, sive sumamus 'omnes' collective sive distributive, sicut postea | videbitur. 56rb I

⟨2⟩ Ad aliam conceditur quod linea *b* transit per tres et per centum et per mille, sed nego illam clausulam 'et sic sine statu'. Quam tamen clausulam ego concederem, si tu procederes per copulativam et non per categoricam de copulato extremo. Et quando tu dicis quod, quantumcumque statum tu posses dicere, adhuc linea *b* transiret ultra, dico quod non, sed alia transiret ultra. Unde tu non potes signare lineam quae transit per omnes, quia nulla est, et quamcumque lineam tu | signaveris per determinatam suppositionem, 59vb p quae non transit per omnes, alia erit quae transit per plures.

⟨3⟩ Ad aliam quae quaerit ad quotam partem erit standum, dico quod una linea stat ad secundam medietatem, alia ad decimam, alia ad centesimam,

---

1 ergo ... rationes] ergo ad rationes respondeo P : ad rationes respondeo p : ad rationes respondendum est I 3 arguebatur] arguitur IPp ‖ est] *om.* P 4 per^2] *om.* IPp ‖ aut per] et per Ip : et P 5 et] quod P ‖ quantumcumque numerum] quaecumque numerorum p 7 procedat] procedit IPp 8 procedit] procederet IPp 9 concederem] concedam IPp ‖ secundum] istam p 10 et^1] *om.* P ‖ decem et] quattuor P 11 vel per] et per p : aliqua per P : vel I 12–13 centum et] *add.* per mille et P : mille et p 13 ego] *om.* Ip 14 protensa^2] *ante* per p ‖ iterum] *om.* P 19 conceditur] concedo IPp ‖ et^1] *om.* IPp ‖ et^2] *om.* p 21 ego] *om.* IP 22 quantumcumque] quemcumque p 23 dicere] discere I ‖ b] *om.* I 24 potes] posses p 25 signaveris] signares P

---

18 Cf. inf., III, q. 18

et sic in infinitum, sed nulla est per omnes. De cavillatione autem quae interponebatur non est vis, quia concedo eam.

⟨4⟩ Ultima ratio bene arguit quod, si esset aliqua protensa per omnes, illa esset infinita et ad nullum terminum terminata; sed nulla est talis etc.

---

1 autem] *om.* p 2 concedo] concedam C 3 ultima] alia P 4 etc.] quare etc. p : et sic est finis quaestionis I

⟨III.17⟩

⟨Utrum omni numero sit numerus maior⟩

Quaeritur septimo decimo utrum omni numero est numerus maior, supposito semper quod nullum continuum sit compositum ex indivisibilibus, sed quod omne continuum est divisibile, ita quod habet partes alias ab invicem et quod quaelibet suarum partium etiam est divisibilis sive habens partes.

Tunc arguitur | quod sic quia: 75ra P

⟨1⟩ In linea *b*, si aliquae partes sunt decem, aliae partes sunt plures, quia centum, et si aliquae sunt centum, aliae sunt mille, et sic in infinitum, quantumcumque numerum volueris accipere. | 86ra C

⟨2⟩ Item omni binario est numerus maior, scilicet ternarius, et omni ternario alius maior, scilicet quaternarius, et sic de omnibus aliis numeris; igitur etc.

⟨3⟩ Item Aristoteles ponit quod omni numero dato contingit maiorem dare. Et tamen secundum ipsum non potest esse tantus numerus, quin de facto sit tantus, scilicet loquendo de numero partium continui. Quamvis enim partes continui non sint ab invicem divisae per discontinuationem, tamen sunt tot sicut si essent divisae. Igitur omni numero dato est numerus maior.

⟨4⟩ Item loquendo syncategorematice infinitae sunt partes in continuo; igitur similiter loquendo syncategorematice infinitus est numerus earum. Et isto modo capiendo 'infinitum' exponatur 'infinitae partes' quia 'non tot quin plures'; et similiter 'infinitus numerus' quia 'non tantus quin maior'. Igitur etc.

---

3 *Inde ab hac quaestione denuo codicem G adhibuimus* ‖ est] *add.* aliquis P 4 sit] est GPp 5 habet] *post* alias P 6 etiam est] etiam C : est G 8 arguitur] *praem.* ergo P : ergo arguitur sic p 9 in] *om.* P ‖ partes[2]] *om.* P 10 et[1]] *om.* P 13 et sic] ita P 15–16 maiorem dare] *inv.* GPp 16 et] *om.* P 17 tantus] *add.* ergo G ‖ continui] *om.* G 18 sint] sunt GP 22 similiter loquendo syncategorematice] simili syncategorematice loquendo G ‖ earum] eorum G ‖ et] in P 23 exponatur] exponitur GPp

---

15 Cf. Aristoteles, *Physica*, III, 7, 207b1–3, 10

 | DOI: 10.1163/9789004322356_020

Oppositum arguitur quia:

⟨1⟩ Si omni numero esset aliquis numerus maior, oportet vel fingere vel dicere quis est iste; sed quicumque dicatur, sequitur inconveniens, scilicet quod iste esset se ipso maior.

70ra G

⟨2⟩ Item si esset omni magnitudine aliqua magnitudo maior, ista esset infinita; igitur similiter, si omni numero esset aliquis maior, iste esset infinitus. Igitur si probatum est quod nullus numerus est | infinitus, sequitur propositum quod non omni numero est numerus maior.

Nunc igitur probo quod nullus numerus est infinitus. Et ad hoc Aristoteles ponit unam rationem, talem scilicet: numerus ex eo dicitur numerus, quia est numeratus vel numerabilis; sed infinitum non est numeratum neque numerabile; igitur nullus numerus est infinitus.

Item confirmo quod nullus numerus est infinitus quia: differt concedere quod est numerus infinitus et quod potest esse numerus infinitus, quia concedere quod est numerus infinitus est concedere quod est infinitum in actu, non solum in potentia; sed Aristoteles semper negavit infinitum in actu, licet concederet infinitum in potentia; igitur etc.

Bene difficile est mihi loqui in ista quaestione de proprietate sermonis. Sed primo excludo aliqua de quibus non est magna difficultas.

Numerus enim secundum Aristotelem vel est ratio animae discretiva qua numeramus alias res intelligendo quot ipsae sunt, vel est ipsae res numeratae, ut quod quattuor homines essent quaternarius.

Item si 'numerum' capiamus pro rebus numeratis vel numerabilibus, hoc potest esse dupliciter. Uno modo quod solum loquimur de rebus ab invicem separatim existentibus, ita quod nec una sit pars alterius nec plures sint

---

2 esset] est GPp 3 quicumque dicatur] quicumque detur p : quiscumque dicatur G : quantumcumque dicamus C 6 similiter] *om.* P ‖ omni numero esset] omni numero est G : est omni numero Pp ‖ esset²] est GPp 7 est¹] fuerit P ‖ sequitur] sequeretur P 8 propositum] *add.* scilicet GP ‖ numerus] aliquis GPp 9 igitur] ego P : *add.* ego p ‖ aristoteles] *ante* ad GPp 10 talem scilicet] *inv.* GPp 11 est²] esset G ‖ neque] nec GPp 13 confirmo] confirmatur GPp ‖ numerus est] numerus sit GP : sit p 14 quia] et P 16 semper negavit] *corr. in* solum negat C : solum negat P 17 licet] sed P ‖ igitur etc.] etc. p : *om.* P 18 bene] unde P ‖ mihi] *om.* GPp ‖ in] de p 19 primo] *add.* ego Pp 20 animae discretiva] *inv.* GPp 21 numeramus] numerat G ‖ quot] quod P 22 essent] *add.* quattuor G : esset C 23 numeratis] numerandis G 24 loquimur] loquamur GPp

---

9–10 Cf. Aristoteles, *Physica*, III, 5, 204b7–10 16–17 Cf. Aristoteles, *Physica*, III, 6, 206a18–21, 206b12–13; 7, 207b10–15 20 Cf. Aristoteles, *Physica*, IV, 11, 219b5–7

partes eius|dem totius, ita quod nihil loqueremur de partibus continui. Et 60ra p tunc statim esset dicendum quod aliquis est numerus rerum quo nullus est maior secundum multitudinem. Non enim est infinita | multitudo rerum 75rb P separatarum a magnitudine. De rebus autem magnitudinem habentibus inter eas quae separatim existunt de facto est aliqua minima, scilicet qua nulla alia sit minor | separatim existens; aliter continuum esset divisum 86rb C in infinitum. Minima autem concessa non est possibile esse infinitatem secundum multitudinem rerum quarum quaelibet esset tanta vel maior, nisi ex eis poneremus resultare magnitudinem actu infinitam, quam non ponimus.

Alio modo possumus 'numerum' accipere pro rebus numerabilibus non ab invicem separatis per discontinuationem, sed tamen ab invicem diversis, sicut sunt partes magnitudinis continuae. Et de istis est difficultas de qua volumus tractare. Et ideo, quandocumque in ista quaestione loquemur de numero sive universaliter sive particulariter, non intelligemus nisi de numero magnitudinum vel continuorum vel partium suarum. Hoc protestemur.

Item notandum est quod simpliciter et proprie loquendo illud dicitur maius alio vel minus, quod est eo maius vel minus secundum magnitudinem corpoream. Alia autem dicuntur maiora vel minora secundum quid, scilicet cum additione, ut maius secundum longitudinem vel latitudinem et maius secundum multitudinem, scilicet quia sunt plura, maius secundum intensionem vel durationem etc.

Hic autem non intelligimus nisi de maiori vel minori simpliciter, scilicet secundum magnitudinem, vel de maiori et minori secundum multitudinem, id est de pluribus vel de paucioribus.

Et tunc de maiori vel minori simpliciter pono conclusiones faciles.

Prima est quod nulla est unitas indivisibilis, immo omnis unitas est divisibilis et etiam quaelibet pars eius est divisibilis. Ego enim non loquor nisi

---

1 loqueremur] loqueremus C 2 est[2]] *praem.* numerus P : *add.* numerus Gp 5 aliqua] *add.* de facto Gp 7 autem concessa] quo concesso p 8 rerum quarum] Ox : quarum ABCGILNQSTUZp : quare P : secundum quam DErFHJKLaMR : quam EXY : qua BrO : quia W : quod ? scilicet ? V : *deest* Pb 9 nisi] nonne P : ubi G ‖ quam] quod GPp 11 numerabilibus] numeralibus p 13 sicut] *om.* p 14 et] *om.* P 15 intelligemus] intelligimus p 16–17 protestemur] protestamur Pp 18 est] *om.* Gp ‖ dicitur] dicimus P 21 scilicet] et Pp : *om.* G ‖ et] *om.* Pp 23 intensionem] intentionem C 24 maiori vel minori] maiore vel minore G : maiori C 25 et] *corr. sup. lin. ex* vel C : vel GP 26 id est] scilicet p ‖ de[2]] *om.* GP 28 prima] *add.* conclusio Gp 29 et] *om.* P ‖ est] *om.* Gp

de magnitudinibus et partibus earum. Unitas autem est idem quod res una, sicut multitudo et numerus sunt res multae et numeratae; ideo omnis uni-
70rb G tas est magnitudo, et omnis magnitudo est sic divisibilis, | ut dicetur in sexto libro.

Secunda conclusio est quod omnis binarius est ternarius et centenarius, quia cum totum sit suae partes, sequitur quod quaelibet unitas est quinquaginta unitates; ideo binarius est centum unitates et sic est centenarius. Sed tamen ista conclusio non convertitur universaliter, quia non omnis ternarius vel centenarius est binarius, quia tres homines sunt ternarius et non binarius; et in magnitudine continua cuius sunt decem decimae non participantes prima decima et tertia decima et sexta decima sunt ternarius, et tamen non sunt binarius, quia sunt tres unitates quarum nullae duae sunt unitas propter partes interiacentes.

Tertia conclusio est quod omnis unitas est binario maior et centenario, quia omnis unitas est magnitudo secundum dicta. Si igitur tu vis ponere instantiam contra dictam universalem conclusionem, sit ista magnitudo
75va P *b* quam non dicis esse ternario | maiorem. Et ista habet decem decimas, quarum tres primae sunt ternarius. Sed isto ternario magnitudo *b* est maior, quia omne totum est maius sua parte et iste ternarius est pars magnitudinis *b*. Et ita posset argui de centenario et millenario.

Et ita manifestum est quod illae sunt concedendae 'binarius est centenario maior' vel 'ternario aequalis' et huiusmodi, quia si magnitudo *a* sit
86va C aequalis magnitudini *b*, | scilicet quod utraque est pedalis, tunc *a* est binarius et *b* est ternarius; sed *a* et *b* sunt aequales; igitur binarius et ternarius sunt aequales (dico: ad invicem). Similiter *a* est binarius et medietas ipsius
60rb p *b* est centenarius; et tamen *a* | est duplum medietati ipsius *b*; igitur binarius est duplus centenario, scilicet maior in duplo. Et sic posset dici de omni proportione numerali.

---

4 libro] huius p 6–7 quinquaginta] quotcumque P 7 ideo binarius] ergo binarius et G : et p 8 ista] *om.* P 9 non] *add.* sunt Pp 11 et^1] *om.* p 11–12 et tamen] cum P : et p 13 unitas] unitates p 17 non dicis] *inv.* Gp : tu dicis non P 19 est^1 ... parte] sua parte est maius P 20 posset] possit P 21 sunt concedendae] *inv.* p : *add.* omnis P 22 vel ternario] *del. et add. in marg.* binarius vel ter†...† centenario C : binarius est centenario vel ternario p : binarius vel ternarius est centenario P ‖ et huiusmodi] *del.* C : *om.* P 23 magnitudini] magnitudine P : magnitudinis p ‖ est^1] sit GPp 24 est ternarius] ternarius G : est centenarius P 25 dico ad] dico ab C : ab P ‖ similiter ... et] a est binarius similiter P : *add.* capitur G ‖ ipsius] *om.* P 26 et] *om.* P 28 proportione] proportionali C ‖ numerali] naturali G

4–5 Cf. Iohannes Buridanus, *Quaestiones super libros Physicorum*, VI, q. 4 (ed. Parisiis 1509, ff. 96rb–98va)

Sed tunc oritur magna difficultas. Cum enim dictum sit quod omnis binarius est ternarius, et tamen omnis binarius est numerus par et omnis ternarius est numerus impar, sequitur quod numerus par est numerus impar. Et quia hoc nomen 'impar' est privativum, sequitur 'impar, igitur non par', sicut sequitur 'caecus, igitur non videns', 'iniustus, igitur non iustus'. Sed iterum sequitur 'par est non par, igitur par non est par' per illam regulam in *Peri hermeneias* 'ab affirmativa de praedicato infinito sequitur negativa de praedicato finito'. Igitur de primo ad ultimum sequitur, si binarius est ternarius, quod par non est par, quod est impossibile.

Propter solutionem huius dubitationis est sciendum quod indubitanter, si hoc nomen 'impar' ponatur privative opponi huic nomini 'par', tunc nihil idem est par et impar, quia sequitur 'impar, igitur non par'. Et si iste modus sumendi 'impar' concedatur, tunc pono quartam conclusionem quod nullus numerus est impar (et dico semper de numeris magnitudinum), quia omnis numerus est par; igitur nullus est impar secundum dictam significationem. Quod autem omnis numerus sit par probatur quia: omnis ternarius est centenarius, qui est par; igitur omnis ternarius est par. Et similiter omnis quinarius est denarius, qui est par; et sic omnis quinarius est numerus par. Et sic de aliis. Ergo omnis numerus est numerus par et sic nullus est impar. Et hoc videtur esse concedendum de virtute sermonis.

Similiter de aequalitate vel inaequalitate possunt poni conclusiones de virtute sermonis.

Et erit quinta conclusio quod nullum aequale est inaequale et nullum aequale alicui est inaequale alicui, sed omne alicui aequale est alicui inaequale et e converso. Et tu vides vel faciliter videre potes causam in hoc; ideo dimitto rationem. Et licet etiam nulla aequalia sint inaequalia, tamen ali-

---

1 cum ... sit] *in marg.* C : *om.* G ‖ quod] quod *vel* quia C : quia G 4 et] *om.* P 5 sequitur] *om.* GPp ‖ iustus] *add.* etc. GPp 6 par[1]] quod P 7 ab affirmativa] *praem.* quod P : quod ad affirmativam p 7–8 sequitur ... finito] sequitur negativa de finito G : ad negativam de praedicato finito valet consequentia P 9 par[2]] *om.* P ‖ est[2]] *om.* P 10 solutionem huius dubitationis] huiusmodi dubitationis solutionem C ‖ est sciendum quod] sciendum est quod GP : sciendum est p 11 nomini] non G 12–13 quia ... impar] *om.* (*hom.*) P 13 sumendi] supponendi C ‖ tunc] *add.* ego Pp 14 et] *om.* GPp ‖ numeris] numero GPp 16–17 ternarius est centenarius] *corr. in marg. ex* numerus est senarius C : ternarius est senarius GPp 17 similiter] sic P 19 et[1] ... par] *in marg.* C ‖ de] *add.* omnibus Gp 20 et] *sup. lin.* C : *om.* Gp 21–22 similiter ... sermonis] †...† aequalitas vel inaequalitas †...† conclusiones de virtute sermonis *in marg.* C 23 erit] est G 24–25 alicui[2] ... inaequale] *om.* P 25 potes] *ante* videre P : posses p 26 licet] *om.* p ‖ sint inaequalia] sunt inaequalia p : sint G

7–8 Cf. Aristoteles, *De interpretatione*, 10, 20a20–21

quibus aequalia sunt aliquibus inaequalia. Et ego bene concederem quod aliqua aequalia ad invicem sunt aliqua inaequalia sive non aequalia ad invicem, 70va G quia duae medietates | lineae *b*, quae sunt ad invicem aequales, sunt una tertia et una duplex tertia, quae sunt eiusdem lineae et quae sunt inaequales 75vb P et non aequales | ad invicem.

Sexta conclusio est quod non sunt plures partes vel pauciores in linea *b* quam in eius medietate vel e converso, quia sicut in linea *b* sunt centum, sic in eius medietate sunt centum; et sic de mille et de quacumque multitudinis quantitate.

Item capiatur circulus parvus circa polum ultimae sphaerae, cuius diameter sit solum pedalis, et capiatur circulus aequinoctialis. Sit igitur parvus circulus *a* et circulus aequinoctialis *b*. Constat quod non sunt plures partes in circulo *b* quam in circulo *a* nec e converso. Probo quia: uterque in 86vb C eodem tempore | movetur et simul incipiunt et desinunt unam circulationem et continue moventur sine interruptione. Tunc ponatur gratia exempli quod praecise in uno die perficiant huiusmodi circulationem et tempus illius diei vocetur *c*. Constat tunc quod non sunt plures partes in circulatione *b* quam in tempore *c*, nec sunt plures partes in spatio pertransito per circulationem *b*, si imaginemur spatium quiescens pertransiri, quam in tempore *c*, quia cuilibet parti circulationis vel spatii correspondet aliqua pars temporis. Dato enim quod sit aliqua pars circulationis vel spatii cui non correspondet aliqua pars temporis, tunc ista pars circulationis vel spatii fieret vel transiretur in non tempore, et sic esset motus in non tempore, quod est impossibile. 60va p Sed etiam probatur quod non sunt plures in tempore *c* quam in circu|latione vel parvo circulo *a*, quia cuilibet parti temporis *c* correspondet aliqua pars circulationis *a* vel spatii pertransiti; aliter esset aliqua pars illius temporis, in qua circulus *a* non moveretur, et ideo non moveretur continue. Igitur de primo ad ultimum non sunt plures partes in magno circulo *b* quam in parvo circulo *a*.

---

1 bene] etiam Pp 3 aequales] aequalia P 4 quae sunt[1]] *in marg.* C : *om.* Gp ‖ et[2]] *sup. lin.* C : *om.* Gp 4–5 inaequales] aequales p 6 sexta] nona C 7 sicut] si Pp ‖ sic] ita GPp 8 et[1]] *om.* P ‖ et[2]] *add.* sic G 8–9 multitudinis quantitate] *inv.* GPp 11 solum] *om.* p 11–12 parvus circulus] *inv.* Pp 13 circulo[1]] circulos p ‖ nec] et sic C ‖ probo] probatur P : probatio p ‖ uterque] utriusque G 14 simul] similiter p 15 moventur] moveretur G ‖ ponatur] ponamus Gp 16 in uno] una P ‖ illius] huius P 17 constat tunc] *inv.* GPp 18 nec] tunc non G 19 imaginemur] imaginetur GPp ‖ quiescens] *add.* et P 21 dato] da Pp ‖ correspondet] correspondeat GPp 22 vel spatii] *om.* G 24 probatur] probabitur GP 25 vel] *add.* in Gp ‖ parvo circulo] *inv.* GPp ‖ cuilibet] *om.* G 26 temporis] *om.* G 27 qua] AT : quo BCGHLMPUp ‖ circulus ... et] non moveretur circulus a GPp 29 a] c G

Septima conclusio est quod nulla tria sunt plura duobus, immo etiam nulla tria sunt aliquibus duobus plura (dico semper: in magnitudinibus), quia si tria sunt tria, ita duo sunt tria, et si tria sunt centum, ita duo sunt centum et mille.

Octava conclusio est quod nullus numerus est alio numero maior secundum multitudinem. Ista sequitur ex praecedente; nam centenarius non est maior secundum multitudinem binario, quia sicut centum sunt centum, ita duo sunt centum.

Item idem non est se ipso plura vel pauciora secundum multitudinem; sed idem est centenarius et binarius; igitur etc.

Istis conclusionibus positis oportet videre quomodo principia supposita in arithmetica et conclusiones ibidem probatae et plurima saepe dicta et supposita ab Aristotele debeant intelligi, quae omnino et simpliciter videntur contrariari conclusionibus praedictis. Supponit enim arithmeticus unitatem esse indivisibilem. Et Aristoteles in isto tertio dicit quod rationabile est in omni numero terminum esse ad minimum, ad plus autem semper excellere omnem multitudinem. Et dicit causam huius esse, quia unum est indivisibile, quodcumque sit; propter quod necesse est divisionem numeri stare ad indivisibile. Et in quinto *Metaphysicae,* capitulo de uno, dicitur quod secundum quantum omnino indivisibile et nullo modo divisibile sunt | punctum 76ra P
et unitas. Item arithmeticus dividit numerum in pares et impares et reputat membra divisionis non coincidere, quia definit 'par' et 'impar' definitionibus contrariis et repugnantibus, scilicet 'par' dividi in partes aequales nulla unitate remanente et 'impar' non sic dividi. Et ita supponit nullum numerum parem esse imparem. Et omnem numerum binarium credit esse parem et

1 septima] alia P : nona C ‖ etiam] et p 3 tria[2]] *om.* G ‖ si[2]] sic P 5 octava] decima C 6 ista] ita G ‖ praecedente] praecedenti G 7 quia] quoniam Gp 10 centenarius et binarius] binarius et centenarius GPp 12 et[1]] *add.* quomodo G 12–13 supposita] sumpta G 13 omnino] omnia p 14 arithmeticus] aristoteles P 15 dicit] *om.* C 16 minimum] *add. in marg.* alias numerum C ‖ semper] *ante* autem G : *om.* P ‖ excellere] *add.* autem G 17 multitudinem] magnitudinem P ‖ quia] quod P 18 numeri] *om.* G 19 quinto] *add.* etiam G 20 indivisibile] *add.* est CG 21 numerum] numeros GPp ‖ et[2]] *add.* in Gp 23 contrariis et] *om.* Pp 23–24 par ... impar] *corr. ex* per dividi in partes aequales nulla unitate remanente et per C : per dividi in partes aequales nulla unitate remanente et per p 24 non sic] *inv.* P 25 binarium] *rep.* G ‖ esse[2]] *add.* numerum GPp

15 Cf. Aristoteles, *Physica*, III, 7, 207b1–3, 7–10 19 Cf. Aristoteles, *Metaphysica*, V, 6, 1016b23–27

omnem ternarium esse imparem, et sic credit nullum binarium esse ternarium. 70vb G Et supponit etiam omnem quaternarium esse binario du|plum et per consequens maiorem. Et omnino perirent omnes propositiones arithmeticae, si binarius esset ternarius et centenarius.

87ra C Et | propter ista notandum est cum diligentia quod iste terminus 'numerus' supponit pro pluribus rebus ab invicem distinctis, etiam licet sumeretur cum pronomine demonstrativo, ut 'iste numerus'. Aliter pro nullo supponeret. Et connotat quod ista multa sunt numerata vel numerabilia, id est sunt scita vel scibilia quot sunt, per rationem sive per conceptum animae discretivum, scilicet quo anima intelligit et intelligere potest ista plura discrete et divisim ab invicem. Et sic iste terminus 'numerus' connotat talem animae rationem. Et ideo bene dicit Aristoteles in quarto *huius* quod, si non posset esse numerans, scilicet anima, non posset esse numerus, quoniam de ratione numeri est quod sit numeratus vel numerabilis. Si enim non posset esse anima numerans, non potest esse illa ratio quam connotat 'numerus'; et deficiente connotatione termini terminus pro nullo supponit; ideo nullus esset numerus.

Sed tu obicies quia: si nulla anima intelligeret, tunc de facto nulla esset talis ratio; et tamen adhuc esset numerus; igitur 'numerus' non connotat talem rationem. Quod autem numerus esset, licet nulla anima intelligeret, manifestum est, quia adhuc Socrates et Plato essent duo homines; et non essent duo sine dualitate; et tamen omnis dualitas est numerus, immo est praedicatio quidditativa.

Ad hoc faciliter respondetur quod iste terminus 'numerus' non connotat 60vb p quod res pro quibus suppo|nit sint actu numeratae et scitae quot sunt, sed quod sunt numerabiles. Ideo 'numerus' non connotat quod res pro quibus supponit sunt numeratae actu, ut dictum est. Et ideo non connotat istam

---

1 omnem] *add.* numerum P : omne p ‖ esse[1]] *om.* GPp ‖ nullum] *add.* numerum P 2 duplum] duplo P 3 propositiones] proportiones Pp 4 centenarius] *add.* etc. Pp 5 et] *om.* Pp 8 sunt[1]] sint Gp ‖ id est] scilicet p ‖ sunt[2]] sint GPp 9 quot sunt] quod sunt P : quod G ‖ per[2]] *om.* P 10 intelligere] intelligi P 11–12 animae rationem] *inv.* P 12–13 posset] possit GP 13 posset] possit P 14 posset] possit P 15 potest] posset Pp 16 nullo] *add.* deficiente p ‖ ideo] *praem.* et p 22 est[1]] esset GPp 24 faciliter] *sup. lin.* C : *post* respondetur Pp : *om.* G 25 quot] quod sic P 26 sunt] sint GPp ‖ numerabiles] *add. sup. lin.* si sint C 27 sunt numeratae actu] sunt numeratae actu (*sup. lin.*) C : sint numeratae G : sint actu numeratae et scitae quot (quod P) sunt Pp ‖ ut dictum est] *om.* p ‖ connotat] connotant G

---

12 Cf. Aristoteles, *Physica*, IV, 14, 223a22–25

rationem animae discretivam tamquam existentem, sed tamquam possibilem existere. Ideo non dicit Aristoteles quod omnis numerus sit numeratus, sed numeratus vel numerabilis. Modo si non potest esse anima numerans, non potest esse ista ratio. Sed licet nulla esset anima numerans, tamen posset esse, quod sufficit ad hoc quod sit numerus.

Deinde etiam considerandum est quod, quamvis idem sit duo et tria, binarius et ternarius, tamen non secundum eandem rationem discretivam dicitur duo et dicitur tria vel dicitur binarius et dicitur ternarius. Ideo oportet intelligere quod omnem rem unam quae est multae partes nos possumus intelligere discernendo inter partes, scilicet intelligendo quod haec est et ista est et haec non est | ista, et sic intelligimus istas partes esse multa. Et (76rb P) cum scimus quod illud est suae partes, concludendo scimus quod illud est multa. Deinde etiam illud possumus intelligere conceptu privativo, scilicet privative opposito conceptui discretivo, scilicet quo intelligimus ipsum aliqua pluralitate non esse plura. Et secundum istam rationem dicimus ipsum esse unum unitate opposita huiusmodi pluralitati, ut hominem unum, quia non est plures homines, sed ipse et alius homo essent plures homines. Et ideo bene dicit Aristoteles decimo *Metaphysicae* quod isti termini 'unum' et 'multa' opponuntur privative. Et sic etiam in quinto *Metaphysicae* dicitur quod unum esse est indivisibile esse et quod quaecumque non habent divisionem, inquantum non habent divisionem, ut sic unum dicuntur. Ideo ad illum sensum omne unum | vel omnis unitas est indivisibilis, quia omne (87rb C) unum dicitur unum secundum istam rationem secundum quam intelligitur esse et tamen non esse plura talia vel tanta vel taliter se habentia, quale ipsum est vel quantum vel qualiter se habens, sicut homo dicitur unus homo, quia non | est divisibilis, sed indivisibilis in plura quorum quodlibet sit (71ra G) homo. Sed Socrates et Plato sunt plures homines, quia secundum rationes

1 discretivam] distinctam P 2 numerus] *add.* non P 3 potest] possit P : posset p 4 potest ... ratio] possit esse ista ratio P : posset illa ratio esse p 4–5 posset] possit P 6 deinde] unde P ‖ sit] sunt P ‖ tria] *add.* et GPp 7 tamen non] *inv.* P 8 et dicitur[2]] vel dicitur P : et G ‖ ideo] ergo GPp 9 nos] *om.* P 10 scilicet] sed G 11 et[2]] *om.* P ‖ multa] multas GP 12 illud[1]] idem P ‖ suae partes] *inv.* p ‖ illud[2]] idem P 13–14 scilicet privative] *om.* p 16 huiusmodi] huius P 17 et[2]] *om.* P 19 et[2]] *om.* P ‖ etiam] *om.* p 20 unum] uni Pp ‖ quod quaecumque] quicumque P 21 habent] *om.* P 22 sensum] *om.* G ‖ vel] ut P 23 quam] *add.* ipsum Pp 24 tanta vel taliter] taliter vel tanta P 25 vel[2]] aut GPp ‖ sicut] sic G : *om.* P 26 est] *om.* p

2 Cf. Aristoteles, *Physica*, IV, 11, 219b6 18 Cf. Aristoteles, *Metaphysica*, X, 3, 1054a23–24 19 Cf. Aristoteles, *Metaphysica*, V, 6, 1016b23–25

discretivas sunt divisibiles in plura quorum quodlibet sit homo. Et linea pedalis dicitur una pedalis, quia non est divisibilis in plures quarum quaelibet sit pedalis, licet sit divisibilis in plures semipedales. Et multa etiam difformia sunt unus exercitus vel unus populus, quia non sunt divisibilia in plura quorum quodlibet sit exercitus, vel saltem, si sint divisibilia in plura quorum quodlibet est exercitus, tamen non in plura quorum quodlibet sit talis vel tantus exercitus.

Et omnino oportet considerare in homogeneis, quorum quaelibet pars recipit denominationem totius, ut quod quaelibet pars quantitativa aquae est aqua. Dicam igitur quod ea ratione aliqua aqua dicitur una aqua, quia est aqua et non est divisibilis in plura quorum quodlibet sit tanta aqua vel taliter se habens. Et ad talem sensum concedendum est quod omnis unitas est indivisibilis, quia non est divisibilis in plura quorum quodlibet sit tale et tantum et taliter se habens sicut ipsa est, quamvis sit divisibilis in plura minora vel sibi dissimilia aut diversimode se habentia. Sic signum in caelo non est divisibile in signa, sed in gradus, nec gradus in gradus, sed bene in minuta. Et haec sint dicta de divisibilitate unitatis.

Deinde ulterius considerandum est quod eandem rem possumus intelligere secundum rationes discretivas suarum partium valde diversas. Nam alia est
76va P ratio secundum | quam discernimus inter medietates lineae *b*, et alia secundum quam discernimus inter eius tertias. Secundum igitur istam rationem
61ra p secundum quam discernimus inter medietates | lineae *b* et non plus, dicimus quod linea *b* est duo vel binarius, et numquam secundum istam rationem dicimus quod ipsa sit ternarius. Sed secundum istam rationem secundum quam discernimus inter tertias eius, dicimus quod ipsa est tria vel ternarius, et numquam secundum istam rationem diceremus quod ipsa sit binarius. Et tamen non obstat, quin idem sit binarius et ternarius, quia diversae ratio-

---

1 et] etiam P 2 quarum] quorum p 3 etiam] et G 4 difformia] *praem.* valde P : valde distincta p ‖ quia] *add.* sunt exercitus et Pp 5 sit] est Pp ‖ sint] sunt GPp 6 non] *add.* est p 8 oportet] esset C 9 denominationem] praedicationem GPp 10 est] sit P ‖ dicam] dico G ‖ ea] eadem p 11 est[2]] *om.* Gp ‖ quodlibet] quaelibet P ‖ tanta] *post* aqua[2] P : tota G 14 et[1]] *om.* Pp ‖ et[2]] vel P 15 aut] *om.* P ‖ sic signum] signum enim GPp 16 sed[1]] *add.* bene Pp 17 minuta] *add.* etc. GPp ‖ sint dicta] sunt dicta P : sunt G ‖ divisibilitate] diversitate C 18 ulterius] *post* est GP 19 alia est] *inv.* P 20 medietates lineae b] eius medietates G ‖ lineae] *sup. lin., post* b C ‖ alia] *add.* est P 21 inter] *add.* medietates P 23 linea b] *sup. lin.* C : *om.* G ‖ numquam] non P 24 dicimus] diceremus GPp ‖ istam] *om.* P 25 eius] omnes p 27 et[1]] *om.* P ‖ quin idem] quod ipsa P ‖ binarius et ternarius] ternarius et binarius G

nes terminorum non impediunt veritatem propositionis, si termini supponunt pro eodem. Dicam igitur quod ad istum sensum nulla duo sunt tria, quia nulla secundum eandem rationem discretivam dicuntur duo secundum quam dicuntur tria vel e converso. Dicam igitur quod quaelibet duo sunt quibuslibet tribus pauciora ad istum sensum quod paucior est discretio secundum quam dicuntur duo, quam secundum quam dicuntur tria. Anima enim dividendo duo non discernit nisi inter hoc et illud, et sic de aliis. Et si dicamus 'ille debet mihi centum poma', non oportet intelligere quodlibet eorum seorsum, sed sufficit intelligere quod ista | possumus discrete et seor- 87[va] C
sum intelligere et numerare secundum rationem tot unitatum. Ad hoc enim quod sunt centum non oportet esse illam rationem discretivam, sed sufficit quod possit esse et quod esset in re correspondentia illi rationi, si esset, sicut satis dictum est prius.

De paritate et imparitate dicendum est quod, licet in quolibet numero infinitae sunt unitates, tamen in ternario non sunt nisi tres unitates inter quas ratio discretiva sit, secundum quam haec dicuntur tria, ut si linea *b* est tres tertiae, quaelibet istarum et nulla alia pars est una; inter quas discernit anima intelligendo et discernendo quod ista linea est tres tertiae. Dicam igitur quod, licet simpliciter et de proprietate sermonis nullus | numerus sit 71[rb] G
impar, quia omnis numerus est par, tamen ad illum sensum dicitur numerus impar, quia rationes inter quas discernit ratio discretiva, secundum quam dicitur talis numerus, sunt impares, quia non possunt dividi sufficienter in duas partes, quin una istarum partium contineat plures rationes istarum unitatum quam alia. Et ad istum sensum diceremus omnem ternarium vel quinarium esse numerum imparem, et omnem binarium vel quaternarium esse numerum parem, et nullum numerum parem esse imparem, et sic de aliis quae supponunt arithmetici.

1 si] *add.* terminus sive p 1–2 supponunt] supponant p 7 dividendo] *corr. in* in dividendo inter C : dicendo p ‖ et[1]] *om.* P 8 mihi] *om.* G ‖ intelligere] *add. in marg.* alias discernere C 10 et numerare] *sup. lin.* C : *om.* G 11 sunt] sint GPp ‖ esse illam] *inv.* P ‖ rationem] *add. sup. lin.* esse C 12 correspondentia] correspondenti p 14 paritate] *add.* ergo p ‖ licet] quodlibet P 14–15 infinitae sunt] infinitae sint p : infinitae G 16 sit] scit C ‖ b] *om.* G 17 tres] tertia G ‖ quaelibet] qualibet G ‖ alia pars] *inv.* G ‖ una] *add.* unitates P : *add.* unitatum p 18 discernendo] HMT : dicendo BLPUp : dividendo ACG ‖ tres] tertia *sed del.* G ‖ dicam] dico GP 19 proprietate] virtute P 21 rationes] *add.* unitatum Pp 24 sensum] *add.* secundum G 25 numerum] *om.* G 26 esse[1]] *om.* GPp ‖ numerum[2]] *om.* p 27 supponunt arithmetici] supponit arithmeticus p

Sed tunc etiam considerandum est ad quid tales modi loquendi secundum diversas rationes proficiunt. Ad hoc dicendum est quod scientia numerandi, scilicet arithmetica, inventa est principaliter et finaliter ad mensurandum 76[vb] P motus et magnitudines et tempora. Unde naturali ordine | geometria supponit arithmeticam secundum Aristotelem. Continuum enim magnum non potest sciri quantum sit, nisi discernendo partes eius seorsum notae quantitatis, ut pedales vel ulnales, et numerando eas, ut dicamus pannum esse trium ulnarum et columnam decem pedum et spatium centum leucarum et pondus decem talentorum et sic de aliis.

Et sciendum est quod per haec nunc dicta ultra ea quae prius dicta fuerunt possunt reddi causae multorum dici consuetorum, videlicet quod omnis binarius omni binario est aequalis et omni ternario minor, et quod omnis quaternarius est omni binario duplus, et nullus numerus par est impar etc. Haec enim omnia intelligunt mathematici et naturales de numeris rerum eiusdem rationis, quia alia non sunt proprie ad invicem comparabilia, sci- 61[rb] p licet secundum cer|tam proportionem numeralem. Et sic debet intelligi de numeris unitatum ad invicem aequalis quantitatis, ut si cuiuslibet unitatis quantitas fuerit unius pedis vel unius ulnae vel unius leucae vel unius diei vel unius sextarii aut unius talenti et sic de aliis. Sic enim binarius non est ternarius, quia duae ulnae non sunt tres ulnae nec duo talenta tria talenta, et omnis ulna una est indivisibilis secundum longitudinem in plures ulnas. Nec mathematici curant de numeris unitatum diversarum rationum vel etiam 87[vb] C inaequalis quantitatis ad invicem nisi forte inquantum illae | unitates essent reducibiles ad unitates aequalis quantitatis vel eis secundum certam proportionem proportionales.

---

1 etiam] *om.* P ‖ secundum] *add.* tales Pp 2 hoc] quod GPp ‖ dicendum est] dicitur P 2–3 numerandi scilicet] *om.* P 3 est] *om.* G 4 motus et magnitudines] magnitudines et motus p ‖ ordine] *add.* doctrinae p 5 enim magnum] *corr. in marg. ex* in magnitudine C : enim P 6 seorsum] *add.* sint C 7 ulnales] ulnares P ‖ numerando] *corr. sup. lin. ex* numerare C : numerare Gp 8 trium] viginti Pp : triginta G 9 pondus ... et] *in marg.* C : *om.* (*hom.*) G 11 multorum] multarum G ‖ videlicet] scilicet P 12 minor] *corr. in* maior C : maior P 13 duplus] duplex p ‖ etc.] *om.* P 14 numeris] numero G 16 numeralem] HMTU : numerabilem B : naturalem LPp (*sup. lin.*) C, *add.* ut habetur septimo huius Pp : *om.* AG ‖ sic debet] etiam debet G : etiam debent Pp 17 quantitatis] qualitatis P ‖ si] *add.* a G 18 quantitas] *om.* P ‖ ulnae] lineae p 18–19 unius diei vel] unius diei aut p : diei vel G : unius dei vel C 19 aut ... aliis] etc. (†...† unius talenti et sic de aliis *in marg.*) C : etc. G 20 nec ... talenta] †...† talenta sunt tria talenta *in marg.* C : *om.* G 21 ulnas] *add.* ut P 25 proportionales] proportionalem G

---

5 Cf. Aristoteles, *Analytica posteriora*, I, 27, 87a34–35

Istis visis bene concedendum est cum Aristotele quod omni numero est numerus maior secundum multitudinem ad istum sensum quod quotarumcumque unitatum potest esse ratio discretiva, plurium potest esse ratio discretiva. Et sic etiam concederetur quod infinitus est numerus capiendo 'infinitum' syncategorematice, quia ad praedictum sensum non est tantus, quin possit esse maior. Sed non sequitur inde quod numerus est infinitus, sicut non sequebatur, si infinita sit linea gyrativa, quod linea gyrativa sit infinita etc.

Et per hoc solvantur rationes hinc inde adductae, quia procedunt viis suis, si aliquis voluerit intendere.

Haec de quaestione etc.

---

2 multitudinem] *add.* scilicet Gp ‖ quod] quia GPp 2–3 quotarumcumque] C ? : quarumcumque Pp 3 unitatum ... discretiva] *om.* G 6 possit esse] sit GPp ‖ inde] *om.* p 7 si] scilicet G ‖ sit[1]] est Pp ‖ linea[1]] *in marg.* C : *om.* G 7–8 quod ... infinita] *in marg.* C : *om.* G 8 etc.] *om.* Pp 9 hoc solvantur] hoc solvuntur G : dicta possunt solvi P : praedicta solvi possunt p ‖ inde] *add.* prius Pp ‖ quia] quae p 10 aliquis] *om.* G 11 haec ... etc.] et sic finitur quaestio G : et sic est finis P : etc. sequitur quaestio decima octava p

⟨III.18⟩

⟨Utrum in quolibet continuo infinitae sint partes⟩

Sed ut magis videamus de ista dictione 'infinitum' syncategorematice sumpta, ideo decimo octavo quaeritur utrum in quolibet continuo infinitae sunt partes, verbi gratia utrum infinitae sunt partes in hac linea pedali, et utrum infinitae sunt partes in hac die vel in hac hora, et similiter utrum in hac linea pedali vel in hac hora infinitae sunt partes.

Arguitur quod non per rationes quae alias fiebant et quae reservabantur ad istam quaestionem quia:

⟨1⟩ Sequeretur quod esset infinita magnitudo secundum extensionem, ita quod esset extensa sine terminis. | 77ra P Consequens reputatum est falsum in praecedentibus quaestionibus. Consequentia probatur quia: ex quibuscumque duabus partibus magnitudinis | 71va G simul appositis resultat maior extensio quam ex una istarum, et sic pari ratione ex tribus quam ex duabus, et sic esset in infinitum; igitur ex infinitis resultaret infinita.

⟨2⟩ Et hoc iterum confirmatur fortius quia: si linea esset composita ex punctis indivisibilibus, ita quod duo puncta aut tria redderent aliquantam extensionem, sequeretur quod infinita redderent infinitam extensionem. Hoc supponatur. Quod ex hoc apparet, quia si duo puncta redderent aliquantam extensionem, quattuor redderent duplam, et sic omnino infiniti binarii punctorum redderent infinitam extensionem. Sed hoc supposito dicemus etiam quod quaecumque duae magnitudines divisibiles redderent extensionem maiorem quam duo puncta indivisibilia. Igitur multo magis

---

3–4 sed ... sumpta] *om.* P 4 ideo ... quaeritur] quaeritur decimo octavo P : potest quaeri decimo octavo G : quaeri potest decimo octavo p ‖ infinitae sunt] sint infinitae G 5 verbi gratia] *om.* G 6 vel] *om.* P ‖ et similiter] similiter P : et universaliter C 7 infinitae sunt] sint infinitae GPp 10 sequeretur] sequitur p 11 terminis] termino G ‖ reputatum est] reputatur esse GP 12 quia] *om.* GPp 13 appositis] compositis p ‖ resultat] *add.* magis sive p 14 quam[1] ... istarum] ex una illarum quam ex alia illarum P 15 resultaret] resultarent Gp 17 aut tria] *sup. lin.* C : vel tria G 18 sequeretur] sequitur p 18–20 sequeretur ... extensionem] *in marg. inf.* C : *om.* (*hom.*) G 18 infinita] *add.* puncta p 19 quod] quia p ‖ redderent] reddunt p 22 etiam] *om.* P ‖ duae] *om.* G 23 maiorem] *in marg.* C : *ante* extensionem GPp

---

8 Cf. sup., III, q. 14, 122^{13-21}

 | DOI: 10.1163/9789004322356_021

infinitae partes magnitudinis divisibiles redderent infinitam extensionem, quod est falsum.

Item ad principale sic arguitur quia:

⟨3⟩ si in linea *b* essent infinitae partes, vel hoc esset actu vel potentia solum. Non est dicendum quod actu, quia numquam concessit Aristoteles quod actu est infinitum. Unde ipse descripsit infinitum per esse in potentia ad ulterius accipiendum, scilicet cuius quantitatem accipientibus semper contingit ultra | accipere. Sed etiam similiter, si non concederet quod in 61va p actu, non erit concedendum quod in potentia. Probo quia: frustra esset potentia quae non aliquando posset esse reducta ad actum. Secundo quia: quaecumque possunt esse partes huius lineae, illae iam sunt partes eius. Da enim quod | alia pars nova adderetur, tunc illa linea *b* non esset illa quae 88ra C resultaret ex tali additione, sed esset pars eius.

⟨4⟩ Item omnes partes continui sunt numerus. Hoc patet per inductionem in qua non reperitur instantia; istae enim sunt duae et istae tres etc. Sed nullus est infinitus numerus, cum omnis numerus sit numeratus vel numerabilis, ut dicitur in isto tertio et in quarto etc.

⟨5⟩ Item infinitis repugnat quod sint omnia accepta, cum dicat Aristoteles quod semper contingit ultra accipere; sed omnes partes huius continui sunt simul acceptae accipiendo hoc continuum totum; igitur non sunt infinitae partes.

⟨6⟩ Item infinitis secundum multitudinem non sunt plura, sicut infinito secundum magnitudinem nihil est maius; igitur non sunt infinitae partes in linea *b*. Consequentia ex hoc patet, quia plures sunt partes in lineis *a* et *b* quam in linea *b*. Probatio quia: sicut maius dicitur secundum quid nominis quod est tantundem et amplius, ita plura dicuntur, si sint illa et quaedam

---

1 magnitudinis divisibiles] magnitudinis divisibilis P : divisibiles G 3 sic] *om.* GPp 5–6 concessit ... est] aristoteles concessit (concessisset P) actu GPp 7 cuius] *add.* secundum GPp ‖ accipientibus] *rep.* P 8 similiter] *om.* GPp ‖ concederet] conceditur GPp 9 erit] *add.* etiam P ‖ probo] *sup. lin.* C : probatio Pp : *om.* G 10 non aliquando posset] aliquando non posset p : aliquando non possit P 12 adderetur] addetur P ‖ illa linea] *corr. ex* ista C : linea G 16 sit] est P 17 et in] et p : vel P ‖ etc.] ergo etc. G : igitur P 18 infinitis] infinito P : infinitus G ‖ sint] sunt G 19 partes huius] huiusmodi partes G 20 totum] *add.* simul GPp 22 non] nulla GPp ‖ infinito] infinitis GPp 24 ex hoc] *om.* P ‖ quia] quod G ‖ lineis] *add.* scilicet P 25 b probatio] *om.* G ‖ dicitur] *om.* p

---

6–8 Cf. Aristoteles, *Physica*, III, 6, 207a7–8 17 Cf. Aristoteles, *Physica*, III, 5, 204b8; IV, 11, 219b5–7 18 Cf. Aristoteles, *Physica*, III, 6, 207a7–8

alia; modo partes linearum *a* et *b* sunt partes lineae *b* et cum hoc aliae, scilicet lineae *a*; igitur sunt plures.

77rb P ⟨7⟩ Item non sunt infinitae | partes eiusdem quantitatis, ut dicit Aristoteles; sed concludo: igitur non eiusdem proportionis. Et si hoc concludatur, tunc nullo modo debet concedi quod sunt infinitae. Consequentia prima et principalis probatur quia: si multiplicas partes proportionales, utputa medietates proportionales, ita multiplicabo partes eiusdem quantitatis ad invicem, quia si tu dividis lineam *b* in duas medietates, istae sunt aequalis quantitatis ad invicem; et si tu vis dividere secundam medietatem et procedas per medietates proportionales, ita volo dividere primam et sic erunt quattuor partes eiusdem quantitatis ad invicem; et si ultra tu dicis quod alia istarum dividitur, ita dicam de aliis tribus et resultabunt adhuc partes aequalis quantitatis; et sic semper.

⟨8⟩ Iterum sequitur quod impossibile esset omnes partes lineae *b* esse pertransitas et per consequens impossibile esset eam esse pertransitam. Consequens est falsum. Consequentia probatur quia: infinitorum non est aliquod ultimum, et tamen numquam est pertransitum aliquod totum, nisi perventum sit ad ultimum eius.

Oppositum arguitur quia:

⟨1⟩ Aliter non esset continuum in infinitum divisibile, et aliter non esset
71vb G divisibile | in semper divisibilia, et aliter esset compositum ex indivisibilibus. Et haec omnia sunt improbata in sexto *huius*.

⟨2⟩ Item non sunt tot quin plures; igitur infinitae. Antecedens patet, quia sicut sunt decem, ita sunt centum et mille, et si sunt mille, ita sunt decem milia, et sic semper in infinitum.

1 linearum] *om.* G 2 lineae] *om.* P ‖ igitur] *add.* non P 3 dicit] tenet p 4 concludatur] *corr. in marg. ex* concedatur C : concedatur Gp 5 sunt] sint GP ‖ infinitae] infinita CGp 6 multiplicas] simul multas P ‖ proportionales] proportionabiles G ‖ utputa] verbi gratia GP : ut p 6–7 utputa medietates proportionales] *in marg.* C 7 multiplicabo] proportionabo C : multitudo P 8 medietates] quantitates C 9 et[2]] *corr. sup. lin. ex* ut C : ut GPp 10 ita] *add.* ego Pp 11 dicis] vis P ‖ alia] altera p 12 dividitur] dividatur GPp ‖ ita] *add.* ego Pp ‖ adhuc partes] *inv.* P 14 sequitur] sequeretur GP 15 pertransitam] *add.* sed Pp 16 consequens est falsum] *in marg.* C : consequens apparet falsum Pp : *om.* G ‖ infinitorum] infinitarum p 17 aliquod[1]] aliquid P ‖ pertransitum aliquod totum] aliquod pertransitum totum Gp : aliquid pertransitum P 18 eius] *add.* ergo etc. p 22 improbata in] improbata Pp : probata G 23 patet] apparet GPp 24 sicut] *corr. in marg. ex* infinite decem (*seq. ras.*) C : si GPp ‖ sunt[4]] *om.* G 25 semper] *om.* p

3–4 Cf. Aristoteles, *Physica*, III, 6, 206b7–12; cf. Aristoteles, *Physica*, VIII, 10, 266b3 22 Cf. Aristoteles, *Physica*, VI, 1, 231a21–b18

⟨3⟩ Item si non essent infinitae partes in linea *b*, sequitur quod essent in duplo plures in duplicata linea, quod est falsum, sicut dictum fuit in quaestione praecedenti.

Ista quaestio continet multas difficultates.

Prima est, cum enim 'infinitum' sumatur categorematice et syncategorematice, quomodo utroque modo exponatur. Et quia nomina sunt ad placitum, multi ponunt expositiones sicut placet eis, et secundum exigentiam istarum oportet dicere consequenter, quia quid nominis est principium | 88rb C omnis doctrinae, ut dicit Aristoteles.

Et videtur mihi quod ca|tegorematice accipiendo 'infinitum' exponeret 61vb p Aristoteles in magnitudinibus per hoc quod est extensum sine terminis vel extensum non terminatum. Et sic non refert dicere 'infinita magnitudo' et 'magnitudo infinita', sicut non refert dicere 'homo albus' et 'albus homo'. Et utrobique iste terminus 'homo' supponit determinate sine aliqua confusione, si pro aliquo supponat, nisi in propositione interveniat aliud confundens. Et de hoc dictum est prius quod magnitudo nec est infinita nec infinita est magnitudo, nec linea gyrativa est infinita secundum longitudinem nec infinita est linea gyrativa | etc. Nec etiam tempus est infinitum nec infini- 77va P tum est tempus.

Et proportionaliter etiam aliqua dicerentur infinita secundum multitudinem categorematice loquendo, quia consequenter numerando non esset ultimum istorum. Et esset aequivalens dicere 'festucae sunt infinitae secundum multitudinem' et 'infinitae sunt festucae secundum multitudinem'. Et tunc ego credo esse ponendum quod nec infinita secundum multitudinem sunt aliqua nec aliqua sunt infinita secundum multitudinem, quia si aliqua

1 sequitur] sequeretur GP 2 duplicata] dupla P : duplicati G : duplici p 3 praecedenti] praecedente G 4 continet multas difficultates] multas continet difficultates P : multas difficultates habet p 5 enim] *om.* Pp ‖ infinitum] finitum G 6 quomodo] quaestio p ‖ quia] si G 7 sicut placet] *praem.* suas G : suas quales placent Pp 8 dicere consequenter] *inv.* GPp 10 accipiendo] sumendo P : capiendo p 10–11 infinitum exponeret aristoteles] infinitum aristoteles exponeret GP : aristoteles infinitum exponeret p 11 terminis] termino Gp : *praem.* termino vel extensum sine P 14 homo] *om.* P 17 est[1]] *om.* G 18 etc. ... tempus] etc. nec etiam quod tempus G : et quod tempus nec Pp 20 etiam] *om.* P 21 categorematice] *add.* sumendo vel P ‖ quia] *del.* C ‖ consequenter] *add. in marg.* alias assumendo (?) C 23 et[1] ... multitudinem[2]] et infinitae (infinita p) secundum multitudinem sunt festucae Pp : *om.* (*hom.*) G 24 tunc] *add.* etiam GPp 25 aliqua[3]] aliquando P

9 Cf. Aristoteles, *Metaphysica*, IV, 7, 1012a21–22; cf. Aristoteles, *Analytica posteriora*, I, 1, 71a12–16 16 Cf. sup., III, qq. 15–16

essent infinita secundum multitudinem, hoc maxime esset verum de partibus continuorum, ita quod haec esset vera 'partes vel aliquae partes lineae *b* sunt infinitae'; sed haec esset falsa, scilicet quod partes lineae *b* sunt infinitae. Probatio igitur quod haec sit falsa 'partes lineae *b* sunt infinitae' quia: quaereres quae sunt illae, et oportet hoc dicere, cum ille terminus 'partes' stat determinate; et hoc non potest dici, nam nec istae tres sunt infinitae, nec istae demonstratis mille, quia numerando veniremus ad ultimam istarum, nec istae decem, nec istae centum, et sic de aliis.

Respondetur quod nullae sic numeri determinati sunt infinitae, sed omnes simul sumptae sunt infinitae.

Contra hoc arguitur quia: si dicis 'omnes' distributive loquendo, totum est falsum, quia sic istae duae essent infinitae et istae tres, quod est falsum, quia est devenire numerando ad ultimam earum. Sed si etiam capiamus 'omnes' collective loquendo, adhuc propositio est falsa, quia nullae sunt omnes collective loquendo. Cum enim dicimus collective quod duodecim apostoli sunt omnes apostoli, sensus est quod isti sunt apostoli et quod nullus est apostolus qui non sit aliquis eorum. Sic autem nullae partes continui sunt omnes, quia quaecumque partes continui habent iterum alias partes in quas sunt divisibiles; et omnino praeter quascumque sunt adhuc aliquae aliae quae non sunt aliquae istarum. Et sic etiam dicendum est de medietatibus proportionalibus columnae *b*, scilicet quod nullae sunt omnes.

Et est notandum quod huius nominis 'infinitum' categorematice sumpti 88va C multae sunt proprietates, quae sibi convenirent, si pro aliquo supponeret. |

Primo hoc nomen 'infinitum' est privative oppositum huic nomini 'finitum', 72ra G sicut opponuntur 'terminatum' et 'non terminatum' vel | 'habens terminos ipsius' et 'carens terminis ipsius'. Et sic numquam possunt simul verifi-

---

2 ita] *om.* p 3–4 infinitae] *add.* igitur P : *add.* ergo etc. p 4 probatio] probo GPp ‖ sit] est P 5 quaereres] quaererem Pp 6 stat] supponit P : supponat p 7 nec ... mille] *om.* Gp ‖ nec ... quia] nam *sed del. et add. in marg.* nec istae sunt infinitae de†...† mille quia C 7–8 istarum] earum GPp 9 respondetur] *praem.* sed p ‖ nullae] *add.* partes G : mille p 11 dicis] dicas GPp ‖ distributive loquendo] *inv.* GPp 12 sic ... quia[2]] *om.* (*hom.*) G 13 devenire] evenire G ‖ etiam] *sup. lin.* C : *om.* GPp 14–15 adhuc ... loquendo] *om.* (*hom.*) P 14 propositio] *post* falsa p 15 collective[2]] *add.* loquendo GP 16 quod[2]] *om.* G 17 sit aliquis eorum] sit aliquis illorum G : est aliquis istorum P : aliquis illorum sit p ‖ sic] si P 18 habent] haberent p 19 sunt adhuc] sunt P : adhuc p ‖ aliquae] *in marg.* C : *om.* Gp 20 sunt] *om.* P ‖ aliquae] aliqua p 22 nominis] nomini P 22–23 categorematice ... sunt] categorematice sumpti (sumptum G) sunt multae Gp : sumpti categorematice sunt multae P 23 supponeret] supponerent GP 24 infinitum] finitum G 25 vel] *rep.* G

cari de eodem et pro eodem, quamvis etiam aliquid esset infinitum, ita quod nihil idem esset finitum et infinitum secundum magnitudinem vel finitum et infinitum secundum multitudinem vel secundum longitudinem vel durationem etc.

Item si esset infinitum secundum magnitudinem, nihil esset eo maius. Et non posset consumi per ablationem finiti quantumcumque multotiens factam, et ipsum esset omni finito maius, nec ex quacumque multitudine finitorum posset resultare. Et quantacumque magnitudo signaretur, infinitum | 77vb P infinitas tantas contineret non | 62ra p participantes. Et similiter, quantumcumque parva magnitudo signaretur, ex infinitis tantis non participantibus resultaret infinitum. Et ideo, si omnis quantitas aequalis grano milii vocetur *a* et omnis aequalis caelo vel mundo vocetur *b*, non essent in infinitum plura *a* non participantia quam *b*, licet quodlibet *b* contineret plus quam mille milia *a*. Nec tempus, si esset infinitum, contineret plures dies quam annos. Et hoc est rationabile, quia sicut infinito secundum magnitudinem non est maius, ita infinitis secundum multitudinem non sunt plura; et tamen tempus infinitum, si esset, contineret annos infinitos secundum multitudinem; ideo non contineret plures dies quam annos.

Et iterum probatur quia: si poneremus tempus praeteritum infinitum, nulli dies praeteriti essent plures quam anni praeteriti, nisi acciperentur a parte ante omnes dies, sic quod nulla dies eos praecessisset. Sed hoc est impossibile, quia nulli essent omnes dies praeteriti, sicut nullae medietates proportionales lineae *b* sunt omnes eius medietates proportionales. Sic bene dicit Commentator quod infinitum infinito non esset comparabile nec secundum maius nec secundum minus nec secundum aequale.

Sed ultra videtur mihi esse concedendum quod in infinito secundum magnitudinem, si esset, essent plura *a* quam *b*, ex quo quodlibet *b* contineret mille milia *a* secundum casum prius dictum; et similiter quod in tempore

---

2 idem esset] idem G : esset P ‖ et infinitum] *om.* p 3 et] vel P 3–4 longitudinem ... etc.] durationem vel secundum longitudinem etc. P : longitudinem durationem G 5 eo] esse p 6 quantumcumque] quamcumque P 7 nec] nam C ‖ quacumque] quantacumque GPp 7–8 finitorum] finitarum p 8 signaretur] assignaretur P 9 quantumcumque] quantacumque P : quamcumque G 12 vel mundo] *om.* P ‖ infinitum] infinito G 14 infinitum] *ante* si Pp 16 infinitis] infinito p ‖ et tamen] cum P 17 secundum] *add.* magnitudinem vel p 19 et] hoc P : *add.* hoc Gp ‖ tempus praeteritum] *inv.* P 21 eos] *ante* nulla Pp 23 sic] sicut P : *praem.* et Gp 24 infinitum] finitum p 26 mihi esse] mihi GP : ideo p 27 b^2] *om.* C

---

24 Cf. Averroes, *In Physicam*, III, comm. 43, f. 104G

infinito, si fuerit infinitum, plures fuerint dies quam anni, cum quilibet annus contineat plus quam trecentos dies.

Et ex hoc concluditur quod impossibile est esse magnitudinem infinitam vel etiam tempus infinitum sumendo 'infinitum' categorematice, quamvis etiam concederemus aeternaliter fuisse tempus et infinitum fuisse tempus sumendo 'infinitum' syncategorematice. Nam impossibile est ad quod sequuntur contradictoria; et tamen ad esse magnitudinem infinitam vel tempus infinitum sequuntur contradictoria, ut dictum est. Nec propter hoc est neganda perfectio Dei infinita, quia nos non loquimur hic nisi de finitate
88vb C vel infinitate quantorum divisibilium. |

Adhuc esset alia proprietas infiniti, si esset, scilicet quod non posset determinate et proprie assignari quantum ipsum esset, nec de multitudine quot essent, quia si quaereretur quantum est vel quot sunt, non potest aliter assignari responsio nisi quod ipsum est infinitum vel ipsa sunt infinita; et haec non est responsio propria et determinata, quia illa non est responsio determinata ad 'quantum est hoc?', quae non deberet mutari, si ab illo auferretur tanta magnitudo quantus est mundus; et tamen adhuc residuum esset infinitum, quia si esset finitum, tunc appositio mundi non reddebat totum infinitum.

72rb G Adhuc ex istis mihi videtur quod non est possibile esse | magnitudinem
78ra P infinitam, quia sequeretur quod totum | non esset maius sua parte, cuius oppositum supponitur specialiter de toto quantitativo. Consequentia patet, quia sit corpus infinitum *a*, cuius una pars finita sit tanta quantus est ille mundus, quae vocetur *b*, et residuum illius infiniti sit *c*. Constat quod *b* et *c* sunt partes illius *a*, ita quod utrumque est pars eius. Et tamen *a* non est maius quam esset *c*, si annihilaretur *b* vel circumscriberetur. Sicut enim *a*

1 fuerit] fuit p ‖ fuerint] fuerunt GP 5 concederemus] consideremus P ‖ et] *add.* in P 7 sequuntur] *add.* duo P ‖ et] *om.* P 9 est neganda] neganda esset P ‖ nos] *om.* GPp ‖ nisi] *om.* G 10 vel] et Pp ‖ quantorum] quantitativorum p 11 infiniti] *praem.* proprietatis G : *om.* P ‖ scilicet] *om.* GPp 12 determinate] determinari C ‖ assignari] signari P 13 quot^{1}] quod P ‖ quaereretur] quaeratur GPp ‖ potest] posset G 14 responsio] *om.* G ‖ responsio nisi quod] *corr. in marg. ex* quam respondendo C 14–15 responsio ... responsio^{1}] respondeo P 15 est^{2}] esset P 16 illo] ipso Gp : ipsa P 17 auferretur] auferetur CGp ‖ quantus est] quantus et p : quanta est CP 18 finitum] infinitum p 20 adhuc] *praem.* et GPp ‖ mihi videtur] *inv.* GPp ‖ est] sit GPp 21 sequeretur] *praem.* si G : *add.* sic P : sequitur p ‖ non] *om.* G 23 cuius ... sit^{2}] et sit una pars eius finita scilicet GPp ‖ quantus est] quantus G : quanta est P 24 infiniti] finiti P 25 partes] *om.* G ‖ illius] *sup. lin.* C : ipsius Gp : *om.* P ‖ ita ... eius] *om.* GPp ‖ est^{2}] esset GP 26 esset] *om.* P ‖ annihilaretur b vel] b annihilaretur vel G : b annihilaretur et p : annihilaretur vel P

esset infinitum, ita et *c*. Esset enim ex infinitis pedalibus non participantibus constitutum. Probatio consequentiae nam: ipsum esset compositum adhuc ex infinitis partibus quarum quaelibet esset tanta quantus est iste mundus; et non posset tali esse maius.

Sed tunc est bene dubitandum circa dicta. Nos enim posuimus secundum dicta quod, | licet aeternaliter fuerit mundus et tempus et motus, tamen 62rb p
nullum fuit tempus infinitum et sic omne tempus praeteritum fuit finitum. Nunc igitur quaeritur utrum fuerint plures dies quam anni.

Arguitur quod non quia: si dicas quod infiniti fuerint dies, ita dicam quod infiniti fuerint anni; et si dicas quod decem fuerunt dies, ita dicam quod decem fuerunt anni; et si dicas quod mille, ita dicam quod mille vel decem milia, et sic sine statu; et dicam verum sicut tu.

Oppositum arguitur quia: in omni tempore finito praeterito in quo fuerunt anni et dies fuerunt plures dies quam anni; sed omne tempus praeteritum est tempus finitum praeteritum quo fuerunt dies et anni; igitur omni tempore praeterito fuerunt plures dies quam anni. Et ad istam universalem de tempore sequitur ista non universalis de tempore, sed indefinita, scilicet quod fuerunt plures dies quam anni. Et si haec est vera, tunc ista sibi contradictoria debet negari, scilicet quod non fuerunt plures dies quam anni.

Respondeo quod haec est concedenda 'fuerunt plures dies quam anni' et similiter ista 'dies fuerunt plures quam anni', quia iste terminus 'fuerunt' connotat tempus praeteritum et nihil praecedit illud verbum quod sit distributivum illius connotationis; igitur illud verbum 'fuerunt' stat pro tempore

---

1 esset[1]] erat GPp ‖ et] *om.* P ‖ enim] *om.* Pp 2 probatio ... compositum] *corr. in marg. ex* immo C ‖ ipsum] *om.* GPp ‖ adhuc] *om.* GPp 3 quaelibet esset] quaelibet est P : quaelibet G : quodlibet esset C ‖ est] esset p 4 posset tali esse] possit esse tali P : posset esse Gp 5 tunc ... dubitandum] nunc bene dubitandum est G : nunc dubitandum est P : bene dubitandum est p ‖ dicta] *add.* nam P : *om.* G 5–6 posuimus secundum dicta] posuimus (secundum dicta *in marg.*) C : secundum dicta ponimus GP : secundum dicta ponamus p 8 nunc] tunc GPp ‖ fuerint] fuerunt GP 9 fuerint] fuerunt G 10 fuerint] fuerunt GPp 10–12 quod[1] ... milia] mille ita dicam mille et si dicas quod decem milia etc. P : quod mille ita dicam mille G : quod fuerunt mille ita dicam quod mille et si dicas quod decem milia ita dicam quod decem milia milia p 12 et[1] ... tu] *om.* G ‖ dicam] dico P ‖ tu] *add.* etc. P 14 anni et dies] dies et anni GPp 15 quo ... anni] *om.* GPp ‖ igitur] *add.* in GPp 16–18 et ... anni] *om.* (*hom.*) p 17 sed indefinita] scilicet indefinita G : sed de infinita P 18–19 sibi contradictoria] sibi contraria G : scilicet illius contradictoria P 19 scilicet quod] *om.* GPp 21 et] *om.* GP ‖ ista] isti Gp ‖ plures] *add.* dies G 23 illius] huius GPp ‖ igitur] ideo GPp 23–174.1 tempore praeterito] *inv.* P

praeterito indefinite. Ideo si pro aliquo tempore praeterito est verificatio, simpliciter est verificatio. Sed ista est vera, scilicet 'aliquo tempore, scilicet 89ra C sequente tempus nativitatis Christi, fuerunt | plures dies quam anni'. Ideo simpliciter est verum quod fuerunt plures dies quam anni. Ideo haec est neganda 'non fuerunt plures dies quam anni'. Et sic etiam haec est concedenda 'quandocumque fuerunt dies et anni, plures fuerunt dies quam anni'. Sed si fuisset tempus infinitum, ista esset falsa 'quandocumque fuerunt dies et anni, plures fuerunt dies quam anni', quia instantia esset pro isto tempore infinito. Et istae etiam essent concedendae 'aliquando fuerunt plures dies quam anni' et 'aliquando non fuerunt plures dies quam anni'. Ideo ista esset neganda 'non fuerunt plures dies quam anni', quia aequivalet isti 'numquam 78rb P fuerunt plures dies quam | anni', contra quam esset instantia pro temporibus finitis.

Haec sint dicta de infinito categorematice sumpto.

Sequitur de infinito syncategorematice sumpto. De quo notandum est quod diversis modis solet exponi illud nomen 'infinitum' syncategorematice sumptum. Uno modo in magnitudinibus quia 'aliquantum et non tantum quin maius'; et in multitudine quia 'aliquota et non tot quin plura'. Et videtur mihi quod aequivalens expositio datur sub verbis manifestioribus et brevioribus, scilicet quod infinitum esse *b* significat quod omni *b* est *b* maius, et infinitum esse *b* secundum longitudinem significat quod omni *b* est *b* longius, et sic de infinito secundum velocitatem vel tarditatem vel parvitatem etc. Et intendo idem per 'infinitum secundum longitudinem' et per 'infinite longum' et per 'in infinitum longum'.

Et de ista expositione ponam conclusiones.

---

1 indefinite] *praem.* et p : indeterminate GP ‖ si pro] illud scilicet quod P ‖ est] sit GPp 2 scilicet[1]] *om.* GPp 3 tempus nativitatis] nativitatem GPp 4 dies] *om.* G 4–5 ideo ... anni] *om.* (*hom.*) G 5 et sic] sic P : ideo p 6 quandocumque] quaecumque G 7 tempus] *om.* P ‖ quandocumque] quaecumque G 8 esset] *om.* G ‖ isto] aliquo GPp 9 infinito] finito G ‖ et] *om.* P ‖ aliquando] *add. sup. lin.* non C 9–10 fuerunt ... aliquando] *om.* (*hom.*) GPp 10 ideo] *add.* etiam G 14 haec] *praem.* et GPp 15 sequitur] *add.* dicere GPp ‖ sumpto] sumptis G ‖ notandum est] nota G 16 diversis modis] diversimode G ‖ illud nomen] hoc nomen p : *om.* P ‖ syncategorematice] categorematice G 17 non] nihil P 18 in] de p ‖ aliquota] aliquanta p ‖ et[3]] *om.* p 19 aequivalens expositio] est aequivalens expositio quae p ‖ sub verbis] *om.* P 20 b[1]] ABCGHLMPTU : *add.* secundum magnitudinem p 21 esse] est G 22 vel[1]] *add.* secundum GPp ‖ vel[2]] aut Pp 23 et[2]] ut G 24 et ... longum[2]] *om.* (*hom.*) P ‖ in] *om.* Gp 25 expositione] *om.* C ‖ ponam] pono GPp

Prima est quod infinita est linea gyrativa secundum longitudinem, quia qualibet data est alia | [72va G] longior et nulla est, quin sit alia longior, secundum dicta prius. Et ita etiam secundum ea quae dicta fuerunt in primo libro, in quaestione de maximo et minimo, manifestum est quod in infinitum longum est corpus etiam secundum rectitudinem, et in infinitum est parvum corpus, et est infinite velox motus, et infinite tardus est motus etc.

Secunda conclusio est quod haec est falsa 'infinitum est corpus', quia non omni corpore est corpus maius, immo etiam si esset corpus aliquod infinitum capiendo 'infinitum' categorematice vel etiam si esset aliqua linea gyrativa infinita secundum longitudinem, adhuc ista esset falsa capiendo 'infinitum' syncategorematice, scilicet 'infini|[62va p]tum est corpus' vel etiam 'in infinitum est linea gyrativa', quia aliquod esset corpus, scilicet illud infinitum, quo nullum corpus esset maius, et aliqua esset linea, scilicet ista infinita, qua nulla esset longior; tamen si esset istud corpus infinitum, ista esset bene vera | [89rb C] 'infinitum est corpus finitum', quia omni corpore finito esset maius corpus, scilicet infinitum.

Tertia conclusio est quod ista est falsa 'infinita est secundum longitudinem haec linea gyrativa' et similiter ista 'infinite velox est iste motus' et sic de aliis huiusmodi, quodcumque demonstretur, quia oportet demonstrare illud idem in exponente et non aliud; modo impossibile est quod hac linea gyrativa sit haec eadem longior et sic de aliis. Et eodem modo manifestum est quod ista est falsa 'linea gyrativa est infinita secundum longitudinem' vel etiam 'motus est infinite velox', quia illi termini 'linea' et 'motus' stant determinate; ideo si essent propositiones verae, oporteret quod pro aliqua determinata linea signata vel signabili et pro aliquo | [78va P] motu determinato

1 prima] *add.* conclusio G : *add.* ergo p 2 data] *sup. lin.* C : *om.* GPp || est alia] *inv.* P || sit alia] sit illa Pp : illa sit G || secundum] per P : bene G 3 secundum] per P || in[2]] *om.* G 4 est] *om.* G 4–5 longum est] *inv.* GPp 5 et in] *om.* p 6 est infinite velox] infinite velox est Gp : in infinitum velox est P 7 falsa] *add.* in G 8 corpus aliquod] *inv.* P 9 capiendo infinitum categorematice] categorematice sumptum GPp || linea] *rep.* P 11 scilicet] *add.* quod P : licet p 13 esset[1]] est G || aliqua] *add.* etiam GPp 13–14 scilicet … longior] *om.* P 14 qua] quo G || tamen] unde P || ista] ita C 14–15 esset bene] *inv.* GPp 16 maius corpus] *inv.* GP || scilicet] *om.* GPp 17 infinita] infinitum P 18 et[1]] *om.* P || iste motus] *inv.* GPp 19 huiusmodi] *om.* P || demonstretur] demonstraretur P : monstretur G || oportet] oporteret Gp 19–20 demonstrare … aliud] illud (*om.* P) idem in exponente et non aliud demonstrare GP : illud idem in exponente demonstrari et non aliud p 20–21 hac … haec] haec linea gyrativa sit hac p 23 motus[2]] velox G 25 signata vel signabili] figurata vel figurabili P

3–4 Cf. Iohannes Buridanus, *Quaestiones super libros Physicorum*, I, qq. 12–13 (ed. Streijger, Bakker, 118–142)

essent verae; et sic ista linea esset longior se et iste motus velocior se, quod est impossibile.

Quarta conclusio de infinito secundum multitudinem, scilicet quod haec est falsa 'infinitae secundum multitudinem sunt partes in continuo', quia si in continuo sunt duae partes, tamen non sunt in eo plures partes quam illae duae, quoniam illae duae sunt et centum et mille, sicut ante dictum est. Et ita etiam concluditur secundum dicta quod haec est falsa 'infinitus est numerus secundum multitudinem'.

Sed quinta conclusio potest poni. De qua difficilius est videre utrum sit vera vel falsa haec propositio, scilicet quod inter infinitas partes lineae *b* est ratio numeralis discretiva, quia inter quotascumque est ratio numeralis discretiva, inter plures est ratio numeralis discretiva, quia non nisi inter duas est ratio binarii discretiva et inter plures est ratio ternarii discretiva, quia inter tres, et sic deinceps. Et ad illum sensum possunt intelligi omnes illae communes auctoritates, scilicet quod infinitae sunt partes in continuo, quod omni numero est numerus maior etc.

Contra istam conclusionem sunt fortes dubitationes. Prima dubitatio quia: cum non sit eadem ratio discretiva duorum et trium, non est possibile quod infinitorum sit ratio discretiva, nisi infinitae sint rationes discretivae, et hoc est impossibile, quia non sunt infiniti homines nec in aliquo homine sunt infinitae rationes sive discretivae sive aliae.

Ad istam primam dubitationem respondetur quod, si aeternaliter mundus duraverit, sicut videtur credidisse Aristoteles, verum est dicere quod infiniti homines sunt mortui, et tamen non est verum quod infiniti sunt homines. Ita etiam dico quod non infinitae sunt rationes, sed tamen infinitae

1 esset] essent G 3 conclusio] *add.* est P ‖ scilicet] *om.* P 4 falsa] *add.* scilicet quod G 5 non] *omnes codd. (deest* Pb); *an omittendum?* ‖ in eo] *post* partes[2] GPp 6 quoniam] quia Gp ‖ et[1]] *om.* GPp ‖ mille] *add.* etc. GPp 7 etiam] *om.* P ‖ dicta] *add.* prius GPp 11–12 numeralis discretiva] *inv.* G 13 quia] scilicet p 14 et[2]] *om.* P ‖ omnes] *om.* GPp 15 communes] *sup. lin.* C : *om.* p ‖ scilicet] *om.* GPp 16 est numerus] *inv.* P ‖ etc.] *om.* P 17 contra] *praem.* sed tamen GPp ‖ istam conclusionem] istas conclusiones P ‖ dubitatio quia] est quia p : est GP 18 cum non sit] non est p ‖ ratio discretiva] discretio p ‖ trium] *add.* et tamen p 20 homine] *om.* GPp 21 sive[1]] sibi P 22–23 mundus duraverit] *inv.* GP : duraret mundus p 23 videtur credidisse aristoteles] videtur aristoteles credidisse p : credidit aristoteles GP 24 homines sunt] *inv.* GPp ‖ et] *om.* P ‖ verum] *add.* dicere P 25 infinitae sunt] *inv.* G

23 Cf. Aristoteles, *Physica*, VIII, 1, 250b10–252a5; cf. Aristoteles, *Metaphysica*, XII, 6, 1071b6–9

sunt rationes discretivae, nam hoc nomen 'discretivum' ampliat ad possibilia sicut haec nomina 'activum' et 'motivum' etc.

Secunda dubitatio | [72vb G] quia: dictum fuit quod tria vel centum non sunt plura duobus; igitur si est aliqua ratio discretiva duarum | [89va C] partium, ipsa est tot partium discretiva, quia nulla potest esse ratio discretiva plurium, cum nulla sunt plura illis duabus partibus. Igitur ratio quae ad istam quintam conclusionem adducebatur non valebat. Falsum enim sumebat dicendo quod, si decem aut centum partium sit ratio discretiva, adhuc plurium erit ratio discretiva; hoc enim est falsum.

Solutio: dicendum est quod, licet duae medietates lineae *b* sint tres tertiae eiusdem lineae, tamen ratio discretiva inter duas medietates non est discretiva inter tres tertias, quia sicut dicebatur in primo libro, quamvis haec sit vera 'partes sunt suum totum' sumendo partes coniunctim, tamen non est vera divisim sumendo. Nulla | [62vb p] enim partium alicuius totius est illud totum. Et etiam, licet duae medietates lineae | [78vb P] sint tres eiusdem tertiae, tamen nulla earum medietatum est aliqua istarum tertiarum. Modo haec dictio 'discernere' et maxime haec dictio 'inter' designant quod accipiantur ista inter quae discernitur divisim et non coniunctim. Si enim dico me discernere inter duas medietates lineae *b*, sensus est quod ego utramque intelligo divisim et distincte contra aliam; ad quod non sequitur quod ego intelligam tertias distincte, scilicet quamlibet contra alias. Immo videtur mihi quod haec dictio 'discernere', licet auferretur haec dictio 'inter', satis connotat sensum divisum. Nam secundum grammaticam 'discernere' dicitur quasi 'distincte cernere'. Ideo si dico me discernere tres vel quattuor partes lineae *b*, sensus est quod unamquamque intelligo distincte contra alias. Quia igitur non tot possumus intelligere distincte contra invicem secundum aliquam rationem,

2 haec nomina] hoc nomen C ‖ et] *om.* GPp **3** quia] est quia p : est quod GP **4** ratio discretiva] *inv.* P **5** quia] quod Pp : sed G **6** sunt] sint GPp ‖ duabus] duobus p ‖ istam] *om.* GPp **8** quod] sed G **8–9** adhuc ... discretiva] adhuc plurium est ratio discretiva G : †...† erit ratio discretiva *in marg.* C **10** lineae ... tertiae] b G ‖ sint] sunt P **12** sit] est P **13** suum] *om.* G ‖ coniunctim] *ante* sumendo GPp **13–14** tamen ... vera] (tamen est vera *sup. lin.*) non C **14** sumendo] *in marg.* C : *om.* GPp ‖ illud totum] eius totum G **15** etiam licet] *inv.* GPp ‖ lineae] *add.* b GPp ‖ eiusdem] eius GPp **16** earum] illarum GPp **16–17** discernere] *add.* discretive p **18** discernitur] discernetur p **19** utramque] *add. sup. lin.* alias unamquamque C : utrumque GP **20** distincte] *add.* et G **21** immo] *add.* etiam Gp **22** licet] si Gp : *om.* P ‖ inter] *add.* adhuc GPp **25–26** tot ... distincte] tot possumus distincte intelligere G : possumus tot distincte intelligere p : tot possemus intelligere distincte P

3 Cf. sup., III, q. 17, 159^{1-8} **12** Cf. Iohannes Buridanus, *Quaestiones super libros Physicorum*, I, q. 9 (ed. Streijger, Bakker, 100^{12-21})

quin plura possumus distincte intelligere secundum aliam rationem, ideo sequitur conclusio quinta, scilicet quod infinitarum partium lineae est ratio numeralis discretiva.

Sed adhuc restat tertia difficultas, scilicet quia si distincte *a* et *b* intelligo, sequitur quod habeo in mente mea duos distinctos conceptus, scilicet unum ipsius *a* et alterum ipsius *b*; ideo etiam, si tria intelligo distincte contra invicem, ego habeo tres conceptus, et ita mille, si mille. Constat autem, cum intellectus noster finitae virtutis sit, quod non potest habere simul infinitos conceptus distinctos. Immo manifestum est nobis per experientiam quod attendentes ad considerationem aliquorum non bene possumus simul attendere ad considerationem aliorum, et quod tot possunt nobis proponi quod confusio est et non possumus inter ista discernere. Igitur quamvis possumus habere plures conceptus simul, tamen status est ad aliquotos, ita quod non possumus habere plures simul. Sit ergo | [89vb C] status in mille vel centum vel decem vel sicut vult dicere adversarius. Manifestum est quod, si est status in mille, tunc impossibile est quod plurium sit aliqua ratio discretiva quam istorum mille.

Videtur mihi quod sit difficile respondere. Sed tamen dicamus concedendo quod non sit possibile in aliquo intellectu humano esse infinitos conceptus simul, immo quod ad aliquotos sit status. Ideo ponamus gratia exempli quod possibile est in intellectu esse simul centum conceptus et non plures. Tunc igitur est dubitatio quomodo potero habere conceptum dis|[73ra G]cretivum plurium quam centum. Hoc enim videtur impossibile.

Ad hoc igitur ego dico quod in mente nostra multipliciter potest esse discretio numeralis. Prima est numeratio qua unamquamque unitatum

---

1 possumus] possimus p ‖ intelligere] *add.* contra invicem GPp 2 conclusio quinta] quinta conclusio prius posita GPp ‖ lineae] *add.* b GPp 4 restat] *om.* C ‖ scilicet] *sup. lin.* C : *om.* Gp ‖ quia] quod P 4–5 intelligo] *ante* distincte (4) P : *ante* a (4) Gp 5 sequitur quod] sicut G ‖ mea] *om.* GPp ‖ distinctos conceptus scilicet] conceptus distinctos GPp 6 et] *om.* G ‖ etiam] *om.* P ‖ intelligo] intelligam G 7 ego] esse p ‖ et] *om.* P 8 cum] quod GP ‖ sit] *ante* finitae GPp 10 considerationem] experientiam G ‖ bene] *om.* G 10–11 simul attendere] *inv.* GPp 11 aliorum et quod] aliquorum G 12 est] *om.* G ‖ possumus inter ista] possumus etiam ista GP : possimus illa etiam p 12–13 possumus] possemus GPp 14 plures] *add.* conceptus G : plura P ‖ sit] sic p 14–15 vel centum] *rep.* p 15 dicere adversarius] *inv.* G 16–17 plurium … quam] sit aliqua ratio discretiva plurium quam (*om.* G) GPp 18 videtur mihi] *inv.* p ‖ sit difficile] hic est difficile P : hic difficile est p : hoc est difficile G ‖ sed tamen] cum P 18–19 concedendo] *om.* GP 20 sit] est GPp 21 in … simul] esse simul in intellectu P : in intellectu esse G 22 plures] *add.* igitur etc. G ‖ igitur] *om.* P 23 enim] *om.* P 24 igitur ego] ego P : enim G 25 numeralis] naturalis G ‖ unitatum] unitatem G : rem unicam P

distincte contra aliam intelligimus et numeramus, ut quia | [79ra P] intelligimus et numeramus hunc equum esse unum equum, deinde illum esse alium equum, et sic dicimus eos esse duos; deinde etiam illum esse alium equum, et sic dicimus illos esse tres; et sic consequenter usque ad decem vel centum. Et manifestum est quod in tali numeratione plurium oportet quodlibet eorum intelligere seorsum et distincte. Et sic etiam ego concederem quod haec est falsa 'inter infinita potest discernere homo' vel 'infinita potest numerare', saltem in sua temporali vita, quia nec potest infinita distincte intelligere simul nec successive, nisi ponatur infinito tempore durare.

Secunda ratio discretiva est quae ex numeratione infertur, ut quod illi equi sint centum. Et ista potest permanere sine hoc quod oporteat permanere singulorum equorum conceptus singulares | [63ra p] et distinctos, sicut si ex praemissis sit conclusio conclusa et scita, possibile est actu considerare de conclusione et ei assentire nihil actu considerando de praemissis.

Adhuc est alia ratio discretiva remotior a numeratione, scilicet per adaequationem vel proportionem numeratorum ad non numerata, ut si per numerationem sciamus hanc virgam esse decem pedum, concluderemus aequalem sibi esse decem pedum et triplam sibi esse triginta pedum, quia quae est proportio trium ad unum, ista est triginta ad decem. Et isto modo aliquorum est ratio numeralis discretiva quorum nullum seorsum et distincte ab aliis intelligitur, ut si ego scio quod cuiuslibet lineae sunt duae medietates et tres tertiae, ita concludam proportionaliter quod sunt centum centesimae, et nullam istarum intelligam seorsum contra aliam. Et tamen hoc non obstante illa ratio dicitur numeralis et discretiva, quia sumpta est secundum proportionem ad aliam discretivam quae erat numeratio vel ex numeratione accepta. Et quamvis non intelligatur quaelibet istarum centesimarum seorsum, tamen illa ratio centenarii dicitur discretiva inter centum et non inter mille, licet illa centum sunt | [90ra C] mille, quia cum dicimus

---

1–2 intelligimus et numeramus] numeramus et intelligimus p : numeramus intelligimus P   2 alium] *add.* unum Gp   4 illos] eos GP : *om.* p   4–5 centum] *add.* vel mille GPp   6 eorum] illorum P ‖ etiam ego concederem] ego concederem G : concederem P : ego crederem p   7 discernere homo] *inv.* GPp ‖ homo] *sup. lin.* C   8 saltem] *om.* P   10 quae] quod p   11 sint] sunt GPp ‖ ista] ita p ‖ oporteat] oportet GP   13 sit] sicut G   14 ei] *om.* GPp ‖ nihil actu] nihil GP : vel p   15 est] *om.* P   16 numeratorum] numeratarum p   17 numerationem] *add.* alicuius CP ‖ numerationem … concluderemus] *corr. in marg. ex* mutationem C ‖ sciamus] scimus Gp : *om.* P ‖ concluderemus] concludemus G   17–18 concluderemus … pedum[1]] *om.* (*hom.*) p   18 esse[1]] *ante* aequalem P ‖ quia] et C   19 est[1]] *post* proportio p : *om.* P ‖ ista] ita p   23 intelligam] intelligo G ‖ contra aliam et] *om.* P   27 discretiva] *add. in marg.* alias discretio seu distinctio C : distinctio P : discretio Gp   28 et] *sup. lin.* C : *om.* Gp ‖ sunt] sint Gp ‖ quia] et GPp

aliqua esse centum, nos intelligimus quod ista sunt aliqua quae, si contra invicem numerantur, concluderentur esse centum et non decem aut mille, licet essent decem et mille. Et isto modo vera est quinta conclusio, scilicet quod infinitarum partium lineae *b* est aliqua ratio discretiva, quia quacumque discretione data, ut centum vel mille partium, possum concludere quod sunt bis centum vel bis mille medietates istarum partium. Et omnino quascumque partes dicas, possum dicere quod bis tot sunt medietates istarum partium.

Sed adhuc est quarta difficultas, quae est cavillosa. Videtur enim quod ratio binarii sit tot partium discretiva sicut ratio ternarii vel quaternarii, quia pono quod non sunt plura entia quam quattuor, puta *a*, *b*, *c*, *d*. Tunc igitur
79rb P patet quod dicendo 'duo' ego | discerno non solum inter *a* et *b*, immo pari ratione inter *c* et *d* et inter *a* et *c* et inter *b* et *d*; igitur inter omnia ita bene
73rb G discerno sicut si dicerem | 'quattuor'.

Solutio: dico quod numquam ratio binarii est discretiva plurium ad invicem quam duorum nec aliorum quam duorum, sed bene multiplicium duorum est discretiva, quia omnium duorum ad invicem indifferenter est ratio discretiva communis et specifica binarii, sicut omnis hominis indifferenter est ratio communis et specifica hominis. Sed non sequitur 'haec duo, scilicet *a* et *b*, ad invicem discerno vel numero, et haec etiam duo, scilicet *c* et *d*, ad invicem discerno; igitur haec tria vel haec quattuor ad invicem discerno vel numero'. Sed requiritur alia ratio discretiva et alterius speciei in genere numeri.

Quinta dubitatio est circa totam praedictam expositionem. Quomodo supponit ille terminus '*b*' in ista propositione 'infinitum est *b*' vel 'infinite longum est *b*' et sic de aliis?

Responsio non solum de ista propositione, sed de omni alia quae indiget exponente vel exponentibus: si aliquis terminus semel tantum capiatur in exposita et indigeat capi pluries in exponente vel exponentibus et in illis pluribus acceptionibus supponit diversis suppositionibus, in tali casu videtur

---

1 quae si contra] quae si G : qua contra P 2 numerantur] numerarentur Gp 3 quinta conclusio] *inv.* Gp 5 ut] vel C ‖ vel] aut P 13 et^{2} … d^{2}] *om.* (*hom.*) G ‖ et^{4}] *add.* sic C ‖ b et d] d et b P ‖ ita] ista GPp 14 discerno] discernerem P ‖ si] *om.* P 15 solutio] secundo modo G 16 nec] vel P ‖ aliorum] aliquorum *sed add. in marg.* aliorum C 17 indifferenter] differenter P 17–18 ratio … specifica] discretiva ratio communis et specifica G : discretiva ratio communis et discretiva sive specifica p : discretiva ratio specifica et P 18 hominis] *corr. ex* homini C 19 haec duo] haec duas P : has duas G 20 etiam duo] *inv.* P ‖ et^{3}] *om.* p 27 responsio] respondetur GPp 28 si] scilicet G 29 indigeat] indiget P : in debeat p 30 supponit] supponat GPp

mihi dicendum quod ille terminus in exposita non supponit unica suppositione, sed illis pluribus. Et sic ego dico quod in ista 'infinite longum est *b*' haec dictio '*b*' supponit suppositione distributiva et suppositione confusa tantum, quia cum dico 'omni *b* est *b* longius', primum '*b*' supponit distributive et secundum confuse tantum.

Haec sunt dicta de ista expositione 'infiniti' syncategorematice sumpti.

Alia expositio 'infiniti' syncategorematice sumpti est per carentiam status in rationibus numeralibus. | Ideo primo exponuntur | 'infinita' secundum multitudinem, scilicet quod infinita esse *b* significat duo esse *b* et tria esse *b* et centum esse *b* et mille et sic sine statu. In aliis autem exponitur certa quantitate accepta per multiplicationem numeralem tantae quantitatis sine statu, ut infinite longum esse *b* significat quod data longitudine alicuius *b*, verbi gratia pedali, tunc est *b* duorum pedum et est *b* trium pedum et est *b* centum pedum et sic sine statu. Ita etiam, si infinitum velocem dicamus esse motum, hoc significat quod dato motu alicuius determinatae velocitatis est motus dupliciter velocior et est motus tripliciter velocior et motus centupliciter velocior et sic sine statu. 63rb p 90rb C

Et secundum istam expositionem ponuntur conclusiones communiter concessae.

Prima est quod infinitae sunt partes continui secundum multitudinem, quia duae, tres, centum et sic sine statu, immo haec linea est infinitae partes, quia duae, tres, centum etc.

Sed aliqui obiciunt dicentes quod omnis | dictio posita a parte praedicati tenetur categorematice et non syncategorematice, et ideo in ista propositione 'linea est infinitae partes' haec dictio 'infinitae' non potest teneri syncategorematice; propter quod male dicebatur quod ista propositio esset vera capiendo 'infinitum' syncategorematice. 79va P

---

1 unica] una G 2 et sic ego] ita P 3 et] etiam P : *add.* etiam Gp ‖ confusa] confuse GP 5 secundum] *add.* b GPp 6 sunt] sint p ‖ ista expositione] ista suppositione p : istis expositione G 7 expositio ... est] expositio infiniti categorematici sumpti est G : est expositio syncategorematice sumpti p 8 exponuntur] exponitur P ‖ infinita] *om.* C 10 mille] *add.* esse b GPp 11 multiplicationem] multitudinem p ‖ numeralem] naturalem G ‖ quantitatis] *add.* et sic p 12 ut] et GPp ‖ longitudine] longitudinem p 13 verbi gratia] ut p : videlicet G ‖ est b^{1}] *inv.* G ‖ est b^{2}] *inv.* G 14 est ... et^{2}] *om.* (*hom.*) p ‖ est b] *inv.* G ‖ etiam] *add.* sic in infinitum GPp ‖ infinitum velocem] infinitam velocitatem p 15 quod] quia p 17 motus ... et] *om.* (*hom.*) GPp 20 prima] *add.* conclusio GPp 22 tres] *om.* G 24 et^{2}] *om.* P 26 male dicebatur] multi dicebant p 27 capiendo] tenendo P ‖ infinitum syncategorematice] *inv.* G

Ad hoc ego respondeo quod dicendo 'Socrates est omnis homo' haec dictio 'omnis' non perdit suam significationem distributivam respectu huius termini 'homo', immo distribuit ipsum sicut si a parte subiecti poneretur. Unde esset bona consequentia sumendo sub isto termino sic 'Socrates est omnis homo; Plato est omnis homo; igitur Socrates est Plato'. Dico igitur quod, licet dicamus hoc totum 'omnis homo' teneri categorematice respectu huius subiecti 'Socrates', quia hoc totum dicimus esse praedicatum de isto subiecto, tamen hoc non obstante ille terminus 'omnis' respectu illius termini 'homo' retinet suam significationem distributivam et syncategorematicam. Et ita est in proposito. Nam licet | 73va G hoc totum 'infinitae partes' teneatur categorematice ad illum sensum quia hoc totum est praedicatum, tamen non debet negari quod haec dictio 'infinitae' respectu illius dictionis 'partes' retineat suam significationem syncategorematicam et quod debeat exponi secundum expositionem praedictam.

Secunda conclusio est quod etiam infinita est linea gyrativa secundum longitudinem, quia si est aliqua pedalis, ita est aliqua bipedalis et alia centum pedum et sic sine statu.

Tertia conclusio est quod haec est falsa 'linea gyrativa est infinita secundum longitudinem', quia haec dictio 'linea' supponit determinate, cum nullum confundens eam praeveniat; ideo si ista propositio est vera, oportet quod hoc sit pro aliqua linea determinate signata vel signabili; ad quod sequitur quod ipsa esset se ipsa longior, quod est impossibile. Unde quamvis omni linea sit aliqua longior, tamen non est aliqua quae sit omni longior.

90va C Sed | dubitatur quia: ego concessi quod aliqua linea est infinitae partes; quare igitur ego non similiter concedo quia aliqua linea est infinitae gyrae?

Respondeo quod hoc est quia eadem linea est duae partes et tres et mille et sic sine statu; ideo illa est infinitae partes. Sed non est eadem linea quae est duae gyrae et quae est tres gyrae; ideo de nulla tu potes inferre quod ipsa sit infinitae gyrae.

---

1 ego] ergo p : *om.* GP ‖ homo] *add.* et G 3 poneretur] teneretur P 4 unde] immo P 5 omnis[2]] *sup. lin.* C : *om.* Gp ‖ socrates est plato] plato est socrates GPp 7 huius] *add.* vel p ‖ totum] *om.* P 8 illius] huius GPp 9 distributivam] *om.* P 10 et] *om.* P ‖ nam] *om.* P 11 hoc] *om.* Gp ‖ tamen] *om.* p 12 negari] notari C ‖ illius dictionis] huius termini GPp 13 syncategorematicam] *om.* P ‖ debeat] debet P 15 secunda] alia P ‖ etiam] *om.* P 16 ita] ista P ‖ alia] aliqua P 18 tertia] secunda P 20 eam praeveniat] *inv.* GPp ‖ ista] haec P 21 signabili] significabili G 22 sequitur] sequeretur P ‖ ipsa[1]] ista P ‖ esset se ipsa] se ipsa est G ‖ est] *om.* G 25 igitur ego] igitur P : ego p ‖ similiter] *om.* GPp ‖ quia] quod GPp ‖ infinitae] finitae G 27 illa est] illae essent C : non est G

Quarta conclusio est quod haec est falsa 'infinite velox est motus', quia signato motu veloci non secundum quamcumque proportionis augmentationem motus est velocior, ut in duplo velocior, in triplo velocior, in centuplo velocior, et sic sine statu. Et ita etiam haec est falsa 'infinite longum est corpus secundum rectitudinem'.

Quinta conclusio est quod haec est vera 'in infinitum parva est magnitudo' et haec etiam 'infinite tardus est motus' et | huiusmodi, quia omni magnitu- 63va p dine secundum quamcumque proportionem minoris inaequalitatis est alia minor, ut in duplo minor, in triplo minor, | in centuplo minor, et sic sine 79vb P statu, vel magis proprie loquendo in subduplo minor, in subtriplo minor, in subcentuplo minor, et sic sine statu. Et quocumque motu tardo dato est in decuplo vel in subdecuplo et in centuplo vel in subcentuplo tardior; ita de aliis huiusmodi. Igitur secundum sensum primae conclusionis sunt vera quae communiter dicuntur, scilicet 'infinitae sunt partes in continuo', 'infinitus est numerus'.

His visis videndum est quae sunt proprietates huius nominis 'infinitum' syncategorematice sumpti. Et est prima proprietas quod isti termini 'finitum' et 'infinitum' non opponuntur ad invicem, sicut nec isti termini 'omnis' et 'homo'. Et hoc etiam apparet per praedicationes affirmativas veras in quibus dicti termini ponuntur, ut quia 'infinitae sunt partes finitae huius continui' et 'infinita est secundum longitudinem linea gyrativa finita' et 'infinite tardus est motus finite tardus'. Hae enim sunt verae, ut patet per exponentes. Alia proprietas est quod, si infinitum est secundum longitudinem, ipso est longius; si infinitum esset corpus, ipso esset maius. Et infinitum secundum longitudinem resecaretur per ablationem finiti multotiens factam.

---

2 quamcumque] quantumcumque C : quamque p 3 motus est] *inv.* GPp ‖ in duplo velocior] velocior in duplo GPp ‖ in triplo velocior] velocior in triplo Pp : in triplo G 3–4 in centuplo velocior] velocior in centuplo p : in centuplo GP 4 et[2]] *om.* P ‖ etiam] *om.* G 6 haec] *om.* p ‖ in] *om.* G ‖ est[3]] *add.* haec G 7 huiusmodi] huius P 8 quamcumque] quantumcumque C ‖ minoris inaequalitatis] numerorum inaequaliter G 9 duplo ... minor[3]] duplo minor G : decuplo minor et C 10 in[1]] ut C 10–11 in[2] ... subcentuplo] et in subtriplo (*corr. ex* subcentuplo) C : in subtriplo G 12 vel[1]] et P ‖ subdecuplo] subduplo C ‖ et ... subcentuplo] tardior etiam est in centuplo vel in subcentuplo p : tardior et etiam in centuplo vel in subcentuplo G : vel subcentuplo P ‖ ita] et sic GPp 13 huiusmodi] huius P 14–15 infinitus est numerus] †...†us est numerus *in marg.* C : *add.* etc. Gp 16 his] *add.* autem GPp ‖ quae sunt] quae sint p : quot sunt C 17 sumpti] sumptum G 18 omnis] *add.* homo Pp 20 ponuntur] praedicantur P ‖ finitae] *om.* G 21 est] *om.* C 22 hae ... verae] haec enim sunt vera p 23 alia] secunda P ‖ quod] *om.* P 24 esset[1]] est P ‖ esset[2]] est P 25 factam] *add.* etc. G

Et ex istis apparet quomodo rationes a principio quaestionis adductae procedunt.

⟨1⟩ Ad primam rationem conceditur quod infinitae sunt partes in linea *b*, quia duae et decem et sic sine statu. Et tamen non resultat extensio infinita, quia non maiorem extensionem reddunt tres tertiae quam duae medietates nec centum centesimae quam duae medietates, quia quanto tu ponis numerum pluris discretionis, tanto unitates inter quas est ratio discretiva 73vb G 90vb C sunt minores. | Unde non oportet quod plura | faciant maiorem extensionem quam pauciora, nisi illorum plurium unumquodque sit aequale vel maius unicuique istorum pauciorum.

⟨2⟩ Ad aliam concedo quod, si duo puncta indivisibilia facerent aliquantam extensionem, infinita facerent infinitam. Ideo concludendum est quod impossibile est ex punctis indivisibilibus resultare aliquantam extensionem. Ideo tandem concludetur in sexto quod impossibile est esse puncta indivisibilia.

⟨3⟩ Ad aliam potest dici quod ad illum sensum Aristoteles negavit actu infinitum et actu infinita et concessit infinitum vel infinita in potentia, quia non est verum quod infinitum est corpus, tamen verum est quod infinitum potest esse corpus, quia non omni corpore est corpus maius, sed omni corpore potest esse corpus maius. Et etiam in multitudine partium ab invicem separatarum, quia non infinitae sunt partes continui ab invicem separatae, scilicet per discontinuationem, sed infinitae possunt esse partes separatae ab invicem; sed haec non sunt contra dicta. Et de hoc etiam dicetur in alia quaestione.

⟨4⟩ Ad aliam concedo et concessum est quod, sicut infinitus est numerus, sic infinitae sunt partes in continuo; et sicut non infinitus est numerus, sic nec infinitae sunt partes in continuo.

---

1 et ex istis] ex dictis autem faciliter P : et etiam ex dictis faciliter p : sic ergo ex dictis faciliter G **1–2** adductae procedunt] adductae procedant G : factae procedunt P **3** rationem] *om.* p **4** decem] *add.* et centum Pp **6** nec] *corr. sup. lin. ex* et C : et G ‖ medietates] *add.* et sic sine statu Pp **7** pluris] pluralis p **8** faciant] faciunt P **10** pauciorum] paucorum p **11** concedo] conceditur Pp **11–12** aliquantam] aliquam G **13** est] *add.* quod G ‖ aliquantam] aliquam Pp **14** ideo] immo Pp ‖ concludetur] concluditur P **16** aristoteles negavit] negat aristoteles P **19** est] *om.* P **20** et] *om.* P ‖ ab] ad C **21** ab] ad C **22** partes] *om.* Pp **23** sed] et GPp ‖ contra dicta] contradictoria C ‖ etiam] *om.* P **23–24** in alia quaestione] *om.* G **26** est] *om.* P **27** sunt] *om.* P

---

14 Cf. Iohannes Buridanus, *Quaestiones super libros Physicorum*, VI, q. 4 (ed. Parisiis 1509, ff. 96rb–98va) **23–24** Cf. inf., III, q. 19

⟨5⟩ Ad aliam concedo quod capiendo 'infinitum' vel 'infinita' categorematice infinitis | [80ra P] repugnat quod sint accepta vel etiam quod sint, ad istum sensum quod impossibile est infinita esse accepta vel etiam esse aut fuisse. Sed tamen syncategorematice capiendo 'infinita' haec est vera, quod infinita possunt esse accepta secundum rationem discretivam. Immo etiam infinitas partes ego capio, quando ego capio librum meum vel tunicam meam, quia tres et centum et mille et sic sine statu. Sed impossibile est quod aliquis infinitas partes capiat successive nume|[63vb p]rando quamlibet contra quamlibet aliarum.

⟨6⟩ Ad aliam dictum fuit prius quod non sunt plures partes in lineis *a* et *b* quam in linea *b*, immo nec quam in una centesima lineae *b*. Unde in continuo bene sequitur 'tantundem et amplius, igitur maius', quia illud amplius reddet maiorem extensionem. Sed non sequitur 'tot et adhuc alia, igitur plura', quia ista non reddunt plurem multitudinem quam essent illa tot, immo nec quam esset unum istorum, quia quodlibet est infinitum secundum multitudinem.

⟨7⟩ Ad aliam concedo quod infinitae sunt partes lineae *b* aequalis quantitatis ad invicem, quia et centum et mille et sic sine statu. Sed signata aliqua parte non sunt infinitae non participantes quarum quaelibet sit aequalis illi signatae. Sed signata una infinitae sunt consequenter se habentes ad invicem secundum eandem proportionem secundum quam se habebat prima signata ad ipsum totum. Et hoc intendebat Aristoteles.

⟨8⟩ Ad aliam dicitur quod omnium partium lineae *b* est aliqua ultima earum, et etiam infinitarum est aliqua ultima earum capiendo 'infinitarum' syncategorematice, et etiam omnium medietatum proportionalium est aliqua ultima earum. | [91ra C] Et tamen nulla est sic ultima medietas proportionalis quod non sit alia ulterior omnium. Igitur est aliqua ultima, sed nulla est ultima omnium etc.

---

1 infinitum] infinite G 1–2 infinitum ... categorematice] categorematice infinitum P 2 accepta ... sint^{2}] *om.* (*hom.*) G 3 infinita esse] quod infinita est G ‖ etiam] *om.* G 4 capiendo] loquendo P : sumendo p 6 ego^{1}] *om.* P ‖ ego^{2}] *om.* GPp ‖ vel tunicam meam] *om.* Pp 7 et^{1}] *om.* G 8 partes] *om.* GPp 10 lineis] linea G 11 quam ... b^{2}] *om.* (*hom.*) GPp 12 bene] b C 12–13 quia ... reddet] quia illud amplius reddit Pp : reddit G 14 multitudinem] p, *in marg.* W : magnitudinem seu maiorem multitudinem T : magnitudinem *reliqui codd.* (*deest* Pb) 15 infinitum] infinita Pp 17 concedo] dico GPp 18 ad invicem] *om.* G ‖ et^{1}] duae p 19 non^{2}] nec G 21 se habebat] *om.* p 22 totum] *om.* C ‖ intendebat] intendit p : *add.* ipse G 23 dicitur] dico GPp ‖ b] *sup. lin.* C : *om.* p 24 et ... earum^{2}] *om.* (*hom.*) G 27 alia] altera Pp ‖ omnium] omni G 28 etc.] et sic est etc. P : et sic est finis quaestionis G : *om.* p

⟨III.19⟩

⟨Utrum possibile sit infinitam esse magnitudinem et in infinitas partes lineam esse divisam⟩

Adhuc restant difficultates de infinito quantum ad propositiones de possi- 74ra G bili. Et ideo formatur ultima quaestio utrum possibile est | infinitam esse magnitudinem et in infinitas partes lineam esse divisam.

Arguitur quod sic quia:

⟨1⟩ In qualibet proportionali parte horae posset Deus creare unum lapidem pedalem. Ponamus igitur quod sic faciat. Nihil sequitur impossibile. Et tamen, cum infinitae sunt partes proportionales horae, sequeretur quod in fine horae essent infiniti lapides pedales. Et tamen ex infinitis partibus pedalibus non participantibus resultaret infinita magnitudo. Igitur magnitudinem infinitam potest Deus facere.

⟨2⟩ Item argumentum formatur sic: quandocumque est possibile quod Deus formet unum lapidem, possibile est quod formet ipsum vel alium pedalem; sed possibile est quod in qualibet proportionali medietate huius horae Deus creet unum lapidem; igitur possibile est quod in qualibet creet unum lapidem pedalem. Ponamus igitur quod hoc possibile sit verum. Sequitur ut prius quod ex infinitis pedalibus resultat magnitudo infinita. Maior huius argumenti videtur nota per se. Minor declaratur quia: non minus posset 80rb P Deus continue partem post partem facere vel creare unum lapidem | per unam totalem horam quam sol posset per unam horam totalem facere unum lumen intensum gradum post gradum; et tamen de aurora usque ad primam sol sic facit lumen continue intendendo; igitur similiter potest Deus continue per unam horam unum lapidem facere partem post partem. Et sic

---

4–5 adhuc … quaestio] quaeritur decimo nono P 4 restant difficultates] restat difficultas p 5 ideo formatur ultima] formetur decima nona Gp ‖ infinitam esse] *inv.* Pp 8 proportionali parte] *inv.* Pp 9 pedalem] *om.* p ‖ nihil] vel P 10 et] *om.* P ‖ sunt partes] sint medietates GPp ‖ sequeretur] sequitur p 11 et] *om.* P ‖ tamen] *add. sup. lin.* alias cum C ‖ partibus] *om.* GPp 12 non] *om.* G 15 formet[2]] format P : formaret G 16 proportionali medietate] *inv.* Pp 17 creet[1]] creat P 17–18 igitur … lapidem] *om.* (*hom.*) p 17 creet[2]] parte creat P 19 huius] istius p 21 continue] *add.* per P 22 unam[1]] *om.* P ‖ sol] *corr. in* solem G ‖ posset] possit p ‖ horam totalem] *inv.* GPp 23 intensum] *add.* per P ‖ aurora] anteriori P 24 sol] *om.* GPp 25 unum] *om.* p

 | DOI: 10.1163/9789004322356_022

faciendo ipse faceret in qualibet medietate proportionali, immo in qualibet parte huius horae, unam partem illius lapidis; et quaelibet pars illius lapidis est lapis, quia loquor de partibus quantitativis; igitur in qualibet parte illius horae faceret unum lapidem. Igitur haec est possibilis et potest esse vera quod in qualibet medietate proportionali horae Deus facit unum lapidem, et hoc volebamus probare.

⟨3⟩ Item Deus cognoscit omnes partes lineae *b* distincte, ita distincte sicut Socratem vel Platonem. Igitur potest inter omnes discernere et omnes numerare. Ex quo sequitur quod omnes actu discernit et numerat, quia non est in eo potentia intelligendi distincta ab actu etiam completo. Sed inter quaecumque ipse discernit et quaecumque numerat, possibile est quod ista dividat. Igitur etc.

⟨4⟩ Item si quaecumque propositiones possibiles et compossibiles ponantur simul verae, nihil debet sequi impossibile. Sed illae sunt possibiles et compossibiles 'in prima medietate facit Deus lapidem pedalem' et 'in secunda facit lapidem pedalem' et sic de omnibus aliis, quia de nulla poterit fieri instantia. Igitur si | omnes tales ponantur simul verae, nihil debet 64ra p sequi | impossibile. Et tamen sequitur quod in qualibet facit Deus lapidem 91rb C pedalem, et sic infiniti sunt lapides pedales, ex quibus resultat infinitum.

⟨5⟩ Vel arguitur sic: omnis propositio universalis est possibilis, cuius omnes singulares sunt possibiles et compossibiles; sed huius universalis 'in qualibet medietate proportionali huius horae Deus facit vel faciet unum lapidem pedalem' omnes singulares sunt possibiles et compossibiles; igitur illa universalis est possibilis. Ponamus igitur quod sit vera vel quod omnino sit ita sicut significat. Nihil debet sequi impossibile, et tamen sequitur infinitos esse lapides pedales etc. Maior huius rationis patet per quid nominis. Nam propositiones | vocantur compossibiles, quarum non solum quaelibet 74rb G est possibilis, sed etiam quod possunt simul esse verae, et ut sic non omnes possibiles sunt compossibiles. Haec enim est possibilis 'Socrates currit' et illa

---

1 ipse] ipsum p **1–2** medietate ... parte] parte proportionali sive medietate p **2** huius] illius GPp ‖ et] *add.* sic p **4–5** igitur ... lapidem] *om.* (*hom.*) p **6** volebamus] voluimus G **7** distincte[1]] *om.* p ‖ distincte[2]] *add.* scilicet GP **8** vel] et Gp **10** potentia] *ante* in GPp ‖ completo sed] complete quod p **11** ista] illas G **15–16** in[2] ... et] *om.* (*hom.*) GPp **16–17** poterit fieri] potest dari GPp **17** simul] *om.* C **18** et] *om.* P ‖ quod] *add.* si P ‖ deus] *om.* GPp **19** pedalem] *om.* G ‖ et] *om.* P ‖ lapides pedales] *om.* GPp **20** arguitur sic] sic P : aliter si G **21** huius] huiusmodi p **22** huius horae] *inv.* G **24** ponamus igitur] *inv.* P : ideo ponamus Gp **25** sicut] *add.* ipsa Pp ‖ nihil] vel P ‖ et] *om.* P ‖ sequitur] sequeretur Gp : *add.* quod P **26** etc.] igitur P : et G : et etiam p **27** solum quaelibet] *inv.* P **28** simul esse] *inv.* Gp ‖ et] *om.* GPp

etiam 'Socrates non currit', et tamen non possunt esse simul verae. Si igitur omnes singulares alicuius universalis possunt simul esse verae, universalis potest esse vera, quia posito quod sint simul verae, universalis est vera. Sed tunc etiam minor dictae rationis apparet, quia haec est possibilis 'in prima medietate creat vel creavit' et haec etiam 'in secunda' et sic de singulis, quia nulla potest dari instantia. Et illae etiam omnes sunt compossibiles, quia possibile est quod istae sint simul verae 'in prima medietate creabit lapidem
80va P pedalem' | et 'in secunda etiam creabit' etc. Et ita etiam possibile esset de tribus et centum et sic de singulis. Unde de nullis potest dari instantia, quin possint esse simul verae.

⟨6⟩ Et similiter etiam arguitur quod possibile est lineam esse divisam in infinitas partes quia: sicut infinitae sunt medietates proportionales horae, ita sunt infinitae partes huius lineae *b*. Et Deus potest in prima medietate horae separare primam medietatem lineae *b* et separatim conservare et in secunda secundam et sic de singulis, et ita in fine horae essent infinitae partes lineae *b* ab invicem divisae et separatae et separatim conservatae.

Oppositum arguitur quia:

⟨1⟩ Aristoteles sic negat infinitam divisionem continui, scilicet quod possibile sit, sicut dixi, ipsum esse in infinitum divisum.

⟨2⟩ Et iterum quia: dictum fuit prius quod implicat contradictionem magnitudinem esse infinitam; et tamen hoc sequeretur ad istam quod Deus in qualibet medietate proportionali horae facit vel faciet unum lapidem pedalem; igitur impossibile est Deum facere sic in qualibet medietate proportionali unum lapidem pedalem.

---

1 etiam] *ante* illa (186.29) G : *om.* P ‖ si] sed C 2 simul esse] *inv.* GPp 3 posito] T : po° U : pono ACGHLMPp : pone B 4 etiam] *om.* P 5 medietate] *add.* deus p ‖ creavit] creabit p ‖ singulis] aliis P : *praem.* aliis p 6 illae] ideo P ‖ omnes] *om.* G 7 sint] sunt P : *om.* G ‖ creabit] *add.* deus Pp : *add.* et deus G 8 in secunda etiam] etiam in secunda p : in secunda P : sed etiam in secunda G ‖ et[2]] *om.* P 9 et[1]] *add.* de Gp ‖ centum] *add.* etc. P ‖ unde] *om.* P ‖ potest] posset GPp 9–10 quin possint] quin possunt P : quin possent G : quoniam possent p 11 et similiter etiam] et consimiliter etiam Gp : consimiliter ergo P 12 quia] et C ‖ medietates proportionales] medietates proportionabiles p : partes proportionales P ‖ horae] *om.* G 13 huius] *om.* P 14 horae] *om.* P ‖ primam] *om.* G ‖ lineae b] lineae p : b P 15 secunda] *add.* parte horae P ‖ et[2]] *om.* P ‖ essent] erunt Pp 16 separatae et] *om.* P 17 quia] quod P 18–19 possibile] impossibile CG 19 sicut dixi] sicut ego dixi G : *om.* Pp ‖ infinitum] *add.* esse p 20 fuit] est P 21 et] *om.* P ‖ sequeretur] *praem.* non p : sequitur GP 22 proportionali] *om.* GPp 23 pedalem] *in marg.* C : *ante* lapidem (22) p : *om.* G

---

18 Cf. Aristoteles, *Physica*, III, 6, 206a18–21, 206b12–13; 7, 207b10–15 20 Cf. sup., III, q. 18

Ista quaestio apparet mihi bene difficilis. Et breviter de ea ponam conclusiones.

Prima est quod haec est impossibilis 'linea *b* est divisa in omnes medietates suas proportionales', et similiter ista est impossibilis 'linea *b* erit hodie divisa in omnes suas medietates proportionales'; loquor de divisione rerum per realem separationem.

Probatur ponendo quod lineae *b* unus conus sit *a* et alter *c* et incipiamus medietates proportionales ab *a* procedendo versus *c* et ponamus lineam *b* tangere lapidem exteriorem secundum conum eius *c*. Tunc igitur, si ponat adversarius | quod linea *b* sit divisa in omnes eius medietates proportionales, 91va C quaeratur utrum aliqua illarum partium sic ab invicem separatarum tangebat illum lapidem vel nulla, et si aliqua, utrum plures vel unica. Necesse esset enim alterum membrorum concedere, quia per contradictionem ponuntur. Et tamen quodlibet eorum esset impossibile stante dicta positione adversarii. Igitur ista positio est impossibilis.

Nunc igitur probo quodlibet illorum membrorum esse impossibile. Primo quidem impossibile est quod sint plures istarum partium ab invicem separatarum quarum quaelibet tangeret illum lapidem, quia hoc non posset esse, nisi una fuisset cum alia sine differentia situs per modum penetrationis, 64rb p quod non | erat verum, antequam ista linea divideretur, licet | forte Deus 74va G posset facere talem penetrationem.

Secundo etiam impossibile est quod istarum partium sic separatarum unica tangebat, quia vel ista est indivisibilis vel ipsa est adhuc divisibilis. Si ipsa est indivisibilis, tunc linea erat composita ex indivisibilibus, quod in sexto probabitur esse falsum. Sed si ista est divisibilis, tunc prima eius medietas erat de medietatibus proportionalibus lineae *b* et sic non erat | 80vb P

1 bene] *om.* p ‖ de ea ponam] de ea ego pono p : ponuntur de ea P 3 prima] *add.* conclusio G 3–4 est^3 ... b] *om.* (*hom.*) p ‖ medietates suas] *inv.* GP 4 et] *om.* P ‖ similiter ista] *inv.* G 5 medietates] partes G ‖ rerum] *om.* GPp 7 ponendo] *om.* Gp ‖ incipiamus] incipimus p 8 procedendo] *post* c p : *om.* P 11 quaeratur] quaeritur p 13 enim] eum *sed add. sup. lin.* enim C ‖ alterum] *add.* illorum GPp 14 et] *om.* P ‖ eorum] illorum Pp 15 positio] propositio P 16 primo] primum G 17 ab] ad C 18 quia] *om.* GPp ‖ posset esse] potest esse p : posset G 20 ista linea] *inv.* P : linea illarum G 22 secundo] secundum G 23 ista est indivisibilis] ipsa est indivisibilis GP : ipsa indivisibilis est p ‖ ipsa est adhuc] ipsa adhuc est G : adhuc ipsa est p 24 ipsa est] est Pp : *om.* G 25 probabitur] *praem.* libro GP : libro probatur p ‖ prima] post p 26 erat^1] *om.* G ‖ proportionalibus] proportionabilibus p

25 Cf. Iohannes Buridanus, *Quaestiones super libros Physicorum*, VI, q. 4 (ed. Parisiis 1509, ff. 96rb–98va)

facta divisio lineae *b* in omnes eius medietates proportionales, quod est contra positionem adversarii.

Tertio etiam ostendo quod impossibile sit quod nulla istarum partium sic ab invicem separatarum tangebat illum lapidem quia: si nulla tangebat, tunc quaelibet distabat, et non est imaginabile, si quaelibet pars lineae *b* distabat a lapide, quod linea *b* tangeret lapidem.

Item nec aliqua medietas proportionalis lineae *b* tangebat lapidem nec plures istarum nec omnes. Nam ultra omnem et omnes semper aliquid relinquitur. Unde dicunt ponentes lineam gyrativam protractam per omnes medietates proportionales esse infinitam quod illa non tangeret lapidem extrinsecum nec esset terminata ultimo termino columnae, nisi forte hoc esset, sicut aliqui dicunt, exclusive, quod non esset tangere lapidem extrinsecum. Ponamus igitur quod omnes medietates proportionales sint resecatae. Adhuc remanet aliquid tangens lapidem, de quo quaeritur utrum sit divisibile vel indivisibile, sicut prius quaerebatur.

Secunda conclusio est quod haec est impossibilis 'in qualibet medietate proportionali huius diei Deus creat unum lapidem pedalem' vel etiam ista 'in qualibet medietate proportionali huius diei Deus creabit unum lapidem pedalem'. Et non loquor hic de creare prout diceremus Deum creare me, quamdiu sum, et angelum vel caelum, quamdiu est, eo quod continue hoc dependet a Deo, et nihil essent, si non dependerent a Deo. Sed accipio in hac die Deum creare lapidem, quia in hac die est ille lapis et a Deo et non erat
91[vb] C ante hanc diem, et sic de qualibet medietate proportionali. |

Et in hoc sensu dicta conclusio probatur quia: non magis est possibile quod in qualibet medietate proportionali huius diei Deus creet unum lapidem pedalem praecise quam quod ipse in qualibet medietate proportionali

1 eius medietates] suas partes G ‖ proportionales] proportionabiles p 4–5 tunc quaelibet distabat] tunc quaelibet distabit C : *om.* G 5 b distabat] distabit C 6 tangeret] tangebat p 7 medietas proportionalis lineae] proportionalis medietas lineae G : medietas proportionabilis p ‖ tangebat] tangit GPp 8 nam] et p ‖ et omnes] *om.* p 10 proportionales] proportionabiles p 11–13 nec ... extrinsecum] *om.* (*hom.*) GPp 14 quaeritur] quaereretur GPp 16 est[1]] *om.* Gp ‖ impossibilis] *add.* quod G 17 vel] et p 17–19 vel ... pedalem] †...† in qualibet medietate †...†us diei deus †...† unum lapidem †...† *in marg.* C 18 deus] *om.* GPp ‖ creabit] creavit P 19 hic] *sup. lin.* C : *om.* GPp ‖ creare[1]] *corr. sup. lin. ex* creatione C : creatione G 20 sum] sim p ‖ eo] *om.* P ‖ continue hoc] *inv.* G 21 dependet] dependent C ‖ et] *add.* quod Pp ‖ essent] esset C ‖ sed] *add.* ego GP : *add.* eo p ‖ hac] *om.* p 22 die[2]] *om.* Pp ‖ et[1]] *om.* G 23 hanc diem] hunc diem G ‖ proportionali] proportionabili p 24 et] *om.* P ‖ quia] sic GPp 25 proportionali] proportionabili p ‖ creet] creat Pp 26 proportionali] proportionabili p

huius diei resecaret unam medietatem proportionalem lineae *b* praecise; sed hoc secundum est impossibile, ut dicebat prima conclusio; igitur similiter primum est impossibile.

Item sicut prius arguebatur, illud est impossibile, ad quod sequitur magnitudinem esse infinitam, quia esse magnitudinem infinitam est impossibile, sicut prius dictum fuit et ut postea dicetur; sed hoc ad illud sequeretur, ut prius dicebatur; igitur etc.

Item non magis hoc est possibile quam facere gnomones pedalis latitudinis circa quadratum pedalem vel sphaeras spissitudinis pedalis circa invicem; et tamen impossibile est quod sic faciat Deus in qualibet medietate proportionali diei unum gnomonem circa quadratum vel unam sphaeram circa praecedentem, quia sequeretur quod in fine diei esset superficies infinita propter infinitarum superficierum latitudinis pedalis appositionem et sequitur quod esset | [74vb G] corpus infinitum propter infinitarum sphaerarum appositionem. Et tamen utrumque horum in dicto casu implicat contradictionem, quia ex quo non ponitur Deus creare nisi corpora sphaerica sive orbicularia, semper oportet totum resultans esse sphaericum; et implicat | [81ra P] contradictionem quod sphaericum sit infinitum. Sphaera enim est figura, et definitio figurae est quod sit clausa termino vel terminis, et sic clausum est terminatum et finitum.

Et similiter etiam dico de gnomonibus quod semper oporteret totalem superficiem esse quadratam, et implicat contradictionem quod quadratus sit infinitus, quia est figura et quia etiam oporteret costas quadrati infiniti esse infinitas (non enim esset dia|[64va p]meter quadrati longior costa), quod est impossibile.

---

1 medietatem proportionalem] medietatem proportionabilem p : partem proportionalem G 2 dicebat] ostendit p 4 illud] hoc GPp ‖ sequitur] sequeretur G 6 prius] *om.* G ‖ ut[1]] *om.* GPp ‖ sequeretur] sequitur p 7 prius ... etc.] praedicebatur P 8 hoc est possibile] esset hoc possibile Gp : hoc esset impossibile P 8–9 pedalis latitudinis] *inv.* Pp : latitudinis pedales G 9 pedalem] pedale G ‖ spissitudinis] sphaericitudinis p 10 et tamen] tamen P : et cum p 12 sequeretur] sequitur p 13 superficierum] *add. in marg.* alias sphaerarum C : sphaerarum P ‖ latitudinis pedalis] *inv.* Gp 14 sequitur] sequeretur P : sequeretur etiam G : *add.* etiam p ‖ esset] essent C ‖ corpus infinitum] *corr. in marg. ex* sphaerae infinitae C : conus infinitus G ‖ infinitarum sphaerarum] *inv.* GPp 15 et] *om.* P 16 ponitur deus] sequitur deum P ‖ nisi] *add.* in P 17 oportet] oporteret GPp 19 est[1]] *om.* G ‖ clausum] clausa p 21 etiam dico] dico quod est P ‖ quod] quia GPp 23 figura] finita P ‖ quadrati] quadratis G

---

7 Cf. sup., III, q. 18

Item si esset possibile quod in qualibet medietate proportionali diei fieret linea pedalis, ita esset possibile quod fieret bipedalis. Ponamus igitur quod pedales fiant et coniungantur ad invicem in longum constituendo lineam *a*, et iuxta fiant bipedales et etiam coniungantur ad invicem constituendo lineam *b*. Constat quod in qualibet medietate proportionali diei erit linea *b* dupla ad lineam *a*. Igitur similiter in toto congregato ex omnibus illis medietatibus proportionalibus erit linea *b* dupla ad lineam *a*, quod est impossibile, quia linea *a* est infinita; ideo non esset aliquid sibi duplum nec maius.

Et iterum in dicto casu linea *a* esset infinita et etiam non esset infinita,
92ra C sed finita, quia semper linea *b* excederet eam secundum longitudinem; | ideo linea *a* esset terminata ad illud in quo linea *b* excederet eam.

Item in dicto casu ego pono quod in qualibet medietate proportionali secundae diei Deus auferat de linea *b* unam pedalem incipiendo ubi inceperat creare. Constat quod in fine diei non erit tota linea *b* remota. Igitur illud quod remotum est erat finitum, quia a parte ante habebat principium et etiam a parte post habebat finem, quia terminabatur ad illud quod adhuc restat de linea *b*. Et tamen oportet dicere quod erat infinitum, quia non erat minus quam linea *a* propter hoc quod non pauciores pedales remotae sunt de linea *b* in die sequente quam creatae fuerunt in die praecedente de linea *a*.

Sed aliquis forte diceret quod nihil remaneret de linea *b*. Contra hoc arguitur quia: non plures pedales removentur de linea *b*, si consequenter sine intervallo removeantur pedales, scilicet in qualibet medietate proportionali una, quam si cum intervallo removerentur, verbi gratia quod in prima medietate proportionali removeatur prima medietas primae lineae pedalis, et in secunda medietate medietas secundae lineae bipedalis, et in tertia medietate medietas tertiae, et sic de singulis aliis. Et tamen sic auferendo remane-

3 pedales] bipedalis GP 3–4 lineam a] a lineam unam G 4 etiam] *om.* P 5 erit] erat P 6 omnibus] *om.* GPp 7 proportionalibus] proportionabilibus p ‖ linea] *om.* G 8 esset aliquid] esset aliquod G : est aliquod P ‖ nec] *corr. sup. lin. ex* vel C : vel G 10 et[1]] nec (*sup. lin.*) etiam C : nec P ‖ a] *sup. lin.* C : *om.* Gp ‖ et[2] ... infinita] et non esset infinita *del.* C : *om.* P 11 semper] *om.* G 14 auferat] auferet G 16 est] erat p ‖ quia] *om.* G 16–17 principium et etiam] principium etiam p : primum P 17 quod adhuc] ad quod p 18 quod erat] *om.* P 20 in die praecedente] *post* a (21) GPp 24 scilicet] si P ‖ proportionali] *add.* parte C 25–26 una ... proportionali] *om.* (*hom.*) p 25 removerentur] *in marg.* C : removeantur G ‖ quod] quia G 26 removeatur ... pedalis] removeretur medietas primae lineae pedalis G : removeretur primae lineae bipedalis medietas P : removerentur medietates primae lineae bipedalis p ‖ lineae] *add. sup. lin.* b C 27 medietas] *om.* p ‖ lineae] *add.* b P 28 et[2]] *om.* P 28–193.1 remaneret] removeret p

ret tantum quantum au|ferretur. 75[ra] G Igitur similiter auferendo pedales continue adhuc remaneret tantum quantum auferretur.

Item si possibile esset Deum sic facere, sequeretur quod esset dare ultimam medietatem proportionalem huius diei, quod est | 81[rb] P falsum, sicut dicebatur quod non est ultima medietas columnae proportionalis. Consequentia probatur imaginando, si Deus sic in qualibet medietate proportionali diei faciat unum lapidem pedalem, quod primum faciat tangentem manum meam semper quiescentem; deinde faciendo secundum elonget primum a manu mea, ut iterum illum secundum faciat tangentem manum; et iterum faciendo tertium elonget primos duos, ut faciet illum tertium tangentem manum, et sic semper ultra. Tunc in fine diei erit unus lapis tangens manum et non plures. Et tunc oporteret dicere quod ille est ultimo factus. Et non potest esse ultimo factus, nisi sit factus in ultima medietate proportionali. Igitur etc.

Et hoc idem apparet secundum alium casum, scilicet quod faciendo secundum lapidem annihilaret primum et faciendo tertium annihilaret secundum et sic deinceps, et quod nullum corrumpat nisi praecise faciendo | 92[rb] C alium. Tunc igitur numquam erit nisi unus lapis pedalis et semper erit unus. Igitur semper in fine diei remanebit unus solus pedalis. Et oportet dicere quod ille erat ultimo factus et in ultima medietate proportionali.

Tertia conclusio, sive sit vera sive sit falsa, videtur apparenter sequi, sive apparentia fuerit probabilis sive sophistica, scilicet quod non est possibile per aliquam potentiam esse magnitudinem infinitam. Probatur primo quia: magna ratio ad probandum quod hoc sit possibile destructa est, scilicet quod possibile est Deum in qualibet medietate proportionali horae vel diei creare lapidem pedalem.

---

3 sequeretur] sequitur p **4** proportionalem] proportionabilem p ‖ huius] *om.* Pp ‖ sicut] ut P **5** est] esset C ‖ proportionalis] *om.* GPp **6** si deus sic] *add.* et C : sic deus si p **8** quiescentem] quiescendo G **9** mea] *in marg.* C : *om.* GP ‖ faciat] faciet p **9–11** et ... manum[1]] meam p **9** iterum[2]] etiam G : *om.* P **10** faciet] faciat GP ‖ tangentem] *add. sup. lin.* aliam C **11** diei] *in marg.* C : *om.* G ‖ manum[2]] *add.* meam p **12** tunc oporteret] tunc oportet Pp : oportet G **15** et hoc] adhuc p **16** annihilaret[1]] annihilet Pp ‖ primum] *add.* lapidem G ‖ annihilaret[2]] annihilet Pp **16–17** secundum] *add.* et faciendo quartum annihilet tertium P **18** et] *corr. sup. lin. ex* igitur C : igitur G **19** solus] *om.* GPp **20** et] igitur GPp **21** sit[2]] *om.* GP ‖ apparenter sequi] *inv.* P **22** fuerit probabilis] fuerat probabilis G : fuerit P **23** esse magnitudinem] *inv.* P ‖ infinitam] *add.* hoc Pp **24** probandum ... possibile] arguendum quod hoc sit possibile Gp : arguendum hoc P ‖ destructa est] *inv.* P ‖ quod[2]] *add.* hoc sit possibile videlicet quod P

Sed diceret aliquis: alia est ratio maior, scilicet quod si Deus fecit mundum de novo, tamen ab aeterno potuisset fecisse eum, et Aristoteles | [64vb p] ponit quod aeternaliter fuit mundus; posito igitur quod aeternaliter fuit mundus constat quod Deus in quolibet die praeterito potuisset creasse unum lapidem pedalem et conservare; igitur nunc essent infiniti lapides pedales, ex quibus esset magnitudo infinita; igitur hoc est possibile.

Respondeo quod, si fuerit mundus aeternus, tunc infiniti fuerunt dies et in quolibet potuit Deus creare unum lapidem pedalem. Tales enim de possibili in sensu diviso concedendae sunt, sicut post dicetur. Tamen in sensu composito non oportet concedere quod haec sit possibilis 'Deus in quolibet die creavit unum lapidem pedalem reservando eos semper'.

Sed adhuc contra hoc est obiectio difficilis, quia sicut natura potest facere aliquid ex aliquo, ita Deus posset illud facere ex nihilo et illud semper conservare. | [75rb G] Sed si mundus fuit aeternus, natura in quolibet die praeterito fecit aquam maiorem quam pedalem, ut | [81va P] apparet ex generatione fluviorum. Igitur haec est possibilis, immo de facto vera, quod omni die praeterito Deus fecit lapidem pedalem et reservavit.

Ad illud potest responderi per illud quod dicitur primo *Caeli*, quod non est potentia ad praeteritum. Ideo dicitur quod non est possibile etiam per potentiam Dei iam praeterita non fuisse vel Aristotelem non fuisse; et ita etiam, quod non fecit heri Deus, non est possibile quod faceret illud heri. Quamvis igitur verum sit quod Deus omni die praeterito potuit facere unum lapidem pedalem, tamen, quia non fecit, non est possibile quod fecerit. Ideo

---

1 aliquis] *add.* quod GPp ‖ scilicet quod] scilicet quia Gp : quia P ‖ fecit] fecisset istum P **2** eum] ipsum GPp **3** fuit[2]] fuerit P : fuisset p **4** quolibet die praeterito] qualibet die praeterita p **5** conservare] conservasse P ‖ pedales] *om.* p **7** fuerit mundus] *inv.* GP **8** potuit] potest P ‖ tales] talem P **9** concedendae sunt sicut] concedo ut P ‖ tamen] *praem.* sed Gp : sed adhuc P **10** possibilis] *add.* possibile est P **11** quolibet die] qualibet parte proportionali P ‖ eos semper] eos P : semper eas p **12** adhuc] *om.* Pp ‖ natura] *om.* G **13** illud[1]] *om.* P **14** si] *om.* p ‖ aeternus] *add.* et p **15** fecit] facit P ‖ maiorem] *add. in marg.* quantitas C : *add.* quantitatem P ‖ ut apparet] prout apparet p : prout patet P **17** fecit] *ante* deus (16) Gp : facit P ‖ lapidem] l (?) *in marg.* (*corr. ex* aquam) C ‖ pedalem] talem G : *add.* igitur similiter haec est possibilis quod omni die praeterito fecit Deus lapidem pedalem C ‖ et] *om.* p **18** illud[1]] hoc GPp ‖ per illud] *om.* P ‖ dicitur] *add.* in P ‖ quod[2]] *praem.* scilicet G : sicut p **19** dicitur] dicit G ‖ etiam] *om.* P **20** potentiam dei iam] divinam potentiam GPp **20–21** ita etiam quod] etiam ita G **21** heri deus] *inv.* P ‖ faceret] fecit GP : fuerit p **22** igitur] iterum P : istud G ‖ potuit] poterit G : poterat P **23** pedalem] *om.* G ‖ fecerit] faceret p

---

2 Cf. Aristoteles, *Physica*, VIII, 1, 250b10–252a5; cf. Aristoteles, *Metaphysica*, XII, 6, 1071b6–9
18 Cf. Aristoteles, *De caelo et mundo*, I, 12, 283b12

haec est impossibilis 'omni die praeterito Deus fecit unum lapidem pedalem semper conservando eum post'. | 92va C

Sed adhuc obicitur quia: ex quo omni die Deus potuit creare unum lapidem pedalem et post semper conservare, quaeritur, si ita fecisset, quid modo esset. Nonne modo essent lapides infiniti?

Respondeo quod, si ita fecisset, lapides nunc essent infiniti. Sed huius antecedens et consequens sunt impossibilia, immo posita mundi et temporis aeternitate semper fuit impossibile quod Deus omni die praeterito creavit unum lapidem semper post conservando, quia semper fuit verum dicere quia ita non fecit, et tamen, cum non sit potentia ad praeteritum, impossibile est, si Deus aliquando fecit *b*, ipsum non fecisse *b*. Et similiter impossibile est quod, si aliquando non fecit *b*, ipsum tunc fecisse *b*, licet ante fuerit possibile quod ipsum faceret.

Adhuc tertia conclusio prius posita potest probari per rationes quae in alia quaestione fuerunt adductae.

Quarta conclusio est quod omnes partes lineae *b* Deus potest separare ab invicem et separatim conservare, quia et istas duas et istas centum et sic de singulis. Nullae enim sunt de quibus posset dari instantia, nisi daretur de omnibus collective; et dictum est prius quod omnes non sunt capiendo 'omnes' collective.

Item aliquas potest separare; et non est ratio quare magis illas quam alias; igitur.

Quinta conclusio est quod in qualibet medietate proportionali huius diei potest Deus facere unum lapidem pedalem conservando ipsum semper post. Probatur per inductionem ut prius, quoniam de nulla medietate potest dari

2 semper conservando] *inv.* GPp 3 deus potuit creare] potuit creare deus P : potuit deus creare p 5 nonne] numquid C ‖ lapides infiniti] *inv.* P 6 huius] hoc G : etiam P : et p 7 antecedens et consequens] consequens et antecedens GPp 8 die] tempore Pp 9 quia] et p 10 quia] quod GPp ‖ praeteritum] praeterita P 11 b^2] *om.* G ‖ similiter] consimiliter GP : consequenter p 12 quod] et P : *om.* Gp ‖ aliquando] aliter p ‖ ante fuerit] *praem.* tamen p : ante fuit G : non fuerit C 13 ipsum] *om.* GPp 14 prius posita] praesupposita P ‖ per] *add.* illas P 15 fuerunt adductae] *add.* ad hoc P : fuerunt ad hoc adductae p : ad hoc adductae sunt G 16 b] *om.* P 17 et^2] *om.* GP 18 posset] possit GPp 19 prius] *om.* P 19–20 capiendo omnes] sumendae GP : sumendo omnes p 22 igitur] *in marg.* C : *add.* etc. G : *om.* Pp 24 potest deus] *inv.* P 25 inductionem] indictionem p

15 Cf. sup., III, q. 18 19 Cf. sup., III, q. 18

instantia. Si enim aliqua posset dari instantia, hoc maxime esset de ultima; sed de ista non potest dari, quia illa non est.

Sexta conclusio quod infinita magnitudo potest esse capiendo 'infinita' syncategorematice, quia non posset esse tanta finita, quin posset esse maior etiam in duplo, in centuplo et sic sine statu. Et cum non posset esse magnitudo nisi finita, sequitur simpliciter loquendo quod omni magnitudine possibili potest esse maior in duplo, in centuplo etc., scilicet per potentiam
75va G divinam. |

81vb P Sed tu | dices quod possibili posito in esse nullum sequitur impossibile. Si igitur in qualibet medietate proportionali Deus potest creare lapidem pedalem, ponatur igitur hoc in esse.

65ra p Dico quod universali de pos|sibili in sensu diviso non oportet correspondere universalem de inesse possibilem, sed sufficit quod quaelibet singularis de inesse sit possibilis. Verbi gratia, haec est vera 'omne astrum possum videre', et sine miraculo, et tamen sic non est ista possibilis 'omne astrum video', sed quaelibet singularis est possibilis.

Sed tu replicabis quia est dissimile de dicto exemplo et de proposito
92vb C nostro, quia licet de astris quaelibet sit | possibilis, tamen non omnes sunt compossibiles; sed in proposito quantum ad quartam conclusionem omnes singulares sunt possibiles et compossibiles; igitur universalis de inesse debet esse possibilis.

Dico quod hoc non sequitur. Sed bene sequitur quod omnes singulares possunt esse simul verae; tamen impossibile est quod omnes sint simul verae. Semper enim in proposito deficit consequentia de divisa de possibili ad compositam stante universalitate. Unde corollarie concludendum est quod aliqua universalis est impossibilis, cuius tamen omnes singulares sunt possibiles et compossibiles. Ad hoc enim quod omnes singulares sint possibiles et compossibiles non requiritur quod universalis sit possibilis, sed

---

1 enim] *add.* de GPp ‖ posset] potest P 2 illa non] illa (*sup. lin.*) non C : non p : nulla G 3 conclusio] *add.* est GPp ‖ capiendo] et capio G 3–4 syncategorematice] categorematice P 4 posset[1]] potest GPp ‖ posset[2]] possit p 5 duplo] *add.* et GPp ‖ cum] tamen G ‖ posset] possit Pp 6 nisi finita] AHLMTUp : nisi infinita BP : nisi infinita *sed corr. in* infinita nisi G : nisi (*del.*) infinita C 7 scilicet] et hoc p 8 divinam] *add.* etc. G 9 dices] dicis P ‖ nullum] nihil GPp 11 igitur hoc] hoc GP : *om.* p 12 quod] *add.* in G 13 singularis] singulares P 14 vera] vere p 15 sic ... ista] sic non est ita P : ista non est G 17 replicabis] replicares P ‖ est dissimile] *inv.* Pp 19 quartam] *in marg.* C : *om.* G 23 omnes sint] *inv.* G 24 in] *om.* P 24–25 de[1] ... possibili] de diviso de possibili CGp, *sed corr. in* de possibili divisa (*in marg.*) C 26 cuius tamen] bene cuius P : cuius p 27–28 sint ... compossibiles] sunt compossibiles et possibiles Pp

requiritur quod omnes possunt esse simul verae, scilicet si proponantur, vel etiam quod de omnibus possibile est, qualitercumque significant, ita simul esse. Et sic est in proposito. Et tamen haec est impossibilis 'qualitercumque omnes significarent, si proponerentur, ita est'.

Sed adhuc est dubitatio de qua inter multos est controversia, scilicet utrum ista sit concedenda 'Deus potest separare et separatim conservare omnes partes lineae *b*', posito quod haec dictio 'omnes' teneatur distributive, quia si teneretur collective, manifestum est quod esset falsa, quia sic nec omnes sunt aliquae nec aliquae sunt omnes, sicut ante dictum est, licet omnes sint aliquae capiendo distributive.

Et haec dubitatio est consimilis sicut, cum omne astrum videre possum, utrum ego possum videre omne astrum. Ad hoc dicunt multi quod haec est falsa quod possum videre omne astrum, quia praedicatum appellat formam vel formalitatem; ideo propositio de possibili, si sit vera, debet poni in esse salvato praedicato secundum eius totam formam, et quod ista propositio de inesse sit possibilis, ita quod ad eam nihil sequatur impossibile. Ideo si haec esset vera 'Deus potest separatim conservare omnes', ista esset possibilis 'Deus separatim conservat omnes', quod est falsum. Et etiam, si haec esset vera 'ego possum videre omne astrum', haec esset possibilis 'ego video omne astrum', quod est falsum.

Sed quamvis ista opinio possit forsan probabiliter | sustineri, tamen magis 82ra P
opinor oppositum, scilicet quod haec sit vera 'ego possum | videre omne 75vb G
astrum' et quod Deus potest separare et separatim conservare omnes partes lineae *b*, quoniam omnes concedunt quod non oportet ponere in esse illam de possibili salvata tota forma subiecti, etiam si propositio sit singularis, particularis vel indefinita. Non enim sequitur, si verum est quod hoc album potest esse nigrum, quod haec est possibilis 'hoc album est nigrum'

---

1 proponantur] ponantur G 2 est] *add.* quod C 3 et sic] ita G 4 proponerentur] ponerentur G 5 est^{2}] *ante* inter Pp 6 ista sit] *inv.* p 7 haec ... teneatur] hae dictiones teneantur P 8 teneretur] tenerentur P ‖ est] esset C ‖ quod] *add.* ipsa GP : *add.* ista p 9 est] fuit P 11 videre possum] *inv.* P : ego possum videre p 12 possum] possim p ‖ hoc] quod Pp 13 quod] ego GPp ‖ appellat] *add.* suam p 14 vel formalitatem] sive formalitatem G : *om.* Pp 15 salvato] servato p ‖ totam] tantam p 16 sequatur] sequitur p 17 omnes] *add.* etc. P ‖ ista] ita haec G 18 separatim] *post* omnes p 19 esset] est p 21 ista] haec GPp ‖ possit forsan probabiliter] forte possit probabiliter G : forte posset probabiliter p : probabiliter potest P 21–22 magis opinor] *inv.* P 25 salvata] servata p ‖ etiam] *add. sup. lin.* praedicati G 27 est^{1}] sit GPp

---

9 Cf. sup., III, q. 18

vel 'album est nigrum', sed sufficit quod ponatur in esse per pronomen demonstrativum demonstrato eo pro quo subiectum supponebat aliis quae 93ra C in subiecto implicabantur circumscriptis. | Et ita credo de praedicato esse, quoniam sicut propositioni de possibili verae debet correspondere propositio de inesse possibilis, ita propositioni de futuro verae debet correspondere propositio de praesenti quae erit vera, ut si haec est vera 'ego curram', sequitur quod haec erit vera 'ego curro', si tunc proponatur; et tamen non oportet quod illi de futuro correspondeat illa de praesenti salvata tota forma praedicati, quoniam haec est vera 'ego bibam cras', et tamen haec numquam erit vera 'ego bibo cras'; ita igitur de possibili. Nam demonstrato puero qui de ventre matris nascitur, haec est vera 'iste puer potest currere in tempore futuro'. Probatio quia: potest currere in aliquo tempore, cum non posset 65rb p esse | cursus nisi in tempore; et tamen non potest currere in hoc tempore praesenti, quia nondum habet membra sufficientia ad currendum; et multo minus potest currere in tempore priori; igitur potest currere in tempore futuro et non nisi in tempore futuro. Et tamen haec numquam potest esse vera nec est possibilis 'ille currit in tempore futuro'. Sufficit igitur ad hoc quod haec sit vera 'ego curram in tempore futuro', quia haec erit vera 'ego curro'. Futuritio enim connotatur in hoc verbo 'erit'. Et ita ad hoc quod haec sit vera 'iste potest currere in tempore futuro' sufficit quod haec sit possibilis 'ille currit in tempore' vel etiam 'iste currit'. Et ita, sicut non oportet illam de possibili quae est de subiecto universali distributo ponere in esse remanente distributione, sed per singulares, ita credo quod nec oportet de praedicato distributo. Et hoc ex alio apparet, quia licet praedicatum appellet suam formam, tamen iste terminus vel haec oratio 'omnes partes lineae *b*' nec significat tempus nec connotat; nomen enim significat sine tempore. Ideo nec significat nec connotat simultatem temporis. Immo etiam nec illud verbum praesentis temporis 'videre' vel 'dividere' vel 'conservare' connotat simulta-

3 credo] crede G ‖ esse] *ante* de GPp 4 possibili] *codd.* (*corr. in* futuro *in marg.* IS, possibile EX, *deest* Pb) ‖ verae] *ante* de P 5–6 de inesse ... propositioni ... propositio] *scripsimus* : de inesse ... propositio ... propositio B : de inesse UW : *om.* (*hom.*) *reliqui codd.* (*deest* Pb) 7 haec erit] haec (hic p) aliquando erit Gp : aliquando est P ‖ et] *om.* P 8 salvata] servata p 9 ego] *add.* vivam vel p 11–12 tempore futuro] *inv.* p 12 cum] tamen G ‖ posset] possit Pp : potest G 13 cursus] currens P ‖ nisi in] in non G ‖ et] *om.* P 14 praesenti] praesente p 15 priori] priore G : praesente priori p 16 futuro[1]] posteriori *sed add. sup. lin.* futuro C 18 haec[1]] *om.* GPp ‖ quia] quod Pp ‖ erit] est C 19 curro] curram C ‖ futuritio] *corr. in marg. ex* futuro C : futurum G ‖ connotatur] connotat C : notatur P ‖ et] *om.* P 19–20 haec sit] esset P 21 ita] *add.* etiam GPp 22 universali distributo] distributo GP : distincte p 24 appellet] appellat p 25 vel] non C 26–27 tempus ... significat] *om.* (*hom.*) GPp 27 etiam] *om.* p 28 connotat] denotat C

tem temporis, immo potius successionem et prius et posterius. Ideo nullo modo sequitur propter praedicatum appellare formam quod, si ego possum videre omne astrum, quod ego possum videre omne astrum simul, sed sufficit quod ego possum videre hoc astrum et quod possum videre illud et sic de aliis, licet | successive unum | post alterum. Ita similiter opinor quod haec 82rb P 76ra G sit vera 'Deus potest separare et separatim conservare omnes partes lineae *b*', quia omnes singulares, quantum ad singularitatem correspondentem isti universalitati 'omnes partes', sunt possibiles et compossibiles et possunt esse simul verae, licet non sit possibile quod omnes sint simul verae.

Tunc respondendum ad rationes.

⟨1⟩ Ad primam apparet manifeste quomodo sit respondendum secundum divisum sensum et sensum compositum.

⟨2⟩ Ad secundum argumentum vel ad secundam formam argumenti ego concedo | quod in qualibet medietate proportionali diei Deus fecit vel natura 93rb C unum lapidem vel unam aquam et quemlibet lapidem factum vel quamlibet aquam factam in hac die potest Deus semper in posterum conservare. Et sic ista manifeste est possibilis, cum de facto sit vera, 'Deus in qualibet medietate etc. facit unum lapidem et conservat', sed hoc facit modo continuo, non per modum discernendi et numerandi inter omnes istas medietates proportionales. Et hoc nullum est impossibile, quia sicut dies est una, continua et finita, ita lapis est unus, finitus et continuus et factio una et continua et finita. Sed non est possibile quod in qualibet medietate proportionali faciat unum lapidem pedalem, quia oporteret hoc esse per modum discretionis et numerationis inter omnes partes non solum capiendo 'omnes' distributive, sed collective, ita scilicet quod praeter illas nulla esset alia, scilicet quae non esset aliqua istarum, et hoc est impossibile, quia sic non omnes sunt aliquae nec aliquae sunt omnes.

---

2 propter] proprie C : semper G ‖ quod] *om.* GPp 3–4 omne[2] ... videre[1]] *om.* (*hom.*) p 4 et quod] quod ego G 5 licet] *add.* possum videre G ‖ similiter opinor] *inv.* G 6 sit] est GPp 7 singularitatem] singularem p 10 respondendum] *add.* est G : *om.* P 12 divisum sensum] *inv.* GPp ‖ sensum[2]] *om.* P 13 secundum ... argumenti] secundam formam argumenti sive ad secundum argumentum p 14 deus fecit] *inv.* GPp 15 et] *add.* quod GPp ‖ quamlibet] *om.* Pp 16 factam] *om.* p 17 manifeste] manifesta p 18 facit[2]] *corr. in* fecit G 20 nullum] non p 20–21 continua et finita] et continua C 21 finitus et continuus] continuus et finitus GPp ‖ factio] factione C : facto P 21–22 et[4] ... finita] continua et finita p : finita et continua G 22 qualibet] *add.* parte vel p 25 scilicet quod] sicut p ‖ scilicet[2]] *sup. lin.* C : *om.* G ‖ quae] *om.* p 26 istarum] earum G ‖ est] esset G ‖ non] nec Pp

⟨3⟩ Ad aliam concedo quod omnes partes lineae *b* Deus cognoscit ita distincte sicut Socratem et Platonem. Sed non cadit in eo proprie actus numerandi, quia ille est apprehendendo successive unitatem post unitatem et talis successio non est in Deo. Sed tamen Deus bene numerat ad talem sensum quod omnia scit quot sunt, sed non scit quot omnia sunt, quia capiendo 'omnia' collective nec aliquota nec tot sunt omnia, et capiendo distributive non est verum quod aliquota sunt omnia, sicut nec aliquoti apostoli sunt omnes apostoli, quamvis omnes apostoli sunt aliquoti apostoli. Refert enim dicere 'omnes sunt aliquot' et 'aliquot sunt omnes', quia in prima 'aliquot' stat confuse et in secunda determinate.

65va p Et tunc etiam concedendum est quod, inter quaecumque Deus discer|nit et quaecumque Deus scit quot sunt, ipse omnia ista potest dividere et separare et separatim conservare. Sed haec est impossibilis 'omnia illa separat vel dividit vel separatim conservat'.

76rb G ⟨4⟩ Ad aliam dictum est quod omnes | propositiones possibiles et compossibiles possunt esse simul verae saltem quantum spectat ad propositum, sed non est possibile omnes esse simul veras.

⟨5–6⟩ Et per hoc cum dictis prius totum solutum est quod restat.

Et sic est finis quaestionis et per consequens quaestionum tertii libri *Physicorum* a magistro Iohanne Buridano etc. | 82va P

---

3 apprehendendo] apprehendendus p 4 deus bene] *inv.* Pp 5 quod] quia GP ‖ quot[1]] quod GP ‖ quot[2]] quod GP 6 aliquota] quod P : quot p 7 verum] *add.* dicere GPp 9 aliquot[1]] aliquod G ‖ aliquot[2]] aliquod G 10 aliquot] aliquod G 11 et] *om.* P ‖ quaecumque] quotcumque p 12 quot sunt] quod sunt GP : quod C ‖ ipse] *post* ista p : *om.* GP 13–14 separat vel dividit] dividit vel separat GPp 14 separatim] *om.* G 18 per] tamen G ‖ totum] *post* est P 19–20 et ... etc.] expliciunt quaestiones tertii libri physicorum P : et sic est finis quaestionum tertii libri physicorum deo gratias ihesus christus maria G : et sic est finis tertii libri expliciunt quaestiones tertii libri physicorum p 20 buridano] *scripsimus* : buridani *vel* buridam C

⟨Tabula quaestionum quarti libri *Physicorum*⟩

Tabula quaestionum quarti libri *Physicorum* 94[ra] C

⟨1⟩ Et est prima quaestio utrum locus est aequalis locato. Quod eiusdem locati sunt infinita loca propria. Idem locatum habet locum proprium maiorem se et locum proprium minorem se.

⟨2⟩ Secunda est utrum locus est terminus corporis continentis. De proprietatibus loci. Quare locus dicitur esse superficies et non corpus, cum omnis superficies sit corpus.

⟨3⟩ Tertia est utrum locus sit immobilis. De alia acceptione 'loci' ab ista quam definit Aristoteles. Quare dicit Aristoteles locum esse immobilem, cum sit mobilis sicut locatum.

⟨4⟩ Quarta est utrum definitio loci sit bona in qua dicitur: 'locus est terminus corporis continentis' etc.

⟨5⟩ Quinta est utrum terra sit in aqua sive in superficie aquae tamquam in loco sibi proprio et naturali. Quod multa sunt corpora quae non habent loca propria, quia saepe unius corporis locus proprius non est una res, sed multae et valde diversae. De diversitate locorum naturalium diversorum corporum. Unde locus dicatur naturalis vel violentus. Utrum, quando de profundo terrae aufertur aliqua pars terrae, aer descendit naturaliter vel violenter ad replendum.

⟨6⟩ Sexta est utrum suprema sphaera est in loco. An ultima sphaera movetur secundum locum vel secundum situm. De opinionibus Commentatoris et sancti Thomae.

⟨7⟩ Septima est utrum possibile est vacuum esse. Ostenditur quod non naturaliter.

⟨8⟩ Octava est utrum possibile est vacuum esse per aliquam potentiam.

---

2 *Tabula quaestionum quarti libri Physicorum deest in G* ‖ tabula] *praem.* sequitur P ‖ physicorum] *om.* p 3 et … quaestio] prima est p ‖ utrum] *add.* omnis Pp ‖ aequalis] *add.* suo P 4 idem] item C 5 et] *add.* habet P ‖ se²] *om.* P 6 est¹] quaestio p : *om.* P 9 est] quaestio p ‖ loci] *om.* p 10 dicit aristoteles] dixit aristoteles P : definit p 12 est¹] *om.* Pp ‖ in qua dicitur] *om.* P 13 corporis] *om.* Pp ‖ etc.] *om.* P 14 est] *om.* Pp 15 sibi] suo Pp 16 quia] qui C : quod p 17 valde] *om.* P ‖ diversitate] diversorum P 18 unde] utrum C ‖ utrum] ultimum C 21 est¹] *om.* Pp ‖ an] aut p 22 secundum situm] situm P : secundum finitum C 23 sancti] beati P : *om.* p 24 est¹] *om.* Pp 26 est¹] *om.* Pp

 DOI: 10.1163/9789004322356_023

⟨9⟩ Nona est utrum in motibus gravium et levium ad sua loca naturalia tota successio proveniat ex resistentia medii. Quid est successio quae est principaliter a motore. Quid est resistens in motu simpliciter gravis deorsum. An partes resistunt sibi invicem tendentes quaelibet ad centrum. Quod una pars aquae respectu alterius non inclinat sursum nec deorsum, sed ratione gravioris vel levioris coniuncti, et sic de terra etc. Quod gravitas vel levitas aquae vel aeris non est composita ex partibus diversarum rationum, sicut esset tepiditas. De resistentia in gravibus mixtis. Unde provenit successio in motu caeli.

⟨10⟩ Decima est utrum, si vacuum esset, grave moveretur in eo. An in [94rb C] vacuo vel ultra supremam sphaeram posset homo extendere bracchium | suum.

⟨11⟩ Undecima est utrum rarefactio et condensatio sunt possibiles. Quod sic et per compressionem etc. Quod omnis rarefactio est per generationem [82vb P] magnitudinis et condensatio | per corruptionem.

⟨12⟩ Duodecima est utrum tempus est motus. Quod tempus est successivum. Quod dicitur diversis intentionibus in tantum quod aliquando vinum aut panis dicitur tempus.

⟨13⟩ Tredecima est utrum definitio temporis est bona in qua dicitur 'tempus' etc. De proprietatibus temporis.

⟨14⟩ Quarta decima est utrum cuiuslibet motus tempus sit mensura. Quid est mensurari aliqua mensura. De quinque diversis modis mensurandi. De duplici magnitudine motus et quae res sit utraque. De mensuratione parvi per magnum.

⟨15⟩ Quinta decima est utrum quies mensuratur tempore. Quod res quae [66vb p] non mutatur non mensuratur tempore; ideo nec quies. | Quomodo esse rei non mutatae mensuratur. Quomodo sine motu et tempore posset esse maior duratio et minor duratio et quomodo esset prius et posterius.

---

1 est] *om.* Pp ‖ et] aut P 2 quid] quae P 2–3 quae est principaliter] quae naturaliter et principaliter est P : quod principaliter est p 3 est] *om.* P 4 quaelibet] quilibet C ‖ centrum] centum C 5 sed] *om.* p 6 gravioris ... coniuncti] levioris vel gravioris coniuncti p : levioris vel gravioris coniunctim P ‖ levitas] *om.* C 8 de] *om.* p 10 est utrum] utrum p : *om.* P ‖ an] vel C : aut P 13 est] *om.* p 14 et] etiam Pp 15 corruptionem] *add.* magnitudinis p 16 est^{1}] *om.* Pp ‖ tempus^{1}] temporis p 17 dicitur diversis intentionibus] diversis rationibus dicitur p ‖ quod^{2}] *praem.* etiam P : *add.* etiam p 18 tempus] tempore C 19 est^{1}] *om.* Pp ‖ est^{2}] sit Pp 19–20 tempus] *om.* Pp 21 est] *om.* Pp 23 res] natura P 25 est] *om.* Pp 25–26 quod ... tempore] *om.* (*hom.*) C 26 quomodo] *add.* igitur Pp 27 mutatae] mutat et p ‖ mensuratur] *add.* tempore Pp ‖ et] ex p ‖ posset] possit P

⟨16⟩ Sexta decima et ultima est utrum tempus est, quamvis non esset aliqua anima intellectiva. Quid iste terminus 'numerus' connotat super illum terminum 'multitudo'. Utrum esset numerus, licet non esset numerans vel licet non posset esse numerans.

Et sic est finis tabulae etc.

---

1 et ultima est] *om.* Pp ‖ est^{2}] esset Pp 2 anima] *om.* P 3 multitudo] magnitudo C 4 posset] possit P 5 et … etc.] explicit tabula quarti P : *om.* p

⟨IV.1⟩

⟨Utrum omnis locus sit aequalis locato suo⟩

77ra G In isto quarto libro tractabitur de loco. Et dicit Aristoteles esse supponendum quod locus est aequalis locato. Ideo primo hic quaeratur utrum omnis locus est aequalis locato suo.

Arguitur quod non quia:

⟨1⟩ Locatum est corpus et locus est superficies; sed nulla superficies est aequalis corpori, nisi linea sit aequalis superficiei et punctum lineae, quod est impossibile; igitur etc.

⟨2⟩ Item dicit Aristoteles quod idem est locus totius et partis etiam, ut totius terrae et unius glaebae; sed non est possibile quod idem sit aequale toti et parti; igitur etc.

⟨3⟩ Item dicit Aristoteles quod ignis est in caelo tamquam in loco eius proprio, et non est ei aequalis, immo caelum est maius.

⟨4⟩ Et iterum continens videtur esse maius contento et totum parte. Et tamen locus debet continere locatum et se habere quasi totum ad partem; dicit enim Aristoteles quod unumquodque natura manet in loco suo proprio rationaliter, quia se habet ad locum sicut pars divisibilis ad totum. Igitur etc.

94va C ⟨5⟩ Item ultima sphaera habet locum, | cum moveatur localiter, et tamen non apparet quod aliquid sit sibi aequale.

⟨6⟩ Item si necesse esset locum esse aequalem suo locato, sequeretur quod ad augmentationem locati oporteret locum augmentari. Consequens est fal-

---

3–4 in … quaeratur] incipiunt quaestiones quarti libri physicorum reverendi magistri iohannis buridani in isto quarto libro tractatur primo de loco et dicit aristoteles esse supponendum quod locus est aequalis locato ideo primo quaeratur de hoc p : quaeritur primo quaestio talis G : quaeritur circa quartum librum ista quaestio consequenter P 3 quarto libro] *rep.* C 4 omnis] *om.* G 5 est] sit GPp ‖ suo] *ante* locato P : *om.* G 8 et punctum] et punctus P : punctum G 10 etiam ut] ut GP : *om.* p 13–14 eius proprio] *sup. lin.* C : *om.* GPp 15 et[1]] *om.* G ‖ esse] *om.* P 16 debet continere] locabit G 17 natura] naturaliter G ‖ suo] *sup. lin.* C : *om.* Pp 18 rationaliter] rationabiliter p ‖ locum] locus *corr. in marg. ex* totum C ‖ igitur etc.] *om.* G 21 suo] *om.* GPp ‖ sequeretur] sequitur p 22 locati] *om.* P ‖ locum] locus C ‖ est] *om.* G

---

3 Aristoteles, *Physica*, IV, 3, 211a2; cf. *AA*, 2: 127 10 Aristoteles, *Physica*, III, 5, 205a10–11 13 Aristoteles, *Physica*, IV, 5, 212b21 17 Cf. Aristoteles, *Physica*, IV, 5, 212b34–213a1

 | DOI: 10.1163/9789004322356_024

sum. Primo, quia omne quod augmentatur est corpus et Aristoteles negat locum esse corpus. Secundo patet falsitas: quod augmentatur movetur, cum augmentatio sit motus, et Aristoteles dicit locum esse immobilem. Tertio patet falsitas, quia locus est continens locatum et non oportet continens augeri, si contentum augeatur, immo potius diminui. Verbi gratia ignis in sphaera sua continet aerem. Ponamus igitur quod magna quantitas ignis convertatur in aerem, | 83ra P quia elementa sunt ex se invicem generabilia. Constat quod ille ignis continens diminuetur et tamen aer contentus augmentabitur. Igitur etc.

Oppositum arguitur per Aristotelem dicentem quod oportet locum proprium non maiorem esse locato nec minorem. Et hoc dicit esse supponendum.

Sciendum est quod multa verba Aristotelis de loco sunt praeter proprietatem sermonis et ideo difficile est proprie loqui de loco.

Primo tamen, sicut bene dicit Aristoteles, distinguendum est de loco, quoniam alius est locus proprius et alius est locus communis. Locus communis Socratis est qui cum Socrate continet alia corpora, ut si dicamus Socratem esse in ecclesia, non dicimus hoc tamquam in loco sibi proprio, sed communi sibi et multis aliis. Sed ille diceretur locus proprius Socratis, qui contineret totum Socratem et cum eo non contineret aliud corpus, saltem quod non sit pars illius. Sic etiam dicit Aristoteles locum esse tibi proprium qui continet te et nihil plus quam te.

Tunc ponuntur conclusiones.

---

1 primo quia] quod P ‖ augmentatur] augetur G 1–2 et ... corpus] *om.* (*hom.*) G 2 patet falsitas] quia GPp ‖ augmentatur] augetur G 4 patet falsitas] *om.* GPp 5 augeri] *add. in marg.* augmentari C : augmentari Pp ‖ augeatur] augmentatur Pp 6 quantitas] *add.* illius GPp 7 convertatur in aerem] contineatur in aere C ‖ ex se] ad G : extra P ‖ generabilia] generalia C 8 diminuetur] diminuitur G 9 igitur etc.] *om.* GP 11 non] neque GPp ‖ esse[1]] *om.* P ‖ nec] neque GPp 13 est] *om.* P ‖ sunt] dicit C 14 et] *om.* P ‖ proprie loqui] *inv.* GPp 15 primo] proprio G ‖ sicut bene] ut P 15–16 quoniam] quia p : *om.* P 16 et] *om.* GPp ‖ est locus[2]] est p : *om.* GP 18 ecclesia] *corr. sup. lin. ex* aere C : aere G ‖ dicimus] dicamus P 19 multis] etiam *sed add. in marg.* multis C ‖ qui] quae p 21 etiam] enim GPp 23 tunc] et tunc statim Pp

---

1 Cf. Aristoteles, *Physica*, IV, 2, 209a5–7 3 Aristoteles, *Physica*, IV, 4, 212a17–18, 20–21 10 Aristoteles, *Physica*, IV, 3, 211a2; cf. *AA*, 2: 127 15 Aristoteles, *Physica*, IV, 2, 209a32–33; cf. *AA*, 2: 120 21 Cf. Aristoteles, *Physica*, IV, 2, 209a35–b1

Prima est quod idem est locus proprius et communis respectu diversorum, quia concavum orbis lunae est locus communis terrae et aliis et est 67ra p locus proprius congregati ex omnibus quattuor elementis et | mixtis.

Secunda conclusio est quod non omne corpus habet locum proprium secundum dictam expositionem huius termini 'locus proprius'. Primo enim est instantia de ultima sphaera, quia nihil continet eam. Secundo excipitur pars in toto continua aliis partibus, quia locus debet esse continens divisum, ut dicit Aristoteles. Tertio excipitur omne corpus continens aliud corpus quod non est pars ipsius, ut ignis in sphaera sua. Nihil enim continet ipsum, quin cum eo contineat alia, scilicet terram, aerem etc. Et nullus etiam est locus continens hominem, quin contineat multa alia quae sunt in homine 94vb C et quae non sunt partes | eius, sicut sunt superfluitates et sanguis et cibus in stomacho et aer inspiratus et cetera huiusmodi. Et propter hoc homo non habet locum proprium, sed congregatum ex homine et contentis in eo bene haberet locum proprium. Verum est quod bene omne corpus haberet locum proprium et vere sibi aequalem, si locus esset spatium separatum. De quo post dicemus, sed ex dicendis post supponimus non esse ita.

Tertia conclusio quod eiusdem locati infinita sunt secundum multitudinem loca propria, ut huius mundi inferioris aggregati ex omnibus elementis et mixtis locus proprius est concava superficies orbis lunae circumdans totum istum mundum inferiorem, sicut supponimus et post probabitur. 83rb P Tunc igitur arguo sic: omnis superficies concava orbis lunae | tangens illum mundum et circumdans ipsum totum est locus proprius huius mundi inferioris; sed infinitae sunt tales superficies; igitur etc. Maior nota, quia aliqua talis ponitur esse locus proprius huius mundi inferioris et non est ratio quare una magis quam alia, si plures sint. Sed minor apparet, quia non ponimus superficiem distinctam a corpore nec terminum corporis distinctum a cor-

---

1 prima] *add.* conclusio P 2 et[1]] *corr. in marg. ex* vel C : vel GPp 4 est] *om.* G 5 huius ... proprius] loci proprii P ‖ enim] *om.* P 6 instantia] *post* sphaera P ‖ secundo] *add.* autem P 7 continua] *ante* in C 10 terram] *add.* et Pp 11 quin] qui non GPp ‖ quae] *add.* non P 12 quae] *om.* p 13 cetera] *om.* GPp ‖ et propter hoc] propter quod GPp 14 congregatum] aggregatum P ‖ contentis] contentum G ‖ bene] unde C 15–16 verum ... proprium] *om.* (*hom.*) C 15 bene] *om.* p 17 dicemus] dicetur p 18 conclusio] *add.* est Gp 19 aggregati] congregati P : et congregati Gp 20 superficies] *in marg.* C : *ante* concava GPp 21 post] postea GPp 22 igitur arguo] arguo G : arguitur P 23 mundum] *add.* inferiorem GPp 24 nota] patet G : apparet Pp 25 inferioris] *om.* P 26 una magis] *inv.* GPp ‖ sint] *add.* tales GPp ‖ apparet] *praem.* ex hoc P : ex hoc patet Gp

---

8 Aristoteles, *Physica*, IV, 4, 211a2–3 17 Cf. inf., IV, q. 2 21 Cf. inf., IV, q. 2

pore, sicut ante dictum fuit et dicetur in sexto. Immo superficies extrema vel terminus corporis est eius ultima pars. Et sicut infinitae sunt ultimae partes lineae, ita et corporis. Igitur debemus imaginari quod, si orbis lunae dividatur orbiculariter in duas medietates vel in tres tertias vel in decem decimas vel centum centesimas et sic deinceps, semper istarum partium erit una ultima versus nos tangens illum mundum inferiorem, scilicet sphaeram ignis. Verbi gratia erit ultima duarum medietatum et ultima decimarum et ultima centesimarum et sic sine statu et quaelibet istarum erit ultima superficies orbis lunae versus nos, quia qua ratione una, eadem ratione alia. Et sic quaelibet istarum partium est locus proprius.

Sed aliquis obiciet quod quaelibet istarum partium | vel superficierum 77[rb] G continet aliquid extra istum mundum inferiorem, quia continet suas partes.

Non valet instantia. Et dicendum est quod hoc non obstat, quin sit locus proprius, quia cum proprius locus non debet esse continuus locato, sed divisus, ut dicetur post, ideo loci proprii debet esse quod nihil a se divisum contineat praeter locatum cuius dicitur esse locus proprius. Potest tamen continere partes suas ab eo non divisas.

Quarta conclusio est quod idem locatum habet locum proprium | aequa- 95[ra] C lem sibi et locum proprium maiorem se et locum proprium minorem se. Magnitudo enim corporis non attenditur solum secundum longitudinem nec solum secundum latitudinem nec solum secundum profunditatem, sed secundum haec tria simul. Unde cera non est maior, si protendatur in longum, quam si etiam contrahatur in unum globum rotundum, quamvis multo sit longior. Ponamus igitur quod vini in dolio locus proprius sit dolium vel superficies tangens vinum. Et tamen illud dolium est minus quam illud

---

1 sicut] ut P 2 corporis] communis P 3 igitur] *praem.* nos GP : nos p 4 orbiculariter] *om.* p 5 vel] *add.* in GPp 6 nos] non P 7 duarum] *om.* GPp ‖ et] vel P : *om.* Gp 8 et[1]] *om.* Gp ‖ ultima centesimarum et] *om.* (*hom.*) P ‖ et[3]] *om.* C ‖ ultima[2]] una G 9 quia] et C ‖ ratione[2]] *om.* GP 10 partium] *om.* GPp ‖ proprius] *add.* etc. Pp 11 sed aliquis obiciet] nec valet si aliquis obiciat GPp 12 aliquid] aliqua GPp ‖ quia] qui C ‖ suas partes] *inv.* GP : partes duas p 13 non[1] ... est] dicendum est enim GPp 14 proprius[2]] *om.* GPp ‖ debet] debeat Pp 15 ideo] ratio p ‖ divisum] *ante* a GPp 19 et[1]] *add.* habet G ‖ locum proprium[2]] *om.* P 21 nec solum secundum[2]] vel G 22 protendatur] *corr. sup. lin. ex* tendatur C : procedatur GP 23 etiam ... rotundum] retrahatur in unum globum rotundum G : retrahatur in globum rotundum et unum P : in unum globum rotundum retrahatur p 23–24 multo sit] *inv.* GPp 24 vini] *post* dolio P ‖ sit[2]] est GPp 25 superficies] *add.* dolii GPp ‖ et] *om.* P 25–208.1 et ... vinum] *om.* (*hom.*) C 25 quam illud] quam sit illud p : quam P

---

1 Cf. sup., III, q. 8; cf. Iohannes Buridanus, *Quaestiones super libros Physicorum*, VI, q. 4 (ed. Parisiis 1509, ff. 96[rb]–98[va]) 15 Cf. inf., IV, q. 4

vinum. Quod apparet, quia asseres si truncentur per medium et congregentur simul in unum cumulum et ligentur, iste cumulus erit minor illo vino et
67rb p secundum longitudinem et secundum latitudinem et profundita|tem; ideo locus ille erat minor simpliciter isto locato vino.

Sed etiam, si superficies magni aeris tangens undique granum milii sit locus proprius illius grani, illa superficies poterit capi tantae profunditatis
83va P quod ipsa erit in centuplo | maior illo grano, ut si ille aer ponatur dividi orbiculariter circa illud granum in tres tertias vel in quattuor quartas, quaelibet istarum partium erit multum magna, licet ultima istarum tertiarum vel quartarum tangens granum sit eius locus proprius. Sed iste aer potest orbiculariter dividi in tot partes quod una est aequalis illi grano aut minor, immo in infinitum potest dari minor; ideo in infinitum est locus proprius illius locati parvus. Immo etiam concludetur quod nullus locus proprius grani milii est ita parvus, quin aliquis locus proprius totius mundi inferioris sit minor.

Quinta conclusio est quod omnis locus proprius aequalis est suo locato tribus modis, qui forte omnes coincidunt secundum quod est expressio per tres modos locutionum.

Primus dicitur aequalitas secundum continentiam, scilicet quod quantumcumque est locatum, tantum locus continet, et quantumcumque locus continet, tantum locatum est. Et non est mirum, quia nihil est locatum aliud ab eo quod locus continet nec locus continet aliud a locato.

Secundus modus dicitur aequalitas secundum mensuras superficiales, videlicet quod, si aliqua mensura circumvolveretur alicui corpori locato extrinsece tangendo semper corpus locatum et alia etiam circumvolveretur loco intrinsece tangendo corpus locans, istae duae mensurae, si in rectum deducerentur, essent sibi invicem aequales. Sic enim mensurant ligatores
95rb C doliorum vini vocatos circulos quibus ligant dolia; mensurant enim eadem | mensura dolium extrinsece et circulum intrinsece. Vel etiam, quia si superfi-

---

1 asseres si] si asseres dolii G : asseres dolii si Pp 2 minor] maior C ‖ et^2] etiam p 3 et^2] *add.* secundum p 4 erat] erit P ‖ minor] maior *sed add. sup. lin.* minor C ‖ vino] *ante* locato Pp : *om.* G 5 etiam] iterum GPp ‖ sit] *add. sup. lin.* sibi C 6 illius grani] grani milii p 7 ipsa] illa G 8 illud granum] *inv.* GPp ‖ tres tertias] *inv.* p ‖ vel in] et in G : vel Pp 9 tertiarum] *om.* P 10 potest] posset sic GPp 11 est] erit G : esset Pp 12 potest] posset GPp 13 concludetur] concluderetur p 14 proprius] *add.* istius GPp 15 est quod] *om.* p ‖ aequalis est] *inv.* GPp ‖ locato] *add.* proprio GPp 16 forte omnes] *inv.* Pp : omnes forte hic G ‖ secundum quod] sed GPp 20 et non est] et neque Pp : neque G ‖ nihil] nec G : neque P : non p 23 videlicet] scilicet G ‖ alicui] *om.* GPp 24 extrinsece] extrinseco p ‖ alia] una Gp 25 intrinsece] intrinseco p ‖ rectum] rectitudine G : rectitudinem p 26 essent] *add. in marg.* simpliciter C ‖ aequales] *om.* G 27 doliorum] *om.* G ‖ enim] in P

cies unius superponatur superficiei alterius tangendo, quaelibet pars superficiei unius tangit aliquam partem alterius et e converso, quantum ad partes divisas contra invicem secundum duplicem mensuram, scilicet longitudinis et latitudinis, circumscripta divisione secundum mensuram tertiam, scilicet profunditatis. Et sic est de loco quantum ad superficiem eius interiorem et de locato quantum ad exteriorem.

Tertius modus dicetur aequalitas secundum suas diametros, videlicet quia quanta esset linea diametralis ducta per ipsum locatum de uno puncto eius extremo ad alterum punctum extremum ipsius terminata per illa puncta inclusive, tanta esset linea diametralis inter puncta loci correspondentia illis punctis locati terminata ad illa puncta loci exclusive, immo illa esset eadem linea, secundum quamcumque distantiam illa diameter duceretur.

Et istos modos aequalitatis intendit Aristoteles, cum dicit locum proprium debere esse aequalem locato.

Respondetur ad rationes.

⟨1⟩ Ad primam dicendum est quod superficies est corpus; ideo est aequalis corpori. Et punctum etiam vel nihil est vel est linea et sic est aequalis lineae. Et adhuc, si essent superficies indivisibiles secundum profunditatem, locus esset aequalis locato | non simpliciter, sed sic quod esset aequalis eius 83vb P
superficiei exteriori.

⟨2⟩ Ad aliam dicendum est quod bene idem est locus naturalis et conveniens toti et parti, sed non esset proprius utrique secundum prius dictam expositionem, immo proprius toti esset communis parti.

⟨3⟩ Ad aliam dico quod Aristoteles glossat se quod non totum caelum est locus ignis proprius, sed eius ultimum.

⟨4⟩ Ad aliam dicitur quod continens potest esse contento minus vel aequale, ut dictum est. Sed de toto et parte dicendum est quod locatum non

1 superponatur] supponatur p 2 partem] *add.* superficiei G || partes] *add.* dictas C 4 latitudinis] *add.* et P || tertiam] tertii p 5 profunditatis] profunditas CG || eius] *in marg.* C : *ante* superficiem GPp || et[2]] *add.* sic C 7 dicetur] diceretur GPp || suas] suos GPp || videlicet] *om.* G 8–9 puncto eius extremo] eius puncto extremo P : extremo puncto G 9 alterum punctum extremum] alterum extremum punctum P : aliud G || ipsius] diximus *sed add. sup. lin.* ipsius C 10–11 correspondentia] correspondens p 12 illa] ille p 14 cum] qui C 16 respondetur] tunc Gp : *om.* P 22 dicendum est] dicitur GPp || est locus] locus esset GPp 23 et] *om.* p 26 eius ultimum] *inv.* G 27 contento] *add.* maius vel P 28 ut] sicut ante GPp

14 Aristoteles, *Physica*, IV, 3, 211a2; cf. *AA*, 2:127 25 Cf. Aristoteles, *Physica*, IV, 5, 212b19

67va p est pars loci, sed se habet se|cundum quandam similitudinem sicut pars, quia continetur in loco et pars in toto.

⟨5⟩ De loco ultimae sphaerae fiet quaestio specialis.

⟨6⟩ Ad ultimam conceditur quod non oportet locus augmentari, si locatum augmentatur, quia potest esse minus suo locato. Sed oportet latera eius elongari ab invicem, ut infra possit esse maius corpus. De mobilitate autem loci fiet postea quaestio specialis.

Et sic quaestio finitur.

---

4 locus] locum GPp ‖ augmentari] augmentare P 5 augmentatur] augetur G 6 elongari] elongare P ‖ mobilitate] immobilitate G 8 et ... finitur] et sic est finis istius quaestionis etc. P : etc. p

---

3 Cf. inf., IV, q. 6 7 Cf. inf., IV, q. 3

⟨IV.2⟩

⟨Utrum locus sit terminus corporis continentis⟩

Quaeritur secundo utrum locus sit terminus corporis continentis.

Et arguitur quod non quia: terminus corporis continentis est eius superficies et locus non est superficies. Hoc probatur multipliciter.

⟨1⟩ Primo quia: in *Praedicamentis* ponitur locus esse species quantitatis | distincta a superficie. 77va G

⟨2⟩ Secundo quia: superficies est de praedicamento quantitatis; et non videtur quod locus sit de praedicamento quantitatis | secundum rei veritatem, sed de praedicamento ubi, cum eaedem sint differentiae loci et ubi, scilicet sursum et deorsum, et cum motus secundum locum vocetur motus ad ubi (quod etiam apparet, quia per motum localem mutatur responsio ad ubi). 95va C

⟨3⟩ Tertio etiam quia: loca sunt sibi invicem contraria, scilicet sursum et deorsum; et superficiei, cum sit quantitas, nihil est contrarium.

⟨4⟩ Quarto quia: sursum et deorsum sunt differentiae naturales locorum, non autem superficierum, immo eiusdem speciei specialissimae in praedicamento quantitatis est superficies aliqua in caelo et superficies aliqua in igne vel terra, sursum et deorsum etc.

⟨5⟩ Quinto quia: dicit Commentator quarto *huius* quod quantitas non est de virtutibus activis; nec igitur superficies. Et tamen locus est de virtutibus activis, quia est principium generationis quemadmodum et pater, ut dicit Porphyrius.

⟨6⟩ Sexto quia: superficies non est aequalis corpori et locus est aequalis locato. Sed de hoc dictum est in alia quaestione.

---

3 quaeritur secundo] secundo consequenter quaeritur G **4** et] *om.* GPp **5** hoc] *om.* P **6** primo] *om.* P ‖ locus] *in marg.* C : *ante* in P : *om.* G **10** cum eaedem sint] cum eodem sint p : sicut G **11** cum] tamen P ‖ vocetur] vocatur P **13** ubi] *add.* est hoc GPp **14** sibi] ad P : *om.* Gp **16** naturales] *om.* G **17** non autem superficierum] *om.* G **18** superficies[2]] *om.* P **19** et] vel p **20** quia] *om.* P ‖ quarto huius] *om.* P **21** virtutibus[2]] *post* activis (22) p : potentiis G **22** quemadmodum] sicut G ‖ dicit] dicitur C

---

6 Cf. Aristoteles, *Praedicamenta*, 6, 4b24–25 **20** Averroes, *In Physicam*, IV, comm. 84, f. 171L; cf. *AA*, 2: 144 **23** Cf. Porphyrius, *Isagoge*, 2, 4–5 (ed. Minio-Paluello, 6) **25** Cf. sup., IV, q. 1

 | DOI: 10.1163/9789004322356_025

⟨7⟩ Septimo quia: locus est immobilis secundum Aristotelem et superficies continentis est mobilis, immo superficies orbis lunae continens ignem movetur continue.

⟨8⟩ Octavo quia: quod continue fit in alio et alio loco, movetur localiter et non quiescit. Quod tamen esset falsum, si locus esset superficies continentis, nam arbor quiescens in campo fieret continue in alio et alio loco, quando ventus flat et movet aerem circumstantem.

⟨9⟩ Nono quia: tunc manens in eodem loco continue moveretur localiter, quod videtur implicare contradictionem. Consequentia patet de pluma 84ra P quae continue | defertur et movetur in eodem aere ad motum illius aeris, et de vino quod existens in navi movetur continue manens semper in eodem dolio.

⟨10⟩ Decimo quia: lapidem vel arborem in campis tu dices anno sequente esse in eodem loco in quo erat anno praeterito, quamvis mutatus sit aer continens et per consequens superficies eius.

Oppositum determinat Aristoteles.

In ista quaestione volo procedere textualiter, scilicet explanando vel recolligendo processum Aristotelis. Et ipse dicit prooemio *De anima* quod accidentia magnam partem conferunt ad cognoscendum quod quid est. Ideo etiam dicit hic Aristoteles quod per proprietates quae loco apparent inesse et convenire poterit inquiri essentia sive quidditas loci. Aristoteles statim enumerat plures proprietates locorum supponendas tamquam manifestas.

Prima est quod locus debet continere locatum, quia omnes communiter dicunt locatum esse in suo loco.

---

1 septimo quia] item septimo P ‖ secundum aristotelem] ut dicit aristoteles GPp 3 movetur continue] *inv.* GPp 4 quia] *om.* P ‖ et[1]] *add.* in P 6 arbor quiescens] aliquod quiescens sicut arbor p ‖ et] *add.* in Pp 7 flat] flaret p ‖ circumstantem] circumdantem GPp 9 pluma] pluvia Cp 10 quae] quia C 11 quod] *om.* G ‖ manens semper] *inv.* P 12 dolio] loco G 13 dices] dicis p 14 erat] erant P ‖ praeterito] praecedente GPp 15 consequens] *add.* etiam G 17 scilicet] videlicet P 18 processum] textum G ‖ et ipse] ipse enim C ‖ dicit] *add.* in p 19–20 ideo etiam] *praem.* et P : ideo p : etiam G 20 dicit hic] hoc dicit C : dicit G 21 et convenire] *om.* P ‖ essentia] *add. in marg.* sive natura C : natura GPp ‖ loci] *add.* et p 21–22 statim enumerat] numerat P 22 supponendas] suppositas P

---

1 Aristoteles, *Physica*, IV, 4, 212a17–18, 20–21 16 Cf. Aristoteles, *Physica*, IV, 4, 212a20–21 18 Aristoteles, *De anima*, I, 1, 402b21–22; cf. *AA*, 6: 7 20 Cf. Aristoteles, *Physica*, IV, 4, 210b32 21–22 Cf. Aristoteles, *Physica*, IV, 4, 210b34–211a6

Secunda est quod locus non est aliquid ipsius locati, scilicet tamquam pars eius vel tamquam accidens sibi inhaerens, | 95vb C quia tale oportet moveri cum locato ad motum eius, locus autem non oportet moveri cum locato, im|mo 67vb p cum locatum movetur, dicimus ipsum mutare locum et recedere ab uno loco et ire ad alium locum.

Tertia est quod locus proprius debet esse aequalis locato. Et de ista dictum est in alia quaestione. Et Commentator arguit ad probandum istam proprietatem quia: locatum si excederet continentiam loci, oporteret esse penetrationem corporum, scilicet locati et locantis, et si locatum deficeret a continentia loci, oporteret esse vacuum inter locum et locatum; et haec sunt impossibilia.

Quarta est non deficere locum corpori, id est omne corpus habere locum. Et hoc non credo esse verum loquendo de loco continente locatum, quia ultima sphaera sic non habet locum. Sed hoc de facto est verum de omni corpore quod est et quod est mobile naturaliter motu recto, de quibus in isto libro loquitur Aristoteles magis quam de aliis, ut dicebatur in tertio libro, capitulo de infinito. Et huic proprietati Aristoteles addit quod locus est separabilis a locato et e converso; et hoc pertinet ad secundam proprietatem et ad eius probationem.

Quinta est quod locorum differentiae sunt sursum et deorsum. Et hoc non est universaliter verum. Aliqua enim loca differunt quorum neutrum est magis sursum vel deorsum quam reliquum, ut si terra sub aequinoctiali esset rotunda et aliquid moveretur super eam tangendo eam, illud moveretur de uno loco ad alium locum sine differentia eorum penes sursum et deorsum. Et haec proprietas est vera de locis naturalibus quattuor elementorum, gravium et levium.

Sexta est gravia naturaliter quiescere deorsum, si ibi sint, et levia sursum, etiam gravia moveri naturaliter deorsum, si sint sursum et non prohibita, et

1 secunda] *add.* proprietas GP : *add.* proprietas autem p 3 locus] locum GPp 6 tertia] *add.* proprietas GPp 8 locatum si] *inv.* GPp 9–10 a continentia] ad continentiam G 10 esse] interesse Pp ‖ haec] sic P 12 quarta] *add.* proprietas GPp ‖ locum[1]] locus C ‖ corpori id est] scilicet p ‖ omne] esse P 14 de facto] *post* verum G 15 et quod est] *sup. lin.* C : *om.* P 16 aristoteles] *ante* in (15) G 17 aristoteles addit] *inv.* GPp 18 secundam] illam G 19 ad] *om.* G 20 quinta] *add.* proprietas Gp ‖ deorsum] *add.* etc. P 22 vel] *om.* P 23 aliquid] aliquis p 24 alium] alterum Gp ‖ eorum] earum GP ‖ et[2]] sed GPp 27 est] *praem.* proprietas G : proprietas Pp ‖ quiescere] quiescent p 28 etiam] et G : *praem.* et p ‖ moveri] moventur Gp ‖ sint] sunt P ‖ et[1]] si sint G

7 Cf. sup., IV, q. 1 7–10 *locus non inventus* 16–17 Cf. sup., III, q. 14

levia e converso. Et hoc apparet ad sensum, quia locus sursum est naturalis 84rb P et conveniens | levibus et locus deorsum gravibus.

Et ex hoc sequitur septima proprietas, scilicet quod locorum sursum et deorsum sunt potentiae diversae naturales, quia aliter non esset ratio quare gravia magis moverentur deorsum naturaliter quam sursum et levia e converso, nec esset ratio quare locus deorsum esset magis naturalis terrae et locus sursum igni quam e converso.

Et ex his debet inferri quod, cui non conveniunt dictae proprietates, hoc non debet poni esse locus vel saltem hoc non est locus proprius et naturalis alicuius corporum gravium et levium.

Istis visis dicit Aristoteles ulterius quod de quidditate loci fuerunt quattuor 96ra C opiniones antiquorum. Una scilicet quod | locus est materia corporis locati, quia utrumque est receptivum. Sed tamen ratio non valet, quia non eiusdem, immo locus corporum actu existentium et localiter motorum, materia vero transmutationum et formarum substantialium. Secundo etiam, quia remotis a corpore formis substantialibus et accidentibus, ut quantitatibus et qualitatibus, nihil videtur remanere nisi materia, et tamen remaneret locus. Sed manifestum est quod, licet nihil remaneret quod esset de intrinseci- 77vb G tate corporis illius, | tamen bene remanet corpus quod extrinsece continebat ipsum.

Secunda opinio fuit quod locus esset forma ipsius corporis locati (et nota quod per 'formam' non intendebat formam substantialem, sed propriam superficiem corporis locati), quia locus est terminus et illa forma est terminus, et quia etiam locus continet locatum et etiam superficies continet ipsum. Sed dicit Aristoteles quod illa forma vel superficies locati est terminus intrinsecus locati et continens intrinsecum, locus autem est continens extrinsecum et terminus corporis extrinseci.

---

1 et] *om.* G ‖ sensum] *add.* et hoc est Gp 2 et conveniens] *om.* G 3 septima] alia P ‖ scilicet] *om.* Pp 4 potentiae diversae] *inv.* G ‖ quia] *sup. lin.* C : *om.* GPp 5 gravia] *om.* G ‖ deorsum naturaliter] deorsum P : seorsum naturaliter p 6 esset[1]] etiam P ‖ esset[2]] est G 9 vel … locus[2]] *om.* (*hom.*) P 10 alicuius] aliquorum G 11 aristoteles ulterius] *inv.* GPp 12 una scilicet] *inv.* P ‖ corporis] *add.* vel p 14 locus] *add. sup. lin.* est C ‖ et] *add.* locatorum p 16 accidentibus] *praem.* etiam p : etiam accidentalibus GP ‖ ut] et P 17 videtur] videretur G ‖ remaneret] remanet Gp 18–19 intrinsecitate] intrinsitate p 19 corporis illius] *inv.* G ‖ remanet] remaneret GP ‖ extrinsece] intrinsece p 21 fuit] *om.* G ‖ nota] notate Pp : forte G 22 substantialem] *add.* corporis locati P 23 illa forma] illa Pp : ita superficies G 24 continet[1]] tenet p ‖ et etiam] et illa G : et p : etiam P 27 extrinseci] intrinseci p

---

11 Cf. Aristoteles, *Physica*, IV, 4, 211b5–9 25 Aristoteles, *Physica*, IV, 4, 211b13–14

Tertia opinio fuit quod locus est spatium separatum aequale secundum omnem dimensionem corpori locato, in quo imaginamur recipi corpus locatum secundum modum penetrationis. Sic enim imaginamur spatium infinitum ultra caelum, in quo | infiniti possunt fieri mundi. Et sic imaginamur (68ra p) quod ante creationem mundi erat spatium tantundem quantus est iste mundus, in quo fiebat iste mundus. Quod quidem spatium occupat iste mundus ita quod de illo spatio quodlibet corpus partibile huius mundi occupat partem sibi aequalem. Et sine dubio, si esset spatium tale, rationabile esset dicere quod ipsum esset locus. Et ideo, quia omnes imaginantur vel imaginari possunt quod sit tale spatium, ideo fuit opinio multorum quod tale spatium sit locus. Ad talem imaginationem apud vulgares iuvat multum insensibilitas aeris, propter quam reputat vulgus vas esse vacuum in quo non est nisi aer, et credit | in illo vase non esse aliquod corpus naturale, sed solum (84va P) tale spatium.

Quarta opinio est Aristotelis, scilicet quod locus est continens extrinsecum ipsi locato secundum magnitudinem et situm. Et hoc dicit esse superficiem corporis continentis.

Ut igitur videatur de veritate istarum opinionum, componuntur | conclusio- (96rb C) nes.

Prima est quod locus non est materia nec forma ipsius locati nec est pars aliqua nec aliquod accidens sibi inhaerens. Ista conclusio patet, quia nulli istorum conveniunt dictae proprietates loci. Nam proprie loquendo pars non continet totum suum nec accidens subiectum suum, sed potius e converso; et locus debet continere locatum. Et iterum locatum potest removeri a loco in quo est et intrare alium per motum localem remanente ipso toto locato et accidentibus sibi inhaerentibus; sic autem non potest separari a suis partibus et accidentibus. Et corpus motum localiter mutat

---

2 imaginamur] imaginatur CG 3 secundum] per P ‖ imaginamur] imaginatur G 4 possunt] possent p ‖ imaginamur] imaginatur G 5 tantundem] mundi G 6 quod quidem] quidquid C ‖ occupat iste mundus] iste mundus occupat G 7 de illo spatio] illud spatium P 8 spatium tale] *inv.* GPp 11 ad talem] et ad illam GPp ‖ iuvat] innuit P 12 quam] quod P ‖ vas] totum C : pottum P 13 aliquod] aliud P 14 tale spatium] vacuum p 15 scilicet] *om.* G ‖ est^{2}] sit GPp 16 ipsi locato] ipsum locatum p ‖ magnitudinem] imaginationem C ‖ dicit] dixit p 18 componuntur] pono GPp 20 prima] *add.* conclusio GPp ‖ non] nec Pp ‖ ipsius] *om.* P ‖ est^{3}] etiam G 21 nec] vel Gp 22 conveniunt] competunt P ‖ nam] quia P 23 subiectum suum] *inv.* P 24 et^{2}] *om.* Gp 26 potest] *add.* locatum p

---

15–17 Cf. Aristoteles, *Physica*, IV, 4, 212a5–6

locum et non mutat materiam vel formam vel accidentia sibi inhaerentia. Et terra aliquando etiam per violentiam removetur a loco suo naturali cum omnibus suis partibus in ipsa manentibus. Nec umquam corpus integratum ex materia et forma movetur ad illam materiam vel formam, quia habet eas; et bene movetur localiter ad suum locum naturalem. Nec umquam materia et forma alicuius corporis sunt sibi violentae et innaturales; et corpora naturalia saepe sunt in locis sibi violentis et innaturalibus.

Secunda conclusio est quod locus non est spatium separatum secundum imaginationem prius positam. Probatio quia: non est aliquod tale spatium, sicut magis dicetur in capitulo de vacuo. Etiam de hoc dictum fuit in tertio libro. Illud enim spatium non esset substantia; et si esset accidens, tunc esset accidens sine subiecto, quod non est possibile naturaliter. Et esset penetratio dimensionum, quia illud spatium non ponitur cedere corpori imposito, et tamen illud spatium esset dimensio corporea habens latum, longum et profundum; sed impossibilis est penetratio dimensionum, ut magis dicetur in capitulo de vacuo.

Iterum illud spatium, ex quo non esset aliud quam dimensio, nihil proficeret ad locandum, quia locatum habet suas dimensiones proprias, quibus dimensionatae sunt eius materia et eius forma et omnes eius qualitates; ideo quantum ad hoc non indiget alia dimensione. Si autem dicatur quod eius propriae dimensiones indigent aliis dimensionibus in quibus recipiantur, hoc est inconveniens, quia pari ratione illae aliae dimensiones iterum indigerent aliis dimensionibus et sic in infinitum, quod est falsum.

Iterum ex quo magnitudo mundi non est aliud quam dimensio et per consequens spatium, et sic etiam magnitudines partium mundi non sunt nisi
96va C spatia et dimensiones, | non apparet quid proficeret aliqua alia dimensio vel
84vb P aliud spatium, quia Deus non indigebat spatio praesupposito | ad creandum

1 materiam] *add.* suam GPp ‖ formam] *add.* suam G 2 aliquando etiam] *inv.* GPp ‖ loco suo] *inv.* GPp ‖ cum] tamen P 3 manentibus] remanentibus G 5 localiter] naturaliter GPp 6 et[3]] sed P 8 secunda] *add.* autem p 9 probatio] primo GPp ‖ aliquod] *sup. lin.* C : aliud P 10 etiam de hoc] et de hoc etiam Gp : et hoc etiam P 11 esset[1]] *add.* sic P 13 imposito et] imposito P : in proposito et p 14 latum longum] *inv.* GP 15 impossibilis] impossibile P ‖ magis dicetur] *inv.* p 16 in] *om.* P 17 esset] est p 18 suas dimensiones] *inv.* GPp 19 dimensionatae] dimensionati P 21 dimensiones] *ante* eius (20) GPp 22–23 iterum indigerent] *inv.* GPp 23 in] *om.* p 24–25 consequens] *add.* sic P 25 et] *om.* P : partium p ‖ etiam] et p ‖ partium] spatia G 26 spatia et dimensiones] dimensiones et spatia G ‖ apparet] *add.* ad p ‖ aliqua] *om.* GPp 27 quia] quoniam GPp

10 Cf. inf., IV, q. 7 10–11 Cf. sup., III, q. 15 16 Cf. inf., IV, q. 7

mundum et eius magnitudinem, quoniam Deus, etiam si erat tale spatium, poterat ipsum annihilare et iterum tale creare. Et si igitur in creando illud spatium non indi|gebat alio in quo crearet, ita nec in creando magnitudinem 68[rb] p mundi indigebat alia magnitudine vel dimensione in qua crearet mundum.

Iterum non imaginamur illud spatium nisi quia imaginamur corpus motum localiter occupare successive aliud et aliud spatium recipiens ipsum. Sed sicut potest moveri hoc corpus, ita Deus posset cum eo movere spatium quod occupat. Ideo illud spatium indigeret iterum alio spatio recipiente ipsum et sic in infinitum.

Et iterum alias persuasiones ponit Aristoteles, videlicet quia sequeretur quod pars quaelibet corporis per se subsistentis ita esset per se in loco sicut illud totum per se subsistens, quod videtur falsum, cum non moveatur per se localiter, sed ad motum totius. Consequentia patet, quia ita quaelibet pars occuparet spatium sibi aequale sicut totum. | 78[ra] G

Alia persuasio est quod quodlibet corpus naturale totale et per se mobile, debet habere locum totalem sibi proprium, quod non esset ita, immo unus solus esset locus totalis, scilicet spatium continuum continens totum mundum. Et nullum corporum particularium, quantumcumque esset, per se mobile haberet locum aliquem totalem, sed solum unam partem loci continui qui esset totius mundi locus.

Iterum alia persuasio est quod loca naturalia debent habere diversas proprietates naturales sursum et deorsum, ut dicebatur in assignando proprietates loci. Sed sic non esset, quia spatium tale separatum esset ubique eiusdem rationis. Et ita non esset aliqua ratio ex parte locorum quare grave magis moveretur sursum quam deorsum et leve e converso.

Et istas ultimas rationes vocavi persuasiones, quia si esset tale spatium separatum, ipsae essent faciliter solubiles; ideo illae rationes quae probant quod non sit tale spatium separatum sunt magis demonstrativae.

---

1 mundum ... magnitudinem] mundum et magnitudinem eius Pp : mundi magnitudinem G ‖ erat] esset p : creat G 2 tale] *add.* spatium G ‖ igitur in] in Pp : *om.* G ‖ illud] tale G 3 crearet] creat G ‖ ita] item P 4 crearet] creat G 5 imaginamur[1]] imaginatur G ‖ imaginamur[2]] imaginatur G 6 motum localiter] *om.* G 7 movere] *ante* cum P 8 illud] non P ‖ iterum] *om.* G 10 et] *om.* GPp ‖ videlicet] scilicet G ‖ sequeretur] sequitur p 11 pars quaelibet] *inv.* p ‖ per[2]] pro C 12 se[1]] *add.* existens sive P 13 localiter] *ante* per[2] (12) P 15 quod] quia G ‖ naturale totale] *inv.* G ‖ et] scilicet C 18 particularium] particularis P 19 aliquem] *add.* locum P 19–20 continui] contineri P 20 locus] *in marg.* C : *ante* totius p : *om.* GP 21 quod] quia Pp 23 spatium tale] *inv.* P 24–25 magis moveretur] *inv.* GPp 25 sursum quam deorsum] deorsum quam sursum p 26 et] *om.* G ‖ ultimas rationes] *inv.* GPp ‖ tale] *om.* p 27 illae rationes] *inv.* GPp

---

10 Cf. Aristoteles, *Physica*, IV, 4, 211b19–30; 8, 214b12–215a28

Tertia conclusio est quod locus est superficies corporis continentis locatum et tangentis ipsum, 96vb C | quia remotis aliis opinionibus tribus non apparet quid aliud posset dici locus et cui conveniant proprietates loci.

Item illud debet poni locus proprius alicuius cui conveniunt omnes loci proprietates; sed illi superficiei conveniunt; igitur etc. Minor probatur per inductionem. Illa enim superficies continet et circumdat locatum nec est pars locati nec accidens sibi inhaerens. Et ab huiusmodi superficie potest locatum recedere sine hoc quod illa superficies moveatur cum eo. Et illa superficies est aequalis locato modis prius determinatis. Et omne corpus excepta ultima sphaera, de qua post dicetur, in tali superficie continetur. Et est superficies orbis lunae continens totum istum mundum inferiorem quae dicitur locus sursum, et est superficies immediate continens corpus medium 85ra P | mundi quae vocatur locus deorsum. Et etiam simpliciter gravia videmus quiescere naturaliter in illo loco deorsum et moveri ad ipsum, si sint extra, et levia videmus ascendere ad illam superficiem magnam orbis lunae. Et illae etiam superficies, si non habeant dissimiles naturas et virtutes ea ratione qua sunt superficies et magnitudines, tamen sunt cum eis virtutes dissimiles propter naturas corporum diversas quorum sunt superficies aut propter diversitatem earum in propinquitate et distantia ad caelum, quod in singula illorum corporum inferiorum influit diversas virtutes secundum diversum situm eorum ad ipsum. Concedendum est igitur quod locus est superficies corporis continentis locatum immediata locato.

Ex hoc sequitur quarta conclusio, scilicet quod locus est terminus corporis continentis, quia superficies sunt termini corporum sicut lineae superficierum et puncta linearum.

Sed difficultas 68va p | restat: cum omnis superficies sit corpus, quare magis dicimus quod locus sit superficies corporis continentis quam quod locus sit corpus continens?

---

2 tangentis] tangens P **3** posset] possit Gp ‖ locus] *om.* G ‖ conveniant] conveniunt P **6** illa] alia P **7** huiusmodi] huius P : hac p **8** sine] absque P ‖ et] quia G **10** post dicetur] prius dicebatur G **11–12** orbis … superficies] *om.* (*hom.*) BCLMP, *et* U, *sed add. in marg. verba inde ab* et est (11) *usque ad* sursum (12) **12** quae] A, *in marg.* U (*vide supra ad lineas 11–12*) : qui HTGp **13** mundi] *om.* p ‖ etiam simpliciter gravia] etiam similiter grave P : similiter gravia Gp **15** magnam] *in marg.* C : *om.* Gp **16** illae] illa G ‖ habeant] habeat G ‖ naturas et virtutes] virtutes et naturas P **17** sunt[1]] *praem.* etiam p : *om.* G **17–18** virtutes dissimiles] *inv.* G **18** diversas] *ante* naturas G ‖ quorum] quarum P **19** quod] quia G **20** inferiorum] *om.* p ‖ virtutes] naturas G **21** diversum situm] diversos situs p ‖ igitur] *ante* concedendum G **22** immediata] immediate CG **23** hoc] quo G ‖ scilicet] *om.* P **27** locus[1] … continentis] superficies corporis continentis sit locus G ‖ sit[1]] est P

---

9 Cf. sup., IV, q. 1 **10** Cf. inf., IV, q. 6

Dico quod secundum veritatem locus est corpus continens locatum, cum omnis superficies sit corpus. Sed quamvis locus sit corpus, tamen non dicitur locus proprius secundum illam rationem secundum quam dicitur corpus, sed secundum illam rationem secundum quam dicitur superficies, quoniam linea dicitur linea secundum quod est divisibilis secundum unam dimensionem, scilicet longitudinis, non considerata divisione | secundum alias 97ra C dimensiones, et superficies dicitur superficies secundum quod intelligitur divisibilis secundum duas dimensiones, scilicet longitudinis et latitudinis, non considerata divisione secundum aliam dimensionem. Sed corpus dicitur corpus secundum quod intelligitur omniquaque esse divisibile, scilicet secundum triplicem dimensionem longitudinis, latitudinis et profunditatis. Et haec dicentur in sexto et supponuntur modo.

Notandum est etiam quod, cum non possit esse penetratio corporum, unum corpus non potest esse proximum alteri corpori et tangens tali modo quod quaelibet pars unius tangat aliquam partem alterius nisi discernendo partes secundum duplicem dimensionem, scilicet longitudinis et latitudinis, et non discernendo partes secundum tertiam dimensionem, scilicet profunditatis. Si enim corpus *a* sit positum supra corpus *b* adaequate et immediate et corpus *a* in infinitum dividatur secundum distantiam de ante ad retro vel de dextro ad sinistrum, adhuc quaelibet pars ipsius *a* tangit aliquam partem ipsius *b*. Sed si corpus *a* dividatur secundum eius altitudinem, in quotascumque partes hoc fuerit, sola pars inferior tanget corpus *b*. Ex quibus concluditur quod contactus corporum dicitur secundum rationes secundum quas dicuntur superficies et non secundum rationes secundum quas dicuntur corpora. Modo | locus ex eo est et dicitur locus proprius, quia tangit undi- 85rb P

---

1 dico] respondeo P : respondetur G : responde p **2** corpus[2]] *corr. in marg. ex* superficies locus C : superficies et locus G : superficies et corpus locus p **5** quod] illam rationem secundum quam G **6** divisione] divisio P : dimensione p **9** considerata] *add. sup. lin.* alia C **10** quod] quam P ‖ omniquaque] undiquaque p : undique P **11** dimensionem] *add.* scilicet Gp **12** haec dicentur] hoc dicetur Pp ‖ sexto] *add.* libro GPp ‖ supponuntur] supponitur CGp **13** notandum est etiam] notandum est G : postea notandum est P : postea etiam notandum p **15** tangat] tangit p ‖ nisi discernendo] non discernendo C : nisi destruendo P **16** dimensionem] divisionem Cp ‖ scilicet] secundum P **17** discernendo … scilicet] *om.* G ‖ dimensionem] divisionem p **19** ad] et p **20** adhuc quaelibet] adhuc aliqua G : et hoc quaelibet P ‖ ipsius] *om.* P ‖ tangit aliquam] tangat aliquam P : tanget G **21** a] *om.* p ‖ eius] *om.* p **22** tanget] tangat P : tangit p ‖ quibus] quo P **23** contactus] cum tactus p **25** est] eo P **25–220.1** quia … proprius] *om.* (*hom.*) P

---

12 Cf. Iohannes Buridanus, *Quaestiones super libros Physicorum*, VI, q. 3 (ed. Parisiis 1509, ff. 95va–96rb)

que locatum exterius. Igitur corpus locans dicitur locus proprius secundum rationem secundum quam dicitur superficies et non secundum rationem secundum quam dicitur corpus. Et hoc intendebat Aristoteles, cum dicit 78rb G locum proprium esse superficiem continentis | et non totum corpus continens.

Ad rationes.

⟨1⟩ Ad primam dicitur quod locus est quantitas et est superficies. Sed credo quod ille terminus 'locus' non sit proprie de praedicamento quantitatis, sed potius de praedicamento ad aliquid, quia locus dicitur locati vel saltem locabilis locus, et locatum vel locabile dicitur loco locatum vel locabile. Et adhuc, si esset de praedicamento quantitatis distincta species contra istum terminum 'superficies', tamen hoc non obstaret, quin illi termini bene supponerent pro eadem re, sicut isti termini 'linea', 'superficies' et 'corpus'.

⟨2⟩ Et eodem modo ad secundam, de quocumque praedicamento ponatur ille terminus 'locus'.

Sed tu quaeres de quo praedicamento est ille terminus 'locus'. Iam dixi quod, ut mihi videtur, est de praedicamento ad aliquid et non quantitatis. Propter quod dicit Commentator quinto *Metaphysicae* quod Aristoteles in 97rb C libro | *Praedicamentorum* saepe locutus est non secundum veram determinationem, sed secundum famositatem. Nec est de praedicamento ubi, quia nec ille terminus 'locus' nec concreta sua, scilicet isti termini 'locans' et 'locatum', respondentur ad quaestionem factam per ubi, immo sic respondentur isti termini 'alicubi', 'in domo', 'prope Parisius' et 'extra domum' etc. Sed tamen, cum illi termini de praedicamento ubi significent loca vel habitudines ad loca, possibile est quod eis attribuantur similes proprietates et differentiae sumptae ex ipsis rebus significatis.

---

1 igitur] *praem.* ex his p **2–3** rationem secundum] *om.* G **3** dicit] dixit P **4** superficiem] *add.* corporis G **6** ad] *praem.* tunc respondendum est p **7** primam] *add.* cum p ‖ sed] *om.* p **9** potius] *add.* forte GPp **10** locabilis] localis CG ‖ et] est C ‖ locabile] locale CG **11** adhuc] hoc P ‖ quantitatis] *add.* adhuc esset p ‖ distincta] distinctas P **13** sicut] *add.* faciunt GPp ‖ corpus] *add.* et quodam modo C **14** et … secundam] ad secundam rationem respondetur eodem modo Gp : ad aliam respondetur eodem modo P ‖ ponatur] *add.* esse G **16** tu quaeres] tu quaeris p : tamen quaeris P ‖ iam] *praem.* et GPp **17** quod] *post* videtur G ‖ et non] non de (*om.* p) praedicamento GPp **19** saepe] *ante* in (18) G **20** famositatem] falsitatem p **21** et] aut GPp **23** domo] *add.* in foro GPp ‖ et] *om.* GPp ‖ etc.] et huiusmodi GP **24** significent] significant G

---

18 Cf. Averroes, *In Metaphysicam*, V, comm. 18, f. 125K

⟨3⟩ Ad tertiam dico quod loca sunt bene invicem contra|ria, non ratione suae essentiae, sed ratione virtutum naturalium sibi coniunctarum. Et cum hoc quantitas est bene quantitati contraria, sed termini de praedicamento quantitatis non dicuntur invicem contrarii. Albedo enim et nigredo sunt quantitates, quia saltem sunt numeri. 68vb p

⟨4⟩ Ad quartam dicendum est eodem modo quod non sunt differentiae specificae locorum secundum suas essentias sursum et deorsum, quia idem locus cum locato suo posset ascendere sursum, cum prius esset deorsum. Sed sunt differentiae locorum, quantum ad sursum et deorsum, per distantiam vel propinquitatem ad caelum et sunt diversae naturales potentiae communicantes, quia aliter caelum influit propinque et remote.

⟨5⟩ Ad aliam dicitur per idem, scilicet quod locus ratione essentiae, quae est superficies vel quantitas, non est potentiae activae vel passivae, sed dicitur esse, quia sibi sunt coniunctae virtutes activae et passivae.

⟨6⟩ Ad sextam dictum est in alia quaestione.

⟨7–10⟩ Ad alias rationes sequentes dicetur in sequenti quaestione etc.

---

1 dico] concedo P ‖ sunt bene] *inv.* Gp 2 suae essentiae] *inv.* p ‖ cum] *corr. sup. lin. ex* secundum C : secundum G 3 bene] *ante* est Gp : *ante* quantitas P 4 dicuntur] *add.* ad p 6 dicendum est] dicitur quod P 7–8 quia … suo] *rep.* G 10 vel] *corr. sup. lin. ex* et C : et G 11 communicantes] ABBrCDEFGIJLLaNOOxPPbQSTXW, Y? : comitantes? Z : *om.* UVp (propinquitatem … communicantes *om.* ErHKMR) ‖ caelum] *om.* P ‖ et] aeternaliter P 12 aliam] quintam p : quintum P ‖ scilicet quod] sicut p ‖ quae] qui P 14 coniunctae] *add.* ut P ‖ et] vel P 16 alias rationes] omnes alias GPp ‖ sequenti] *post* quaestione p : alia P ‖ etc.] *om.* GP : *add.* sequitur quaestio tertia p

---

15 Cf. sup., IV, q. 1

⟨IV.3⟩

⟨Utrum locus sit immobilis⟩

85va P Quaeritur tertio utrum locus | sit immobilis.

Arguitur quod non quia:

⟨1⟩ Omne corpus est mobile et locus est corpus, ut dictum est ante.

⟨2⟩ Item cum locus sit superficies corporis, quamvis ista superficies non esset corpus, tamen moveretur ad motum corporis cuius est superficies.

⟨3⟩ Item locus ignis est orbis lunae vel superficies eius et utrumque continue movetur.

⟨4⟩ Item locus augmentatur, si locatum augeatur; aliter non maneret aequalis. Sed augeri est moveri, quia augmentatio est motus. Igitur etc.

⟨5⟩ Item in istis inferioribus nihil apparet incorruptibile nisi materia; et locus non est materia; igitur est corruptibilis. Et corruptio non est sine motu vel praecedente vel concomitante. Igitur etc.

⟨1⟩ Oppositum dicit Aristoteles definiendo locum. Et ad hoc arguitur per differentiam eius a vase quia: licet sit de ratione utriusque continere, tamen de ratione vasis est quod sit mobile cum contento. Ad hoc enim res ponitur in vase, ut cum quiete vasis quiescat et cum eius motu moveatur. Nesciremus enim bene portare vinum sine expansione, nisi ipsum poneretur in
97va C vase; cum quo vase | portaretur. Locum autem non attribuimus corpori, ut cum locato moveatur, sed ut ex permanentia locorum assignemus locatum moveri per hoc quod prius erat in hoc loco et post in alio.

---

3 quaeritur tertio] tertio quaeritur consequenter G 5 et] *sup. lin.* C : *om.* G ‖ ante] *ante* dictum Pp : *add.* igitur G 6 item] *praem.* et Pp ‖ quamvis ista] quam ille P ‖ superficies^{2}] *add.* corporis p 7 tamen] *add.* illa P 8 eius] *in marg.* C : *ante* superficies GPp 10 locus] *post* augmentatur P : *om.* G ‖ augeatur] augmentatur Pp 11 aequalis] *add. in marg.* eo G ‖ augeri] augmentari Pp 13 est^{2}] *om.* P 14 concomitante] communicante CG 15 definiendo] definiens P ‖ locum] motum p : motum *sed add. in marg.* alias locum P ‖ arguitur] arguit Pp 16 sit] *post* utriusque GPp 17 ad] ex P 18 eius motu] *inv.* P 18–19 nesciremus] nescirem p 19 sine expansione] sine expansione (*corr. in marg. ex* extra ponere C) : sive exponere G ‖ ipsum poneretur] poneremus ipsum GPp 21 locato] *corr. sup. lin. ex* loco C : loco Gp ‖ ut] *om.* p 22 post] postea P

---

5 Cf. sup., IV, q. 2 15 Cf. Aristoteles, *Physica*, IV, 4, 212a20–21 16–22 Cf. Aristoteles, *Physica*, IV, 4, 212a14–19

 | DOI: 10.1163/9789004322356_026

⟨2⟩ Item si locus esset mobilis, sequeretur quod idem simul moveretur localiter et continue maneret in eodem loco, sicut arguebatur in alia quaestione de pluma. Sequeretur etiam quod idem continue et continue esset in alio et in alio loco et tamen non moveretur localiter.

Commentator respondet ponendo tres conclusiones.

Prima est quod omnis locus est mobilis, quia est superficies corporis continentis; et omne corpus est mobile et ad motum eius movetur simul eius superficies.

Secunda conclusio est quod locus non est mobilis per se, scilicet solitarie, quia locus est superficies corporis, quae non movetur sine corpore cuius est superficies, sed cum eo. Et hoc est verum, nisi dividatur ab eo. Sed ea ratione qua divideretur et moveretur per se, non diceretur superficies, sed corpus.

Tertia conclusio est quod locus non de necessitate movetur ad motum locati, quia est continens extrinsecum a quo locatum potest recedere. Et dicit Commentator quod per hoc differt locus a materia et forma locati et accidentibus sibi inhaerentibus.

Et vult Commentator | quod ita intendit Aristoteles locum esse immobilem, (78va G) scilicet quia non per se movetur et quia non de necessitate movetur ad motum locati, sed potest remanere locato remoto.

Sed licet istae tres conclusiones sunt verae, tamen istae non sufficiunt in proposito, quia non | amplius possemus assignare rationem quod aliquod (69ra p) corpus moveretur localiter vel quiesceret, quia non ex alio dicitur | aliquid (85vb P) moveri localiter nisi quia mutat locum, nec quiescere localiter nisi quia manet in eodem loco. Et hoc esset totum falsum, quia si contingeret locum moveri et non cum locato, tunc locatum quiesceret et tamen fieret in alio loco; et si etiam contingeret locum moveri cum locato, licet non de necessitate, tunc contingeret quod locatum localiter moveretur et non mutaret locum, sicut dicebatur de pluma.

---

1 sequeretur] sequitur p ‖ simul] *sup. lin.* C : *om.* G 3 pluma] plu[a] C : pluvia p ‖ sequeretur] sequitur Pp ‖ idem ... continue[2]] illud corpus P : idem continue p ‖ esset] fieret GPp 4 et in alio] et alio P : *om.* G 5 respondet] respondit G 7 et[2]] *om.* p ‖ movetur] moventur P ‖ eius[2]] eiusdem C : omnes P 9 quod] quia p 11 nisi] si p 12 divideretur] *add.* ab eo GPp ‖ sed] vel p 14 locati] localem P 15 differt locus] *inv.* P ‖ et[2]] *add.* ab Gp : ab P 17 intendit] intendat Gp 18 per se] *om.* P 19 locato] loco p 20 licet] *om.* G ‖ conclusiones sunt] conclusiones sint p : condiciones sunt C 22 aliquid] *om.* GPp 23 locum] locus C 24 esset totum] *inv.* p 25 et[2]] *om.* P 27 tunc contingeret] *rep.* G 28 sicut] ut P ‖ pluma] pluvia CGp

---

2–3 Cf. sup., IV, q. 2, 212[8–12] 5 Cf. Averroes, *In Physicam*, IV, comm. 41, ff. 139L–140B 28 Cf. sup., IV, q. 2, 212[8–12]

Item turres Beatae Mariae dicuntur permansisse in eodem loco a tempore quo fuerunt factae. Quod non esset ita, si contigisset loca earum moveri et recedere ab eis.

Et ad hoc solvendum dicunt multi expositores Aristotelis quod ad loca concurrunt duo, scilicet materiale, quod est superficies corporis continentis (et quantum ad illud materiale bene dicit Commentator et bene tenendae sunt conclusiones ipsius); aliud concurrit formale, videlicet distantia vel propinquitas ad caelum et terram | 97vb C et partes mundi quiescentes. Caelum enim quodam modo capitur tamquam quiescens, quia non movetur motu recto; propter quod in habitudine ad ipsum possemus iudicare motus rectos aliorum corporum. Et quantum ad tale formale locus dicitur immobilis pro tanto, quia quamdiu aliquod corpus quiesceret, tamdiu diceremus ipsum manere in eadem propinquitate vel distantia ad caelum et ad terram et ad omnia corpora mundi quiescentia, quantumcumque corpus ipsum continens moveretur et mutaretur, ut si aliquando esset aer et aliquando esset aqua.

Sed adhuc restat dubitatio: cum isti dicunt locum pro materiali esse mobilem et locum pro formali esse immobilem, videretur quod magis deberent dicere e converso, quoniam distantia illa vel propinquitas magis est mobilis per se quam superficies, quia superficies corporis non est per se mobilis, scilicet solitarie, quin moveatur corpus cuius est superficies, nisi fiat illius corporis dissolutio; distantia autem magis esset per se mobilis, quia distantia huius lapidis a caelo vel a terra non est aliud quam ille lapis vel non est aliud quam corpus medium per quod distat, et utrumque horum est per se mobile.

Iterum non apparet quare illa distantia debeat magis dici formale quam illa superficies, immo magis e converso. Nam illa distantia forte est unum corpus totale et per se mobile et superficies corporis non est aliquod corpus totale, sed est pars vel terminus alicuius corporis totalis; et magis videtur quod pars vel terminus debeat dici formale eius cuius est pars vel terminus, quam corpus totale et per se subsistens diceretur forma vel formale alicuius.

1 dicuntur] dicimus GP : diximus p 2 fuerunt factae] *inv.* GPp 4 et] *om.* Pp ‖ dicunt] respondent GPp ‖ multi expositores] *inv.* P ‖ loca] locum GPp 6 ad] *om.* P ‖ bene^{2}] vere p 7 videlicet] scilicet GPp 9 quodam] quod p 11 immobilis] *ante* dicitur P : mobilis G 12 quia] quod G 13 manere] esse P ‖ vel distantia] *om.* P 14 omnia] *om.* P 14–15 ipsum continens] *inv.* GPp 17 restat] restaret Pp ‖ dubitatio] *add.* quia GPp ‖ dicunt] dicant Pp 18 locum] *om.* P ‖ videretur] videtur Gp 18–19 deberent dicere] deberet dici G 22 magis esset] *inv.* Pp 23 est aliud] *inv.* G 24 horum] eorum p 28 est] *om.* p

Verum est tamen quod haec opinio appropinquat | ad veritatem, sed ponitur sub verbis valde impropriis et non solventibus clare difficultatem. 86ra P

Ideo pono conclusiones quae mihi videntur verae de proprietate sermonis.

Prima est quod omnis locus est mobilis, quia omnis locus est corpus et omne corpus est mobile. Et ista propositio, scilicet quod omnis locus est corpus, apparet, quia dictum est quod omnis superficies est corpus et quod omnis locus est superficies. Et adhuc, si locus poneretur esse distantia vel ad caelum vel terram vel ad aliquid aliud, adhuc locus esset dimensio vel habens dimensionem et omne tale est corporeum et mobile.

Secunda conclusio quod omnis locus, saltem si sit aliquid unum, est mobilis | per se, scilicet solitarie, absque hoc quod aliquid aliud moveatur 98ra C
cum eo vel ipsum cum alio (dico 'aliud' vel 'alio' quod non sit pars eius), quia omnis superficies alicuius corporis, cum sit pars, posset saltem per potentiam divinam separari et separatim conservari ab aliis partibus et tunc posset Deus illud movere solitarie.

Tertia conclusio est quod non est possibile naturaliter locatum exire a totali loco pro|prio in quo est sine motu loci illius vel alicuius partis eius, 69rb p
quia locus ille undique circumdat locatum; ideo ex nullo latere potest aliquid de locato exire, nisi in illo latere penetret dimensionem loci, quod est impossibile, vel nisi illi corpori exeunti aliqua pars loci cedat per quam exibit, et sic illa pars cedens movetur.

Quarta conclusio est quod ita possibile est locum moveri locato quiescente sicut e converso, quia si ignis quiesceret in sphaera sua, non minus moveretur orbis lunae, et turribus Beatae Mariae quiescentibus movetur et mutatur aer circumdans per ventum.

Quinta conclusio quod idem corpus manet in eodem loco proprio continue per quoddam totum tempus, et tamen continue movetur per illud

3 videntur] *add.* esse G 4 prima] *add.* conclusio GPp 5 scilicet] *in marg.* C : *om.* GPp 6 quod[2]] *om.* G 8 vel[1]] *add.* ad GPp || aliquid] aliquod Gp || adhuc] *om.* p 10 conclusio] *add.* est GPp || aliquid] aliquod G || unum] *add. sup. lin.* per se G 11 aliquid] aliquod G 12 ipsum] *om.* p || aliud vel alio] aliquid vel aliud P : aliquod aliud G : aliud p 15 posset ... solitarie] posset deus movere illud solitarie G : illud posset deus solitarie movere P 16 naturaliter] *om.* P 17 loci] *om.* p || alicuius partis eius] partes alicuius ipsius P 18 circumdat] circumstat P 19 illo] *add.* tempore G 20 exeunti] exienti p 22 locato] loco p 23 e converso] *add. in marg.* scilicet locatum loco quiescente C : locatum loco quiescente GPp 25 circumdans] circumstans GPp 26 conclusio] *add.* est Gp || proprio] *om.* P 27 quoddam] aliquod G || totum tempus] *inv.* P

6 Cf. sup., IV, q. 1, 206[26–27]; cf. sup., IV, q. 2, 218[1–2]

totum tempus motu quem consuevimus vocare motum localem. Sic enim sol movetur et luna et unumquodque | [78vb G] astrorum. Et sic moveretur pluma delata in aere continente eam ad motum illius aeris. Sic etiam forte movetur ignis continue circulariter cum orbe lunae continente ipsum et locante.

Sexta conclusio est quod possibile est aliquod corpus per aliquod totum tempus quiescere et tamen ipsum continue vel multotiens fieri in illo tempore in alio et alio loco. Sic enim est de turre Beatae Mariae Parisiensis vel de ligno fixo in litore maris, quod cotidie bis successive continetur, modo ab aqua per accessum maris, modo ab aere per | [86rb P] recessum maris.

Ex his concluditur septima conclusio quod possibile est sine loci mutatione, immo sine loco, fieri motum quem communiter vocamus localem vel sibi similem quantum ad essentiam et proprietates intrinsecas motus. Et hoc dictum fuit in tertio libro.

Octava conclusio est: impossibile est nos percipere saltem sensu aliquid moveri motu quem vocamus localem, nisi percipiamus ipsum aliter et aliter se habere ad aliquod corpus aliud secundum situm distantiae vel propinquitatis (dico: aliter et aliter se habere totum ad totum vel partes ad partes). Et hoc experiuntur illi qui sunt in fundo navis velociter motae, non aspicientes extra. Non enim percipiunt quando navis | [98rb C] movetur vel quiescit.

Nona conclusio est quod iudicamus motum fieri quem vocamus localem, si percipiamus corpora diversa continue se habere aliter et aliter ad invicem secundum situm. Immo hoc non debet poni conclusio, immo communis animi conceptio, quia omnes sic iudicant. Sed non possunt cum certitudine iudicare quid illorum movetur, nisi sciant alterum illorum non moveri vel nisi percipiant illud aliter se habere secundum situm ad illud alterum quod sciunt non moveri vel saltem non tali motu moveri vel non ita velociter qualiter aut quomodo velociter percipiunt illud ad illud aliter et aliter se habere. Et haec possunt experiri et dicta fuerunt in tertio libro.

---

2 sic moveretur] sic movetur p : similiter movetur G ‖ pluma] plu[a] C : pluvia GPp 3 sic] *om.* P 4 orbe] orbi P ‖ locante] locantem P 5 totum] *om.* G 6 ipsum] *om.* p 7 turre] turri GPp ‖ parisiensis] *om.* GPp 8 quod] et p ‖ cotidie bis] continue bis *in marg.* C 9 accessum] assessum C ‖ maris[2]] aeris P 10 septima] alia P 11 motum] *add.* continue p 12 et[2]] *om.* P 13 tertio] secundo P 14 est[1]] *add.* quod GPp ‖ saltem] *post* sensu G : *om.* P 16 corpus aliud] *inv.* GPp ‖ distantiae] substantiae P 17 dico] dicitur G ‖ et[2]] *om.* P 18 fundo] profundo P ‖ aspicientes] *add.* ad aliquod G 20 quod] *add.* nos GPp ‖ motum fieri] *inv.* P 22 hoc] haec GPp ‖ immo[2]] sed GPp 24 quid] quod p ‖ illorum non] illorum P : non G 25 aliter] *post* habere G ‖ illud[2]] aliquod GPp 26 sciunt non] sciant non p : sciunt CP 27 qualiter] *add.* autem p ‖ illud[2]] aliquod G 28 experiri] experire P

---

13 Cf. sup., III, q. 7, 76 28 Cf. sup., III, q. 7

Decima conclusio vel forte propositio per se nota per quid nominis est quod omne quod continue movetur localiter per aliquod totum tempus, ipsum continue per illud tempus se habet aliter et aliter secundum locum vel ad locum, quia moveri vel mutari est aliter et aliter se habere prius et posterius secundum quid nominis, ut dicebatur in tertio libro; et sicut est mutari, ita est aliter et aliter se habere; ideo si est mutari secundum locum vel localiter, hoc est se habere aliter et aliter secundum locum vel localiter.

Undecima conclusio quod saepe utimur hoc nomine 'locus' secundum aliam significationem quam secundum illam quam prius diximus, scilicet prout diximus omnem locum esse superficiem continentis corporis et secundum quam dicimus omnem locum alicuius corporis continere illud corpus. Illa conclusio probatur quia: omnes communi conceptione concedunt stellas moveri localiter et ultimam | sphaeram; igitur secundum praecedentem conclusionem ipsi habent concedere ultimam sphaeram se habere 69va p aliter et aliter secundum locum et ad locum; et tamen hoc non debet concedi de | loco continente illam sphaeram, quia non est, nec de loco continente 86va P ignem vel aerem vel aquam vel etiam terram, quia quamvis omnia circa terram simul volverentur uniformiter terra quiescente, nos existentes in terra, percipientes caelum se habere aliter et aliter ad nos et ad terram, nihil considerantes de loco nostro vel terrae, dicemus caelum et astra moveri localiter.

Item omnes ex communi animi conceptione concedunt quaestionem per ubi quaerere de loco rei. Ideo etiam respondentes convenienter ad illam quaestionem debent assignare locum de quo quaeritur. Sed saepius non responderemus locum continentem illud de quo quaeritur, sed aliquando contentum ab eo, aliquando aliud quod nec est continens ipsum nec contentum ab eo. Quod apparet, quia omnia ista nomina 'intra' et 'extra', 'infra'

---

1 vel forte] *praem.* est p : est vel P : est quod forte G ‖ per²] ex GPp ‖ quid] quod p 2 quod²] quid p 3 secundum] per P 4 locum quia] locum quod G : locatum quia p 5 dicebatur] dicitur p 6 ita] sic G 7 secundum ... localiter²] secundum locum vel ad locum G : *om.* P 8 quod] *praem.* est p : est quia P ‖ locus] loci GPp 10–11 esse ... locum] *om.* (*hom.*) C 10 continentis corporis] *inv.* Pp 11 quam dicimus] quam diximus P : quem diximus p 12 conceptione] *add.* animae P : acceptione animi G 13 moveri] movere P ‖ et] ad P 13–14 praecedentem conclusionem] *inv.* P 14 concedere] *post* sphaeram C 15 et²] vel G 16 illam ... continente²] *om.* (*hom.*) C ‖ nec] *om.* G 18 volverentur] moverentur P ‖ terra quiescente] *inv.* P 19 et²] vel G 20 dicemus] diceremus GPp ‖ moveri] movere P 21 animi conceptione] conceptione animae P 22 rei ideo] ideo etc. p ‖ convenienter] communiter p 23 locum] locus C 23–24 non responderemus] non respondemus P : respondemus non G 24 locum] locus C 25 eo] *add. sup. lin.* et G ‖ aliud] illud p ‖ ipsum] illud G 26–228.1 infra et supra] supra et infra Gp

---

5 Cf. sup., III, q. 7, 75[13–24]

et 'supra', 'ante' et 'retro' et huiusmodi et multa alia respondentur ad quaestionem quaerentem ubi est, ut 'ubi est sphaera stellata?'—'supra sphaeras
98va C planetarum'; et non | dicemus quod sit in alia aliqua sphaera, si non credamus esse sphaeram aliquam superiorem. Et si credamus esse spatium infinitum vel vacuum infinitum et quaeramus ubi est, dicemus 'ultra caelum'. Et 'ubi est sol?'—dicemus 'in oriente' vel 'in meridie'; et erit responsio secundum habitudinem ad nos, quia quando erit nobis in oriente, erit quibusdam aliis in meridie. Et 'ubi est Robertus?'—dicimus quod extra villam, versus talem arborem aut supra talem domum. Et 'ubi est villa sancti Dionysii?'—dicimus quod ad duas leucas prope Parisius versus septentrionem. Et 'ubi est talis puer?'—'ipse est cum matre sua'. Et 'quo ivit mater?'—'ad puteum ivit'. Et sic de multis aliis.

Igitur secundum haec dicta videtur mihi quod oportet dicere de loco sicut de sano et tempore et quasi de omnibus aliis nominibus. Sicut enim sanum dicitur multipliciter, ut de animali, de urina, de cibo, de aere etc., sed tamen est unus primus modus ad quem alii habent attributionem, ita manifeste apparet quod multipliciter dicitur tempus, ut quia tempus proprie est motus caeli, et aliquando dicimus tempus esse panem aut vinum, quia dicimus tempus esse carum; aliquando dicimus aerem esse tempus, licet sit in quiete et permanentia, ut si dicamus tempus esse serenum, et aliquando etiam esse aerem motum et alteratum diversis alterationibus, ut cum dicimus tempus esse ventosum, pluviosum, frigidum, calidum etc.

---

1 et[2]] *om.* P ‖ huiusmodi] huius P ‖ et multa alia] *om.* G ‖ respondentur] responderentur p : respondetur CG 2 quaerentem ubi est] ubi est quaerentem G : per ubi P ‖ stellata] stellarum G 3 et] *om.* p ‖ dicemus] dicimus P ‖ alia aliqua] *inv.* GPp 4 esse sphaeram aliquam] aliquam esse P ‖ si] *om.* P 5 infinitum] *om.* P ‖ dicemus] *add.* quod Pp : diceremus quod G 6 in[2]] *om.* p 7–8 oriente ... meridie] meridie erit quibusdam aliis in oriente G : oriente erit aliquibus in meridie P 8 versus] vel supra GP : vel super p 9 talem[1]] *om.* P ‖ aut ... domum] vel domum P : etc. G 9–11 et ... ivit[2]] †...†la sancti dionysii †...† ad duas leucas †...† versus septentrionem †...† iuvencula illa †...† suo et quo ivit †...† ivit ad filium *in marg.* C 9 sancti] *om.* P 10 dicimus] dicemus p ‖ quod] *om.* P ‖ ad] per G 11 puer ... ivit[2]] socius cum amica sua et quo ivit socius ad amicam P ‖ ad] *praem.* ipsa p ‖ ivit[2]] *add.* ad habendum aquam p 12 et ... aliis] et sic de multis p : etc. G 13 igitur] ideo G 14 et[1]] *add.* de Gp ‖ nominibus] *om.* G 15 ut] sicut P ‖ urina ... aere] cibo et de urina de aere G : cibo de urina de aere p : cibo aere urina P 16 ita] item P 18 dicimus tempus] *inv.* GPp 19 tempus esse] *inv.* G ‖ carum] *add.* quia vinum est carum vel panis et P 19–20 carum ... esse[1]] *om.* (*hom.*) p 19 aerem esse tempus] esse aerem G 20 serenum] sanum G ‖ aliquando etiam] *inv.* p 20–21 esse[2] ... ut] aerem motum et alteratum diversis alterationibus dicimus esse tempus et G 21 dicimus] dicamus P 22 frigidum] *add.* et P : *om.* G

Ita igitur rationabile est dicere de loco. Prima enim acceptio huius termini | 'locus' est prout Aristoteles definit ipsum in isto libro. Sed consequenter secundum attributionem ad istam primam acceptionem, quia si locus proprie dictus non moveretur et sciretur non moveri et locatum perciperetur moveri et aliter et aliter se habere secundum situm ad illum locum, nos iudicaremus locatum moveri motu quem vocamus localem, ideo propter veritatem illius condicionalis, nos omne illud vocamus locum alicuius corporis vel respectu alicuius corporis, per quod propter mutationem eius ad ipsum secundum situm iudicamus illud corpus moveri. Et similiter de quiete, quia si sciremus locum proprie dictum non moveri et locatum continue se habere similiter ad illum locum secundum situm et totius ad totum et singularium partium ad singulas partes, iudicamus locatum quiescere. Ideo consequenter omne per quod propter non mutationem alicuius corporis ad ipsum secundum situm nos iudicamus illud quiescere, nos vocamus locum respectu ipsius. Et sic valde bene dictum est quod omne quod cognoscimus moveri motu quem vocamus localem, habet locum; habet enim aliud ad quod apparet aliter et aliter se habere secundum situm. Unde sic non est inconveniens terram dicere locum respectu | caeli, quamvis non sit proprie locus eius. 79ra G 86vb P 69vb p

Sed iterum, quia si quaeramus de aliquo ubi est, si habeat locum proprium nobis manifestum, propriissima et finalis responsio esset assignando illum locum proprium (ut si | quaeramus 'ubi est Socrates?' et dicamus 'ipse est Parisius', hoc non sufficeret, sed iterum quaeremus 'et ubi inveniam ego eum Parisius?'—diceretur 'in tali vico' et 'in tali domo'; et adhuc, cum venies ad domum, tu quaeres 'ubi ille est?', donec assignabitur tibi vel apparebit tibi locus proprius ubi ipse est, et tunc cessabit quaestio), ideo secundum quandam similitudinem et attributionem ad talem locum proprie dictum nos vocamus omne illud locum respectu alicuius, quod respondemus ad quaestionem factam per ubi de illo. Et ita non est mirum, si multis modis impropriis dicatur locus, sicut improprie tempus diceretur panis aut vinum. 98vb C

1–2 termini] nominis GPp **2** prout] quod ut G ‖ ipsum] *post* libro P ‖ sed] *om.* p **5** moveri et] movere et P : moveri p : *om.* G **5–6** locum nos] non P **7** locum] locatum G **8** per quod propter] propter quam P **9** et] *om.* P ‖ similiter] *add.* est p **10** locatum] locum p **12** iudicamus] *praem.* nos GPp **13** omne] esse P ‖ propter] *om.* P **14** ad ipsum] *post* situm G ‖ nos^2] et G **20** sed] *om.* G ‖ quia] *om.* G **20–22** si^2 ... est^1] *om.* (*hom.*) P **21** esset] est p **22** et] *om.* C ‖ dicamus] *add.* quod P **23** sufficeret] sufficit Pp ‖ et] *om.* GP ‖ inveniam ego] invenirem ego p : inveniam G **24** diceretur] dicitur P : dicetur p ‖ vico] loco Pp ‖ et^1 ... domo] vel tali domo P : *om.* G **27** proprie] *add.* locum P **30** tempus diceretur] *inv.* P

Sed adhuc restat dubitatio: cum locus sit mobilis et per se mobilis sine locato quemadmodum locatum sine loco, quare Aristoteles in definitione loci apposuit hanc particulam 'immobilis'?

Ad hoc respondet Aristoteles quod hoc fecit ad differentiam vasis, non quia realiter locus differat a vase vel vas a loco, sed quia licet forte sit idem vas et locus vel tamen conveniunt, quia utrumque est continens, tamen secundum diversas rationes sumptas a diversis proprietatibus continentis imposita fuerunt haec nomina 'vas' et 'locus'. Ex eo enim continens dicitur vas respectu contenti, quia cum contentum sit per se fluxibile et dispergi- 87ra P bile secundum partes | eius, prohibetur a vase, ne sic defluat et dispergatur, et sine huiusmodi dispersione portatur de loco ad locum cum vase, scilicet ad motum vasis. Igitur hoc nomen 'vas' imponitur ad significandum continens secundum quandam rationem mobilitatis, scilicet secundum quam ad motum eius portatur cum eo contentum de loco ad locum. Sed sicut bene dicit Aristoteles, non cognovissemus locum, scilicet secundum illam rationem secundum quam imposuimus sibi nomen loci, nisi percepissemus motum localem secundum quem videmus unum corpus exire ab uno loco et moveri ad alium locum. Hoc autem non debetur loco nisi secundum illam rationem secundum quam non movetur cum locato. Et ideo nomen loci impositum fuit ad significandum continens secundum quandam rationem immobilitatis respectu locati. Et non dico 'immobilitatis' simpliciter, sed sic quod non convenit sibi secundum rationem secundum quam dicitur locus quod moveatur cum locato, licet hoc aliquando bene sibi conveniat secundum aliam rationem, scilicet secundum quam dicitur vas. Et hanc differentiam secundum rationem inter nomina ista 'locus' et 'vas' intendebat Aristoteles per istam clausulam 'immobilis' positam in definitione loci, sicut 99ra C ipsemet dicit. | Et non intendebat per hoc quod locus differret realiter a vase,

---

2 quemadmodum] sicut G : *add.* aliquod P : *add.* et p 2–3 aristoteles ... apposuit] in definitione loci apponit aristoteles G 4 respondet aristoteles] *inv.* G 5 locus differat] *inv.* p 6 tamen[1]] cum P 7 secundum] propter P 8 enim] *om.* p 9 cum] *om.* C 10 defluat] *ante* sic P : fluat p 11 dispersione] dispergatione P ‖ portatur] *add. in marg.* continens C ‖ locum] *add.* continens P 13 secundum quam] secundum quod G : quam p 14 ad[1]] *add.* talem G 15 aristoteles] *om.* P ‖ cognovissemus] cognoscimus G ‖ scilicet] *om.* P 16–19 imposuimus ... quam] *om.* (*hom.*) p 17 quem] quam G ‖ unum] idem GP 18 alium locum] alterum locum P : alium G 19 cum] a G ‖ ideo] *add.* motui loci G 21 et] *om.* G ‖ dico] *add.* respectu P 22 quod] quia Pp 23 aliquando] *post* bene G : aliter p 24 scilicet] *post* quam p : *om.* P 25 nomina ista] haec nomina G 26 loci] *om.* p 27 differret realiter] *inv.* G

---

4 Cf. Aristoteles, *Physica*, IV, 4, 212a14–19 15 Cf. Aristoteles, *Physica*, IV, 4, 211a12–13

immo notavit quod possunt esse idem dicens: 'est autem, sicut vas locus transmutabilis, sic locus vas immobile'.

Sed tamen sciendum est quod rationes illae differentes secundum quas imponuntur nomina vasis et loci non repugnant, quia tunc esset impossibile quod idem simul esset vas et locus, sed sunt rationes disparatae. Ratio enim vasis est quod sit continens ad cuius motum contentum est natum moveri delatum in eo; ratio autem loci est quod sit continens non necessarium moveri cum locato, sed quod locatum possit moveri, non moto loco cum eo.

Per haec dicta apparet quod rationes quae in principio quaestionis fiebant concludebant verum.

⟨1⟩ Sed quando arguebatur ad oppositum per auctoritatem Aristotelis et per rationem eius de differentia vasis et loci, manifestum est ex dictis quomodo illa ratio et illa auctoritas debeant intelligi.

⟨2⟩ Ad aliam dicendum est quod saepe dicimus aliquid moveri localiter vel secundum locum non capiendo 'locum' proprie, quia sit manens in eodem loco et se habens eodem modo ad ipsum, sed capiendo 'locum' improprie pro eo ad quod percipimus illud quod movetur | 79rb G aliter se habere secundum situm, quamvis accipiamus illud tamquam quiescens. Et de hoc dictum est prius.

Sed tamen adhuc restat una dubitatio, quia omnes concedunt tamquam communem animi conceptionem turres Beatae Mariae esse in eisdem locis | 70ra p suis in quibus erant, quando primo factae fuerunt, non obstante | 87rb P quod saepe

1 immo] et P 2 transmutabilis] transibilis *sed add. in marg.* transmutabilis C ‖ sic] sicut p ‖ immobile] *add.* unde cum quidem in eo quod movetur moveatur et mutatur quod intus ire in flumine navis utitur magis quam loco continenti vult autem immobilis locus esse etc. P : *add. in marg. sup.* (*supplevimus litteras rescissas*) unde in eo quod quod mo⟨vetur⟩ movea⟨tur⟩ et muta⟨tur⟩ quod intus ut in flu⟨m⟩i⟨n⟩e navis tamquam vase utitur magis quam loco continente vult autem immobiliter esse locus etc. C 3 sed tamen] *om.* P ‖ est] *om.* Pp ‖ rationes illae] *inv.* GPp 5 sed] *add.* non P ‖ rationes] *add.* sic C 6 natum] *ante* est² P : innatum p 7 delatum] dilatum C 8 possit] potest P ‖ moto] motu P 9 per] *praem.* sed p ‖ in] a Pp 10 verum] veritatem GPp 11 sed] et G ‖ arguebatur] arguitur Gp ‖ auctoritatem aristotelis] *inv.* p 12 vasis et loci] loci et vasis P 13 illa²] *om.* P ‖ debeant] debent G 14 aliam] *add.* rationem Gp ‖ est] *add.* breviter GPp 15 sit manens] sic manet Gp 16 habens] *add.* in P 17 percipimus] *add.* ad p 18 et] *om.* P 19 prius] *om.* P 20 sed tamen adhuc] adhuc G : sed tunc P ‖ una] *om.* P 21 communem animi] *inv.* G 22 suis] *om.* P

1 Aristoteles, *Physica*, IV, 4, 212a14–16

fuerit alius et alius aer continens, et etiam non obstante quod corpora media per quae distabant a caelo fuerunt saepius facta alia et alia. Quid igitur est ille locus manens idem secundum quem dicimus illam turrim esse et fuisse in eodem loco continue?

Hoc videtur difficile et est facile, quia non debet hic accipi 'idem' pro eodem simpliciter et essentialiter, sed debet accipi 'idem' pro aequalitate distantiae vel propinquitatis ad caelum vel ad terram vel ad aliud corpus per quod iudicamus locatum moveri aut quiescere, scilicet moveri, quia aliter et aliter se habet ad illud secundum situm, et quiescere, quia non aliter et aliter se habet ad ipsum. Sic enim lapidem vel arborem in campis dicimus permanere in eodem loco, quamvis continue habeat alium et alium locum proprie dictum, ex eo quod continue manet in aequali distantia ad caelum
99[rb] C vel ad terram vel ad aliquid aliud quod imaginatur | tamquam non motum. Et cum etiam dicimus plumam delatam ab aere continente moveri localiter et esse in alio et alio loco prius et posterius, non intendimus de alietate essentiali loci proprie dicti, sed capimus 'alietatem loci' pro inaequalitate distantiae vel propinquitatis plumae ad caelum vel terram vel aliud per quod iudicamus etc., sicut ante dictum est. Constat enim, ut in *Metaphysica* dicitur in pluribus locis, quod unum vel idem dicitur multipliciter, vel quia idem simpliciter vel quia aequale aut quia simile et aliis pluribus modis.

Et sic est finis quaestionis.

---

1 fuerit] fuit p ‖ et alius] continuus *sed add. sup. lin.* et alius C ‖ et etiam] vel etiam G : et P ‖ media] mota G 2 fuerunt] fuerint Gp ‖ igitur est] *inv.* p 3 quem] quam C ‖ et] vel G 4 in … continue] continue in eodem loco P 5 debet hic] debent hic p : debet ita capi vel P 6 debet] *add.* hic G ‖ idem] *om.* P 7 distantiae vel propinquitatis] distantia vel propinquitate G ‖ ad[2]] *om.* G ‖ ad[3]] *om.* G 9–10 ad … habet] ad illud secundum situm et quiescere quia non aliter et quiescere quia non aliter et aliter se habet ad ipsum *in marg.* C 10 sic] sicut G 11 habeat] habet P ‖ et alium] *om.* G 12 continue manet] *inv.* P 13 vel ad[1]] et ad p : vel G ‖ ad aliquid] ad aliquod p : aliquod G ‖ non motum] unus motus G 14 cum] *add. sup. lin.* hoc C : *add.* hoc P ‖ plumam] pluviam Pp ‖ continente] continue p 15 alietate] alienitate G 17 plumae] pluviae CPp ‖ vel[2]] *add.* ad p ‖ vel[3]] *add.* ad p 19 unum] *om.* P 20 aut] vel G 21 et … quaestionis] et sic patet quaestio G : etc. Pp

---

18 Cf. e.g. Aristoteles, *Metaphysica*, V, 6, 1015b16–1017a3; 9, 1017b27–1018a9; X, 1, 1052a15–1052b1

⟨IV.4⟩

⟨Utrum definitio loci quam assignat Aristoteles sit bona, qua dicitur 'locus est terminus corporis continentis immobilis primum'⟩

Quaeritur quarto utrum definitio loci quam assignat Aristoteles sit bona, qua dicitur 'locus est terminus corporis continentis immobilis primum'.

Arguitur quod non quia:

⟨1⟩ Iam dictum est quod locus alicuius corporis saepe accipitur non pro continente, sed pro contento vel aliquo alio quod nec continet illud corpus nec continetur ab eo; et illa definitio non conveniret tali loco; ideo non conveniret omni contento sub definito; igitur non esset bona.

⟨2⟩ Item prima pars definitionis debet esse genus definiti; sed sic non est hic, quia illud nomen 'terminus' est de praedicamento ad aliquid et ille terminus 'locus' non, sed de praedicamento quantitatis vel ubi. Sed de hac ratione dictum fuit prius.

⟨3⟩ Item posset argui contra istas clausulas 'terminus continentis' et 'immobilis' per duas quaestiones praecedentes. Sed etiam de hoc dictum est prius. Ideo haec ratio et praecedens dimittantur.

⟨4⟩ Item quaerendum est ad quid ista dictio 'primum' refertur. Et non poterit bene dici. Nam si refertur ad terminum continentem, tunc non deberet poni cum ista dictione 'immobilis', sed cum ista dictione 'terminus' vel cum ista dictione 'continentis'. Et etiam continens primum non est nisi ultima sphaera; et multa alia sunt loca; ideo illa dictio 'primum' non debet

---

5 quaeritur quarto] quarto quaeritur consequenter G ‖ sit bona] *ante* quam P ‖ qua] *praem.* in Gp 6 corporis] *om.* p 7 non] *add.* primo G 8 quod] quia G ‖ non] *add.* solum P 10 conveniret] convenit p 11 conveniret] convenit p ‖ igitur ... bona] igitur non est bona Gp : *om.* P 12 definitionis] *om.* P 13 illud] hoc P 13–14 ille terminus] hoc nomen P 14 non sed] est P 15 fuit] est P 16 posset] potest P 17 quaestiones] conclusiones CGp ‖ etiam] *om.* GP 18 haec ... dimittantur] postponatur modo P 19 est ad] ad (*rep.*) p ‖ ista ... refertur] refertur haec dictio primum P ‖ et] quia G 20 poterit] potest P ‖ nam] quod G 22 cum ista dictione] *om.* P ‖ est] *om.* p 23 ideo] *praem.* et P

---

6 Aristoteles, *Physica*, IV, 4, 212a20–21 8 Cf. sup., IV, q. 3, 227^{21}–228^{12} 15 Cf. sup., IV, q. 2, 220^{7-13}

 | DOI: 10.1163/9789004322356_027

87va P referri ad 'continentis'. Si autem dicatur quod refertur ad 'immobilis', | tunc in hac definitione non ponitur differentia inter superficiem concavam corporis continentis et superficiem convexam. Et ita locus proprius ignis ita bene esset superficies convexa orbis lunae sicut superficies concava, quod est falsum, vel definitio conveniret alteri a definito.

⟨5⟩ Item sequeretur quod pars media corporis continui haberet locum proprium, scilicet superficiem partis extremae continentis illam mediam. Et hoc negat Aristoteles rationabiliter, quia locus non debet moveri de necessitate cum locato; et partes continuae ad invicem in toto de necessitate
99va C moventur, nisi fiat | dissolutio et diruptio illius totius.

Oppositum videtur dicere Aristoteles.

70rb p Respondeo breviter quod ista definitio, sicut posita est, non | est bona definitio loci, sed indiget quibusdam supplementis.

Notandum est igitur quod hic intendit Aristoteles definire locum proprium corporis locati et proprie dictum. Et dictum est prius quod de ratione talis loci est quod contineat totum locatum et cum eo non contineat aliud corpus quod non sit pars illius locati vel contenti in eo. Propter quod necesse est illud continens quod est locus proprius esse immediatum ipsi locato, scilicet tangens ipsum. Sed iterum oportet quod illud continens quod est locus sit divisum a locato et non sibi continuum, quia secundum illam rationem secundum quam esset sibi continuum necesse esset quod moveretur simul cum contento, nisi fieret diruptio et discontinuatio, et de ratione loci est quod sic de necessitate non moveatur cum locato, ut dictum est ante.

Ideo concluditur quod in definitione loci, si debeat esse perfecta, debet apponi cum eo quod dicebatur, videlicet quod continens sit immediatum locato et divisum ab eo, ita quod sit definitio completa talis: 'locus proprius

---

1 si autem] sed si G ‖ ad[2]] *add.* li P **2** concavam corporis] *inv.* P **3** et[1]] *add.* inter p ‖ et[2]] *om.* P ‖ locus proprius] *inv.* P ‖ ignis] *om.* p **4** concava] *ante* superficies[2] P : convexa p **4–5** quod est falsum] *om.* G **6** sequeretur] sequitur p ‖ media corporis] *inv.* P **8–9** moveri de necessitate] de necessitate moveri Gp : de necessitate movere P **9** ad] ab C **14** igitur] *om.* P ‖ hic intendit aristoteles] aristoteles intendit hic P **16** contineat[1]] continet P ‖ cum eo] quod p **17** sit] est p ‖ contenti] contentum C **20** illam] unam p **21** esset sibi] *inv.* P **22** diruptio et] diruptio vel G : descriptio et p ‖ et[2]] *om.* p ‖ de … est] sequitur de ratione loci P **23** non] *ante* sic GPp **24** debeat] debet Pp **25** apponi] opponi p ‖ videlicet] *om.* G

---

8 Cf. Aristoteles, *Physica*, IV, 4, 211a29–31, 212a18–19 **11** Cf. Aristoteles, *Physica*, IV, 4, 212a20–21 **15** Cf. sup., IV, q. 1, 205[15–22] **23** Cf. sup., IV, q. 3, 223[13–14]

est superficies corporis continentis locatum, immediata locato et divisa ab eo, immobilis primum'. Et istas clausulas notavit Aristoteles expresse esse ponendas, licet non congregavisset eas simul. Dicit enim sic Aristoteles: 'cum quidem igitur non divisum sit continens, sed continuum, non dicimus esse in illo sicut in loco, sed sicut pars in toto; cum vero divisum sit et contactum, in primo quodam ultimo continentis etc'. Ecce quod manifeste dicit quod oportet continens esse divisum et immediatum, quia contactum. Et etiam cum dicit Aristoteles 'in primo quodam ultimo continentis', per 'ultimum | 79va G continentis' intendit superficiem continentis et per 'primaevitatem' intendit immediationem. Unde 'in primo ultimo continentis', id est in superficie continentis immediata ipsi locato.

Tunc igitur dico quod ista definitio loci est valde bona. Propter cuius expositionem et manifestationem debemus dicere quod Aristoteles accepit in definiendo locum istam dictionem 'terminus' loco illius dictionis 'superficies' propter hoc quod superficies est terminus | corporis continentis. Nunc 87vb P autem isti termini 'superficies' et 'locus' se habent sicut subiectum et passio, quia supponunt pro eadem re, tamen ille terminus 'locus proprius' super significationem huius termini 'superficies' addit quasdam connotationes, scilicet continentiam locati et divisionem a locato et immediationem et quod non sit de ratione eius, inquantum dicitur locus, quod moveatur cum locato, licet hoc non repugnet, sicut ante dictum fuit. Et ideo manifestum est quod ille terminus 'locus' se habet ad istum 'superficies' sicut passio ad subiectum. Modo ita est quod passio debet definiri per subiectum suum et per terminos explicantes connotationes quas addit | super signifi- 99vb C cationem subiecti. Sic enim definitur simum: 'simum est nasus cavus'. Igitur bene definitur locus per superficiem, tamquam per subiectum suum, et per

1 immediata] immediate G 2 primum] primo P 3 enim] *om.* P ‖ sic aristoteles] *inv.* GPp 4 sit] et P 5 illo] eo P 6 quodam ultimo continentis] quodam continente C : quodam continentis P : quod continentis p ‖ quod[1]] quam P ‖ dicit] *add.* aristoteles G 7 et[2]] *om.* P 8 cum] *om.* C ‖ quodam] quidem G : quod p ‖ continentis] continente C 9 primaevitatem] praemeditatio p 10 id est] non C ‖ in[2]] *om.* p 12 igitur] ego p 13 accepit] accipit p : cepit G 15 est terminus] *om.* p ‖ corporis continentis] corporis G : continentis P 19 et[1]] *om.* G 20 et quod non] *om.* P 21 licet ... repugnet] sed hoc repugnat G ‖ fuit] est P 22 istum] *add.* terminum GPp 23 modo] *om.* p 25 simum[1]] *add.* quod G ‖ nasus cavus] nasus curvus Pp : nasi curvitas G 26 definitur] dicitur p ‖ superficiem] superficies p

3 Aristoteles, *Physica*, IV, 4, 211a29–32 21 Cf. sup., IV, q. 3

particulas sequentes quae exprimunt praedictas connotationes. Ideo descriptio est bona.

Unde potest sic argui: ista descriptio est bona passionis, quae est data per subiectum illius passionis et per terminos explicantes connotationes ipsius, et etiam quae est convertibilis cum ista passione definita; sed ista descriptio loci est huiusmodi, ut apparuit. Quod enim sit convertibilis apparet, quia bene convenit omni loco proprie dicto et nulli alteri. Per hoc enim quod dicitur 'superficies' vel 'terminus' differt ab omni eo quod non est superficies. Et quamvis locus sit superficies et corpus, tamen dicitur superficies descriptive melius quam corpus, quia potius sibi convenit quod dicatur locus proprius ea ratione qua dicitur superficies quam ea ratione qua dicitur corpus. Et per hoc quod dicitur 'corporis continentis locatum' differt a superficiebus
70va p aliorum corporum. | Et per hoc quod dicitur 'immediata locato' differt a superficie remota; verbi gratia superficies convexa orbis lunae non est locus proprius ignis, sed superficies concava. Et per hoc quod dicitur 'divisa' differt a superficie partis in aliquo toto continuo continentis aliam partem. Et per hoc quod dicitur 'immobilis primum', id est non de necessitate mobilis cum locato, differt a vase, non realiter, sed secundum rationem, sicut ante expositum est.

Tunc igitur respondendum est ad rationes.

⟨1⟩ Ad primam conceditur quod hic non definitur locus quantum ad omnem acceptionem huius termini 'locus', sed solum quantum ad primam et propriam eius acceptionem.

⟨2⟩ Ad aliam dico quod in definitione passionis debet poni subiectum quoad modum generis.

⟨3⟩ De tertia ratione dictum est prius.

⟨4⟩ Ad aliam dictum est quod ista dictio 'primum' est determinatio huius termini 'immobilis'. Unde per hoc notat Aristoteles quod locus non est simpliciter immobilis, sed immobilis primum, id est non per se et de necessitate mobilis cum motu locati. Et iam dictum est quod ad ponendum differentiam

---

2 est] *add.* manifestum G : *add.* multum Pp **3** descriptio] *post* bona GPp **4** explicantes] *add.* omnes G **5** etiam] *om.* P ‖ convertibilis] principalis p **7–8** dicitur] dico GPp **16** continentis] *add.* continue P **17** id est] scilicet p **18–19** sicut ante expositum] ut ante expositum P : sicut iam dictum p **20** tunc … rationes] tunc igitur ad rationes respondetur P : ad rationes G **21** conceditur] concedendum est p **25** quoad] primo ad Pp **26** prius] *om.* p **28** unde] et G **29** primum] primo P ‖ et] sed G **30** iam] ideo p ‖ differentiam] differentias G

---

19 Cf. sup., IV, q. 3, 230–231

inter superficiem concavam locantis et convexam oportebat apponere istam clausulam 'immediata | locato'. 88ra P

⟨5⟩ Propter ultimam rationem dictum est quod oportuit apponere istam clausulam 'divisa a locato'.

Et sic est finis quaestionis etc.

1–2 oportebat ... clausulam] oportebit clausulam illam ponere G 3 oportuit] oportebat P 4 divisa] divisam GP 5 et ... quaestionis] *om.* GPp

⟨IV.5⟩

⟨Utrum terra sit in aqua sive in superficie aquae tamquam in loco suo proprio et naturali⟩

Quaeritur quinto utrum terra sit in aqua sive in superficie aquae tamquam in loco suo proprio et naturali.

Arguitur quod non quia:

⟨1⟩ Terra pro magna eius parte non continetur ab aqua; et locus proprius
100ra C debet continere | totum locatum; igitur etc.

⟨2⟩ Item licet aqua circumdaret totam terram, tamen cum ipsa terra continet multa alia corpora, ut lapides et radices plantarum; et locus proprius praeter corpus cuius est locus proprius non debet continere alia corpora.

⟨3⟩ Item corpora naturalia, cum fuerunt in locis suis naturalibus, debent naturaliter quiescere, et cum fuerunt extra, debent naturaliter moveri ad ea. Sed sic non est de terra respectu aquae. Nam glaeba terrae, si fuerit omnino circumvoluta aqua, non quiescit, sed movetur, donec venit ad fundum aquae; et glaeba etiam terrae posita extra aquam, ut in litore maris, non movetur ad illam aquam.

⟨4⟩ Item videtur esse e converso, scilicet quod aqua naturaliter locatur in terra. Quod patet primo, quia aqua pluta vel expansa super terram descendit naturaliter in terram. Secundo etiam, quia aqua naturaliter generatur in terra, ut patet de fontibus; et ille est locus naturalis rei in quo illa naturaliter generatur, propter quod dicebat Porphyrius quod locus est principium generationis. Tertio, quia in profunditate terrae aqua naturaliter conservatur; et ideo, ubicumque fodias profunde, invenies aquam. Et si ibi intra terram non

---

4 quaeritur quinto] quinto quaeritur consequenter G 7 et] sed G 8 totum locatum] *inv.* G 9 totam] totum p 9–10 continet] contineret C : commaneret P 11 continere] coninere (!) P ‖ corpora] *add.* multa ut lapides etc. G 12 fuerunt] fuerint Gp : erunt P ‖ locis suis] *inv.* GPp 13 fuerunt] fuerint Gp 15 aqua] aquae P ‖ venit] venerit Pp 15–16 fundum] profundum p 18 e converso] *ante* videtur GPp 19 patet] apparet G ‖ vel expansa] et expansa G : vel expensa C 20 terram] terra G 22–23 generationis] *add.* etc. GP 23 quia] *add.* etiam G ‖ profunditate] profundis GPp ‖ et] *om.* GPp 24 invenies] invenias P ‖ intra] infra p

---

22 Porphyrius, *Isagoge*, 2, 4 (ed. Minio-Paluello, 6)

 | DOI: 10.1163/9789004322356_028

esset locus naturalis eius, ipsa ibi existens deberet naturaliter moveri sursum, donec esset supra terram, quod apparet falsum.

⟨5⟩ Item non aqua, sed centrum mundi debet poni locus naturalis terrae. Quod apparet primo, quia glaeba terrae moveretur, donec esset ad | centrum, si non esset prohibita, scilicet si terra perforaretur usque ad centrum. Secundo, quia terra est simpliciter gravis; ideo locus eius naturalis debet esse simpliciter deorsum et hoc est centrum mundi. Tertio, quia primo *Meteororum* dicitur quod locus terrae est medium lationis circularis, et hoc est centrum. 79vb G

⟨6⟩ Item si superficies aquae esset locus naturalis terrae et non centrum, sequeretur quod partes terrae quae sunt in centro vel iuxta centrum distarent multum a loco suo naturali; ideo ibi non quiescerent naturaliter, quod est falsum. | 70vb p

Oppositum dicit Aristoteles. Et arguitur ratione quia: elementa quattuor debent esse ordinata secundum exigentiam suarum gravitatum et levitatum; ideo terra, quae est simpliciter gravis, debet esse medium vel in medio mundi; et aqua consequenter circa terram undique, quia est minus gravis, ideo debet continere et locare terram; et consequenter aer aquam etc.

Ista quaestio est duplex quaestio. Quaerit enim de loco proprio terrae et de loco naturali terrae. Et non secundum eandem rationem dicitur locus proprius | et locus naturalis. Primo igitur respondebo de proprietate, secundo de naturalitate. 100rb C 88rb P

De proprietate dictum est quantum ad quid nominis quod ille dicitur locus proprius alicuius corporis, qui continet ipsum totum et non continet aliud corpus, scilicet quod non sit pars illius.

Et tunc faciliter ponuntur conclusiones.

Prima est quod terrae locus proprius est aer vel superficies eius, quia quaelibet glaeba terrae est terra, et tamen alicuius glaebae terrae locus proprius est aer, scilicet glaeba quae proiecta est sursum in aere. Ipsa undique cir-

---

1 esset] esse G ‖ eius ipsa] aquae ipsa Pp : aquae aqua G 4 apparet] patet GPp 5 si[1]] sed P 6 locus eius] *inv.* p 8 circularis] circulationis G : et circulationis p 11 sequeretur] sequitur Gp 12 loco suo] *inv.* p 14 arguitur] arguit P ‖ elementa quattuor] *inv.* G 16 est simpliciter] *inv.* p 20 dicitur] *add.* enim p 23 quod] quia p 27 prima] *add.* conclusio P ‖ quia] et P 28–29 terrae[2] ... glaeba] *om.* CPp 29 ipsa] *add.* enim G

7–8 Cf. Aristoteles, *Meteora*, I, 7, 344a11–13 14 Cf. Aristoteles, *De caelo et mundo*, IV, 4
23 Cf. sup., IV, q. 1, 205$^{15-22}$

cumdata est et contenta aere tangente eam, sive moveatur, sive quiescat in termino reflexionis, ita quod ad praesens non discutio quomodo continue motum de uno loco ad alium locum habeat vel non habeat locum proprium; et de hoc dicetur alias.

Secunda conclusio quod terrae locus proprius est aqua. Probatur consimili modo sicut de aere. Possibile enim est glaebam terrae esse totam aqua circumdatam.

Tertia conclusio est quod terrae locus proprius nec est aer nec est aqua nec superficies aquae nec superficies aeris, quia est quaedam terra quae nec tota est contenta aqua nec tota est contenta aere, verbi gratia totalis terra a centro mundi imaginato continua usque ad nos et usque ad aquam et usque ad aerem.

Et illae conclusiones sunt indefinitae; ideo stant simul.

Sed tunc videtur quod quaestio intendebatur de illa totali terra quae modo dicta est; ideo quaeritur quid sit locus eius proprius.

Et ego respondeo ponendo quartam conclusionem quod nullus est locus eius proprius, quia omnis superficies quae assereretur vere continere totam illam terram contineret cum ea multa alia corpora, ut lapides multos et mineralia, aquas visceribus terrae inclusas, radices plantarum etc. Sic etiam aeris totalis continui de sphaera ignis usque ad sphaeram terrae vel aquae nullus est locus proprius, quia quicumque assignaretur continens totum illum aerem, ille contineret simul aquam et terram, quae non sunt partes illius aeris. Et etiam sphaerae totalis ignis nec orbis lunae vel solis est aliquis locus proprius.

---

1 est] *om.* G ‖ contenta] condensata G ‖ eam] ipsam P 2 ita] *om.* G 3 locum[1]] *om.* G 4 et] sed G ‖ dicetur alias] *inv.* G 5 conclusio] *add.* est GPp ‖ terrae locus] *inv.* G 5–6 consimili modo] consimiliter p 6–7 glaebam ... circumdatam] quod glaeba terrae sit totaliter aqua circumdata P 7 circumdatam] circumdata G 8 locus proprius] *inv.* P ‖ est[3]] *om.* GP 9 aquae ... aeris] aeris nec superficies aquae G 10 est contenta[2]] *om.* p 10–11 a centro] ad centrum P 11 imaginato] *add. in marg.* alias imaginata C ‖ imaginato continua] imaginata continue GP 13 et] *add.* sic p ‖ ideo] igitur G 14 intendebatur] intendebat p : non intendebat nisi G 15 locus eius] *inv.* GPp 16–17 locus eius] *inv.* GPp 17 assereretur] asseretur P : appareret G : assignetur p 18 contineret] continet p : continere P 18–19 multos et mineralia] multos mineralia P : multos G : multa mirabilia p 19 aquas] *add.* in GPp 20 aeris totalis] totalis aer p ‖ terrae vel aquae] aquae vel terrae G 21 quicumque assignaretur] quocumque assignarentur P 22 illum] alium C ‖ quae] quia G 23 etiam] ita GPp ‖ vel] nec p ‖ est] esset p

---

4 *locus non inventus*

Quinta conclusio ponitur quod congregati ex orbe totali lunae et ex omnibus contentis sub illo est locus proprius, scilicet superficies orbis immediate superioris, scilicet orbis Mercurii, quia continet totum etc. Et sic superficies concava orbis lunae est locus proprius congregati ex igne immediato illi orbi et ex contentis ab illo igne.

Sexta conclusio est quod nulla res est locus proprius congregati ex terra totali, quae prius dicebatur, et ex contentis in ea, quia locus | 100va C proprius debet undique tangere locatum et esse immediatum locato; et nulla res sic se habet ad illud congregatum, quia ex una parte illud congregatum tangitur ab aqua et ex alia | 88va P parte ab aere et nulla res est aer et aqua vel etiam superficies aquae et aeris, immo illae sunt diversae res numero et specie et ad invicem discontinuae et secundum situm separatae, et talia non sunt aliquid unum simpliciter nec per consequens aliquod ens vel aliqua res. Et consimiliter dicetur quod hominis in balneo sedentis vel arboris aut plantae vel lapidis, cuius una pars est in aere et alia in aqua, nulla res est locus proprius, quia congregatum ex aere et aqua non est aliqua res, sed aliquae res.

Septima conclusio est quod multa sunt corpora totalia et per se mobilia etiam motu recto quae sic se habent quod uniuscuiusque eorum locus proprius non est aliqua | 71ra p res una, sed est congregatum ex multis rebus ex quibus non est aliquod unum loquendo simpliciter nec aliquod ens. Sic etiam exercitus non est aliqua res, sed multae res, vel etiam | 80ra G populus. Nec plantae habentis radices in terra et ramos in aere nulla res est locus proprius, sed congregatum ex quadam superficie terrae et quadam superficie aeris est locus eius proprius. Et ita de baculo habente unam partem in terra, aliam in aqua et aliam in aere, locus proprius esset congregatum ex quadam superficie terrae et quadam superficie aquae et quadam superficie aeris, quia huiusmodi congregato et nulli alteri vel etiam nullis aliis quam tali congregato conveniret continere undique totam plantam et esset divisum ab illa et immediatum sibi.

---

1 congregati] congregatum p **1–2** omnibus] *om.* G **2–3** immediate … orbis] *om.* (*hom.*) P **3** superioris] superiorum p ‖ totum] *om.* P **6** est^{1}] ponitur Pp **8** esse] est G ‖ immediatum] *add.* illi p **11** et^{3}] *om.* G **12** separatae] separatum p ‖ aliquid] aliquod GP **13** simpliciter nec] singulariter videtur G **14** dicetur] diceretur Gp : dicitur P ‖ aut] vel P ‖ vel^{2}] aut P **15** et] *om.* P **16** sed aliquae] sed est aliquae G : vel aliqua P **17** totalia] *om.* G **19** est^{2}] *om.* Pp **20** loquendo] *om.* P ‖ sic] sicut GPp **21–22** nec plantae habentis] vel plantae habentes CPp **22** nulla res] nullus *sed add. sup. lin.* nulla res C **23** et] *add.* ex P **24** eius proprius] *inv.* G ‖ ita] sic P ‖ terra] *add.* et Gp **25** congregatum] aggregatum p **26** quia huiusmodi] et huius P **27** congregato^{1}] congregatio C ‖ vel] nec C ‖ nullis aliis] nulli alio G **28** continere] quod contineret G ‖ esset divisum] esse divisim P : esse divisam p

Octava conclusio ponitur de totali terra, id est de maxima terra, quae scilicet est continua de centro terrae imaginato usque ad aquam et aerem nobis apparentes. Et est ista conclusio quod congregati ex huiusmodi terra maxima et contentis in ea locus proprius est congregatus ex valde pluribus corporibus diversis et ab invicem discontinuis. Verbi gratia illud congregatum vocetur *b*. Tunc oporteret locum proprium ipsius *b* circumdare undique totum *b* et tangere ipsum. Sed constat quod alicubi superficies aquae tangit ipsum *b* extrinsece et alicubi superficies aeris et alicubi superficies ignis facti super terram ex straminibus et alicubi pes tuus et pes formicae et sic de aliis quoad nos infinitis. Et tamen cuiuslibet sic extrinsece tangentis ipsum *b* superficies, secundum quam tangit ipsum, est pars loci totalis proprii ipsius *b*. Ideo congregatum ex omnium talium superficiebus est locus proprius
100vb C ipsius *b*. Et secundum exigentiam istorum dictorum | aliquis posset multiplicare conclusiones, sicut sibi videretur expedire.

Ultimo igitur concludendum quod non est de ratione huius termini 'locus proprius' quod supponat pro una re continua, sed aliquando pro una re continua, aliquando pro pluribus non ad invicem continuis. Et multa sunt
88vb P talia nomina, ut aliquando onus quod tu portas est | una res continua, scilicet si sit totum de ferro, et aliquando est multae res discontinuae, ut onus straminum. Et possibile est quod cultellus est una res continua, scilicet si sit totus de ferro, et aliquando non est ita, sed est congregatum ex ferro et ligno, si manubrium sit de ligno. Ideo non sequeretur 'locus est unus locus, igitur locus est unum ens vel una res', sicut non sequitur 'domus est una domus, exercitus est unus exercitus, igitur est una res sive unum ens'.

Nunc igitur dicendum est de naturalitate loci. Et videtur mihi secundum dicta auctorum quod locus dicitur ex eo locus naturalis alicui corpori, quia

---

1 conclusio] *add.* est et p ‖ id est de] scilicet de p : id est P 2 aquam et aerem] aerem et aquam G 3 congregati] congregatum C 4 est congregatus] *inv.* P 5 et] *om.* G ‖ discontinuis] disconvenientes P 6 b[1]] *add.* et tangeret P ‖ oporteret] oportet G ‖ circumdare undique] *inv.* P 7–8 aquae ... superficies[2]] *om.* (*hom.*) CP 9 facti] factae P ‖ et[1]] vel p ‖ et[2]] vel G 10 infinitis] *ante* quoad P : *om.* G ‖ tangentis] tangentes P 11 ipsum] *add.* b P 12 omnium talium] omnibus talibus P 13 istorum dictorum] *inv.* p 14 conclusiones] quaestiones C ‖ sicut ... expedire] sicut sibi viderentur expedire p : *om.* G 15 concludendum] *add.* est GPp ‖ huius] huiusmodi p 16 supponat] supponatur P 19 si ... ferro] magnum plumbum G 20 straminum] stramineum p ‖ scilicet] *om.* p 21 et[1]] *om.* P ‖ et[2]] *add.* ex P 22 sequeretur] sequitur GPp 23 ens] *om.* p ‖ domus[2]] *add.* vel Pp 24 igitur] *add.* exercitus P ‖ una ... ens] una res vel unum ens p : unum ens vel una res G : una res vel una domus est unum ens P 26 auctorum] aristotelis G ‖ dicitur] *post* eo Pp ‖ locus[2]] *om.* GP

est sibi conveniens tamquam naturaliter conservativus ipsius, et dicitur locus innaturalis sive violentus alicui corpori, quia est sibi disconveniens tamquam non naturaliter conservativus ipsius, sed potius destructivus. Etiam ob hoc dicimus quod locatum est innatum naturaliter quiescere et permanere in loco sibi naturali, si fuerit in eo, et moveri ad ipsum, si fuerit extra vel remotum ab eo. Unumquodque enim naturaliter appetit esse et bene esse et conservari in suo bene esse; ideo appetit esse ubi bene sit et conservetur in suo bene esse.

Ex istis dictis manifestum est iam quid debeamus dicere de locis naturalibus multorum mixtorum. Locus enim naturalis plantae nec est terra nec aqua nec est aer aut ignis vel superficies alicuius istorum, quia in nullo istorum, si esset in eo totaliter et immediate circumdata, posset diu durare et conservari; sed locus eius naturalis est congregatum ex superficie terrae et superficie aeris, quia indiget esse in terra quantum ad radices et etiam in aere quantum ad ramos. Locus naturalis piscis est bene aqua. Locus naturalis lapidis est bene terra. Locus na|turalis hominis indiget bene quod sit compo- 71rb P
situs ex aere, ut possit respirare, et ex terra vel aliquo solido ad sustinendum ipsum; caderet enim, si esset undique circumdatus aere. Locus naturalis hirundinis est in ista patria in aestate, sed non in hieme. Et sic forte non esset locus naturalis hominis sub aliquo polorum caeli propter nimiam frigiditatem nec sub | circulo aequinoctiali, ut aliqui dicunt, propter nimiam 101ra C
caliditatem; et alii dicunt oppositum. Sed de hoc non est hic discutiendum. Et haec sint dicta de mixtis pro prima conclusione. Potest enim dici suo modo de aliis mixtis et locis, sicut dictum est de praecedentibus.

Sed de terra et aliis elementis, scilicet aqua, aere et igne, dicendum est pro secunda conclusione quod locus naturalis ignis est superficies orbis lunae immediata huic mundo inferiori, quia ille est locus naturalis ignis ad quem movetur naturaliter ignis, si fuerit extra; sed nobis est manifestum | 80rb G
quod flamma et exhalationes calidae participantes multum de natura ignis

4 etiam] ipsius et G : et P 5 sibi] *ante* in P 7–8 ideo ... esse] *om.* (*hom.*) p 7 et^{2}] ut C 8 conservetur] conservare P 9 ex] *praem.* et G ‖ istis ... dicere] his patet quid dicendum p 10 mixtorum] illorum G ‖ nec^{2}] *add.* est p 11 est] *om.* GP ‖ aut] nec P ‖ vel superficies alicuius] aut aliquis G ‖ alicuius] aliquorum *sed add. sup. lin.* aut alicuius C ‖ quia] qui G 11–12 quia ... istorum] *om.* (*hom.*) P 13 eius] est G 14 superficie] *om.* p ‖ radices] radicem P ‖ et etiam] etiam C : et G 15 locus^{1}] *add.* enim G 16 sit] est locus P 17 ut] et P ‖ ex^{2}] *om.* G ‖ sustinendum] sustentandum GPp 18 esset undique circumdatus] undique circumdatus esset P ‖ naturalis] proprius *sed add. in marg.* naturalis C 19 in^{1}] *om.* P 22 de hoc] *om.* P ‖ est hic] *inv.* p 23 sint] sunt P 23–24 suo ... locis] *add.* suis Pp : de aliis suo modo G 24 sicut] ut P 28 quem] hoc quod P ‖ naturaliter] *om.* G 29 et] vel G ‖ participantes] percipientes P

89ra P moventur | naturaliter ad illum locum, si non prohibeantur. Unde flammam videmus velociter ascendere, tamquam regio aeris non sit locus naturalis ipsius ignis, nec etiam infra, sed supra. Modo supra regionem et locum aeris non est locus debitus istis generabilibus nisi illa superficies orbis lunae; igitur illic est locus naturalis ipsius ignis. Et hoc est rationabile, quia nobiliori elemento debet esse locus nobilior naturalis; ignis autem est nobilissimum quattuor elementorum, quod apparet ex eius maxima activitate; et non posset assignari locus nobilior tangens ista generabilia quam caelum; igitur etc.

Et iterum, cum primo debeatur igni caliditas, ut apparet secundo *De generatione*, ille locus debet igni esse naturalis, qui maxime est effectivus et conservativus caliditatis; et ille est caelum vel superficies caeli propter ipsius velocem motum, quoniam de natura motus localis est calefacere, ut habetur primo *Meteororum* et secundo *Caeli*.

Sed contra hanc conclusionem obicitur quia: cum locatum fuerit in suo loco naturali, debet in eo naturaliter quiescere; ignis autem in concavo orbis lunae non quiescit, sed movetur motu diurno cum ipso caelo, ut habetur primo *Meteororum* et scitur per motum stellae comatae; igitur etc.

Item Aristoteles ponit quod stella sit de natura orbis. Igitur cum luna ponatur esse frigida et humida, videtur etiam quod orbis lunae sit frigidus et humidus et virtualiter et active. Et tale non est conveniens conservativum ignis, sed potius destructivum, cum ignis debeat esse calidus et siccus.

Ad primam respondetur quod non universaliter est verum quod locatum in suo loco naturali debeat quiescere, quia possibile est, si locus ille naturaliter moveatur, quod locatum naturaliter cum eo moveatur. Sed quantum ad
101rb C elementa et loca eorum naturalia locatum in suo loco naturali | non debet moveri naturaliter motu per quem exeat et recedat a loco illo. Et sic impro-

1 naturaliter] *om.* p 2 tamquam] *add.* si G 4 est locus] sit locus naturalis p ‖ generabilibus] generalibus p ‖ orbis] *om.* P 5 illic] illis p ‖ est[2]] *om.* p 6 nobilior naturalis] *inv.* P ‖ autem est] *inv.* P 7 apparet] patet P 7–8 non posset] non potest GP : potest p 8 locus] *om.* G 10 et] *om.* G ‖ apparet] patet G 13 habetur] patet p 14 et secundo caeli] *om.* G 15 obicitur] *add.* primo GP 16 debet ... quiescere] quiescit G 19 igitur] *post* cum P : ideo G 20 ponatur esse] ponitur P ‖ etiam] *post* quod p : *om.* P 21 et[2]] scilicet GPp 22 debeat] debet P 23 primam] *add.* obiectionem GPp ‖ universaliter est] *inv.* Gp : est naturaliter P 25 moveatur[1]] movetur Pp ‖ naturaliter] *post* eo GPp 26 eorum] *om.* P 27 moveri] movere P ‖ a loco illo] ab illo loco GPp

10–11 Cf. Aristoteles, *De generatione et corruptione*, II, 3, 331a3–6 14 Cf. Aristoteles, *Meteora*, I, 3, 340b10–14, 341a19–36; cf. *AA*, 5: 5; cf. Aristoteles, *De caelo et mundo*, II, 7, 289a21–22; cf. *AA*, 3: 63 18 Cf. Aristoteles, *Meteora*, I, 3, 341a1–3 19 Cf. Aristoteles, *De caelo et mundo*, II, 7, 289a13–15

prie loquendo quiescit ignis in concavo orbis lunae, scilicet quia non movetur ibi naturaliter motu recto quo recedat ab illo concavo. Et hoc sufficit.

Ad aliam obiectionem solet dici quod caelum in corpus sibi propinquum fortius agit per suum motum calefaciendo, propter hoc quod ille motus est velocissimus, quam per alias suas virtutes vel influentias infrigidando. Et ob hoc locat naturaliter ignem, cuius est esse primo calidum.

Tertia conclusio ponitur quod superficies concava illius ignis totalis qui est immediatus orbi lunae est locus naturalis totalis aeris (et voco totalem aerem maximum, scilicet qui est continuus ab illo igne usque ad terras et aquas hic apud nos existentes), quia cum aer sit levior aqua et terra, debet locari naturaliter supra aquam et terram; et ideo, cum non sint elemen|ta 71[va] p
nisi quattuor, scilicet terra, aqua, ignis et aer, non potest inter aerem et ignem esse aliud elementum; | propter quod necessario attingit usque ad sphaeram 89[rb] P
ignis et est ibi situs eius naturalis. Ideo naturale est quod sphaera ignis contineat aerem immediate; ideo eius superficies est locus eius naturalis.

Et hoc etiam consonat qualitatibus naturalibus ipsorum. Cum enim aer sit naturaliter calidus, ignis, qui etiam est naturaliter calidus, est naturaliter conservativus aeris quoad illam qualitatem, licet quantum ad siccitatem et humiditatem sint ad invicem activa et passiva et sibi invicem corruptiva. Talem enim contrarietatem inter elementa se invicem locantia ingeniavit natura, ut ipsa elementa invicem possent misceri et quod generarentur ex eis mixta, plantae et animalia ad quae illa elementa finaliter ordinantur.

Quarta conclusio est quod aer est locus naturalis aquae (vel totalis locus vel partialis), quia consequenter se habet aqua post aerem quantum ad gravitatem et levitatem; ideo naturale est quod immediate sub aere contineatur aqua. Et huic etiam consonat quod aer in una qualitate communicat cum aqua, scilicet in humiditate, sicut aer cum igne in caliditate.

---

1 loquendo] *rep.* G ‖ scilicet] *om.* p 2 naturaliter] *om.* G ‖ et hoc sufficit] *om.* G 3 obiectionem] *om.* P 4 suum] *om.* p ‖ calefaciendo] calefaciendi G 4–5 calefaciendo ... velocissimus] *om.* p 5 alias ... vel] *om.* p ‖ influentias infrigidando] intelligentias frigefaciendo G 6 ignem] *rep.* C ‖ primo] primum P 7 ignis totalis] *inv.* P 9 aerem] *om.* G 10 debet] debeat P 11 locari] *add.* ibi C ‖ supra] super P 12 nisi] *add.* illa G ‖ potest] *add. in marg.* communiter C 13 quod] *add.* aer GPp ‖ attingit] pertingit GPp 14 sphaera] superficies G 15 eius superficies] *om.* G 16 et ... consonat] et haec etiam consonant G : hoc etiam consonat et P 17 qui etiam] *inv.* Cp ‖ est naturaliter[1]] *inv.* G ‖ naturaliter[2]] *om.* P 18 licet] sed G 19 sint] sunt G : sicut C ‖ sibi] sui Gp 21 possent] possunt P 22 ad ... ordinantur] quae illa elementa finaliter ordinant G 23 quarta] *add.* etiam p ‖ locus[2]] *om.* P 25 et] vel p 27 scilicet in humiditate] *om.* G ‖ aer ... in[2]] ignis cum aere in Pp : ignis in aere cum G

80va G Quinta conclusio quod aqua est | locus naturalis terrae (dico: locus sive totalis sive partialis), quia secundum exigentiam gravitatis aut levitatis terra debet contineri sub aqua immediate. Et est etiam convenientia aquae cum terra in una qualitate elementari, scilicet in frigiditate.

Sexta conclusio quod locus naturalis terrae est aer vel superficies aeris (dico ut prius: locus totalis vel partialis), quia quod perpetuo invenitur in rebus naturalibus et ordine earum ad invicem debet dici naturale, non 101va C violentum nec contra | naturam; sed perpetuo secundum Aristotelem terra locata est ab aere quantum ad partem deputatam habitationi animalium et plantarum; igitur etc.

Item videmus totalem terram naturaliter quiescere sub aere quantum ad partem habitabilem, sicut sub aqua quantum ad partem coopertam aquis; et tamen ista elementa non quiescunt naturaliter nisi in locis suis naturalibus.

Etiam videmus glaebam terrae elevatam in aere descendere, donec contineatur sub totali aere cum totali terra, sicut glaeba descenderet in aqua, donec contineretur sub totali aqua; locus autem est naturalis, in quo locatum naturaliter quiescit, si sit in eo, et ad quod movetur, si sit extra, non prohibitum; igitur etc.

Septima conclusio quod nullus locus dicitur naturalis vel violentus alicui corpori secundum praecisam essentiam loci aliis circumscriptis, quia huiusmodi essentia praecisa loci est dimensio, quae non esset alterius rationis in caelo quam in terra, circumscriptis substantiis subiectis illis dimensionibus et etiam circumscriptis qualitatibus activis et passivis, immo sicut dicit Aristoteles quarto *huius*, dimensio vacui, si esset, non esset alterius rationis a dimensione pleni, quantum est ex praecisa ratione et essentia dimensionis. 89va P Sed tamen oportet eiusdem elementi locum naturalem et | locum violentum esse aliarum rationum et diversarum naturarum vel potentiarum natura-

---

1 quinta] alia P : *add.* etiam p ‖ conclusio] *add.* est G ‖ locus naturalis] *inv.* GPp ‖ dico locus] *om.* P 2 aut] et G 4 in^{2}] *om.* G 5 sexta] alia P ‖ conclusio] *add.* est GPp 7 earum] eorum p ‖ naturale] *add.* et P 9 habitationi] habitativam G 10 igitur] *om.* G 12 et] *om.* G 13 locis suis] *inv.* P 14 etiam] *add.* ut P : et etiam nos Gp ‖ elevatam] elementatam p 15–16 cum ... aqua] *om.* P 15 terra] aqua Cp 16 autem] *om.* C 17 extra] *add.* et GPp 19 septima] alia P ‖ conclusio] *add.* est Gp 20 quia] et G 20–21 huiusmodi] huius P 21 dimensio] demonstratio P 22–23 subiectis ... et^{1}] vel dimensionibus vel G : subiectis illis demonstrationibus P : et illis dimensionibus et p 23 sicut] ut P 23–24 aristoteles] *add.* in GPp 24 dimensio] demonstratio P 25 dimensione] demonstratione P ‖ dimensionis] demonstrationis P

---

8 *locus non inventus* 24 Cf. Aristoteles, *Physica*, IV, 8, 216a27–b2

lium, quia aliter non esset ratio quare ille diceretur naturalis et ille violentus, nec quare esset naturalis uni et violentus alteri. Igitur non dicitur naturalis vel violentus ratione simplicis dimensionis quae est locus, sed ratione naturae subiectae dimensioni quae est locus, vel ratione naturae aut qualitatum naturalium cum illa dimensione quae est locus existentium in subiecto illius dimensionis quae est locus.

Octava conclusio quod aqua est locus (sive totalis sive partialis) magis naturalis terrae quam aer, quia | ceteris paribus est magis conveniens terrae, 71vb p scilicet quantum ad qualitates naturales. Primo, quia magis convenit ei in gravitate et levitate. Secundo, quia quantum ad primas qualitates tangibiles aqua et terra conveniunt in una qualitate, scilicet in frigiditate; aer autem et terra non, sed contrariantur invicem et secundum caliditatem et frigiditatem et etiam secundum humiditatem et siccitatem.

Sed contra praedicta obicitur primo quia: videtur quod aer nullo modo sit locus naturalis terrae (nec totalis nec partialis), quia omnino est sibi disconveniens et contrarius quantum ad eorum qualitates naturales, scilicet calidum, frigidum, humidum et siccum, immo etiam quantum ad | grave vel 101vb C leve, rarum et densum (dico vel 'contrarius' vel 'prope contrarius').

Item nullo modo ponitur ignis locus naturalis aquae; igitur pari ratione nec aer terrae.

Propter istas rationes considerandum est quod, cum nullus locus dicatur naturalis vel violentus ea ratione qua est superficies vel corpus vel dimensio aliis circumscriptis, sed ratione naturae vel dispositionis naturalis activae vel passivae, convenientis vel disconvenientis naturae ipsius locati, quae est in illo loco sive in illa superficie vel cum ea, sicut dicebat septima conclusio, aer vel aqua respectu terrae potest dupliciter dici locus naturalis vel conveniens. Uno modo quantum ad virtutes vel qualitates suas elementares,

---

1 quia] *sup. lin.* C : *om.* GPp 2–3 nec ... violentus] *om.* (*hom.*) G 2 naturalis[1] ... violentus] violentus uni et naturalis P 3 dimensionis] demonstrationis P 4 subiectae dimensioni] quae subest dimensioni p : subest dimensioni G : subiectae demonstrationi P ‖ aut] vel G 5 cum] sed G : quam P ‖ dimensione] demonstratione P 6 dimensionis] demonstrationis P 7 conclusio] *add.* est GPp 8 magis] *om.* p 9–10 ei ... levitate] in gravitate et levitate ipsi p 10 quantum] *om.* p 11 in[2]] *om.* G 12 et[1]] *om.* p ‖ et[2]] *add.* secundum C 13 etiam] *om.* G 14 contra praedicta obicitur] tunc obicitur contra praedicta (dicta G) GPp 15 totalis nec partialis] partialis nec totalis G 16 quantum] tantum P ‖ eorum] earum Pp 17 calidum] *add.* et Gp : *add.* vel P ‖ ad] *rep.* C ‖ vel] et GPp 18 dico vel] vel P : dicitur p ‖ prope] proprie p 21 quod cum] quod G : cum p 22 vel[3]] et P 23 sed] vel P 24 convenientis vel disconvenientis] convenienter vel disconvenienter CP 25 dicebat] dicit p ‖ septima] alia P 26 dupliciter dici] *inv.* G 27 vel] seu Gp : sive P

quae sunt caliditas, frigiditas etc. et qualitates secundae ex istis consequentes, alio modo quantum ad virtutes et influentias quas ista elementa recipiunt a corporibus caelestibus alias et alias, secundum quod difformiter se habent secundum situm ad caelum, et maxime, quantum spectat ad propo-
80[vb] G situm, secundum quod | difformiter se habent secundum propinquitatem vel distantiam ad caelum. Aliter enim influit caelum propinque et aliter remote et est motus influentiae de propinquo conveniens uni elemento et de remoto alteri. Unde sicut dicitur primo *Meteororum*, ille mundus inferior
89[vb] P continuus est lationibus superioribus, ut omnis virtus eius inde gubernetur. | Caelum igitur in corpus sibi immediatum influit virtutem igni convenientem et in corpus ab eo remotissimum influit virtutem convenientem terrae et in corporibus intermediis virtutes convenientes aeri et aquae.

Notandum quod, sicut Deus et natura dederunt elementis qualitates ad invicem activas et passivas, secundum quas possent ad invicem transmutari et misceri propter generationes animalium et plantarum, ad quae elementa finaliter ordinantur, ita dederunt eis naturales gravitates et levitates, secundum quas movent se ad loca sibi convenientia ratione propinquitatis vel distantiae ad ipsum caelum, scilicet ignis super unumquodque aliorum elementorum, et terra sub unoquoque aliorum, et aer sub igne et super aquam et terram, et aqua super terram et sub igne et aere. Ideo concludit Commentator quod, si terra elevata moveatur ad superficiem concavam aquae, non movetur ad eam inquantum est superficies aquae, sed inquantum est in tali
102[ra] C distantia ab orbe. Unde, si ubi | est aqua, esset aer aut ignis, ita illa terra elevata moveretur ad superficiem illius aeris vel ignis tangentem et continentem terram, sicut movetur ad illam superficiem aquae.

Per hoc solvuntur igitur obiectiones. Aer enim continuus terrae non est locus naturalis terrae inquantum est aer, sed inquantum eius superficies

---

1 etc.] humiditas et siccitas GPp 4–5 situm ... secundum[2]] *om.* (*hom.*) p 4 ad[1]] *add.* ipsum GP 5 secundum[2]] *add. sup. lin.* alias ad C : ad P 6 vel] et secundum P ‖ aliter enim] aliquando p ‖ et] *om.* G 8 inferior] *om.* p 9 inde gubernetur] *inv.* Pp 10 in[1]] *sup. lin.* C : et P ‖ igni convenientem] *inv.* G 11 et] etiam C 13 notandum] *praem.* et est GPp 13–14 ad invicem] *post* passivas P : *om.* G 14 possent] possunt P 15 elementa] *om.* G 17 movent] moverent GPp 18 aliorum] *praem.* illorum p : illorum P 20 terram[1] ... sub] aquam super terram et terram et aquam sub aquam P ‖ super] supra p ‖ igne] ignem (*add. sed del.* igne) C ‖ ideo] *praem.* et GPp 21 superficiem] superiorem P 22 est[1]] *om.* P ‖ est[2]] *om.* P 23 orbe] orbi P 24 et] vel GP 26 per] *praem.* et p ‖ solvuntur igitur obiectiones] igitur solvuntur obiectiones Gp : igitur solvuntur rationes et obiectiones P ‖ continuus] contiguus GP 27 eius] *om.* p

---

8 Aristoteles, *Meteora*, I, 2, 339a21–23; cf. *AA*, 5: 2 20–21 *locus non inventus*

concava est in tali distantia ad caelum, scilicet quia inter superficies elementorum continentium terram ipsa est remotissima a caelo versus partes terrae habitabiles et discoopertas aquis. Immo etiam sic esset ignis locus naturalis terrae, si esset sic contigue expansus super terram, sicut est ille aer.

Ad ratio|nes principales. 72ra p

⟨1–2⟩ Duae enim primae procedunt secundum ea quae dicta fuerunt de loco proprio.

⟨3⟩ Tertia solvitur per hoc quod dictum est, quod terra non quaerit superficiem aquae secundum quod est superficies aquae etc.

⟨4⟩ Ad quartam dicendum est quod, si terra esset continua, scilicet sine poris et concavitatibus, aqua non descenderet in eam. Sed intra terram sunt pori vel concavitates repleti vel repletae aere, quia natura non potest permittere vacuum; et terra etiam solida existens, non fluxibilis, non potest defluere in illos poros vel concavitates ad replendum eas; ideo necesse est corpus subtilius circumstans fluere in illas concavitates. Tunc adveniente aqua supra terram, quia aqua appetit esse sub aere et aer supra aquam, aqua illa descendit in illos poros et in illas concavitates et aer ibi existens elevatur extra superius.

Sed tunc aliquis dubitaret utrum sic descendat naturaliter in terram vel | 90ra P praeter naturam. Et adhuc magis dubitatur de aere descendente in terram, quando de profundo terrae aufertur aliqua pars terrae, ut si fiat puteus, utrum igitur ille aer descendit naturaliter in fundum putei vel violenter. Et sic etiam est de aqua quae ascendit in phialam, quando aer qui est in phiala infrigidatur et condensatur, qui fuerat prius calefactus, utrum igitur illa aqua sic ascendit naturaliter vel violenter.

Dico, sicut mihi videtur, quod illi motus sunt naturales. Ab ipso enim Deo entia sic habent ordinem communem secundum situm quod necesse

1 concava] *add. sup. lin.* quae G ‖ scilicet] *sup. lin.* C : *om.* p ‖ quia] quod G 2 continentium] continentes P 2–3 partes ... discoopertas] terras habitabiles et discoopertas G : partem terrae habitabilis et discoopertis P 3 sic esset] si esset p : si essent P 4 sic] sibi Cp : *om.* P ‖ contigue expansus] contiguus expansus p : extensus contiguus G ‖ super] supra P 5 ad] *praem.* tunc ergo respondendum est Pp 8 non] *om.* p 9 est] *om.* P 11 et] vel p 12 repleti vel repletae] repletae (*et add. sup. lin.* vel repleti) C ‖ aere] *add.* nam GPp 13 existens] *add.* et GPp 14 vel] et P ‖ ideo] *om.* G 15 subtilius] *om.* G ‖ concavitates] *add.* et GPp 16 aqua supra] aqua super Pp : aere supra G ‖ quia] *om.* G ‖ supra^{2}] super Pp ‖ aqua^{3}] *om.* p 17 in^{1}] ad P ‖ ibi] in P 22 fundum putei] puteum G : *add. sup. lin.* naturaliter C 24 fuerat prius] *inv.* Gp : prius fuit P 25 sic ascendit] *inv.* G 26 dico sicut] respondeo sicut Gp : respondeo ut P

102rb C est semper totum mundum esse plenum, ita quod inter | nulla corpora mundi sit vacuum interceptum. Ideo hoc est omni corpori naturali, saltem infra caelum, naturale, quod si aliquod corpus removeretur a suo loco vel situ, corpus ei proximum movetur ad replendum illum locum. Et si sint diversa corpora proxima, tunc illud movetur ad illum locum replendum, quod est magis innatum moveri ad illum locum vel minus prohibitum. Unde
81ra G cum fiat puteus profundus, verum est | quod terra de eo removetur per violentiam, tamen ad huiusmodi remotionem consequitur ex natura quod corpus propinquum descendat in illum locum unde terra removetur; terra autem circumstans propter eius soliditatem minus est innata moveri ad illum locum quam aer; ideo aer ibi naturaliter descendit. Et si aqua esset circumstans, ipsa ibi descenderet, et non aer, quia appetit esse sub aere et aer super eam ratione suarum gravitatum vel levitatum.

Et est bene considerandum quod, licet aeri vel aquae bene conveniat naturaliter moveri ad replendum locum a quo illud corpus removetur, tamen hoc non convenit aeri secundum quod aer nec aquae secundum quod aqua et sic de aliis, immo hoc convenit eis secundum rationem communem, scilicet secundum quod sunt corpora naturalia. Hoc enim conveniret terrae solidissimae, si esset contigua corpori quod ab aliquo loco auferretur. Si enim non esset aliquod aliud corpus fluxibile quod posset in illum locum intrare, terra illa solidissima, etiam si esset lapis durissimus, dissolveretur ad replendum locum a quo illud corpus removetur; alioquin illud corpus ab illo loco nulla virtute naturali posset removeri.

Et sic debet intelligi illud quod antiqui dixerunt, videlicet quod corpora sic moveri, ne sit vacuum, est per naturam universalem sive communem, non per naturam specialem aeris et aquae etc. Hoc enim sic debet intelligi quod non convenit aeri sic moveri secundum quod aer nec aquae secundum quod aqua etc., sed inquantum corpus naturale et naturaliter mobile ad

---

1 totum] *post* esse G 2 interceptum] *add.* et G : et p 3 removeretur] removetur GP ‖ suo] *om.* P 4 movetur] moveretur p ‖ illum locum] *inv.* GPp 5 illud] *om.* G 8 ad huiusmodi] ad huius p : per huiusmodi P ‖ consequitur ex natura] sequitur P 9 descendat in] descendet in P : descendit in p : descendat ad G ‖ unde] *add.* cum P 10 ad] in P 11 ibi naturaliter] *inv.* GPp 12 ibi descenderet] *inv.* G ‖ quia] qui P 13 vel] et G 14 bene conveniat] conveniat Gp : convenit P 15 replendum locum] *inv.* G ‖ illud] aliud G 16 nec aquae] et aquae non convenit G 17 et … aliis] *om.* G 18 corpora naturalia] corporalia p 19 loco auferretur] auferetur p 20 aliquod] aliud P : *om.* Gp ‖ posset] possit P 21 solidissima] solidissime p 22 removetur] removeretur GP 23 posset] possit P 24 et] *om.* P ‖ videlicet] scilicet G 25 est] et C 26 et] aut GPp 27 quod[1]] pro p 28 inquantum] secundum quod GPp

replendum locum a quo corpus proximum ei separatur; hoc est | dictu quod 90rb P haec non est per se et primo vera 'aer est naturaliter mobilis ad locum a quo separatur corpus sibi proximum' nec ista etiam 'aqua vel ignis' etc., sed ista 'corpus proximum' etc. cum circumstantiis prius dictis.

Sed ultra, quando dicitur quod aqua naturaliter gene|ratur in visceribus 72rb p terrae et conservatur, dicendum est quod hoc est verum, quia cum aer ibi in concavitates terrae descendit, tamen quantum ad qualitates elementares contiguas habet qualitates contrarias et corruptivas ipsius (ille enim aer ibi descendens circumdatur undique frigidis); ideo ex eo | virtute illorum 102va C circumstantium generatur aqua, quia magis naturale saltem respective est ibi inferius esse aquam quam aerem. Et ob hoc etiam semper est aqua in visceribus terrae, immo etiam et aer. Et haec omnia apparent rationabilia ex praedictis.

⟨5⟩ Ad aliam, quae arguit quod non superficies aquae, sed centrum mundi sit locus proprius et naturalis terrae, dicitur primo quod centrum mundi est totalis terra et ipsa non est locus sui ipsius. Et si capiamus 'centrum' pro puncto indivisibili secundum imaginationem mathematicam in medio mundi, tamen illud centrum non esset locus, quia nihil contineret nec ad ipsum moveretur terra, postquam esset cum alia terra sub aliis elementis.

Nec valet quod dicitur, quod si terra esset perforata etc. Constat enim quod oporteret secundum naturam illud foramen aliquo repleri, immo etiam ignis potius descenderet de caelo quam quod ibi remaneret aliquod vacuum.

Quod etiam dicitur, 'cum terra sit simpliciter gravis, debet esse simpliciter deorsum', dico quod hoc conceditur respectu aliorum elementorum, sed non oportet, quin una pars sit magis deorsum quam alia secundum situm distantiae vel propinquitatis ad caelum.

---

1 corpus proximum ei] proximum corpus eius P ‖ dictu] dictum GP 3 sibi proximum] proximum ei P 3–4 nec ... proximum] *om.* (*hom.*) CPp 4 prius dictis] praedictis p 7 descendit] *praem.* naturaliter G : descenderet et p ‖ quantum ad] adhuc p 8 contiguas habet] habet sibi contiguas Gp : habent sibi contiguas P ‖ ille] ipse G 9 circumdatur undique frigidis] circumdat undique frigiditatem p 10 circumstantium generatur] circumdantium generatur GP : circumstantium generat p ‖ est] *ante* naturale GPp 11 ibi] *sup. lin.* C : id est P 12 apparent rationabilia] rationabilia sequuntur p 14 aliam] *add.* rationem GPp 15 et] *om.* G ‖ dicitur] respondetur GPp 16 est[1]] etiam P ‖ et ipsa] et etiam ipsa p : ipsa C : *om.* P 18 nihil contineret] nec contineret terram G 20 etc.] *om.* G 21 illud] *om.* GPp ‖ aliquo] *add.* impleri vel P : aliquod G 22 ibi remaneret aliquod] aliquid ibi remaneret P : aliquid remaneret ibi p : aut quid ibi remaneret G 24 quod] *om.* p ‖ cum terra] quod terra cum G 25 conceditur] *add.* in GP

Et de causa quare sic in profundis terrae generatur et conservatur aqua, dictum est satis.

⟨6⟩ De ratione ultima, quando dicitur quod partes iuxta centrum terrae essent multum remotae et distantes a suo loco naturali etc., dico quod, quamvis essent remotae et distantes a suo loco, quia non habent locum nisi locum totalis terrae, tamen essent continuae in illo toto quod esset immediatum illi loco; et pars adhuc magis naturaliter est in suo toto quam
81rb G totum locatum in suo loco; ideo magis naturaliter quiescunt partes terrae | existentes iuxta centrum quam partes terrae contiguae aquae vel aeri etc.

---

1 et[1]] *add.* ideo p ‖ generatur … aqua] conservatur aqua et generatur iam G : conservatur et generatur aqua iam Pp 3 de ratione ultima] de ultima ratione Pp : ad ultimam G 4 et distantes] *om.* Gp 4–5 naturali … loco] *om.* (*hom.*) P 4 etc.] *om.* p 5 et distantes] *sup. lin.* C 6 locum totalis] locum totalem P : locus totalis C ‖ essent] esset C 8 ideo] immo Gp 9 contiguae] continuae C ‖ etc.] et sic est finis istius quaestionis P

⟨IV.6⟩

⟨Utrum ultima sphaera, scilicet suprema, sit in loco⟩

Quaeritur sexto utrum ultima sphaera, scilicet suprema, sit in loco.

⟨1⟩ Et arguitur quod non auctoritate Aristotelis dicentis: 'cui quidem igitur corpori inest aliquid extra continens ipsum, hoc est in loco, cui | vero non, (90va P) minime'. Et ad hoc est ratio demonstrativa quia: locus est terminus corporis continentis.

⟨2⟩ Item dicit Aristoteles: 'caelum autem amplius in alio non est', quia nullum est continens ipsum. Et hoc est verum de ultima sphaera.

⟨3⟩ Item dicas: quis est ille locus? Apparet enim quod non poteris dicere, cum nihil sit continens ipsum.

⟨1⟩ Oppositum arguitur per Aristotelem quia: movetur localiter, igitur habet locum. Consequentia est de se manifesta. Et antecedens est communiter concessum ab omnibus et ab Aristotele octavo *huius*, qui omnino determinat primum motum esse loci mutationem circularem etc.

⟨2⟩ Iterum dicit Aristoteles in isto quarto quod quaedam sunt in loco per accidens, ut anima et caelum. Et non intendit de caelo pro sphaeris inferioribus, quia illae sunt per se in locis suis; igitur intendit de caelo | pro (102vb C) ultima sphaera. Ideo illa est in loco, licet per accidens.

---

4 quaeritur sexto] sexto quaeritur consequenter G : quaeritur consequenter sexto p ‖ scilicet suprema] quaecumque sit illa G 5 et] *om.* GPp ‖ igitur] *om.* p 6 est] *post* loco GPp ‖ non] est G 9 dicit] *post* aristoteles p : *om.* GP ‖ amplius ... est] non est amplius in alio p : non amplius in alio G : non in alio P ‖ quia] *praem.* et hoc dicit Gp : et hoc dicit quod P 11 quis] quid GPp 12 cum] quod p ‖ ipsum] ipsam p 13 arguitur per aristotelem] dicit aristoteles Gp : est aristoteles P ‖ movetur localiter] ultima sphaera movetur motu locali P 14 est de] apparet per GPp 15 concessum] *post* omnibus GPp 16 circularem etc.] certiorem G 17 isto] *om.* P 18 ut ... caelum] *om.* G 19 per se] *om.* p

---

5 Aristoteles, *Physica*, IV, 5, 212a31–32 9 Aristoteles, *Physica*, IV, 5, 212b22 15 Cf. Aristoteles, *Physica*, VIII, 9, 265a13–266a9 17 Aristoteles, *Physica*, IV, 5, 212b11–12

 | DOI: 10.1163/9789004322356_029

Ista quaestio reputata fuit difficillima. Et puto quod fuit propter non distinguere aequivocationem huius termini 'locus'. Dictum est enim prius quod uno modo dicitur locus proprie, scilicet continens locatum divisum ab eo et immediatum sibi; alio modo improprie vel minus proprie secundum attri-
72[va] p butionem, aut quocumque modo loqui volueris, | pro eo per quod aliquod corpus iudicatur moveri ex eo quod aliter et aliter se habet ad ipsum secundum situm, aut totum ad totum aut partes ad partes. Si igitur haec distinctio data sit concessa, quaestio est valde facilis.

Tunc sit prima conclusio quod capiendo 'locum' proprie, scilicet pro continente, tunc ultima sphaera nec est in loco nec habet locum qui sit locus ipsius, quia supponimus nullum corpus eam continere, sed ipsam omnia alia corpora continere.

Secunda conclusio quod sic capiendo 'locum' ultima sphaera non movetur localiter sive secundum locum, quia nullo modo se habet aliter et aliter secundum locum vel ad locum, saltem qui sit locus ipsius, cum nullus sit talis.

Tertia conclusio quod capiendo 'locum alicuius corporis' pro illo per quod illud corpus apparet moveri vel quiescere ex hoc quod aliter et aliter vel eodem modo secundum situm apparet se habere ad ipsum prius et posterius, suprema sphaera habet locum, scilicet terram vel lapidem aut murum etc. Et hoc est per se notum.

Quarta conclusio est Commentatoris, scilicet quod ultima sphaera non habet locum per se, sed per accidens. Hoc enim est verum ad illum sensum quod per 'locum per se' intelligimus locum proprie dictum eius, scilicet

---

1–2 reputata … aequivocationem] est difficilis sine distinctione G 1 puto] credo Pp ‖ quod] *add.* hoc Pp 2 enim] *ante* est P : *om.* G 3 proprie] proprius C 4 proprie] *add.* vel G 5 quocumque] quoque p ‖ loqui volueris] *add.* scilicet Gp : volueris loqui scilicet P ‖ per] *om.* Cp ‖ aliquod] aliud P 6 moveri] *add.* et P 7 aut[1]] ut p ‖ igitur] autem G ‖ distinctio] diffinitio p 8 data sit] *add.* et P : sit data et G 9 tunc … conclusio] prima conclusio est ista G : et est prima conclusio Pp 10 tunc] etc. Pp : *om.* G ‖ qui] quid G 11 quia] et C ‖ corpus eam continere] esse corpus continens ipsam GPp ‖ ipsam] ipsa C 13 conclusio] *add.* est GPp 15 vel] et G ‖ qui] quod p ‖ nullus] non P 17 conclusio] *add.* est Gp ‖ locum] locus C ‖ illo] eo GPp 18 hoc] eo GPp 20 suprema] *corr. in marg. ex* sic supra C : *post* sphaera P : ultima G ‖ vel] *add.* me vel Pp 21 etc.] aut quadratum Gp : aut quadrantum P ‖ et] *om.* P ‖ est] esset P 22 scilicet] *om.* P 23 locum] *post* se G ‖ sed] *add.* habet locum Pp 24 intelligimus locum] intelligamus locum GP : intelligimus locus C ‖ eius] *ante* proprie GPp

---

2 Cf. sup., IV, q. 3 22 Cf. Averroes, *In Physicam*, IV, comm. 43, f. 142G–H

continentem eam etc., et quod per 'locum per accidens dictum' intelligimus locum improprie dictum modo prius exposito.

Sed tamen mihi videtur quod Commentator in ponendo | (90vb P) hanc conclusionem ponit falsas circumstantias. Cum enim diceret ultimam sphaeram esse in loco ratione centri, quod ponitur esse terra, dicit quod locatum dicitur habere fixionem et permanentiam ex fixione et permanentia loci. Ideo dicit ultimam sphaeram habere fixionem ex fixione centri, scilicet terrae. Quod reputo omnino absurdum. Haec enim inferiora ordinem habent ex superioribus et non e converso, immo si Deus moveret caelum motu recto, oporteret centrum moveri vel continue esse aliud et aliud centrum.

Quinta conclusio est Avicennae dicentis quod ultima sphaera non movetur secundum locum, sed secundum situm, quia non habet locum proprie dictum, sed bene habet situm, id est determinatam distantiam vel propinquitatem ad alia corpora, secundum quas aliter et aliter se habet ad alia corpora vel partes eius ad partes illorum, scilicet quiescentes.

Sed mirandum est quomodo Averroes et sanctus Thomas ausi fuerunt | (103ra C) arguere contra istam conclusionem Avicennae, quia nihil probabile dixerunt contra eum. Commentator enim non arguit nisi auctoritate Aristotelis, qui illum motum vocavit localem et qui etiam quinto *huius* non posuit | (81va G) esse motum nisi secundum tria genera, scilicet secundum quantitatem, secundum qualitatem, secundum locum sive secundum ubi, non secundum situm. Et sanctus Thomas arguit quia: situs, ut dicit, est ad aliquid et in quinto *huius* dicitur quod non est per se motus in ad aliquid. Et iterum quia: situs, ut dicit, consistit in indivisibili et ad tale non est motus.

---

1 dictum intelligimus] intelligimus p : intelligamus GP 3 tamen] *om.* P ‖ mihi videtur] *inv.* GPp 4 ponit] posuit Pp 5 dicit] dixit P : dicitur G ‖ dicitur] debet GPp 6 dicit] dixit P 8 reputo omnino] *inv.* p : ego omnino reputo GP 9 immo] ideo P 12 locum^{2}] *add.* scilicet G 13 id est] scilicet Gp : et est P 14 quas] quae G ‖ habet] habent C ‖ alia^{2}] illa G 15 vel] *praem.* saltem Gp : *add.* saltem P ‖ quiescentes] quiescentis C 16 sanctus] beatus GPp ‖ ausi] visi P 17 arguere ... conclusionem] contra istam conclusionem arguere G 18 eum] eam p 19 qui^{1}] quia C ‖ vocavit] *ante* motum p : *post* localem P ‖ et qui] quem C 20–21 quantitatem secundum qualitatem] quantitatem qualitatem et GP : qualitatem secundum quantitatem et p 21 non] *add.* igitur GPp 22 sanctus] beatus GPp ‖ est] *add.* in P 23 huius] *om.* p 24 tale] talem (*corr. ex* tale) C : talem P

4–7 Cf. Averroes, *In Physicam*, IV, comm. 43, f. 142G ?; cf. *AA*, 2: 143 11 Cf. Averroes, *In Physicam*, IV, comm. 45, f. 144E–F 18 Cf. Averroes, *In Physicam*, IV, comm. 45, f. 144F 22–24 Thomas Aquinas, *In octo libros Physicorum expositio*, IV, lect. 7, 475 (4) (ed. Maggiòlo, 232) 23 Cf. Aristoteles, *Physica*, V, 2, 225b11

Sed faciliter respondetur quod, cum Aristoteles vocavit illos motus locales secundum quos mobile variat situm propinquitatis vel distantiae ad res quiescentes, ipse non accepit 'locum' secundum eius propriam acceptionem, sed secundum impropriam et communem, de qua dictum est prius. Et tamen dicit etiam talem motum esse secundum ubi. Unde dixit quod responsio ad ubi non solum fit de loco proprie dicto, sed saepe respondetur de situ rei ad alias res; secundum cuius situs variationem percipitur talis motus et aliter non perciperetur, ut praedictum fuit in tertio libro.

De hoc autem quod dicit beatus Thomas ego dico quod termini de praedicamento situs non magis sunt ad aliquid quam termini de praedicamento
72[vb] p ubi. Et bene dictum fuit in tertio libro quod de essentia illius mo|tus non est quod sit secundum locum nec secundum situm nec secundum ubi, quia haec omnia significant habitudines unius corporis ad aliud corpus sibi extrinsecum et ille motus posset esse sine variatione habitudinis ad aliud extrinsecum, et quia etiam aliquo corpore non moto potest variari habitudo
91[ra] P ipsius ad aliud extrinsecum per motum illum alterius, sicut est de aliis quae | ad aliquid dicuntur.

Sed ultra, mirabile est quomodo dicit sanctus Thomas quod situs consistit in indivisibili, quoniam cum sedere et stare sint de praedicamento situs, tamen mutatio fit de sedere ad stare secundum motum continuum et temporalem secundum quem membra hominis se habent continue aliter et aliter secundum situm ipsorum ad invicem et ad locum. Et ita negandum est omnino quod situs magis consistat in indivisibili quam ubi.

Tunc igitur faciliter respondetur ad rationes principales.

Auctoritates enim Aristotelis quod illa sphaera non sit in loco sunt verissimae capiendo 'locum' proprie.

---

1 faciliter respondetur] *inv.* GPp ‖ cum] *om.* G 2 quos] quod C ‖ mobile variat] *inv.* G 2–3 vel ... quiescentes] *rep.* G 3 accepit] accipit Pp ‖ eius propriam] *inv.* GPp 4 secundum] *om.* p ‖ et] vel Pp ‖ est] fuit GPp 5 et tamen] et cum p : tamen P ‖ dicit etiam] *inv.* Gp : etiam dixit P ‖ secundum] per P : *om.* G ‖ dixit] dicit Gp ‖ quod] quia Pp 6 ad] per P ‖ dicto] *om.* P 7 variationem] variatione p 8 ut praedictum] prout dictum GPp 10 magis sunt] *inv.* P 11 et] ut P 14 aliud] aliquod Gp 15 etiam] in p : *add.* non C : *om.* G ‖ aliquo corpore] *inv.* GPp ‖ variari] variare P 16 aliud] aliquod p ‖ illum] illius p ‖ quae] qui G 18 quomodo ... thomas] quomodo dicit beatus thomas Gp : quod beatus thomas dicit P 19 in] *om.* C ‖ quoniam] quando p ‖ sint] sit C 20 situs] *add.* et C ‖ ad] et P 21 quem] quam C ‖ se habent continue] continue habent se G 22 et[1]] *om.* G 23 consistat in] consistit in G : consistat ex P 25 auctoritates enim] ad auctoritates C 25–26 quod ... verissimae] concludunt quod illa sphaera non sit in loco p

---

8 Cf. sup., III, qq. 7–8 11 Cf. sup., III, q. 7

Sed de hoc quod Aristoteles motum ultimae sphaerae vocat localem, dicendum est quod in hoc dicendo ipse non capit 'locum' proprie, sed communiter pro habitudine illius sphaerae vel suarum partium ad alia corpora per quae percipitur quod illa sphaera movetur, vel pro illis corporibus, sicut dictum est. Unde quia per illum motum mutatur situs illius sphaerae vel partium suarum ad illa corpora, ideo illum motum Aristoteles vocat localem.

De hoc autem quod Aristoteles dicit, caelum esse in loco per accidens, potest dici quod per 'caelum' intendit | ultimam sphaeram, quia est in loco 103rb C
non per se, sed per accidens, ad illum sensum secundum quem exposuimus quartam conclusionem, scilicet opinionem Commentatoris. Vel forte per 'caelum' intendit totum mundum, qui non est in loco secundum se totum, sed ratione aliquarum partium. Modo saepe istud quod dicitur ratione partis vocatur per accidens, non prout 'per accidens' distinguitur contra 'esse per se', sed prout distinguitur contra 'primo'.

Haec de quaestione etc.

---

2 capit] accipit P 3 suarum partium] *inv.* G 4 quod] motus et G ‖ sicut] *add.* ante Gp : ut ante P 5 quia per] per C : secundum P ‖ situs illius] situs p : motus illius C 11 totum mundum] *inv.* GPp 12 dicitur] dicit p 13 esse] omne p 14 contra primo] contra per se G : contra per se primo p : primo P 15 haec … etc.] etc. G : *om.* Pp

⟨IV.7⟩

⟨Utrum possibile sit vacuum esse⟩

Quaeritur septimo utrum possibile est vacuum esse.

⟨1⟩ Et arguitur quod sic rationibus quas facit Aristoteles quia: si motus localis rectus est, vacuum est; sed ille est, ut omnes concedunt; igitur etc. Maior probatur quia: quod motu recto movetur, vel recipitur in plenum vel in vacuum. Si in vacuum, habetur propositum. Si in plenum, tunc vel illud plenum cedit vel non. Si non, tunc esset penetratio; quod reputamus impossibile. Si cedit, tunc iterum cedendo vel recipitur in plenum vel in vacuum. Si in plenum, quaeretur ut prius. Et sic tandem oporteret uno motu omnia ad ante moveri et caelum tumultuari et cedere et expelli de loco suo, quod est absurdum.

⟨2⟩ Item quia aliquis respondendo rationi factae posset fingere quod, si ego moveor ad ante, non oportet quod aer mihi cedens moveatur motu recto ad ante, sed movetur lateraliter et quodam modo circulariter (aer enim qui
81vb G erat mihi lateralis movetur ad replendum locum | a quo ego exeo, et aer mihi anterior dispergitur lateraliter ad replendum locum quem occupabat ille aer
91rb P qui | prius erat mihi lateralis), ideo contra imaginationem istam ponitur alia ratio, scilicet supponendo quod per calefactionem vel frigefactionem et elementorum ad invicem transmutationem fit condensatio vel rarefactio et maioris vel minoris loci occupatio, ut ex straminibus parvae quantitatis generatur magna flamma et magnus fumus occupans valde maiorem locum

---

3 septimo] *add.* consequenter G : *add.* circa secundum tractatum istius quarti scilicet circa tractatum de vacuo P   4 et] *om.* Pp   5 est³] *add.* motus G   6 movetur] *ante* motu GPp   7 tunc vel illud] vel illud P : igitur G   8 esset] est p ‖ quod] dimensionum quam P ‖ cedit] cedat p   9 vacuum] *add.* et Gp   10 quaeretur] quaereretur P ‖ motu] moto Cp ‖ ad ante] alia ante C : ante p : *om.* G   11 tumultuari et] *om.* P ‖ et³ … suo] et expelli de loco s† *in marg.* C : *om.* Gp ‖ absurdum] inconveniens P   12 respondendo] movendo P   13 moveor] movear P ‖ mihi] *om.* P   14 lateraliter] localiter CPp ‖ quodam modo] quomodo P ‖ qui] quantum p   15 lateralis movetur] naturalis movetur P : localis moveretur G ‖ aer mihi] aer tamen p : *rep.* P   16 lateraliter] *corr. ex* localiter G : localiter CP ‖ occupabat] occupat GP   17 prius] *post* mihi P ‖ lateralis] localis GP ‖ istam] *in marg.* C : *ante* imaginationem GPp   20 et] ad P ‖ ut] et P ‖ quantitatis] calliditatis G   21 magna] maxima P ‖ magnus] *om.* P ‖ valde] *om.* G ‖ maiorem] magnum p

---

4 Cf. Aristoteles, *Physica*, IV, 6, 213b4–29

 | DOI: 10.1163/9789004322356_030

quam occuparent illa stramina. Tunc igitur arguitur sic: cum ille ignis genitus occupet multo maiorem locum quam facerent stra|mina, igitur oportet [73ra p] corpora circumstantia cedere, nisi fiat penetratio corporum. Et illis cedentibus oportet alia iterum cedere. Et sic tandem oportet caelum cedere, nisi sint in corporibus aliquae vacuitates in quas corpora sic cedentia recipiantur, vel nisi dicatur quod necesse est quod, quantumcumque generatur hic de denso rarum et ex straminibus ignis, tantundem alibi oportet generari ex raro densum et simul horum utrumque fieri, | ut semper remaneat totum aequale. Et [103va C] sic dicere est omnino ficticium, quia ego experior quod ego libere possum, quando placet mihi, comburere ista stramina, quamvis alibi alia non sunt in potestate mea. Igitur potius ponendum est quod sit vacuum.

⟨3⟩ Item sicut de rarefactione argutum est, ita de condensatione arguitur quia: si condensatio est, vacuum est; sed condensatio est (hoc communiter conceditur); igitur vacuum est. Maior probabitur quia: si aliquod corpus condensatur, partes eius extremae undique moventur appropinquando ad centrum et recipiuntur sic movendo in plenum vel in vacuum. Si in vacuum, habetur propositum. Si in plenum, hoc est impossibile, nisi sit penetratio, quia non est possibile partes medias, scilicet versus centrum, cedere, quoniam qua ratione cederent ad unam partem, eadem ratione ad aliam; ideo vel ad neutram vel undiquaque; quod est impossibile, et maxime quia undique partes exteriores moventur versus centrum; ideo ex nulla parte possibile est partes centrales cedere, quia exteriores obviarent eis.

⟨4⟩ Item si augmentatio est possibilis in viventibus per nutritionem, vacuum est; sed illa est possibilis, prout hic supponitur; igitur etc. Maior probatur quia: supponitur quia nutriti quaelibet pars sit nutrita et aucti aucta et quod nutritio et augmentatio fiant adveniente ab extrinseco aliquo cor-

---

1 igitur] *om.* P **2** occupet multo] occupat multum Pp ‖ facerent] *om.* P ‖ igitur] *om.* GPp **3** nisi] ne Pp **4** oportet[1]] *om.* GPp ‖ nisi] ne P **5** sic cedentia] sic tendentia sic C : sic cedenda p : cedentia G **6** dicatur] dicant P ‖ quod[2]] JOOx : *om. reliqui codd.* (*deest* Pb) ‖ generatur ... denso] hic generatur de denso p : generatur extensio vel G **7** et] ut p ‖ oportet] *omnes codd.* (*ante* alibi P) ‖ raro] aere p **8** et[1]] vel P ‖ horum utrumque] *inv.* GPp **9** ficticium] fictivum p ‖ ego[2]] *om.* P **10** comburere] *om.* P ‖ alibi alia] alibi ista alia P : alia alicubi G ‖ sunt] sint Gp **13** est hoc] est ut P : in hoc p **14** probabitur] probatur P **16** movendo] vel G ‖ plenum ... vacuum[1]] vacuum vel in plenum G **18–19** cedere quoniam] cadere quia P **19** cederent] caderet P ‖ ratione[2]] *om.* GP ‖ aliam] alteram G **20** vel[1]] et G ‖ neutram] neutrum p ‖ undiquaque] undique P **20–21** quia undique] quia undiquaque G : quod undique P **21** nulla] una p **22** obviarent] obviant P **25** quia[2]] quod GPp ‖ nutriti ... sit] quaelibet pars nutriti est P ‖ aucti aucta] aucti est aucta P : augmentata p **26** augmentatio] auctio P ‖ adveniente] ad invicem G ‖ ab] *del.* C **26–260.1** corporeo] corpore Pp

poreo. Quod si recipitur in vacuum, habetur propositum. Et si in plenum quod non cedat, erit penetratio, quae reputatur impossibilis. Et si illud plenum cedat, tunc non nutritur nec augetur, quia non recipit nutrimentum; 91va P et hoc est contra suppositum, quia supponitur | quod quaelibet pars nutriti nutritur et aucti augetur.

⟨5⟩ Item experimentum est quod pottus repletus cineribus, sicut repleri potest, tantum recipit de aqua, quantum reciperet, si non essent cineres ibi. Et hoc non esset possibile, nisi essent inter partes cinerum multae vacuitates in quas illa aqua reciperetur, vel nisi esset penetratio corporum, quae est impossibilis; igitur etc.

⟨6⟩ Item non apparet quare rarum esset minus ponderosum et magis transparens quam densum, si totum utrobique esset solidum et continuum. Ideo posuerunt aliqui antiquorum vacuum immixtum corporibus, propter quod est maior raritas et levitas etc.

⟨7⟩ Aliqui etiam, ut dicit Aristoteles, posuerunt vacuum inter corpora contigua et dixerunt esse prohibens eorum continuationem, tamquam omnia corpora essent ad invicem innata continuari, nisi essent ab invicem divisa per vacuum intermedium. Et ita auctoritate illorum arguitur quod vacuum sit.

⟨8⟩ Item etiam auctoritate vulgari quod omnes communiter dicunt do-
82ra G lium aut pottum esse vacuum, quando ab eo extractum est vinum. |

Oppositum determinat Aristoteles.

3 augetur] augmentatur G 5 augetur] augeatur P 6 est] *om.* P ‖ quod] *add.* si G 6–7 sicut … recipit] sicut eis repleri potest tantundem recipit P : sic eis repleri potest tantundem recipit G : potest tantundem recipi p 7 aqua] qua C ‖ quantum] *corr. sup. lin. ex* quam C : quam G : sicut P ‖ cineres ibi] *inv.* GPp 8 et] *add.* sic C 8–9 nisi … quas] nisi inter partes cineris multae essent (*in marg.*) vacuitas in quas C : si inter partes cineris non essent multae vacuitates in quas P : nisi esset inter partes cineris magna vacuitas in qua p 9–10 quae est impossibilis] quod est impossibile P : *om.* p 11 quare] *add.* tibi P ‖ magis] *om.* P 13 aliqui antiquorum] antiqui aliqui p 15 aliqui] *praem.* et GPp ‖ posuerunt] *om.* p 16 et] *om.* p 17 ad invicem innata] innata ad invicem G ‖ ad … essent^2] *om.* (*hom.*) p ‖ essent ab invicem] ab invicem essent P ‖ divisa] diversa C 18 per vacuum intermedium] *om.* Pp ‖ et ita] ita P : ideo p ‖ arguitur] arguit G 20 etiam] in G ‖ quod omnes] qui omnes G : quia omnes p : quia homines P 20–21 dolium aut pottum] pottum aut dolium P 22 oppositum] *add.* enim p

7 Cf. Aristoteles, *Physica*, IV, 6, 213b20–22 15 Cf. Aristoteles, *Physica*, IV, 6, 213a32–b2 22 Aristoteles, *Physica*, IV, 7–9, 213b30–217b28

Primo intendo loqui de possibilitate essendi vacuum quantum ad potentias naturales, post|ea autem modicum de potentia supernaturali dicetur. [103vb C]

Et oportet, sicut dicit Aristoteles, praemittere quid nominis. Aliqui enim descripserunt vacuum quod vacuum est in quo nihil est, vel in quo non est corpus sensibile, vel etiam in quo non est corpus. Et illae non sunt bonae descriptiones, quia sic, si essent puncta indivisibilia, punctum esset vacuum. Immo etiam sic Deus esset vacuum, quia in eo non est corpus nec aliquid aliud, saltem secundum illum modum essendi in aliquo secundum quem | [73rb p] illi intendebant de vacuo, licet sint in eo omnia sicut in efficiente vel in fine. Sic etiam una species intelligibilis esset vacuum, cum nihil esset in ea.

Isti igitur deficiunt, quia dimittunt genus ponendum in definitione vacui, scilicet 'locus'. Oportet igitur dicere quod vacuum est locus non plenus. Nam si esset possibile vacuum esse, ista nomina 'plenum', 'vacuum' essent passiones huius nominis 'locus'. Supponerent enim pro loco; et ultra 'plenum' connotat quod in illo loco sit corpus contentum ab eo, et 'vacuum' connotaret quod non sit corpus contentum in eo. Et ista nomina 'vacuum' et 'plenum' essent ad invicem privative opposita. Quorum subiectum esset iste terminus 'locus', de quo essent innata dici successive | [91vb P] et non simul. Privatio autem est innata describi per suum subiectum et per habitum sibi oppositum cum dictione negativa, ut quod caecum est oculus non habens visum vel carens visu. Ideo rationabiliter potest dici quod vacuum est locus non plenus. Ex eo autem dicitur locus non plenus, quia non est in eo corpus contentum ab eo; ideo vacuum dicitur locus in quo non est corpus contentum ab eo.

Postea notandum est quod, sicut dupliciter potest imaginari locus, ita etiam dupliciter potest imaginari vacuum. Nam si esset spatium praeter magnitudines corporum naturalium, in quo non cedente reciperentur corpora naturalia, de quo spatio unumquodque corpus naturale occuparet partem sibi aequalem, sicut multi imaginati sunt, illud spatium sine dubio debe-

---

1 possibilitate essendi] possibilitate esse p : potentia essendi G 2 postea autem] postea aliquid G : et post aliquid p ‖ de … dicetur] dicetur de potentia supernaturali p : dicere de potentia supernaturali G : de potentia supernaturali directe P 3 sicut] ut P ‖ enim] ergo Pp 4 in quo non] in quo nullum G : nullum P 5 non[1]] nullum G 6 si essent] se habent p 9 illi] *om.* p ‖ vel in fine] *om.* G 10 etiam] enim C ‖ ea] eo P 11 dimittunt] mittunt p 12 locus[1]] locum p ‖ plenus] repletus corpore C 13 plenum] *add.* et GP 16 et[2]] *om.* P 17 subiectum] substractum Gp 18 de] in G 19 subiectum] substractum Gp 20 caecum est] caecus est (*rep.* p) Gp 21 plenus] *corr. in marg. ex* repletus corpore C 22 dicitur] *post* locus Gp : *post* plenus P 23 ideo] igitur P 24 dupliciter potest] *inv.* p ‖ locus] plenum CG 25 etiam] *om.* P 26 cedente] sedente non C

---

2 Cf. inf., IV, q. 8 3 Cf. Aristoteles, *Physica*, IV, 7, 213b30–31

ret poni esse locus. Et esset locus plenus, quando in eo et cum eo esset corpus naturale adaequate, et diceretur vacuus, quando cum eo vel in eo non esset corpus naturale. Et sic apparet quod vacuum esset dimensio corporea tanta secundum longitudinem, latitudinem et profunditatem, quantum esset corpus naturale per quod illud vacuum repleretur, si poneretur in eo. Et videtur quod secundum istam imaginationem, quae est satis vulgaris, locutus est Aristoteles in libro *Praedicamentorum* de loco, ubi dicit quod loci particulae
104ra C quae obtinent singulas | corporis particulas ad eundem terminum copulantur, ad quem et corporis particulae. Et ideo bene dicit Commentator quinto *Metaphysicae* quod saepe Aristoteles in *Praedicamentis* locutus est secundum famositatem, non secundum veram determinationem, scilicet de illis quorum propria perscrutatio pertinebat ad alias partes philosophiae.

Nunc igitur dicendum est quod nec est vacuum isto modo nec potest esse naturaliter, quia non est locus tale spatium, ut prius determinatum est; igitur nec vacuum est tale spatium, quia vacuum, si est, est locus, ut dictum est. Et etiam in tractatu de loco diximus non posse esse tale spatium naturaliter,
82rb G quia esset | penetratio dimensionum et accidens sine substantia et non proficeret, immo etiam frustra esset. Et haec dicta fuerunt prius.

Alio modo secundum Aristotelem ponitur locus esse superficies corporis continentis locatum. Et tunc, si vacuum esset, deberet imaginari sic quod ex loco pleno auferretur corpus contentum vel annihilaretur loco remanente in sua figura, videlicet quod latera loci non approximarentur ad invicem,
92ra P verbi gratia imaginando quod ille mundus inferior annihilaretur totaliter | caelo remanente in sua magnitudine et figura, sicut est nunc. Si enim sic esset, tunc superficies orbis lunae, quae modo est locus repletus isto mundo inferiori, esset locus vacuus, quia non esset in eo aliquod corpus contentum ab eo, immo nec aliquod spatium nec aliqua dimensio, immo nihil. Non

---

1–2 esset[2] ... adaequate] adaequate esset corpus naturale GP : adaequate corpus naturale esset p **2–3** et ... et] *om.* P **2** cum ... in] in eo vel cum p : in eo et cum G **4** longitudinem] *add.* et P ‖ latitudinem] *om.* G **8** corporis particulas] *inv.* p ‖ terminum] *add.* communem p **9** et[1]] *om.* P ‖ dicit] dixit P **10** saepe aristoteles] *inv.* GPp **12** quorum] *add.* determinatio et P ‖ alias] illas G **13** est vacuum] vacuum potest esse G **14** est[2]] *om.* P **15** est[3]] *om.* p ‖ et] *om.* P **16** diximus] dicemus G **18** etiam frustra esset] esset frustra GPp ‖ esset] *sup. lin.* C **19** locus esse superficies] locum esse superficiem P **20** quod] *praem.* esse G : *add.* si P **21** vel] et G **22** videlicet] videndum est C : ita P ‖ loci] corporis C **24** figura] *add.* sic quod P **25** locus] *om.* P ‖ isto] ipso p **26** eo aliquod] ea aliud P **27** non] *om.* p

---

7 Aristoteles, *Praedicamenta*, 6, 5a11–13 **9–10** Averroes, *In Metaphysicam*, V, comm. 18, f. 125K **14** Cf. sup., IV, q. 2, 215–217 **16** Cf. sup., IV, q. 2, 216 **19** Cf. Aristoteles, *Physica*, IV, 4, 212a20–21

igitur deberet imaginari quod intra latera orbis lunae esset vacuum secundum distentionem rectam de uno polo ad alium sibi oppositum secundum situm vel de uno latere ad aliud latus sibi oppositum secundum | situm, quia 73[va] p omnino nihil esset intra huiusmodi latera. Ideo affirmativa esset falsa, quae diceret quod intra illa esset vacuum. Sed vacuum esset mobilis res, quia esset orbis lunae vel pars eius, scilicet superficies concava ipsius.

Secundum istam imaginationem dicendum est quod nec est vacuum nec potest esse naturaliter. Quod probatur primo quia: sequeretur quod essent aliqua duo corpora extra invicem secundum situm, quae nec tangerent se nec distarent a se invicem, saltem secundum rectitudinem; quod non est possibile per naturam, licet non sit simpliciter impossibile, scilicet per potentiam divinam. Consequentia patet, quia poli in orbe lunae non essent ad invicem proximi sive tangentes se, et tamen etiam non distarent ab invicem secundum rectitudinem, quia distantia est per dimensionem intermediam sive per spatium intermedium et nullum esset tale.

Item secundum naturam nulla distant ab invicem inter quae nihil est medium; et sic esset in proposito.

Et si aliquis | diceret quod adhuc illi poli distarent ab invicem per latera 104[rb] C orbis lunae intermedia, quae non esset distantia secundum rectitudinem, sed secundum curvitatem, tunc sequeretur aliud quod non est naturale, scilicet quod poli magis distarent quam modo distant, quia modo non distant nisi in tanto quanta est longitudo diametri rectae protensae de polo ad polum, et illa distantia esset ablata et non remaneret distantia nisi longior, scilicet illa curva. Modo non est naturale quod duae partes corporis continui distent aliquando plus, aliquando minus, illo corpore remanente semper in sua magnitudine non mutata et similiter in sua figura et secundum se totum et quamlibet sui partem, et etiam quod quaelibet pars illius corporis rema-

1 intra] inter P 2 distentionem] distantiam P : extensionem p 2–3 polo ... uno] *om.* (*hom.*) G 2 alium] *add.* polum Pp 3 latere ... latus] latere ad aliud G : loco ad alium locum P 4 huiusmodi] huius P 5 intra] inter P ‖ illa] *sup. lin.* C : *om.* GPp ‖ mobilis] *add. in marg.* alias notabilis C 7 secundum] *praem.* et GPp 8 sequeretur] sequitur Pp 10 invicem] *om.* GPp 11 simpliciter impossibile] *inv.* GPp ‖ scilicet] sed P 13 ad] *om.* C ‖ etiam] *om.* G 14 rectitudinem] *add.* ideo G ‖ est] *om.* P ‖ dimensionem] divisionem P 15 tale] *add.* igitur G : *om.* Pp 16 nulla] illa non Gp : illa P ‖ nihil est] nihil esset Gp : idem esset P 18 et] *om.* p 20 sequeretur aliud] sequeretur aliquod P : sequitur p ‖ est] esset P 21 distarent] *add.* tunc p ‖ distant[1]] *in marg.* C : *om.* Gp 22 tanto quanta] tantum quanta Gp : quantum P ‖ protensae] protensi p 23 nisi] illa P 24–25 continui distent] continui distantes G : continentis distent p 25 semper] *om.* G 27 et[1]] *add.* secundum GPp 27–264.1 remanet] remaneat GPp

net proxima et immediata omni parti cui prius erat proxima et immediata. Et tamen sic esset in proposito de sphaera lunae et de suis partibus.

Item omnis propositio universalis in scientia naturali debet concedi tamquam principium, quae potest probari per experimentalem inductionem sic quod in pluribus singularibus ipsius invenitur manifeste ita esse et in nullo apparet umquam instantia. Sic enim bene dicit Aristoteles quod oportet multa principia esse accepta et scita sensu et memoria et experientia, immo 82va G aliter non potuimus scire quod omnis ignis | est calidus. Sed per talem induc- 92rb P tionem experimentalem | apparet nobis quod nullus locus est vacuus, quia ubique invenimus aliquod corpus naturale, scilicet vel aerem vel aquam vel aliud. Et iterum experimur quod non possumus unum corpus ab alio separare, quin interveniat aliud corpus. Unde si perfecte obstruantur omnia foramina follis, ita quod non possit aer subintrare, numquam possemus latera follis ab invicem elevare, immo nec viginti equi hoc possent, si decem traherent ad unam partem et decem ad aliam. Numquam separarent latera follis ab invicem, nisi aliquid rumperetur vel perforaretur, per quod aliud corpus posset intrare. Et per calamum cuius unum conum ponis in vino et alium in ore tuo, tu attrahendo aerem existentem in calamo attrahis vinum movendo ipsum superius, licet sit grave, propter hoc quod aerem quem tu attrahis necesse est sequi aliud corpus semper immediate existens, ut non sit vacuum. Et sic sunt multae aliae experientiae manifestae. Ideo debemus concedere quod non potest esse naturaliter vacuum, tamquam scitum per illum modum qui est sufficiens ad ponendum et concedendum principia in scientia naturali. Et per hanc inductionem habetur quod non sit vacuum per

---

1 omni parti cui] cum parte cui Pp : cum parte quae G **3** universalis] naturalis (?) P **5** pluribus] plurimis Gp ‖ invenitur manifeste] *inv.* Pp ‖ ita esse] *inv.* G **6** apparet umquam] *inv.* GP : numquam apparet p ‖ sic] sicut p **7** esse] *om.* G ‖ et scita] inscita G ‖ et[2]] *om.* GPp **8** aliter] *praem.* et P : aliquando p ‖ potuimus] possumus P : poterimus G ‖ omnis] *om.* P ‖ est] sit P **11** iterum] *add.* nos GPp **12** aliud] aliquod G ‖ obstruantur] obstruerentur GPp **13** follis] *add. in marg.* ad invicem C : *add.* ad (*rep.*) invicem p : forili ab invicem P ‖ possit] posset GPp **14** follis] forilli P ‖ elevare] levare G ‖ viginti ... decem] decem equi si G **15** decem] tot G ‖ separarent] *praem.* enim Gp : enim separarentur P **16** follis] *om.* P **17** intrare] *corr. sup. lin. ex* intercidere C : intercidere Gp : *add.* et intercidere P ‖ cuius] eius P **18** alium] alterum Gp ‖ tuo] *om.* P ‖ aerem existentem] aerem exeunte P : vinum exeunte aere G **19** movendo ipsum] *inv.* P ‖ licet] sed G **20** aliud corpus] aliquod corpus p : *om.* G ‖ existens] *om.* GPp **21** sit] *add.* corpus P ‖ manifestae ideo] manifestae igitur G : mathematicae ideo p **22** esse naturaliter] *inv.* p ‖ scitum] situm P **24** scientia naturali et] naturali scientia P **24–265.1** per aliquem istorum] per aliquod istorum p : aliquo G

---

6 Cf. Aristoteles, *Analytica posteriora*, II, 19, 100a3–9, 100b3–5; cf. *AA*, 35: 125

aliquem istorum duorum modorum praedictorum. Semper enim | [104va C] videmus corpora naturalia consequi ad invicem tangendo nec inter ea manere spatium sine corpore naturali, ut sine aere vel aqua vel huiusmodi. | [73vb p]

Tunc ad rationes.

⟨1–2⟩ Ad primam, quae est de motu locali recto, et ad secundam, quae est de hoc quod ex denso fit rarum, dicendum est quod rationes illae bene procederent, nisi corpora circumstantia condensarentur. Sed quia condensantur, ideo non oportet ultra ea alia corpora cedere vel moveri, ita quod non dico quod oportet omnem condensationem fieri per generationem unius substantiae ex alia, immo nec per infrigidationem, sicut post magis dicetur.

⟨3⟩ Et illa etiam ratio quae arguit quod, si condensatio est, vacuum est, solvetur, quando videbitur de motu rarefactionis et condensationis.

⟨4⟩ Ad rationem quae est de augmentatione dicendum est in libro *De generatione et corruptione*, ubi apparet quod non quaelibet pars eius quod augetur augetur.

⟨5⟩ Ad rationem de cineribus Commentator bene dat causam propter quam ita est. Cineres enim, maxime si sint novi, sunt calidi et sicci virtualiter; ideo agunt in aquam infusam evaporando magnam quantitatem ex ea. Et etiam virtute aquae subintrantis plures partes subtiles cinerum vel etiam inter cineres inclusae exeunt aqua intrante; non enim essent partes cinerum continuae ad invicem, sed erat multus aer interclusus. Et sic tandem possibile est quod ille pottus plus reciperet | [92va P] de aqua quam si non essent ibi cineres, sicut etiam si in illo potto essent frustra ferri igniti et candentis.

---

1 duorum] *om.* P ‖ praedictorum] prius dictorum GPp 2 manere] movere p 3 ut] et GP 4 tunc] *praem.* et P : *add.* respondendum est p : *om.* G ‖ rationes] *add.* principales GPp 5 secundam] illam G 6 rationes illae] *inv.* P 7 procederent] procedunt P ‖ circumstantia] circumscripta C 9 oportet] oporteat GPp 10 sicut] ut P 11 illa] ita p 12 videbitur] videtur P ‖ motu] modo Gp ‖ rarefactionis et condensationis] condensationis et rarefactionis G 13 est[1]] arguebat P ‖ augmentatione] auctione P 14 et corruptione] *om.* GPp ‖ non] *om.* P 15 augetur[2]] augeatur G 17 cineres enim] scilicet enim cineres p ‖ maxime … novi] si sint novi maxime P : si sint novi et maxime p ‖ calidi et sicci] sicci et calidi P : *add.* et activi p 18 magnam] *post* quantitatem G : *add.* partem sive p 19 et] *rep.* p ‖ cinerum] cineris Pp ‖ etiam[2]] *om.* G 20 essent] erant GPp 20–21 cinerum] cineris Pp 21 interclusus] inclusus P 22 est] esset GPp 23 frustra] *i.e. 'frusta'*

---

10 Cf. inf., IV, q. 11 12 Cf. inf., IV, q. 11 13–15 Cf. Iohannes Buridanus, *Quaestiones super libros De generatione et corruptione*, I, q. 14 (ed. Streijger, Bakker, Thijssen, 117–119) 16 Cf. Averroes, *In Physicam*, IV, comm. 56, f. 150A

⟨6⟩ De alia ratione dicetur, quando dicetur de rarefactione et condensatione. Omnino enim negatur quod rarum vel leve sit per vacuum vel mixtionem vacui cum pleno.

Et sic est finis quaestionis.

2 vel[1]] et P 3 pleno] *add.* etc. GPp 4 et … quaestionis] *om.* Gp

1 Cf. inf., IV, q. 11

⟨IV.8⟩

⟨Utrum possibile sit esse vacuum per aliquam potentiam⟩

Quaeritur octavo utrum possibile est esse vacuum per aliquam potentiam.

Arguitur quod non quia:

⟨1⟩ Per nullam potentiam possibile est quod implicat | contradictionem 82vb G
vel ad quod sequuntur contradictoria; sed ad vacuum esse sequuntur contradictoria. Probatio quia: sequitur, si vacuum est, quod ipsum sit locus per eius descriptionem dicentem quid nominis; sed etiam sequitur, si vacuum est, quod ipsum non sit locus, quia non conveniret sibi descriptio loci, scilicet 'superficies corporis continentis', quia nihil contineret.

⟨2⟩ Item sequeretur alia contradictio. Verbi gratia, si orbis lunae maneret in sua magnitudine et figura isto mundo inferiori annihilato, sequeretur quod latera caeli essent ad invicem proxima et tangentia, quia nihil esset intermedium, et non essent tangentia neque proxima, quia hoc non posset compati illa figura orbicularis.

⟨3⟩ Item sequeretur alia contradictio, scilicet quod intra latera caeli esset vacuum et quod intra latera caeli nihil esset. Hoc enim implicat contradictionem, quia cum prima sit affirmativa, requiritur ad veritatem eius quod
termini supponant pro aliquo; ideo sequitur | 'intra latera caeli est vacuum, 104vb C
ideo intra latera caeli est aliquid', et haec contradicit isti 'intra latera caeli nihil est'.

Oppositum arguitur quia: Deus posset annihilare omne quod est sub orbe lunae manente orbe lunae in magnitudine et figura in qua est; et tunc

---

4 quaeritur octavo] octavo quaeritur consequenter G ‖ esse vacuum] *inv.* GPp 5 arguitur] *add.* primo P 6 est] *add.* esse P 7 vel] et G 8 sequitur] *add.* quod P ‖ quod] *om.* G ‖ sit] est GPp ‖ locus] *add.* ut patet G 10 sit] est GPp ‖ descriptio] definitio G 11 corporis continentis] *inv.* P 12 sequeretur] sequitur Pp 13 mundo] *post* inferiori P : *om.* C ‖ sequeretur] sequitur Pp : sequitur enim G 14 ad] *om.* C 15 neque] aeque P 16 orbicularis] *om.* P 17 sequeretur] sequitur p ‖ intra] inter GPp 18 intra] inter Pp ‖ hoc enim implicat] haec enim implicant G 19 prima] *om.* P 20 intra] inter GP 21 ideo intra] igitur inter G : ergo inter P : ergo intra p ‖ intra[2]] inter GPp 23–24 est … lunae[1]] sub orbe lunae est G 24 orbe lunae] illo orbe G

---

11 Cf. Aristoteles, *Physica*, IV, 4, 212a20–21

 | DOI: 10.1163/9789004322356_031

concavum orbis lunae, quod modo est plenum isto mundo inferiori, esset vacuum, sicut si dolio remanente Deus annihilaret vinum quod est in eo absque hoc quod intraret vel fieret in ipso aliquod aliud corpus, dolium illud esset vacuum.

Iam aliqui dominorum et magistrorum in theologia improperaverunt mihi de hoc quod aliquando in quaestionibus meis physicis intermisceo aliqua theologicalia, cum hoc non pertineat ad artistas. Sed ego cum humilitate respondeo quod ego bene vellem non esse ad hoc astrictus, sed omnes magistri, cum incipiunt in artibus, iurant quod nullam quaestionem pure theologicam disputabunt, utpote de trinitate vel in|[74ra p]carnatione. Et ultra iurant quod, si contingat eos disputare vel determinare aliquam quaestionem quae tangat fidem et philosophiam, eam pro fide determinabunt et rationes in oppositum dissolvent prout eis videbuntur dissolvendae. |[92vb P] Constat autem, si aliqua quaestio tangit fidem et philosophiam, ista est una de illis, scilicet utrum possibile sit esse vacuum. Ideo si eam volo disputare, oportet me dicere quid de ea apparet mihi esse dicendum secundum theologiam, vel esse periurum, et evadere rationes ad oppositum prout apparebit mihi possibile. Et non possum solvere eas, nisi moverem eas; igitur sum ad hoc faciendum coactus.

Dico igitur quod, cum duplici modo possemus imaginari vacuum, sicut dictum est in alia quaestione, possibile est utroque modo vacuum esse per potentiam divinam. Et hoc est mihi creditum et non ratione naturali pro-

---

2 remanente] manente G 3 intraret ... aliquod] in ipso fieret vel intraret G 3–4 dolium ... vacuum] *in marg.* C : iam dolium esset vacuum p : *om.* P 5 iam] ideo G : et ideo p ‖ et magistrorum] *praem.* meorum G : *add.* meorum p 6 de] in G ‖ physicis intermisceo] physicalibus intermisceo p : philosophicis intermisceo aliquando P : physicis interposui G 7 theologicalia] theologica p 8 non ... hoc] de hoc non est G 9 pure] puram P : *om.* G 10 de trinitate vel] de trinitate de G : determinate vel de p 11 contingat] contingit P 13 in oppositum] *om.* p 14 si] quod G ‖ tangit] tangant G ‖ philosophiam] theologiam Pp ‖ ista] *add.* quaestio P : *add.* enim G 15 sit esse vacuum] est esse vacuum Gp : est vacuum esse P 16 ea] illa G ‖ apparet mihi] *inv.* P ‖ esse] *om.* Gp 17 vel] *add.* etiam G 17–18 evadere ... possibile] eiusdem rationes in oppositum solvere G 18 possum] possem p 18–19 sum ... coactus] ego divisione (?) ad hoc faciendum G : ego sum de hoc faciendo coactus P : sum ad haec facienda coactus p 20 igitur] *om.* G ‖ cum] *om.* Gp ‖ possemus] possumus P ‖ sicut] ut P 21 quaestione] *add.* et p ‖ esse] *om.* G 22 hoc] *om.* P

---

21 Cf. sup., IV, q. 7, 261–262

batum; ideo nec istud intendo probare, sed solum dicere modum secundum quem hoc apparet mihi possibile.

Primo igitur quantum ad primum modum imaginandi vacuum esse, ego suppono quod Deus potest facere accidens sine subiecto et potest accidentia separare a subiectis suis et separatim conservare; ideo potest simplicem dimensionem creare absque hoc quod cum ea sit aliqua substantia vel etiam aliquod accidens distinctum ab ea. | 83ra G

Secundo videtur quod non est impossibile apud Deum dimensionum penetratio, immo ipse potest plura corpora facere simul esse in eodem loco absque hoc quod differant ab invicem secundum situm, scilicet absque hoc quod unum sit extra alterum secundum situm. Igitur Deus potest facere simplicem dimensionem sive spatium ab | 105ra C omni substantia naturali separatum, in quo vel cum quo absque hoc quod cedat recipi possunt corpora naturalia; et hoc vocabitur vacuum secundum primam imaginationem prius narratam.

Deinde de secundo modo imaginandi vacuum credo, sicut prius arguebatur, quia Deus posset annihilare istum mundum inferiorem conservando caelum et magnitudines et figuras quales et quantas nunc habet; et tunc concavum orbis lunae esset vacuum. Et de hoc et de dubitationibus quae circa hoc accidunt dictum fuit satis in quinta decima quaestione tertii libri.

Et nunc ultra concludo corollarium, de quo aliquando quaeritur, scilicet quod possibile esset per vacuum vel partes vacui videre et audire, quia Deus posset aerem conservare in magnitudine et figura in qua nunc est circa aquam et terram, et annihilare aquam et terram et omnia in eis contenta, et sic ille aer esset vacuus. Et si Deus in illo aere sustentaret duos homines prope invicem, ipsi viderent se invicem per illum aerem et possent loqui ad

1 dicere modum secundum] modum dicere per G 2 hoc apparet mihi] apparet hoc mihi P : apparet mihi hoc G 3 imaginandi] imaginando P ‖ esse] *om.* G 4 suppono] pono p ‖ subiecto] substantia G 5 a ... conservare] et separatim conservare a substantiis suis G 7 distinctum] determinatum G 8 videtur] *add.* mihi GPp 8–9 impossibile ... penetratio] impossibilis apud deum penetratio dimensionum G : apud deum impossibile penetratio dimensionum P : apud deum impossibilis penetratio dimensionum p 9 simul esse] *inv.* p : simul G ‖ eodem] *add.* subiecto vel G : *add.* subiecto vel in eodem Pp 10 differant] differunt P ‖ scilicet] sed G 11 extra] ad P 14 vocabitur] vocabatur p ‖ narratam] enarratam G 15 de] *om.* G ‖ vacuum] *om.* GPp 16 quia] quod GPp ‖ posset] potest G 17 et[1] ... figuras] magnitudines et figuras p : in magnitudine et figura G ‖ habet] habent C ‖ tunc] *om.* p 18–19 quae ... accidunt] circa hoc accidentibus Pp 19 quinta decima] decima nona C : quinta G ‖ libri] *om.* G 20 concludo corollarium] concludendum corollarium G : concludo de corpore P ‖ aliquando] aliter C : ante G 21 per] *om.* P ‖ vel] *add.* per p ‖ videre et audire] videre vel audire G : audire vel videre P 22 posset] potest G ‖ in[2]] *om.* P 23 in eis contenta] contenta in eis G : quae in eis contenta sunt p 24 et[1]] *om.* P

invicem, sicut faciunt nunc. Et dictum fuit prius quod illa superficies aeris, quae est vel esset locus, non est indivisibilis secundum profunditatem, immo est bene eius tertia pars vel quarta pars aeris secundum eius divisionem 93[ra] P orbicularem. Ideo in illa superficie | aeris, quae esset locus vacuus, possent poni homines et animalia et a Deo sustineri, scilicet in illa profunditate aeris.

Tunc ergo ad rationes respondeo.

⟨1⟩ Ad primam dico quod illa descriptio loci quam dat Aristoteles non est simpliciter bona descriptio huius termini 'locus', quia propter quemcumque casum possibilem non debet propositio universalis falsificari, in qua definitio affirmatur de definito; et tamen, si esset vacuum, non omnis locus esset continens locatum. Sed Aristoteles dedit istam descriptionem, quia credi- 74[rb] p dit quod non posset esse locus va|cuus. Et cum hoc ego etiam dico quod ista est valde bona descriptio huius totalis termini 'locus proprius corporis', ita quod si haec oratio 'locus proprius corporis' aequivalet in significando huic termino *b*, illa descriptio esset bona descriptio huius termini *b*. Omnis enim locus proprius alicuius corporis est superficies corporis continentis illud corpus immediata ei et divisa etc. Et de hoc etiam descripto intendebat Aristoteles dare istam descriptionem. Ideo valde bene dedit eam. Et cum hoc 105[rb] C etiam Aristoteles non intendit dare et verificare istam descriptionem | nisi secundum casus naturaliter possibiles. Ideo ad hoc bene dedit eam secundum exigentiam suae intentionis.

⟨2–3⟩ De aliis duabus rationibus dictum est satis in quinta decima quaestione libri tertii.

Et sic est finis quaestionis etc.

---

1 faciunt nunc] *inv.* G ‖ et dictum] dictum enim G **3** est bene eius] bene est G ‖ pars[2]] *om.* G ‖ divisionem] dimensionem G **5** sustineri] sustentari P ‖ profunditate] *add.* illius G **6** tunc ... respondeo] tunc respondeo breviter ad rationes p : ad rationes GP **7** primam] *add.* ego p ‖ dat] dicit C **8** locus] *add.* et P **8–9** quemcumque casum] quamcumque causam P **10** de] *add.* suo G ‖ tamen] tunc p **12** posset] potest G ‖ locus] *om.* P ‖ hoc ego etiam] hoc etiam ego p : eo etiam P **13** totalis termini] *inv.* p **14** oratio] omnia p ‖ aequivalet] aequivaleret GPp ‖ significando] significatione GPp **15** esset ... b[2]] huius termini b esset bona descriptio ipsius P **16** proprius] proprie C **17** etiam] et P ‖ intendebat] intendit p **18** ideo] *add.* bene p ‖ hoc] *om.* G **20** ad hoc] adhuc C **22** de] *praem.* et P ‖ rationibus] quaestionibus G **22–23** satis ... tertii] in tertio p **22** quinta decima] decima nona C **23** libri tertii] *inv.* GP **24** et ... etc.] etc. GP : *om.* p

---

1 Cf. sup., IV, q. 1

⟨IV.9⟩

⟨Utrum in motibus gravium et levium ad sua loca naturalia tota successio proveniat ex resistentia medii⟩

Consequenter quaeritur nono | utrum in motibus gravium et levium ad sua loca naturalia tota successio proveniat ex resistentia medii. 83rb G

Arguitur quod non quia:

⟨1⟩ Dicit Commentator quod universaliter in omni motu oportet esse resistentiam mobilis ad motorem; igitur ex illa proveniret aliqua successio, quamvis medium nihil resisteret.

⟨2⟩ Item ipse dicit quod in omni motu mobile est quodam modo contrarium motori, et contrarium resistit contrario.

⟨3⟩ Item successio est in motu caeli; et tamen ibi non est resistentia medii, quia ibi non est medium aliud quam ipsum mobile; igitur in motu caeli ipsum mobile resistit. Et pari ratione videtur esse de aliis ita.

⟨4⟩ Item sequeretur quod tota illa successio esset violenta. Consequens est falsum; igitur et antecedens. Falsitas consequentis apparet, quia illa successio non est aliud quam ille motus; ideo si illa esset violenta, ille motus esset violentus, quod est falsum. Sed consequentia prima manifesta est ex descriptione violenti tertio *Ethicorum*. Violentum enim est quod est a principio extrinseco, cum passum ad hoc nullam vim conferat. Et sic esset in proposito.

---

5 consequenter] *post* quaeritur p : *post* nono G : *om.* P 8 quod universaliter] *inv.* GPp 9 illa] illis G 11 ipse] ipsemet P : met G ‖ in omni motu] *om.* G ‖ est] *post* contrarium (11–12) P 13 in] *praem.* motus caeli C : *add.* omni G ‖ et] *om.* P ‖ medii] *om.* P 14 ibi] *post* est Pp ‖ aliud quam] aliquod quantum P ‖ in] *add.* illo G 15 ita] *ante* esse GP : *ante* de p 16 sequeretur] sequitur p ‖ tota illa] *inv.* P ‖ consequens] quod p 17 igitur et antecedens] *om.* p ‖ apparet] patet p 18 ille[1]] ipse p ‖ motus[2]] *post* violentus (19) P 20 a] *om.* G 21 ad] et P ‖ et sic esset] et sic est p : sic etiam esset G

---

8 Cf. Averroes, *In Physicam*, IV, comm. 71, f. 161H 11 Cf. Averroes, *In Physicam*, IV, comm. 71, f. 161I 20 Cf. Aristoteles, *Ethica Nicomachea*, III, 1, 1110a1–3

 | DOI: 10.1163/9789004322356_032

⟨5⟩ Item motus ille non est solum ex medio vel ex resistentia medii, immo magis et principaliter a motore; igitur similiter est de successione, cum non sit aliud illa successio quam ille | [93rb P] motus, ut dictum est.

⟨6⟩ Item oporteret assignare in quo genere causandi illa successio esset ex illa resistentia, quod non est facile.

Oppositum dicit Commentator, quia aliter nihil valeret processus Aristotelis ad probandum quod in vacuo, si esset, grave moveretur in instanti, vel etiam ad probandum quod aequali velocitate moveretur in pleno et in vacuo.

Notandum est quod successio non est aliud quam motus neque etiam tarditas vel velocitas est aliud quam motus. Sed tamen hoc nomen 'successio' manifeste connotat quod continue pars post partem acquiratur dispositio secundum quam est motus, et non tota simul. Et si hoc significat vel connotat hoc nomen 'motus', tamen hoc nomen 'successio' manifestius hoc connotat vel significat.

Ex hoc statim sequitur quod successio non solum provenit a resistentia medii, immo principalius a motore, sicut prius arguebatur.

Deinde notandum est quod resistentia vocatur inclinatio mobilis ad oppositam dispositionem ei quam motor intendit. Et si potentia resistens | [105va C] superet in resistendo potentiam motoris in movendo, tunc ab illo motore non fit motus; immo etiam non fieret motus, si essent aequales ad invicem, haec in resistendo et illa in movendo. Sed si motor superet, tunc fit motus, et quanto in maiori proportione superat, tanto fit motus velocior. Et si nulla esset resistentia, tunc fieret mutatio instantanea, si movens instanter applicaretur mobili et non successive. Verbi gratia, si Deus in instanti crearet unum magnum lucidum in aere tenebroso, instanter fieret lumen | [74va p] inten-

---

1 motus ille] *inv.* G ‖ ex[2]] *om.* G 2 et principaliter] et (*om.* G) principalius est GPp 4 oporteret] oportet p ‖ genere] *add.* causae seu p 5 est] apparet GPp 6 quia] quod p ‖ quia … valeret] *scr. sed loco maculis corrupto del. et add. in marg.* dicit commentato †…† nihil valeret †…† C 7 etiam] *om.* P 8 moveretur] movetur p ‖ et in] vel in G : et P 9–10 neque … motus] *om.* (*hom.*) G 10 vel … aliud] aut velocitas est aliud p : est aliud aut velocitas P ‖ sed] et G 11 manifeste] intrinsece G ‖ continue] *om.* G 13 hoc nomen motus] *ante* hoc (12) GPp 13–14 tamen … significat] *om.* p 13 hoc nomen[2]] videtur quod nulla G ‖ manifestius] maius G 13–14 connotat vel significat] significat vel connotat P : connotat G 15 a] ex G 16 sicut] ut P 17 est] *om.* Gp 18–19 resistens superet] resistentiae superat P 20 fieret] fierent G 21 sed] et G ‖ superet] superat P 24 in instanti] instanter GPp

---

6 Cf. Averroes, *In Physicam*, IV, comm. 71, f. 160G

sum et non prius remissum quam intensum, et instanter etiam ad tantam distantiam ad quantam posset illuminare, et non prius prope quam longe.

Et haec dicta sunt vera de motoribus naturalibus, prout naturale distinguitur contra voluntarium. | Sed non | oportet quod sint vera de motoribus 106ra C 83va G per voluntatem liberam, quia non oportet quod moveant maxima velocitate qua movere possunt, sed maxima qua simul possunt et volunt. Et ideo quidquid ego dicam, donec ego loquar de motu caeli, ego volo quod intelligatur de motibus qui fiunt a potentiis naturalibus, scilicet non voluntariis. Immo nihil intendo ad praesens dicere nisi de motibus inanimatis.

Tunc pono secundam conclusionem, scilicet quod impossibile est motore sufficienter applicato mobili esse motum sine resistentia, quia si non est successio, scilicet quod pars post partem et non tota simul acquiritur dispositio secundum quam innatus est esse motus, | non est motus, sed mutatio instan- 93va P tanea; sed non est talis successio, nisi sit resistentia, ut dictum est; igitur etc. Nec obstat, si Deus sine resistentia moveat successive, quia non esset determinatio ad talem successionem nisi per voluntatem liberam, scilicet Dei, et hoc est exclusum a proposito.

Tertia conclusio sequitur: necesse est in omni motu naturali gravis deorsum esse resistentiam motori. Haec sequitur ex praecedenti conclusione, quia omnis talis motus est successivus et non est mutatio instantanea.

Quarta conclusio est quod in talibus motibus materia prima non resistit motori, quia materia prima vel ad nullum locum inclinat et ad nullam dispositionem, vel si passive dicatur habere inclinationem et appetitum, tamen indifferenter ita habet inclinationem ad illud quod motor intendit sicut ad oppositum. Ideo quantacumque esset eius inclinatio, illa non prohiberet, quin motor moveret. Et hoc non vocamus resistentiam. Resistentia enim est inclinatio per modum activum, scilicet determinata ad unum ita quod non ad oppositum, et ad illud cuius oppositum motor intendit. Et talis inclina-

1 prius] *om.* G 4 voluntarium] violentum G : *corr. in marg. ex* violentum, *add.* sed non oportet etc. nihil deficit; *quod folii restat vacat* C ‖ sint] sunt P 6 qua[1]] quam P ‖ possunt[1]] possent P ‖ qua[2]] quam p ‖ possunt et volunt] volunt et possunt GPp 7 loquar] loquor GP 9 intendo] *post* praesens G ‖ motibus] motoribus GPp 10 secundam] primam Pp ‖ scilicet] *om.* GP 11 esse] est G ‖ est] esset P 12 acquiritur] acquiratur G 14 sit resistentia] *inv.* p 15 esset] est p 16 dei] *om.* G 17 est exclusum] exclusi P 18 tertia] secunda Pp ‖ sequitur] *add.* quod GPp 19 haec] et G ‖ praecedenti] praecedente p 20 quia] quod G 21 quarta] tertia Pp 22 et] vel G 24 ita] *om.* p 25 quantacumque] quanta Pp 26 vocamus ... enim] vocatur resistentia quia resistentia G 27 modum] ALMp : mo^{m} B : motum CGHPTU ‖ activum] accidentium G ‖ non] *add.* est P 28 oppositum[1]] aliud (*sup. lin.*) oppositum C : alium G ‖ illud cuius] aliud cuius p : illud P

tio non convenit materiae, quia indifferens est omni formae vel dispositioni possibili inesse rebus materialibus.

Quinta conclusio est quod in pure et simpliciter gravi nihil est quod resistit motori moventi ipsum grave naturaliter deorsum. Et voco illud pure et simpliciter grave, quod nullum habet gradum levitatis. Conclusio patet, quia motor illius gravis est sua gravitas et cum hoc forte sua forma substantialis intendentes locum deorsum, et non resistunt sibi ipsis; nec materia etiam resistit, ut dictum est; nec sunt accidentia quae etiam resistant, quia nullum est accidens quod inclinat ad oppositum locum, nisi illud sit levitas vel aliquis gradus levitatis.

Sexta conclusio est quod in motu naturali simpliciter gravis deorsum medium per quod ipsum movetur resistit motori. Hoc apparet, quia quanto medium est densius, tanto tardior est motus; et si est tardior, hoc est propter
106rb C maiorem | resistentiam; igitur medium resistit.

Item terrae pulverizatae videmus partes ita parvas existere sursum in aere quod non moventur deorsum vel valde tarde moventur, quamvis essent simpliciter graves et quod gravitas sua sit motor intendens locum deorsum.
83vb G Et non potest reddi causa quare sic tarde moventur vel deorsum forte non |
93vb P moventur, nisi ex eo quod aer | in quo sunt vel per quem innatae sunt moveri resistit, et quod potentia motiva earum vel non superat resistentiam
74vb p vel in valde parva proportione superat propter illarum par|tium nimiam parvitatem.

Iterum in tali motu est resistentia, ut dicit secunda conclusio, et non ex aliquo quod sit in ipso gravi, ut dicit quinta conclusio; et non apparet quid extrinsecum possit magis resistere quam medium quod oportet dividi; igitur ipsum resistit.

---

2 materialibus] naturalibus GPp 3 quinta] quarta Pp ‖ in ... gravi] *corr. in* in motu simpliciter gravis C : simpliciter gravi p : simpliciter grave P ‖ est[2]] *add.* intrinsecum p 3–4 resistit] resistat GPp 4 naturaliter] natura p 5 conclusio] *praem.* ista G : consequentia p 6 et] *add.* igitur C ‖ forte] *ante* cum GPp 7–8 etiam resistit] *inv.* P 8 resistant] resistent P 9 inclinat] inclinet P ‖ oppositum locum] oppositum locus C : quartum locum P 11 sexta] quinta Pp ‖ deorsum] *add.* sive P 12 ipsum] ipse P 13 tardior est] *inv.* GPp ‖ tardior[2]] *add.* motus P 15 partes] *post* parvas G 16 moventur[1]] moveretur C : moverentur p ‖ moventur[2]] moverentur p 17 sit] sive P ‖ locum] motum G 18 moventur] moverentur p ‖ deorsum] *om.* GPp 19 innatae] innata P 20 earum] eorum G 21 in] *om.* G ‖ proportione] portione p ‖ nimiam] nimiarum P 24 aliquo] *add.* nisi G 24–25 quid ... magis] quid extrinsecum magis posset P : quod extrinsecum possit magis p : quod extrinsece magis possit G

Sed contra hanc conclusionem obicitur quia: medium non inclinat ad oppositum eius quod motor intendit; igitur non resistit. Consequentia patet ex prius dictis. Antecedens probatur quia: si aqua descendit per aerem, sicut gravitas movens aquam inclinat ad esse sub aere, ita levitas aeris inclinat ad esse super aquam; ideo aer in aqua existens ascenderet et aqua in aere descenderet; et hae inclinationes non repugnant sibi invicem, sed consonant.

Solutio: dico quod, licet dictae inclinationes non repugnent, sed consonent, tamen alia est inclinatio, scilicet quod unumquodque corpus naturale appetit suam continuationem, quia virtus unita est fortior se ipsa dispersa. Grave autem non potest descendere nisi dividendo medium; ideo inclinat ad divisionem medii. Et huic inclinationi resistit inclinatio medii ad sui continuationem.

Et iterum, si aer est sub aqua et aqua est supra aerem, aqua inclinat ad descensum per viam rectam, quia illa est brevior, et aer inclinat ad ascensum per viam rectam; et haec non possunt fieri, quia non est possibilis penetratio corporum; ideo sic resistunt sibi invicem. Et oportet unum eorum dividi vel lateraliter moveri per viam obliquam et sic est resistentia.

Unde circa hoc est notandum quod aliqui propter modum nunc tactum ponunt in motu naturali gravis deorsum resistentiam intrinsecam bene subtiliter, videlicet quod (ponatur grossus lapis descendens) omnes partes eius tendunt ad centrum secundum lineam rectam, et extremae partes laterales non possunt incedere ad centrum secundum viam rectam partibus mediis prohibentibus; ideo videtur quod partes gravis habeant sibi invicem quandam prohibitionem vel resistentiam contra inclinationem earum ad centrum. Et hoc videtur esse contra conclusionem quintam prius positam.

---

1 hanc] istam G ‖ quia] quod P ‖ medium non] motum non G : medium P 3 probatur quia] declaratur quia G : probatur P 4 aquam] *om.* C ‖ ad] *om.* P 5 ad] *om.* C ‖ super] supra GPp ‖ in aqua existens] in aqua G : aquae inexistit P 5–6 ascenderet ... descenderet] ascenderet C : *om.* P 5 aere] *add.* existens p 6–7 consonant] repugnant (*seq. spat.*) G 8 dico] *om.* G 8–9 non ... consonent] non repugnant sed consonent P : non repugnant sed consonant p : consonant G 9 alia est] *inv.* Gp ‖ unumquodque] *in marg.* C, *et add.* est 10 est fortior] *inv.* GP 11 inclinat] inclinet P 14 est[1]] *sup. lin.* C : *ante* aer Gp : *om.* P ‖ est[2]] *sup. lin.* C : *post* aerem P : *om.* Gp ‖ inclinat] inclinet P 15–16 quia ... rectam] *om.* (*hom.*) G 15 illa] iste P ‖ ascensum] *add.* etiam p : descensum P 16 possibilis] possibile P 17 sic resistunt] si resistant C ‖ eorum] *om.* G ‖ vel] et G 18 lateraliter moveri] naturaliter P 19 unde] *add.* sic p : *om.* G ‖ est] *post* notandum G : *om.* p 20–21 bene subtiliter] *om.* p 22 lineam rectam] rectam viam GPp 23 possunt incedere] possent cedere G : possent recedere p ‖ viam rectam] *inv.* GPp 24 habeant] habent G 25 earum] eorum p 26 conclusionem quintam] *inv.* GPp ‖ prius positam] praesuppositam G

106va C Sed mihi videtur respondendum supponendo quod nihil est centrum | vel medium mundi quod sit res indivisibilis, sicut imaginaretur punctus 94ra P in linea. Immo centrum vel medium mundi est magnum, longum, | latum et profundum, ut totalis terra vel aqua vel pars quantitativa ipsius. Nec locus qui est infimus et summe deorsum est medium mundi, immo est continens medium mundi. Et ideo etiam lapis motus deorsum non intendit nec inclinatur ad medium mundi indivisibile, immo si non esset aliquod grave nisi ille lapis, sed quod totum esset aer, ubi nunc sunt terra et aqua, iste lapis inclinaretur ad hoc quod fieret medium mundi. Et ad hoc et non ad aliud omnes partes eius simul tenderent et inclinarentur et tandem iste 84ra G lapis | fieret medium mundi nec ad hoc partes impedient se invicem.

Item totus ille lapis simul movetur multo velocius quam moveretur una pars eius seorsum; igitur non impediunt nec retardant se invicem, sed potius iuvant et velocitant.

Item adhuc oportet imaginari quod alicuius magnae aquae continuae una pars respectu alterius partis non appetit esse inferius, si sint aequalis gradus in levitate et gravitate. Et ideo, si nauta descendit ad fundum maris, ut habeat super umeros suos centum dolia aquae, ipse non sentit gravedinem illius 75ra p aquae, quia illa aqua quae est super ipsum non inclinat | ad amplius esse deorsum. Sed respectu aeris inclinaret, si esset aer inferior.

Item quamvis aqua esset non in suo loco naturali, sed multum alte in vase, ut in cacumine turris Beatae Mariae, tamen una pars respectu alterius non inclinaret ad esse deorsum, ut si aliquis ibi esset in balneo et haberet tibiam suam in fundo, ita quod supra eam esset magna aquae quantitas quam ipse in aere non posset portare, tamen non sentiret pondus illius, scilicet quia illa aqua in respectu aquae inferioris vel circumstantis non trahit nec inclinat ad esse inferius, licet totalis aqua cum vase respectu

---

2 punctus] punctum G 3 vel medium] in medio G 3–4 magnum … profundum] res magna longa lata et profunda Gp 3 longum] *add.* et P 4 aqua vel] *om.* P : aqua G : aliqua p ‖ nec] ut P 5 infimus] inferius p 6 et ideo etiam] immo etiam P : et ideo G 6–7 intendit nec] tendit vel G 8 sunt] est G 9 inclinaretur] *add.* vel cederet G : *add.* et moveretur Pp ‖ non] *om.* G 10 inclinarentur] inclinarent GPp 11 impedient] eius impedirent GPp 12 movetur] moveretur p 13 se] *add.* ad CPp 15 magnae] magnitudine P 16 partis] *om.* GP 17 levitate et gravitate] gravitate et levitate GP 17–18 habeat … suos] super umeros habeat G : habeat super umeros p 18 gravedinem] gravitatem G 19 quae est super] quae est supra p: quae est superius G : super P 20 esset aer] *inv.* GPp 21 item] et iterum GPp ‖ aqua] *om.* C ‖ esset non] *inv.* GPp ‖ in suo loco] in loco suo G : suo loco P ‖ sed] *add.* et P 23 ibi] *post* esset p : *om.* P 24 aquae quantitas] *inv.* GPp 26 scilicet quia] *praem.* aquae p : scilicet quod C : aquae quia G 27 ad] *om.* GP

aeris inferioris vel circumstantis inclinaret ad esse inferius. Sic igitur dico de totali terra, quae modo est medium mundi, quod non solum pars eius media quiescit naturaliter, immo etiam partes eius extremae, nec amplius habent inclinationem ad punctum medium quod imaginatur esse centrum. Et ita etiam est credendum quod, si totalis terra esset nunc elevata simul usque ad orbem lunae, ipsa non tenderet ad punctum quod imaginatur esse centrum, immo ipsa et omnes partes eius simul tenderent ad hoc quod illa totalis terra esset medium mundi. Et ita ipsa et omnes partes eius una inclinatione continua tenderent et moverentur per viam | rectam ad 106vb C occupandum tantum locum quantum nunc occupant, absque hoc quod pars media | et partes extremae inclinarent aliquo modo vel resisterent contra 94rb P invicem.

Nunc igitur pono septimam conclusionem, scilicet quod aliquando in motu naturali gravis deorsum est aliud resistens a medio per quod ipsum movetur. Et hoc primo apparet rudi exemplo, scilicet quod plumbum in horologio descendit continue et naturaliter per suam gravitatem intrinsecam, et tamen plus resistat chorda ad quam pendet quam medium per quod descendit.

Sic etiam patet sine artificio. Si enim lapis descendat in aere, sicut oportet aerem inferiorem dividi et cedere et fieri quandam violentam condensationem, ut post videbitur, ita oportet aerem superiorem consequi ad replendum locum a quo ille lapis recedit, quod non potest esse sine aliqua divisione vel distractione partium aeris superioris vel circumstantis; contra quam distractionem ille aer superior habet inclinationem, quia dictum est quod naturaliter habet inclinationem ad permanendum | in sua continuitate. Et 84rb G oportet etiam in aere superiori fieri quandam violentam rarefactionem, ut dicetur post; ad cuius oppositum aer naturaliter inclinatur. Et ita aer superior per huiusmodi inclinationes resistit aliqualiter.

1 inclinaret] inclinat G || igitur] *add.* ego Gp 2 modo] non p || non] nondum p 3 immo] *om.* G || eius extremae] *inv.* P 5–7 et … centrum] A (credendum est) Tp : et ita credendum est quod si totalis terra est nunc elevata simul usque ad orbem lunae ipsa non tenderet quod imaginatur esse centrum G : *om.* (*hom.*) BCHLMUP 7 et omnes] *rep.* G 8 terra] *om.* Cp || et[1] *om.* P 9 viam] lineam P 10 quantum nunc occupant] *om.* p 13 nunc igitur] nunc iterum C : tunc iterum p || pono septimam] ponam septimam G : pono sextam p || scilicet] *om.* P 14 est] etiam G 15 et] *om.* P || primo apparet] *inv.* P 16 descendit continue] *inv.* G || naturaliter] *add.* quia G 17 resistat] resistit GPp || pendet] pandet C 19 sic] sed GPp || descendat] descendit GPp || sicut] *om.* p 20 quandam] aliquam G 24 ille aer] *inv.* Pp 25 naturaliter habet] *praem.* nihil C : *inv.* Pp || continuitate] continuatione p 27 post] *om.* G

21 Cf. inf., IV, q. 11 27 Cf. inf., IV, q. 11

Dicto autem de pure et simpliciter gravi et levi dicendum est de aliis gravibus et levibus. Et oportet hic aliqua supponere ex libro *Caeli et mundi*. Nam ex quarto et ex his quae in illo debent declarari et probari ego suppono quod, si esset aqua pura et in dispositione sibi convenientissima, qualitas secundum quam inclinaretur ad essendum sub aere et supra terram esset ita simplex, sicut qualitas terrae purae qua inclinaretur ad esse in infimo loco elementorum. Et hoc est ita intelligendum quod, licet in tepido sit gradus alterius rationis, quo tepidum remitteret frigidissimum, et alterius rationis, quo remitteret calidissimum, quia ille est gradus caliditatis et ille est gradus frigiditatis, tamen non sic esset in proposito. Immo eadem qualitas omnino et secundum omnem eundem gradum ipsius moveret aquam deorsum, si esset in aere, et sursum, si esset in terra, et resisteret motui sursum vel deor-
107ra C sum, si esset intermedia, scilicet in loco naturali. | Et ita de aere quantum ad
75rb p esse in intermedio aquae et ignis et quantum ad moveri superius, si esset in aqua, et inferius, si esset in igne. Hoc non probo hic, sed suppono tamquam probatum in quarto *Caeli et mundi*.

Deinde etiam suppono ex eodem quarto *Caeli et mundi* et etiam ex primo et ex libro *De generatione* quia mixtum participat aliquo modo qualitates naturales elementorum vel aliquos gradus earum. Participat enim aliquid de caliditate ratione ignis vel aeris et aliquid de frigiditate ratione terrae
94va P vel aquae. Et ita etiam participat aliquid de levitate | et aliquid de gravitate. Et ita qualitas gravis vel levis mixti motiva ipsius secundum locum non est simplex, sicut erant qualitates elementorum, sed composita ex partibus et gradibus diversarum rationum et inclinantibus ad diversa loca.

---

1 autem] *om.* GPp ‖ pure et] puro et p : potentia C 2 hic] licet P 3 nam] *om.* p ‖ his] illis p ‖ illo] *add.* quarto GPp 5 inclinaretur] *om.* P 6 qua] quae p 7 ita] *om.* P 8 quo] quod p 8–9 tepidum … quo] *om.* (*hom.*) P 11 omnem eundem gradum] eundem gradum omnem p : omnem gradum G 12 vel] et G 13 intermedia] *add.* sibi C ‖ et] *om.* P 14 in[1]] *om.* P ‖ intermedio] *add. sup. lin.* alias intermedium C : medio Gp ‖ ignis] aeris Pp 15 hoc] haec GPp 16 probatum] probanda GPp ‖ quarto] secundo p 17 et] *om.* P 17–18 etiam[2] … et] ex primo et etiam G 18 ex] in P ‖ quia] quod GPp ‖ participat] *praem.* aliquo modo G : *praem.* prout p : participiet P 19 earum] L : eorum ABCGHMPTUp ‖ participat enim] participantium C 20 vel] et Gp 21 vel] et Gp ‖ etiam] *om.* G ‖ participat] *post* aliquid[1] p : participet P ‖ gravitate] frigiditate P 22 vel] et G 24 inclinantibus] inclinationibus C

---

3 Cf. Iohannes Buridanus, *Quaestiones super libros De caelo et mundo*, IV, qq. 6–7 (ed. Moody, 261–269) 17 Cf. Iohannes Buridanus, *Quaestiones super libros De caelo et mundo*, I, q. 7 (ed. Moody, 33); IV, q. 5 (ed. Moody, 258) 18 Cf. Iohannes Buridanus, *Quaestiones super libros De generatione et corruptione*, I, q. 22 (ed. Streijger, Bakker, Thijssen, 168)

Istis visis pono conclusiones. Prima est quod, cum aer existens in aqua ascendit et existens in igne descendit, ille aer in tali motu nullam habet resistentiam intrinsecam, quia non per materiam, ut dictum fuit; nec per formam, quia illa inclinat ad esse supra aquam, et sic ad ascendendum, et ad esse sub igne, et sic ad descendendum; nec per qualitatem motivam ipsius pari ratione (ipsa enim tota et quilibet gradus eius inclinat ad esse sub igne et supra aquam); et nullus dicit quod sit resistentia per alias qualitates naturales ipsius aeris. Et similiter diceretur de aqua, quae naturaliter ascenderet in terra et descenderet in aere. Dico enim bene quod aqua, si esset in terra, ascenderet naturaliter, si terra circumstans esset fluxibilis, ut faciliter posset moveri ad replendum locum a quo illa aqua ascenderet.

Secunda conclusio est | quod grave mixtum, si esset in igne, descenderet 84va G
naturaliter et non haberet resistentiam intrinsecam, quia non haberet eam ratione aeris, aquae aut terrae vel participationis suarum qualitatum, quia haec omnia inclinant ad esse sub igne et sic inclinant ad descendendum; non igitur resistunt descensui. Sed etiam nec esset resistentia ratione ignis vel qualitatis quam ratione ignis illud mixtum participat, quia illa qualitas in igne et respectu partium ignis circumstantium nec inclinat ad esse superius nec ad esse inferius. Unde omnino satiatus est appetitus ipsius ignis ad esse
in loco naturali, | sive sit superius sive inferius, dum tamen sit supra aerem et 107rb C
quod non habeat supra aliquid gravius se. Ideo nulla est resistentia ex parte illius, quamdiu est in sphaera ignis, licet ex parte eius bene esset resistentia, quando exiret a sphaera ignis.

Et proportionaliter etiam debet poni quod grave vel leve mixtum, cum existens in terra ascenderet, nulla esset resistentia intrinseca, donec exiret a terra. Et hoc proportionali modo apparet posito quod terra esset bene et faciliter fluxibilis ad replendum locum a quo illud ascenderet. Sed quia non sic est fluxibilis, sed solida, ideo non solum prohibetur ascensus talis mixti,

---

1 istis] his P ‖ conclusiones prima est] conclusiones prima conclusio est p : conclusionem primam P ‖ cum] *om.* C 3 resistentiam intrinsecam] *inv.* P 4 supra] super G ‖ ascendendum] descendendum P 6 tota] *add.* ex p 8 diceretur de] dicetur de P : diceretur quod G ‖ ascenderet] *add.* existens G 9 terra[1]] terram C ‖ aqua] *post* esset p 11 illa] *om.* G ‖ ascenderet] descenderet P 12 quod] *om.* G 13 eam] ex p 14 aut] vel GPp 15 descendendum] essentiam P 16 esset] esse G 17 quam] qua G 18 et] est G ‖ circumstantium] circumstantis G 19 esse[1]] eius C 21 aliquid] aliquod P ‖ ideo] omnino p 22 licet] sed G 24 et proportionaliter etiam] et proportionabiliter etiam p : etiam proportionaliter P 25 existens] *om.* P ‖ ascenderet] ascenderent P 26 proportionali modo apparet] apparet proportionali modo P : apparet proportionabili ratione G : apparet proportionabiliter modo p ‖ esset] est P 27 ascenderet] exivit ascendendo P 28 prohibetur] prohiberetur p

immo etiam prohiberetur ascensus levis simplicis, scilicet aeris existentis in profundis cavernis terrae.

Tamen contra istam conclusionem obicitur quia: pure et simpliciter grave, si esset in sphaera ignis, velocius descenderet per illum ignem quam grave mixtum habens aliquos gradus levitatis; et hoc non videretur esse nisi
94vb P quia illi gradus levitatis resisterent. |

Solutio: concedo quod velocius descenderet, sed hoc non esset quia gradus levitatis resistunt, sed quia non iuvant ad descendendum. Verbi gratia
75va p sit globus puri gra|vis et alius globus mixti aequaliter ex quattuor elementis et sint similes in magnitudine et figura. Et habeat globus gravis, scilicet terrae, octo gradus gravitatis. Ita globus mixti habebit octo gradus proportionales illis octo, scilicet duos gravitatis ratione terrae et duos etiam gravitatis ratione aquae et duos levitatis ratione aeris et duos ratione ignis. Modo in descendendo per ignem omnes octo moverent simul ad descensum illius terrae et ad descensum illius mixti non moverent nisi sex, quia duo ex parte ignis nec moverent nec resisterent; ideo terra descenderet velocius propter maiorem virtutem moventem, licet resistentia non sit maior vel minor.

Tertia conclusio videtur esse ponenda quod grave vel leve mixtum, quando naturaliter movetur in aere vel in aqua sursum vel deorsum, habet intrinsecam resistentiam supponendo quod sit mixtum ex quattuor elementis, ita quod de cuiuslibet elementi virtute aliquid participet. Conclusio pro-
107va C batur quia: gradus gravitatis quos habet ratione | terrae inclinant ad locum
84vb G deorsum, | scilicet ad esse sub aqua et aere; ideo si movetur sursum per virtutes aliorum elementorum dominantes, tamen illi gradus ex parte terrae resistunt propter inclinationem ad oppositum. Et similiter gradus levitatis quos habet ratione ignis inclinant ad locum ignis, scilicet ad esse supra aerem et aquam; ideo si descenderet per gravitatem terrae vel aquae domi-

1 prohiberetur] prohibetur G 2 cavernis] nervis G 3 conclusionem] *praem.* secundam p : *add.* secundam G 6 levitatis] *add.* aliquid P ‖ resisterent] *add.* solutio conceditur quod in tepido gradus caliditatis et gradus frigiditatis non agunt ad invicem nec in aliud consimiliter tepidum sed illud tepidum agit in calidius ratione suae frigiditatis vel resisteret ei et etiam ageret ratione suae caliditatis in frigidius vel resisteret ei et ita etiam dico quod mixtum ex gravitate aquae et terrae C 7 velocius descenderet] velocius desc†...† *in marg.* C 8 resistunt] resistit G ‖ iuvant] iuvat G 9 sit] si G ‖ puri gravis] pure gravis p : puri ignis G 11 octo[1]] a G ‖ octo[2]] a G 12 octo] *sup. lin.* C : *om.* p ‖ duos[1]] *add.* gradus G ‖ et] *om.* Pp ‖ etiam gravitatis] *inv.* P 12–13 etiam ... duos[1]] *om.* (*hom.*) G 13 et[1]] *om.* Pp ‖ in] *corr. sup. lin. ex* ad C : *om.* P 16 terra] *add.* moveretur sive p 17 vel] aut GPp 18 videtur esse] videtur mihi GP : est et videtur mihi p 19 aqua] aliquo P 21 participet] participat Gp 22 ad locum] solum ad G 24 dominantes] dividentes G 25 et] *om.* P 26 inclinant] inclinat G ‖ supra] super p 27 descenderet] *post* terrae G : descendit Pp

nantem, tamen illi gradus ex parte ignis resistunt propter inclinationem ad oppositum.

Sed aliqui volunt istam rationem solvere dicentes quod in tepido gradus caliditatis et gradus frigiditatis simul existentes non habent ad invicem contrarietatem nec agunt vel patiuntur ad invicem; ita igitur in gravi vel levi mixto gradus levitatis non oppugnant gradibus gravitatis vel e converso; ideo nec sibi invicem resistunt.

Solutio: dico quod in tepido gradus caliditatis et frigiditatis non agunt in invicem nec in aliud consimiliter tepidum, sed illud tepidum agit in calidius ratione suae frigiditatis vel resisteret ei et etiam ageret ratione suae caliditatis in frigidius vel resisteret ei. Et ita etiam dico quod mixtum ex gravitate aquae et terrae et levitate aeris et ignis ita se habet quod nec gravitas in levitatem nec levitas in gravitatem agit. Nec etiam in respectu corporis consimiliter gravis et levis ageret inclinando ad esse supra vel infra illud. Sed tamen in respectu gravioris existentis superius illud mixtum ratione levitatis moveret se superius, nisi nimia resistentia esset extrinseca vel nisi | esset 95ra P
tanta virtus aut maior inclinans ad inferius.

Deinde magis difficile est de motu caeli, de quo aliae rationes tangebant. Et de illo pono conclusiones.

Prima est quod primum mobile nullam resistentiam intrinsecam motui suo vel motori suo habet, quia primi mobilis perfectio naturalis est continue moveri et non in aliquo termino vel loco vel situ quiescere; sed omne ens naturale naturaliter inclinatur in suam perfectionem et non in oppositum; igitur primum mobile naturaliter inclinatur ad semper moveri et numquam quiescere; igitur non resistit motui vel motori. Et pari ratione diceretur quod nullae naturales dispositiones illius primi mobilis resisterent; omnia enim disposita sunt et ordinata ad movendum.

---

4 caliditatis ... frigitidatis] frigiditatis et caliditatis G ‖ habent] haberent G 5 vel[1]] nec G ‖ ad invicem] *om.* P 6 non] nec P ‖ oppugnant] opponuntur G : repugnant p ‖ vel] et P 8 caliditatis et frigiditatis] caliditatis gradus frigiditatis C : et frigiditatis et caliditatis p ‖ in[2]] AHTMp : ad L : *om.* BCGPU 9 in[1]] ad P ‖ consimiliter] similiter P ‖ agit] ageret GPp 10 suae[1]] *om.* P ‖ et] *corr. sup. lin. ex* vel C : vel G : *add.* sic P 11 in frigidius] *ante* ratione[2] (10) G 12 aquae et] aliqua Pp ‖ et[2]] vel P ‖ aeris et] aqua C : aliqua Pp ‖ habet] habent C ‖ nec] *om.* G 13 in[2]] *om.* P 14 ageret] agerent p 15 illud] aliud p 16 nimia resistentia esset] esset nimia resistentia Gp : esset resistentia nimia P ‖ extrinseca] *add. in marg.* alias intrinseca C 18 deinde] etiam et G : nec p ‖ difficile] dissimile p ‖ aliae] aliquae GPp ‖ tangebant] tangebantur P 19 illo] *add.* motu GPp 20–21 resistentiam ... habet] habet resistentiam intrinsecam motui suo vel motori GPp 22 loco vel] *om.* (*hom.*) p 23 in[1]] ad G 25 motori] moventi GPp 26 primi mobilis] mobilis primae P

Secunda conclusio quod etiam quaelibet sphaera caelestis movetur sine resistentia intrinseca. Et haec conclusio apparet sicut prior.

Tertia conclusio est difficilis, scilicet quod nulla sphaera caelestis in motu
75vb p suo vel in motibus suis habet aliquam resistentiam. Proba|tur quia: non
107vb C habet | resistentiam intrinsecam, ut dictum est, nec extrinsecam, quia non
85ra G ex parte Dei vel intelligentiarum, quia illae, ut dicit Aristoteles, nullo modo | adversantur ad invicem nec aliquid potest resistere potentiae divinae propter eius infinitatem; nec ex parte caelorum et motuum potest esse quod unum alteri resistat, quia non sunt invicem continua nec colligata, propter quod debeat unus orbis alteri resistere vel alterum trahere aut pellere.

Sed tamen contra istam conclusionem sunt difficiles rationes.

⟨1⟩ Prima: ex quo non est ibi resistentia, deberet fieri mutatio instantanea, non autem temporalis, secundum dicta prius.

⟨2⟩ Secunda ratio est quia: sequeretur quod musca vel saltem intelligentia quae non esset fortioris potentiae quam musca, posset movere caelum etiam velociori motu quam nunc movetur motu diurno, quia ad movendum vel etiam ad movendum velocius non requiritur maior potentia nisi ad magis superandum resistentiam.

⟨3⟩ Item nos videmus manifeste quod in eodem mobili, si sint motus plures diversi et diversae inclinationes, una resistit alteri et retardat vel impedit alterum motum. Verbi gratia, si lapis proiciatur lateraliter et velociter, non poterit cadere deorsum per longum tempus, quia motus ille lateralis et velox impedit vel resistit inclinationi quam ille lapis habet per suam gravitatem ad movendum deorsum. Cum igitur eadem sphaera duplici motu moveatur a duplici motore, scilicet motu diurno et in obliquo circulo, oportet quod motus unus resistat alteri motui et retardet ipsum et e converso.

95rb P ⟨4⟩ Item oportet ibi imaginari aliam | causam resistentiae et retardationis. Quia scilicet, in quocumque loco vel ubi caelum fuerit, illud est sibi naturale

1 conclusio] *add.* est GPp ‖ etiam] *om.* Pp 2 prior] *add.* quare etc. p 6 vel] et GPp ‖ ut dicit aristoteles] secundum aristotelem GPp 8 caelorum] *add.* inferiorum P 9 sunt] *post* continua p 10 debeat] *post* orbis Gp : debet (*post* resistere) P ‖ vel alterum] nec alterum G : vel alium P 11 tamen] *om.* P 12 prima] *add.* est p : *add.* est quod GP 13 autem] *om.* G 14 secunda ratio est] item secundo P ‖ sequeretur] sequitur p 16 movetur] moveatur GPp 17 etiam] *om.* P 19 eodem] caelo p : *om.* G 20 et retardat] vel excedit G 21 proiciatur] proicitur p 22 et velox] est velox P : qui est velocior G 25 in] *om.* P ‖ oportet] apparet P 26 resistat] resistit G ‖ retardet] retardat GP 27 item] et iterum p ‖ ibi imaginari] *inv.* p 28 quia scilicet] *inv.* P : scilicet quod G : quia p ‖ illud] *om.* Pp

6 Cf. Aristoteles, *Physica*, VIII, 10, 267b2–5; cf. *AA*, 3: 224; cf. *De caelo et mundo*, II, 1, 284a11–18

et conveniens, ideo appetit ibi esse, sicut materia appetit formam quam habet per modum delectationis, ut dicebatur in primo libro, quamvis etiam per modum desiderii appeteret aliud, sicut materia aliam formam. Et ita appetitus ad ubi quem habet est quaedam resistentia motui ad alterum ubi.

⟨5⟩ Item beatus Thomas adhuc in omni motu locali imaginatur aliam resistentiam, scilicet incompossibilitatem terminorum. Non enim est possibile naturaliter quod idem lapis sit simul sursum in sphaera ignis et in terra et in locis intermediis, scilicet in aqua et aere, quia oporteret ipsum distare a se ipso, quod est impossibile. Ideo necesse est, si sit | in sphaera ignis et post 108ra C in loco terrae, quod hoc sit successive, prius in aere et post in aqua et post in terra.

⟨6⟩ Et ad hoc iuvat auctoritas Aristotelis quarto *huius* dicentis quod prius et posterius provenit in motu ex priori et posteriori in magnitudine, scilicet in spatio in quo est motus. Unde in sexto *huius* dicitur quod oportet motum, tempus et spatium dividi proportionaliter in partes priores et posteriores et quod in nullo eorum est dare primum. Hoc enim provenit ex incompossibilitate essendi simul terminos magnitudinis. Ita igitur, cum impossibile sit quod sol simul sit in oriente et in occidente, oportet quod sit successio, | licet 85rb G non esset aliunde resistentia.

⟨7⟩ Iterum imaginatur adhuc aliter resistentia ex eo quod caelum secum trahit ignem in sphaera sua et supremam regionem aeris, prout hoc arguitur de stella comata primo *Meteororum*; et in hoc tractu ignis et aer resistunt inclinationem habentes ad ibi quiescendum, sicut lapis, si deberes ipsum trahere post te, resisteret et retardaret ambulationem tuam.

---

1 ibi esse] *inv.* P 2 dicebatur] dicitur p 3 aliud] *add.* ubi GP : ad ubi p ‖ ita] ille G 4 ad[1]] et P ‖ quem] quod Gp 5–6 aliam resistentiam] alia resistentia C 6–7 enim est possibile] est possibile est G 8 et[2]] *add.* in P ‖ oporteret] oportet G 9 in] de P ‖ post] postea P 10 hoc sit successive] sit hic successio P ‖ post[1]] postea P ‖ post[2]] tandem GPp 12 iuvat] vadit p ‖ auctoritas aristotelis] *inv.* P 13 provenit] *post* motu GPp ‖ priori et posteriori] primeitate et posterioritate P ‖ in[2]] *om.* G 14 motum] *add.* et G 16 enim] *add.* totum GPp 16–17 incompossibilitate] compossibilitate P 17 terminos] terminus P ‖ cum] quia P : quod G : *om.* p ‖ sit] est GPp 18 sol simul] *inv.* Gp ‖ in[2]] *om.* GP ‖ oportet] *add.* enim G 20 imaginatur adhuc aliter] adhuc aliter imaginatur Pp : aliter adhuc imaginatur G ‖ resistentia] resistentiam p 21 hoc arguitur] arguitur hic P 22 de] ex G ‖ comata] *add.* in GPp ‖ in hoc] quinto G ‖ tractu] tractatu p 23 ibi] ubi G ‖ ipsum] eum G 24 resisteret] *add.* tibi GP

---

2 Cf. Iohannes Buridanus, *Quaestiones super libros Physicorum*, I, q. 24 (ed. Streijger, Bakker, 234–239) 5 Thomas Aquinas, *In octo libros Physicorum expositio*, IV, lect. 12, 534 (8)–536 (10) (ed. Maggiòlo, 257–258) 12 Cf. Aristoteles, *Physica*, IV, 11, 219a16–18 14 Cf. Aristoteles, *Physica*, VI, 4, 235a11 sqq. 22 Cf. Aristoteles, *Meteora*, I, 4, 341b1–342a33

Ista sunt bene difficilia, sicut mihi videtur. Quibus non obstantibus apparet 76ra p mihi quod motibus corporum caelestium vel eorum motoribus nihil | resistit.

⟨1⟩ Ideo ad primam rationem dicitur primo quod intelligentia movet voluntarie; ideo non movet quantacumque velocitate potest, sed quanta vult.

Et iterum intelligentia non intendit aliquem terminum finalem in quo caelum debeat quiescere, quia si intenderet, ipsa forte faceret instanter caelum esse in illo termino eo quod non haberet resistentiam. Sed ipsa intendit movere secundum se, non propter esse in termino, quia perfectio caeli consistit in continue moveri, non in esse motum nec in esse alicubi in quiete; ideo successive et continue et perpetuo movet qua velocitate intendit et vult movere.

Contra istas solutiones obicitur quia: licet intelligentia non moveat forte 95va P quacumque velocitate potest, sed | quacumque vult, tamen possibile est de intelligentia virtutis finitae quod vellet movere quantumcumque potest velocissime, et tunc, cum non sit resistentia, videretur sequi infinita velocitas.

Ad hoc respondetur quod non est dare maximam velocitatem qua potest movere, quia infinita velocitate potest movere capiendo 'infinita' syncategorematice; sed non potest movere infinita velocitate capiendo 'infinita' categorematice, quia illa non est possibilis etiam per potentiam divinam. Ideo negatur quod possit movere vel etiam velle movere maxima velocitate 108rb C qua potest movere. | Unde etiam ista concederetur quod quantumcumque ipsa potest movere velociter, ipsa potest velle movere ita velociter. Sed haec reputaretur impossibilis 'quantumcumque ipsa potest movere velociter, ipsa movet vel vult movere velociter'.

⟨2⟩ Verum est quod ista solutio non solveret argumentum secundum, quod erat de musca. Ideo alia solutio datur ad principalem rationem satis apparens, videlicet quod ex terminatione potentiae activae provenit termi-

1 bene] *om.* G ‖ sicut … obstantibus] sed G 2 motoribus] motibus p 4 primo] *om.* G 5 quantacumque velocitate] quacumque voluntate p 8 caelum debeat quiescere] debeat quiescere caelum P : caelum debet quiescere G ‖ si] *add.* ipsa p 10 propter] *add.* motum G 11 caeli] *add.* continue p 12 perpetuo] perpetue GP 14 contra istas solutiones] sed contra istam solutionem P ‖ quia] quod P 15 quacumque velocitate potest] quacumque potest velocitatis G : quantumcumque potest velociter Pp ‖ quacumque[2]] quantum GPp 16 vellet] velit Pp : debet C ‖ potest] *om.* G 17 videretur] videtur GP 17–18 velocitas] voluntas G 19 velocitatem] voluntatem G 20 velocitate] *om.* P 22 illa] ita G 23 negatur] negetur P ‖ movere[1]] *om.* G 25 ipsa[1]] *om.* GPp 27 movere] *add.* ita G 28 verum] *add.* etiam p 29 alia] aliter G : illa p

natio in effectu, licet nulla sit resistentia. Unde maiorem effectum, intensiorem et perfectiorem faceret maior potentia et minorem minor, licet nulla esset resistentia. Verbi gratia magis lucidum vel lucidius facit lumen intensius ad maiorem distantiam quam minus lucidum. Et ita etiam debile calefactivum, si approximaretur calefactibili non habenti aliquem gradum frigiditatis nec habenti aliquam resistentiam, tamen non faceret in eo caliditatem intensissimam nec multum intensam, sed proportionatam suae virtuti. Motus igitur qui est effectus intelligentiae virtutis finitae non fieret infinitae velocitatis sive infinite intensus, licet intelligentia illa | moveret secundum 85va G
extremum suae potentiae. Unde Aristoteles et Commentator, ut puto, crediderunt quod intelligentiae, saltem aliae a Deo, movent quantumcumque velociter possunt movere. Unde dicunt quod, si in caelo adderetur una stella, intelligentia non moveret amplius ipsum vel tardius moveret.

Sed licet ista solutio sit satis subtilis, tamen videtur esse contra dicta prius, quia prius dictum est quod, si non esset resistentia, fieret mutatio instantanea.

Ad hoc respondetur quod hoc est verum de virtutibus naturalibus inanimatis, de quibus tunc erat sermo, propter hoc quod illae non intendunt motum secundum se, sed intendunt terminum, ut lapis existens sursum intendit esse deorsum; ideo si non esset resistentia, faceret se instanter deorsum et non indigeret facere motum nec faceret, sed propter resistentiam se non potest facere deorsum instanter, immo indiget quod per motum auferat resistentiam, ideo facit motum.

Sed adhuc haec omnia non videntur sufficere, quia secundum istam solutionem sequeretur quod virtus movens lapidem deorsum esset fortior quam virtus movens sphaeram lunae aut solis, quod omnino videtur ficticium

1 effectu] effectum G 1–2 intensiorem] intentiorem C 3 magis] maius G 3–4 vel … lucidum] *rep.* C 4 ita etiam] etiam p : *om.* P 4–5 calefactivum] calefactum P 5 habenti] *add.* autem p 7 intensissimam … intensam] intensiorem G ‖ proportionatam] proportionate P 8 motus] in illo G 9 sive] sed C ‖ intensus] intensius CG 11 movent] *om.* P 12 possunt movere] *inv.* G 13 moveret amplius] *inv.* GPp 14 sed] *om.* p ‖ esse] *om.* P 15 prius[1]] *om.* p 16 instantanea] *add.* etc. G 17 ad] *praem.* et p ‖ respondetur] est dicendum p ‖ est verum] *inv.* p ‖ naturalibus] *add.* et G 19 motum] *om.* G 21–22 se … instanter] non potest instanter facere se deorsum p : non potest se facere instanter deorsum G : se non potest in facere deorsum P 22 immo indiget] quin primo G 24 sufficere] satisfacere P 25 sequeretur] sequitur p

12 Cf. Aristoteles, *De caelo et mundo*, II, t. 71 (*secundum translationem Michaelis Scoti*): 'Et si in aliquo orbe orbium erraticorum essent plures stelle, esset orbis ultimus in labore' (ed. Carmody, Arnzen, 407); cf. Averroes, *In De caelo et mundo*, II, comm. 71 (ed. Carmody, Arnzen, 407–408)

76rb p 95vb P 108va C dicere et absurdum. Consequentia probatur supponendo quod omnis vir|tus activa vel motiva est finita excepta illa virtute quae Deus est. Ideo virtus | intelligentiae | moventis lunam in obliquo circulo est finita, cum ipsa secundum Aristotelem sit intelligentia alia a Deo. Et quocumque mobili dato ipsa non posset infinita velocitate movere ipsum, sed etiam esset determinata ad certam velocitatem sic quod non posset ipsum movere velocius. Hoc oportet concedere secundum praedictam solutionem. Tunc igitur ego arguo quod virtus lapidis *b* sit fortior quia: illa virtus est maior, quae aliquod mobile certum datum infinita velocitate potest movere ipsum, quam illa quae nullum mobile certum datum potest infinita velocitate movere ipsum (et capio semper 'infinita' syncategorematice); sed nullum mobile datum intelligentia posset infinita velocitate movere, ut dictum est; virtus autem lapidis illum lapidem infinita velocitate posset movere, quia in qua proportione minoretur resistentia, in illa vel in conversa proportione maioraretur velocitas; sed saltem per potentiam divinam in infinitum, scilicet in subduplo, in subcentuplo et sic sine statu, posset minorari resistentia; igitur in duplo et in centuplo et sic sine statu posset intendi vel augeri velocitas. Et sic illa virtus motiva lapidis, licet sit finita, non esset terminata ad effectum finitum, quod est contra praedictam solutionem. Et non oportet hic recurrere ad potentiam 85vb G divinam et supernaturalem in dicendo | quod resistentia posset diminui in subduplo etc., immo Aristoteles ponit hic in isto loco quod, quantacumque subtilitate medii data, potest in quacumque proportione dari subtilior et minus resistens. Verum est tamen quod hoc dictum Aristotelis non credo esse verum nisi per potentiam supernaturalem, sed tamen credo ipsum esse verum sine dubio.

Non apparet mihi quod illud argumentum posset bene solvi sustinendo solutionem contra quam arguit, nisi concedendo resistentiam in caelo vel nisi recurrendo ad opinionem Avempace subtilissimi philosophi in omni-

---

1 quod] quia p 2 illa virtute] *inv.* GPp 3 ipsa] illa G 4 et] *add.* sic GPp ‖ dato] data G 5 posset] potest G ‖ etiam] *om.* G 6 oportet] *om.* P 7 praedictam] dictam G ‖ ego] *om.* P 9–10 quam ... ipsum] *om.* (*hom.*) P 12 est] *om.* G 13 posset] potest p ‖ quia] *om.* p 13–14 minoretur] minoraretur GPp 15 subduplo] *add.* et P : *add.* in centuplo G 16 et in] in G : et p 17 intendi vel augeri] augeri vel intendi G 18 esset] est p : tamen G 19 non] *om.* P 20 in dicendo] dicendo P : inducendo G ‖ posset] possit GPp 21 subduplo] subtilitate G ‖ ponit hic] *inv.* GPp ‖ loco] *add.* scilicet G ‖ quod] *om.* P 22 quacumque] quantacumque G ‖ et] etiam G 23 hoc] illud P ‖ credo] intendo P 24 credo ipsum] credo G : credendum est ipsum P 24–25 esse verum] *inv.* p 25 dubio] *add.* et G 26 posset] possit G

---

21 Cf. Aristoteles, *Physica*, IV, 8, 215a31–216a7

bus in quibus Commentator recitat eum. Erat autem eius opinio, ut credo, quod omni resistentia circumscripta, tamen sive in faciendo motum sive in faciendo aliam rem, ex determinatione potentiae moventis provenit determinatio effectus, et quod tanta potentia non posset effectum maiorem producere vel intensiorem. Et ita oportet dicere quod, licet in motu gravis deorsum non resisteret medium vel aliquid aliud, tamen si gravitas vel aliud movens inclinaret ad movendum illud grave inferius, motus esset determinatae velocitatis. Ideo neganda essent quae dicunt Aristoteles et Commentator, scilicet quod si sit idem movens vel aequale in movendo idem grave vel consimile per diversa media et dissimilia, motus ad motum se habebit in consimili proportione in velocitate et tarditate sicut medium ad medium in subtilitate et grossitate sive in magis vel minus | 108vb C resistendo. Hoc enim non esset verum, quia imaginando quod sit aliquis gradus tarditatis ex determinatione moventis omni resistentia circumscripta, tamen resistentia medii addit alios gradus tarditatis; et tunc non quantum ad totam tarditatem, sed quantum ad tarditatem additam | 96ra P ex resistentia medii valent illae proportiones quas ponit Aristoteles de velocitate et tarditate motus ad subtilitatem et grossitatem medii.

Et secundum hoc etiam oportet corrigere quod ante dictum est, scilicet quod quanto est maior proportio secundum quam virtus motiva superaret resistentiam, tanto esset motus velocior, et quanto minor, tanto tardior, et quod si non esset resistentia, non esset successio. Hoc enim totum non esset verum nisi quantum ad tarditatem additam, defalcando aliam quae esset | 76va p ex determinatione potentiae. Verbi gratia in spatio *a* sint iam duo lapides, in spatio *b* nulli. Tunc utrobique ponantur lapides alii, et quandocumque in spatio *a* ponitur unus, in spatio *b* ponantur duo. Constat quod mutabitur proportio numeralis lapidum spatii *b* ad lapides spatii *a*. Nam in prima

1 ut] sicut GPp 2 tamen] *om.* p ‖ faciendo] patiendo G 2–3 motum … faciendo] *om.* (*hom.*) p 3 faciendo] patiendo G 4 posset] potest G 5 oportet] oporteret G 6 vel[1]] nec GPp 7 motus] movens G 8–9 et commentator] *om.* P 9 si] *om.* P 10 motum] medium G ‖ habebit] habebunt Gp 11 consimili] simili Pp ‖ et] *om.* P 12 magis] maius CG ‖ vel] aut Gp : sive in P 13 esset] esse P 14 tamen] *corr. sup. lin. ex* cum C : cum P 15 totam] *om.* CPp ‖ tarditatem] *add.* moventis p 16 valent] valerent Pp 16–17 proportiones] proportionalitates p 17 motus ad] in motu ad Gp : in motu et P 19 est] fuit Pp : *om.* G 21 tanto[2]] tanta C 22 esset[3]] esse P 23 defalcando] defalcerando P 24 sint] sunt GPp ‖ lapides] *add.* et GPp 25 lapides] *in marg.* C 26 ponantur] ponuntur Gp ‖ quod] *add.* continue G 27 numeralis] *post* lapidum Pp : naturalis (*post* lapidum) C ‖ spatii[1]] spatio P ‖ lapides] lapidem P

1–5 Cf. Averroes, *In Physicam*, IV, comm. 71, f. 160D–G 8–9 Cf. Aristoteles, *Physica*, IV, 8, 215a31–b12; cf. Averroes, *In Physicam*, IV, comm. 71, f. 160H–L appositione erunt in *a* tres lapides et in *b* duo et est proportio sesquialtera, 86ra G scilicet | trium ad duo. Et in secunda appositione erunt in *a* quattuor lapides et in *b* quattuor et erit proportio aequalitatis. Sed tamen defalcando duos primos lapides praesuppositos semper quantum ad alios manebit proportio eadem, quia semper lapides *b* erunt dupli ad lapides *a*. Et non apparet mihi quod ista imaginatio Avempace possit demonstrative improbari.

Sed adhuc, si haec imaginatio, quam non nego, non concederetur, apparet mihi alia imaginatio, quae etiam, sicut mihi videtur, non possit demonstrative improbari, licet non sit secundum opinionem Aristotelis, scilicet quod non quaelibet virtus activa potest in quodlibet passivum agere, sed determinata in determinatum. Et hoc bene dicit Aristoteles in primo *Physicorum*. Caliditas enim ageret caliditatem in corpus opacum sibi approximatum, lucidum autem non posset agere in ipsum lumen nec color in suam speciem; et sol non calefacit corpora caelestia, calefacit tamen ista inferiora; nec gravitas aut levitas, caliditas aut frigiditas possent movere caelum. Diceretur igitur quod nulla est virtus creata quae moveat caelos nec quae posset 109ra C movere caelos, nisi Deus daret | ei ad hoc virtutem, sed Deus movet eos quanta velocitate vult et sicut vult. Nec sequitur quod potentia gravis sit maior aut aequalis potentiae divinae, licet infinita posset movere velocitate, quia si aliqua virtus movet tali velocitate talem resistentiam, non erit maior virtus, sed aequalis, quae dupla velocitate movebit subduplam resistentiam.

Adhuc est alia imaginatio, quam nescirem demonstrative improbare, scilicet quod a creatione mundi Deus movit caelos tot et talibus motibus, sicut nunc moventur, et movendo eos impressit eis impetus per quos postea movebantur uniformiter, propter hoc quod illi impetus, cum non habeant resistentiam contrariam, numquam corrumpuntur nec diminuuntur, sicut 96rb P nos dicimus lapidem proiectum post recessum a proiciente moveri per | impetum sibi impressum, sed tamen propter magnam resistentiam tam ex

1 erunt] erant G ‖ in[1]] *add.* spatio p 4 lapides] *add.* prius p ‖ quantum] inquantum P ‖ alios] alias G 4–5 proportio eadem] *inv.* GPp 6 possit] posset P ‖ improbari] reprobari vel probari p 7 concederetur] concedetur P : consedetur C 8 possit] posset GPp 10 potest] *om.* P 11 determinatum] determinatam G ‖ in[2]] *om.* Pp 12 ageret] *add.* in P ‖ sibi] *add.* oppositum seu P 13 posset agere] ageret GPp ‖ color] calor CP ‖ in[2]] *om.* CG 14 ista] corpora p 15 aut[2]] vel GPp ‖ possent] possunt P : posset G 16 est] esset G ‖ creata] danda G ‖ nec quae] vel quae G : neque P 17 ei ad hoc] adhuc ei Cp ‖ eos] *om.* P 19 posset movere velocitate] velocitate possit movere GPp 21 dupla] duplici GPp 22 est alia] *inv.* P 23 quod] *om.* C ‖ movit] movet P 24 eos] *om.* GPp ‖ quos] *add.* spiritus (?) G 25 movebantur] movebatur G 26 contrariam] *om.* p ‖ nec] et p

11 Cf. Aristoteles, *Physica*, I, 5, 188a31–34 27 Cf. Iohannes Buridanus, *Quaestiones super libros Physicorum*, VIII, q. 12 (ed. Parisiis 1509, ff. 120rb–121rb)

medio quam ex inclinatione ad alium locum ille impetus continue diminuitur et tandem cessat.

Et secundum istam imaginationem non oporteret ponere intelligentias appropriate moventes corpora caelestia, immo etiam non oporteret quod Deus moveret ea nisi per modum generalis influentiae, sicut nos dicimus quod ad omne quod fit ipse cooperatur.

Et etiam cum ista imaginatione posset salvari quod dicunt Aristoteles et Commentator, scilicet quod ubi est resistentia, proportio motus ad motum in velocitate vel tarditate est sicut proportio ad proportionem moventium ad suas resistentias, et quod si non esset resistentia, non esset successio. Et hoc etiam potest salvare secunda imaginatio, quae non ponit caelos moveri | 86rb G
per tales impetus, sed ab ipso Deo.

Imaginatio autem Avempace non aufert opinionem Aristotelis de intelligentiis, sed aufert imaginationem Aristotelis et Commentatoris de hoc quod motus ad motum in velocitate et tarditate sit sicut medii ad medium in subtilitate et grossitate.

Item per istas tres imaginationes solutum est secundum argumentum, quod erat de musca. Omnino enim absurdum est dicere quod in infinitum parva potentia posset movere caelos.

⟨3⟩ Ad tertiam rationem dicitur quod pila per hoc quod volvitur super terram movetur recte de termi|no ad terminum melius quam si non volveretur. 76vb p
Motus enim plures in eodem mobili non retardant vel impediunt se invicem, nisi sint secundum inclinationes contrarias vel ad terminos incompossibiles. | Sic enim gravitas et impetus in proiectione sibi invicem resistunt. Sed 109rb C
hoc in caelo non est, quia omnino possunt simul stare motus plures eiusdem sphaerae super diversis polis absque impedimento unius ab altero.

⟨4⟩ Ad quartam rationem dicitur quod caelum, saltem ultima sphaera, nec est in ubi nec in loco; et si esset in loco vel ubi, tamen non esset sibi naturale

1 ex] *om.* p 3 et] *om.* G ‖ imaginationem] opinionem G 5 ea] eas G ‖ modum] motum GP 6 ipse] *om.* G 7 etiam cum] tamen P 7–8 aristoteles et commentator] commentator et aristoteles P 8 scilicet] *om.* G 9 ad proportionem] *om.* P 11 potest] posset P ‖ secunda] ista Gp 13 aufert] auferat P 13–14 de … aristotelis] *om.* (*hom.*) p ‖ intelligentiis] intelligentis C 14 imaginationem] opinionem G 16 grossitate] grossitie etc. G 17 per] *add.* omnes GPp 18 quod[1] … musca] *om.* p ‖ absurdum est] *inv.* G ‖ in infinitum] infinita P 20 rationem] *om.* G ‖ dicitur … quod[2]] dico per hoc quod pila p ‖ super] supra P 21 movetur] non C 22 non] nec G 24 sibi] se P ‖ resistunt] *om.* P 26 diversis polis] diversos polos GPp 27 rationem] *om.* G ‖ saltem] *add.* vel C 28 nec] *add.* est GPp

7–8 Cf. Aristoteles, *Physica*, IV, 8, 215a31–b12; cf. Averroes, *In Physicam*, IV, comm. 71, f. 160H–L vel conveniens esse in quiete ibi, sed in motu, qui est perfectio quaedam ipsius caeli.

⟨5⟩ Ad quintam dico quod nullus terminus per quem possit esse resistentia vel inclinatio vel qui aliquod operetur ad motum caeli, est assignandus in caelo nisi partes fluxus, sicut prius dicebatur in tertio libro. Nec etiam ibi est incompossibilitas terminorum quantum ad ea quae sphaerae motae sunt intrinseca. Et iterum incompossibilitas terminorum non sufficit ad hoc quod mutatio sit successiva, quia in mutatione instantanea esset incompossibilitas terminorum, immo etiam corpus quod est in caelo Deus posset facere instanter esse in terra.

⟨6⟩ Ad aliam conceditur bene quod prius et posterius in motu est propter prius et posterius in magnitudine vel in spatio una cum resistentia, ut
96va P lucidum | non prius illuminaret prope quam longe, sicut dicebatur.

⟨7⟩ Ad ultimam dico quod non debet imaginari quod caelum trahat secum ignem vel aerem, quia esset motus violentus et oporteret quod caelum esset illi igni colligatum, quod non est ita. Sed ignis naturaliter insequitur locum suum et naturaliter inclinatur voluntati primi moventis, quia omnia sunt naturaliter gratia ipsius et gratia ipsius operantur. Nec ignis illic habet inclinationem ad quiescendum nisi quiete opposita motui recto secundum quem recederet ab illo loco suo naturali. Et si ignis ibi ex toto quiesceret, adhuc non impediret motum caeli, cum sibi non sit continuus nec colligatus.

Tunc igitur respondendum est ad rationes principales.

⟨1⟩ Prima auctoritas Commentatoris est concedenda de motibus gravium; sed illud mobile quod resistit gravi descendenti non est illud grave, sed est
86va G medium, quod movetur, | quia dividitur.

⟨2⟩ Et eodem modo procedit secunda auctoritas. Medium enim quod movetur et dividitur quodam modo habet inclinationem contrariam motori vel eius inclinationi, per quam resistit ei.

1 ibi] *ante* in[1] p : *om.* GP 4 vel[2]] *add.* etiam G ‖ aliquod] aliquid GPp ‖ motum] motus p 5 nisi] nec Pp ‖ prius] *om.* G ‖ libro] huius p 5–6 etiam ibi est] est ibi aliter G 8 quia] quae P 12 resistentia] *add.* aliter non G 13 non prius illuminaret] non illuminaret prius P : *rep.* G ‖ sicut] ut GP 14 trahat secum] *inv.* G 17 locum suum] *inv.* G : motum suum *in marg.* C ‖ voluntati] velocitati G 18 ignis illic] *inv.* P 20 quem] quam G ‖ loco suo] *inv.* GP 21 adhuc] ex hoc P ‖ sibi] *post* sit G ‖ nec] neque G 22 tunc … est] *om.* G 23 gravium] *add.* et levium G 26 et] *om.* G ‖ secunda] ista P ‖ enim] *add.* per p 27 modo] *add.* etiam G 28 inclinationi] inclinatione P

5 Cf. sup., III, q. 8

De aliis rationibus apparet ex dictis nisi de ultima, quae quaerit in quo genere causae resistentia se habet ad successionem vel motum.

⟨6⟩ Ad quod ego dico quod resistentia est causa activa non quia agat motum cui resistit, sed quia innata est agere ad oppositum eius quod motor agit. Multis enim modis dicitur hoc nomen 'causa activa vel agens', sicut dictum fuit in secundo libro.

Et videtur mihi quod ista quaestio est bene longa et | difficilis. Et conve- 109va C
nienter potuisset fuisse divisa in tres quaestiones, scilicet unam de pure et simpliciter gravibus et levibus, aliam de gravibus et levibus non simpliciter et pure et tertiam de caelo. Et sic dividat eam qui voluerit.

Et sic est finis quaestionis.

---

2 resistentia] resistentiae P ‖ habet] habeat P 3 causa activa] *inv.* p 4 est] esset G ‖ motor] *add.* eius G 5 enim] *om.* p ‖ activa vel agens] agens vel activa GPp 7 bene] *om.* G 8 divisa] *post* tres G ‖ quaestiones] *sup. lin.* C : *om.* Gp ‖ unam] una GPp ‖ pure] puris G 9 et[1]] aut G ‖ aliam] alia GPp 9–10 simpliciter et pure] pure et simpliciter GPp 10 et tertiam] et tertia G : tertia P : alia p ‖ dividat] dividet P 11 et … quaestionis] et sic dictum sit ad quaestionem etc. G : *om.* Pp

---

6 Cf. Iohannes Buridanus, *Quaestiones super libros Physicorum*, II, q. 8 (ed. Streijger, Bakker, 303 306)

⟨IV.10⟩

⟨Utrum, si vacuum esset, grave moveretur in eo⟩

Quaeritur decimo utrum, si vacuum esset, grave moveretur in eo.

Arguitur quod sic quia:

⟨1⟩ Aristoteles probat quod in instanti moveretur; igitur moveretur.

⟨2⟩ Probat etiam quod aeque velociter moveretur in pleno et vacuo; igitur moveretur in vacuo.

Oppositum tamen intendit et dicit Aristoteles.

77[ra] p Ista quaestio, sicut formatur, est una condicionalis, | quae aequivalet uni consequentiae, scilicet isti consequentiae 'vacuum est, igitur grave movetur in eo'. Ideo quaestio facta non quaerit nisi utrum ista sit bona consequentia 'vacuum est, igitur grave movetur in eo'.

Et statim ponitur prima conclusio, scilicet quod Aristoteles concessisset istam consequentiam tamquam bonam, et similiter istam 'vacuum est, igitur nullum grave movetur in eo', et quod esset motus in eo in instanti et quod non esset motus in eo in instanti, quia ipse credidit quod simpliciter esset impossibile vacuum esse et ad impossibile sequitur quodlibet. Igitur
96[vb] P concessisset illas 'si vacuum esset, grave moveretur in eo', 'si vacuum esset, | nullum grave moveretur in eo', et quod esset motus in instanti et quod non esset motus in instanti, et sic de aliis. Nec tales condicionales sunt contra-

---

3 quaeritur ... esset] decimo quaeritur consequenter si vacuum esset utrum G : quaeritur consequenter utrum si vacuum esset p ‖ eo] ipso G **6–7** probat ... moveretur] *om.* (*hom.*) p **6** moveretur] movetur G ‖ et] *add.* in G **8** intendit] *add.* aristoteles G ‖ et dicit] ut dicit P : et dicit ipse G **9** sicut formatur] sic formata G : sicut formata est P : sicut formata p **10–11** grave ... eo] in ipso movetur aliquid grave G **11–12** ideo ... eo] *om.* G **11** quaestio] *add.* quae est P **12** grave] gravetur C **14** et similiter istam] *om.* P **15** in eo[1]] *om.* P **15–16** et[1] ... instanti] *om.* GPp **17** esset] esse G ‖ igitur] ideo GPp **18** eo] *add.* et GPp ‖ esset[2]] *add.* grave non moveretur in eo et P **19** nullum grave] grave non P ‖ motus in instanti] motus in eo in instanti Pp : in eo in instanti motus G **20** motus in instanti] *om.* GPp **20–293.1** sunt contradictoriae] contradicunt GPp

---

5 Cf. Aristoteles, *Physica*, IV, 8, 215b1–216a8 6 Cf. Aristoteles, *Physica*, IV, 8, 215b1–216a8 8 Cf. Aristoteles, *Physica*, IV, 8, 214b12–216b21

 | DOI: 10.1163/9789004322356_033

dictoriae sibi invicem, sicut illae non contradicunt, sed sunt simul verae, 'si tantum pater est, pater est' et 'si tantum pater est, nullus pater est'. Sed oporteret contradictoriam condicionalis accipere praeponendo toti propositioni negationem cadentem super totam propositionem.

Sed non possumus sic dicere, quia concedimus quod vacuum esse est possibile, scilicet per potentiam divinam; ideo non debemus dicere quod ad hoc sequuntur contradictoria.

Ideo dico pro secunda conclusione quod ista non est bona consequentia 'vacuum est, igitur grave movetur in eo', quia posito quod vacuum esset, cum hoc sit possibile, tamen forte nullum grave esset in eo, vel licet esset grave in eo, tamen forte quiesceret aut per potentiam divinam aut aliter.

Tertia conclusio quod etiam ista non est bona consequentia 'vacuum est, igitur grave non movetur in eo', quia possibile est quod moveretur saltem per potentiam divinam, sicut dicetur. Ideo istae condicionales sunt negandae 'si vacuum esset, grave moveretur in eo' et 'si vacuum esset, grave non | 86vb G moveretur in eo'.

Quarta conclusio est quod possibile est grave moveri in vacuo, scilicet per potentiam divinam. Hoc enim non minus est possibile quam totum mundum moveri motu recto, et de hoc dictum fuit in quinta decima quaestione tertii libri.

Sed magis eundo ad intentionem Aristotelis et quaerentium ponamus | 109vb C casum quod vacuum sit, verbi gratia quod aere circumdante sphaeras aquae et terrae remanente in sua quantitate et figura orbiculari, sicut nunc est, et quod omnia quae sunt infra illum aerem essent annihilata, ita quod iste aer esset vacuus secundum imaginationem dudum positam, et lapis esset positus in illo aere vacuo, utrum moveretur naturaliter descendendo.

Respondeo quod dupliciter potest imaginari quod lapis esset in illo aere: uno modo quod esset infra latera concavitatis illius aeris, sicut nunc sunt terra et aqua; alio modo quod esset inter superficiem concavam et superficiem convexam illius aeris, ut si esset in media regione aeris.

2 est[1]] *om.* G 3 contradictoriam] contradictionem P || accipere] capere G || propositioni] orationi P 4 super] supra P 5 quia] quod C : *add.* nos Gp : eo quod nos P || esse est] esset G 6 scilicet] *om.* P 8 ideo] *add.* secundo P 10 esset[1]] *add.* etiam G 12 conclusio] *add.* est GPp 14 dicetur] dicebatur G 15 moveretur ... grave[2]] *om.* (*hom.*) P || et] *om.* p 15–16 et ... eo] *om.* (*hom.*) C 17 vacuo] vacuum P 18 hoc enim] nam hoc G 19 fuit] *add.* satis G || quinta decima] quinta G 22 casum] tamen P 23 remanente ... orbiculari] in figura et quantitate sua G 25 esset vacuus] *post* imaginationem P 27 esset] *add.* positus p 28 nunc sunt] *inv.* p 29–30 superficiem] *om.* P

Et tunc erit quinta conclusio quod, si lapis esset in aere vacuo secundo modo, ita quod haberet aerem sub se, ipse moveretur deorsum naturaliter per suam gravitatem, donec esset sub aere et quod non haberet aerem sub se, quia inclinationes gravium et levium ad movendum superius et inferius sunt secundum exigentiam corporum sibi proximorum. Verbi gratia ita lignum existens in aqua ascenderet ad esse supra aquam, si illa aqua esset in vase detenta in loco altissimo, sicut si esset in profundo putei. Ubicumque igitur gravius et levius essent sibi proxima ad invicem, si gravius | [97ra P] esset supra levius, grave inclinaretur ad descendendum et leve ad ascendendum, donec levius esset supra gravius et gravius infra levius. Ideo lapis ille descenderet per illum aerem, donec esset sub eo. Sic igitur concederetur quod grave per suam gravitatem moveretur naturaliter in vacuo et vacuum etiam vel pars vacui moveretur naturaliter ascendendo, quia quantus esset lapis descendens, tantus aer ascenderet naturaliter de illo lo|[77rb p]co vel situ in quem lapis descenderet ad replendum locum a quo lapis ille descenderet.

Sexta conclusio est de priori modo imaginandi quod lapis sit in vacuo, scilicet quod si ille lapis esset omnino sub illo aere tangens ex uno latere superficiem concavam illius aeris, ille lapis non moveretur nec amplius descenderet per suam gravitatem. Haec conclusio probatur quia: ille lapis nihil haberet sub se levius; ideo nullam inclinationem ad esse sub aliquo alio haberet quam sub illo sub quo esset.

Item ille lapis si inclinaretur ad motum deorsum, hoc esset vel propter recedere a caelo vel propter accedere ad medium mundi. Sed non propter recedere | [87ra G] a caelo, quia quocumque moveretur, | [110ra C] tamen non magis distaret a caelo quam ante, quia nulla distantia esset nisi per dimensionem illius lapidis et dimensionem aeris et ignis, quae omnes manent. Immo forte, si ille lapis non amplius tangeret sphaeram aeris nec aliud corpus, ipse nec esset proximus caelo nec distaret a caelo nec secundum aliquem situm se haberet ad caelum. Et de hoc sufficiat quod dictum fuit in quinta decima quaestione tertii libri.

---

1 et tunc erit] tunc esset ista G ‖ lapis] *sup. lin.* C : *om.* GPp 4 et[1] … movendum] *om.* G ‖ et[2]] vel GPp 5 exigentiam] exigentias GPp ‖ ita] *om.* p 7 si] *om.* G ‖ profundo] loco G 8 sibi] *sup. lin.* C : *om.* Gp 9 ascendendum] ascendum C 10 gravius[1]] *om.* G ‖ levius[2]] *om.* G 11 per illum aerem] sub aere G 12 vel] aut G 13 quia] *om.* GP 14 situ] *add. sup. lin.* et C 15 a] in G 16 sexta conclusio] et tunc sequitur sexta conclusio et P ‖ est] *add.* quod p ‖ imaginandi] concludo p 17 ex] in G 18 aeris] *om.* G 19 probatur] *add.* primo G 20 ideo] igitur G 21 haberet] *ante* ad (20) GPp ‖ esset] *praem.* iam P : iam est G : iam erat p 22 si] *ante* ille p 23 sed] *om.* p 24 recedere a caelo] primum p ‖ quocumque] *add.* motu G ‖ tamen] *add.* inde GP 26 manent immo] manerent immo Pp : manerent ideo G 27 ipse] ille G 28 distaret] distans GPp 29 de] *om.* p ‖ sufficiat] sufficiet G

Sed etiam nec ille lapis descenderet vel moveretur propter accedere ad medium mundi, quia nihil magis esset medium mundi quam ipse lapis esset vel quam aer esset cui ille lapis esset contiguus. Et si moveretur motu recto, tamen nec ad aliquod corpus accederet nec ab aliquo recederet etc., sicut dictum est in illo tertio libro.

Sed aliquis poterit dicere quod vacuum imaginabitur dimensio simplex sine substantia naturali et qualitatibus naturalibus, existens commensurative ubi nunc sunt terra et aqua, scilicet infra latera illius aeris.

Et tunc est septima conclusio quod adhuc ille lapis non moveretur naturaliter per suam gravitatem, quia non esset supra gravius nec supra levius nec recedendo ab aere haberet aliquod corpus superius vel inferius sibi proximum grave vel leve, gravius vel levius. Ideo sicut bene dicit | Aristoteles, non 97rb P esset ratio quare magis deberet inclinari ad superius vel inferius, ad unum latus quam ad alterum.

Nec valeret ratio de maiori recessu a caelo vel de maiori distantia, quia hoc non facit ad inclinationem naturalem ratione simplicis dimensionis, sed si facit hoc, est ratione qua caelum influit aliter propinque quam remote, et non esset influentia virtutis naturalis, si non esset substantia naturalis receptiva illius. Et nos supponimus ac si illa simplex dimensio esset sine aliqua alia virtute naturali, quoniam posito quod esset cum ea gravitas vel levitas tunc forte aliud esset dicendum.

Et iterum, si vacuum poneretur esse talis simplex dimensio separata et immobilis, tamen illa non esset naturaliter penetrabilis; ideo non posset lapis moveri per eam.

---

1 vel] nec G 2 magis esset medium] esset magis medium P : esset magis in medio p : esset maius medio C 2–3 ipse … vel] ille lapis vel p : ipsemet esset et G 3 lapis] *sup. lin.* C : *om.* GPp ‖ contiguus] continuus p 4 ad aliquod] ad aliud G : aliquod P ‖ accederet] ascenderet G ‖ aliquo] alio G ‖ etc.] *om.* GP 5 est] fuit p ‖ illo] *om.* Pp ‖ libro] huius p 7 substantia] subiecto p 8 infra] inter P 9 et] *om.* G ‖ septima] alia Pp 10 supra … supra] *omnes codd.; sed, si intelligas praepositiones, exspectes 'infra' primo loco (quod fortasse sup. lin. in W), si adverbia, altero loco* 11–12 sibi proximum] *inv.* P 12 vel[2]] aut GPp 13 esset] est P ‖ magis deberet] *inv.* P 14 quam ad] vel ad Gp : vel P 15 vel] et p 16 non] *om.* p 17 influit aliter] *inv.* GP : aliquando influit p ‖ quam] et GPp 19 nos] non P 20 aliqua] *in marg.* C : *om.* G ‖ quoniam] quia p : quando dicitur G ‖ cum ea] cum eo P : in eo G ‖ vel] aut P 22 et[1]] *om.* G 23 illa non] *om.* P

---

5 Cf. sup., III, q. 15 12 Cf. Aristoteles, *Physica*, IV, 8, 214b32–215a1

Octava conclusio est quod, si grave simplex moveretur in vacuo, moveretur in instanti et aeque velociter in pleno sicut in vacuo propter nullam esse resistentiam.

Et Aristoteles ad hoc format duas rationes in textu. Prima probat quod in instanti quia: secundum diversas proportiones moventium ad resistentiam 110rb C sunt motus differenter tardi | aut veloces; sed nulla esset proportio vacui 87rb G ad plenum in | resistendo; igitur nec velocitatis ad velocitatem; et tamen esset proportio velocitatis ad velocitatem, si fieret in tempore, quia omnis temporis finiti ad omne tempus finitum est proportio; igitur non fieret in tempore; et tamen, si fieret motus, ipse fieret vel in tempore vel in instanti; igitur fieret in instanti.

Secunda ratio Aristotelis est ad probandum quod aeque velociter moveretur in pleno et in vacuo quia: si moveretur in vacuo, moveretur in aliquo tempore, cum oportet omnem motum fieri in tempore; et in pleno etiam move- 77va p retur in tempore, licet longiori; et illorum temporum esset ad invicem | certa proportio, quia utrumque tempus est finitum. Sit igitur gratia exempli proportio centupla, scilicet quod tempus in quo movetur per plenum sit centuplum ad tempus in quo movetur per vacuum. Deinde ponatur quod illud plenum sive corpus quo locus est repletus subtilietur in centuplo. Tunc in centuplo velocitabitur motus et in subcentuplo minorabitur tempus. Et sic erit velocitas aequalis velocitati in vacuo et tempus aequale tempori et tamen adhuc erit plenum illud in quo erit tale corpus in centuplo subtilius etc.

97va P Sed contra istam rationem obicitur quia: Aristoteles supponit | quod corpore subtili dato posset dari, in quacumque proportione voluerimus, subtilius; et hoc est falsum, sicut non est dare quocumque calido in infinitum calidius.

1 est] *om.* p ‖ moveretur] *add.* per suam gravitatem GPp 1–2 moveretur] *praem.* ipsum GPp 2 sicut] et G ‖ esse] *om.* G 4 in textu] *om.* G ‖ probat] est probans P 5 resistentiam] resistentias GPp 6 differenter] differentes G ‖ nulla esset proportio] nullam possemus invenire proportionem G 7 nec] neque GPp 9 omne] *add. in marg.* alias esse C 10 et … tempore^{2}] *om.* (*hom.*) C ‖ vel^{1}] *om.* P 12–13 moveretur] *post* pleno G 13 et] sicut GPp 14 oportet] *add. sup. lin.* alias necesse sit C : necesse sit GPp ‖ et … etiam] etiam in pleno p 16 est] esset Gp : sic esset P 17 movetur] moveretur G 21 velocitas] *add.* in pleno Gp 22 etc.] *praem.* igitur G : igitur P 23 rationem] conclusionem P 24 corpore] corpori P ‖ posset dari] possit dari p : possit dare G ‖ voluerimus] volueris P 25 est^{1}] esset G ‖ est^{2}] esset G ‖ in] *om.* P

4 Cf. Aristoteles, *Physica*, IV, 8, 215b1–216a8

Respondetur quod, licet non per potentiam naturalem, tamen per potentiam divinam, quocumque subtili dato vel raro vel calido, in infinitum est dare subtilius, rarius et calidius.

Notandum est tamen quod haec octava conclusio et rationes eius positae sunt ex suppositione quod non sit vera opinio Avempace posita in praecedenti quaestione; quam tamen ego nescirem improbare et cui magis ego consentio quam opinioni oppositae. Et quae opinio Avempace si concederetur, ista octava conclusio non esset concedenda nec valent rationes Aristotelis, sicut ex dictis posset probari.

Sed dimittendo haec omnia aliqui quaerunt et dubitant, posito quod in aere sic vacuo esset homo in inferiori et concavo aeris et ibi per potentiam divinam salvaretur, utrum ille homo posset infra illum aerem extendere vel movere suas tibias vel sua bracchia, cum tamen illic nullum sit spatium etc. Et est quaestio similis sicut utrum, si homo esset supra ultimam sphaeram, ipse posset movere ultra illam bracchia sua etc.

Et de hoc pono ultimam conclusionem quod homo sic posset movere membra, quia nihil resisteret ei extrinsece. Nec | valet dicere quod non 110va C posset illic bracchium ponere vel elevare, quia nullum esset ibi spatium in quo posset manum | suam extendere. Dico enim quod spatium non est nisi 87va G dimensio corporis, ut spatium tuum est dimensio corporis tui; et antequam elevares bracchium tuum ultra illam sphaeram, nihil esset ultra illam, sed bracchio elevato esset ibi spatium, scilicet dimensio bracchii tui.

Tunc igitur dico ad auctoritates Aristotelis in principio quaestionis positas quod Aristoteles non intendebat dicere quod in vacuo fieret motus in

---

1 respondetur quod] dicitur p 4 est tamen] tamen p : *om.* P ‖ octava conclusio] *inv.* P ‖ rationes eius] *inv.* p 5–6 in praecedenti quaestione] prius p 6 ego[1]] *om.* Gp 6–7 magis ego consentio] magis ego assentio G : magis consentio p : ego consentio P 8 concedenda] *add.* et G ‖ valent] valerent GPp 10 dimittendo] dimitto p ‖ omnia] *add.* adhuc Gp ‖ quaerunt et dubitant] dubitant et quaerunt P : quaerunt p 11 homo] *add.* et G ‖ et[1]] *add.* in G ‖ et ibi] ille G 12 infra] extra p 13 vel] et p ‖ etc.] *om.* p 14 sicut] scilicet p : *om.* P ‖ homo esset] *inv.* Gp ‖ supra ultimam sphaeram] ultra ultimam sphaeram P : ultra sphaeram ultimam p 15 bracchia sua] *ante* ultra P : sua membra scilicet bracchia p ‖ etc.] *om.* GPp 16 pono ultimam conclusionem] ponitur ultima conclusio scilicet G 17 resisteret ei extrinsece] extrinsece resisteret ei GP : extrinsece ei resisteret p ‖ valet] oportet P 18 posset] possit P ‖ vel] et G ‖ esset ibi] *inv.* P 19 suam] *om.* G ‖ dico] da G ‖ nisi] *om.* G 20 ut] *corr. sup. lin. ex* etiam C : et GPp ‖ est] *sup. lin.* C : *om.* GPp 21 tuum] *om.* GPp ‖ ultra[1]] *add.* ultimam P ‖ ultra illam[2]] ibi p 23 igitur dico] *inv.* p 23–24 positas] *ante* in (23) G

instanti etc., sed intendebat istam condicionalem, scilicet quod si sic esset in vacuo grave, ipsum moveretur per suam gravitatem in instanti etc. Et hoc est concessum, si non concedatur opinio Avempace.

Sed quia consequens est impossibile simpliciter, ideo volebat ex hoc Aristoteles concludere quod impossibile esset grave per suam gravitatem moveri in vacuo. Et hoc etiam concessum est, licet dictum sit quod posset in eo moveri per potentiam divinam.

Et sic finitur quaestio ista.

---

1 scilicet quod] quod G : scilicet P 1–2 sic ... grave] esset in vacuo grave P : in vacuo grave esset p : sic in vacuo grave moveretur G 2 ipsum moveretur] *post* gravitatem G ‖ etc.] *om.* P 3 concedatur] conceditur p 4 sed] *om.* p ‖ volebat] *post* hoc G 5 aristoteles concludere] *inv.* P 6 etiam concessum est] est concessum etiam P ‖ posset] possit G 8 et ... ista] etc. GPp

⟨IV.11⟩

⟨Utrum rarefactio et condensatio sint possibiles vel utrum possibile sit aliquid rarefieri vel condensari⟩

Quaeritur undecimo utrum rarefactio et condensatio sunt possibiles vel utrum possibile est aliquid rarefieri vel condensari.

Arguitur quod non quia:

⟨1⟩ Si esset condensatio, singulae partes | circumferentiales corporis quod 97vb P
condensaretur fierent ad invicem proximiores movendo se versus centrum illius corporis, et sic partes dextrae non cederent sinistris, sed moverentur contra eas, nec partes ante partibus retro nec partes infra partibus supra; sed impossibile est sic eas moveri contra invicem, quia vel reciperentur in plenum non cedens et sic esset penetratio dimensionum, quae reputatur impossibilis, vel reciperentur in vacuum, quod etiam est impossibile naturaliter; igitur etc.

⟨2⟩ Similiter argueretur de rarefactione quia: oporteret partes circumferentiales undique elongari ab invicem; et tunc etiam inter eas re|manerent 77vb p
vacuitates vel oporteret ab extrinseco intrare alia corpora inter illas partes sic ab invicem recedentes, quod non apparet possibile.

Oppositum apparet in multis per multas experientias.

Notandum est quod aliquando capiuntur 'rarefactio' et 'condensatio' improprie, scilicet quando inter corpora grossa sunt corpora subtilia interclusa, ut quod inter partes lanae sunt multae partes aeris. Et tunc pondus lanae videtur minorem locum occupare et esse densius, si comprimantur partes lanae simul, quia exeunt partes aeris. Et iterum remissa compressione lanae partes

---

5 quaeritur undecimo] undecimo quaerendum erit G ‖ sunt] sint G ‖ vel] sive Pp 6 vel] aut Pp 8–9 quod condensaretur fierent] condensarentur fierent enim G 9 proximiores] propinquiores G 10 cederent] *add.* partibus P 11 partes¹ … retro] partibus retro partes ante P 14 vacuum] vacuo P 15 etc.] *om.* p 16 argueretur] arguitur GP : etiam arguitur p 17 undique] undiquaque Pp ‖ remanerent] reciperent G 19 sic] *om.* P ‖ possibile] *om.* P 20 oppositum] *add.* tamen p 21 est] *om.* P 24 minorem] maiorem G ‖ esse] esset G : exinde p ‖ si] dum P 25 exeunt] exirent G

 | DOI: 10.1163/9789004322356_034

lanae elongantur ab invicem et subintrant inter eas partes aeris circumstantis et sic videtur cumulus lanae locum maiorem (110vb C) | occupare et esse rarior. Et iste modus est bene possibilis, sed non est nisi metaphorice dicta rarefactio vel condensatio, de quibus non intelligimus nunc.

Sed rarefactio dicitur proprie, si corpus prius existens minus fiat maius nullo corpore extrinseco subintrante inter partes eius. Et condensatio etiam proprie dicitur, si corpus prius existens maius fiat minus (87vb G) | nullo corpore exeunte ab eo, quod ante esset inclusum inter partes eius. Et est proprietas huiusmodi rarefactionis aut condensationis quod eius quod rarefit quaelibet pars quantitativa rarefit et eius quod condensatur quaelibet pars condensatur, ita quod quaelibet pars quantitativa rari est rara et densi densa. Quod non est in rarefactione et condensatione improprie dicta, immo in prius dicta rarefactione essent partes densae, scilicet lanae, et partes rarae, scilicet aeris.

De huiusmodi autem proprie dictis rarefactione et condensatione ponitur prima conclusio quod rarefactio et condensatio sunt possibiles per calefactionem et frigefactionem. Ista conclusio patet primo, si concedamus elementa ex invicem generari. Aqua enim est densior aere et aer rarior aqua, et ita fit ex denso rarum et ex raro densum; et hoc est condensatio et rarefactio.

Et hoc etiam apparet de musto novo posito in dolio bene obstructo, quod parando calefit et tumescit et sic augetur quod oporteret rumpi dolium, si non fieret apertura.

Hoc etiam apparet de phiala vitrea. Quae si calefiat super carbones, aer interior calefit et rarefit, ita quod (98ra P) | si verso culo phialae os eius ponatur in aqua, tunc quando aer qui est in phiala refrigerabitur, apparebit eum ita

---

2 locum maiorem] *inv.* Pp ‖ esse] esset C **3** bene] *om.* G **4** condensatio] *add.* improprie dicta P ‖ intelligimus] intelligemus P **5** rarefactio dicitur proprie] proprie capitur P **6** corpore extrinseco] *inv.* G ‖ et] sed C ‖ etiam] *om.* P **7** proprie dicitur] *inv.* GPp **8** exeunte] existente C ‖ esset] erat P **9** huiusmodi] huius Gp ‖ aut condensationis] vel condensationis Pp : *om.* C **10** pars[1]] *add.* eius C **10–11** condensatur] *add.* et p **11** quantitativa] *om.* p **12** rarefactione et condensatione] condensatione et rarefactione GPp ‖ dicta immo] dictis p **13** rarefactione] *add.* cumuli lanae GPp **15** autem] ergo GPp **18** ex invicem generari] generari ex invicem G : generari ex se invicem p **19–20** condensatio et rarefactio] rarefactio condensatio p **21** et] *om.* G ‖ etiam apparet] *inv.* P ‖ posito] *om.* P **21–22** quod parando calefit] praeparando calefacit G **24** super carbones] supra carbones G : super carbonos C **25** et rarefit] *om.* p ‖ culo] collo P ‖ ponatur] ponetur G **26** aqua] aquam GP ‖ refrigerabitur] refrigidabitur G ‖ eum ita] ita Pp : ibi C

condensari et fieri minorem quod oportebit aquam ascendere in phialam ad replendum, ne sit vacuum.

Et omnino manifestum est pluribus experientiis et concessum quod rarefactio est possibilis et similiter condensatio illo modo.

Secunda conclusio ponitur quod condensatio est possibilis per compressionem absque hoc quod illud corpus quod condensatur frigefiat vel alteretur secundum primas qualitates. Hoc probatur primo per motum localem rectum. Aliter enim corpore recte moto oporteret concedere vel vacuum vel penetrationem corporum vel quod omne corpus ad ante cederet et tandem caelum, ut arguebatur in principio septimae quaestionis huius quarti libri. Et si respondeatur sicut ibi notatum fuit, tunc arguitur ratione sequente quae ibidem posita fuit et deducta.

Tertia conclusio quod etiam rarefactio est | [111ra C] possibilis absque alteratione illius quod rarefit secundum primas qualitates, quia non apparet quare magis sine alteratione secundum primas qualitates debeat esse possibilis condensatio quam rarefactio. Et sicut argutum est de condensatione, ita rationibus conversis argueretur de rarefactione. Nam si corpore recte moto oportet ad ante fieri condensationem vel tandem caelum cedere, ita retro oportet corpora sequi, ut non sit vacuum. Et sic | [78ra p] oportet ea rarefieri vel etiam caelum sequi, ut non sit vacuum. Et etiam, sicut si ex denso generatur rarum, oportet corpora circumdantia et tandem caelum cedere, nisi fiat condensatio vel nisi simul oporteat tantundem generari | [88ra G] ex raro densum alibi, ita si ex raro generatur densum, oportet corpora circumdantia et tandem caelum consequi, nisi fiat rarefactio vel nisi oporteat alibi simul fieri tantundem ex denso rarum, puta ex aqua aerem. Rationes enim hinc inde sunt proportionaliter se habentes.

---

1 minorem] minor p ‖ phialam] phiala p **3** et omnino] ideo P ‖ pluribus] plurimis Pp : multis G **4** est ... modo] et condensatio sunt isto modo possibiles GPp **6** frigefiat] calefiat G **8** enim] *om.* P ‖ recte] recto CG **9** ad] *om.* G ‖ et] vel p **10** ut] sicut GPp ‖ libri] *om.* p **11** si respondeatur] si respondetur G : sic respondeatur p ‖ ibi] ibidem GPp **12** et deducta] *in marg.* C : et adducta P **13** conclusio] *add.* sequitur GPp ‖ etiam] *om.* P **14** illius] eius GPp ‖ non] nec G **16** et] *add.* quia GPp **17** recte] recto C **18** oportet ad ante] oporteret ad ante G : ad ante oportet P **18–19** vel ... sequi] ita retro oportet corpora sequi vel tandem caelum p **18** ita retro] in tantum (?) G **19** oportet[2]] oporteret p ‖ etiam] *add.* sic P **20** ut ... vacuum] *om.* GPp ‖ si] *om.* p **22** tantundem] tandem C ‖ si] sicut p : sicut hic G **23** generatur] *add.* hic Pp ‖ circumdantia et] circumdantia vel P : caelestia circumdantia sequi et G **24** alibi] *add.* sicut p ‖ fieri tantundem] *inv.* Gp **25–26** proportionaliter] proportionabiliter p : probabiliter C

8–12 Cf. sup., IV, q. 7, 258–259

Item apparet per experientiam. Videtur enim mihi quod aer sit magis rarefactibilis et condensabilis quam aliquod aliud corpus magis grossum. Unde videmus quod, si dolium plenum vino sit perfectissime bene ligatum et obstructum et perforetur inferius ad trahendum vinum, vinum non exibit vel valde modicum exibit, quia non potest multum tali rarefactione rarefieri nec potest aer subintrare ad replendum pro eo quod exiret. Sed cum dolium fuerit semiplenum aere, tunc quamvis sit bene obstructum, multum de vino exit per foramen, quia aer ille potest ad multam quantitatem rarefieri ad replendum pro eo quod exit. Et nihil posset exire, nisi fieret infra dolium rarefactio vini vel aeris, ex quo dolium est bene obstructum et quod est forte et non faciliter plicabile.

98rb P Et est notandum, ut mihi videtur, quod talis condensatio | vel rarefactio est quasi violenta corporibus quae sic rarefiunt vel condensantur. Data enim dispositione aeris quantum ad raritatem et densitatem sibi convenientissima, si ultra sine alteratione secundum primas qualitates rarefiat vel condensetur per hoc quod ab extrinsecis comprimitur vel quod post extrinseca trahitur ad replendum, ne sit vacuum, hoc est praeter eius propriam inclinationem; et ideo intendit et inclinatur naturaliter ad revertendum ad statum priorem sibi convenientissimum et revertitur naturaliter comprimente remoto, sicut aqua calefacta moveret se ad refrigerationem.

Et forte quod in violentis incurvationibus lignorum habent locum huiusmodi condensatio et rarefactio. Nam cum arcus quasi est rectificatus et
111rb C habens superficiem concavam quasi aequalem | secundum longitudinem superficiei convexae, tamen quando multum incurvatur, oportet superficiem concavam fieri multo breviorem et superficiem convexam longiorem. Quod forte est per violentam condensationem partium interiorum et violentam rarefactionem exteriorum; ideo remoto incurvante revertitur velocissime et impetuosissime ad naturalem rectitudinem.

---

1 item] *add.* hoc GPp ‖ sit magis] isto modo magis sit faciliter GPp 2 et] aut GPp ‖ aliquod aliud corpus] aqua vel aliud Gp : aqua vel aliquid P 3 vino] *om.* C 4 perforetur] perforaretur G ‖ inferius] interius G ‖ vinum[2]] *om.* G 5 multum] *om.* P 7 sit bene] *inv.* P 8 exit] exibit p 9 et] *add.* sic p 11 et non] vel G 12 et est] *om.* G 13 sic] *ante* violenta G ‖ enim] tamen C 14–15 convenientissima] Bp : convenientissimam ACGHMPTU : convenientes L 16 post] per p 17–18 inclinationem] *ante* propriam (17) P : *add.* est C 18 et[1]] *om.* P ‖ intendit] tendit GPp 21 habent locum] habet locum Pp : habet locus C 21–22 huiusmodi] huius P 22 condensatio et rarefactio] rarefactio et condensatio p ‖ est] ErK, *et* F, *ubi postea* habet *pro* et habens : *om. reliqui codd.* 24 convexae] convexi P ‖ multum] *post* oportet C 26 quod] et G ‖ per] propter G

Quarta conclusio apparet mihi probabilis, quod in omni rarefactione generatur magnitudo sive dimensio et in omni condensatione corrumpitur etiam aliqua magnitudo vel dimensio. Et hoc argutum fuit in primo libro, quando quaerebatur de distinctione magnitudinis a substantia et qualitate. Non enim per solum motum localem partium fit condensatio, quia tunc cum possim velociter movere aerem localiter, ego ita possem sine alteratione condensare aerem, quantum natura posset per alterationem, quod est falsum. Et haec ratio fuit deducta ubi dictum est.

Et iterum possunt adhuc ad hoc poni persuasiones. | Aliter sequeretur 88rb G quod rarefactio esset motus vilior vel minus nobilis quam condensatio, quod videtur falsum, cum elementa rariora ponantur nobiliora. Consequentia patet, quia ubi nihil generatur vel corrumpitur, videtur nobilius quod corpus naturale uniatur quam quod dispergatur, cum virtus unita sit fortior se ipsa dispersa; et condensatio esset tamquam unio et congregatio partium corporis ad invicem, rarefactio autem esset quasi dispersio.

Iterum si non sit ibi principaliter nisi motus localis, tunc illud corpus naturaliter movebitur motibus contrariis secundum singulas partes, quia contra invicem haec ad dextram, alia ad sinistram. Et hoc non videtur convenire naturaliter et principaliter aliis | ab animatis. 78rb p

Iterum si in naturali rarefactione non generatur alia dimensio faciens distare, non apparet quo appetitu vel qua inclinatione partes elongabuntur ab invicem. Non enim apparet ratio quare partes caliditatis quae generantur appeterent elongari ab invicem vel etiam quare partes aeris vel formae eius vel | materiae eius appeterent elongari ab invicem; non enim debent se odire. 98va P Ideo non apparet quomodo et unde proveniret naturalis rarefactio.

Iterum dimensio reddit extensum sicut caliditas calidum vel lumen luminosum. Ideo videtur rationabile quod, sicut plus de caliditate vel lumine reddit subiectum magis calidum vel luminosum et minus de caliditate vel lumine reddit ipsum minus calidum vel luminosum, ita plus de dimensione

---

2–3 et ... dimensio] *om.* (*hom.*) CPp **3** et hoc] ut prius p ‖ libro] huius p **6** possim] *praem.* ego p : ego possum P : ego possumus G ‖ ita possem] possum ita P **8** deducta ubi] deducta sicut G : adducta ubi P ‖ est] *om.* p **9** et] *om.* G ‖ adhuc ad hoc] ad hoc Gp : adhuc P ‖ poni] apponi GPp ‖ aliter] *add.* enim Gp **10** vel] et p **12** ubi ... corrumpitur] *om.* p **14** congregatio] approximatio GPp **16** ibi principaliter nisi] igitur nisi principaliter P ‖ illud corpus] *inv.* Gp **17** movebitur] movetur G : movebatur p **18** contra invicem] *om.* P ‖ haec] hae G ‖ alia] aliae G ‖ sinistram] *add.* etiam G : *add.* etc. Pp **18–19** convenire] *post* principaliter (19) P **20** alia] aliqua G **25** quomodo et unde] unde et quomodo Gp : quomodo et bene C **27** vel] *add.* de p **28–29** et ... ipsum] vel P **29** vel] et minus P

3–8 Cf. Iohannes Buridanus, *Quaestiones super libros Physicorum*, I, q. 8 (ed. Streijger, Bakker, 87–90)

reddit magis extensum et minus minus. Unde sicut plus de qualitate reddit intensius, ita plus de dimensione reddit extensius.

Sed iterum, quia in argumentis tangitur de penetratione dimensionum, restat aliquid dicendum, scilicet quod penetratio de qua loquimur non est sicut sagitta penetrat ostium vel corpus hominis dividendo partes eius et intrando inter eas cedentes sagittae. Sed intelligimus de penetratione prout duo corpora quantum ad eorum dimensiones essent simul non distincta ab invicem secundum situm. Et tunc dico quod plures dimensiones sic esse 111va C vel fieri simul potest intelligi dupliciter. Uno modo quod | una sit subiectum alterius, sicut materia formae vel etiam substantia accidentis, aut etiam quod idem sit subiectum earum, ut substantia plurium accidentium. Et sic concedunt esse simul plures dimensiones omnes qui ponunt quod omnis res extensa sit magnitudo et dimensio. Alio modo quod plures dimensiones possibiles naturaliter extra invicem existere vel etiam extra invicem existentes fiunt simul secundum situm. Et hoc illi dicunt esse impossibile. Sed haec opinio improbata fuit primo *huius*; ideo dimitto eam hic. Et ideo ego do aliam distinctionem, quod plures dimensiones esse simul et non reddere 88va G subiectum | extensius et maius quam faceret una illarum est impossibile, sicut plures gradus albedinis in eodem subiecto reddunt illud albius quam faceret unus illorum. Sed plures dimensiones esse simul et facere subiectum extensius non est impossibile. Et ita est in rarefactione, quia cum dimensione praecedente generatur in eadem materia alia dimensio, sicut cum gradu caliditatis praecedente generatur alius gradus, et inde materia redditur extensior et maior.

Et ita finaliter concluditur quod rarefactio vere est motus ad quantitatem, secundum quam generatur magnitudo, sicut secundum calefactionem caliditas et secundum illuminationem lumen. Et accidit quod sit motus

---

1 minus[2]] *om.* P **4** restat] ideo de hoc est (*rep.* G) GPp ‖ scilicet] de quo notandum est Gp : de quo notandum P ‖ qua] *add.* hic P **5** sicut] prout GPp ‖ dividendo] dimittendo p **9** vel] nihil G ‖ simul] *om.* G ‖ quod] quia G ‖ una] unum P **10** etiam[1]] *om.* G **11** earum] eorum p **12** simul plures dimensiones] *add.* et plura corpora Pp : plures dimensiones simul et plura corpora G **15** fiunt] fiant G ‖ secundum] *add.* eundem GPp ‖ impossibile] *add.* naturaliter sic enim ponit aristoteles penetrationem dimensionum esse impossibilem G **16** improbata] reprobata GPp ‖ primo huius] *praem.* in p : in primo physicorum GP ‖ dimitto] *post* hic G **17** distinctionem] definitionem C **19** subiecto] *sup. lin.* C : *om.* Pp **20** illorum] eorum P ‖ facere] reddere GPp **22–23** in … generatur] *om.* (*hom.*) C **23** generatur alius gradus] et alius gradus generatur p **24** et maior] *om.* p **25** vere] *post* est P : non p **27** lumen] luminis G

9–16 Cf. Iohannes Buridanus, *Quaestiones super libros Physicorum*, I, q. 8 (ed. Streijger, Bakker, 85)

localis, quia si per potentiam divinam esset aer sine alio corpore continente ipsum, posset rarefieri et non mutaret locum, quia non haberet locum. Tamen verum est quod aliquando, ut in septimo *Physicorum*, Aristoteles dicit non esse motum per se | ad illas dispositiones quae acquiruntur per 98vb P modum sequelae ad alios motus; et sic, scilicet per modum sequelae ad calefactionem vel frigefactionem vel compressionem vel huiusmodi, acquiritur vel corrumpitur dimensio in rarefactione vel condensatione. Ideo solet dici quod non est per se motus ad quantitatem in condensatione vel rarefactione.

Ex dictis bene apparet quomodo sit dicendum ad rationes quae in principio quaestionis fiebant. Non enim est penetratio corporum in condensatione, sed corruptio dimensionis, nec in rarefactione advenit dimensio ab extrinseco, sed generatur et educitur de potentia materiae sicut et aliae formae.

Et sic est finis quaestionis etc. | 78va p

---

1 alio] aliquo p 2 posset] possit P ‖ locum[1]] locus C 3 physicorum] huius p 4 esse] *add.* per C 5 modum[1]] motum C 6 vel[1]] et p ‖ vel[3]] et G 8 est] fit G : sit Pp 9 bene apparet] *inv.* GPp ‖ quomodo] quo G 10 in condensatione] nec condensatio P 12 potentia] *add.* scilicet p ‖ formae] *add.* etc. G 13 et … etc.] et sic est finis istius quaestionis P : *om.* p

---

3 Cf. Aristoteles, *Physica*, VII, 3, 246a6–9, b10–17, 247a5–7

⟨IV.12⟩

⟨Utrum tempus sit motus⟩

Circa tractatum de tempore quaeritur duodecimo utrum tempus est motus.

Arguitur quod non rationibus Aristotelis:

⟨1⟩ Primo quia: motus non invenitur nisi in eo quod movetur; tempus autem in omnibus invenitur motis et quiescentibus. Ita enim dicimus in tempore quiescere, sicut in tempore moveri.

⟨2⟩ Item non est idem motus in diversis corporibus motis, quamvis simul moveantur; sed est idem tempus omnium quae simul sunt et simul moventur; igitur etc.

⟨3⟩ Item omnis motus est velox vel tardus; sed tempus non est velox vel tardum. Probatio quia: velox et tardum definiuntur per tempus (velox enim dicitur quod parvo tempore transit multum, et tardum quod multo paucum); sed tempus non definitur per tempus (idem enim non definitur per se ipsum); igitur etc.

⟨4⟩ Item Aristoteles concludit tempus non esse motum, quamvis non sine motu.

111vb C ⟨5⟩ Item si tempus esset motus, ipsum esset primus motus, | prout concedunt Aristoteles et Commentator. Sed tunc probatur quod non sit ille motus primus. Primo quia: incarcerati percipiunt tempus et tamen non percipiunt primum motum.

---

3 circa … duodecimo] utrum tempus est motus duodecima quaestio G : consequenter quaeritur circa tertium tractatum P : quaeritur duodecimo circa capitulum de tempore primo quaeritur p ‖ est] sit P 4 arguitur] *praem.* et P ‖ rationibus aristotelis] *om.* P 5 primo] *om.* G ‖ nisi] *om.* p 6 invenitur] *add.* scilicet G 7 tempore[1]] *add.* rem G ‖ quiescere … moveri] quiescere sicut moveri Gp : moveri sicut quiescere P 8 item] *add.* etiam P ‖ diversis] omnibus *sed add. sup. lin.* diversis C 12 vel] nec p ‖ et] vel p ‖ per tempus] tempore GPp 13 multo] *add.* tempore P 14 non[1]] *sup. lin.* C ‖ non[1] … tempus[2]] non definitur tempore P : tempore non definitur Gp 15 per se ipsum] se ipso P 16 concludit] *add.* in quarto huius G ‖ non[2]] *add. sup. lin.* nec C : nec P : nec sit p 19 quod] *add.* tempus Gp : quia P 20 primus] *ante* motus (19) Gp : *add.* motus P ‖ et] qui GPp

---

4 Cf. Aristoteles, *Physica*, IV, 10, 218b10–18 16 Cf. Aristoteles, *Physica*, IV, 11, 219a1 19 Cf. Aristoteles, *Physica*, IV, 14, 223b21–23; cf. Averroes, *In Physicam*, IV, comm. 133, f. 205A–B

 | DOI: 10.1163/9789004322356_035

Secundo quia: ille motus magis deberet dici tempus, quo magis | mensuramus motus nostros et alios; sed hoc non est primo motu, immo motu solis composito ex motu diurno et proprio (tali enim mensuramus dies et annos, menses et horas); igitur etc. 88vb G

Tertio quia: sicut arguit Aristoteles, si tempus esset motus primus, sequeretur, si essent plures primi motus, scilicet plures mundi, quod essent plura tempora simul; quod tamen est impossibile.

⟨6⟩ Item tempus est numerus; sed motus non est numerus, sed continuus; igitur etc.

Oppositum arguitur quia:

⟨1⟩ Prius et posterius in motu sunt partes motus; igitur sunt motus. Et tamen prius et posterius in motu sunt tempus secundum quod numerabilia sunt, ut dicit Aristoteles. Igitur etc.

⟨2⟩ Item motus mensuratur tempore; igitur est tempus. Consequentia patet, quia decimo *Metaphysicae* dicitur quod mensura et mensuratum debent esse unigenea, sic quod magnitudo mensuratur magnitudine, pondus pondere, tempus | tempore, motus motu, unitas unitate. Ideo etiam 99ra P dicitur quarto *huius* quod extra metrum nihil aliud videtur esse quod mensuratur, sed autem metra multa totum.

⟨3⟩ Iterum dicit Aristoteles quod tempus est numerus motus non quo numeramus, sed quod numeratur. Et sic tempus est motus numeratus.

Notandum est quod omnes concedunt tempus esse quo numeramus temporalia quantum ad suas durationes, ut 'quanta est vita naturalis hominis?'—

1 ille] *om.* GPp ‖ magis deberet] *inv.* P ‖ quo magis] cum magis P : quo p 2 primo] *add.* in p ‖ immo] nec p 2–3 motu solis composito] motus solis compositus P 3–4 dies ... horas] annos et dies et horas G : dies ad nos et horas P : annos dies p 4 igitur etc.] etc. Gp : ergo P 5 quia] *om.* P 5–6 sequeretur] sequitur p ‖ sequeretur ... motus] *om.* C 6 mundi] *add. in marg.* sequitur C 8 sed[1]] et GPp ‖ sed[2]] cum sit G 12 numerabilia] naturalia p 13 igitur etc.] *om.* P 16 sic quod] secundum quod C : sicut p ‖ magnitudo mensuratur] *inv.* G 16–17 pondus pondere] *om.* G 17 tempus ... motu] motus mobili tempus tempore p ‖ unitas] unitates GPp 18 dicitur] *add.* in G ‖ metrum] motum CPp ‖ aliud] *om.* P 18–19 mensuratur] *add.* tempore C 19 sed] *add. in marg.* apparet C : apparent P ‖ autem] ante p ‖ multa] mota p 21 quod] quo GP ‖ motus numeratus] *inv.* p 22 est] *om.* GP

5 Cf. Aristoteles, *Physica*, IV, 10, 218b3–5 13 Cf. Aristoteles, *Physica*, IV, 14, 223a28 15 Cf. Aristoteles, *Metaphysica*, X, 1, 1053a24–27; cf. *AA*, 1: 241 18 Aristoteles, *Physica*, IV, 14, 223b33–224a2 20 Aristoteles, *Physica*, IV, 12, 220b8–9

dicimus 'septuaginta annorum'. Et 'quanta canis?'—'decem annorum'. Et 'quanta via est de Parisius ad Avignionem?'—dicimus 'duodecim dierum'. Et omnino quaecumque aeque velociter moventur in aliquo spatio, dicimus aequali tempore aequale spatium pertransire et longiori longius et breviori brevius.

Ex hoc sequitur quod tempus est res successiva habens partes priores et posteriores ad invicem et quod illae partes non sunt simul. Unde alio et alio tempore dicimus moveri *a* et *b*, si *a* movetur hora primae et *b* hora meridiei et non in eodem tempore adaequate. Vel si dicimus in eodem tempore communi, quia in eodem die, tamen dicimus hoc esse ratione diversarum partium illius diei et non ratione eiusdem, et dicimus eos non simul moveri, sed *a* prius et *b* posterius. Unde etiam et qui dicunt et tempus et motum esse idem cum mobili, tamen ex eodem dicunt tempus esse et motum, quia aliter et aliter et non similiter se habet prius et posterius aut ad locum aut ad aliquid aliud aut in se ipso. Et sic ratione successionis in se habendo aliter et aliter prius et posterius dicimus esse motum et tempus.

78vb p Item secundum quod coexistimus maiori vel minori tempo|ri, plus dicimur aut minus durare; et hoc non esset ita, si tempus esset res permanens
112ra C secundum se totum et omnes partes eius | simul, quia ita sic coexisto magnae rei sicut parvae simul, sicut caelo et stellae vel turri et grano milii.

89ra G Item si esset tempus | res permanens secundum se et suas partes simul, tunc in eodem tempore adaequate Aristoteles loquebatur et Antichristus ambulabat et non hic prius tempore quam ille, cum rei simpliciter permanentis sint omnes partes simul et non una prius quam alia.

Ideo omnino haec prima conclusio est concedenda, scilicet quod tempus est res successiva cuius partes non sunt simul, sed alia prius et alia posterius,

---

1 septuaginta] sexaginta G 2 de] a G ‖ duodecim] viginti duorum G 3 omnino] omnia G ‖ moventur] moveretur C 7 alio[1]] *add.* modo P 8 movetur] moveatur p : moveretur G ‖ hora primae] hora prima p : horae primo P 9 adaequate] *om.* p ‖ vel] et G 10 communi] *om.* P ‖ eodem] eadem p ‖ ratione diversarum] diversarum rationum P 11 eos non] a et b non P : eos C 12 sed] si Cp ‖ etiam et qui] qui P : etiam aliqui p ‖ et[3]] *om.* GPp ‖ motum] motus p 13 eodem] eo Pp ‖ et] *om.* G 14 et[3]] *add. sup. lin.* seu ad C : aut p 15 aliquid] aliquod P 16 et[2]] vel P 17 tempori] tempore Pp 17–18 plus dicimur] *inv.* GPp 19 partes eius] *inv.* G ‖ ita] *post* sic Gp : *om.* P ‖ magnae] magnitudine P 20 sicut[1]] et G ‖ simul] *praem.* et GP : et simili p ‖ sicut[2]] ut G ‖ et[1]] vel P ‖ vel] et G ‖ et[2]] vel Pp 21 si] sicut G ‖ secundum se] *om.* P 22 in eodem tempore] eodem modo CPp ‖ adaequate] *add.* quare P ‖ et] *om.* p 23 ambulabat] ambulavit G : ambulat p ‖ non] *om.* P 24 sint omnes partes] partes sint Gp : partes sunt P 25 ideo omnino haec] ideo haec p : igitur et G 26 successiva] *add.* quia P ‖ alia[1]] una GPp

et quod priori existente nondum est posterior et posteriori existente non amplius prior est.

Secunda conclusio infertur quod tempus est motus, quia omne continuum sic successivum est motus vel mutatio, quia secundum tale successivum est continue aliter et aliter se habere, quod est mutari.

Item tali successivo convenit definitio motus. Est enim actus entis, quia est, licet in successione, et remanet semper potentia ad illud quod succedit | 99[rb] P
ei quod prius est; ideo est actus entis in potentia secundum quod in potentia.

Item haec nomina 'annus', 'dies', 'mensis', 'hora', quae significant tempora, sumpta sunt ex motu solis; et tamen in illo motu non est fluxus vel successio alia a motu, ut dictum est alias in tertio libro; igitur oportet tempus esse motum.

Tertia conclusio est quod tempus propriissime acceptum est motus primus, quia de ratione temporis est quod sit mensura motuum; ideo magis proprie tempus debet esse ille motus qui magis proprie dicitur esse mensura aliorum; sed ille est primus motus propter hoc quod in unoquoque genere rationabilius est quod primum sit mensura aliorum quam e converso. Per mensuram enim cognoscimus mensuratum quantum est; magis autem est cognitio perfecta et proprie dicta quae est posteriorum per primum quam quae est e converso, sicut scientia propter quid est potior quam scientia quia. Et etiam mensura debet esse regularis et primus motus est regularissimus in succedendo. Primum enim mobile non velocius aut tardius movetur hodie quam heri, licet aliter sit irregularis velocitatis, scilicet quantum ad partes simul existentes, prout motus dividitur secundum divisionem mobilis. Partes enim iuxta polos tardius moventur et aliae velocius. Sed secundum talem divisionem motus non habet rationem temporis. Et hoc etiam manifestum est per astrologos qui in mensuratione motuum recurrunt finaliter ad pri-

---

1 priori] priore G : *praem.* ex p ‖ est posterior] *inv.* GPp **3** infertur] *add.* scilicet G **4–5** successivum] successum P **5** continue] *add.* et continue p **6** successivo] successioni C : successione P ‖ entis] *sup. lin.* C : *om.* Gp **7–8** succedit ei] succedet ei illud G : succedet oportet p **8** prius] plus G ‖ entis] et C **9** annus dies] *inv.* G ‖ mensis] *om.* GPp **10** et] *om.* P **11** ut] quod G ‖ dictum ... libro] alias in tertio libro (huius p) dicebatur GPp **14** motuum] temporum C **15** tempus debet esse] tempus debet dici p : debet tempus dici G ‖ tempus ... proprie²] *om.* (*hom.*) P ‖ esse²] *om.* GPp **16** sed] et p : si P ‖ primus] proprius P ‖ propter] per P **20** sicut] sic p ‖ potior] posterior G **21** et etiam] *praem.* est p : et G : etiam P **23** licet] sed G ‖ sit] esset G ‖ irregularis velocitatis] enim regulari (*corr. ex* regularis) velocitate C : irregularis velocitas G ‖ scilicet] licet G **24** dividitur] dicitur P **24–25** partes ... moventur] pars enim iuxta polos tardius movetur C **27** qui] *add.* enim G : quae CP

---

11 Cf. sup., III, q. 2

mum motum tamquam ad primam et maxime proprie dictam mensuram omnium aliorum.

Quarta conclusio est quod apud vulgares motus solis compositus ex diurno et proprio magis est tempus quam aliquis alius motus, quia sicut praenotatum est, hoc nomen 'tempus' ultra hoc nomen 'motus' connotat quod sit mensura aliorum motuum; ideo unusquisque magis habet illum
89rb G motum pro tempore quo magis mensurat alios | motus; sed maxime vulgares mensurant per dictum motum solis: per annos, per dies, per horas. Causa
112rb C huius est, quia | ille motus est eis notissimus, quia maxime sensui apparens, et non est eis notus motus simplex diurnus, scilicet distincte a motu proprio; ideo non possunt illo mensurare. Mensura enim debet esse notior mensurato, cum per mensuram debeamus cognoscere mensuratum quantum ipsum sit.

Quinta conclusio est quod saepe operatores mechanici utuntur sua operatione per modum temporis, quia ex consuetudine quantitas suae operationis est multum nota eis. Ideo saepe ex ea mensurant alios motus, immo motum solis. Cum enim non videant solem, tamen ex quantitate operatio-
79ra p nis concludunt quod est hora tertia et tempus | comedendi etc. Et etiam ecclesiastici horologio utuntur per modum temporis; tamen non est proprie
99va P tempus, quia primo indiguit horologium | quod motus eius motu solis mensuraretur.

Sexta conclusio est quod adhuc intentione remotiori et secundum locutionem attributivam hoc nomen 'tempus' aliquando ponitur supponere pro rebus temporalibus, quia esse earum mensuratur tempore. Sic enim dicimus tempus esse serenum vel pluviosum vel carum vel aegritudinale vel pacificum vel bellicosum vel frigidum vel calidum etc. Et sic tempus est aer aut panis aut vinum vel homines etc.

---

3 solis compositus] *inv.* C ‖ ex] *add.* motu G **4** alius] terminus P **7** magis] *om.* G ‖ alios] *rep.* G **8** mensurant] mensurantur P ‖ dictum motum] dictos motus C ‖ solis] *add.* ut Pp ‖ per horas] *add.* etc. p : et horas etc. P : etc. G **8–9** causa huius est] *praem.* et p : causa est P : *om.* G **10** motus simplex] *inv.* GPp ‖ diurnus] *om.* G **13** ipsum sit] *inv.* G **14–15** sua operatione] *inv.* P **15** modum] motum Pp ‖ quantitas] quantitatis p **16** eis] *ante* nota p : *ante* multum P **17** quantitate] *add.* suae GPp **18** et tempus comedendi] *sup. lin.* C : *om.* G ‖ etc. et] etc. G : et p : *om.* P **19** modum] motum P **20** indiguit] indiguerit G **22–23** locutionem attributivam] attributionem GPp **23** ponitur supponere] supponit p **24** temporalibus] *rep.* G ‖ earum] eorum C **24–25** sic ... esse] si enim dicimus tempus est G **25** carum vel] *om.* P **25–26** vel^3 ... calidum] vel aegritudinale vel pa†...† vel bellicosum vel †...† vel calidum *in marg.* C : *om.* G **26–27** et ... etc.] *om.* (*hom.*) Pp **26** aut] vel G

Ad rationes principales.

⟨1⟩ Ad primam dicendum est quod, licet idem dicatur esse tempus et motus, tamen hoc est secundum diversas rationes. Dicitur enim motus secundum quod eo aliquid movetur; ideo non dicitur motus nisi eius quod movetur. Sed dicitur tempus secundum quod eo alii motus mensurantur. Et quia eodem motu possunt omnes motus simul existentes mensurari tamquam mensura extrinseca, ideo omnium dicitur esse idem tempus. Tempus igitur idem est motus unius solius mobilis et est mensura multorum aliorum. Et postea dicetur quomodo quies est in tempore et mensuratur tempore.

⟨2⟩ Consimiliter soluta est secunda ratio.

⟨3⟩ Ad aliam dicendum est quod tempus bene est velox. Sed hoc praedicatum non attribuitur sibi secundum illam rationem secundum quam dicitur tempus, sed secundum illam rationem secundum quam dicitur motus. Nec valet ratio qua arguitur quod tempus non sit velox vel tardum, quia simili ratione argueretur quod nasus nec esset simus nec aquilus, quia simum et aquilum naso definiuntur et nasus non definitur naso. Dicam igitur quod argumentum non valet, quia praemissae, si sunt verae, sumunt terminos secundum suppositionem materialem; ideo non deberet inferri nisi quod ille terminus 'tempus' non est ille terminus 'velox' vel ille terminus 'tardum'. Praemissae enim sunt quod iste terminus 'velox' et 'tardum' definiuntur per | 89va G
illum terminum 'tempus' et iste terminus 'tempus' non definitur per istum terminum 'tempus'. Igitur etc.

⟨4⟩ Ad auctoritatem Aristotelis dico quod non intendit negare veritatem illius propositionis 'tempus est motus', sed vult dicere quod non est quidditativa praedicatio, immo denominativa, quia ille terminus 'tempus' est passio huius termini 'motus'.

---

1 ad] *praem.* tunc P : *praem.* tunc ergo p 2 ad … est] ad primum dicendum p : dicitur P ‖ esse] *om.* GPp 5 eo … mensurantur] alii (aliquando p) motus mensurantur eo Pp 7 omnium] omni C ‖ tempus^{2}] *om.* GPp 9 postea] post GPp ‖ et mensuratur tempore] *om.* (*hom.*) P 10 consimiliter] *praem.* et p : et similiter G ‖ soluta est] *post* ratio P 11 aliam dicendum est] tertium dicitur P 13 sed … motus] *om.* G 14 qua arguitur] quae arguit G ‖ vel] nec Pp 15 ratione] modo G ‖ nec esset] nec est p : non esset GP ‖ aquilus] aquilinus p 16 aquilum] aquilinum p ‖ dicam] dico G : dicatur p 17 si sunt verae] sunt verae si P : sicut sunt p 18 deberet inferri] debent inferre G 19 ille terminus^{3}] *om.* P 20 praemissae … tardum] praemissae enim sunt quod †…† velox et tard†…† *in marg.* C ‖ et] vel P 21–22 et … tempus] *om.* (*hom.*) C 22 igitur] *om.* G 23 ad] *add.* aliam G 24 illius] huius GPp

---

9 Cf. inf., IV, q. 15

⟨5⟩ Ad aliam dicendum est quod capiendo propriissime hoc nomen 'tempus' tempus est motus primus. Sed dictum est quod hac acceptione non utuntur vulgares, sed astrologi, qui non per notitiam sensitivam, sed per ratiocinationem intellectualem utuntur illo motu tamquam tempore in suis computationibus ad sciendum situs stellarum ad invicem et ad nos, quamvis non videant illum motum. Alii | 112va C autem motibus sibi notis per sensum vel imaginationem utuntur per modum temporis.

Et bene concederem, si essent plures mundi, unus motus esset tempus illis de uno mundo et alius illis de alio mundo. Nec sic esset impossibile simul esse plura tempora etiam propriissime dicta.

99vb P ⟨6⟩ Ad aliam dicitur quod | numerus continuorum bene est continuus, sed ad numerandum oportet discernere inter partes eius.

Et sic est finis quaestionis praesentis etc.

1 dicendum] dictum GPp ‖ capiendo] *om.* p 2 motus primus] *inv.* GPp ‖ hac acceptione] illa acceptione G : acceptione ista p 4 ratiocinationem] rationem P 5 computationibus] computatibus p 6 non videant] non videat G : vero videant p 8 concederem] *add.* quod G 9 illis de^{2}] aliis de p : *om.* P ‖ simul] *om.* P 10 etiam] et G 11 aliam dicitur] ultimam dicitur (dico P) GPp ‖ bene est] *inv.* P 13 et … etc.] etc. et sic finitur quaestio P : etc. G : *om.* p

⟨IV.13⟩

⟨Utrum definitio temporis in qua dicitur 'tempus est numerus motus secundum prius et posterius' sit bona⟩

Quaeritur tredecimo utrum definitio temporis in qua dicitur 'tempus est numerus motus secundum prius et posterius' sit bona.

Arguitur quod non quia:

⟨1⟩ Si tempus esset numerus, tunc magis deberet poni species quantitatis discretae quam continuae, quod est contra Aristotelem in *Praedicamentis*.

⟨2⟩ Item numerus dividitur in indivisibilia, scilicet in unitates, quas Aristoteles quinto *Metaphysicae* dicit omnino esse indivisibiles secundum quantitatem; tempus autem non dividitur in indivisibilia, ut apparet sexto *huius*; igitur etc.

⟨3⟩ Item cum omnis numerus sit numeratus vel numerabilis, nullus numerus est infinitus; et tamen tempus est infinitum et perpetuum, ut habetur octavo *huius*; igitur etc.

⟨4⟩ Item cum tem|pus sit quo mensuramus et numeramus motus, si ipsum 79rb p esset numerus, esset numerus quo numeramus et mensuramus; sed hoc est falsum; igitur etc.

⟨5⟩ Item termini de praedicamento quantitatis, cum sint absoluti, non debent definiri per terminos respectivos; igitur tempus, cum sit de praedicamento quantitatis, non debet definiri per 'prius et posterius'.

⟨6⟩ Item prius et posterius definiuntur per tempus, quia per distantiam ad praesens nunc, ut vult Aristoteles; igitur non debet fieri e converso.

---

5 utrum] *rep.* P 6 sit bona] *ante* in (5) GPp 8 species] quantitatis discreta sicut P 9 continuae] tempore P 10 dividitur] divideretur p 11 omnino esse] *inv.* G 12 quantitatem] quantitates p ‖ apparet] patet G 15 et tamen] tamen G : et P 16 octavo] *praem.* in G : tertio p 17 et numeramus] *sup. lin.* C : *om.* GPp ‖ ipsum] iterum P 18 esset numerus^2] *praem.* ipsum GP : *om.* p ‖ et] vel p ‖ hoc] *om.* P 19 falsum] *add.* ut dicit Aristoteles GPp 21 tempus cum] *inv.* P 23–24 item … converso] *om.* G 24 fieri] definiri Pp

---

9 Cf. Aristoteles, *Praedicamenta*, 6, 4b20–25 11 Cf. Aristoteles, *Metaphysica*, V, 6, 1016b3–5 12–13 Cf. Aristoteles, *Physica*, VI, 1, 232a18–19 16 Cf. Aristoteles, *Physica*, VIII, 1, 251b10–28 24 Cf. Aristoteles, *Physica*, IV, 14, 223a5–6

 DOI: 10.1163/9789004322356_036

⟨7⟩ Item prius et posterius vel sunt partes quantitativae temporis vel sunt instantia. Si sunt partes quantitativae, tunc male est, quia totum non debet definiri per partes quantitativas suas, immo potius e converso, ut habetur septimo *Metaphysicae*, sicut acutus angulus definitur per rectum et semicir- 89vb G culus per circulum, ut ibi dicitur. Si vero per 'prius' et 'posterius' intelligimus | instantia, adhuc male est, quia haec nomina 'instans', 'punctum', 'momentum' sunt nomina privativa et minus nota quam 'tempus', 'linea' et 'motus'. Unde habitus non definitur per privationem, sed e converso.

Oppositum intendit Aristoteles.

Sicut ex proprietatibus manifestis et communiter concessis loci debet inquiri definitio loci, ita debet esse de tempore. Sunt autem proprietates temporis concessae quod tempore mensuramus motus finitos et quietes ad sciendum quanti sunt, cum hoc fuerit nobis dubium. Deinde quod tempus est successivum habens prius et posterius, sicut dictum est in alia quaestione. Unde dicit Aristoteles quod non est possibile plures partes temporis esse simul quarum una non includit aliam.

Deinde etiam ex praecedente quaestione supponitur quod tempus est motus et quod est motus primus vel secundum vulgares motus solis. Et 112vb C tamen | in octavo *huius* habetur quod infinitus et perpetuus et continuus est primus motus; quod quomodo debeat intelligi dicetur in illo octavo. Ideo etiam sequitur quod infinitum et perpetuum est tempus et continuum.

Deinde etiam quod tempus est continuum. Hoc non solum supponitur, sed probatur quia: magnitudo, motus et tempus proportionaliter dividuntur,

---

3 suas] ipsius P **4** septimo] nono G **6** est] *om.* p **7** sunt] cum sint G ‖ et^1] ut p **8** unde] inde p **10** loci] *om.* CP ‖ debet] *corr. sup. lin. ex* debebat C : debeat G : debebat p **13** sunt] sint GP **14** successivum] successio G ‖ habens ... est^2] *om.* p **15** unde] ut G **15–16** unde ... aliam] *om.* P **16** quarum] quorum p ‖ includit] includat G **17–21** deinde ... continuum] *transp. post* continuum (315.3) GPp **17** etiam] *om.* G ‖ praecedente] praecedenti Gp **18** et^2] *om.* P **19** habetur] habebitur P **19–20** et^2 ... motus] est (et p) primus motus et continuus GPp **20** quod quomodo] secundum quam p ‖ illo octavo] octavo huius G **21** et continuum] *om.* p **22** deinde etiam] et P ‖ hoc] *praem.* et GPp **23** quia] quod P ‖ magnitudo] *add.* et G ‖ proportionaliter] proportionabiliter p

---

4 Aristoteles, *Metaphysica*, VII, 10, 1035b5–10 **9** Cf. Aristoteles, *Physica*, IV, 11, 219b1–2, 220a24–26; cf. *AA*, 2: 137 **14–15** Cf. sup., IV, q. 12, 308 **15** Aristoteles, *Physica*, IV, 10, 218a11–16 **19** Aristoteles, *Physica*, VIII, 1, 251b10–28 **20** Cf. Iohannes Buridanus, *Quaestiones super libros Physicorum*, VIII, q. 3 (ed. Parisiis 1509, ff. 110vb–112va)

ut apparet in sexto; et tamen magnitudo est continua; et nihil | proportio- 100ra P
naliter dividitur continuo nisi continuum. Et quia: etiam ibidem ostenditur quod tempus in infinitum est divisibile, et tale oportet esse continuum.

Deinde etiam ex dictis sequitur quod iste terminus 'tempus' et iste terminus 'motus' se habent sicut passio et subiectum. Supponunt enim pro eodem, cum tempus sit motus, et cum hoc ille terminus 'tempus' ultra significationem motus connotat quod sit mensura, scilicet aliorum motuum.

Ex istis autem suppositis sic inferuntur conclusiones ad propositum.

Prima est quod de ratione temporis sive huius termini 'tempus' est discretio inter partes illius motus qui est tempus. Hoc probatur quia: tempus debet esse mensura motuum finitorum, cum nihil sit mensurabile nisi finitum; et oportet mensuram adaequari vel proportionari mensurato; et infinitum non est proportionale finito. Igitur ad hoc quod ille primus motus sit mensura aliorum motuum finitorum, necesse est discernere inter partes eius ad accipiendum partes motibus mensurandis proportionales. Et hoc etiam manifestum est per usum communem, quia motum caeli, cum per ipsum volumus temporalia mensurare, dividimus secundum rationem discretivam in annos, dies, horas, minuta etc.

Secunda conclusio est quod rationabile est dicere quod tempus est nume-
rus, quia est | mensura, sicut dictum est, et non sine discretione inter partes 90ra G
eius quod est tempus; mensura autem cum discretione suarum partium dicitur numerus. Hoc enim est de ratione numeri, quod sit mensura discreta secundum partes eius.

Tertia conclusio est quod tempus est numerus motus, quia cum sit de
ratione temporis quod sit mensura | discreta, sicut dictum est, tamen non 79va p

---

1 in] *om.* p ‖ sexto] *add.* huius GPp ‖ et tamen] cum P 1–2 proportionaliter] proportionabiliter p 2 continuo] continue G ‖ nisi] *corr. sup. lin. in* quam C : quam Pp ‖ etiam ibidem] ibi G : etiam idem P 3 est] *ante* in G 4 deinde ... tempus] *om.* p 5 passio et subiectum] subiectum et passio P 7 motuum] *add.* etc. G 8 autem] nunc P ‖ sic] *ante* suppositis Gp : *om.* P 11 nisi finitum] nisi sit finitum p : infinitum P : in infinitum C 13 est proportionale] esset proportionale C : est proportionabile p ‖ primus motus] *inv.* Pp 13–14 mensura] *add.* omnium G 14 eius] *om.* G 15 ad accipiendum] et accipiendum P : ad accedendum p ‖ mensurandis proportionales] mensurandae proportionales P : mensurandis proportionabiles p : proportionaliter commensurabiles G 16 caeli] *sup. lin.* C : *om.* G 18 annos] *add.* menses GPp ‖ minuta] momenta (*corr. ex* minuta) C : *praem.* et P 20 sicut] ut GP 22 sit mensura] sic mensuratur P 25 sicut] ut GPp ‖ est] *add.* et P ‖ non] *om.* p

---

1 Aristoteles, *Physica*, VI, 1, 231b18–20 2 Aristoteles, *Physica*, VI, 1, 232a18–19 13 Cf. Aristoteles, *Physica*, VI, 7, 238a29–30

est mensura permanentium secundum quod permanentia, quia dictum est quod tempus est res successiva; et successivum non est mensura rei permanentis ea ratione qua est permanens. Et iterum maioris mensurabilis est maior mensura vel pluries replicata, et minoris minor vel paucius replicata; 100rb P sed rei pure permanentis idem est tempus | parvae et magnae, scilicet totius et partis. Igitur ea ratione qua dicitur tempus non est mensura rei permanentis secundum rationem suae permanentiae. Igitur est mensura successivi vel 113ra C ratione | successionis sibi attributae; et successio est motus; igitur debet dici quod tempus est mensura motus.

Quarta conclusio est quod ad rationem temporis exprimendam non sufficit dicere quod tempus est numerus motus, sed oportet addere 'secundum prius et posterius', quia motus dupliciter est divisibilis, ut apparet in sexto *huius*. Uno modo secundum divisionem mobilis, quia extensus est in omnes partes ipsius mobilis, sicut et alia accidentia. Et omnes partes motus secundum huiusmodi divisionem sunt simul ad invicem, non una prius quam alia; simul enim moventur omnes partes ipsius mobilis, quando totum mobile movetur. Et isto modo tempus non est mensura motus, quia sic in eodem tempore adaequate est maior motus et minor motus; nam sic motus totius est in duplo maior quam motus suae medietatis, et tamen sunt in eodem tempore adaequate. Alio modo dividitur motus in priores partes et posteriores, ut secundum divisionem spatii cuius una pars prius pertransitur et consequenter alia continue, vel secundum divisionem graduum qualitatis quae acquiritur, quorum unus prius acquiritur et consequenter alius continue. Et secundum huiusmodi divisionem et quantitatem motus tempus mensurat motum. In duplici tempore duplex est motus, si sit aeque velox, quia duplex spatium pertransitur vel dupliciter intensa qualitas acquiritur. Igitur ad removendum mensuram motus, prout motus divi-

---

1 mensura permanentium] *inv.* G ‖ quod] *add.* sunt G **3** et] *om.* G **3–4** est maior] *inv.* P **4** replicata[1]] duplicata p **5** parvae et magnae] parvi et magni GPp **6** igitur] *add.* tempus GPp **7–8** vel ratione] *om.* C **9** motus] *add.* etc. G **10** est] *om.* P ‖ exprimendam] exprimendum Pp : *add.* recte G **11** motus] *add.* etc. P **12** quia] quod P ‖ apparet in] apparet G : patet Pp **13** uno] *praem.* est enim divisibilis GPp **15** huiusmodi] hanc p **16** alia] altera Gp ‖ moventur] moverentur p **17** et] *om.* P **18** sic] sicut G : *om.* p ‖ motus[2]] *om.* Pp **19** sic] si p ‖ quam motus] motu GPp ‖ et tamen] et G : tamen p **20–21** priores partes] *inv.* GPp **21** et] *add.* partes Gp **22** pertransitur] transitur P **24** huiusmodi] huius P **25** tempus mensurat motum] mensurat motum P : mensuratur motu G ‖ duplici] *add.* enim Gp : duplo enim P **26** pertransitur] pertransit G **26–27** qualitas] qualitatis p **27–317.1** dividitur] divideretur Pp

---

13 Aristoteles, *Physica*, VI, 4, 234b21–23

ditur secundum divisionem mobilis in partes simul ad invicem existentes, oportuit dicere quod tempus est numerus motus secundum prius et posterius.

Quinta conclusio est quod ista est bona descriptio temporis, quia explicat significationem et omnes connotationes huius termini 'tempus', et est convertibilis cum definito. Ibi enim ponitur 'numerus' per modum generis et glossatur 'numerus': id est mensura requirens | discretionem in suis 90rb G partibus; et per hoc differt ab aliis quae non sunt sic mensurae. Et dicitur 'motus' ad differentiam mensurarum magnitudinum vel rerum permanentium. Et dicitur 'secundum prius et posterius' ad differentiam mensurae motus penes quantitatem sibi debitam ex quantitate mobilis, sicut dicebatur. Igitur etc.

Tunc ad rationes.

⟨1⟩ Ad primam dicit Commentator quinto *Metaphysicae* quod Aristoteles in libro *Praedicamentorum* locutus est multa secundum famositatem, non secundum veram determinationem, scilicet de illis de quibus | spectat 100va P determinatio ad principalem philosophiam, scilicet naturalem vel metaphysicam. Unde potest dici quod, licet tempus sit continuum, tamen magis proprie hoc nomen 'tempus' pertinet ad genus quantitatis discretae quam ad genus quantitatis continuae. Terminus enim non reponitur sub tali genere vel sub tali specie ratione huius pro quo supponit, sed magis ratione formalis | connotationis. Isti enim termini 'album' et 'pater' supponunt pro 113rb C substantiis et non sunt de praedicamento substantiae. Licet ille terminus 'tempus' supponat pro motu continuo, tamen quia connotat quod sit mensura aliorum motuum, quod non potest esse sine discretione partium illius

1 divisionem mobilis] mensuram mobilis sive divisionem G 2 oportuit] *corr. in marg. ex* oporteret C : oportet p || dicere] *add.* et etiam ponere G 4 quinta conclusio est] *om.* C || temporis] *add.* tempus est numerus motus secundum prius et posterius G 4–5 explicat] explicant p 6 cum definito] cum eo G : *om.* Pp || ibi enim] et G 7 et] id est P || id est mensura] scilicet mensura p : id est mensuratur P 8 aliis] illis G || sunt sic] *inv.* GPp 9 magnitudinum vel] magnitudinum et G : magnitudine vel p 11 quantitatem] quantitates P 12 igitur etc.] *om.* GPp 13 tunc] *om.* G 15 multa] *ante* locutus GPp 16 non] nec P || spectat] spectabat G 17 determinatio ad principalem] principalis determinatio ad P || scilicet] *om.* P 19–20 discretae ... quantitatis] discretae quam ad genus p : discretae quam P : *om.* (*hom.*) G 20 sub] in GPp 21 sub] in GP : *om.* p || huius] eius G 23 substantiis] subiectis p || et] *add.* tamen G || ille terminus] igitur hoc nomen GPp 25 potest] possit P

14 Averroes, *In Metaphysicam*, V, comm. 18, f. 125K

79vb p continui motus, ideo ratione discretionis connotatae ille terminus 'tempus' | pertinet ad genus quantitatis discretae. Et ad hoc designandum dicit bene Aristoteles quod tempus esset numerus.

⟨2⟩ Ad aliam dictum fuit in septima decima quaestione tertii libri quod omnis unitas in continuis est in infinitum divisibilis. Et ibi dictum fuit quomodo exponi debeant auctoritates quae videntur esse ad contrarium.

⟨3⟩ Ad aliam dictum fuit in tertio quomodo infinitus sit numerus et infinitum sit tempus sumendo li 'infinitum' syncategorematice. Et tamen nec est numerus infinitus nec tempus infinitum sumendo 'infinitum' categorematice.

⟨4⟩ Ad aliam dicitur quod dupliciter dicitur 'numerus quo numeramus'. Uno modo ratio discretiva animae qua discernimus inter unitates numerandas dicitur numerus quo numeramus. Et sic bene dicit Aristoteles quod tempus non est numerus quo numeramus, sed est motus primus numeratus vel numerabilis quantum ad partes eius et quantum ad rationem discretivam animae inter illas partes discernentis. Alio modo numerus numeratus per talem rationem discretivam dicitur respectu aliorum numerus quo numeramus. Cum per rationem discretivam tu scis digitos tuos esse decem, tu per illos aliquando numeras equos aut alia, unicuique digito applicando unum equum. Et sic tempus, cum fuerit discrete notum per rationem animae, est quo numeramus alios motus. Et sic est numerus numeratus respectu rationis discernentis inter partes primi motus, et est quo iterum numeramus et mensuramus alios motus.

⟨5⟩ Ad aliam potest dici quod termini relativi bene sunt aliquando proprietates vel passiones terminorum absolutorum. Et quamvis subiectum non debet definiri definitione pure quidditativa per passiones suas vel proprietates, tamen bene aliquando describitur per eas. Unde conceditur quod

---

1 continui motus] *inv.* GPp ‖ ratione] *add.* huius G : rationis P ‖ connotatae] connotat p 2 pertinet] partium p ‖ dicit bene] bene dixit P 3 esset] est G 4 septima decima] alia G ‖ libri] huius p 5 est] *ante* in^{1} G 6 exponi debeant] *inv.* Pp : debeat exponi G 7 ad] *praem.* et G ‖ fuit] *add.* etiam GPp ‖ tertio] *add.* libro G ‖ et] *add.* in p 8 li] *om.* GPp 9 sumendo infinitum] sumendo p : *om.* G 12 ratio] ratione P 15 et quantum ad] per GPp 16 discernentis] *ante* inter GPp ‖ numerus numeratus] numeratus P : numeratum C 17–18 numeramus] *add.* ut G 18 cum] tamen P ‖ rationem] *add.* animae G 19 alia] canes GPp 21 quo] *add.* numerus G ‖ sic] *om.* P ‖ numeratus] *om.* P 21–22 respectu rationis] respectu ratione P : et unitas p 22–23 iterum … motus] mensuramus sive numeramus etc. G 24 relativi] connotativi G ‖ sunt aliquando] *inv.* G 25 vel] et G 26 debet] debeat GPp ‖ pure] simpliciter GPp 27 describitur] describetur G ‖ unde] *add.* bene G

---

4 Cf. sup., III, q. 17, 155^{28}–156^{4} 7 Cf. sup., III, q. 16

haec non est | definitio quidditativa; talis enim non convenit terminis connotativis, ut habetur septimo *Metaphysicae*. Sed est bona descriptio. 90va G

⟨6⟩ Ad aliam conceditur quod haec nomina 'prius' et 'posterius' secundum tempus praeteritum vel futurum definiuntur bene per hoc nomen 'tempus praesens' vel per hoc nomen 'nunc'. Sed hoc nomen 'tempus' | per illa nomina non definitur, sed per istos terminos 'prius' et 'posterius' in motu. Est enim tempus mensura motus secundum quod dividitur in partes priores et posteriores, sicut dictum fuit. Igitur etc. 100vb P

⟨7⟩ Ad aliam dictum est quod tempus non definitur per prius et posterius prout sunt partes ipsius temporis, sed prout dicuntur partes ipsius motus cuius tempus est mensura.

Et sic finitur quaestio etc.

---

1 haec] hoc p 1–2 talis ... connotativis] *om.* G 4 vel] et P ‖ definiuntur bene] *inv.* G 5 praesens] *om.* G ‖ illa nomina] istos terminos GPp 7 dividitur] dividere p ‖ et] *add.* partes p 8 sicut] ut P ‖ igitur etc.] *om.* GPp 9 dictum est] dictum fuit G : sit dictum P **10** sunt] dicuntur GPp ‖ ipsius[2]] *om.* GPp **12** et ... etc.] etc. et sic est finis quaestionis P : etc. G : *om.* p

---

2 Cf. Aristoteles, *Metaphysica*, VII, 5, 1030b14–1031a14

⟨IV.14⟩

⟨Utrum cuiuslibet motus tempus sit mensura⟩

Quaeritur quarto decimo utrum cuiuslibet motus tempus est mensura.

Arguitur quod non quia:

113va C ⟨1⟩ Aliqui sunt motus sempiterni et infiniti et illi non sunt mensurabiles. | Unde dicit Aristoteles quod quaecumque sunt semper non sunt in tempore.

⟨2⟩ Item idem non mensuratur se ipso; igitur ille motus qui est tempus non mensuratur tempore.

⟨3⟩ Tertio mensura debet esse notior mensurato, quia mensurare est ad cognoscendum quantitatem eius quod mensuratur per quantitatem mensurae. Sed motus est notior tempore, saltem aliquis qui est nobis manifeste sensibilis. Et etiam tempus non definiretur motu, nisi motus esset notior. Igitur saltem ille motus qui sic est notior tempore non mensuratur tempore.

⟨4⟩ Item mensura debet esse aequalis mensurato vel minor, immo debet esse minima, ut dicitur decimo *Metaphysicae*. Sed nullum est tempus quod sit omni motu minus; immo tempus, cum sit primus motus, est omnium motuum maximum et secundum magnitudinem subiecti et secundum successionem, quia infinitum.

⟨5⟩ Item mensura debet esse prior mensurato, ut dicitur decimo *Meta-*
80ra p *physicae*. Et tempus non est omni motu prius, immo consequitur | motum sicut passio subiectum. Unde etiam dicit Aristoteles quod in magnitudine est primo continuitas et prius et posterius, deinde in motu per magnitudinem et consequenter in tempore per motum. Igitur tempus non est mensura motus vel saltem non cuiuslibet.

---

3 quaeritur] *add.* consequenter G ‖ est] sit GPp 6 unde] ideo G ‖ sunt semper] *inv.* Pp : quae semper sunt G 7–8 ille … tempore] etc. G 9 tertio] item GPp ‖ quia mensurare] quoniam mensurare G : quia mensura p 12 et] *om.* P ‖ motu] per motum GPp 13 sic] satis p ‖ non] *add.* sic C 15 est] *om.* p 16–17 omnium motuum maximum] maximum omnium motuum GP : maximus omnium motuum p 17 maximum] *corr. sup. lin. ex* maximus C 18 quia] *add.* in p 20 est] *rep.* p ‖ consequitur] quod sequitur P 22 primo] prima P ‖ per] propter P 24 non] *om.* G

---

6 Cf. Aristoteles, *Physica*, IV, 12, 221b3–4; cf. *AA*, 2: 141 15 Cf. Aristoteles, *Metaphysica*, X, 1, 1052b31–32; cf. *AA*, 1: 239 19–20 Cf. Aristoteles, *Metaphysica*, X, 1, 1052b18–19; cf. *AA*, 1: 239 21 Cf. Aristoteles, *Physica*, IV, 11, 219a14–19

 | DOI: 10.1163/9789004322356_037

⟨6⟩ Item nihil est mensurabile vel mensura nisi secundum quod est quantum, ut habetur decimo *Metaphysicae* et quarto *huius*. Sed alteratio, cum sit qualitas, non habet quantitatem seu divisibilitatem nisi ex divisione et quantitate subiecti. Et dictum est in alia quaestione quod motus non mensuratur tempore secundum quantitatem vel divisionem quam habet ex divisione subiecti. Igitur alteratio non mensuratur tempore.

⟨7⟩ Item mensura et mensuratum debent esse de eodem genere, ut habetur decimo *Metaphysicae*. Et hoc est, quia mensura est comparabilis mensurato et proportionalis; et tamen quae sunt diversorum generum non sunt ad invicem comparabilia, ut dicitur in septimo *huius*. Sed alteratio et tempus sunt diversorum generum; igitur etc. Et hoc confirmatur quia: in septimo *huius* dicitur quod alteratio non est comparabilis loci mutationi; et tamen tempus est loci mutatio; igitur etc.

⟨8⟩ Item si ambulatio equi mensuraretur tempore, aut hoc esset secundum quantitatem quam habet de quantitate subiecti; quod est falsum, ut dictum est | 101ra P in alia quaestione. Aut hoc esset secundum quantitatem et divisionem quam habet ex quantitate et divisione spatii; quod est falsum, quia tunc oporteret illos motus esse invicem aequales secundum huiusmodi quantitatem, qui fierent in tempore aequali; quod manifeste est falsum de ambulatione equi et formicae, nam in aequali et eodem tempore ambulatio equi est valde maior et longior ambulatione formicae quantum ad longitudinem et magnitudinem spatii. Vel ille motus mensuratur tempore quantum ad fluxum secundum se; quod etiam apparet falsum, quia fluxus non potest dici maior nisi propter maius spatium vel propter subiectum vel | 90vb G propter velocitatem maiorem, sed fluxus non mensuratur tempore quantum

1 nihil] vel p ‖ secundum quod] secundum quid C : quod P 2 ut habetur] quod G 3 seu] sive P ‖ et] in P 5–6 secundum … tempore] *rep.* C 8–9 mensurato et proportionalis] et proportionabilis mensurato Gp : ac proportionalis mensurato P 11 et] *om.* P 11–12 septimo huius] illo septimo G : septimo Pp 12 comparabilis] proportionalis C ‖ tamen] *om.* p 14 aut hoc esset] hoc esse vel P 15 quantitatem] *add.* et divisionem p ‖ de] ex GPp 16 esset] est P ‖ et] vel p 17 quod] *add.* etiam GPp 18 secundum huiusmodi] secundum huius P : per hanc p 19 qui] quia G ‖ manifeste] *om.* G 20 in] *om.* GPp 21 maior et] *om.* G ‖ quantum] *om.* G 21–22 longitudinem et magnitudinem] magnitudinem et longitudinem GPp 22 vel] et G ‖ mensuratur] mensuraretur GPp

2 Cf. Aristoteles, *Metaphysica*, X, 1, 1052b20–25; cf. Aristoteles, *Physica*, IV, 12, 220b26; cf. *AA*, 1: 240 4 Cf. sup., IV, q. 13, 316^{10-20} 7–8 Cf. Aristoteles, *Metaphysica*, X, 1, 1053a24–25; cf. *AA*, 1: 241 10 Aristoteles, *Physica*, VII, 4, 249a3–5 11–12 Cf. Aristoteles, *Physica*, VII, 4, 248a17–18 16 Cf. sup., IV, q. 13, 316^{10-20}

113vb C ad magnitudinem vel intensionem | velocitatis, quia in eodem tempore vel aequali est motus velocior et motus tardior.

⟨9⟩ Eodem modo arguitur de alteratione. Nam alteratio non mensuratur tempore quantum ad magnitudinem quam habet ex magnitudine alterabilis, quia sic minor et maior sunt in aequali tempore; nec quantum ad magnitudinem intensionis vel graduum qualitatis quae acquiritur, quia in aequali tempore acquiritur qualitas intensior et maiorum graduum vel plurium in hoc alterabili quam in illo, secundum quod motus ille est velocior illo. Igitur non apparet secundum quam quantitatem alteratio possit mensurari tempore.

Oppositum arguitur quia:

⟨1⟩ Aristoteles dicit et nititur probare in isto quarto et sexto quod omnis motus et omnis mutatio est in tempore. Et tamen ipse Aristoteles determinat quod 'esse in tempore' non solum significat esse, quando tempus est, quia tunc non essent aliter temporalia in tempore quam ego essem in lapide, cum ego sim, cum lapis est, ita bene sicut quando tempus est. Sed 'esse in tempore' non potest aliud addere quantum ad significationem et connotationem ultra esse, quando tempus est, nisi hoc quod 'esse in tempore' connotet mensurari tempore. Igitur cum omnis motus sit in tempore, sequitur quod omnis motus mensuratur tempore.

80rb p ⟨2⟩ Item Ari|stoteles dicit quod quae semper sunt, secundum quod semper sunt, non sunt in tempore, quia non mensuratur esse eorum tempore. Igitur ad hoc quod aliquid sit in tempore requiritur quod ipsum vel esse eius mensuretur tempore. Sed cum omnis motus vel mutatio est in tempore, omnis motus vel mutatio mensuratur tempore.

⟨3⟩ Item si primus motus, qui est tempus, mensuratur tempore, hoc magis debet concedi de aliis motibus; sed primus motus mensuratur tempore, quia

---

1 vel[1]] et P 2 motus[2]] *om.* P 3 nam] *om.* p 5 quia sic] quia sicut C : et sic G 6 vel] *rep.* G ‖ qualitatis] qualitas C ‖ in] etiam GP : *praem.* etiam p 7 et] *add.* sic p 8 illo[2]] quam ille G 9 quam] *om.* P 12 et[1]] quod P ‖ isto] *om.* G ‖ et[2]] *add.* in GPp 13 tamen] cum tempore aristotelis *del.* C : causam p 15 quia] et p 16 ego sim] ergo sum p ‖ cum[2]] quando GPp 17 et] vel G 18 quod] *add.* est GPp 19 connotet] connotat P ‖ mensurari] *add.* in P 22 esse eorum] eorum esse sub G 24 cum] tamen p ‖ tempore[2]] *add.* ergo GPp 26 mensuratur] mensuraretur G

---

12 Cf. Aristoteles, *Physica*, IV, 14, 222b30–31; cf. Aristoteles, *Physica*, VI, 2, 232b20; cf. *AA*, 2: 173
13 Cf. Aristoteles, *Physica*, IV, 12, 221a2–26 21 Aristoteles, *Physica*, IV, 12, 221b3–5; cf. *AA*, 2: 141

totum mensuratur per partes quae sunt eaedem toti. Unde dicit Aristoteles quod non est aliud metrum et quod mensuratur, sed multa metra totum.

Ista quaestio est iudicio meo bene difficilis et multas implicans in se difficultates. Quas aggrediendo notandum | est primo quantum ad quid nomi- 101rb P nis quod mensurabile mensurari aliqua mensura est scire quantitatem eius prius dubiam per quantitatem illius mensurae prius notam. Hoc apparet inductive. Sic enim ulna notae quantitatis mensuramus quantus sit pannus, et talento vel marca quantum sit pondus, et diebus septimanam, et motu horologii quota sit hora diei etc.

Secundo etiam notandum quod plures et diversi sunt modi mensurandi. Primus modus mensurandi est per mensuram intrinsecam, scilicet totius per partes, numerando unamquamque illarum partium contra invicem et concludendo quotus sit numerus, ut quod illa sunt decem aut viginti. Et sic dicimus numerum mensurari per | unitates. Sic enim numerando seorsum 114ra C unum hominem et consequenter alium et alium scimus quot sunt illi homines. Et ideo dicitur decimo *Metaphysicae* quod primo invenitur metrum in numeris et consequenter in aliis. Et primum metrum unum est. Nam sicut ibi dicitur, 'metrum enim est quo quantitas cognoscitur; cognoscitur vero aut uno aut numero quantitas inquantum quantitas, numerus autem omnis uno; quare omnis quantitas inquantum quantitas cognoscitur uno.'

Secundus modus mensurandi reperitur iam in continuis supponens praecedentem modum. Et est etiam modus mensurandi per intrinsecam mensuram, scilicet totius per partes, scilicet quando continuum magnum de quo dubitamus quantum sit, distinguimus in plures partes ad invicem aequales,

---

2 metrum] motum p **3** est ... bene] iudicio meo est bene P : est G ‖ implicans] implicat p **4** primo] *om.* GPp **5** aliqua] aut G **6** dubiam ... prius[2]] *rep.* p **7** ulna notae] una nocte G ‖ quantus] quamvis G **8** talento vel] talenta vel p : talenta et G ‖ diebus septimanam] dies in septimana p **9** quota] quanta P ‖ etc.] *om.* P **10** secundo etiam] *om.* P ‖ notandum] *add.* est p : nota G **11** mensurandi] *om.* P **12–13** et ... numerus] *in marg.* C : *add.* illarum partium GPp **13** sit] est G ‖ ut quod illa] et quod illa P : ut quia illa p : ut quia G ‖ viginti] duodecim p **14** numerum] *om.* G ‖ unitates] unitate G ‖ enim] *om.* P **15** unum] illum p ‖ alium et alium] alterum et alterum GPp ‖ quot] quod p **16** quod] *add.* in p **17** est] *ante* primum GPp **18** ibi] ibidem GPp ‖ est] *om.* G ‖ cognoscitur[2]] ALp: *om.* BCGHMPTU **18–19** vero ... aut[2]] vere aut uno aut P : una specie aut una G : una tempore ab alia p **19** quantitas[1]] *add.* enim P ‖ autem] aut p **20** uno[1]] una p ‖ uno[2]] numero P **21** mensurandi] commensurandi p ‖ iam in continuis] ideo in continuis G : in continuis iam p : in continuis P **23** per] ad CP

---

1 Cf. Aristoteles, *Physica*, IV, 14, 224a1–2 **16** Cf. Aristoteles, *Metaphysica*, X, 1, 1052b18–19 **18–20** Aristoteles, *Metaphysica*, X, 1, 1052b20–23

et si sciverimus numerando primo quot sunt illae partes et cum hoc sciverimus quanta est quaeque earum, concludemus totum totiens, quot erant illae partes, esse tantum quanta erat unaquaeque illarum partium, ut si sint decem partes et quaelibet sit decem pedum, concludemus totum esse decies decem pedum, scilicet centum pedum.

Sed adhuc tertius modus mensurandi in magnitudinibus invenitur per superpositionem magnitudinis notae quantitatis ad magnitudinem ignotae quantitatis. Tunc si una non transcendat aliam nec transcendatur ab alia, iudicamus magnitudinem mensuratam esse aequalem magnitudini mensu-
91ra G ranti. Sic enim ulna ligni nota men|suramus ulnam panni et modo proportionali quarta aerea quartam vini et libra plumbea examinata libram cerae et huiusmodi. Et est iam ille modus mensurandi per mensuram extrinsecam. Et est possibile quod eadem mensurentur diversa mensurabilia, ut hac ulna ligni possem mensurare ulnam panni, chordae aut terrae.

Quartus modus mensurandi est iam compositus ex tertio et primo. Cum enim per ulnam ligni mensuraverimus per superpositionem singulas partes panni non communicantes, illi ulnae ligni aequales, et illas partes panni sic
101va P mensuratas numeraverimus, sciemus quantus est totalis pannus, ut quia | est decem ulnarum etc.

Et adhuc alii modi mensurandi sunt, ut secundum ductum numeri in numerum, costae superficiei in costam quadrati vel orthogonii et secundum
80va p reductionem aliarum figu|rarum ad quadrangulum orthogonium, et sic de aliis. Qui non sunt multum ad propositum; ideo dimitto illos.

---

1 si] *om.* Gp ‖ sciverimus] *post* numerando P : scimus G ‖ quot] *praem.* modo P : modo quod p 2 quaeque earum] unaquaque earum p : unaquaeque illarum GP ‖ quot] quod p 3 unaquaeque] unaquaque p ‖ sint] essent P 4 et quaelibet] in qualibet P 6 invenitur] *ante* in GPp 7 superpositionem] U : sup^{em} A : suppositionem BCGHMPTp : suppositio L ‖ quantitatis] *om.* C 8 tunc si] tunc enim si P : si enim p ‖ transcendat] transcendit P 10 sic] sicut p ‖ mensuramus] mensuremus G ‖ ulnam panni] pannum P 10–11 proportionali] proportionabili Gp 11 quarta aerea quartam] quartam aeream quartam C : quarta p ‖ libram] libra G ‖ cerae] *seq. spat.* G 12 huiusmodi] huius P 13 eadem] *add.* mensura GPp ‖ mensurabilia] monstrabilia P ‖ ut] *add.* quod GPp 14 possem] possum P : possent p ‖ chordae] *praem.* aut G : cerae p 16 enim] *om.* P ‖ per^{1} ... mensuraverimus] mensuremus per ulnam ligni G ‖ superpositionem] BHMU : suprapositionem T : sup^{em} A : suppositionem CGLPp 17 illi] cum illae C ‖ ligni] *add.* sunt C 18 mensuratas numeraverimus] mensuratas numeravimus p : numeravimus P ‖ sciemus] scimus G 19 decem] viginti G ‖ etc.] *om.* GPp 20 alii] *add.* ulteriores GPp ‖ ductum numeri] dictum numerum p 21 orthogonii] orthogani G : orgonii p 22 figurarum] significativarum P ‖ orthogonium] orthoganium G ‖ de] *add.* multis Pp 23 qui] quae GPp ‖ ideo] non C ‖ illos] *add.* modos G : illas Pp

Sed alius est modus ad propositum. Qui fit secundum proportionalem divisionem mensurae et mensurati, quamvis in partes valde inaequales et quamvis res quae sic dividuntur sint valde diversarum rationum. Sic | enim (114rb C) mensuramus circuitum terrae, quot leucas contineat, per ipsum caelum dividendo caelum in trecentos et sexaginta gradus, deinde proportionaliter dividendo terram et considerando in terra quot leucae correspondent uni gradui in caelo, ut quia forte triginta, quia cum per triginta leucas in plana patria ambulemus de austro ad septentrionem, inveniemus polum arcticum elevatum nobis in uno gradu. Tunc concludemus circuitum terrae esse totiens trecentas et sexaginta leucas, quot leucae correspondent uni gradui in caelo, scilicet tricesies trecentas et sexaginta.

Et apparet manifeste quod ille modus mensurandi non fit secundum adaequationem mensurae ad mensurabile nec alicuius partis mensurae ad aliquam partem mensurabilis, quoniam unus gradus in caelo non est aequalis terrae sibi correspondenti, immo valde maior, et non est vis in qua proportione sit maior.

Unde sic etiam per horologium mensuramus motum solis diurnum sciendo quot gradus in horologio signati correspondent uni horae et quot nocti et quot diei. Sic enim per gradus pertransitos in horologio scimus quota sit hora diei vel noctis et quando sol est in oriente, in meridie, in occidente vel in angulo noctis, licet nullus gradus signatus in horologio sit aequalis parti motus solis sibi correspondenti. Nec est cura quae proportio sit huius ad illum secundum magnitudinem. Ita enim bene per parvum horologium hoc fieret sicut per magnum, licet secundum magnitudinem non esset eadem proportio magni horologii et parvi ad caelum et ad motum solis.

---

1 est] *post* modus G : *om.* p ‖ ad propositum] *praem.* magis Gp : *add.* magis P **2** divisionem mensurae] *inv.* P **3** quamvis] *add.* in p **4** quot] quod G ‖ ipsum] totum G **5** et] *om.* Gp ‖ gradus] *add.* dividendo G **5–6** proportionaliter] proportionabiliter p **6** in ... leucae] quot leucae in terra Pp : quod leucae in terra G ‖ correspondent] correspondeant P **7** quia cum] nam si GPp **8** ambulemus] ambulamus p ‖ inveniemus] invenimus Gp **9** arcticum] antarcticum G **10** et] *om.* Gp ‖ quot] et sic P ‖ leucae] *add.* quae P **11** caelo] *add.* tot sunt P ‖ trecentas et] *inv.* p : trecentae G : centum P ‖ sexaginta] *add.* leucas p **12** apparet manifeste] *inv.* P ‖ secundum] quantum P : per p **14** quoniam unus] unus enim G **15–16** et ... maior] *om.* (*hom.*) G **16** sit] *sup. lin.* C : *om.* Pp ‖ maior] maioris P **17** unde] *om.* P **18** sciendo quot] sciendum quod P ‖ signati] figurati P **19** per] *add.* in G **20** quota] quod P ‖ in[3]] *om.* p **22** parti motus] motu G ‖ sibi] sive P ‖ correspondenti] correspondentis p **23** illum] illud P ‖ magnitudinem] imaginationem C ‖ enim] *om.* P ‖ bene] *sup. lin.* C : *om.* Gp **24** hoc fieret] *om.* G **25** eadem] *om.* p ‖ horologii] *om.* p ‖ et[2]] *om.* Gp

Tunc descendo ad ponendum conclusiones ad propositum.

Prima est quod aliquando tempus mensuramus per tempora per simplicem numerationem, ut numerando dies concludimus quod sunt septem vel decem, sicut concluderemus quod sunt decem canes.

Secunda conclusio est quod etiam aliquando tempus mensuramus per tempora secundo modo prius dicto, quia scimus quantum tempus est una
101vb P dies et numeramus dies mensis et concludimus | quod mensis est totiens tantum tempus quanta erat una dies.

Tertia conclusio est quod non mensuramus tempus per tempus nec motum per tempus per superpositionem, scilicet tertio modo. Partes enim temporis fluunt et non permanent; ideo non possemus unam ponere super aliam. Et si dicamus aliquam partem primi motus coexistere alicui alteri motui, hoc non est secundum adaequationem, immo ita magnum coexistit parvo sicut magnum magno vel parvum parvo, quantum ad illam coexistentiam quae est in coexistendo hoc, quando illud est.

Quarta conclusio etiam sequitur quod secundum quartum modum prius dictum mensurandi non mensuramus tempus per tempus nec motum per
91rb G tempus, quia modus ille continet vel praesupponit | tertium modum men-
114va C surandi, qui non convenit | tempori, sicut praedictum est in conclusione praecedente.

Quinta conclusio est quod per tempus et per motum qui est tempus bene mensurantur alii motus, sive sint locales sive sint alterationes vel aliae continuae mutationes, secundum ultimum modum mensurandi praedictum, scilicet per proportionales divisiones temporis mensurantis et motus mensurati.

Et primo hoc declaratur de motu locali, postea de motu alterationis. De
80vb p motu locali hoc | primo declaratur quantum ad magnitudinem et divisionem motus secundum magnitudinem et divisionem spatii pertransiti quia:

---

1 tunc] nunc GPp 2 prima] *add.* conclusio G 2–3 per simplicem] secundum simplicem P : per simplicitatem C 3 ut] et C : aut P ‖ concludimus] *om.* p 5 conclusio] *om.* G ‖ etiam aliquando] *inv.* GPp 7 mensis[1]] menses Gp ‖ totiens] *add. in marg.* alias continens C 8 quanta] quantum p 10 superpositionem] suppositionem CP ‖ scilicet] id est p 11 possemus] possumus GPp 11–12 unam … aliam] unum ponere super alium p 12 si] sic G 12–13 alteri motui] *inv.* p 14 quantum] *om.* p ‖ illam] aliam G 15 coexistendo] existendo G 16 etiam] *om.* GP 16–17 prius dictum] *om.* G 18 modus] *post* ille G : motus Pp ‖ praesupponit] supponit Gp ‖ modum] motum P 19 tempori] tempore P ‖ praedictum] dictum GPp 19–20 in conclusione praecedente] *sup. lin.* C : in praecedente conclusione GP : in praecedenti conclusione ideo etc. p 21–22 bene mensurantur] *inv.* GPp 22 sint[2]] *om.* P ‖ vel] et G : sive sint p 24 per proportionales] secundum proportionales P : secundum proportionabiles Gp 26 locali] *add.* et P

scire quantus est motus tali quantitate est scire quantum est spatium pertransitum; sed per tempus hoc scimus. Verbi gratia, cum viderimus quod ambulatio de Parisius ad Romam perficitur in viginti diebus, posito quod fuerit illa ambulatio regularis, et sciverimus quod in primo die transivimus decem leucas, sciemus spatium totale pertransitum usque ad Romam esse decies viginti leucarum, scilicet ducentarum leucarum. Et ita, si motus continuus et regularis durat praecise per diem naturalem et appareat quod in una hora transitur spatium trium pedum, sciemus totale spatium pertransitum per talem motum esse ter viginti quattuor pedum, scilicet septuaginta duorum pedum. Et si motus esset difformis et sciretur modus et quantitas difformitatis, ut si in secunda die poneretur dupliciter velox et in tertia tripliciter, adhuc multiplicando per invicem concluderetur quantitas spatii.

Secundo etiam declarabitur conclusio quantum ad magnitudinem et divisionem motus secundum magnitudinem et divisionem mobilis, scilicet signando aliquod signum in spatio quod pertransitur et tunc illud mobile pars eius post partem continue pertransit illud signum. Et sic prima medietas navis prius pertransit pontem quam secunda medietas et quam navis, et iterum prius prima medietas medietatis | quam secunda medietas, et sic 102ra P in infinitum secundum infinitam divisionem. Si igitur signaverimus tempus in quo mobile pertransit illud signum, scilicet tempus imaginatum inter instans primum in quo mobile attingit illud signum et instans primum in quo totum mobile est ultra signum, et sit illud tempus una dies naturalis viginti quattuor horarum, si appareat quod pars pedalis illius mobilis transit illud signum in una hora et motus sit regularis, concludemus quod mobile est longitudinis viginti quattuor pedum.

---

3 posito] supposito p **4** illa ambulatio] *in marg.* C : *om.* Gp ‖ primo] prima p ‖ transivimus] *add. in marg.* alias iverimus C : pertransivimus p **6** leucarum[2]] leucas p ‖ et] *om.* P **8** transitur] transit P ‖ trium pedum] *inv.* p **9** per talem motum] per totalem motum Gp : *om.* P **9–10** septuaginta duorum] septuaginta et duorum p : sexaginta octo et quattuor per totalem motum P **10** modus] motus GP **11** ut] et P ‖ secunda] secundo P ‖ velox] velocitas G ‖ tertia] tertio P **12** tripliciter] *add.* et C **14** declarabitur] declaratur GPp **15** motus … divisionem[2]] *om.* (*hom.*) C ‖ secundum magnitudinem] secundum quantitatem p : ad magnitudinem P **17** et] *sup. lin.* C ‖ et sic] sicut Gp **18** pertransit] pertransibit P ‖ et quam] vel quam p : *om.* G **19** et iterum] item G ‖ prima] *om.* GPp ‖ secunda] illa GPp **20** infinitam divisionem] *inv.* P **21** pertransit] transit G ‖ scilicet] *om.* P **22–23** mobile … ultra] totum (*del.*) mobile est ultra signum et instans primum in quo totum mobile attingit C **23** totum mobile] *inv.* P ‖ sit] sic C **24** viginti] et p ‖ si] *om.* G ‖ transit] pertransit p **26** longitudinis] *post* pedum G

Sed tamen per hoc nesciremus latitudinem vel profunditatem mobilis, quia partes mobilis diversae secundum latitudinem eius et profunditatem simul transeunt signum, non prius una quam alia; et quantum ad tales partes quae simul sunt, inquantum sunt simul, tempus non est mensura motus, sicut dicebatur in alia quaestione. Unde etiam tempus numquam esset mensura longitudinis mobilis considerando simpliciter quod in tali tempore tale mobile movetur, propter hoc quod non prius movetur pars prior quam pars posterior. Sed bene esset mensura eius considerando tempus transitionis signi | [114vb C] signati, quia sic non simul transeunt pars prior et posterior.

Deinde etiam hoc declaratur de alteratione. Primo quantum ad magnitudinem intensionis qualitatis in eodem subiecto et eadem parte eius; quae quidem magnitudo solet vocari magnitudo gradualis. Nam si tempore decem horarum intenditur qualitas in aliquo subiecto continue et regulariter, quanta intensione est qualitas acquisita in una hora, decies tanta intensione est acquisita in totali tempore decem horarum.

Et similiter etiam de quantitate et divisione alterationis secundum quantitatem et divisionem alterabilis. Si ponatur casus quod calefaciens sit ex una parte contiguum calefactibili et quod non simul multiplicetur calefactio in totum calefactibile, sed prius in partem propinquiorem calefactibilis et posterius in posteriorem continue et regulariter, et sit tempus decem horarum a principio calefactionis, donec primo erit multiplicata per totum alterabile, et in prima hora multiplicetur per quantitatem alterabilis | [81ra p] pedalem, sequitur quod alterabile erit decem pedum et quod tanta erit alteratio quantum ad quantitatem et divisionem alterabilis.

---

1 sed] *om.* P ‖ nesciremus] *ante* per P ‖ profunditatem] profunditudinem G 2 diversae] divisae P ‖ latitudinem eius] *inv.* Pp ‖ profunditatem] profunditudinem G 3 transeunt signum] transierint signum p : transeunt P ‖ una] *add.* pars G ‖ alia] altera p 4 quae] qui p ‖ sunt simul] *inv.* GPp ‖ tempus] tempore C ‖ motus] *om.* P 5 sicut … quaestione] sicut dicebatur in alia conclusione p : *om.* G ‖ etiam tempus] et tempus G : *om.* p 6 longitudinis] *om.* P 7 propter] per P ‖ prius movetur] prius moveretur p : primo movetur G ‖ pars[2]] *om.* P 9 et] *add.* pars Gp 10 etiam] *om.* P 11 qualitatis] quantitatis G ‖ eius] *om.* P 11–12 quae quidem] quicquid p 12 si] *add.* in GPp 13–14 continue et regulariter] regulariter et continue P 14 intensione] intensive Gp ‖ una] illa P 14–15 intensione] intensive GPp 16 et[1]] *om.* P ‖ etiam] *add.* est G : est P 17 alterabilis] alterationis C ‖ ponatur casus] ponamus casum G ‖ ex] in G 19 calefactibilis] calefactibili G 20 sit] sic Gp 22 et] vel p 23 sequitur] sequeretur Pp 23–24 quantum ad] quoad GPp 24 divisionem] *add.* eius secundum quantitatem et divisionem G

---

5 Cf. sup., IV, q. 13, 316[10–20]

Sed hic est dubitatio posito | casu quod esset motus quem solemus vocare (91va G) localem absque hoc quod mobile pertransiret aliquod spatium, ut si Deus moveret mundum motu recto vel lapidem posito quod omnia alia essent annihilata, utrum iste motus haberet magnitudinem aliquam nisi ex magnitudine mobilis.

Et videtur quod non, quia motus localis non videtur habere magnitudinem nisi secundum divisionem mobilis vel spatii. Cum | enim tempus sit (102rb P) extrinsecum motui, non est magnitudo motus ex magnitudine temporis nisi arguitive propter proportionalem divisionem, sicut dictum fuit.

Ad hoc ego respondeo quod motus ille nec ex loco nec ex spatio pertransito nec ex tempore habet aliquam magnitudinem nisi arguitive. Sed tamen dupliciter oportet dicere motum localem habere magnitudinem sicut et alterationem. Primus modus est, quia quantitative extensus est ad extensionem eius quod movetur; ideo continue est tantus secundum longitudinem, latitudinem et profunditatem, | quantum est illud quod movetur. Et isto modo non (92ra G) maioraretur motus propter longiorem successionem, sicut caliditas intensa non est longior, latior vel profundior quam caliditas remissa. Sed alio modo sic est magnitudo successionis, sicut est in qualitate magnitudo intensionis; hoc enim modo dicimus caliditatem in parva scintilla ignis esse in duplo vel in decuplo maiorem | intensive quam in magno aere. Et sic in motu quem (115ra C) vocamus localem est magnitudo successionis et fluxus, et sic aliquando motus parvi corporis est motu alterius magni corporis maior.

Sed quia dictum est alias quod ille motus non perciperetur, nisi mobile perciperetur aliter se habere ad locum vel spatium vel ad quiescens, ideo

---

1 est] erit G : esset P ‖ esset] *post* motus Pp : *add.* hic G **2** hoc] *om.* G ‖ aliquod] illud P **6** et] *om.* P **9** proportionalem] proportionabilem Gp ‖ sicut] ut P **10** ego] *om.* P ‖ ille] *om.* C ‖ ex[2]] *om.* G **12** et] *om.* p **13** quia] quod GP ‖ ad extensionem] extensione Gp : extensio P **14** continue] *praem.* sic GP : *om.* p **15** profunditatem] profunditudinem G, *add. sed del.* quantum est illud quod movetur et isto modo non maioratur motus propter longiorem successionem sicut caliditas intensa non est longior latior et profundior quam qualitas remissa sed alio modo sicut est magnitudo successionis sicut etiam, *add.* hic est defectus pecuniae non scripturae hic nihil deficit nisi pecunia; *quod folii restat vacat* G ‖ est] ad P **16** maioraretur] *corr. sup. lin. ex* maioretur C : maioratur Gp ‖ propter longiorem] propter maiorem G : per longiorem P **17** vel] et GP ‖ caliditas] qualitas p **18** sic] *corr. in marg. ex* sicut C ‖ sic ... sicut] sicut est magnitudo successionis sic Gp ‖ est[2] ... magnitudo[2]] in qualitate est magnitudo p : in qualitate est magnitudo in G : in qualitate et magnitudine P **19** parva scintilla] *inv.* GPp **19–20** duplo ... decuplo] duplo vel decuplo Pp : decuplo in duplo G **20** et] *sup. lin.* C: *om.* GPp **22** magni corporis] *inv.* p **24** ad[2]] aliquod GPp

---

23 Cf. sup., III, q. 7; cf. sup., IV, q. 8

etiam istam quantitatem fluxus oportet cognoscere et arguere ex quantitate spatii vel eius quod pertransitur vel eius quod imaginatur pertransiri, quoniam motu maiori secundum quantitatem successionis maius spatium pertransiretur, si ille motus fieret commensurative spatio; in pertranseundo spatium tamen non esset maior aut minor secundum huiusmodi successionem, licet nullum esset spatium aliud a magnitudine mobilis.

Et tunc rationabiliter dubitatur utrum sit eadem res quae est quantitas motus secundum quantitatem et divisionem mobilis, et quae est quantitas motus secundum successionem sive fluxum.

Ad quod ego respondeo quod non, immo quantitas quam motus habet secundum divisionem mobilis est dimensio qua ipsum mobile et omnia accidentia sibi inhaerentia extenduntur. Quantitas autem intensionis formae, scilicet quantitas gradualis qualitatis, est ipsamet qualitas quae ex infinitis gradibus seu partibus simul in eodem subiecto et in eadem parte eius sine distinctione earum ab invicem secundum situm composita est, sicut magnitudo ex infinitis partibus secundum situm extra invicem existentibus. Magnitudo autem successionis in ipso motu quem vocamus localem est illemet motus localis, quia ex infinitis partibus | [102va P] non simul, sed successive una post alteram est compositus.

Et ideo notandum est quod, si dicamus ambulationem formicae et cursum equi esse ad invicem aequales secundum durationem, non est per hoc intelligendum quod sunt aequales ad invicem secundum quantitates ipsorum, sed hoc solum significat quod tempus est idem coexistens utrique adaequate, vel si non | [81rb p] idem, tunc significat quod tempus coexistens uni est aequale tempori coexistenti alteri. Duratio enim rei non est aliud quam res illa durans, sed duratio eius extrinseca bene dicitur tempus sibi coexistens.

Sic igitur dictum est de modo per quem motus mensuratur tempore.

---

2 eius[1]] *om.* G ‖ eius quod] quod P : *om.* G 4 si] sed P 5 aut] vel P ‖ huiusmodi] huius P 5–6 successionem] commensurationem p 7 et] *om.* G 8 quantitatem] qualitatem p 8–9 quantitatem ... secundum] *om.* (*hom.*) C 9 sive] seu p : et G 10 ego] *om.* P 11 ipsum] *om.* GPp 12 extenduntur] exceduntur G 13 quantitas gradualis] gradualis qualitas P ‖ qualitatis] qualitas C : qualiter P ‖ qualitas] *add.* gradualis p 14 seu] sive P 15 distinctione] disiunctione C ‖ ab invicem] *om.* GPp 16 magnitudo] *add.* est p 17 autem] enim C 18 quia] qui P : quo p ‖ una] unam Cp 19 alteram] aliam GP 20 et[1]] *om.* P ‖ quod] *om.* p 22 sunt] sint GPp ‖ secundum] *add.* aliquas GPp 23 idem] illud P 24 vel] *om.* P ‖ si non] *add.* est G : signum p ‖ uni] *om.* G 28 sic igitur] sic enim G : sicut igitur p ‖ mensuratur] commensuratur C : mensurantur p

Nunc est magis directe dicendum de quaesito. Et sic est ibi duplex dubitatio. Prima est utrum omnis motus sit mensurabilis tempore; de qua prius argutum fuit. Secunda est utrum omne tempus est mensura motus; de qua argutum est prius quod non. Sed | [92rb G] arguitur modo quod sic quia: aliter definitio temporis non valeret, quia non conveniret omni contento sub definito; non enim conveniret illi tempori quod non esset mensura motus. | [115rb C]

Et de his ponitur sexta conclusio quod nec omni tempore mensuratur motus actuali mensuratione nec omnis motus mensuratur tempore, quia cum in infinitum sit tempus magnum secundum dicta prius, constat quod omni tempore quo aliquis mensurat aliquem motum est aliud tempus maius. Et ideo nullus mensurat nisi forte ratione partis, ut si diceremus anno mensurare, quia die mensuramus, quae est pars anni; sed haec est impropria locutio. Similiter etiam multae sunt alterationes quarum quantitas est nobis ignota et non potest fieri nobis nota, ut alterationes quae modo sunt piscium in profundo maris; ideo istas alterationes nullus mensurat.

Septima conclusio ponitur quod nulli motui repugnat secundum rationem eius specificam vel generalem, quin possit mensurari tempore, quia suppono quod de omni genere vel specie motus aut mutationis potest aliquis esse nobis notus aut per sensum aut per ratiocinationem; et tunc illum, si sit ignotae quantitatis propter magnitudinem, poterimus mensurare per tempus.

Sciendum est quod multis motibus repugnat propter circumstantias particulares quod possint a nobis mensurari, ut quia sunt in tali loco ad quem homo non potest naturaliter pertingere; ideo nullus potest naturaliter istos motus percipere ad mensurandum eos, licet alibi possint similes in specie percipere. Et non dico, quin | [102vb P] per potentiam divinam posset homo istos

1 est[1] ... dicendum] dicendum est directe P ‖ sic] *om.* GPp 2 qua] quo G 3 fuit] est G ‖ secunda] *add.* etiam Gp ‖ est[2]] sit GPp 4 quia] *om.* p 5 temporis] motus C 8 mensuratur] mensuratus C 9 in] *om.* GPp 10 omni] aliquo G ‖ mensurat] mensuratur G ‖ aliquem] *sup. lin.* C : *om.* G ‖ motum] *add.* non P 11 ideo] illo P ‖ anno] *post* mensurare (12) G : annum p 12 die] diem p ‖ haec] hoc p 13 locutio similiter] locutione et p 14 potest fieri] *inv.* Gp ‖ piscium] *ante* quae G 16 motui] motus tantum P 17 possit] posset P 18 genere vel specie] specie vel genere p : genere P 18–19 aliquis ... notus] esse nobis aliquid notius P 20 poterimus mensurare] potest posterius mensurari G 22 sciendum] notandum P 24 non ... pertingere] naturaliter non potest attingere P ‖ naturaliter[2]] *om.* p 25 eos] *om.* p ‖ alibi possint] alibi possit G : alibi posset p : aliquis possit alibi P 26 posset homo istos] homo posset istos G : homo possit istos P : homo posset p

2–3 Cf. sup., IV, q. 14, 320–322 3–4 Cf. sup., IV, q. 13, 315[9–18]

percipere et sic eos tempore mensurare, sed quod hic dicimus intendimus dicere quantum ad potentias naturales.

Notandum est etiam quod non possumus motus naturales omnino praecise et punctualiter mensurare, scilicet secundum modum mathematicae considerationis. Non enim possumus per stateram scire si praecise libra cerae sit librae plumbi aequalis. Potest enim esse excessus in ita parva quantitate quod non perciperemus excessum. Sed sufficit saepe mensuratio ad prope, iuxta illud quod de modico non est curandum.

Octava conclusio ponitur probabilis quod nulli motui repugnat ratione suae magnitudinis vel parvitatis esse mensurabilem tempore; et etiam nulli tempori ratione suae magnitudinis vel parvitatis repugnat esse mensurativum motus. Haec conclusio cum difficultate deducitur, sed tamen sic potest deduci supponendo quod, licet infinitum sit tempus syncategorematice loquendo et etiam infinitus sit motus, | [92va G] tamen nec aliquod tempus est infinitum nec aliquis motus est infinitus; dico secundum extrema et non dico secundum divisionem, ita quod nullum est tempus et nullus est motus carens partibus extremis. Hoc apparet ex prioribus dictis. Ideo ex hoc sequitur quod omni tempore est tempus maius, si perpetuus sit mundus, et omni motu locali est motus maior secundum quantitatem successionis. Tunc apparet quod propter temporis aut motus exce|[81va p]dentem magnitudinem non obstat, quin illud tempus sit mensurabile per partes. Si enim sit centum annorum vel mille | [115va C] vel millesies mille et sic sine statu, tamen adhuc mensuraretur per annos et per dies aut tibi per sensum notos aut per ratiocinationem quantum ad annos praeteritos et futuros qui ante te fuerunt et post te erunt.

Deinde etiam quantum ad parvitatem notandum est quod, sicut quantumcumque magnum possumus numerare congregando partes, donec red-

---

1 quod] *sup. lin.* C : hoc quod nos G ‖ intendimus] *praem.* hoc G 3 notandum] *praem.* et p ‖ quod] *add.* nos G ‖ motus naturales] *om.* G 4 modum] motum P 5 si praecise] *inv.* P 6 excessus] *corr. in marg. ex* extensus C : *ante* esse G : extensio P 7 non perciperemus] nos non possemus percipere nec etiam perciperemus G : non percipiemus P ‖ sufficit] *om.* P 8 prope] proprie P ‖ curandum] *add.* et sequitur statim post hoc conclusio octava sicut videre poteris et dicit quod nulli motui ratione magnitudinis suae vel parvitatis repugnat esse mensurabilem tempore etc. G 9 ponitur probabilis] ponitur p : est P ‖ repugnat] *post* parvitatis (10) G 10 et] *om.* P 12 sed] *om.* P 13 supponendo] supponitur GP : simpliciter p 14 etiam] *om.* P 17 prioribus dictis] *inv.* P : dictis in prioribus Gp 18 est tempus] *om.* G 19 omni motu] tam motui P 20 temporis] tempus G 21 sit²] sint p 21–22 centum annorum] decem annorum vel centum G 22 millesies] *om.* G ‖ statu] statum p 23 et] *om.* Pp ‖ tibi] etiam C 24 et²] aut p : *add.* qui P 26 etiam] *om.* p ‖ est] *om.* G 27 possumus] possimus p 27–333.1 reddant] reddunt p : reddat G

dant quantitatem illius magni, et numerando eas et sciendo de unaquaque quanta sit, ita etiam quantumcumque parvum possumus mensurare quantum sit per proportionalem divisionem magnae mensurae et magni mensurati. Et est ille modus mensurandi ultra prius dictos et diversus ab eis. Verbi gratia, si sciverimus per ratiocinationem quod motus Saturni in obliquo circulo perficiendo unam circulationem est triginta annorum et quod etiam in triginta annis est tantus motus, scilicet perficiens talem circulationem, et si supposuerimus quod ille motus sit regularis, et quaeratur quantus est iste motus in uno anno, dividemus zodiacum in | triginta partes aequales; qui 103ra P
quidem zodiacus cum sciatur esse trecentorum et sexaginta graduum, dividemus trecentos et sexaginta per triginta et concludemus motum unius anni esse per duodecim gradus zodiaci.

Deinde quaeritur forte quantus erit motus in die. Et tunc videbimus quot sunt dies in anno, ut quia sunt trecenti et sexaginta quinque, et dividemus istos duodecim gradus per fracturas in trecentas et sexaginta quinque partes aequales. Et erit motus unius diei per unam istarum, et millesima parte diei erit motus per unam millesimam illius partis, et sic in infinitum secundum minorationem. Ideo ratione parvitatis non obstat, quin omnis motus sit tempore mensurabilis et quin etiam omne tempus possit esse mensura motus.

Et tunc possunt solvi rationes quae fiebant. Apparet enim bene quod omni tempori convenit definitio temporis, quia omne tempus est mensura motus, non sic quod actu eo mensuretur motus, sed quia eo potest mensurari motus. Et omnis motus etiam est in tempore, non solum quia est, quando tempus est, sed etiam quia potest mensurari tempore, capiendo 'posse' pro non

---

1 sciendo] considerando P 3 proportionalem] *post* divisionem G : proportionabilem p ‖ magnae] magni P ‖ et] *add. in marg.* †…†agnitudinis mensurae C ‖ magni] *om.* P 4 est] *om.* G ‖ ab] *rep.* G 5 sciverimus] scimus G 6 unam] integram p ‖ quod etiam] *inv.* G 8 est] sit p 9 uno] nono G ‖ dividemus] divideremus P 10 cum … graduum] constituitur ex trecentis sexaginta gradibus G ‖ sciatur] scitur p 10–11 dividemus] divideremus P 11 trecentos] trecenta P : trecentum p ‖ et^{1}] *om.* G ‖ concludemus] tunc videmus p ‖ motum] *om.* G 12 esse] *om.* P ‖ duodecim] decem P : viginti duo p 13 quaeritur forte] quaeretur forte P : quaeretur G : quaeritur p ‖ videbimus] videmus P 14 ut] et p : *om.* P ‖ et^{1}] *om.* G ‖ dividemus] divideremus P 15 in trecentas et] in trecentas G : et trecentas et P : in trecentum et p 16 unam istarum] unum illorum P 16–17 millesima … millesimam] *add. in marg.* †…†a parte diei erit †…† unam centesimam C : in centesima parte diei erit motus per unam centesimam GPp 19 etiam omne] etiam esse P : omne G ‖ possit] potest p 21 et … fiebant] ad rationes ad primam p 21–22 et … motus] *om.* (*hom.*) P 22 tempori] tempore G 23 eo^{2}] iterum p 24 est^{1}] *om.* p

repugnantia secundum modos praedictos, videlicet quod nulli motui repugnat ratione suae speciei vel sui generis aut etiam ratione suae magnitudinis aut parvitatis, quin sit tempore mensurabilis, licet forte aliunde sibi repugnet.

92vb G ⟨1⟩ Ad primam igitur rationem dictum est | alias quod nullum tempus est infinitum sive perpetuum capiendo 'infinitum' categorematice, licet infinitum sit tempus syncategorematice capiendo 'infinitum'.

⟨2⟩ Ad aliam dicendum est quod motus qui est tempus mensuratur bene aliis motibus qui sunt etiam tempora. Et sic etiam bene idem mensuratur se ipso, quia per suas partes, quae collective sumptae sunt ipsum totum.

⟨3⟩ Ad aliam dicendum est quod in omni genere vel specie motuum est invenire aliquos motus quorum quantitas non est nobis vel astrologo notior quam quantitas temporis.

115vb C ⟨4⟩ Ad aliam | dicendum est quod in modo mensurandi per proportionales divisiones mensurae et mensurabilis non oportet quod mensura sit aequalis mensurabili vel minor, immo non oportet quod aliqua pars mensurae sit aequalis alicui parti mensurabilis. Sed sufficit quod sint proportionaliter
81vb p divisibilia, ita quod si mensura fuerit uni|formis et mensurabile uniforme, et medietati mensurae correspondeat medietas mensurabilis et tertiae parti tertia pars, millesimae millesima; et si alterum eorum non fuerit regulare, tunc oportet bene scire modum et quantitatem irregularitatis. Unde per parvum horologium mensuramus motum solis et per motum solis horologium.

⟨5⟩ Ad aliam dicendum est quod, sicut aliquando demonstramus prius per posterius et aliquando e converso, quia aliquando posterius est nobis notius

---

1 praedictos videlicet quod] prius dictos scilicet quia G 1–2 repugnat] repugnant p 1–3 repugnat ... parvitatis] ratione suae speciei vel sui generis aut etiam ratione suae magnitudinis aut parvitatis repugnat G 3–4 repugnet] repugnat P : repugnent p 5 igitur] *om.* G ‖ quod] quia G 7 syncategorematice] *post* infinitum G 8 est^{1}] *om.* P 9 sunt etiam tempora] etiam sunt tempus p : sunt etiam tempore P : etiam sunt in tempore G ‖ etiam bene idem] etiam idem G : idem bene P 10 sumptae] *om.* G ‖ totum] *om.* G 11 dicendum est] dicitur P 12 invenire] invenires P ‖ astrologo] astrologis G 14 mensurandi] *om.* p ‖ per] vel G : *om.* (*spat.*) C 14–15 proportionales] proportionabiles p 15 mensurae] mensuretur P ‖ mensurabilis] mensurabiles G ‖ mensura] mensuratur P 16 minor] minoris P ‖ quod] *add.* mensura sit aequalis mensurabili nec quod G : *add.* mensura sit aequalis mensurabili quod p 16–17 mensurae] mensuretur P 17 alicui] *om.* G 17–18 proportionaliter] proportionabiliter p 19 uniforme] uniformiter p 20 millesimae millesima] *add. in marg.* et centesimae centesima C : et centesimae centesima GPp 21 tunc] *om.* p ‖ oportet bene] *inv.* G ‖ modum] motum GP 22 motum^{1}] motus P 24 per] et P

---

5 Cf. sup., III, q. 16

quantum ad quia est, et e converso | quantum ad propter quid est, ita non est inconveniens quod aliquando mensuremus prius per posterius et aliquando e converso; vel etiam unum per alterum, licet non sit naturaliter ordo inter ea secundum prius et posterius, dum tamen sciatur quod sint proportionaliter divisibilia. Tunc enim quodcumque acciderit esse notius altero quantum sit, illo mensurabimus alterum. Et saepe hoc est isto notius secundum sensum et e converso secundum rationem, vel etiam hoc est illo notius notitia confusa et e converso notitia determinata, vel etiam hoc est isto notius tibi et e converso mihi aut per sensum aut propter consuetudinem aut propter aliam causam. Ideo etiam bene dicit Aristoteles quod aliquando motum mensuramus tempore et aliquando tempus motu et etiam aliquando tempus aut motum magnitudine et aliquando e converso. Tamen magis proprie dicta est mensuratio posterioris per prius, sicut scientia propter quid est magis proprie dicta seu potior scientia quam scientia quia. 103rb P

⟨6⟩ Ad aliam dictum est quod alteratio vel qualitas dupliciter dicitur quanta: aut quantitate extensionis partis extra partem aut quantitate graduum seu intensionis. Et dictum est quomodo utroque modo mensuretur tempore.

⟨7⟩ Ad aliam dicendum est quod quaecumque sunt mensurabilia unum per alterum secundum adaequationem unius ad alterum, oportet illa esse eiusdem generis, vel etiam si sint | comparabilia ad invicem secundum certam et determinatam proportionem numeralem excessus unius ad alterum, ut quod unum sit alteri duplum aut sesquialterum. Sed de mensura secundum proportionales divisiones hoc non est necessarium. 93ra G

⟨8⟩ Ad aliam dictum est quod cursus equi et ambulatio formicae sunt eodem tempore mensurabiles, licet nullo modo sint invicem aequales, quia eidem tempori maius et minus sunt proportionaliter divisibilia.

---

1 ita] aliter p 2 et] *om.* G 3 sit naturaliter] sit naturalis p : stat naturaliter G 4 proportionaliter] proportionabiliter p 5 quodcumque] quemcumque P ‖ quantum] quanto p 6 illo mensurabimus] illud mensurabimus G : illo mensurabilius p ‖ mensurabimus … isto] *om.* P 7 et] vel etiam P ‖ est] *om.* G ‖ notius] *add.* et C 8 et[1]] vel G ‖ est] *om.* G 9 per sensum] propter sensum Pp : propter sensum aut propter sensum G 10 etiam] *om.* P 11 etiam aliquando] *inv.* Gp 12 et aliquando] aliquando etiam P 14 seu potior scientia] scientia sive potior P 15 vel] et p 16 quantitate[1]] quantae P 17 seu] sive P ‖ quomodo] quoniam G 19 sunt] etiam P 22 numeralem] ABHMTUp : naturalem CP : numerorum G : *om.* L 24 proportionales] proportionabiles p ‖ divisiones] *add.* dicens C 25 est] *om.* P ‖ formicae] forte p 27 eidem tempori] eodem tempore p ‖ maius et] maius aut p : magis et P ‖ proportionaliter] proportionabiliter p

---

10 Cf. Aristoteles, *Physica*, IV, 12, 220b15–32

⟨9⟩ Et eodem modo dicendum est de ultima ratione, quae erat de alteratione.

116ra C Et haec pro nunc de quaestione etc. |

1 dicendum est] *inv.* G 3 et ... etc.] etc. G : etc. sequitur P : *om.* p

⟨IV.15⟩

⟨Utrum quies mensuretur tempore⟩

Quaeritur quinto decimo utrum quies mensuratur tempore.

Arguitur quod non quia:

⟨1⟩ Si lapis quiescit, quies eius non est aliud quam ipse lapis; et tamen ipse secundum se non mensuratur tempore; igitur etc.

⟨2⟩ Item cum omnis successio sit motus vel mutatio et non oportet in quiete esse aliquem motum vel aliquam mutationem, sequitur quod non oportet in quiete esse aliquam successionem, sed possibile est quod sit simplex permanentia. Sed dictum est quod nullius tempus est mensura nisi ratione successionis. Igitur etc.

⟨3⟩ Item si quies mensuraretur tempore, sequeretur quod in definitione temporis deberemus apponere quod tempus est numerus motus et quietis, sicut in definitione | naturae dicimus quod natura est principium motus et 103va P
quietis; quod tamen non fecit Aristoteles; igitur etc.

⟨4⟩ Item omne quod mensuratur tempore est divisibile in prius et posterius aut secundum se aut secundum aliquam eius dispositionem. Sed quiescens non est huiusmodi, immo est totum simul, et quidquid est eius hodie, erit etiam cras. Nec etiam quies est sic divisibilis, quia quies non est aliud quam res quiescens. | 82ra p

Oppositum tamen determinat hic Aristoteles et in sexto libro.

Prima conclusio ponitur quod res simpliciter permanens, scilicet quae nulla mutatione mutatur, non mensuratur tempore, quia nihil mensuratur tempore nisi ratione alicuius successionis; et in tali re nihil est successivum.

---

3 quaeritur quinto decimo] quinto decimo quaeritur G : quaestio non minus difficilis p ‖ utrum … tempore] *rep.* G **5** lapis^{2}] *sup. lin.* C : *om.* Gp ‖ et] *om.* P **7–8** in quiete] quietem P **8** sequitur] *om.* p **9–10** sit simplex] est simplex GP : est simpliciter p **10** sed] *add.* tamen G ‖ tempus] temporis GP **12** sequeretur] sequitur p **13** numerus] mensura C **15** fecit] facit P ‖ igitur] *om.* p **16** tempore] *om.* G **18** huiusmodi] huius P ‖ est eius] *inv.* GPp **19** etiam cras] cras sed p **21** determinat] intendit P ‖ hic aristoteles] *inv.* GPp ‖ libro] *om.* G **22** permanens] *corr. in marg. ex* quiescens C

---

10 Cf. sup. IV, qq. 12–13 **21** Cf. Aristoteles, *Physica*, IV, 12, 221b7–8; VI, 3, 234a33–b9

 | DOI: 10.1163/9789004322356_038

Item in tali re nihil est divisibile proportionaliter tempori; et tamen omne aliud a tempore quod est mensurabile tempore est proportionaliter divisibile tempori.

Item omnis eius quod mensuratur tempore aliquo, maius est illud quod est in isto toto tempore eo quod est in eius medietate; sed in sic permanente nihil est maius in uno die quam in una hora; igitur etc.

Secunda conclusio est quod quies non mensuratur tempore, quia quies non est aliud quam res quiescens; et si res sit omnino quiescens, ipsa est permanens sine aliqua mutatione; et talis non mensuratur tempore; igitur etc.

Tertia conclusio est quod omnis res quae non semper est mensuratur tempore quantum ad esse ipsius, ad istum sensum qui post dicetur. Hoc apparet per Aristotelem in quarto *huius* dicentem quod motum mensurari
93rb G tempore est ipsummet et esse eius mensurari tempore, alia autem a motu | esse in tempore est esse eorum mensurari tempore, sed non ipsa. Et hoc est manifestum ex communi locutione. Dicimus enim Socratem esse vel fuisse per sexaginta annos et per unam diem mansisse in domo et parietem fuisse album per duos annos et sic de aliis rebus et earum dispositionibus permanentis naturae et certo tempore permanentibus. Manifestum enim est quod tempore mensurantur quantum ad esse, fore vel fuisse.

116rb C Sed tunc oportet videre ad quem sensum hoc sit | verum. Nam in quarto *Metaphysicae* debet ostendi quod esse rei non differt ab eius essentia. Et in primo *huius* diximus quod non est aliud hominem esse vel hominem currere quam homo. Quomodo igitur poterit esse verum quod homo vel lapis non mensuratur tempore et tamen esse hominis et esse lapidis mensurantur tempore? Hoc enim videtur contradictionem implicare.

1 proportionaliter] proportionabiliter p 2 tempore[1]] caliditate p ‖ proportionaliter] *post* divisibile (2–3) G : proportionabiliter p 4 omnis eius] omne p 5 toto] *om.* G ‖ eo] *om.* P ‖ in sic] *inv.* p ‖ permanente] permanentibus G 6 uno] una GPp ‖ una hora] *inv.* P ‖ igitur etc.] *om.* G 7 est] *om.* p 9 mensuratur] mensuretur p 12 post dicetur] prius dicebatur p 13 quod] *om.* p 14 est ... tempore[2]] *om.* (*hom.*) p ‖ ipsummet] ipsum GP 15 est] aut p 15–16 et ... manifestum] et hoc etiam manifestum est GP : adhuc est etiam manifestum p 17 unam] unum P 18 duos] quattuor p ‖ rebus] *om.* G ‖ earum] eorum G 19 enim] *om.* G 20 mensurantur] mensurari P ‖ fore] fuere P 21 tunc] sic p 22 essentia] existentia p 23 hominem esse] *inv.* Pp 25–26 et[1] ... tempore] *om.* (*hom.*) C 25 esse lapidis] *inv.* P 26 enim] *om.* GPp

13 Cf. Aristoteles, *Physica*, IV, 12, 221a4–5, 7–9 21–22 Cf. Iohannes Buridanus, *Quaestiones super libros Metaphysicorum*, IV, q. 8 (ed. Parisiis 1518, f. 18vb) 23 Cf. Iohannes Buridanus, *Quaestiones super libros Physicorum*, I, q. 18 (ed. Streijger, Bakker, 184)

Dicendum est ad hoc quod proprie loquendo de virtute sermonis nec homo nec esse hominis mensurantur tempore, nisi hoc sit ratione motus vel mutationis alterius ab | homine et hominis esse. Sed haec verba 'esse', 103vb P 'fore' et 'fuisse' consignificant tempora. Et tunc verum est quod esse hominis mensuratur tempore, non ad illum sensum quod illud pro quo supponit haec oratio 'esse hominis' mensuretur tempore, sed ad illum sensum quod tempora connotata per istas orationes 'hominem esse', 'hominem fore' vel 'hominem fuisse' mensurantur tempore. Et ita dico de hoc quod dicimus Socratem fuisse album vel in domo etc. Unde si dico 'Socrates fuit tribus annis Parisius', hoc non est dicere quantus est vel fuit Socrates nec quantum est esse eius nec quanta est essentia eius, sed hoc est dicere quod tempus in quo fuit Parisius est vel fuit tempus trium annorum; et hoc est mensurare tempus et non Socratem.

Quarta conclusio est quod quies mensuratur tempore etiam quantum ad esse eius, hoc est dictu quod quiescens est mensurabile tempore quantum ad quiescere. Et oportet totum intelligere ad sensum praedictum, videlicet quod tempus connotatum per hoc verbum 'quiescere' vel 'quievisse' aut huiusmodi mensuratur tempore, ut si sit septimana, mensuratur per septem dies.

Dico igitur finaliter quod, si dicamus 'hominem esse vel hominem esse album vel hominem quiescere aut huiusmodi mensuratur tempore', hoc non debet nobis aliud significare nisi quod tempus coexistens homini, dum est vel dum est albus aut dum quiescit, mensuratur tempore; et hoc est dictu quod nos scimus vel possumus scire quantum est istud tempus quod sic coexistit. Etiam nec tempus est mensura quietis sic quod quies sit proportionaliter divisibilis tempori sicut motus, sed modo praedicto solum.

Sed adhuc est fortis dubitatio, | quia quamvis nullus esset motus, tamen 82rb p adhuc res aliquae nullam habentes quantitatem nisi durationis viderentur

---

2 mensurantur] mensuratur GP **3** esse[2]] *post* fore (4) G : *add.* et p **4** et[1]] vel P **6** mensuretur] mensuratur p **7** tempora] tempore P ‖ vel] *om.* G **8** mensurantur] mensuratur G ‖ dico] *add.* quod C **9** vel] *add.* socratem fuisse P ‖ unde] *praem.* et G **10** quantus est] quantum est vel quantus p **11** eius[1]] *om.* p **12** tempus] *om.* Pp **13** et] *om.* GP **15** dictu] dictum P **16** oportet totum] hoc totum oportet P ‖ intelligere] *add.* quantum P ‖ videlicet] scilicet G **17** vel] aut p **17–18** aut huiusmodi] vel huiusmodi G : aut huius P : *om.* p **19** dies] *add.* etc. Gp **20** dico igitur] *inv.* P ‖ dicamus] *add.* quod G **21** vel hominem quiescere] *om.* p ‖ huiusmodi] huius P ‖ hoc] hic p **23** quiescit] *om.* P ‖ dictu] dictum GP **24** vel] aut GPp ‖ sic] sibi P **25** etiam] etc. p ‖ mensura] *om.* p **25–26** proportionaliter] proportionabiliter p **26** sicut] *add.* est GPp **27** quia] quod G ‖ esset] est P **28** viderentur] videntur P

esse mensurabiles quantum ad suam durationem; et cum mensura rei quantum ad suam durationem non sit nisi tempus, videtur quod tempus esset, licet non esset motus.

Et hoc sic declaratur. Ponamus quod, antequam esset mundus creatus et per consequens antequam esset aliquis motus, Deus creasset tres angelos *a*,
93va G *b*, *c*, et quod creavit prius *a*, deinde creavit simul *b* et *c* | et postea simul illos
tres annihilavit. Illo casu posito oportet concedere quod angelus *a* plus duravit quam angelus *b* vel *c*, etiam in certa proportione, quia non minus duravit
116va C quam si motus caeli fuisset coexistens eis; et tamen, si fuisset | motus caeli
104ra P coexistens eis, duravisset plus in | certa proportione, ut forte plus in duplo,
scilicet si motus coexistens ei fuisset duplus ad motum coexistentem aliis; igitur etiam plus duravit in duplo. Sed angeli *b* et *c* aequaliter duraverunt, non unus plus vel minus quam alter. Cum igitur non sit aequale vel duplum, nisi sit certa mensura qua illa sunt commensurabilia vel unum commensurabile per alterum, ideo licet non esset aliquis motus, tamen erat certa mensura eorum quantum ad suam durationem.

Sed tunc, si hoc concedatur, sequuntur alia magna inconvenientia, quia duratio angeli *a* non erat aliud quam angelus *a*, nec extrinsece nec intrinsece, nisi diceretur quod ista duratio erat Deus, quia ponimus quod nihil erat nisi Deus et iste angelus. Et non potest dici quod ista duratio maior angeli unius et duratio minor alterius sit Deus, cum ipse semper sit indivisibilis et numquam maior vel minor. Si igitur duratio angeli *a* non erat aliud quam angelus *a* et sic de angelis *b* et *c*, tunc illi angeli nec essent aequales nec unus maior altero, cum omnes essent indivisibiles; vel si loquamur quod unus est maior altero, id est perfectior, tamen sic erat possibile quod angelus *a* erat minor, id est minus perfectus. Et sic non apparet quod angelus *a* durasset plus vel quod eius duratio fuisset maior duratione angeli *b* et *c*.

---

1 mensura rei] mensurari G 2 tempus videtur] potest videri P ‖ quod tempus esset] esse tempus G 5 creasset] creavisset P ‖ angelos] angulos C 6 creavit prius] *inv.* G ‖ creavit^{2}] *om.* P ‖ et^{2}] *om.* P ‖ postea] deinde P ‖ simul^{2}] *post* tres (7) GPp 7 oportet] oporteret p ‖ angelus] angulus C 8 angelus] angulus C ‖ etiam] et G ‖ duravit] duraverit P 9 fuisset^{1}] *ante* motus^{1} G ‖ et] *om.* P ‖ caeli^{2}] ex *sed del.* C : *om.* Pp 10 duravisset] durasset p 11 scilicet] *om.* p ‖ duplus] duplex G 13 cum] quia P 14 sunt] sint Gp 14–15 commensurabile] mensurabile G 15 esset] erat (*sed corr. in* esset ?) C 16 eorum] *om.* p ‖ ad] *add.* certam p 17 tunc] *om.* p ‖ hoc concedatur] haec concedantur P ‖ sequuntur] sequunt G ‖ alia magna inconvenientia] aliae magnae inconvenientiae p 18 angelus] angulus C 18–19 extrinsece nec intrinsece] intrinsece nec extrinsece Gp 19 duratio] *add.* esset vel G ‖ nihil] *add.* tunc GPp 20 angelus] angulus C 20–21 angeli unius] *inv.* GPp 22 erat] est G 23 tunc] tamen CPp ‖ essent] erant Gp : erunt P 24 loquamur] loquamus C ‖ est] sit GPp 25 altero] alio G ‖ tamen sic] si G 27 c] *om.* P

Istud est difficile, quia sensus vel imaginatio non cadit super talia, sed solum intellectus. Et adhuc apud intellectum non posset forte convinci per rationes ex sensibus deductas quod dicti casus sint possibiles, sed hoc simpliciter fide credendum est. Ideo quasi mendicare oportet intellectum humanum.

Et de hoc videtur mihi probabile dicere quod non erat aliud duratio angeli *a* vel angeli *b* aut *c* quam ille angelus *a* vel *b* vel *c*. Et ideo, cum angelus *a* non esset maior angelo *b*, duratio eius non fuit maior duratione illius; nec durationes angelorum *b* et *c* fuerunt aequales, cum illi angeli non essent aequales. Et tamen apparet mihi quod angelus *a* erat prius angelo *b*, non quia in priori tempore, sed quia erat illo non existente. Et potest dici quod illa prioritas erat secundum tempus ad sensum condicionalem, scilicet quia si illis angelis coextitisset tempus, sicut coexistit his quae nunc sunt, tempus coexistens angelo *a* fuisset prius in succedendo tempore quod coextitisset angelo *b* vel angelo *c*. Ita similiter ad sensum praedictum condicionalem | 93vb G et extrinseca denominatione | 104rb P diceremus angelum *a* plus durasse quam angelus *b* duravit, et etiam in certa proportione, ut in duplo vel in triplo, scilicet quia in duplo vel in triplo fuisset maius tempus coexistens angelo *a* quam | 116vb C coexistens angelo *b*, si fuissent eis tempora coexistentia. Et aequalis fuisset duratio angeli *b* et angeli *c* tali modo, quia aequalis vel eadem fuisset du|82va p ratio temporis coexistentis angelo *b* et coexistentis angelo *c*, si fuisset eis tempus coexistens. Et aliter non possumus percipere nec exprimere quanto prius fuit vel quanto plus duravit angelus *a* quam angelus *b*. Nec aliter etiam exprimimus quanto pater tuus fuit prior te vel quanto plus duravit quam tu. Sicut etiam motum quem vocamus localem non possemus dicere nec

---

1 est difficile] esset divisibile P ‖ cadit] cadunt G 2 non posset forte] non potest forte P : forte non potest Gp 3 rationes] *add.* et p ‖ quod] ut G ‖ sint] sunt P ‖ hoc] *om.* P 4 fide] *om.* p 7 aut] et P ‖ vel^{3}] aut Gp 8 duratio ... maior^{2}] non fuit maior duratio eius vel G : duratio non erat maior p 9 angelorum] *om.* P 11 existente] coexistente P 12 ad] et P 13 coextitisset] *corr. sup. lin. ex* coexistisset C : coexistisset P ‖ coexistit] *praem.* nunc G : coexistat P ‖ quae] qui Gp 14 in succedendo tempore] tempore in succedendo G : tempore in succedendo tempus p : in succendendo tempori P ‖ quod coextitisset] quo coextitisset C : quo extitisset p : quam quod (*sup. lin.*) coextitisset G 15 angelo^{2}] *om.* P 17 angelus] *corr. ex* angelo C : angelum GP ‖ duravit] *om.* G ‖ et etiam] *om.* G ‖ certa] incerta G ‖ ut] et P ‖ in^{3}] *om.* p 18 scilicet quia] quia scilicet vel p 18–19 a ... angelo] *om.* (*hom.*) G 20 fuisset^{1}] *corr. in marg. ex* fiat C : fuit Gp : fuit duratio angeli a et aequalis fuit P ‖ tali] *om.* P ‖ vel eadem fuisset] fuit P 22 coexistens] existens C 23 prius] plus G ‖ b] *om.* P ‖ nec aliter etiam] ideo etiam aliter p 24 duravit] duraverit P 25 motum] motus P 25–342.1 dicere nec percipere] percipere nec dicere P : dicere G : dicere nos percipere p

percipere quantus esset nisi per habitudinem mobilis ad aliquid aliud quod tamen nihil operatur ad motum istum.

Ex dictis apparet manifeste quomodo rationes quae fiebant in principio quaestionis procedunt etc.

---

1 ad aliquid] vel aliquod P 3 ex dictis apparet] et ideo apparet ex dictis P 4 etc.] *om.* Pp

⟨IV.16⟩

⟨Utrum tempus esset, quamvis non esset aliqua anima intellectiva⟩

Quaeritur sexto decimo et ultimo utrum tempus esset, quamvis non esset aliqua anima intellectiva.

Arguitur quod sic quia:

⟨1⟩ Tempus est ens reale et non ens rationis, quia aliter non pertineret ad scientiam naturalem, sed ad logicam.

⟨2⟩ Item quamvis annihilaretur omnis anima intellectiva, tamen adhuc posset Deus salvare motus caelestes; igitur adhuc esset tempus. Consequentia probatur primo quia: tempus est primus motus, qui salvaretur. Secundo quia: adhuc esset unus motus in caelo velocior alio et alter tardior; et tamen remanente veloci et tardo oportet tempus remanere, quia velox et tardum definiuntur tempore, et tamen remanente eo pro quo definitum supponit oportet remanere ea pro quibus termini definitionis supponunt, aliter definitio non esset convertibilis cum definito.

⟨1⟩ Oppositum arguitur, sicut arguit Aristoteles, quia: si impossibile est numerantem esse, impossibile est numerum esse, propter hoc quod numerus ex eo dicitur numerus quod est numeratus vel numerabilis; et tamen nihil esset numerabile, si impossibile esset esse numerantem, quia numerabile non dicitur nisi quia potest numerari, et nihil posset numerari, si nihil esset aptum natum numerare. Sed tempus est numerus et nihil est aptum numerare nisi anima, non adhuc quaecumque, sed solum intellectiva, ut dicit Aristoteles. Igitur si non posset esse anima intellectiva, non posset esse tempus.

---

4 et ultimo] *om.* GPp 6 arguitur] *praem.* et P ‖ quia] per C 7 non[1]] *add.* est P 8 logicam] musicam P 9 tamen] *om.* G 10 posset deus] *inv.* G 12 alio] altero Gp 15 termini definitionis supponunt] definitio supponit G 17 arguitur] *add.* sic P ‖ arguit aristoteles] *inv.* p 19 dicitur] *rep.* C ‖ quod] quia GPp 20 esse numerantem] *inv.* GP 21 et] *add.* cum p 22 esset aptum natum] esset aptum Pp : posset G ‖ aptum[2]] *ante* est[2] p : *add.* natum G 23 intellectiva] *add.* solum G 24 igitur] *om.* P ‖ posset[1]] possit P ‖ esse[1]] *post* intellectiva G ‖ posset[2]] possit P

---

17 Cf. Aristoteles, *Physica*, IV, 14, 223a21–26 24 Cf. Aristoteles, *Physica*, IV, 14, 223a25–26

 | DOI: 10.1163/9789004322356_039

⟨2⟩ Item si terminus pro aliquo supponit et aliquid etiam appellat, deficiente appellato | 104va P ille terminus amplius pro nullo supponit. Verbi gratia, si 'album' supponit pro homine vel lapide et appellet albedinem, si non sit albedo, ille terminus 'album' pro nullo supponit. Modo verum est quod iste terminus 'tempus' supponit pro motu primo, sed appellat numerationem vel numerabilitatem. Et iam hoc appellatum deficeret, si non esset anima | 94ra G intellectiva, ut dicebatur in alia ratione. Igitur ille terminus 'tempus' pro nullo supponit; ideo nihil esset tempus. | 117ra C

⟨3⟩ Item Commentator dicit quod tempus est bene praeter animam in potentia, sed non in actu, propter hoc quod prius et posterius in motu sunt tempus secundum quod numerata sunt; et non sunt numerata praeter animam nisi in potentia.

Prima conclusio ponitur manifesta quod, si nullus esset intellectus, nullum esset tempus nec posset esse, immo nihil esset nec posset esse, quia Deus est intellectus, et si Deus non esset, nihil posset esse, quia ad impossibile sequitur quodlibet, et maxime ad Deum non esse.

Secunda conclusio est quod ad Deum non spectat aliquid mensurare, saltem capiendo 'mensurare' et 'mensurari' ad istum sensum ad quem dicimus quod nos mensuramus unam quantitatem per aliam, quia sic mensurare est cognoscere quantitatem mensurabilis prius ignotam per quantitatem mensurae notam; nihil autem est Deo ignotum vel dubium vel imperfecte notum, immo de quolibet mensura|82vb p bili Deus scit evidenter quantum est et quantum esset, licet nullum esset aliud quantum per quod mensuraretur.

Deinde notandum est quod, licet isti termini 'numerus' et 'multitudo' pro eisdem supponant, tamen ultra significationem multitudinis iste terminus 'numerus' connotat quod multitudo sit numerata vel numerabilis. Unde si esset multitudo infinita capiendo 'infinita' categorematice, illa non esset

---

1 terminus] tempus p ‖ aliquid etiam] *inv.* p : etiam aliquod P 1–2 deficiente] si deficiat aliquid de G 3 appellet] appellat GPp 4 album] *add.* amplius GPp 5 motu primo] *inv.* GPp 6 iam] ideo Gp 6–7 intellectiva] *add.* et P 7 dicebatur] dicitur p ‖ ratione] quaestione G 8 supponit] supponeret GPp ‖ nihil] nullum G 10 in[2]] *rep.* G 11 tempus] *add.* et C 13 quod] quia G 14 nec[1] ... immo] probo G 15 posset esse] *inv.* GPp 17 est] *om.* p 18–19 dicimus ... mensuramus] nos mensuramus ut dicimus G 21 notam] notae G 22 immo] ergo G ‖ scit evidenter] *inv.* GPp ‖ est et quantum] *om.* (*hom.*) p 24 deinde ... quod] notandum P ‖ isti termini] iste terminus P 25 supponant] supponunt P 25–26 iste terminus numerus] *om.* G 26 unde] ut G 27 infinita[2]] infinitum p

---

9 Cf. Averroes, *In Physicam*, IV, comm. 131, f. 202E

numerus nec esset aliquota, nec etiam magnitudo esset aliquota, si esset infinita. Et hoc notavit Aristoteles quinto *Metaphysicae* dicens: 'multitudo autem quantum quoddam est, si numerabilis fuerit, magnitudo autem, si fuerit mensurabilis'.

Sed apparet mihi quod iste terminus 'numerus' non connotat actualem numerationem, sed numerabilitatem. Quod potest declarari quia: si connotaret actualem numerationem, sequeretur quod nullus esset numerus nisi quando res actu numerarentur; et hoc non videtur esse verum, quia si omnes dormirent et quod nullus numeraret, tamen ego et tu essemus duo homines; et non essemus duo sine dualitate vel binario; igitur adhuc esset dualitas sive binarius et per consequens numerus, quia necesse est, si est aliquis binarius, quod omnis binarius sit numerus. Ibi enim est praedicatio | generis de spe- 104vb P
cie, quae semper est universaliter vera, si species pro aliquo supponit.

Tunc sequitur tertia conclusio, scilicet quod quamvis omnia cognoscentia praeter Deum essent annihilata manentibus aliis, scilicet non cognoscentibus, adhuc res essent numerus, quia quamvis non essent numeratae, tamen adhuc essent numerabiles et possent numerari, quia Deus posset statim creare homines et animas quae possent illas res numerare; et hoc sufficit ad hoc quod illae res sint numerus secundum praedicta.

Sed posito quod ille terminus 'numerus' ex impositione, quae est ad placitum, connotaret actualem numerationem, quid esset dicendum, si nullum esset cognoscens nisi Deus?

Dico quod per illam | actualem numerationem connotatam tu | posses 117rb C 94rb G
intelligere uno modo illam divinam cognitionem qua Deus distincte cognoscit omnes res alias ab invicem et qua cognoscit distincte personas divinas. Et tunc statim certum est quod adhuc res essent numerus, quia non deficeret connotatum. Sed alio modo per illam actualem numerationem tu posses intelligere rationem secundum quam numerans cognoscit res quot

1 etiam] *om.* GP ‖ magnitudo] *add.* et sic P ‖ aliquota²] aliquanta GPp 2 aristoteles] *om.* G 3 quantum quoddam est] quantum quodam est C : quantum quidem est P : quantumcumque p 3–4 autem² … mensurabilis] autem si mensurabilis fuerit Gp : aliter P 5 sed] item G 7 sequeretur] sequitur p 8 numerarentur] numeraretur G ‖ omnes] *add.* homines G 10 vel] et P ‖ sive] *add.* numerus p 11 quia] et sciendum est quod P 12 sit] est P 13 est] *post* vera G 16 res essent numerus] esset numerus res illae P ‖ quia] et G : *om.* p 17 numerari] numerare P 19 illae res sint] illae sint GP : ille sit p 20 posito] pono GP : pone p 21 connotaret actualem] connotaret G : connotet aliqualem p ‖ esset] esse P ‖ nullum] nullus p 22 nisi deus] praeter deum P 24 uno modo] *om.* G 26 statim] *om.* G ‖ adhuc] *ante* certum C ‖ essent] esset G 27 illam] *om.* G 28 posses] potes GP ‖ quot] quod P

2 Aristoteles, *Metaphysica*, V, 13, 1020a8–10

sunt, cum ante hoc esset sibi dubium vel ignotum. Talis enim est numeratio qua numeramus loquendo proprie de nostra numeratione, sicut etiam prius dicebamus de mensurare nostro. Et tunc ego dicerem quod, si Deus cras annihilaret omnia cognoscentia praeter ipsum, adhuc numerus esset, sed non esset numerus, quia non deficerent isti duo lapides qui modo sunt numerus, sed deficeret connotatum per hoc nomen 'numerus'. Ideo numerus esset, quia isti duo lapides qui modo sunt numerus essent, sed non essent numerus nec essent duo, licet adhuc essent bene alii ab invicem, sicut album esset, sed non esset album, si cessaret esse albedo.

Et istae conclusiones possent numerari cum aliis. Sed non curavi, quia credo quod 'numerus' de facto connotet non actualem numerationem, sed numerabilitatem, sicut dixi prius.

Quarta conclusio est quod, si non posset esse numerans, non posset esse numerus, sicut bene arguebat ratio Aristotelis prius narrata. Nihil enim esset possibile numerari, si nihil esset potens numerare. Et haec etiam conclusio
83[ra] p patet per istam regulam com|munem quia ad impossibile sequitur quodlibet; et antecedens dictae conclusionis condicionalis est impossibile simpliciter. Haec enim est impossibilis, scilicet quod non potest esse numerans.

Quinta etiam conclusio sequitur quod, si non posset esse intellectus, non posset esse numerus. Hoc probatur statim per dictam communem regulam. Sed etiam hoc probatur speciali probatione quia: nulla anima est
105[ra] P innata numerare nisi intellectiva, ut | dicit Aristoteles. Et hoc persuadetur quasi experimento. Gallina enim habens multos pullos percipit bene eorum multitudinem, sed non numerat eos; ideo si de decem auferantur tres vel quattuor, ipsa videns alios non curabit de ablatis; nec quaereret eos, nisi audiat eos vel videat. Et eodem modo esset de sue quae haberet decem porcellos et de cane habente quinque vel sex catulos. Et ita etiam gallina

---

1 est] *om.* G 3 mensurare nostro] *inv.* GPp ‖ et] *om.* P 7 modo] *om.* G 8 licet] sed p ‖ adhuc … alii] bene adhuc essent G 9 sed] si P 10 possent] possunt P ‖ curavi] curari P 11 numerus] *post* facto GPp ‖ connotet] connotat GP 12 sicut] ut P 13 posset[1]] possit P ‖ posset[2]] possit P 14 arguebat] arguit p 16 regulam] rationem G ‖ quia] quod GPp 17 conclusionis] *om.* p 18 haec enim] *inv.* GPp ‖ potest] posset G 19 etiam conclusio sequitur] conclusio sequitur etiam G : conclusio etiam sequitur P : conclusio sequitur scilicet p 20 hoc] *om.* P ‖ statim] *om.* p 20–21 communem regulam] *inv.* Pp : rationem communem G 21 etiam hoc] *inv.* P ‖ speciali] *add.* ratione vel G 24 non] *om.* P ‖ ideo si] si enim p ‖ de decem] decem habeat G 25 curabit] curat P ‖ nec quaereret] nec quaeret Gp : non quaeret P 26 eos] *post* videat G 26–347.4 quae … numeravit] et similiter de aneta si amoveantur ova de nido non amplius ponet in nido nisi remaneant aliqua ova G 27 porcellos] porcos P ‖ catulos] canes Pp

---

3 Cf. sup., IV, q. 14 22 Cf. Aristoteles, *Physica*, IV, 14, 223a25–26

ponens ova in uno nido, si auferantur omnia, non amplius in isto nido ponet, si inveniat ubi alibi ponat; et si multa auferantur et remaneant aliqua, non dimittet ponere in nido, quia non percipiet quod aliquid sit ablatum, eo quod non numeravit. Et si haec persuasio non valeat, inveniant alii aliam.

Deinde iuxta praedicta de numero | 117va C dicendum est proportionaliter de tempore. Quamvis enim idem sit tempus et motus, tamen hoc nomen 'tempus' ultra significationem huius nominis 'motus' connotat quod sit mensura etiam. Sic in praecedentibus dictum fuit. Sed non credo quod connotet actualem numerationem, sed potentialem.

Et sic ponuntur conclusiones quod, si nullus esset intellectus, nullum esset tempus, quia non esset Deus; ad quod sequitur nihil esse. Sed si | 94va G nullus esset intellectus praeter Deum et esset motus caeli, adhuc esset tempus, quia motu caeli posset aliquis, qui forte cras crearetur, mensurare istos motus, quamvis adhuc hoc nomen 'tempus' connotet possibilitatem ad mensurandum secundum nostrum modum mensurandi. Sed si non posset esse intellectus, non posset esse tempus. Etiam enumeret conclusiones qui voluerit secundum dicta de numero.

Ad rationes igitur dicendum est quod:

⟨1⟩ Tempus est res extra animam et praeter operationem animae nostrae existens, quamvis hoc nomen 'tempus' connotet operationem animae possibilem. Nec 'tempus' est nomen secundae intentionis. Et cum hoc etiam entia quae sunt ab anima et per animam bene pertinent ad scientiam naturalem vel ad metaphysicam. Ideo ista ratio prima undique nihil valet.

⟨2⟩ Secunda ratio et eius confirmatio conceduntur.

Ad rationes in oppositum.

---

1 in^2 ... ponet] ponet in illo nido p : ponit in illo nido P   2 si ... ponat] et si inveniret ea alibi ponit ibi P   3 percipiet] percipit P   5 proportionaliter] proportionabiliter p   7 motus] *om.* p   8 etiam sic] etiam sicut G : etc. sicut p ‖ fuit] fuerit GP ‖ connotet] connotat G   9 numerationem] mensurationem G   11 esse] est P   12 adhuc] et hoc P   13 qui forte] *inv.* P ‖ istos] alios GPp   14 connotet] connotaret G : connotat Pp   15 posset] possit P   16 posset] possit P ‖ etiam enumeret] etc. et enumeret p : etc. enumerat P ‖ qui voluerit] *post* numero (17) p : *om.* G   20 connotet] connotat G ‖ operationem animae] *inv.* P   21 secundae] *add.* impositionis vel G ‖ hoc etiam entia] etiam G   23 ista ratio prima] ratio ista prima GP : ratio illa p ‖ undique nihil] nihil utique G   24 eius confirmatio conceduntur] eius conclusio conceduntur Pp : aliae faciliter solvuntur per dicta C   25 rationes in] rationes post P : rationem in C

⟨1⟩ Dico quod bene prima ratio concludit quod, si non posset esse numerans vel etiam si non posset esse intellectus, non posset esse numerus neque tempus.

⟨2⟩ Ad aliam dicitur quod, licet omnes creaturae annihilarentur, tamen non propter hoc deficeret numerabilitas.

⟨3⟩ Ad auctoritatem Commentatoris potest dici quod forte ipse opinabatur quod hoc nomen 'tempus' connotaret actualem numerationem; quod non credo. Si tamen connotaret actualem numerationem, dicerem sicut dixi
105rb P de | numero, si etiam iste terminus 'numerus' connotaret actualem numerationem.

Et sic est finis quarti libri etc. Expliciunt quaestiones quarti libri *Physicorum* editae a reverendo philosopho magistro Iohanne Buridano etc.

---

1 bene] *post* ratio GPp ‖ posset] possit Pp 2 etiam] *om.* P ‖ posset[1]] possit GP ‖ non posset[2]] nec possit P ‖ numerus] numerans G 5 non propter hoc] propter hoc non G : non per hoc P 6 auctoritatem] auctoritates p ‖ ipse] ille p 8 actualem] *om.* G 9–10 actualem numerationem] *add.* etc. G : actualitatem P 11–12 est ... etc.] *add.* istae quaestiones sunt fratris iohannis teobaldi ordinis praedicatorum conventus C : sunt completae quaestiones huius quarti incipiunt consequenter supra quintum eiusdem G : est finis quarti explicit quartus liber physicorum p : sit finis quarti libri P 12 buridano] *scripsimus* : buridani C

Index locorum

Index codicum manu scriptorum

Index nominum

All names mentioned in the book, with the exception of Aristotle and John Buridan, are included in this index. Medieval authors (before ca. 1500) are alphabetically listed according to their first names, modern authors according to their last names. Medieval authors are generally mentioned under their English names, except for lesser known authors and authors whose foreign names are currently used in English scholarly literature.

www.ingramcontent.com/pod-product-compliance
Lightning Source LLC
LaVergne TN
LVHW010555300926
849162LV00002B/334